16th International Conference on Micro and Nanotechnology for Power Generation and Energy Conversion Applications (PowerMEMS 2016)

Journal of Physics: Conference Series Volume 773

Paris, France
6 - 9 November 2016

ISBN: 978-1-5108-3530-6
ISSN: 1742-6588

Printed from e-media with permission by:

Curran Associates, Inc.
57 Morehouse Lane
Red Hook, NY 12571

Some format issues inherent in the e-media version may also appear in this print version.

Printed by Curran Associates, Inc. (2017)

For permission requests, please contact the Institute of Physics
at the address below.

Institute of Physics
Dirac House, Temple Back
Bristol BS1 6BE UK

Phone: 44 1 17 929 7481
Fax: 44 1 17 920 0979

techtracking@iop.org

Additional copies of this publication are available from:

Curran Associates, Inc.
57 Morehouse Lane
Red Hook, NY 12571 USA
Phone: 845-758-0400
Fax: 845-758-2633
Email: curran@proceedings.com
Web: www.proceedings.com

Table of contents

Volume 773

16th International Conference on Micro and Nanotechnology for Power Generation and Energy Conversion Applications (PowerMEMS 2016)

6–9 December 2016, Paris, France

Accepted papers received: 9 November 2016
Published online: 14 December 2016

Preface

Oral presentations

PowerMEMS in Action

Poster session A

Late-news posters

Poster session B

Oral presentation

WELCOME TO POWERMEMS 2016

It is our great pleasure to welcome you to PowerMEMS'16 - the 16th International Conference on Micro- and Nano-Technology for Power Generation and Energy Conversion Applications - in Paris at the UIC Espace Congrès, a few meters away from the Eiffel Tower.

The objective of the PowerMEMS conference is to catalyse innovation in micro- and nano- scale technologies for the energy domain. The scope of the meeting ranges from basic principles, to materials and fabrication, to devices and systems, to applications. The many applications of Power MEMS concern the harvesting, storage, conversion and conditioning of energy, to integrated systems that manage these processes, to actuation, pumping, and propulsion.

Our Conference aims to stimulate the exchange of insights and information, as well as the development of new ideas, in the PowerMEMS field. Our goal is to allow the attendees to interact and network within our multidisciplinary community that includes professionals from many branches of science and engineering, as well as energy, policy, and entrepreneurial specialists interested in the commercialization of Power MEMS technologies.

This year's technical program is highlighted by four plenary talks from prominent experts on M/NEMS for ultra-low power in electronics, advanced nanomaterial for solar cells, bioinspired devices and thermal transistor. The contributed program received 159 abstract submissions this year. After careful review by the 33-members of the Technical Program Committee, a total of 123 papers will be presented. The 40 contributed oral presentations are arranged in two parallel sessions. The 83 posters are arranged in a "two-in-one" poster session in which the poster session time is divided in two; half the posters will be presented during each half-session, allowing the poster presenters to also browse the posters during the poster session. Posters will remain up during the meeting, so please feel free to peruse them at your leisure. The Proceedings will be visible and accessible through IOP after conclusion of the Conference.

For the first time this year we have plan a "Power MEMS in Action" special session that is specifically dedicated to table-top demonstrations of micro power sources and small scale energy harvesting systems. With the growing interest in wireless sensors for the Internet of Things and other distributed or portable devices, there is no better time to bring power MEMS out of the lab and into applications. The Power MEMS in Action session will provide this opportunity by allowing researchers to demonstrate their technologies at the conference.

As every year, this meeting is made possible by many generous contributions of time, effort, and financial support. Many thanks are due to the Technical Program Committee led by Pr Einar Halvorsen for their intensive efforts in reviewing abstract submissions, and to the International Steering Committee for their advice and support. Thanks to The PowerMEMS School chair Mickaël Lallart and the expert speakers that made the School possible. The local organizing committee, led by Dimitri Galayko, has provided us with invaluable assistance in making PowerMEMS 2016 happens.

We wish you a productive and enjoyable conference and a wonderful stay in Paris.

Philippe Basset

Skandar Basrour

CONFERENCE OFFICIALS

Conference Chairs
Philippe Basset Université Paris-Est, FRANCE
Skandar Basrour Grenoble Alpes Université, FRANCE

Technical Program Chair
Einar Halvorsen University College of Southeast Norway, NORWAY

PowerMEMS School Chair
Mickaël Lallart Université de Lyon, FRANCE

PowerMEMS in Action Chair
Luc Frechette University of Sherbrooke, CANADA

International Steering Committee
Mark G. Allen – University of Pennsylvania, USA
Philippe Basset – Université Paris-Est, FRANCE
Skandar Basrour – Grenoble Alpes Université, FRANCE
Steve Beeby – University of Southampton, UK
Luc Fréchette – University of Sherbrooke, CANADA
Takayuki Fujita – University of Hyogo, JAPAN
Reza Ghodssi – University of Maryland,
Einar Halvorsen – Grenoble Alpes Université, FRANCE
Hiroki Kuwano – Tohoku University, JAPAN
Jeff Lang – Massachusetts Institute of Technology, USA
Carol Livermore – Northeastern University, USA
Ryutaro Maeda – National Institute of Advanced Industrial Science and Technology, JAPAN
Kazusuke Maenaka – University of Hyogo, JAPAN
Paul Mitcheson – Imperial College London, UK
Yuji Suzuki – University of Tokyo, JAPAN
Shuji Tanaka – Tohoku University, JAPAN
Luis Velásquez-García – Massachusetts Institute of Technology, USA
Peter Woias – University Freiburg-IMTEK, GERMANY
Eric Yeatman – Imperial College London, UK

Technical Program Committee
Mahmoud Almasri, University of Missouri, USA
David Arnold, University of Florida, USA
Mustafa Beyaz, Antalya International University, TURKEY
Danick Briand, EPFL, SWITZERLAND
Stephen Burrow, Univesrity of Bristol, UK
Francesco Cottone, University of Perugia, ITALY
Alper Erturk, Georgia Institute of Technology, USA
Luis Fonseca, IMB-CNM (CSIC)Micro-Nanosytems, SPAIN
Takayuki Fujita, University of Hyogo, JAPAN
Dimitri Galayko, UPMC-Sorbonne, FRANCE
Tzeno Galchev, Analog Devices Inc., USA
Gideon Gouws, Victoria University of Wellington, NEW ZEALAND

Andrew Holmes, Imperial College London, UK
Yoshihiro Kawahara, The University of Tokyo, JAPAN
Sang-Woo Kim, Sungkyunkwan University, SOUTH KOREA
Jeffrey Lang, Massachusetts Institute of Technology, USA
Janet Ledesma-García, Universidad Autónoma de Querétaro, MEXICO
Carol Livermore, Northeastern University, USA
Yiannos Manoli, University of Freiburg, GERMANY
Jianmin Miao, Nanyang Technological University, SINGAPORE
Paul Mitcheson, Imperial College London, UK
Koji Miyazaki, Kyushu Institute of Technology, JAPAN
Jaeyeong Park, Kwangwoon University, KOREA
Michael Renaud, Holst Centre-Imec NL, NETHERLANDS
Paul Ronney, University of Southern California, USA
Shad Roundy, University of Utah, USA
Tomonori Seki, OMRON Corporation, JAPAN
Yuji Suzuki, The University of Tokyo, JAPAN
Shuji Tanaka, Tohoku University, JAPAN
Luis Velasquez-Garcia, MIT, USA
Xiaohong Wang, Tsinghua University, CHINA
Peter Woias, University Freiburg, GERMANY

International Advisory Board

Young-Ho Cho – Korea Advanced Institute of Science and Technology, KOREA
Alan Epstein – Massachusetts Institute of Technology, USA
Masayoshi Esashi – Tohoku University, JAPAN
Albert Pisano – University of California, San Diego, USA
Susumu Sugiyama – Ritsumeikan University, JAPAN
Miwako Waga – Susano Berkeley LLC, JAPAN

Local Organizing Committee

Dimitri Galayko – UPMC-Sorbonne University, France
Elie Lefeuvre – Univ. Paris-Sud, France

Peer review statement

All papers published in this volume of *Journal of Physics: Conference Series* have been peer reviewed through processes administered by the proceedings Editors. Reviews were conducted by expert referees to the professional and scientific standards expected of a proceedings journal published by IOP Publishing.

High temperature performance of a piezoelectric micro cantilever for vibration energy harvesting

E Arroyo[1], Y Jia[1,2], S Du[1], ST Chen[1] and A Seshia[1]

[1] Nanoscience Centre, University of Cambridge, Cambridge, CB3 0FF, UK
[2] Department of Mechanical Engineering, University of Chester, Chester CH1 4BJ, UK

E-mail: emmanuelle.arroyo@gmail.com

Abstract. Energy harvesters withstanding high temperatures could provide potentially unlimited energy to sensor nodes placed in harsh environments, where manual maintenance is difficult and costly. Experimental results on a classical microcantilever show a 67% drop of the maximum power when the temperature is increased up to 160 °C. This decrease is investigated using a lumped-parameters model which takes into account variations in material parameters with temperature, damping increase and thermal stresses induced by mismatched thermal coefficients in a composite cantilever. The model allows a description of the maximum power evolution as a function of temperature and input acceleration. Simulation results further show that an increase in damping and the apparition of thermal stresses are contributing to the power drop at 59% and 13% respectively.

1. Introduction

Energy harvesting has become over the recent years the main focus to supply low-power miniaturized sensor nodes from on-site available energy. For industrial machine surveillance, smart infrastructure or aerospace and automotive structures monitoring, the applications of sensor networks often involve harsh environment conditions, such as high g accelerations or high temperatures, that the energy harvester has to withstand as well. Understanding and predicting the effects of high temperatures on a harvester structure could help designing more robust devices and anticipate potential negative effects on the harvested power. In that objective, this work focuses on experimentally studying the effects of temperatures up to 160 °C on the power harvested by a classical MEMS energy harvester, and in a second part proposes a model to explain the harvester behavior.

2. Experimental study

2.1. Structure of the MEMS harvester

A classical cantilever type energy harvester described in a previous publication [1] is studied. A beam is clamped at one end and a proof mass is suspended at the other end to adjust the resonance frequency of the device. The harvester is fabricated using a MEMS process which involves a 400 μm thick silicon substrate used as a proof mass, with a 10 μm doped silicon device layer on which a 0.5 μm aluminum nitride (AlN) piezoelectric layer is patterned, and finally a 1 μm aluminum layer is used as a top electrode (Figure 1). Figure 2 shows a picture of the die composed of 4 such cantilevers. While most piezoelectric thin films lose piezoelectric properties when heated up, AlN has been shown to be stable at high temperatures [2]. However, when it integrated as a thin-film piezoelectric layer on a resonating silicon micro-

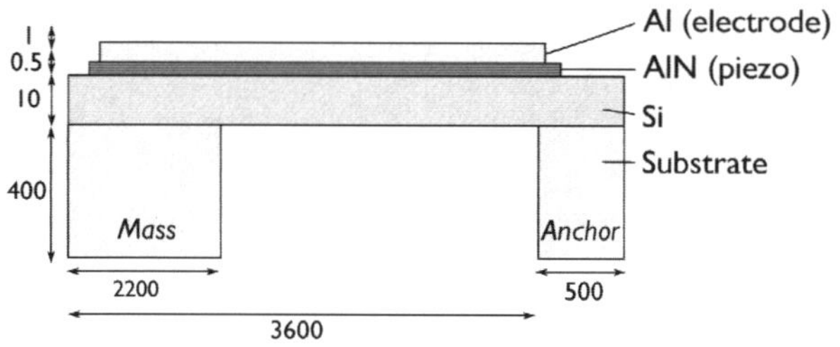

Figure 1. Schematics of the cantilever, dimensions in μm.

Figure 2. Picture of the die.

cantilever, other mechanical or electrical effects induced by temperature can affect the harvested power, which are investigated in this study.

2.2. *Experimental setup*

The die is glued to a ring, itself placed inside a socket (shown in Figure 2). The chip is soldered to a PCB and fixed onto an electromagnetic shaker driven by a function generator and amplifier. The experimental setup is described in Figure 3. The shaker is placed inside a vacuum chamber regulated at 0.33 mbar. Heaters pads are used to heat up the device and a thermocouple provides feedback on the temperature close to the harvester. Three frequency sweeps are performed at each temperature and fixed input acceleration: with the device in open circuit, in short circuit and on a matched 300 kΩ resistive load. At each step the RMS voltages are recorded.

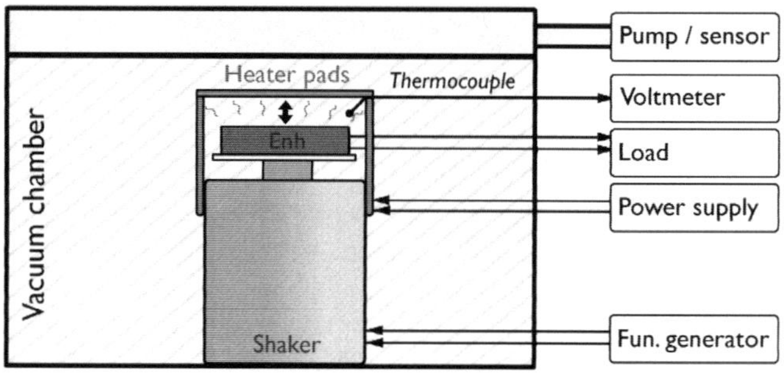

Figure 3. Schematic diagram of the experimental setup.

2.3. *Power as a function of temperature*

The frequency sweeps allow for the calculation of the coupling coefficient, the resonance frequency, the mechanical quality factor and the maximum power as a function of the input temperature. The results at low acceleration are plotted in Figures 4 and 5. The harvested power is linearly decreasing from 9.6 μW to 3.2 μW, showing a decrease of around 67%. The resonance frequency is decreasing from 217.7 Hz to 216.1 Hz, corresponding to a 0.7% drop and the mechanical quality factor is going from 726 to around 475 over the tested range, or a 35% decrease. Other parameters, like the coupling coefficient or the optimal load show very small variations over the tested range. For various acceleration levels the resonance frequency and quality factor show the same rate of decrease, whereas the power slope gets steeper for higher accelerations.

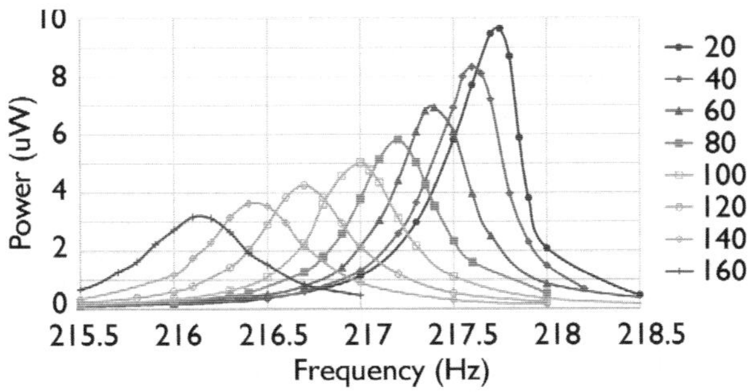

Figure 4. Experimental power as a function of input frequency, for several temperatures.

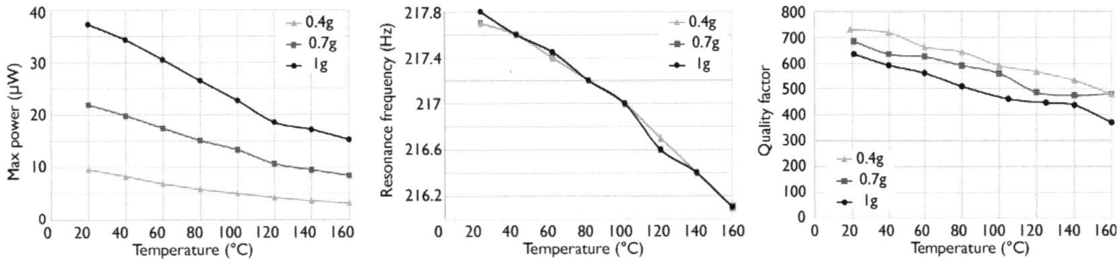

Figure 5. Experimental maximum power, resonance frequency and quality factor as a function of input temperature.

3. Modeling of a micro cantilever energy harvester with temperature

3.1. Thermal-dependent parameters

Experiments show that the resonance frequency and the mechanical quality factor have both a linear and significant decrease versus temperature. The resonance frequency is defined as the square root of the beam stiffness over the equivalent mass, and those two parameters can in turn be defined as a function of the material parameters and cantilever dimensions. The decrease of the silicon Young modulus as a function of temperature has been experimentally measured by several studies reporting temperature coefficients at around: $\frac{1}{E}\frac{dE}{dT} = -80e^{-6}\ {}^{\circ}\mathrm{C}^{-1}$[3, 4]. Another phenomenon contributing to the change in stiffness and mass values, is the thermal expansion of the materials, described by their linear thermal expansion coefficient. The expansion notably leads to a slightly larger mass at higher temperature. Both those phenomenon are contributing to the decrease in resonance frequency.

The mechanical quality factor is primarily limited by air damping, thermoelastic dissipation and support loss. All those terms are linked to the temperature, therefore inducing that the mechanical quality factor is a strong function of temperature [5]. A further article will be dedicated to calculating those parameters from the geometric and material parameters and include them into the model. For the present study, the experimental values of the quality factor drop are used and input into the model.

3.2. Thermal stress

Each material composing the cantilever has a different thermal coefficient. When subjected to increasing temperature, the mismatched dilatations induce stresses on the surfaces. The Finite elements software Comsol is used to simulate those thermal stresses. The cantilever is simulated in 2D, with coupled thermal and mechanical study. Figure 6 shows the resulting stress distribution when the cantilever is submitted to 150 °C. The thermal expansion coefficient of the metal layer being the largest, followed by the AlN one and finally the silicon one which is the smallest, an increase in temperature induces a downward curling of the cantilever and large stresses at the border between the layers. The average stress σ induced in the piezoelectric layer is evaluated for several input temperatures; its variation is shown in the graphic in the right in Figure 6.

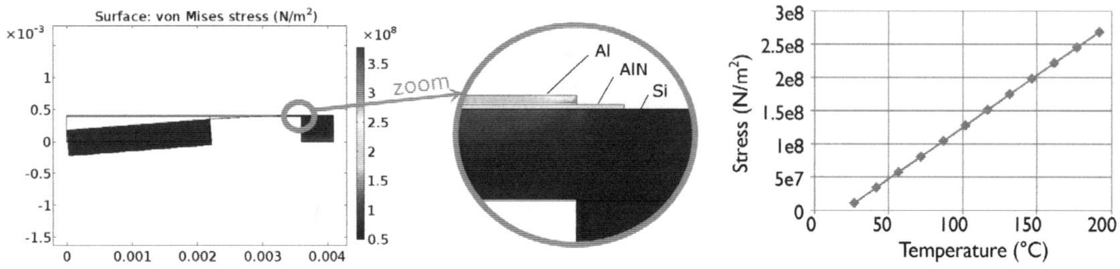

Figure 6. Comsol simulation of the thermal stresses when the cantilever is subjected to 150 °C, and average stress as a function of the temperature.

3.3. Model taking into account temperature

Around its resonance frequency and at small acceleration, the cantilever can be modeled as an equivalent mass-spring-damper system, and its behavior is described by two coupled mechanical and electrical equations involving the mechanical quality factor Q_m, the coupling coefficient k_m^2, and the load coefficient ξ_c. The details of the normalization and parameters definitions can be found in [6].

The young modulus variation with temperature and the materials dilatation are included in the model, as well as the damping variation experimentally measured.

The thermal stresses are added to the model by means of a global force $\sigma^*(T) = \sigma(T)A$, where A is the active piezoelectric area, and $\sigma(T)$, is the stress defined by the Comsol simulation shown in Figure 6. The term is normalized by the mass $m(T)$ and the natural frequency of cantilever $\omega_0^2(T)$, both dependent on the temperature through material parameters.

The model can finally be written as the coupled equations (1) and (2), with y the input vibration displacement, u the cantilever tip displacement and $\underline{v}$ the normalised generated voltage.

$$y'' + \frac{\sigma^*(T)}{m(T)\omega_0^2(T)} = u'' + u'/Q_m(T) + u + k_m(T)^2\underline{v} \quad (1)$$

$$u' = \underline{v}' + 2\xi_c\underline{v} \quad (2)$$

Those two equations are translated into complex variables and solved by first extracting the displacement amplitude u_M to then calculate the power, function of u_M^2 [6]. The resolution is the same as the classical model except for the added term $\frac{\sigma^*(T)}{m(T)\omega_0^2(T)}$ to the displacement expression.

The final calculated power expression is a function of the temperature, the input vibration characteristics like the acceleration and frequency, and the load resistance.

When separately studying the power evolution as a function of the frequency, the quality factor diminution or the thermal stress, the model shows that the quality factor contributes the most to the power drop. The damping alone contributes to a power decrease of 49%, compared to the almost 67% experimentally measured. This result can also been seen as:

the degradation of damping when increasing the temperature accounts for 59% of the power drop. The thermal stresses contributes to 13% of that same decrease. All in all, the damping and thermal stresses explain 72% of the power drop.

Those results are shown in Figure 7 for 3 different input accelerations. The normalized power drop experimental and theoretical is plotted over the range of temperature. The dotted line corresponds to the model taking into account the damping drop only, in which case the normalized power does not depend on the acceleration. The other theoretical curves show the result for the global model, taking into account thermal stresses. In that case, the power slope depends on the input acceleration, and its evolution matches well the experimental results, which confirms the importance of both the thermally induced stresses and the increase in damping as the two main causes behind the power drop at elevated temperatures up to 160 °C. The remaining unexplained difference between the experimental and theoretical power drop is going to be investigated in further work.

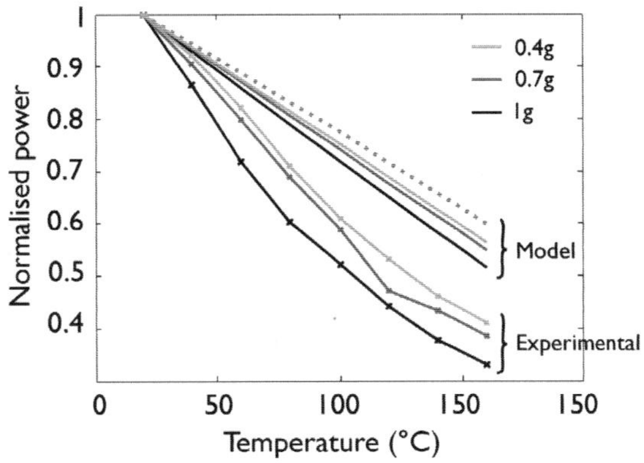

Figure 7. Normalized power as a function of temperature, modeled and experimental data for three different input acceleration.

4. Conclusion

This study experimentally and theoretically studies the evolution of the harvested power of a classical MEMS cantilever under vacuum, as a function of the ambient temperature. When the working temperature is increased to 160 °C, a power drop of almost 70% is experimentally measured. Simulations with a classical lumped parameters model confirm that the quality factor causes the biggest decrease in power, accounting for 59% of this decrease. Thermal stresses appearing in the composite cantilever are the second largest cause for power drop. Those stresses could be further reduced by optimizing the electrode shape, which has the largest thermal expansion coefficient among the cantilever materials. Future work aims at completely understanding and modeling the effect behind the quality factor drop in order to anticipate and reduce its effect on the power decrease at elevated temperatures.

References

[1] Jia Y and Seshia A A 2016 *Journal of Microelectromechanical Systems* ISSN 10577157
[2] Darnjanovic D 1998 *Current Opinion in Solid State & Materials Science* **3** 469–473
[3] Cho C H 2009 *Current Applied Physics* ISSN 15671739
[4] Jeong J H, Chung S H, Lee S H and Kwon D 2003 *Journal of Microelectromechanical Systems* **12**
[5] Kim B, Hopcroft M A, Candler R N, Jha C M, Agarwal M, Melamud R, Chandorkar S A, Yama G and Kenny T W 2008 *Journal of Microelectromechanical Systems* ISSN 10577157
[6] Arroyo E, Badel A, Formosa F, Wu Y and Qiu J 2012 *Sensors and Actuators A: Physical* ISSN 09244247

Multi-Axis Inertial Energy Harvester Based on Piezoelectric Crab-Legs with Partitioned Electrodes

E E Aktakka, K Najafi

Center for Wireless Integrated MicroSensing & Systems (WIMS²)
EECS Department, University of Michigan, Ann Arbor, MI 48104, USA

E-mail: aktakka@umich.edu

Abstract. This paper reports a microfabricated piezoelectric MEMS inertial energy harvester that can scavenge high-frequency mechanical vibrations in all three dimensions with an active inner volume of only 55 mm^3. The device can generate 0.1 to 10.5 μW from out-of-plane vibrations (365-465 Hz), and 0.1 to 3.2 μW from in-plane vibrations (680-930 Hz) at 0.5 g acceleration input. The reported harvester is demonstrated to be effective for multi-axis operation, with a favorable power density of 0.2-2.2 $mW/cm^3/g^2$ (at 0.1-0.5 g) with respect to previously reported microfabricated multi-axis vibration harvesters.

1. Introduction

Harvesting vibrational energy from all spatial directions can both enable higher power output from an energy scavenger, and also extend its practical applications in real-life situations. Previously, three-degrees-of-freedom energy harvesting in a single transducer has been reported only with electrostatic and electromagnetic devices [4-5], but with limited power density (<125 $nW/cm^3/g^2$). On the other hand, piezoelectric inertial energy harvesters have been reported with only up to 2-axis operation, based on asymmetric inertial mass [1], coupled mass-spring combinations [2], and three-dimensional beam designs [3]. In addition, the employed device architectures mostly require unconventional three-dimensional structures with manual assembly, which hinder further device miniaturization. Recently, we reported for the first time a meso-scale piezoelectric device (6500 mm^3) that can be used for 3-axis inertial energy harvesting [6]. In this paper, a highly miniaturized (100×) and batch-mode microfabricated version of this structure and its detailed experimental results are demonstrated.

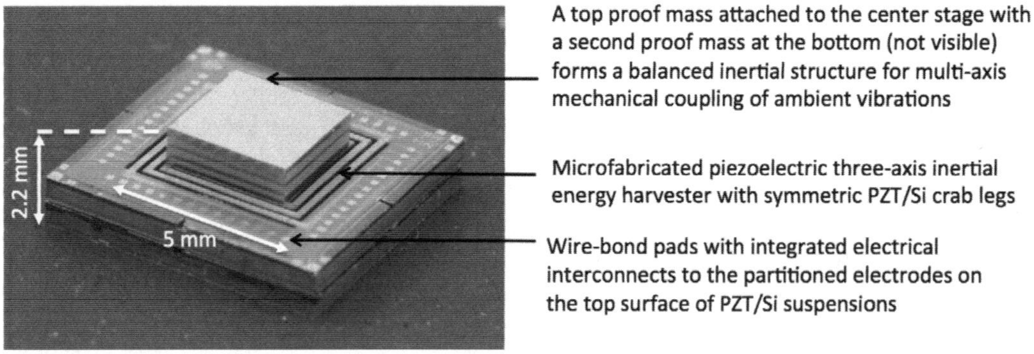

Figure 1. Micro-fabricated piezoelectric energy scavenger for multi-axis vibration harvesting.

2. Design of the Three-Axis Harvester

The harvester design is based on three key aspects (Figure 1). First of all, in order to create a balanced inertial force during both out-of plane and in-plane excitations, two proof masses with equal sizes and weights are attached to the top and bottom surfaces of a center platform. Secondly, this center platform is suspended by four piezoelectric unimorph crab-legs in a symmetric layout (Figure 1). As the inertial proof mass is coupled to ambient vibrational accelerations, this crab-leg beam structure allows its motion in both in-plane and out-of-plane directions (Figure 2a-b). Finally, depending on the direction of applied acceleration, the mechanical strain created on the piezoelectric layer of a crab-leg beam has different polarities at different sections of the beam. The conventional method of using a single monolithic electrode on top of the piezoelectric layer would cause charge cancellation on the surface, and thus lower the power output substantially. Instead, the presented design utilizes partitioned surface electrodes (Figure 2c), such that charge outputs with varying phases from these sections can be harvested efficiently with respect to a common ground electrode in the transverse piezoelectric mode.

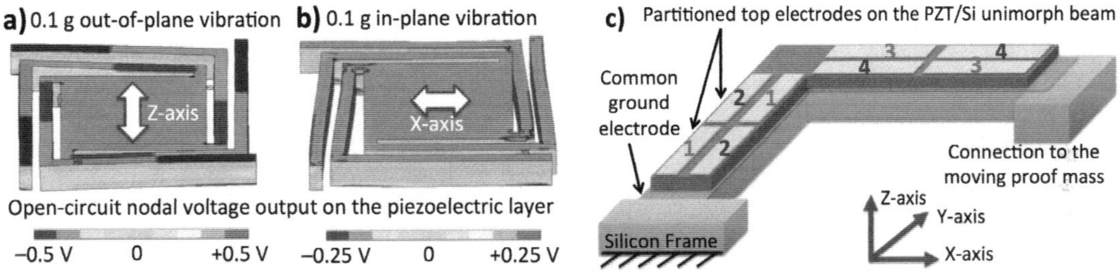

Figure 2. a-b) Simulated electrical output and polarity on the piezoelectric layer at different vibration modes, **c)** Partitioned top electrodes and the common ground electrode on one arm of the crab-stage.

3. Microfabrication

A 10-mask wafer-level fabrication process is developed, based on a previously reported and further advanced "*bulk-PZT on Si*" technology [7]. The process involves diffusion solder bonding, precision lapping, and dry-etch patterning of a bulk-PZT substrate on a Si wafer. The fabricated piezoelectric unimorph beams have a cross-section of 35-μm thick high-quality PZT film on top of 160-μm thick Si. The piezoelectric layer is covered with two layers of metallization buried in between electrical isolation layers, of which the first layer forms the energy-harvesting partitioned top electrodes, and the second layer forms the electrical interconnects. The overall harvester structure, excluding the wire-bonding pads and support frame but including empty space for packaging, occupies only 55 mm^3.

4. Experimental Setup and Results

The harvester is wire-bonded to a dual in-line package with electrical connections to the 16 partitioned top electrodes and the common ground, and all electrodes are connected to individual loads of equal resistivity for electrical damping (Figure 3a). A custom-made jig is utilized to attach the device to a shaker table in different spatial orientations, and test its response for each vibration axis (Figure 3b-d).

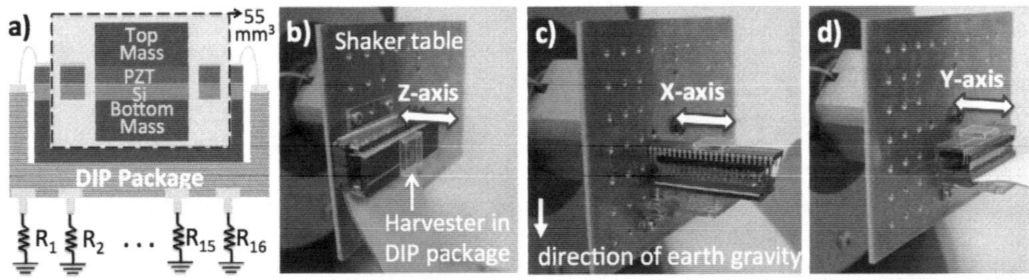

Figure 3. Experimental setup to test the harvester for sinusoidal vibrations in the X, Y, and Z axes.

In order to obtain maximum power transfer, the electrical load on the harvester should be optimized, such that the induced electrical damping equals to the existing mechanical damping in the system. When the device is tested for sinusoidal vibration at its first resonance frequency (433 Hz), the optimum resistive load on a single electrode is measured to be ~800 kΩ (Figure 4a). However, as the excitation frequency is swept from 300 Hz to 850 Hz covering different vibration modes, the optimum load is observed to range between 400 kΩ and 1.5 MΩ (Figure 4b). This suggests that, an impedance matching network will be required to be included in the power management circuitry of this transducer in order to obtain high efficiency in all possible operation conditions.

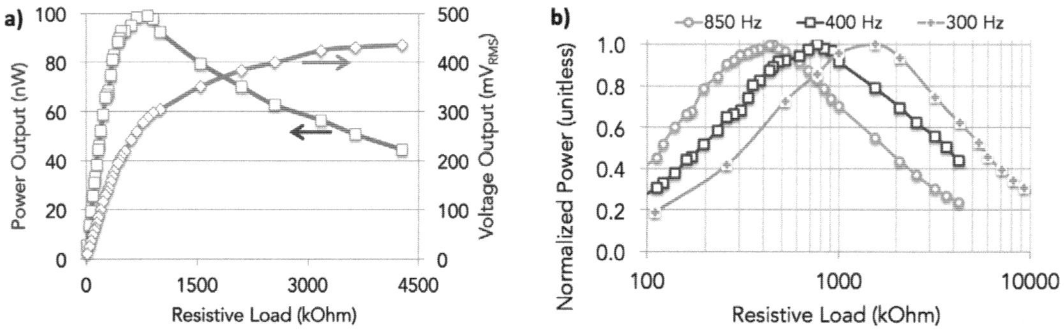

Figure 4. a) Output from a single electrode of the harvester with respect to a resistive load for a sinusoidal vibration at 433 Hz, 0.1 g **b)** Variance of the optimum resistive load for maximum power output when the device is excited at different frequencies and at constant peak acceleration of 0.1 g.

The amplitude and phase of voltage outputs from the 16 partitioned electrodes depend on the direction and frequency of excitation. For instance, the voltage outputs recorded from the electrode pairs 1-2 and 3-4 on a crab-leg, as pictured in Figure 2c, are measured to have similar phase and amplitudes, when the device is tested with an out-of-plane excitation at its first mode at 433 Hz (Figure 5a). However, the phase differences between these electrodes shift, as the direction of input vibration is alternated and an in-plane mode is excited. In this case, electrode pairs 1-4 and 2-3 have similar phase in their outputs (Figure 5b). These results match well with the FEA simulations presented in Figure 2a-2b. The variable phase difference in the outputs demonstrates the effectiveness of using partitioned surface electrodes to prevent charge cancellation across the piezoelectric layer compared to using a single monolithic electrode in a conventional design. One drawback of using multiple harvesting electrodes could be the required number of input channels in a possible power management circuitry connected to this device, and thus a potential increase in the power consumption necessary for active rectification of multiple low-voltage signals. In order to minimize this power consumption and generate a net energy output, a smart circuit architecture could be employed, such that the outputs of only certain electrodes with sufficient voltage levels are actively rectified depending on the vibration mode and amplitude.

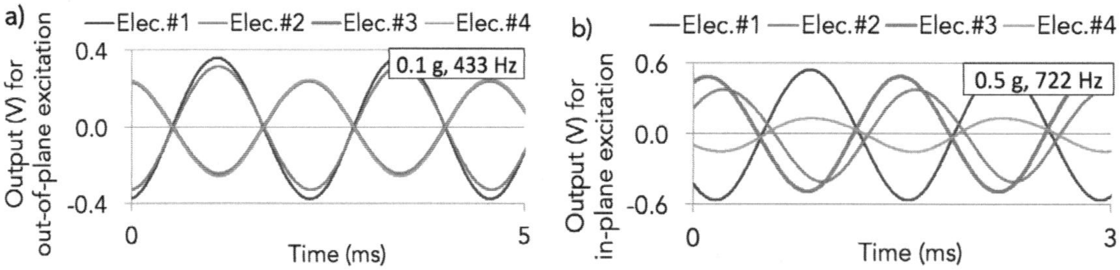

Figure 5. Comparison of voltage outputs and phases of partitioned electrodes in a single arm, when the device is excited at resonance in the **a)** out-of-plane (Z-axis), and **b)** in-plane (X-axis) directions.

When the harvester is tested for its maximum total power output for an input vibration with 0.1 g acceleration amplitude, it is found to generate 1.2 μW and 0.3 μW from out-of-plane and in-plane excitations, respectively (Figure 6-7). At 0.5 g acceleration, the power output increases about 10×, to 10.5 μW and 3.2 μW for the same excitation modes, respectively. For an unpackaged device, these values correspond to a normalized power density of 0.2 to 2.2 $mW/cm^3/g^2$. This result compares favorably with respect to previously reported three-axis microfabricated energy harvesters based on electrostatic and electromagnetic transduction (<0.1 $mW/cm^3/g^2$) [4-5].

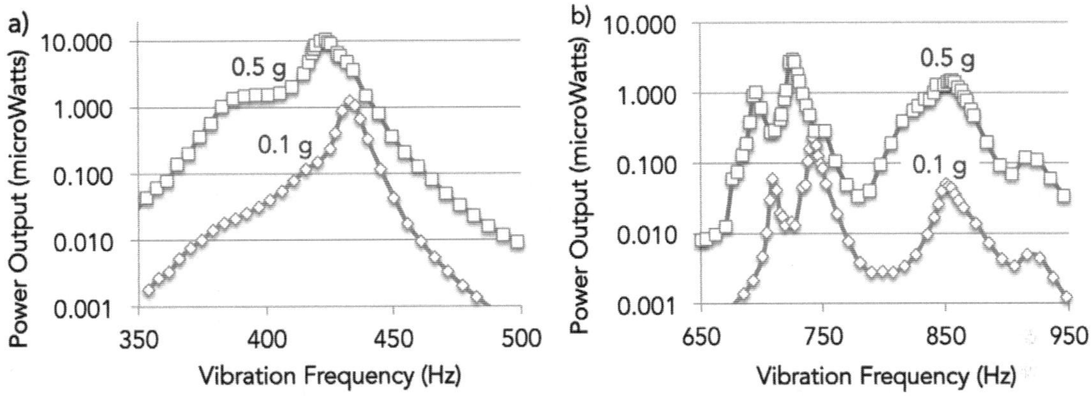

Figure 6. Total power output from all 16 partitioned electrodes of the multi-axis harvester as it is excited in the **a)** out-of-plane mode (Z-axis) with a resistive load of 767 kΩ on each electrode, and **b)** in-plane mode (X-axis) with a resistive load of 450 kΩ, at 0.1 g and 0.5 g acceleration amplitudes.

5. Conclusions

A three-axis vibration energy harvester is developed based on a novel architecture with piezoelectric crab-legs and partitioned surface electrodes. The presented device is microfabricated with an advanced piezoelectric MEMS technology, and experimentally verified with results matching FEA simulations. The multi-axis MEMS harvester can harvest 0.3 to 10.5 μW from input vibrations in the X, Y or Z spatial directions with 0.1 to 0.5 g acceleration amplitude at its modal frequencies (430-850 Hz). Although the power density (0.2-2.2 $mW/cm^3/g^2$) is lower than the values reported for high-performance single-axis microfabricated cantilever beam energy harvesters (6-10 $mW/cm^3/g^2$) [8], the presented device is advantageous for its multi-axis and wide bandwidth operation capability. Future work will focus on decreasing the resonance frequency of in-plane modes, and increasing the power output via optimization of the beam structure for larger in-plane structural flexibility.

Acknowledgement

This research project is partially supported by DARPA PASCAL award #W31P4Q-12-1-0002. The fabrication process was performed at the University of Michigan's Lurie Nanofabrication Facility.

References

[1] H. Wu, L. Tang, Y. Yang, C. K. Soh, *SPIE Active and Passive Smart Structures and Integrated Systems*, vol. 9057, 905703, 2014.
[2] S. Hashimoto, et al., *IEEE Int. Symp. Industrial Electronics (ISIE)*, pp. 1-6, 2013.
[3] J. C. Park, J. Y. Park, *Proc. PowerMEMS*, pp. 281-284, 2011.
[4] A. G. Fowler, S. O. R. Moheimani, S. Behrens, *Proc. PowerMEMS*, vol. 476, 012002, 2013.
[5] H. Liu, et al., *J. Micromech. Microeng.*, vol. 22, 125020, pp. 1-11, 2012.
[6] E. E. Aktakka, K. Najafi, *IEEE Int. Conference on MEMS*, pp. 1141-1144, 2015,
[7] E. E. Aktakka, K. Najafi, *IEEE Int. Conference on MEMS*, pp. 243-246, 2016.
[8] E. E. Aktakka, R. L. Peterson, K. Najafi, *Proc. PowerMEMS*, pp. 139-142, 2011.

Vibrational micro-energy harvesters utilizing Nb-doped $Pb(Zr,Ti)O_3$ films on stainless steel substrates

L Van Minh[1], T Sano[2], T Fujii[2] and H Kuwano[1]

[1]Graduate School of Engineering, Tohoku University, Sendai, Japan
[2]Fujifilm Corp., 577 Ushijima, Kaisei-machi, Kanagawa 258-877, Japan

minhlv@nanosys.mech.tohoku.ac.jp

Abstract. This work presents the micromachined energy harvesters using Nb-doped $Pb(Zr,Ti)O_3$ (PNZT) films grown directly on the stainless steel substrates (SUS430). Piezoelectric materials on metallic substrates have been attracted to practical and robust energy harvesters. Nb-doped PZT films with (001)-preferred orientation grown on SUS substrates provided excellent properties for energy harvesting – high piezoelectric coefficient (e_{31} = -10.6 C/m^2) and low dielectric permittivity (ε_r = 373). The PNZT-based micro-energy harvester comprising a cantilever of 1.7 mm × 5 mm × 0.05 mm and a proof mass of 3 mm × 5 mm × 0.47 mm achieved the normalized power density (NPD) of 2.87 $mW.g^{-2}.cm^{-3}$. It is the highest performance among the published SUS-based energy harvesters, being closer to the best Si-based energy harvesters.

1. Introduction

Vibrational energy harvesters providing electrical power supply to wireless sensor nodes have become compelling issues. Energy harvesters have been developed from piezoelectric materials grown on silicon substrates [1-3]. The silicon-based energy harvesters have achieved substantial progress in term of output power density [1,2]. Minh *et al.* developed the high performance piezoelectric micro-energy harvesters that piezoelectric films grown directly on Si substrates [1,2]. The lead-free (Mg,Zr)-doped AlN based micro-energy harvesters provided the NPD at level of 4 $mW.g^{-2}.cm^{-3}$[1]. However, from practical perspectives of many applications, silicon-based energy harvesters have to face the robustness when exploiting ambient vibrations in low frequency region. Working in low frequency and magnitude vibrations requires that structural materials are capable of carrying heavy weight. Metallic substrates that excel in mechanical properties such as large fracture toughness and high elastic strain limit provide viable solution to the issue. Unfortunately, low piezoelectricity on metallic substrates and difficulties in micromachining contribute to low performance of metal-based energy harvesters [4-7]. Cao *et al.* had used AlN grown on SUS to fabricate piezoelectric macro-energy harvesters with the NPD of 40-400 $\mu W.g^{-2}.cm^{-3}$ [5,6]. These performances are much lower than those of Si-based harvesters. To address the problems, in this work, we developed the piezoelectric micro-energy harvesters that exploited the highly piezoelectric Nb-doped PZT films grown directly on SUS substrate and utilized electrochemical technique for micro-machining the metallic substrates.

2. Fabrication of Nb-doped PZT/SUS based micro-energy harvesters

PNZT films were deposited on SUS substrates with Ir/Ti seed layer by an RF sputter (Model STV4320, Shinko Seiki) with Ar/O_2 gasses and $Pb_{1.3}(Zr_{0.46}Ti_{0.42}Nb_{0.12})O_3$ target. PNZT films were synthesized at the condition – RF power of 200 W, gas pressure of 0.5 Pa with Ar + 2.5% O_2, and substrate temperature of 525 °C [8]. Figure 1 shows X-ray diffraction patterns of the as-deposited PNZT films. The films had highly preferred (001)-PNZT orientation. PNZT films provided high piezoelectric coefficient of 10.6 [8] and low dielectric permittivity of 373. Therefore, we exploited these attractive properties to develop energy harvesters.

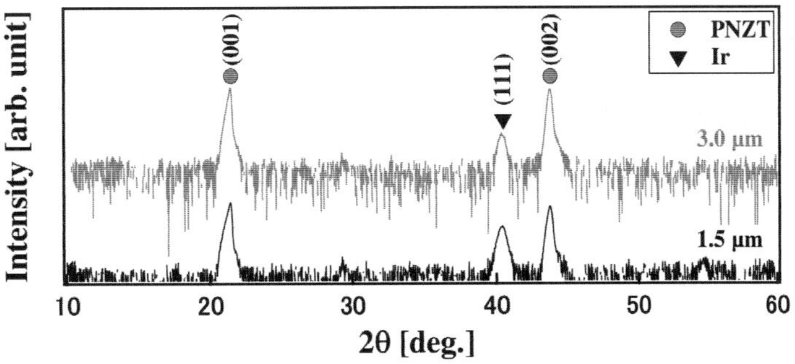

Figure 1. X-ray diffraction patterns: PNZT films on SUS of 1.5 µm (black), and 3 µm (red).

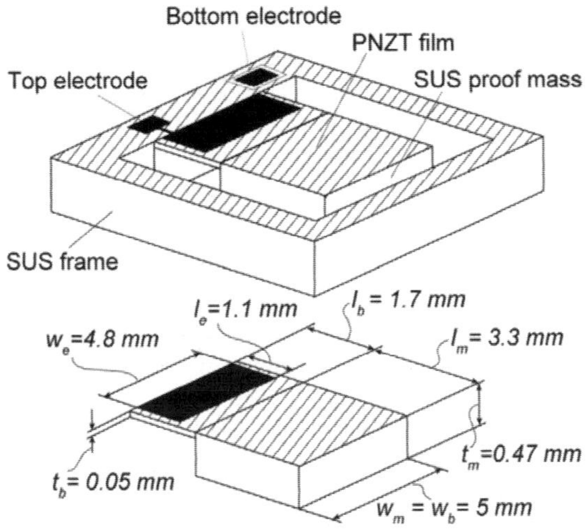

Figure 2. Illustration of PNZT/SUS based micro energy harvester with the geometric parameters.

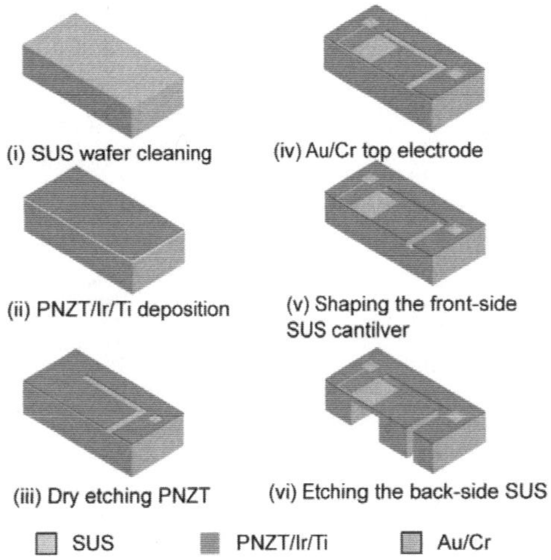

Figure 3. Fabrication flow of PNZT/SUS micro-energy harvesters.

The 3-µm PNZT films on SUS430 substrates of 470 µm were utilized for fabricating the energy harvesters that were composed of a cantilever and a proof mass. The geometric parameters of the harvesters are presented in figure 2. The fabrication flow is presented in figure 3. At first, PNZT film was etched to form a piezoelectric element on the cantilever by using fast atomic beam (FAB) etching technique with SF_6 gas (Figure 3 (ii)). The PNZT etching rate was 14 nm/min on average under the same condition as that in our previous study [2]. Next, Au(200 nm)/Cr(10 nm) thin films were deposited on the wafer and were etched by Au and Cr wet etchants, respectively, to form the top electrode and the open contact window (Figure 3 (iii)). In order to form the SUS cantilever, we etched SUS from the front-side wafers by H_3PO_4-based electrochemical etching (Figure 3(iv)). The electrochemical setup included an electrolyte of H_3PO_4 (99.8%), a stainless steel cathode, and an anode attached by our wafer. The SUS etching rate was approximately 1.00 µm/min at the applied voltage of 4.5 V and the electrolyte current of 0.4 A. The SUS etching depth was the same as the desired cantilever's thickness. Finally, the backside of the SUS wafer was also electrochemically etched to form the proof mass and the desired SUS cantilever (Figure 3(v)). The photographic and scanning-electron-microscopic images of the fabricated PNZT/SUS energy harvester are exhibited in figure 4.

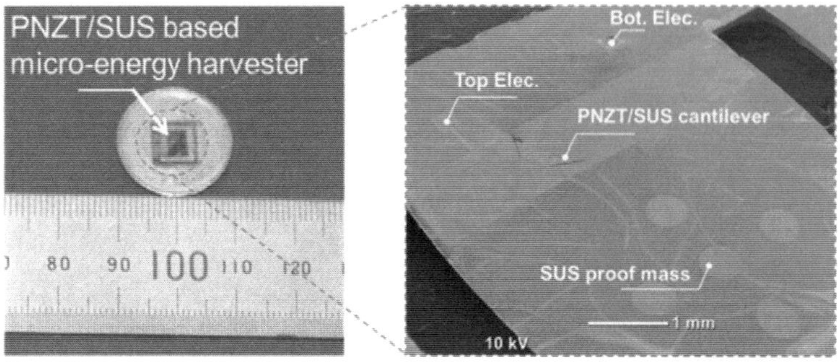

Figure 4. The photographic and scanning electron microscopic (SEM) images of the PNZT/SUS based micro-energy harvester.

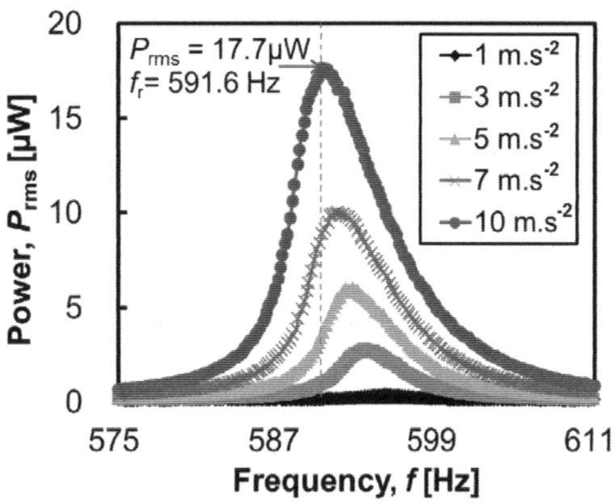

Figure 5. Output power spectra of the PNZT/SUS based micro-energy harvester at the optimal resistance of 49.7 kΩ.

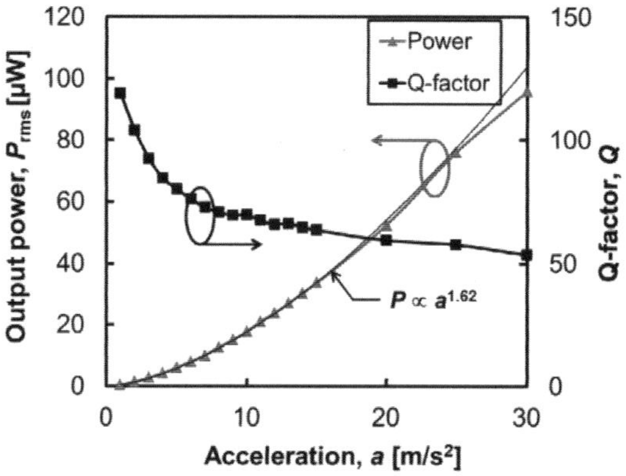

Figure 6. Output power and Q-factor ($Q = f_r/\Delta f$, *Δf is a bandwidth of a power spectrum*) as function of the input acceleration.

3. Results and Discussions

The fabricated PNZT/SUS based energy harvester was evaluated power characteristics using the resistive connection for matching impedance and the shaker for controlling the harmonic acceleration. Figure 5 displays the power spectra of the PNZT/SUS based energy harvester under various accelerations. The harvester had the weakly-softening effect, resonating at 591.6 Hz with the maximum output power of 17.7 μW at the optimal resistance of 49.7 kΩ and the acceleration of 10 $m.s^{-2}$. For ideal linear harvester, output power is proportional to squared acceleration, a^2. Figure 6 shows that the PNZT/SUS based harvester depended on $a^{1.63}$. When increasing the acceleration, the device had the softening effect and the degraded Q-factor led to lowering the output power. The Q-factor decreased from 120 to 54 when the input acceleration increased from 1 $m.s^{-2}$ to 30 $m.s^{-2}$. However, the output power achieved up to 96 μW at the acceleration of 30 $m.s^{-2}$ (Figure 6).

For comparison, performance of the state-of-the-art piezoelectric micro-energy harvesters has been expressed on the NPD-frequency graph including the iso-energy contours (Figure 7). The device achieved the NPD of 2.87 $mW.g^{-2}.cm^{-3}$ that was 7-fold those of KNN/SUS harvester and AlN/SUS harvesters [5-7]. And the NPD of PNZT/SUS based harvester is comparable to those of AlN/Si and

KNN/Si based energy harvesters [1,2,9]. This result of this work is superior to those of the metallic energy harvesters and approaches closely towards the best Si-based energy harvesters.

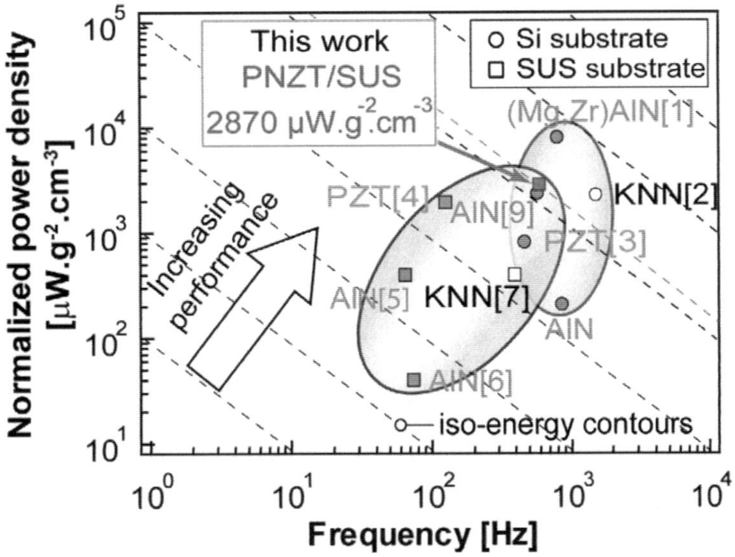

Figure 7. Comparison of the state-of-the-art piezoelectric energy harvesters for both Si- and SUS-based development ($g = 9.89\ m.s^{-2}$).

4. Conclusions

This work presents the Nb-doped PZT/SUS based micro-energy harvester. PNZT films were grown directly on SUS substrates with highly preferred (001) orientation. The micromachining process was established for fabricating PNZT/SUS micro energy harvester. The output power was up to 96 μW at the acceleration of 30 $m.s^{-2}$. The micromachining harvester was confirmed with the high normalized power density of 2.87 $mW.g^{-2}.cm^{-3}$. This achievement was 7-fold increase compared to those of KNN/SUS and AlN/SUS based energy harvesters, being comparable to those of KNN/Si and AlN/Si based energy harvesters. The achievement opens a door for high performance micro energy harvesters exploiting piezoelectric films on metallic SUS substrates and leading to practical applications in which energy harvesters based on Si-platforms are impractical.

References

[1] Van Minh L, Hara M, Yokoyama T, Nishihara T, Ueda M and Kuwano H 2015 *IEEE Trans. Ultrason., Ferroelect., Freq. Control* 62 11 2005-2008
[2] Van Minh L, Hara M, Horikiri F, Shibata K, Mishima T and Kuwano H 2013 *J. Micromech. Microeng.* **23** 035029
[3] Shen D, Park J H, Ajitsaria J, Choe S Y, Wikle H C and Kim D J 2008 *J. Micromech. Microeng.* **18** 055017
[4] Morimoto K, Kanno I, Wasa K and Kotera H 2010 *Sensors and Actuators A* **163** 428-432
[5] Cao Z, He J, Wang Q, Hara M, Oguchi H and Kuwano H 2013 *J. Phys.: Conf. Ser.* **476** 012034
[6] Cao Z, Zhang J and Kuwano H 2012 *Sensors and Actuators A* **179** 178-184
[7] Tsujiura Y, Suwa E, Kurokawa F, Hida H, Suenaga K, Shibata K and Kanno I 2013 *Jpn. J. Appl. Phys.* **52** 09KD13
[8] Fujii T, Hishinuma Y, Mita T and Naono T 2010 *Sensors and Actuators A* **163** 220-225
[9] Elfrink R, Kamel T M, Goedbloed M, Matova S, Hohlfeld D, van Andel Y and van Schaijk R 2009 *J. Micromech. Microeng.* **19** 094005

A handy motion driven hybrid energy harvester: dual Halbach array based electromagnetic and triboelectric generators

M Salauddin, J Y Park[1]
Micro/Nano Devices and Packaging Lab., Dept. of Electronic Engineering, Kwangwoon University, 447-1 Wolgye-dong, Nowon-gu, Seoul, 139-701 Korea.

E-mail: jaepark@kw.ac.kr

Abstract. In this work, we have proposed and experimentally validated of hybrid electromagnetic and triboelectric energy harvester using dual Halbach magnets array excited by human handy motion. Hybrid electromagnetic (EM) and triboelectric (TE) generator that can deliver an output performance much higher than that of the individual energy-harvesting unit due to the combination operation of EM and TE mechanisms under the same mechanical movements. A Halbach array concentrates the magnetic flux lines on one side of the array while suppressing the flux lines on the other side. Dual Halbach array allows the concentrated magnetic flux lines to interact with the same coil in a way where maximum flux linkage occurs. When an external mechanical vibration is applied to the hybrid structure in the axial direction of the harvester, the suspended mass (two sided dual-Halbach-array frame) starts to oscillate within the magnetic springs and TEG part. Therefore, the TEG part, the Al film and microstructure PDMS film are collected into full contact with each other, generating triboelectric charges due to the various triboelectricities between them. A prototype of the hybrid harvester has been fabricated and tested. The EMG is capable of delivering maximum 11.5mW peak power at 32.5Ω matching load resistance and the TEG delivering 88μW peak power at 10MΩ load resistance.

1. Introduction

Extracting energy from mechanical vibration has drawn much remark over the last few decades due to its sufficiency in nature and unlimited lifetime. Various vibration sources, such as water and wind flow, rotary motion, human and machine motion generate vibrations of different frequencies and amplitudes, but most vibrations are of low frequencies and large amplitudes with various cyclic movements in different directions. Widely used techniques for harvesting energy from mechanical vibrations are electromagnetic [1], electrostatic [2] and piezoelectric [3] mechanisms. The electromagnetic generator can also work from the relative movement between the magnet and the coil. Also use of Halbach magnet array in an electromagnetic energy harvester, instead of a single magnet, increases the magnetic flux density which, in turn, partially addresses the power generation issues at low frequencies [4-5]. Behaviours of handy motion show that it has very low frequency (~5Hz) and high acceleration (~2g peak) [6]. The electrostatic mechanism is based on repeated charge pumping with variable capacitors, while the piezoelectric mechanism is based on the reconfiguration of unbalanced dipole moments. Newly, the triboelectric generator (TEG) has been highlighted because of its high power and low fabrication cost. The TENG can scavenge the mechanical energy from the contact/separation between two triboelectric materials. In spite of the continuous improvement of small-scale mechanical energy harvesters, the level

[1] To whom any correspondence should be addressed.

of output energy still need to be improved to meet the requirements of commercial electronic systems and to further expand their fields of application. By combining two types of mechanical-energy-harvesting units, more electricity can be extracted from one mechanical motion, which may meet the power needs of some portable electronic devices. A hybrid mechanical energy harvesting technology may enhance the overall power output [7]. Zhang et al. proposed contact-separation mode hybrid energy harvester with implanted planar coil, which is proper to generate strong instantaneous power [8]. In this work, we have proposed and experimentally validated of hybrid electromagnetic and triboelectric energy harvester using dual Halbach magnets array excited by human handy motion. This paper is mainly focused on Halbach array based hybrid electromagnetic and triboelectric energy harvester. A Halbach array is an arrangement of magnets which increases the magnetic field in one side of the array, minimizing the magnetic field to nearly zero on the other side of the array. Hybrid electromagnetic (EM) and triboelectric (TE) generator that can deliver an output performance much higher than that of the individual energy-harvesting unit due to the combination operation of EM and TE mechanisms under the same mechanical movements.

2. Hybrid energy harvester design and fabrication

In this work, with a reasonable design, the electromagnetic generator (EMG) and triboelectric generator (TEG) are integrated in a suspended structure by magnetic springs as a hybrid cell in Figure 1. The EMG consists of the two dual Halbach array and bobbin containing two coils (each 200 turns) connected in series, whereby the structure of the dual Halbach array is suspended by magnetic springs (Ø6×2 mm^2). Each Halbach array is formed using 3 block (10×5×2mm^3) magnets. Dual array structure is made by attaching the Halbach array to a hollow rectangular shape polycarbonate structure. Each TEG made of two Al films and a microstructure PDMS film. The micro-structured PDMS films were fabricated by using a Si master with the surface of pyramid structures. The mixture of PDMS elastomer and cross-linker with a 10:1 ratio (w/w) was spin-coated on to the Si master. The PDMS film with the surface pyramid micro-structures were peeled off from the Si master after it was cured at 353K for 1 hour in an oven. Dual Halbach array allows the concentrated magnetic flux lines to interact with the same coil in a way where maximum flux linkage occurs. When an external mechanical vibration is applied to the hybrid structure in the axial direction of the harvester, the suspended mass (two sided dual-Halbach-array frame) starts to oscillate within the magnetic springs and TEG part. Therefore, the TEG part, the Al film and microstructure PDMS film are collected into full contact with each other, generating triboelectric charges due to the various triboelectricities between them. During experiment, vibration was applied in horizontal direction in order to reduce the gravity effect on the Halbach array frame structure. The FEM simulation allows us to determine the exact positions of both Halbach arrays, coil positions and the spring magnets in figure 2. Figure 2 shows the results of FEMM simulation of the corresponding dual Halbach array generator structure of figure 1, showing magnetic flux density at 1.5mm distance is 0.82T.

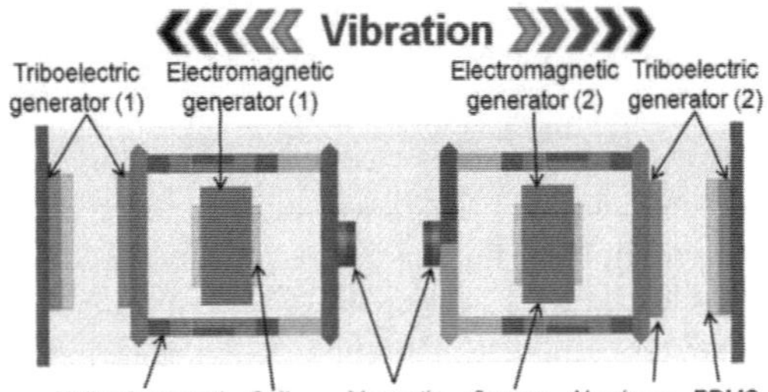

Figure 1. Schematic structure of the proposed hybrid energy harvester.

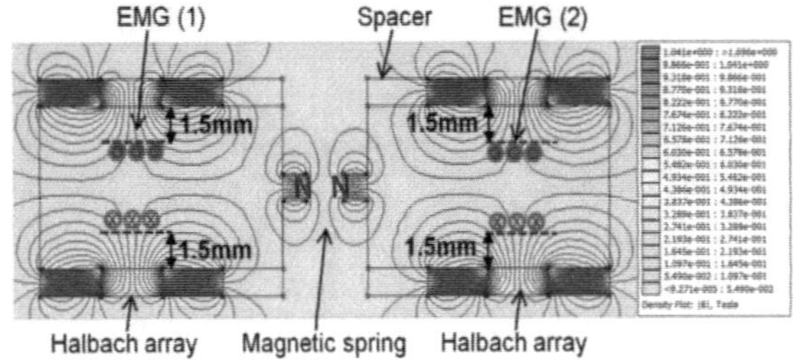

Figure 2. Finite element simulation of magnetic flux lines for both Halbach array and spring magnets by FEMM.

Journal of Physics: Conference Series **773** (2016) 012004 doi:10.1088/1742-6596/773/1/012004

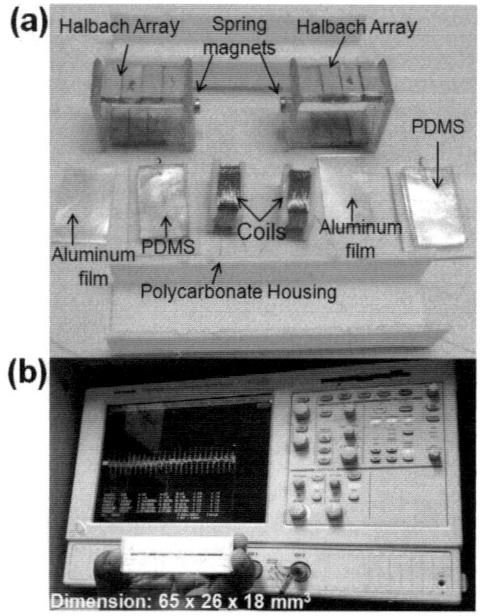

Figure 3. Photographs of the hybrid energy harvester: (a) Components before assembling and (b) fabricated prototype and handy motion test setup.

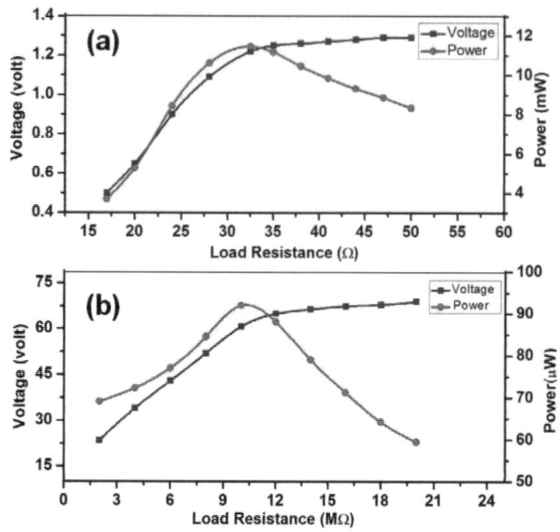

Figure 4. Load voltages (peak-peak) and powers (peak) versus load resistances generated by (a) electromagnetic (series) (b) triboelectric generator (series).

The overall dimensions of the harvester prototype are 6.5 cm ×2.6 cm × 1.8 cm. Figure 3(b) shows that the fabricated prototype after assembling the components; the prototype is small, compact in size, and convenient as it can be held between two fingers.

3. Experimental results and discussions

We proposed testing our hybrid energy harvester prototype under a very low frequency range that is generated by a human-body-induced vibration such as human hand motion. The output voltage signals of the electromagnetic generator were measured by oscilloscope and triboelectric generator were measured a low-noise voltage preamplifier (Keithley 6514 system electrometer). All the components were assembled to make the prototype hybrid energy harvester and tested by handy motion to observe its output performance in figure 3. The hybrid EMG delivers 11.5mW peak power to a 32.5Ω matched load resistance and TEG delivers 88μW peak power at 10MΩ load resistance in figure 4. Figure 5 shows the instantaneous voltage waveform across the optimum load and the peak voltages of EMG and TEG are 1.22V and 60V, respectively. Figure 6 shows instantaneous peak power waveform across the optimum load resistance of electromagnetic generator and triboelectric generator. However, slight variations were observed between the values for the generated open-circuit voltage, and the power and value predictions of the simulation; these variations might be due to the calculation variations because of magnetic-field values, magnetic-flux densities, coil position, effect of coil inductance, phase difference, fabrication of PDMS film etc. and the improper assembly of the harvester components. The dual-Halbach-array based hybrid generator presented here showed an ability to generate meaningful power from low-frequency fitful behavior. The benefits of using magnetic springs in this dual Halbach-array structure include reliability, a simple structure, and an ease of use at low-frequency vibrations. Besides, the EMG component generated low voltage and high current while the TENG component generated high voltage and low current, according to their inherent output resistances. By combining both components in the hybrid one-body device, the advantages of both could be fully utilized without energy loss. The proposed hybrid harvester performance can be more improved by further optimization.

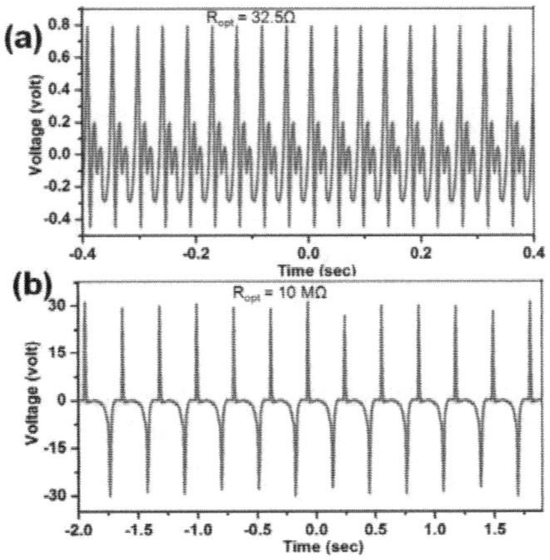

Figure 5. Instantaneous voltage waveform across the optimum load resistance: (a) electromagnetic (series) (b) triboelectric generator (series).

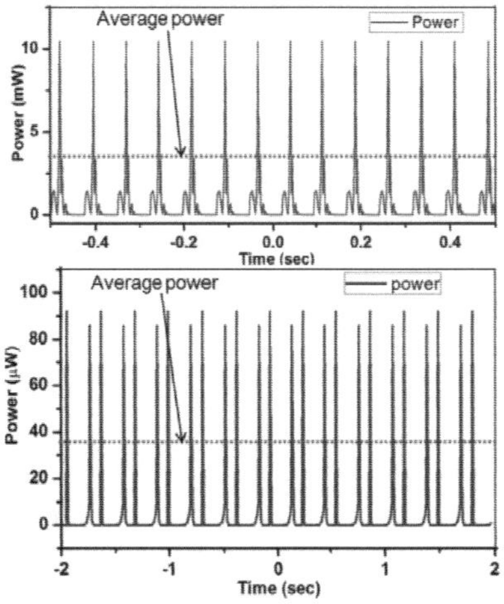

Figure 6. Instantaneous peak power waveform across the optimum load resistance: (a) electromagnetic (series) (b) triboelectric generator (series).

4. Conclusions

A Halbach array based hybrid energy harvester was designed, fabricated, and tested for scavenging significant power from low-frequency vibrations. The main goal of the work was to develop, analyze, and implement a suitable and reliable hybrid energy harvester to scavenge significant power from human hand motions such as hand-shaking which would be able to supply power to portable and wearable smart devices. The use of the dual Halbach array allows the concentrated magnetic flux to interact with the same coil, resulting in the increment of the electromotive force, which in turn increases the power generation. The proposed dual Halbach arrays based hybrid energy harvester device has been designed, fabricated, and tested. The electromagnetic generator is capable of delivering maximum 11.5mW peak power at 32.5Ω matching load resistance and the triboelectric generator delivering 88μW peak power at 10MΩ load resistance. The next stage of this work will include further optimization of the prototype to be used in energy harvesting from different human-body-induced motions such as walking, running, cycling etc. The advantages of our proposed device include its simple structure, low operating conditions and low fabrication cost.

Acknowledgements

The authors are grateful to acknowledge the Basic Science Research Program (2013R1A1A2A 10064810) through the National Research Foundation of Korea (NRF) funded by the Ministry of Science, ICT and Future Planning, Korea.

References

[1] Salauddin M, Halim M A, and Park J Y, 2016 *Smart Mater. Struct.* **25** 095017
[2] Nguyen S D and Halvorsen E, 2011 *J. Microelectromech. Syst.* **20** 1225-1227
[3] Halim M A, Khym S, and Park J Y, 2011 *J. appl. Phys* **114** 044902
[4] Liu H, Qian Y, Wang N, and Lee C, 2013 *Proc. MEMS* 829-832
[5] Wang Y J, Chen C D, Sung C K, and Li C, 2012 *J. Intel. Mat. Syst. Struct.* **23** 933-943
[6] Halim M A and Park J Y, J. Phys. 2013 *Conf. Ser.* **476** 012119
[7] Hu Y, Yang J, Niu S, and Wu W, 2014 *ACS Nano* **8** 7442-7450
[8] Zhang K, Wang X, Yang Y, Wang Z L, 2015 *ACS Nano* **9** 3521

Electrically conductive fabric based stretchable triboelectric energy harvester

Rubaiyet I. Haque[1,2], Pierre-André Farine[2] and Danick Briand[1]

[1] Microsystems for Space Technologies Laboratory, École Polytechnique Fédérale de Lausanne, Neuchâtel, Switzerland
[2] Electronics and Signal Processing Laboratory, École Polytechnique Fédérale de Lausanne, Neuchâtel, Switzerland

E-mail: rubaiyet.haque@epfl.ch

Abstract. Stretchable conductive fabric-based triboelectric generator (TENG), to develop breathing/chest band for harvesting energy at low frequency has been developed. Stretchable conductive nylon-fabric and carbon-based elastomer composites were used as electrodes. During this work, film casting technique was implemented and combination of different materials, such as, polydimethylsiloxane (PDMS) and polytetrafluoroethylene (PTFE)/ polyurethane (PU) were tested as triboelectric layers. The process was compatible with large scale fabrication. At low operation frequency of 1.0±0.1 Hz for the strain of 13±1.5%, developed TENGs provide output power densities of 0.06 μW/cm^2 and 0.11 μW/cm^2 for the load resistance of 100 MΩ, and energy density of 0.19±0.03 nJ/cm^2/cycle and 0.08±0.01 nJ/cm^2/cycle for the use of capacitor of 2.2 μF, for PDMS-PTFE and PDMS-PU based TENGs respectively.

1. Introduction

Triboelectric effect is known for long time mainly for its negative effects in electronics industries. In recent time, triboelectric energy harvesting, which uses the contact electrification effect and electrostatic induction method [1], have attracted extensive attention to harvest energy from mechanical movements or vibrations. In addition, with the fast development of wearable electronics, there is a strong need for fully compliant wearable, stretchable energy harvesters for users' comfort.

Several advanced next generation applications, such as smart textiles, wearable biomedical systems, and smart sensory skin for robotics, require energy harvester with adaptability to different surface topology and mechanical durability to be highly sustainable and to have capability to withstand high deformations, various movements, twisting and stretching. Flexibility and stretchability provide conformability and great deformability without compromising device performance and higher thus the reliability. Such triboelectric generators (TENGs) can be used to harness energy from human elbow, knee and breathing movements. Working principle of stretchable TENG is mainly based on sliding mode, whose output depends heavily on operational frequency and change in overlapping areas of two dielectric surfaces during sliding. Different stretchable TENGs have been reported [2]–[5], but their geometry and fabrication processes were relatively complex and not scalable, as in most cases individual fibre structures are fabricated, nanotextured and then weaved to obtain TENGs.

Here, we are presenting stretchable lateral-sliding mode TENGs manufactured using a film casting approach and having simpler design. Stretchable conductive composite fabric, dielectric elastomers and conductive elastomer composites were implemented during this work to develop the stretchable TENGs with an easier process compatible with large scale fabrication techniques.

2. Experimental procedure

2.1. Materials and design

The proposed design of the stretchable lateral-sliding mode TENGs is composed of two triboelectric layers, assembled in contact, facing each other, as illustrated in Figure 1. Outer layer is stretchable, whereas the inner layer is flexible but non-stretchable. Stretchable conductive fabric and carbon-based elastomer composites were used as electrodes. In this work, Polydimethylsiloxane (PDMS), polyurethane (PU) and polytetrafluoroethylene (PTFE) were used as triboelectric layers. The geometrical dimensions of the sliding mode TENG-bands are mentioned in Table 1.

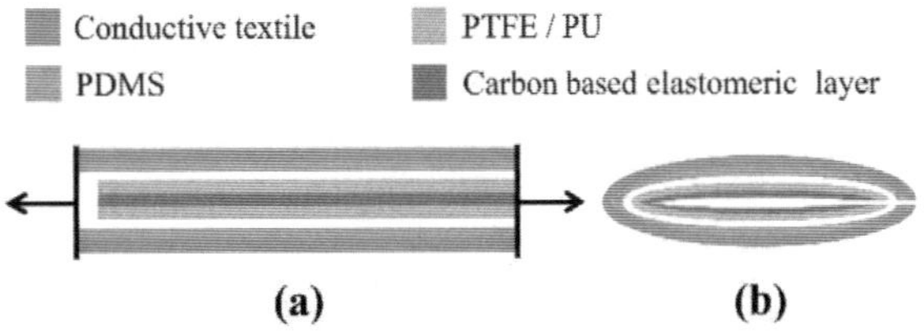

Figure 1. Schematic diagram of stretchable triboelectric generator along the (a) length and (b) width directions.

Table 1. Dimensions of the stretchable TENG-band

Active area	:	72 cm^2
Thickness of PDMS layer	:	80 μm
Thickness of PU/PTFE layer	:	50 μm

Conductive silver-plated nylon fabric (MedTex180) was supplied by Shieldex productions & Vertriebs GmbH. PDMS (Sylgard 186) and PU (MM4520) were purchased from DowCorning and SMP Technology Inc respectively, and PTFE film having thickness of 50 μm was collected from online supplier (Angst+Pfister). Carbon black pellets (Ketjenblack EC-300J & Ketjenblack EC-600JD) that has been used to prepare electrodes were purchased from Akzo-Nobel. Solvents were purchased from Sigma-Aldrich. All products were used as received. Film casting of elastomers were performed on commercially available polyethylene terephthalate (PET) film.

2.2. Process flow

The fabrication process of stretchable TENG-bands was comprised of preparation of two triboelectric layers, one was stretchable and other was flexible, along with electrodes and then assembly. Starting with the casting and curing of PDMS layer on PET substrate along with sacrificial layer; the PDMS layer was then transferred to electrically conductive fabric using elastomeric adhesive to form stretchable triboelectric layer with electrode.

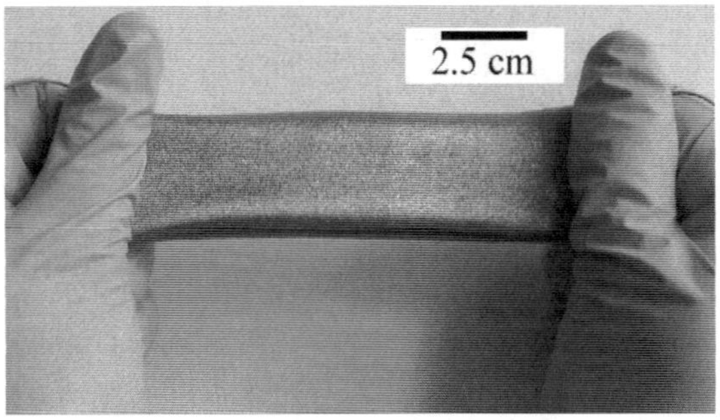

Figure 2. Stretchable triboelectric generator.

In addition, two different flexible but non-stretchable second triboelectric layers were prepared using PTFE and PU materials. In case of PTFE, conductive carbon-based PDMS layer was film casted on PTFE film and cured. Triboelectric PU layer is produced by film casting of dissolved PU layer on casted and cured carbon-based conductive PU layer. At the final stage, layers were assembled by placing triboelectric layers facing each other in surface contact, to form two different stretchable sliding-mode based TENGs with PDMS-PFTE and PDMS-PU triboelectric layer combinations (Figure 2).

The electrical responses of the stretchable TENG-bands were performed via oscilloscope through electronics circuit presented in Figure 3(a). Amount of generated energy of the TENGs were also measured using electronics circuit (Figure 3(b)), where a capacitor (C_L) of 2.2 μF was used. All the tests were performed for the strain of 13±1.5% and the frequency of 1.0±0.1 Hz.

3. Results and discussion

Lateral sliding-mode based stretchable TENGs were fabricated, where PDMS, PU and PTFE were tested as triboelectric layers and stretchable conductive fabric and carbon-elastomer composites were used as electrodes. During stretching, PDMS layer elongated and sliding between PDMS and other triboelectric layer take place. The change of overlapping surface area generates the electrical energy due to triboelectric effect. The performances of PDMS-PTFE and PDMS-PU based TENGs were tested using electronics mentioned in Figure 3(a) and 3(b).

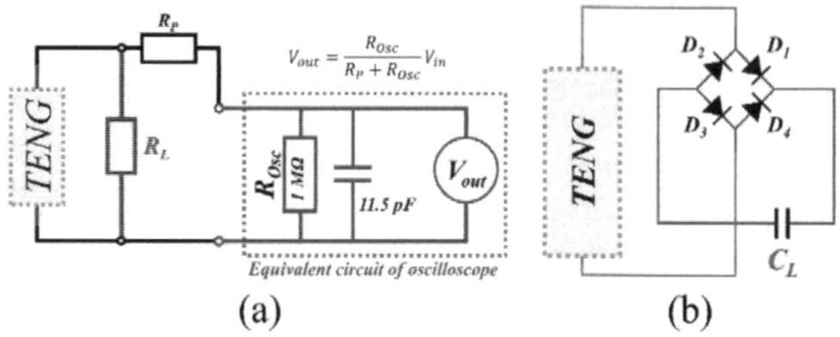

Figure 3. Schematic diagram of the measurement technique (a) to characterize the electrical response and (b) circuit used to store energy/charge generated by TENG based shoe insole during walking.

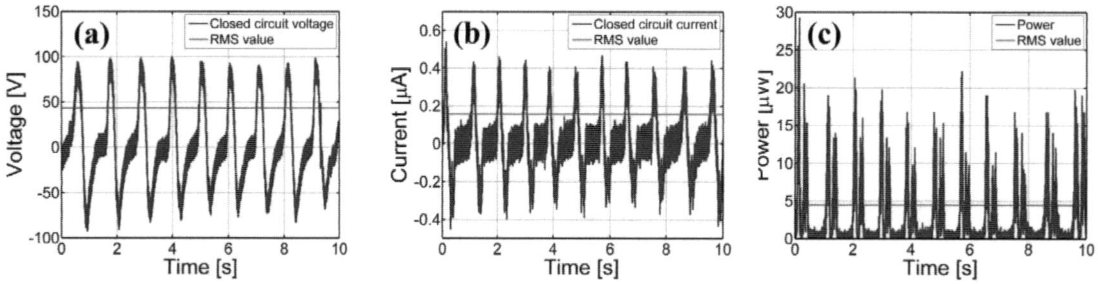

Figure 4. Electrical response of the PDMS-PTFE based stretchable TENGs. (a) open circuit voltage; (b) closed circuit current for the load resistance of 100 MΩ; (c) power generated for closed circuit measurement for the load resistance of 100 MΩ.

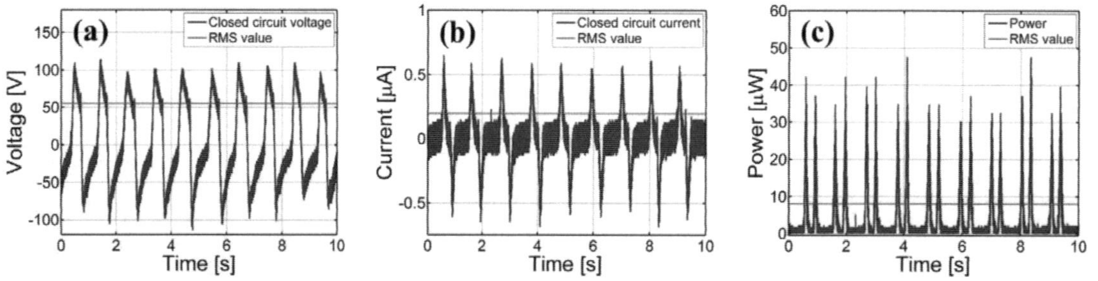

Figure 5. Electrical response of the PDMS-PU based stretchable TENGs. (a) open circuit voltage; (b) closed circuit current for the load resistance of 100 MΩ; (c) power generated for closed circuit measurement for the load resistance of 100 MΩ.

Figures 4 and 5 present the electrical response due to cyclic stretching and relaxing by PDMS-PTFE and PDMS-PU TENG-bands at 1.0±0.1 Hz. According to the observation, for the load resistance of 100 MΩ, PDMS-PTFE based TENG generated open circuit voltage (V_{OC}, rms value) of 43.5 V and closed circuit current (I_{CC}, rms value) of 0.16 μA, which corresponded to the power and power density of 4.5 μW and 0.06 μW/cm^2 respectively. Whereas, PDMS-PU based TENG provided V_{OC} (rms value) of 55 V and I_{CC} (rms value) of 0.19 μA for the load resistance of 100 MΩ, that represented, respectively, the power generation and power density of 8.1 μW and 0.11 μW/cm^2.

Moreover, the amount of charge transfer that occurred during stretching and amount of energy of TENGs were measured using capacitor of 2.2 μF. For PDMS-PTFE based TENG, the achieved amount of electrons transfer in a single charge-transfer process was 0.24±0.02 μC, that corresponded to the surface charge density of 3.35±0.31 nC/cm^2 and energy density of 0.19±0.03 nJ/cm^2. Regarding the other typology, PDMS-PU based TENG provided the surface charge density of 1.65±0.15 nC/cm^2 and energy density of 0.08±0.01 nJ/cm^2 per cycle.

It has been observed that, PDMS-PTFE based TENG provided better performance as lateral-sliding mode based triboelectric energy harvester. The high surface stickiness between PDMS and PU triboelectric layers induced higher friction between layers and thus performing sliding-mode of PDMS-PU based TENG was relatively difficult, which might have led to the reduced performance of PDMS-PU based TENG in comparison with PDMS-PTFE based TENG.

4. Conclusions

In summary, the stretchable sliding-mode based TENGs, with simple design, have been reported. These TENGs can be used to generate energy due to stretching while breathing, keen and elbow movements, and can be developed using simple, economic and scalable fabrication method, like, film casting method. As demonstrated, the developed TENGs were capable of generating energy at low stretching frequency of 1±0.1 Hz for the strain of 13±1.5%.

In comparison to the PDMS-PU based TENG, PDMS-PTFE based TENG perform smoothly due to less friction between triboelectric layers, and provided power density of 0.06 μW/cm^2 for load resistance of 100 MΩ, and charge density of 3.35±0.31 nC/cm^2 and energy density of 0.19±0.03 nJ/cm^2 for capacitor of 2.2 μF during single stretching operation. On the contrary, PDMS-PU based TENG delivered power density of 0.11 μW/cm^2 for load resistance of 100 MΩ, and charge density of 1.65±0.15 nC/cm^2/cycle and energy density of 0.0.8±0.01 nJ/cm^2/cycle for the use of capacitor of 2.2 μF in the circuit during stretching.

However, further investigating is required to explore suitable materials and designs to reduce the friction among the triboelectric layers to ease the sliding mechanism, and in the process, improve the capability and fabrication method. In addition, to protect the TENG from humidity and dust, encapsulation procedures will be studied.

Acknowledgements

The authors would like to acknowledge Swiss National Science Foundation (SNSF) and NanoTera.ch for the funding and support of this work within the framework of BodyPowerSenSE project.

References

[1] Wang Z L, Chen J and Lin L 2015 Progress in triboelectric nanogenerators as a new energy technology and self-powered sensors *Energy & Environmental Science* **8(8)** 2250

[2] Lee K Y, Gupta M K and Kim S -W 2015 Transparent flexible stretchable piezoelectric and triboelectric nanogenerators for powering portable electronics *Nano Energy* **14** 139

[3] Lee J -H, Lee K Y, Gupta M K, Kim T Y, Lee D -Y, Oh J, Ryu C, Yoo W J, Kang C -Y, Yoon S -J, Yoo J -B and Kim S -W 2014 Highly stretchable piezoelectric-pyroelectric hybrid nanogenerator *Advanced Materials* **26(5)** 765

[4] Yi F, Lin L, Niu S, Yang P K, Wang Z, Chen J, Zhou Y, Zi Y, Wang J, Liao Q, Zhang Y and Wang Z L 2015 Stretchable-rubber-based triboelectric nanogenerator and its application as self-powered body motion sensors *Advanced Functional Materials* **25(24)** 3688

[5] Kim K N, Chun J, Kim J W, Lee K Y, Park J -U, Kim S -W, Wang Z L and Baik J M 2015 Highly stretchable 2D fabrics for wearable triboelectric nanogenerator under harsh environments *ACS Nano* **9(6)** 6394

PowerMEMS 2016 IOP Publishing
Journal of Physics: Conference Series **773** (2016) 012006 doi:10.1088/1742-6596/773/1/012006

Magnetically levitated autoparametric broadband vibration energy harvesting

L. Kurmann[1], Y. Jia[2], Y. Manoli[3] and P. Woias[3]

[1]University of Applied Sciences and Arts Northwestern Switzerland, Switzerland
[2]Department of Mechanical Engineering, Thornton Science Park, University of Chester, UK
[3]Department of Microsystems Engineering, IMTEK, University of Freiburg, Germany

E-mail: `lukas.kurmann@fhnw.ch; yu.jia.gb@ieee.org`

Abstract. Some of the lingering challenges within the current paradigm of vibration energy harvesting (VEH) involve narrow operational frequency range and the inevitable non-resonant response from broadband noise excitations. Such VEHs are only suitable for limited applications with fixed sinusoidal vibration, and fail to capture a large spectrum of the real world vibration. Various arraying designs, frequency tuning schemes and nonlinear vibratory approaches have only yielded modest enhancements. To fundamentally address this, the paper proposes and explores the potentials in using highly nonlinear magnetic spring force to activate an autoparametric oscillator, in order to realize an inherently broadband resonant system. Analytical and numerical modelling illustrate that high spring nonlinearity derived from magnetic levitation helps to promote the 2:1 internal frequency matching required to activate parametric resonance. At the right internal parameters, the resulting system can intrinsically exhibit semi-resonant response regardless of the bandwidth of the input vibration, including broadband white noise excitation.

1. Introduction

The mainstream of vibration energy harvesting (VEH) has often relied on oscillators operating in direct resonance in order to accumulate mechanical energy [1]. Such an approach leads to a system that only exhibits resonant response within a specific bandwidth of excitation frequencies. By adjusting the quality factor, widening of this bandwidth would inevitably sacrifice the power peak. Various frequency tuning and frequency broadening strategies have been extensively investigated [2], however, the enhancements to date are still relatively modest and proposed systems are still confined to a particular operational frequency range.

Arraying of either coupled or uncoupled oscillators with varying resonant frequencies can help to accumulate to a relatively wide frequency range, though the power density of the overall system compares unfavorable to a single large oscillator that takes up the whole volume [3]. A multitude of frequency tuning mechanisms, either mechanical or electrical, typically require power input; therefore, it is still limited to single frequency sources that only require intermittent tuning [4]. Duffing nonlinearity has been extensively explored [2, 3], but bandwidth broadening is modest and proposed enhancements towards noise excitations are disputable [5].

Parametric resonance on the other hand is a fundamentally different nonlinear vibratory phenomenon [6] invoked through Mathieu instability and has the potential to exhibit nonlinear broadband behavior [7]. Parametrically excited VEH has been previously demonstrated enhancements in both power peak and the frequency bandwidth [8] over conventional direct resonant VEH. Furthermore, unlike direct resonance, it is possible to simultaneously enhance both power peak and frequency bandwidth with lower damping [9]. However, its activation is non-trivial due to stringent stability bounds and initiation amplitude thresholds [10].

One design route is to establish an autoparametric system where external direct excitation is internally coupled to parametric excitation through a strict 2:1 internal frequency ratio. This has been numerically and experimentally validated [11] with a coupled two degrees of freedom (DOF) oscillatory system at its most basic form, yielding substantially more favorable results than a direct oscillator. Furthermore, the use of strong spring nonlinearity in the first DOF has been suggested as a means to more readily access Mathieu instability [12], while the use of magnetic spring through magnetic levitation [13] is a potential source of intrinsic nonlinearity.

2. Method

2.1. Design

The proposed design (figure 1) consisting of a pair of magnets carrying a pendulum oscillator is magnetically levitated between repulsive magnetic fields from both top and bottom. Figure 2 illustrates the system model where external excitation is directly applied to a strongly nonlinear magnetic spring, which is in turn coupled to the secondary pendulum oscillator.

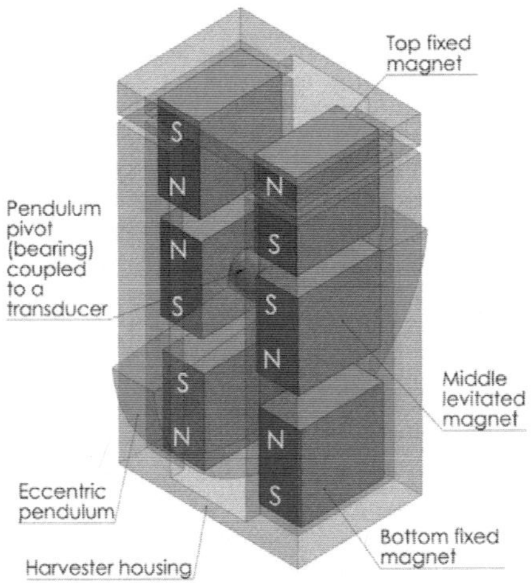

Figure 1. Design of the system with three opposite directed permanent magnet pairs.

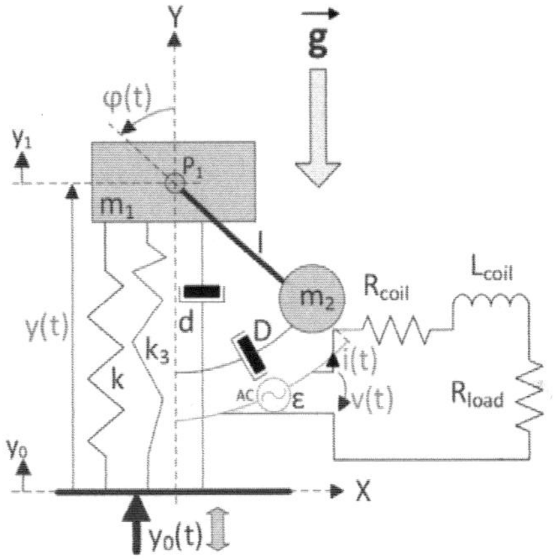

Figure 2. Lumped parameter model with attached transducer shown on the right hand side. k and k_3 are linear and nonlinear springs.

2.2. Model

While magnetic spring restoring force is often reduced to a simple cubic Duffing type nonlinearity added to a linear spring term [13], a more complicated polynomial expression term can be derived from the magnetic dipole interactions [14] as shown in in equation 1.

$$F(t) = -\frac{6\mu_0 M^2}{\pi(r-|y|)^5}y \tag{1}$$

where, r is the total spacing between the magnets at zero displacement, y is the vertical displacement of the magnetically levitated mass, $(r-y)$ is thus the spacing between the magnet dipoles, M is the dipole moment and μ_0 is the relative permeability. Figure 3 compares the restoring force derived from these two approaches as well as a linear spring force. It can be noted that the Duffing approximation holds for small displacements. Therefore, the governing equation of the primary oscillator alone can be approximated by equation 2.

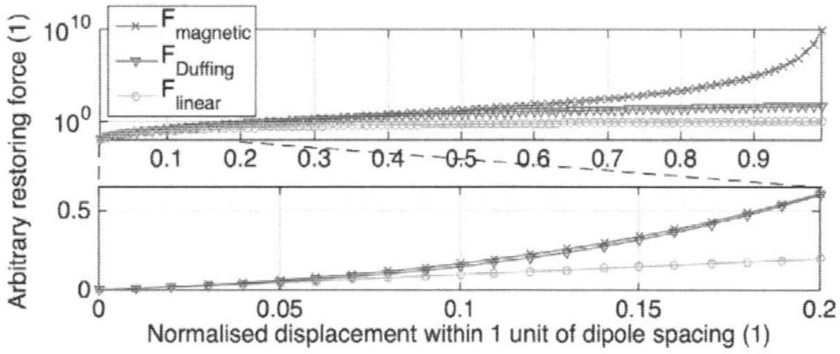

Figure 3. Comparing spring restoring force described by dipole interaction, Duffing nonlinearity and linear spring. Bottom plot is the zoom of the top plot.

$$\ddot{y} + 2\zeta\Omega_0\dot{y} + \Omega_0^2 y + \beta y^3 = A_y\omega^2\cos(\omega t) \tag{2}$$

where, ζ is the damping ratio for this primary oscillator in the y axis, Ω_0 is the natural frequency for this primary oscillator, β is the Duffing coefficient, A_y is the excitation displacement amplitude in the vertical direction, ω is the excitation frequency and t is the time domain.

The response $y(t)$ feeds into the pendulum oscillator, of which the response can thus be modelled by equation 3 (for direct oscillations) and/or equation 4 (for parametric oscillations).

$$\ddot{\varphi} + \gamma\dot{\varphi} + \omega_0^2\sin\varphi = \frac{A_x}{l}\omega^2\cos(\omega t)\cos\varphi \tag{3}$$

$$\ddot{\varphi} + \gamma\dot{\varphi} + \xi\dot{\varphi}|\dot{\varphi}| + [\omega_0^2 - \frac{\ddot{y}(t)}{l}]\sin\varphi = 0 \tag{4}$$

where, φ is the angular displacement of the pendulum, γ is the viscous damping coefficient, ξ is nonlinear damping coefficient, ω_0 is the natural frequency of the pendulum, l is the effective pendulum length and A_x is the horizontal excitation displacement where applicable.

With the onset of parametric resonance in the subsidiary pendulum oscillator, the resulting unidirectional flow of energy limits the oscillatory amplitude growth of the primary levitated oscillator. This is described with the inclusion of a mode coupling term ψ [11] in equation 5.

$$\ddot{y} + 2\zeta\Omega_0\dot{y} + \Omega_0^2 y + \beta y^3 - \psi[\cos(\varphi)\dot{\varphi}^2 + \sin(\varphi)\ddot{\varphi}] = A_y\omega^2\cos(\omega t) \tag{5}$$

Extensive numerical simulations in MATLAB were carried out to better understand the derived systems. The ODE45 Dormand-Prince solver was employed.

3. Results and Discussion

3.1. Spring nonlinearity and frequency response

The internal frequency ratio of the 2-DOF system was matched to 2:1. However, the presence of the displacement-dependent nonlinearity results in the modulation of this ratio during vibration (figure 4), thus promoting the chance of operating around the 2:1 frequency ratio even if for slightly detuned scenarios (figure 5) common as a result of manufacturing tolerance.

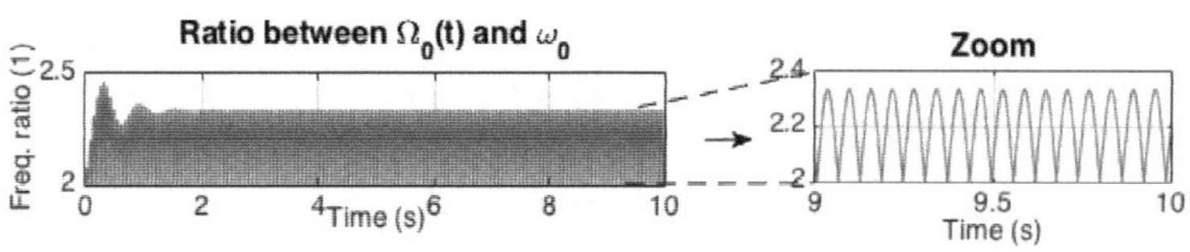

Figure 4. Modulation of internal frequency ratio during oscillation.

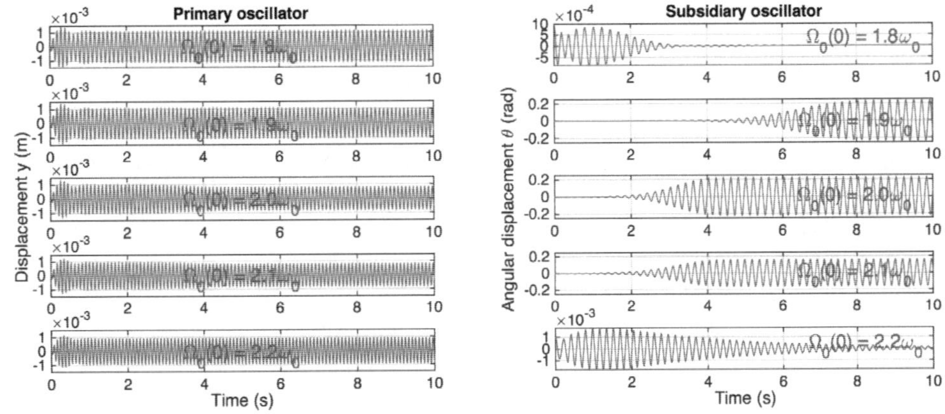

Figure 5. The activation of parametric resonance for varying initial internal frequency ratios.

Figure 6 presents the frequency domain characteristics of the two coupled oscillators for a given acceleration. In addition to the typical spring hardening peak bending for the primary oscillator, there is also a dip where energy is pumped to the secondary oscillator as parametric resonance onsets; thus, creating two peaks for the same resonant regime. The normalised −3dB bandwidth (shaded region) of the directly excited primary oscillator is 0.134, in contrast to 0.195 (∼45% enhancement) of the parametrically excited secondary oscillator. However, a sine wave frequency sweep does not reveal the full frequency responsiveness of the autoparametric system.

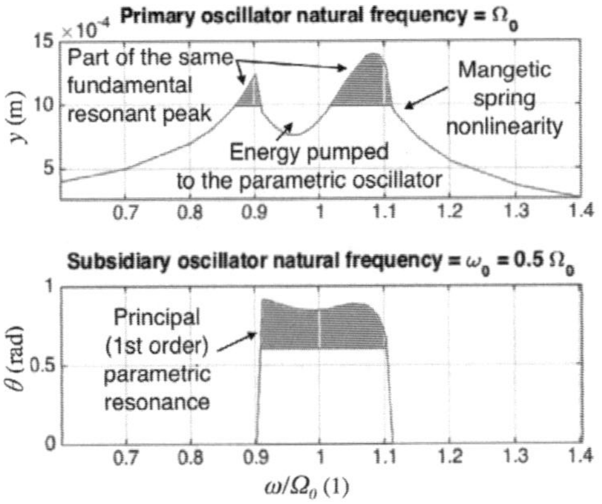

Figure 6. Frequency domain response of directly excited primary oscillator and parametrically excited oscillator.

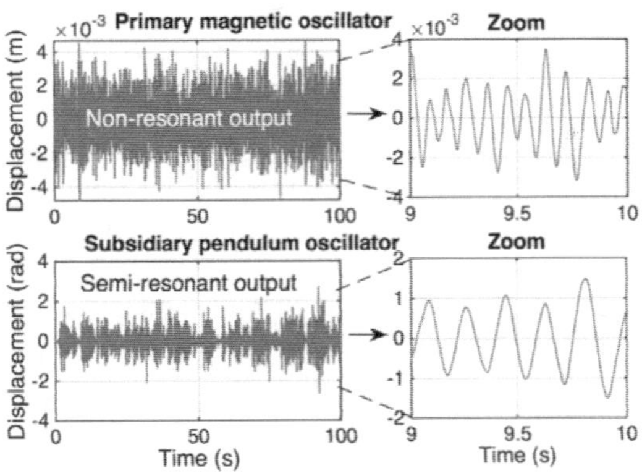

Figure 7. Time domain response when subjected to band-limited (0 Hz - 1 kHz) white noise vibration.

3.2. Noise response

When subjected to noise excitation, the primary magnetic oscillator exhibits non-resonant impact-induced output, which rapidly decays away at around its natural frequency. This is then internally coupled through the 2:1 frequency ratio to parametrically activate semi-resonant response as shown in figure 7. The semi-resonant response can be seen across varying bandwidth as illustrated with 100 Hz band-limited noise (figure 8) and 1 kHz band-limited noise (figure 9).

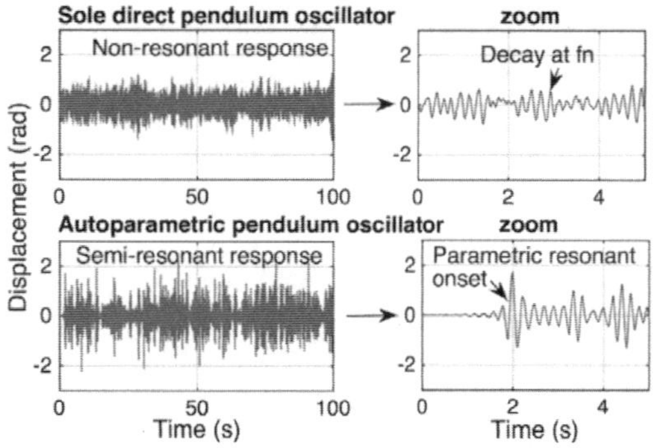

Figure 8. Response from band-limited (0 Hz -100 Hz) white noise at 6.6×10^4 g^2/Hz.

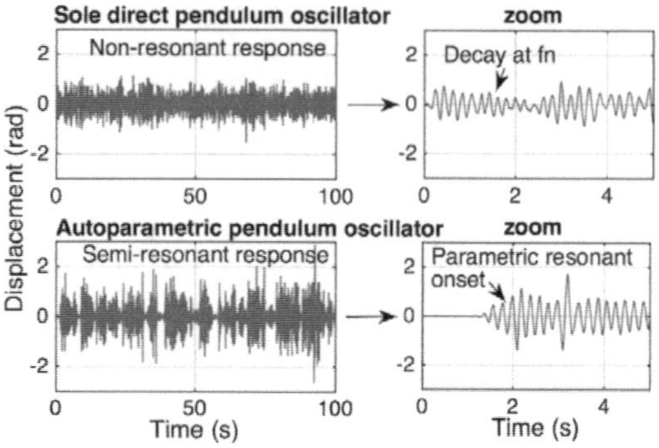

Figure 9. Response from band-limited (0 Hz - 1 kHz) white noise at 6.6×10^4 g^2/Hz.

Figure 10 compares the proposed system with a comparable directly excited pendulum oscillator under broadband noise excitation. It shows that beyond a certain noise threshold, the autoparametric system increasingly performs better than its direct counterpart due to its inherent ability to attain semi-resonant. This gap further widens at higher noise levels.

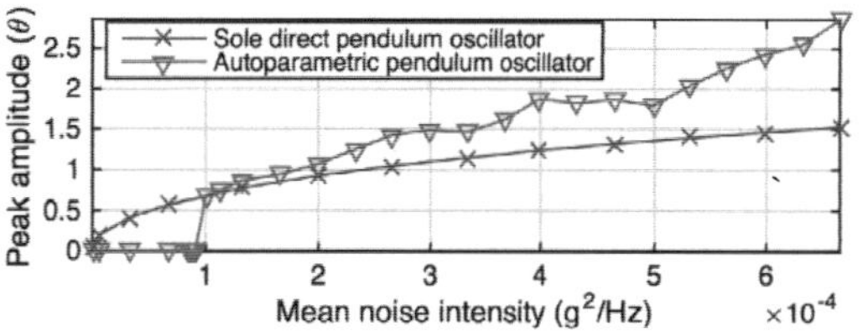

Figure 10. Peak amplitude of the proposed auto-parametric system (semi-resonant output) compared to a conventional direct oscillator when subjected to band-limited (0 Hz - 1 kHz) white noise.

4. Conclusion

This paper explored a pathway towards a highly nonlinear autoparametric VEH solution that has broader operational frequency bandwidth than linear direct oscillators and can inherently exhibit semi-resonant response under broadband noise excitations. The harvester solution would thus be suitable for capturing a broad range of vibration sources. Future work involves experimental investigation, incorporation of transduction mechanisms, and validation with various measured vibration profiles from various applications.

References

[1] S.P. Beeby, M.J. Tudor and N.M. White, 2006, *Meas. Sci. Technol.*, **17**, R175-R195
[2] D. Zhu, M.J. Tudor and S.P. Beeby, 2010, *Meas. Sci. Technol.*, **21**, 29pp
[3] L. Tang, Y. Yang C.K. Soh, 2010, *J. Intel. Mater. Syst. Struct.*, **21**(18), pp. 1867-97
[4] D. Zhu, S. Roberts, M.J. Tudor and S.P. Beeby, 2010, *Sens. Actuators A*, **158**(2), pp. 284-293
[5] R.S. Langley, 2015, *J. Sound Vib.*, **339**, pp. 247-61
[6] A. Nayfeh and D. Mook, *Nonlinear Oscillation*, (Wiley, 1979)
[7] L. Kurmann *et al.*, 2015, *J. Phys. Conf. Ser.*, **660**(1), p 012070
[8] Y. Jia, J. Yan, K. Soga and A.A. Seshia, 2014, *J. Intel. Mater. Syst. Str.*, **25**(3), pp 278-89
[9] Y. Jia, J. Yan, K. Soga and A.A. Seshia, 2013, *J. Phys. Conf. Ser.*, **476**(1), p 012126
[10] Y. Jia, J. Yan, K. Soga and A.A. Seshia, 2014, *Smart Mater. Struct.*, **23**(6), p 065011
[11] Y. Jia, J. Yan, K. Soga and A.A. Seshia, 2014, *Sens. Actuators A*, **220**, pp 69-75
[12] B. Vazquez-Gonzalez and G. Silva-Navarro, 2008, *Shock Vibration*, **15**(3-4), pp 355-68
[13] B. P. Mann, N. D. Sims, 2009, *J. Sound Vib.*, **319**(1-2), pp. 515?30
[14] K. W. Yung, P. B. Landecker, and D. D. Villani, 1998, *Mag. Elect. Sep.*, **9**(1), pp 39?52

Rotary bistable and Parametrically Excited Vibration Energy Harvesting

L. Kurmann[1], Y. Jia[2], D. Hoffmann[3], Y. Manoli[3,4], P. Woias[5]

[1]University of Applied Sciences and Arts Northwestern Switzerland, School of Engineering, Inst. of Automation, Klosterzelgstrasse 2, 5210 Windisch, Switzerland
[2]University of Chester, Department of Mechanical Engineering, Thornton Science Park, University of Chester, Pool Lane, Chester CH2 4PU, United Kingdom
[3]Hahn-Schickard, W.-Schickard-Str. 10, 78052 Villingen-Schwenningen, Germany
[4]Fritz Huettinger Chair of Microelectronics, [5]Lab. for the Design of Microsystems, Department of Microsystems, Engineering – IMTEK, University of Freiburg, Georges-Köhler-Allee 102, 79110 Freiburg, Germany
E-mail: lukas.kurmann@fhnw.ch

Abstract. Parametric resonance is a type of nonlinear vibration phenomenon [1], [2] induced from the periodic modulation of at least one of the system parameters and has the potential to exhibit interesting higher order nonlinear behaviour [3]. Parametrically excited vibration energy harvesters have been previously shown to enhance both the power amplitude [4] and the frequency bandwidth [5] when compared to the conventional direct resonant approach. However, to practically activate the more profitable regions of parametric resonance, additional design mechanisms [6], [7] are required to overcome a critical initiation threshold amplitude. One route is to establish an autoparametric system where external direct excitation is internally coupled to parametric excitation [8]. For a coupled two degrees of freedom (DoF) oscillatory system, principal autoparametric resonance can be achieved when the natural frequency of the first DoF f_1 is twice that of the second DoF f_2 and the external excitation is in the vicinity of f_1. This paper looks at combining rotary and translatory motion and use autoparametric resonance phenomena.

1. Introduction

Energy Harvesting is a technology for capturing non-electrical energy from ambient energy sources, converting it into electrical energy and storing it to power wireless electronic devices. The process of capturing mechanical energy such as shocks and vibrations is a particular field of energy harvesting requiring specific types of energy harvesting devices, so called kinetic energy harvesters (KEH). There are many types of KEH's, but all of those systems have one common goal: an ideal KEH can keep the kinetic proof mass in resonance over an infinite large excitation bandwidth. Conventional, first generation types of such transducers can harvest mechanical vibration energy effectively only in a narrow frequency window. Over time many different types of systems have been analytically characterized, designed and tested. Most of these systems show only small improvements with respect to their bandwidth. None of those systems can transfer mechanical vibration power into electrical energy over a wide frequency band. The ideal kinetic harvester system will have a simple mechanical structure as well as a wide vibration frequency range for which the system can transfer effectively environmental mechanical vibrations into electrical energy. In this paper a new KEH system is analytically and numerically examined, assuming that the basepoint excitation source is infinite

Journal of Physics: Conference Series **773** (2016) 012007 doi:10.1088/1742-6596/773/1/012007

(which for such small energies you can safely assume). In chapter 2.1. a mathematical system model is derived and in chapter 2.2. numerical simulations are presented.

2. Design of 2DoF bistable rotatory-translatory KEH

The lumped parameter model is depicted in Figure 1 having a rotary $\varphi(t)$ and a translatory $b(t)$ degree of freedom. The proof mass m_1 on cantilever l_1 with a 1st DoF $\varphi(t)$ can rotate on the pivot P_1 in the bounded region $\alpha_0 > |\varphi(t)|$, otherwise it will hit lobe N_1 or N_2 and might have hard or soft impact with a similar behavior treated in [3]. Perpendicularly attached ($\gamma_0 = 0°$) to l_1 at P_1 is a passive cantilever beam l_2. This cantilever has a translatory 2nd DoF $b(t)$ and carries an integrated piezoelectric (PE) transducer. On the tip of l_2 is a second proof mass m_2 carrying an electromagnetic (EM) transducer. The system is harmonically basepoint excited via y_0 and the mass m_1 at the end of cantilever l_1 is permanent-magnetically suspended (in a similar fashion as described in [9]) to introduce a linear k and nonlinear stiffness k_3 plus a translatory viscous damping d for the 1st DoF φ. The 2nd DoF b is similarly structured, having a linear stiffness k_b and a nonlinear stiffness k_{3b} (not depicted) plus a viscous damping d_b and an additional electromagnetic damping d_e. Depending on the angle γ_0, the cantilever length $l_{1,2}$ and the proof masses $m_{1,2}$, the system is in a monostable or bistable configuration. In case of the latter, the system has two stable energy wells, depicted in Figure 2, bottom diagram with $\varphi(t) = +\varphi_{01}$ and $b(t) = +b_{01}$. Its unstable equilibrium point is shaped by the mass ratio $\lambda_m = m_2{:}m_1$ and given length ratio's $\lambda_l = l_2{:}l_1$. In case both masses and lengths are equal, the unstable equilibrium position would be 45° (see also $\lambda_m = 1$ crossing the blue line ($\lambda_l = 1$) in top diagram of Figure 2).

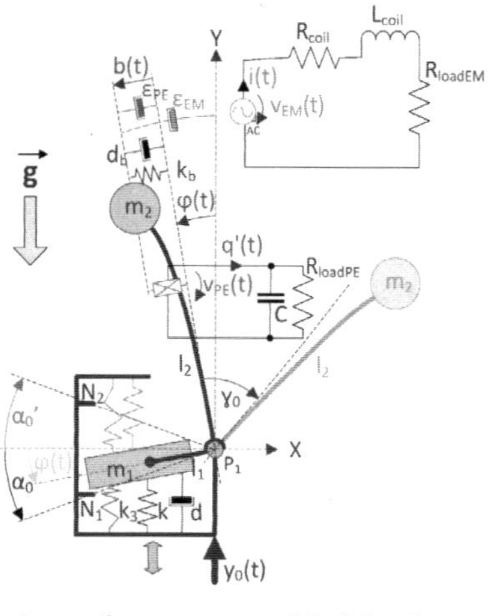

Figure 1. Lumped parameter model of the electromagnetic piezoelectric 2DoF KEH with 1. rotational DoF represented by $\varphi(t)$ and 2. translatory DoF represented by $b(t)$.

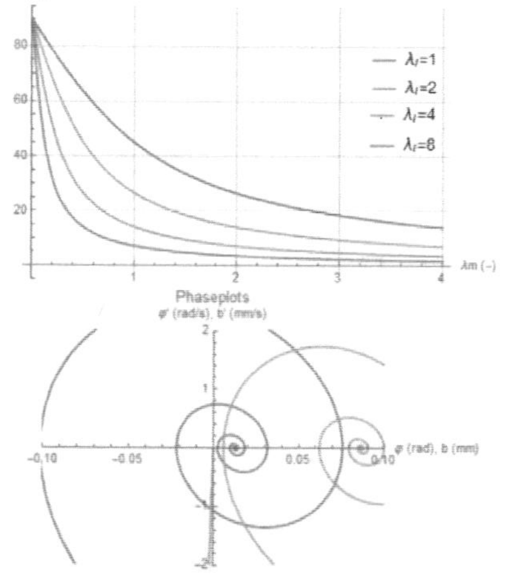

Figure 2. Initial position of φ_0 for the L-cantilever to achieve an unstable equilibrium (top) and the phase plots of $\varphi(t)$ and $b(t)$ with initial displacement and no excitation.

Cantilevers l_1, l_2 are assumed to have no mass and the quasi magnetically levitated mass m_1 on cantilever l_1 attached to the pivot P_1 transforms the translatory basepoint excitation y_0 into a rotary oscillation. The electromagnetic transducer damping (transduction factor ε_{EM} in [Vs/m]) is present in the mechanical domain in the resulting torque DE of $\varphi(t)$ and the force DE $b(t)$ whereas the piezoelectric damping (transduction factor ε_{PE} in [As/m]) is only present in the force DE $b(t)$. In this lumped parameter model the electrical circuit has an inductance L_{coil} and resistor R_{coil} and in series the resistive load R_{loadEM} attached. The piezoelectric transducer is drawn directly on the passive cantilever beam l_2 and the piezo-ceramic is modeled as capacitor C. It is a separately uncoupled circuit in the electrical domain with resistive load R_{loadPE}.

Figure 3 shows studies I-IV with different system angles γ_0. Several studies were conducted for achieving resonance over a large frequency range using following four principle parameters: (1) angle γ_0, (2) mass proportion factor λ_M, (3) proportionality of natural frequencies ω_1, ω_2 and (4) range of Ω

for analysis of sub-resonant and over-resonant response; in this paper only configuration I with system angle $\gamma_0 = 0°$ is presented.

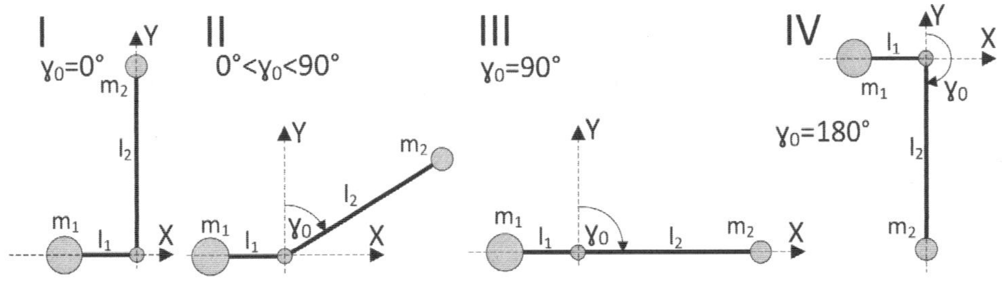

Figure 3. Configurations I-IV of the cantilever system; cantilever l_2 is a passive beam with stiffness k_b; in configuration I $\gamma_0 = 0°$, e.g. beams are perpendicular to each other.

2.1. Lumped parameter model of the 2DoF rotary-translatory KEH system

The equation of motion can be derived using the Lagrangian-Euler method, considering the system at first conservative. The Lagrangian total energy of this 2. DoF system is

$$L(\varphi, b) = T(\varphi, b) - V(\varphi, b) = 0 \tag{1}$$

The nonlinear magnetic spring shall be introduced directly via its energy (2).

$$F_S(y) = k\,y + k_3\,y^3 \;\rightarrow E_S = \int F_S(y)\,dy = \frac{1}{2}k\,y^2 + \frac{1}{4}\,k_3\,y^4 \tag{2}$$

Summing all kinetic T and potential V energies of this system and adding E_S of (2) to V yields

$$T(\varphi, b) = \frac{1}{2}m_2\dot{b}^2 + \frac{1}{2}m_2 l_2\cos\gamma_0\,\dot{b}\,\dot{\varphi} + \frac{1}{2}{l_1}^2 m_1\dot{\varphi}^2 + \frac{1}{2}{l_2}^2 m_2\dot{\varphi}^2 + \frac{1}{2}m_2 b^2\dot{\varphi}^2 + l_2 m_2\sin\gamma_0\,b\dot{\varphi}^2 \tag{3}$$

$$V(\varphi, b) = \frac{1}{2}k_b b^2 + \frac{1}{4}k_{3b}b^4 + gm_2(l_2\cos(\gamma_0 - \varphi) - b\sin\varphi) - gm_1 l_1\sin\varphi + \frac{1}{2}k{l_1}^2\sin^2\varphi + \frac{1}{4}k_3{l_1}^4\sin^4\varphi \tag{4}$$

In (4) also a nonlinear spring for the cantilever beam is introduced in the same fashion as the magnetic spring using linear beam stiffness k_b and nonlinear stiffness k_{3b}. Following the Lagrangian formalisms and writing the set of DE dimensionless using ω_1 as reference will lead to following set:

$$\begin{gathered}\theta''(1 - \lambda_m + \lambda_m\,{\lambda_l}^2 - 2\lambda_l\lambda_m\sin\gamma_0\,u + \lambda_m u^2) + 2\xi_1\sin\theta\,\theta' + \cos\theta\sin\theta\,(1 + \beta\sin^2\theta) + 2\lambda_m uu'\theta' \\ +\varrho\cos\theta\,(-1 + \lambda_m - \lambda_m u) + \varrho\lambda_l\lambda_m\sin(\gamma_0 - \theta) - 2\lambda_l\lambda_m\sin\gamma_0\,u'\theta' + \lambda_l\lambda_m\cos\gamma_0\,u'' + \kappa_E\zeta = -\lambda\,\Omega^2\cos(\Omega\tau)\sin\theta\end{gathered} \tag{5}$$

$$u'' + 2\xi_2\,\Omega_0\,u' + {\Omega_0}^2\,u(1 + \beta_b u^2) + \lambda_l\cos\gamma_0\,\theta'' - u\theta'^2 + \lambda_l\sin\gamma_0\,\theta'^2 - \varrho\sin\theta + \kappa_E\lambda_l{\lambda_m}^{-1}\,\zeta + \kappa_P\lambda_P\lambda_l\,{\Omega_0}^2 \upsilon = 0 \tag{6}$$

$$\lambda_E\,{\lambda_l}^{-1}\Omega_0\,u' + \lambda_E\,\theta' = \zeta' + \lambda_E\,\zeta \tag{7}$$

$$u + \lambda_P\upsilon = \rho \tag{8}$$

$$\rho' = -\upsilon \tag{9}$$

Using following parameters for nondimensionalization (using ω_1 as reference):

$$\lambda = \frac{A}{l_1};\; \lambda_l = \frac{l_2}{l_1};\; \lambda_M = \frac{m_2}{m_1 + m_2} = \frac{m_2}{m};\; \Omega = \frac{\omega}{\omega_1};\; \tau = t\omega_1;\; \xi_1 = \frac{d}{2m_1\omega_1};\xi_2 = \frac{d_b}{2m_2\omega_2} \tag{10a}$$

$${\omega_1}^2 = \frac{k}{m_1 + m_2};{\omega_2}^2 = \frac{k_b}{m_2};\; {\Omega_0}^2 = \frac{{\omega_2}^2}{{\omega_1}^2};\; \varrho = \frac{g}{l_1{\omega_1}^2};\beta = \frac{k_3}{k}\,{l_1}^2;\beta_b = \frac{k_{3b}}{k_b}\,{l_1}^2 \tag{10b}$$

$$i_{e0} = \frac{\varepsilon_{EM}\,l_2\,\omega_1}{R};\upsilon_{p0} = R\,\omega_2 q_{p0};\; q_{p0} = \varepsilon_{PE}\,l_2;\; \kappa_E = \frac{{\varepsilon_{EM}}^2}{m\,R\,\omega_1};\; \kappa_P = \frac{{\varepsilon_{PE}}^2}{C\,k_b};\lambda_E = \frac{R}{L\,\omega_1};\lambda_P = RC\omega_2 \tag{10c}$$

$$angle\,\theta(\tau) = \frac{\varphi(t)}{\varphi_0},\, path\,u(\tau) = \frac{b(t)}{l_1},\, EM\,current\;\zeta(\tau) = \frac{i_e(t)}{i_{e0}},\, PE\,voltage\;\upsilon(\tau) = \frac{\upsilon_p(t)}{\upsilon_{p0}},\, charge\,\rho(\tau) = \frac{q_p(t)}{q_{p0}} \tag{10d}$$

Note also that above set of equation is highly nonlinear and not reflected in above set of DE is the impact when θ reaches $\pm\alpha_0$; see also [3] for a possible nomenclature of such conditions.

2.2. Mechanical domain (θ and u) and electrical domain (ζ, ν, ρ) response

Extensive dimensioned and dimensionless simulations were conducted with Mathematica and Matlab Simulink (solvers ODE23 of Bogacki-Shampine and ODE45 of Dormand-Price) using the coupled DE system (5)-(9) and corresponding parameters for nondimensionalization (10a)-(10d) to investigate the dynamics of this highly nonlinear system. In the following simulations an impact of m_1 into the supporter structure at $N_{1,2}$ (shown in Figure 1) will result in a light elastic impact and its impact velocity is reversed to 40% of its impact velocity.

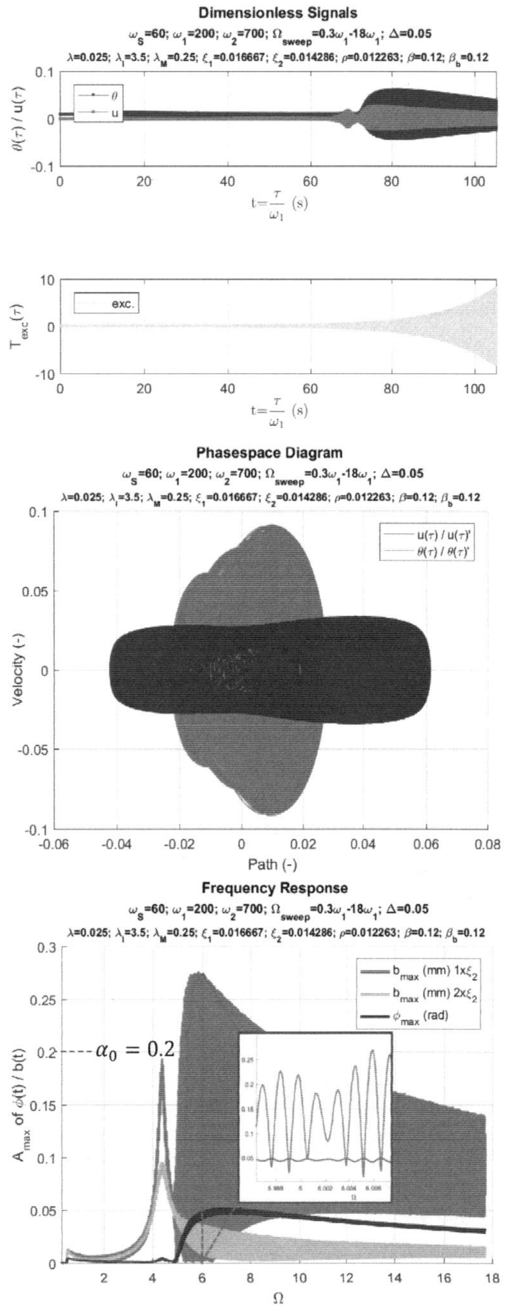

Figure 4. Dimensionless response and input (top); phase space behavior on upward sweep (middle) and its corresponding frequency-response (bottom)

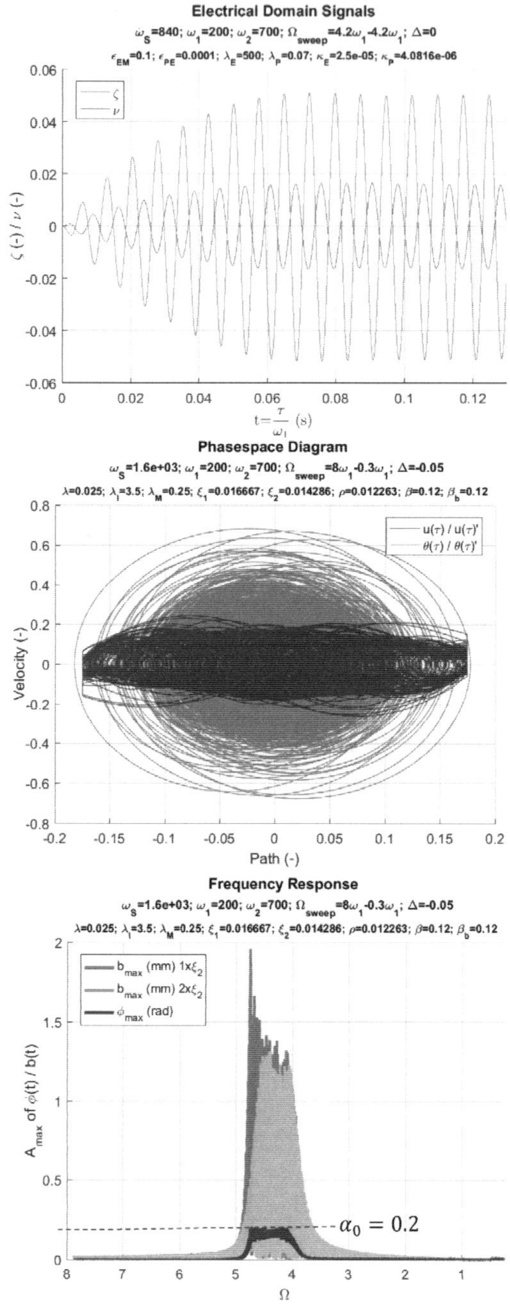

Figure 5. Dim. less response of PE voltage and EM current (top); phase space behavior of downward sweep and its corresponding frequency-response (bottom)

Particular attention was given to the initial positions of the system θ, θ' and u, u', as depending on its configuration, an initial torque is created, which puts the system not in rest. For example, if the starting condition of θ will be set to α_0 using also an according weak stiffness – the angle θ might stay at the oscillating lobe N_1 (depending heavily also of all system parameters) and the 2nd DoF u will show in such a case only a classic linear frequency response. For having a broadband response, m_1 needs to be able to exert a slight oscillation (for configuration I with $\lambda_M = 0.25$).

Figure 4 depicts mechanical domain diagrams, top and middle diagrams show responses dimensionless, bottom diagram depicts the frequency response of $\varphi \equiv \phi$ and $u \equiv b$. Top and middle diagram of Figure 5 show dimensionless electrical response and phase space response, bottom diagram depicts the frequency response of ϕ and b. Top diagram of Figure 4 shows excitation signal of an upward frequency sweep ($\Omega = 0.3 - 18$) in light blue; for each frequency step-cycle 40 periods are used and an amplitude step of $\Delta = 0.05\Omega$ is applied. The system in this configuration is set to have its natural frequencies at $\omega_1 : \omega_2 = 1:3.5$ (like its λ_l relation; other frequency ratios were tested, $\omega_1 : \omega_2 = 1:3$ exerts similar results, but not $\omega_1 : \omega_2 = 1:2$). The damping of θ and u is set to $d = 0.1Ns/m$; in case both damping factors are doubled, resulting amplitudes are at least reduced by factor 6 with exception of the subharmonic response at $\Omega = 0.5$ on an upward seep (depicted in orange, bottom diagram in Figure 4). On a downward sweep, the amplitudes are only reduced by ca. 20% (depicted in orange, bottom diagram in Figure 5). The excitation amplitude is set to $A = 0.5mm$. If a smaller amplitude is used, the system will not behave as depicted in Figure 4 and Figure 5. The response of b and ϕ, particularly ϕ is showing a beat frequency. In the electrical domain, the advantage of having a PE and EM transducer coupled is the resulting 180° phase shift of current and voltage (see also Figure 5, top diagram) which is particularly useful for a DC-DC converter next in line.

3. Conclusions

Presented KEH system is a new approach for the realization of a wideband KEH. Such a device can also be tuned to different environmental frequency bands by simply setting the system cantilever angle γ_0 (see Figure 3). This KEH system is discussed only for bi-stable configuration I, but additional bi-stable (II) and mono-stable (III and IV) configurations are available. It shows a particularly large and long resonance on an upward sweep ($\Omega = 5 - 18$), above the detuned system resonance of $\Omega \approx 4.4$ (which is always accessible, e.g. independent of the frequency sweep direction). Like in all nonlinear systems, the hysteresis causes to access this large frequency band only by tuning into this region with an upward frequency sweep. Having a downward frequency sweep (Figure 5, middle and bottom), also a resonance region is found with ca. 6x larger amplitudes for b and ca. 3.5x larger for φ resulting in an impact behavior where φ reaches α_0 (Figure 5, bottom/middle) in the region of $\Omega = 5 - 3.8$.

References

[1] N. Minorsky, „Nonlinear Oscillations“, Princeton, N.J., Van Nostrand, 1962
[2] A. Nayfeh, D. Mook, „Nonlinear Oscillation“, Weinheim, Wiley, 2007
[3] L. Kurmann, D. Hoffmann, B. Folkmer, Y. Manoli, P. Woias, R. Anderegg, "Autoparametric resonance systems for vibration-based energy harvesters", J. Phys. Conf. Ser., vol. 660, p. 012070, 2015
[4] Y. Jia, J. Yan, K. Soga, and A. Seshia, "A parametrically excited vibration energy harvester", J. Intel. Mat. Syst. Str., 2013.
[5] Y. Jia, J. Yan, K. Soga and A. A. Seshia, "Multi-frequency operation of a mems vibration energy harvester by accessing five orders of parametric resonance", in J. Phys. Conf. Ser., ser. 1, vol. 476, 2013, pp. pp. 607–611
[6] Y. Jia, J. Yan, K. Soga, and A. A. Seshia, "Parametric resonance for vibration energy harvesting with design techniques to passively reduce the initiation threshold amplitude", Smart Mater. Struct., vol. 23, no. 6, p. 13, 2014.
[7] Y. Jia, J. Yan, K. Soga, and A. Seshia, "Parametrically excited mems vibration energy harvesters with design approaches to overcome initiation threshold amplitude", J. Micromech. Microeng., vol. 23, no. 11, p. 10pp., 2013.
[8] Y. Jia and A. A. Seshia, "An auto-parametrically excited vibration energy harvester", Sens. Actuators A, vol. 220, pp. 69–75, 2014.
[9] B. P. Mann, N. D. Sims, "Energy harvesting from the nonlinear oscillations of magnetic levitation", J. Sounds and Vibration, 319, pp. 515–530, 2009.

Broadband Rotational Energy Harvesting with Non-linear Oscillator and Piezoelectric Transduction

H Fu and E M Yeatman

Department of Electrical and Electronic Engineering, Imperial College London, Exhibition Road, London, SW7 2AZ, UK

E-mail: `h.fu14@imperial.ac.uk`

Abstract. Rotational energy is widely distributed in many industrial and domestic applications, such as ventilation systems, moving vehicles and miniature turbines. This paper reports the design and implementation of a bi-stable rotational energy harvester with wide bandwidth and low operating frequency. The rotational energy is converted into electricity by magnetic plucking of a piezoelectric cantilever using a driving magnet mounted on a rotating host. The bistable condition is achieved by introducing a fixed magnet above the tip magnet at the cantilever's free end. The repulsive magnetic force between the magnets creates two equilibrium positions for the piezoelectric beam. The harvester is designed to operate in the high energy orbit (interwell vibration mode) to extract more energy from the rotational energy source. Harvesters with and without bistability are compared experimentally, showing the difference of power extraction on both the output power and bandwidth. The method proposed in this paper provides a simple and efficient way to extract rotational energy from the ambient environment.

1. Introduction

Compared with vibrational energy, rotational energy is also readily harnessed for powering low-power electronics. This energy source also widely exists in many scenarios, including heating, ventilating and air conditioning (HVAC) systems, vehicles and wrist watches [1, 2]. However, rotational energy harvesting is less investigated and developed than vibration energy harvesting. Similar to vibrational energy, rotational energy in daily life also has the characteristic of low frequency and wide bandwidth [3]. Therefore, the harvesters should be low-frequency and broadband to harness rotational energy efficiently.

Generally, for large-scale rotational harvesters, such as wind turbines, gear trains are adopted to increase the rotation frequency of the turbine rotor, but for micro-scale energy harvesters, it is costly and complicated to fabricate gear trains. We developed an airflow energy harvester using miniature turbines and piezoelectric transduction [4]. In this design, a mechanism called frequency up-conversion was employed to replace gear trains. This mechanism was achieved by plucking the piezoelectric beam using a driving magnet mounted on the turbine rotor. The beam vibrates freely at its resonant frequency after each plucking. However, the output power has a significant drop when the driving frequency is high due to the interference between the vibrating beam and the coming driving magnet [5]. This phenomenon was also observed in other vibration energy harvesters using frequency up-conversion [6, 7].

Bi-stability is widely used in vibrational energy harvesting to increase bandwidth. This mechanism is generally achieved by applying magnetic force or external mechanical load to the motion direction of proof masses [8]. Zhou *et al.* presented a nonlinear piezoelectric harvester for vibration energy harvesting [9]. Two rotatable magnets and one tip magnet were used to generate the bistable condition. The harvester had a wide low-frequency operating range by adjusting the magnet orientation. In order to enhance the performance of the rotational energy harvester over a wide bandwidth, we combine the bistable method with frequency up-conversion in this paper. The operating principle is illustrated firstly and experiments were carried out to verify the idea. An improved output power was obtained over a wide bandwidth.

2. Operating principle of the harvester

The schematic of the harvester is shown in Fig. 1. It consists of a driving magnet mounted on a rotating host, a piezoelectric beam, a tip magnet and a fixed magnet. The fixed magnet is above the tip magnet, and the driving magnet is below the tip magnet. The driving magnet is mounted on a rotating host. The beam is plucked once per rotation cycle by the repulsive magnetic force between the driving magnet and the tip magnet.

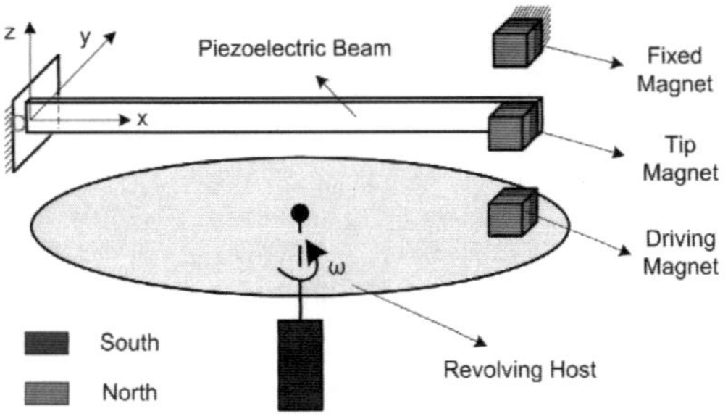

Figure 1. Schematic of the nonlinear piezoelectric rotational energy harvester.

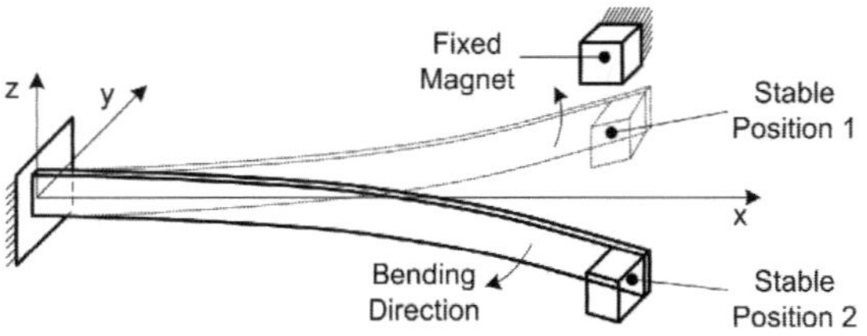

Figure 2. Bi-stability of the energy harvester, showing the nonlinear characteristic of the design. Two equilibrium positions are created by the repulsive magnetic force.

In terms of bi-stability, it is achieved by the repulsive force provided by the fixed magnet. The force pushes the beam to bend towards either side of its original position, introducing two equilibrium positions, as shown in Fig. 2. There are three modes of vibration for the bistable beam, i.e. intrawell vibration, chaotic interwell vibration and periodic interwell vibration. These modes are determined by the driving force provided by the rotating magnet and the restoring force provided by the fixed magnet and the bended beam.

3. Experimental set-up

The experimental set-up is shown in Fig. 3. The piezoelectric beam was clamped on a beam holder at one end. The holder was fixed on an adjustable platform which is capable of precisely controlling the position of the piezoelectric beam in three directions. A high speed stepper motor (Phidgets 3303) with a step angle of 1.8° was placed underneath the beam with a revolving plate mounting on the motor's shaft. The motor is driven by a bipolar motor control circuit (Phidgets 1067) which has a position resolution of $^1/_{16}$ step. The acceleration is also programmable to achieve any desired rotational speed.

A magnet was mounted on the revolving plate as the driving magnet. Magnet plucking is formed by the driving magnet and the tip magnet at the beam's free end. For bistability, a fixed magnet was also employed and placed on a magnet holder above the piezoelectric beam. The magnet holder was also fixed on an adjustable platform. Therefore, the gap between the tip magnet and the fixed magnet can be adjusted as well. A laser displacement sensor was

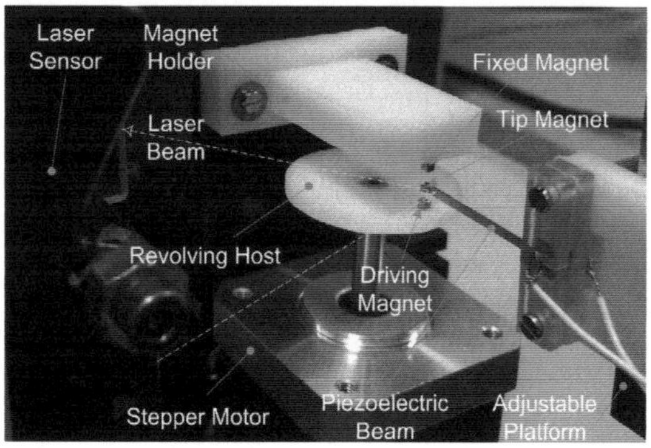

Figure 3. Experimental set-up. The rotational motion is provided by a stepper motor. The rotational frequency can be accurately controlled by the management circuit.

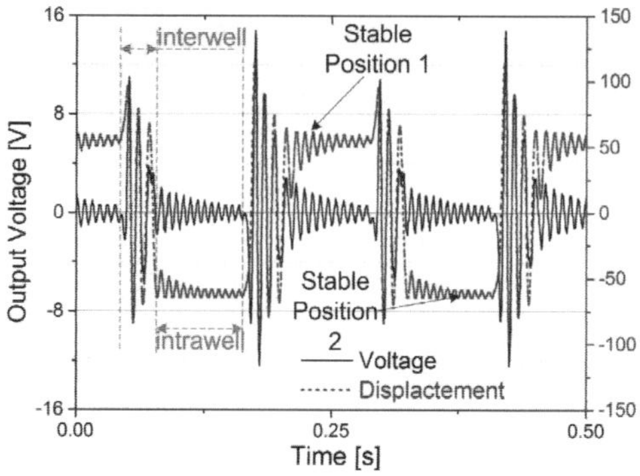

Figure 4. Output voltage and tip displacement of the nonlinear energy harvester, showing the bi-stability of the system under tip excitation (Rotation speed: 9.8 Hz; load: 180 kΩ).

adopted in the experiment to measure the vibration of the piezoelectric beam. The output of the piezoelectric beam was directly connected with a 180 kΩ resistive load.

4. Results and discussion

Fig. 4 illustrates the output voltage and tip displacement of the harvester operating in the high energy orbit (periodic interwell vibration). Due to the limited input energy from each plucking, the beam does not purely vibrate in the interwell mode during one cycle. The beam vibrates in the interwell vibration mode first for several oscillations and then settles into the intrawell mode during one excitation cycle. The output voltage in the interwell mode is much higher than that in the intrawell mode. Two equilibrium positions are clearly observed on the tip displacement curve.

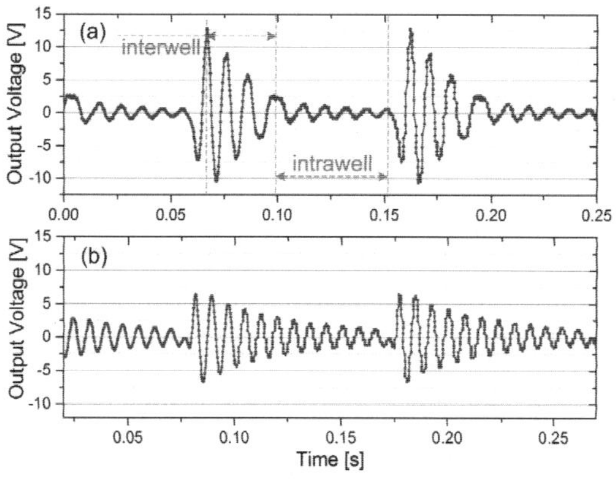

Figure 5. Output voltage of the harvester at 11.2 Hz with and without the nonlinear characteristics. (a) with the fixed magnet (Gap in z-axis: 1.2 mm) and (b) without the fixed magnet.

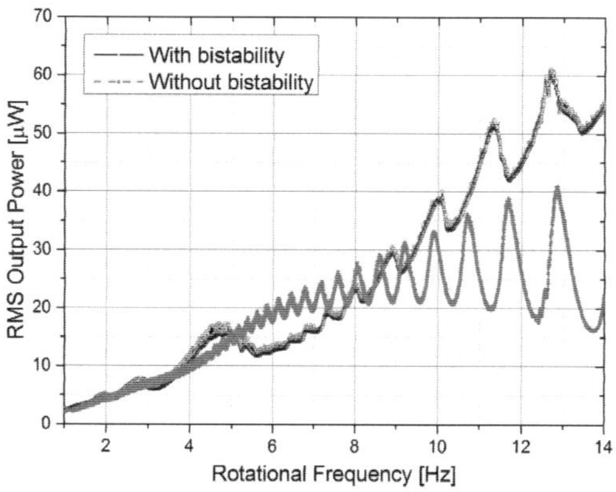

Figure 6. Frequency response of the harvester with and without bistability, showing the wide bandwidth of the bistable device and the improved output power at high frequencies.

Fig. 5 depicts the output voltage of the harvester with and without the nonlinear characteristics. Without bistability, the maximum output voltage is much lower than the maximum voltage from the bistable harvester. The improved voltage in the bistable case will add even greater benefit after rectification. Meanwhile, the output voltage is damped more slowly than that in the bistable case. Therefore, the beam is still in motion when the driving magnet returns in a new cycle, causing interference between plucking cycles and loss of power.

Fig. 6 shows the frequency response of the harvester with and without bistability. The nonlinear harvester has similar performance to the linear harvester at low frequency, but the output power is much higher at high frequency. The improved output power at high frequency is achieved by avoiding the interference between the vibrating cantilever beam and the coming driving magnet. As shown in Fig. 4, for each excitation, the beam finally gets into the intrawell vibration mode with limited oscillating amplitude. Therefore, the beam does not have significant impact on the driving magnet, whereas in the harvester without bistablity, the beam is not damped quickly enough. At high frequency, the beam is still vibrating with a quite large amplitude when the driving magnet comes. Negative work is done on the piezoelectric beam, when the beam vibrating direction and the rotation direction of the driving magnet are opposed.

The quickly damped motion in the bistable case provides a possibility of using 2 or more driving magnets on the rotor to improve the power generation capability at even lower frequencies.

5. Conclusion

This paper presents a bistable rotational energy harvester with low operating frequency and wide bandwidth. A piezoelectric cantilever beam with a tip magnet was employed as the transducer. A bistable beam was achieved by adding a fixed magnet above the tip magnet. Two equilibrium positions were created by the repulsive magnetic force. In terms of the excitation mechanism, the beam was plucked by a rotating driving magnet mounted on a revolving host. The beam vibrated at its natural frequency after each plucking.

An experimental validation was carried out. The bistable beam had higher output voltage and tip displacement, when it vibrated in the interwell vibration mode. Compared with the harvester without bistability, the bistable harvester has a wider bandwidth and improved output power. The proposed nonlinear rotational energy harvester provides a practical solution to harness ambient rotation energy for low-power electronics.

Acknowledgments

The work is supported by the China Scholarship Council (CSC) and the Department of Electrical and Electronic Engineering, Imperial College London. We would also like to thank Prof. Andrew Holmes for use of apparatus.

References

[1] Briand D, Yeatman E and Roundy S 2015 *Micro energy harvesting* (John Wiley & Sons)
[2] Mitcheson P D, Yeatman E M, Rao G K, Holmes A S and Green T C 2008 *Proceedings of the IEEE* **96** 1457–1486
[3] Yeatman E M 2008 *Proceedings of the Institution of Mechanical Engineers, Part C: Journal of Mechanical Engineering Science* **222** 27–36
[4] Fu H and Yeatman E M 2015 *Applied Physics Letters* **107** 243905
[5] Fu H and Yeatman E M 2015 *Journal of Physics: Conference Series* **660** 012058
[6] Tang Q C, Yang Y L and Li X 2011 *Smart Materials and Structures* **20** 125011
[7] Jung S M and Yun K S 2010 *Applied Physics Letters* **96** 111906
[8] Harne R L and Wang K W 2013 *Smart Materials and Structures* **22** 023001
[9] Zhou S, Cao J, Erturk A and Lin J 2013 *Applied Physics Letters* **102** 173901

Organic microcantilevers based on reduced graphene oxide composite for electrostrictive energy harvesting

H. Nesser[1], H. Debéda[1], J. Yuan[2], A. Colin[2], P. Poulin[2], I. Dufour[1], C. Ayela[1]
[1]Université de Bordeaux, IMS, UMR 5218, 351 Cours de la Libération, 33405 Talence Cedex, France
[2]CRPP, CNRS, UPR8641, 115 Avenue Schweitzer, 33600 Pessac, France

Email: hussein.nesser@ims-bordeaux.fr

Abstract. Electrostrictive materials are promising for mechanical energy harvesting applications because of their high power density, low cost and scalability. In this paper, strain sensitive nanocomposite materials based on reduced graphene (rGO) and PDMS are used for energy harvesting; they are characterized by a high electrostrictive coefficient (2.4×10^{-15} m^2/V^2) and a giant dielectric constant (ranging from 100 to 1000 at 100 Hz, depending of rGO concentration). Using these nanocomposite materials, electrostrictive MEMS microgenerators are fabricated with an innovative low-cost and environment friendly process in an all-organic approach. The fabricated microcantilevers exhibit excellent mechanoelectrical performances in dynamic mode. With an acceleration of 1 g of the microcantilever base using a shaker, experiment at the first resonant mode (≈ 16 Hz) generates an electrical power density of 8.15 μW/cm^3.

1. Introduction

To develop autonomous and self-powered electronic devices, increasing attention is focused on harvesting energy from ambient energy sources such as light, heat, and mechanical vibrations [1]. Among them, vibration energy has been widely and intensively studied because it exists almost everywhere in our living environment. In most cases, it can be converted to electricity by three different electromechanical transduction schemes: electrostatic, electromagnetic, and piezoelectric [2]. Nanocomposite electrostrictive materials have been recently introduced since they are suitable for low frequency, high-strain excitation owing to their flexibility. Because of the capacitive nature of nanocomposite electrostrictive polymers, it is possible to directly apply electrostatic cycle for energy harvesting when the materials are submitted to strain [3]. In this paper, we use nanocomposite electrostrictive materials made of near-percolated networks of reduced graphene oxide (rGO) as conducting fillers dispersed in a polydimethylsiloxane (PDMS) matrix which is a soft and deformable dielectric elastomer [4]. A MEMS microgenerator integrating the reduced graphene oxide nanocomposite is fabricated with an innovative low-cost and environment friendly process in an all-organic approach. Then, this original nanocomposite integrated into the MEMS is characterized by measuring the variation of the capacity under mechanical strain. The microdevices are finally mounted in a primitive continuous conditioning circuit to evaluate the electrical energy which can be harvested with these strain-sensitive nanomaterials.

2. MEMS fabrication

The micromachining process used in this work does not require the high-cost semiconductor manufacturing machines used for silicon microfabrication and thus reduces drastically the fabrication cost of the MEMS devices. The electrostrictive organic microcantilevers are fabricated in an all-organic approach thanks to a low-cost method allowing a rapid production rate. The fabricated microdevice is a simple microcantilever composed of the electrostrictive material sandwiched between two electrodes and deposited on a flexible substrate. Simple rectangular microcantilevers without seismic mass are first fabricated and tested in static mode to evaluate the change in capacitance of the electrostrictive material. Then, trapezoidal shaped micro-cantilevers incorporating a seismic mass at their free-end are of particular interest for dynamic mode tests for their relatively low resonant frequency and high uniform strain. The cross-section and top view of the fabricated organic microcantilevers are given in figure 1.a an d 1b. The first step of fabrication consists in the screen-printing of PEDOT:PSS conductive polymer on top of a 50 µm thick PolyEthylene Naphthalate (PEN) sheet used as supporting layer. This printed PEDOT:PSS layer is afterwards cured for 20 min at 120 °C and serves as bottom electrode. Subsequently, the electrostrictive rGO/PDMS layer is spin-coated and soft baked at 200 °C for 3 h. The average thickness of the deposited layer is about 50 µm. An additional PEDOT:PSS top electrode is printed by doctor-blade through a shadow-mask for patterning. Finally, the microcantilever shape is simply defined by xurography using a vinyl cutting machine (Graphtec Craft ROBO Pro) [5]. Note that the mask used for the top electrode printing is also fabricated by xurography. The seismic mass is based on epoxy loaded with Ag particles for higher density. A paste (ESL1901-SD from ElectroScience Laboratory) is used for this purpose and is deposited with a syringe on the tip-end of the trapezoidal microcantilever; it is afterwards dried for 30 min at 120 °C. The deposited mass is measured with a microbalance (≈ 86 mg). Photographs of the fabricated multilayered microstructures are given in figure 1.c. The final dimensions of the trapezoidal microcantilevers are 10 mm long, 5 mm large base width, 2 mm small base width and 100 µm thick.

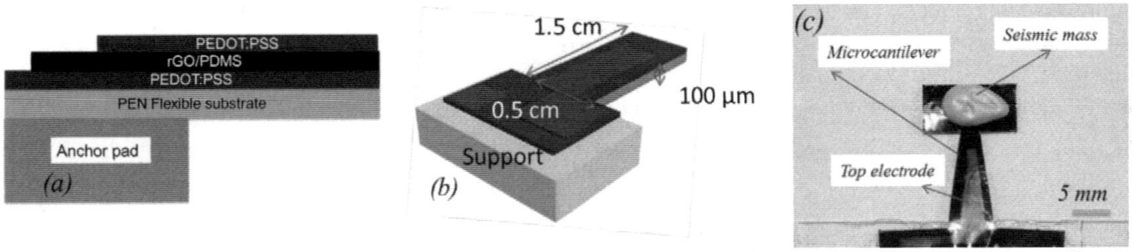

Figure 1. (a) (b) Cross-section and top-view scheme of a rectangular electrostrictive microcantilever (c) Top-view photograph of a trapezoidal electrostrictive microcantilever.

3. Results and discussion

3.1. Static mode

In this part, the organic rectangular microcantilevers are tested in static mode to evaluate the change in capacitance of the electrostrictive materials. Following that, the gauge factor, which gives an image of the electromechanical sensitivity of the material relatively to the strain, is calculated. Concretely, the rectangular microcantilever is bent intermittently by applying a force at its end using a microprobe while monitoring the evolution of both the capacitance and the resistance with an impedance analyser. Several graphene oxide concentrations are tested while pure PDMS is used as reference. The electrostriction coefficient M of the electrostrictive material as well as the capacitive and resistive gauge factor (FG_c and FG_R respectively) are defined by the following equations:

$$FG_C = \frac{\Delta C/C_0}{\Delta\varepsilon} \qquad FG_R = \frac{\Delta R/R}{\Delta\varepsilon} \qquad M = \frac{\Delta e}{2E\Delta\varepsilon} \qquad (1)$$

With ε being the mean top surface strain, and C, R, e and E being the capacitance, resistance, permittivity and Young's modulus of the electrostrictive material, respectively.

As shown in figure 2, an increase of the capacitance as a function of strain is observed for the sample prepared with pure PDMS (geometric effect, with $FG_C = 1$) while a decrease is observed for the rGO/PDMS microstructures with a maximum relative variation of 4 % for a strain of 0.55 % for the material composed of 3.7 wt% of rGO. In addition, it is worth noting a difference in samples response depending on their rGO concentration, as observed in nanocomposite electromechanical gauges. These results confirm that changes of capacitance are dominated by the strain dependence of the permittivity of the nanocomposite and not to geometrical variations. The best gauge factor of the MEMS capacitive sensor is obtained for the 3.7 wt% sample with a value of (-) 6.7 at low strain level ($\varepsilon < 0.6$ %). Using the equation 1, the static electrostriction coefficient M of the best sample of rGO/PDMS is 2.4×10^{-15} m^2/V^2, with dielectric constant of 80 for 3.7 wt% sample and E= 1 MPa.

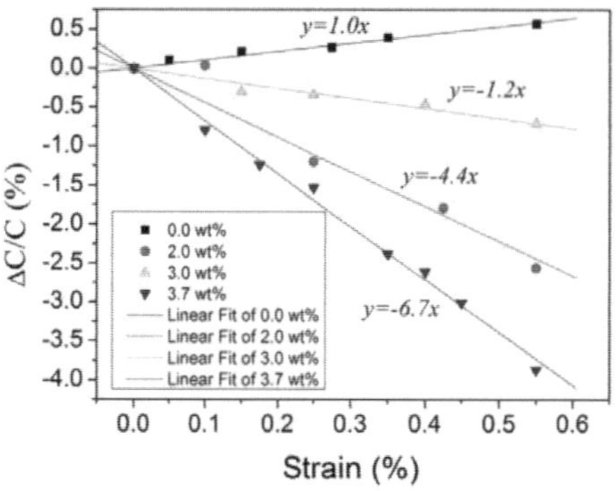

Figure 2. Relative capacitance variation as a function of the mean strain applied to the top surface for 4 concentrations of rGO dispersed in PDMS (rectangular microcantilevers).

3.2. Dynamic mode

Here, the trapezoidal microcantilevers are tested in dynamic mode to evaluate the variation of the admittance of the electrostrictive material at the mechanical resonant frequency of the microcantilevers. A shaker drives the devices into resonance for this purpose. The admittance of the device is measured by impedance analysis, with an applied support acceleration of 1 g. During the measurement, a constant external biais voltage (8 V) is injected to polarize the material.

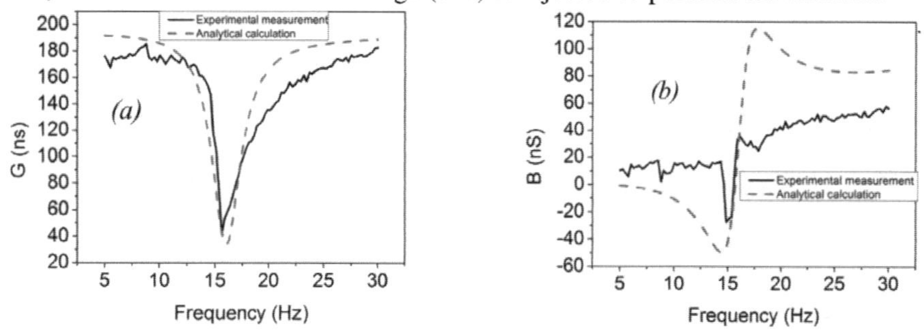

Figure 3. Experimental measurement and analytical calculation of the conductance (a) and susceptance (b) of the electrostrictive MEMS as function of the frequency for 1 g acceleration.

A resonance peak is observed electrically at ~16 Hz and demonstrates the existence of a phenomenon of electrostriction at the resonant frequency of the microcantilever (figure 3). Near the resonant frequency, the equivalent electrical circuit of the device is a 'capacitance' C in parallel with a 'resistance' R, the expression of C and R being expressed as follows (note that R and C are not real numbers but complex numbers due to the resonance phenomenon):

$$C = C_0\left(1 + \frac{K_C}{1 + \frac{j}{Q}\frac{f}{f_0} - \frac{f^2}{f_0^2}}a_s\right) \qquad R = R_0\left(1 + \frac{K_R}{1 + \frac{j}{Q}\frac{f}{f_0} - \frac{f^2}{f_0^2}}a_s\right) \tag{2}$$

Where f_0 is the resonant frequency, Q the quality factor, C_0 and R_0 the capacitance and resistance at rest, a_s the support acceleration and K_C and K_R two constants expressing the dependence of the capacitance and resistance to the applied acceleration (these constants depend on the electrostrictive material and on the cantilever geometry and size).

With these two equations (2) it is possible to express both the conductance and susceptance. A fit, presented in figure 3 with $K_C = 0.1$ s^2/m and $K_R = 0.001$ s^2/m, gives a good adequation with measurement. The high K_c value reflect the electrostrictive effect. As proved by the measurement, the fabricated microdevices exhibit excellent electromechanical performances in dynamic mode. Moreover, they are characterized by a low resonant frequency (≈ 16Hz) which can be advantageous for use in human motion energy harvesting for instance.

The electrostrictive device is then connected to a resistance load R_{Load} with an external voltage source (20V), to evaluate the harvested energy of the resonant MEMS devices. Figure 4.a shows the existence of an optimal resistance R = 166 kΩ, defined as R_{load} in parallel with $R_{oscilloscope}$ (1 MΩ). For this value, a maximum harvested power of ~ 59 nW at 1 g is observed. In addition, the output power increases with the amplitude of the displacement of the microstructure with the largest power measured at the resonance, as expected (figure 4.b). Last, a linear increase of the maximal output power at resonance with the support acceleration (figure 4.c) is observed. It should be noted that a significant part of the harvested power is lost because of the quite low resistance of the electrostrictive layer (few MHz).

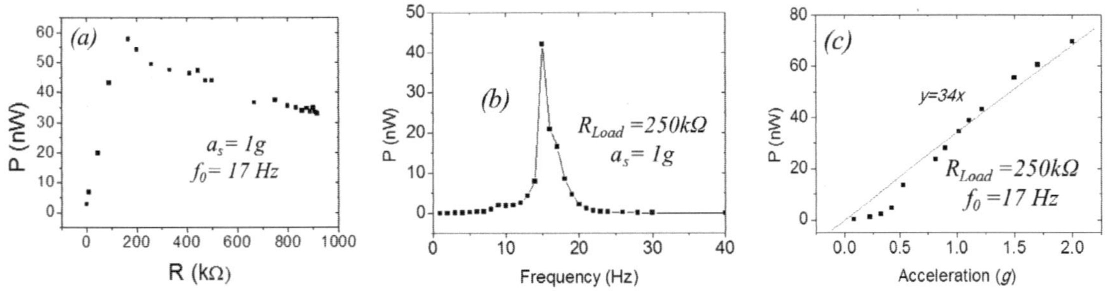

Figure 4: *Output power of the electrostrictive micro-generator (a) as function of the resistance load at the resonant frequency; (b) around the resonant frequency (c) as function of the support acceleration*

4. Conclusion

The present work shows a new organic electrostrictive MEMS for energy harvesting. A novel nanocomposite electrostrictive material based on reduced graphene oxide and PDMS has been integrated in all organic MEMS microcantilevers characterized by a low resonant frequency and a high strain at the resonance. This original electrostrictive micro-generator is an excellent candidate for energy harvesting applications, providing a power density of 8.15 μW/cm^3 at 1 g and f_0 ~17Hz. This output power density is in the same order of magnitude as the one extracted from the standard P(VDF-TrFE-CFE) electrostrictive material [6]. The next step will be to use charge pump with flyback return circuit to manage the converted energy in an optimal way.

5. Acknowledgment

This work is funded by the French National Agency for Research (ELENA project no ANR-12-NANO-0002-03).

6. References

[1] S. Basagni, M. Y. Naderi, C. Petrioli, D. Spenza, "Wireless sensor networks with energy harvesting" Mobile Ad Hoc Networking: Cutting Edge Directions, S. Basagni, M. Conti, S. Giordano, and I. Stojmenovic, Eds. Hoboken, NJ: John Wiley & Sons, Inc, 703-736 (2013).
[2] S. P. Beeby, M. J. Tudor, N. M. White, "Energy harvesting vibration sources for microsystems applications". Measurement science and technology, 17(12), R175 (2006).
[3] Y. Liu, K. L. Ren, H. F. Hofmann, Q. Zhang, "Investigation of electrostrictive polymers for energy harvesting" Ultrasonics, Ferroelectrics, and Frequency Control, IEEE Transactions on, 52(12), 2411-2417 (2005)
[4] J. Yuan, A. Luna, W. Neri, C. Zakri, T. Schilling, A. Colin, P. Poulin, "Graphene liquid crystal retarded percolation for new high-k materials", Nature communications 6, 8700 (2015).
[5] E. Lemaire, D. Thuau, B. Caillard, I. Dufour, "Fast fabrication process of low environmental impact MEMS" J. of Cleaner Production 108, 207-216, (2015)
[6] M. Meddad, A. Eddiai, A. Hajjaji, D. Guyomar, S. Belkhiat, Y. Boughaleb, A. Chérif. "Lowest of AC–DC power output for electrostrictive polymers energy harvesting systems". Optical Materials, 36(1), 80-85 (2013).

Nitride Nanowires: From Rigid to Flexible Piezo-generators

L. Lu[1], N. Jamond[2], P. Chrétien[3], F. Houzé[3], L. Travers[2], J. C. Harmand[2], F.Glas[2], F. H. Julien[1], E. Lefeuvre[1], N. Gogneau[2], M. Tchernycheva[1]

1 Centre de Nanosciences et de Nanotechnologies – site Orsay, Université Paris Saclay, Bât. 220 - Centre Scientifique d'Orsay, F-91405 Orsay, France
2 Centre de Nanosciences et de Nanotechnologies – site Marcoussis, Université Paris-Saclay, Route de Nozay, F-91460 Marcoussis, France
3 GeePs, UMR 8507 CNRS-Centrale Supélec, Paris-Sud et UPMC Université, 91192 Gif-sur-Yvette, France

lu.lu@u-psud.fr

Abstract. Here we employ self-assembled Mg-doped GaN nanowires (NWs) grown by plasma assisted molecular beam epitaxy on Si(111) substrates to fabricate piezogenerators. We first discuss the fabrication and testing of rigid nanowire-based generators and then a flexible generator prototype is shown.

1. Introduction

The development of autonomous electronic equipment requires efficient and portable power supplies. Besides the solar energy, another renewable resource for small-scale energy harvesting is the mechanical vibrations. Thanks to their superior mechanical properties and high piezoelectric coefficient with respect to their bulk materials [1,2], piezoelectric semiconductor nanowires (NWs), in particular GaN NWs, have emerged as promising piezo-materials alternative to conventional piezo-ceramics to fabricate novel ultra-compact power sources.

2. GaN nanowires and their piezoelectric properties

2.1. Growth of GaN nanowires

Self-assembled GaN NWs were grown by Plasma-Assisted Molecular Beam Epitaxy (PA-MBE) on a thin n-doped AlN buffer layer [3] deposited on n-type doped Si(111) substrate (the NW morphology is illustrated in Figure 1). The growth was performed at 760°C, under N-rich conditions. NWs have a cylindrical shape with a hexagonal cross section delimited by {10-10} planes [4]. They are characterized by a height and diameter of the order of 1 µm ± 120 nm and 45 nm ± 20 nm respectively and by a density of the order of 10^9 NWs/cm^2.

2.2. Piezoelectric potential of individual GaN nanowires

The piezoelectric properties of single GaN NWs were investigated by atomic force microscopy (AFM) equipped with an adapted home-made Resiscope module [5]. By scanning over the array of vertical NWs, the conductive AFM tip bends the NWs and at the same time records the output voltage piezo-generated by the nanostructures in response to their deformation [6,7]. Figure 2 shows a 3D representation of the output voltage generated by GaN NWs and collected by the AFM for normal constant force of 149 nN. The average and maximum outputs are respectively 228±120 mV and 350 mV.

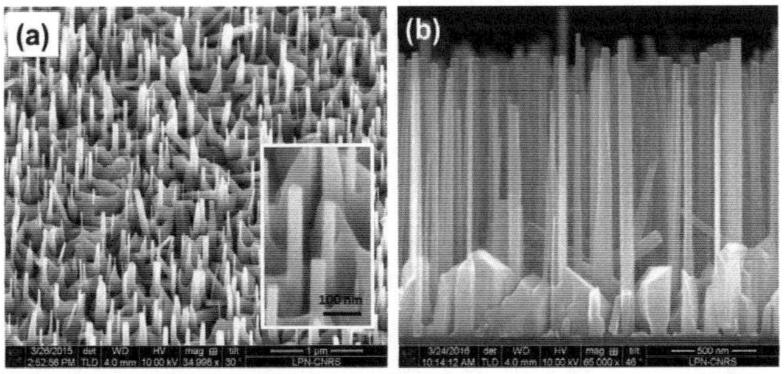

Figure 1. *Tilted (a) and cross section (b) SEM images of Mg-doped GaN NWs grown by PA-MBE on a Si(111) substrate. Inset of (a) displays a high magnification image showing NWs with a diameter as low as 50 nm.*

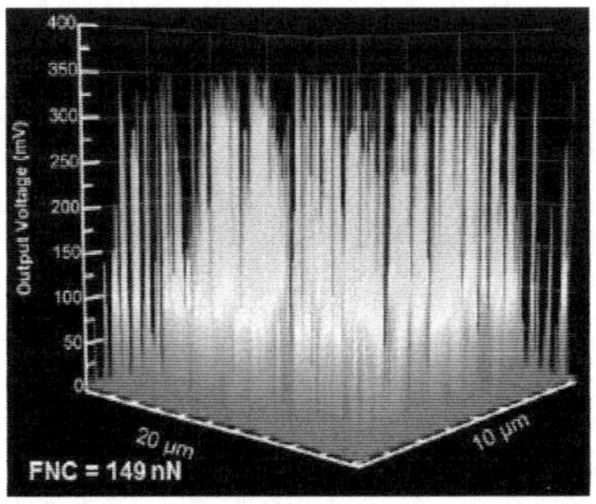

Figure 2.: *3D representation of output voltage generated by GaN NW array under the constant bending force (149 nN) of the AFM tip.*

3. Fabrication and characterization of rigid piezogenerators

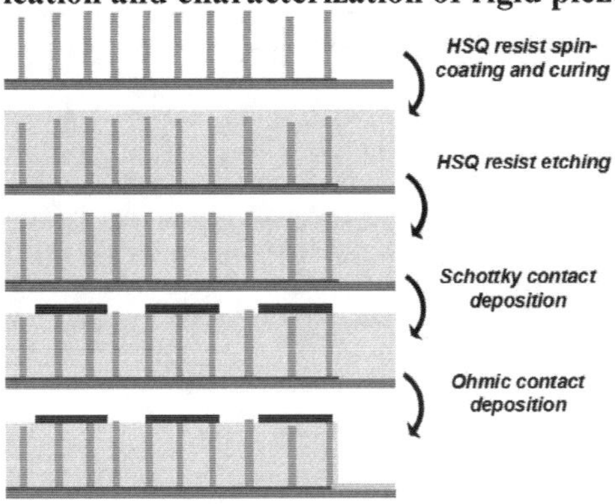

Figure 3. *Process steps to realize rigid piezogenerator device from self-assembled NWs.*

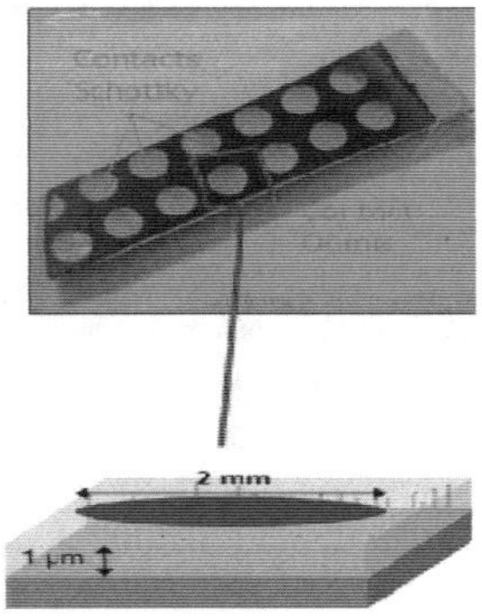

(a)

(b)

Figure 4. *(a) Optical image of a piezogenerator based on a GaN NWs array (b) Schematics of the device.*

3.1 Fabrication of rigid piezogenerators

To fabricate rigid devices, the as grown NW arrays are encapsulated into spin-on glass (HSQ), which is annealed at 400 °C. Then the NW tops are uncovered by doing reactive ion etching and the Schottky contact is deposited, though a shadow-mask, on the NW top side. Finally, an ohmic contact is directly deposited on the substrate (Figure 3).

3.2 Characterization of rigid piezogenerators

The device (Figure 4(a)) is tested by using a cyclic flexion strain setup showed in Figure 5(a). The measured device is fixed at two extremities, and is deformed by the applied force in the middle. Due to the difference of thickness between the substrate (350 μm) and the active layer (1 μm), and thanks to the strong mechanical adherence of the HSQ matrix to the NWs, the lateral bending of the device can induce a stretching of the matrix, thus a thinning of the active layer as a consequence of the volume conservation. This matrix deformation is transmitted to the embedded GaN NWs leading to their homogeneous compression, as illustrated on Figure 5(b).

Journal of Physics: Conference Series **773** (2016) 012010 doi:10.1088/1742-6596/773/1/012010

Figure 6 presents the output voltage generated and synchronized with the magnitude of the force applied to bend the device. We can observe that the generation follows the applied deformation. The solicitation is conducted at low frequency, 3 Hz and 6 Hz separately. By considering the measuring circuit and the device shape, we estimate an average output power density of 6.35 mW/cm^3, and a maximum output power density of about 12.7 mW/cm^3 [8].

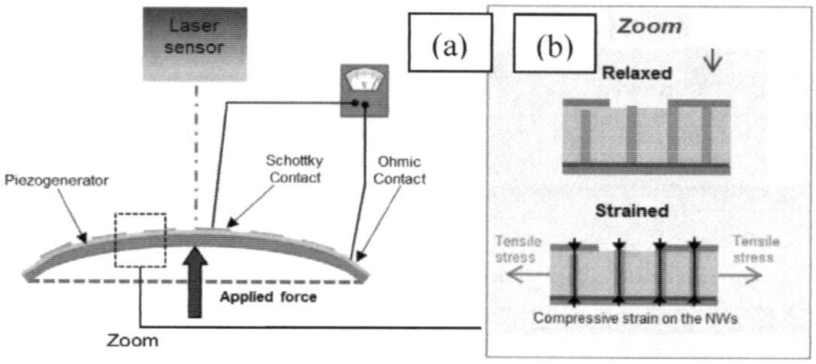

Figure 5. *(a)Schematic of the setup used to characterize the GaN NW-based piezo-generator; The lateral bending of the generator induces a thinning and thus a compression of the active layer as illustrated in (b).*

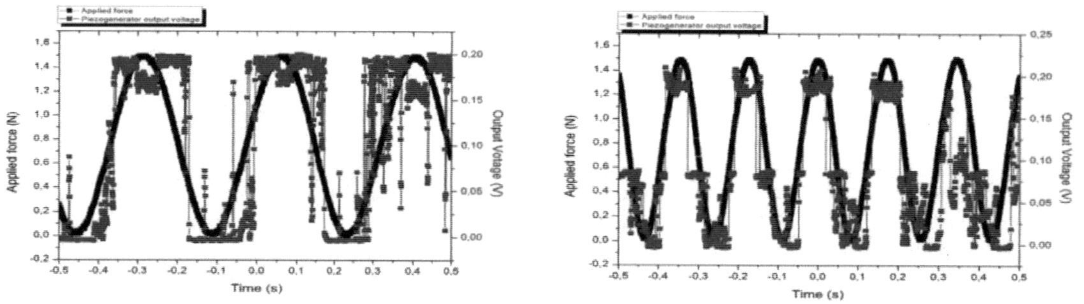

Figure 6. *Output voltages generated by the piezo-generator as a function of its deformation for a bending cycle rate of 3 Hz (left) and 6 Hz (right).*

4. Flexible piezogenerator

The flexibility enables the integration of devices on different objects with arbitrary shape and offers prospects for medical applications. To achieve flexible nanowire generators while keeping the advantages of vertical control-by-design NW arrays, we develop a fabrication method based on PDMS NW encapsulation, membrane lift-off and transfer to a flexible substrate.

First, we embedded the root part of the NWs in a very thin layer of photoresist. Then, diluted PDMS is spin-coated on the NWs to fully encapsulate the NWs array. By doing reactive ion etching, we uncover the top part of the NWs from PDMS layer with a free part length of about 100 nm. Then, a Schottky contact is deposited to fully cover the summits of the NWs, followed by another thick layer of PDMS in order to thicken the structure and facilitate its manipulation. After a short annealing and a submersion of the whole sample in acetone, we use a razor blade to peel the sandwich structure off the substrate and fix the membrane up-side down on another rigid substrate. Making sure that the entire membrane is generally flat over the substrate, we deposit ohmic contact on the NW base part. Finally, the membrane is again flipped and fixed on a flexible substrate. It is easy to get rid of the PDMS thick layer on the other side by simple mechanic lift-off after the sticking step. In this way, we can obtain a vertical-NWs-in-PDMS membrane with metallic contacts on both top and bottom sides of the layer

(Figure 7). A SEM tilted image of this membrane is shown in Figure 8 and Figure 9 spresents an optical image of the final device. The characterization of these flexible piezo-generators is in progress.

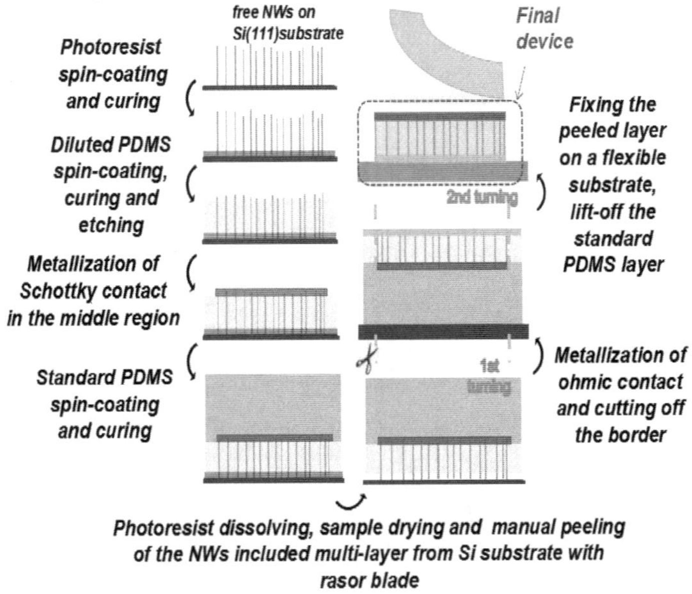

***Figure** 7. Process flow of the nanowires lift-off and the realization of flexible device*

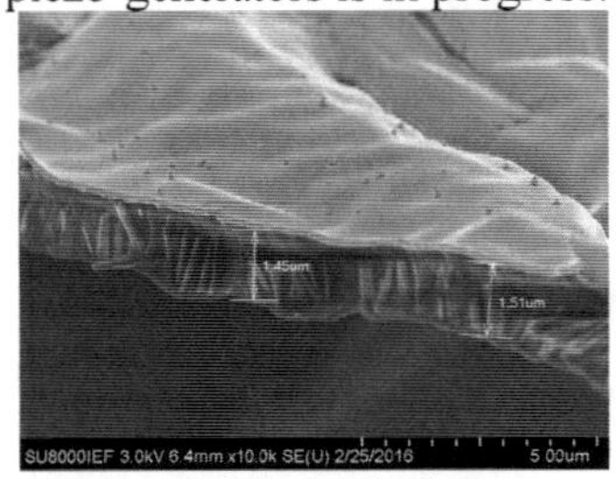

***Figure 8.** SEM titled image of the vertical-GaN-NWs -in-PDMS membrane with top and bottom metallic contacts.*

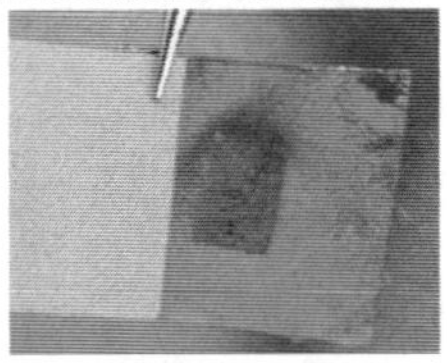

***Figure 9.** An optical image of the as-processed flexible piezo-generator(deposited on flexible copper tape)*

5. Conclusion

Here we demonstrate the fabrication of rigid and flexible piezo-generators integrating vertical array of GaN NWs. The measured results of rigid device set the new state of the art of piezogenerators based on GaN NWs. The generation of the flexible device is promising and is currently under measurement.

References

[1] M. Minary-Jolandan, R. A. Bernal, I. Kuljanishvili, V. Parpoil, H. D. Espinosa, "Individual GaN nanowires exhibit strong piezoelectricity in 3D", Lett. 12, 970 (2012)

[2] R. Agrawal and H. D. Espinosa, "Giant Piezoelectric Size Effects in Zinc Oxide and Gallium Nitride Nanowires. A First Principles Investigation", Nano Lett. 11, 786 (2011)

[3] L. Largeau, E. Galopin, N. Gogneau, L. Travers, F. Glas, J.-C. Harmand, "N-Polar GaN Nanowires Seeded by Al Droplets on Si(111)", Cryst. Growth Des. 12, 2724 (2012)

[4] O. Schneegans, P. Chrétien, F. Houzé, French Patent FR 10 01940, 05/05/2010, international PCT WO 2011/138738, 10/11/20

[5] L. Largeau, D. L. Dheeraj, M. Tchernycheva, G. E. Cirlin, J.-C. Harmand, "Facet and in-plane crystallographic orientations of GaN nanowires grown on Si(111)", Nanotechnology 19, 155704 (2008)

[6] N. Gogneau, P. Chrétien, E. Galopin, S. Guilet, L. Travers, J.-C. Harmand, F. Houzé, "GaN nanowires for piezoelectric generators", Phys. Stat. Sol. RRL, DOI 10.1002/pss.201409105

[7] N. Gogneau, P. Chrétien, E. Galopin, S. Guilet, L. Travers, J.-C. Harmand, F. Houzé, "Impact of the GaN nanowire polarity on energy harvesting", Appl. Phys. Lett. 104, 213105 (2014)

[8] N. Jamond, P. Chrétien, F. Houzé, L. Lu, L. Largeau, O. Maugain, L. Travers, J. C. Harmand, F. Glas, E. Lefeuvre, M. Tchernycheva, N. Gogneau,"Piezo-Generator Integrating Vertical Array of GaN Nanowires",Nanotechnology 27, 325403(2016)

High-performance energy harvester fabricated with aerosol deposited PMN-PT material

C.T. Chen[1], S. C. Lin[1], T. K. Lin[1], W.J. Wu[1*]

[1] Department of Engineering Science and Ocean Engineering Science, National Taiwan University, Taipei, Taiwan

*Email: wjwu@ntu.edu.tw

Abstract. This paper reports a high-performance piezoelectric energy harvester (EH) fabricated with $xPb(Mg_{1/3}Nb_{2/3})$-(1-x)$PbTiO_3$ (PMN-PT) by aerosol deposition method. The result indicates that PMN-PT based EH owns 1.8 times output power which is higher than traditional $PbZr_xTi_{1-x}O_3$ (PZT) based EH. In order to compare the output performance of EH fabricated with PMN-PT compared with PZT, the similar thickness of PMN-PT and PZT thin film is deposited on stainless steel subtracted. The experimental results show that PZT-based EH had a maximum output power of 4.65 μW with 1.11 V_{P-P} output voltage excited at 94.4 Hz under 0.5g base excitation, while the PMN-PT based device has a maximum output power of 8.42 μW with 1.49 V_{p-p} output voltage at a vibration frequency of 94.8 Hz and the same base excitation level. The volumetric power density was 82.95 μW/mm^3 and 48.05 μW/mm^3 for the device based on PMN-PT and PZT materials, respectively. All the results demonstrate that PMN-PT has better output performance than PZT.

1. Introduction

In past years, researchers keep trying to improve the output performance of EH from the optimization of the design of device and the improvement of process. The performance of EH power output has gradually reached the limit of the chosen materials. In our previous research results [1,2], the significant increase of the power output and the improvements of durability were achieved by the change of the EH substrate material. In this study, we tried to improve the overall EH performance by replacing PZT with PMN-PT which has better piezoelectric performance. In several investigations, PMN-PT exhibits high piezoelectric coefficient and electromechanical coupling factor compared with PZT. However, there are still very few studies related to PMN-PT based EH [3,4]. Therefore, this study focuses on intruducing PMN-PT material as the active material. And PMN-PT are deposited on stainless steel substrate with aerosol deposition method with the same design and fabrication process as the PZT that we have done before. The comparison of output performance of the devices which is based on two different piezoelectric materials will be completely analyzed.

2. Device Fabrication Process

Figure 1 shows a schematic illustration of the piezoelectric energy harvester device. Based on microelectromechanical systems (MEMS) fabrication process [5], the lithography and etching are processed as the flow chart shown in Figure 2 .

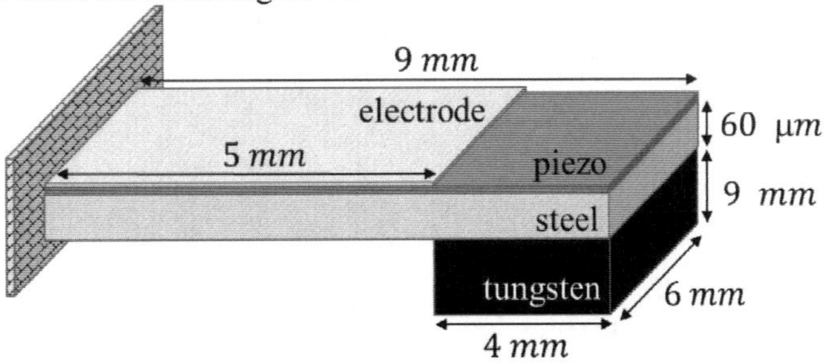

Figure 1. Schematic illustration of the piezoelectric energy harvester device

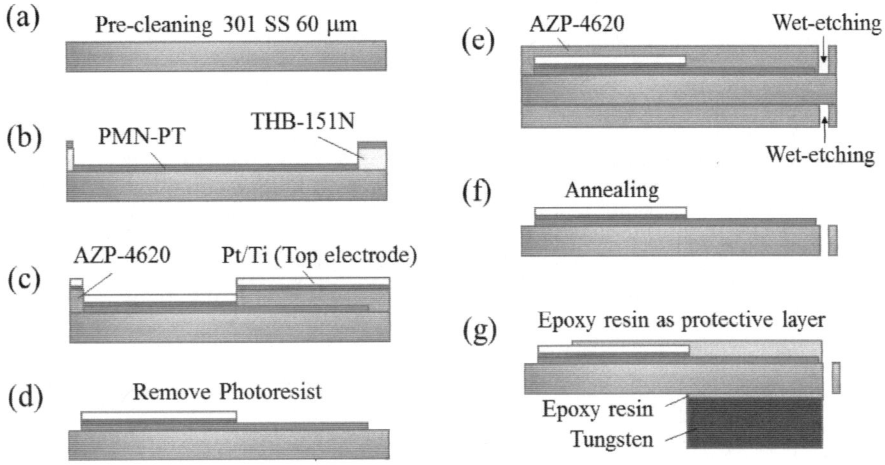

Figure 2. The flow chart of device fabrication process

For better mechanical ductile property [6], 301 stainless steel is chosen as the substrate for device fabrication and the metal substrate also acted as the bottom electrode of the fabricate EH device. The fabrication process steps are described as follows. First of all, metallic oxide and organic residues on the surface of substrate are removed with acid solution. After the cleaning process, a piezoelectric layer is then deposited onto the stainless steel substrate using the aerosol deposition method [7]. The schematic diagram of aerosol deposition equipment is shown in Fig. 3. In order to compare the output performance of the piezoelectric EH based on different materials, PMN-PT and PZT with the thickness of 2.8 um are both deposited on stainless steel substrates. THB-151N photoresist is used for pattering the piezoelectric layer with lift-off process which is shown in Fig 2(c). After patterning the piezoelectric layer, an annealing process with 600 °C is then applied. The upper electrode is then deposited with a 20 nm titanium and 200 nm platinum onto the piezoelectric layer by using E-gun evaporation. Lift-off process is again adopted for patterning the electrode shapes. The beam shape is defined and release with wet etching by using aqua regia. Afterwards, thin epoxy layer is applied on the upper electrode as the protective layer shown in Fig. 2(f). Finally, a tungsten proof mass is glued on the beam tip with epoxy. The tungsten material is selected here because of high density so as to increase the power output at low base excitation level and lower the resonance frequency (Fig. 2(g)).

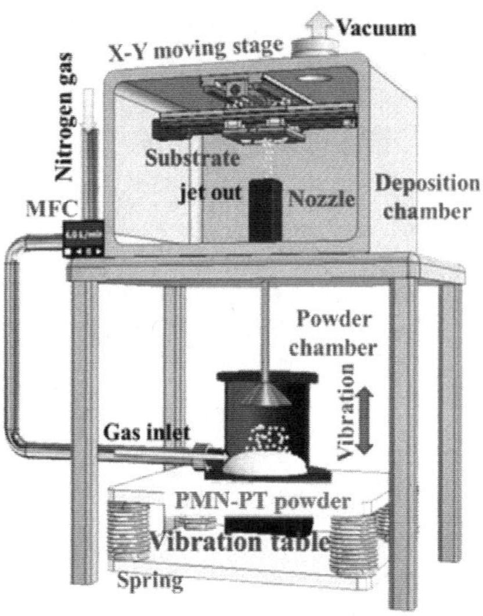

Figure 3. Schematic diagram of aerosol deposition equipment

3. Experimental Setup

Fig. 4 shows a schematic diagram of experimental setup for measuring EH output performance. In order to compare the output performance between PZT and PMN-PT based EH, the thickness of film is analyzed by Probe-Type Surface Analyzer. Then, both EH were mounted on the shaker which was driven by a function generator through a power amplifier to simulation ambient vibration sources with different frequency and amplitude. An accelerometer (B&K Type 4381) is mounted on the shaker with the piezoelectric EH to measure the base excitation acceleration level. The output signal of the device is connected to different load resistance in order to investigate the output characteristic. However, the corresponding output average powers are calculated using the following equation,

$$P = \frac{\left(V_{PP}/_{2\sqrt{2}}\right)^2}{R} \tag{1}$$

,where $V_{p\text{-}p}$ is the peak-to-peak output voltage, and R is the load resistance.

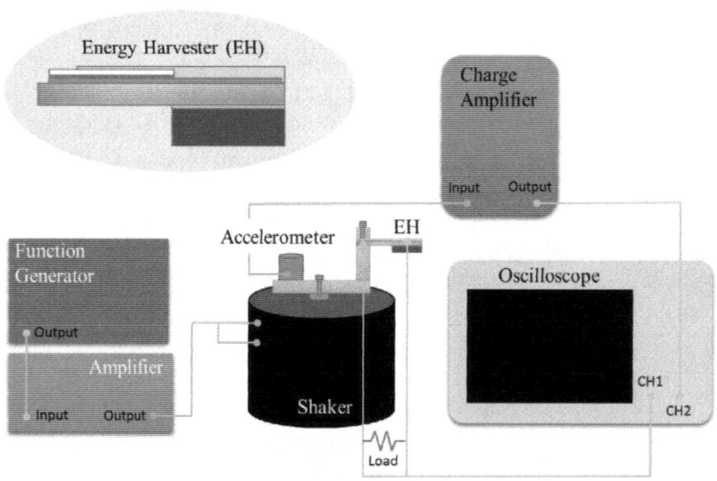

Figure 4. Schematic diagram of experimental setup

4. Results and Discussion

Fig. 5(a)(b) shows the peak to peak voltage ($V_{p\text{-}p}$) versus the vibration frequency under an open circuit condition at different base excitation acceleration levels. The resonant frequency increases when excited acceleration increases. The resonance frequency of the d_{31} mode PZT based EH shifts significantly from 93.4 Hz to 94.4 Hz under 0.3 g to 0.5 g base excitation acceleration levels. Due to the similarity of dimension, thickness and mechanical properties of two piezoelectric layer materials, the resonant frequency of PMN-PT based EH is very closed to PZT based EH. From the experiment, we found out the resonant frequency shifts from 93.6 Hz to 94.8 Hz. The results indicate that the PMN-PT based EH output voltage under an open circuit condition is greater than PZT based EH.

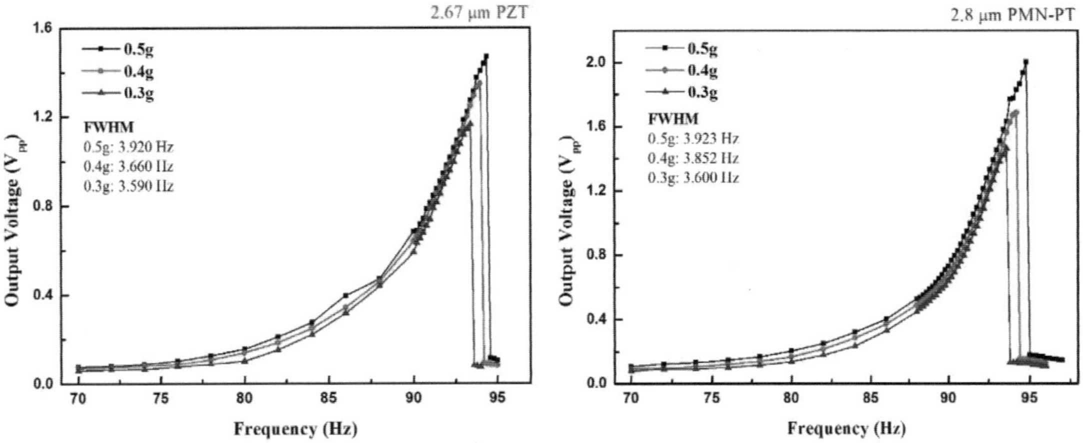

Figure 5. (a) PZT (b) PMN-PT device output voltage vs. frequency under the open circuit condition at 0.3 g to 0.5 g.

In order to measure the output power from EH, the devices are connected to resistors with different resistance values in series. Therefore, the output power could be calculated by equation (1). Fig. 6(a)(b) shows the comparison of PZT and PMN-PT based EH output performance. When both excited at a 0.5 g acceleration level, the PMN-PT based EH could provide 8.42 μW output power which is significantly greater than the PZT based EH with 4.65 μW output power.

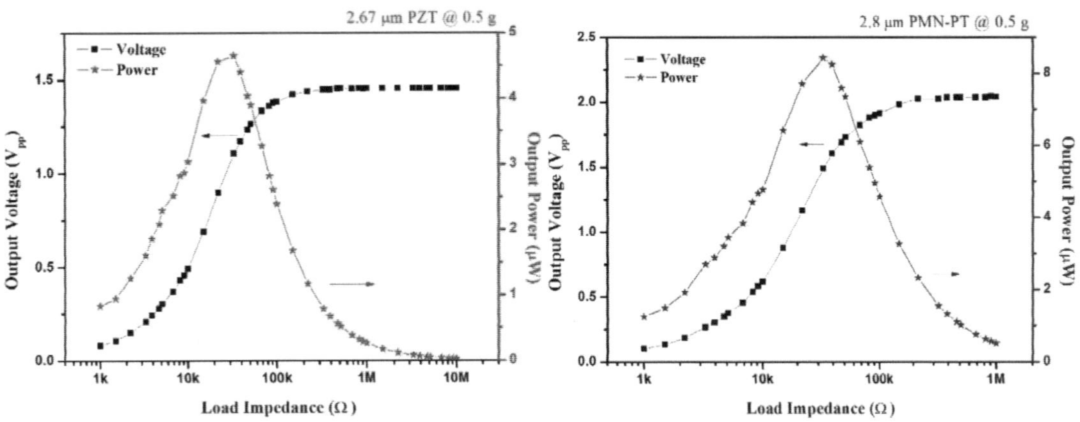

Figure 6. (a) PZT (b) PMN-PT device electrical output vs. load impedance at 0.5 g acceleration.

Table 1. Dimension and Properties of PZT and PMN-PT device

Material	**PZT**	**0.6PMN-0.4PT**
Thickness (μm)	2.67	2.81
Resonant Frequency (Hz)	94.4	94.8
ε_r **(@1 kHz)**	455.08	413.69
d_{33} **(pC/N)**	13.5	18.5
FOM (10^{-24} C/N^2)	0.400(1)	0.8273(2.065)
Voltage (with load) ($V_{p\text{-}p}$)	1.108 @ 0.5g(1)	1.491 @ 0.5g(1.346)
Power (with load) (μW)	4.651 @ 0.5g(1)	8.423 @ 0.5g(1.811)
Power density (μW/mm^3)	48.05 @ 0.5g(1)	82.99 @ 0.5g(1.727)

5. Conclusion

We summarized the characteristics of PZT and PMN-PT based EH in Table 1. With the same exact specification; the voltage output, power output and power density of the PMN-PT device excel the PZT device performance. We concluded that the figure of merit (FOM) of PMN-PT material is better than PZT material. When both devices excited under a 0.5 g base excitation acceleration level, PMN-PT based EH shows 1.8 times maximum power output of the PZT based EH. The volumetric power density of PMN-PT based EH is 1.72 times larger than PZT based EH. According to the results, it indicates that PMN-PT based EH exhibits better output performance than PZT based EH.

References

[1] Wu, W. J., Chen, C. T., Lin, S. C., Kuo, C. L., Wang, Y. J., & Yeh, S. P. 2014 *Journal of Physics: Conference Series* **557** 012027
[2] Lin, S. C., & Wu, W. J. 2013 *Journal of Micromechanics and Microengineering* **23** 125028
[3] Moon, S. E., Lee, S. K., Lee, Y. G., Kim, K. M., Yang, Y. S., Yang, W. S., & Kim, J. 2012 *Journal of the Korean Physical Society* **60** 230
[4] Song, H. J., Choi, Y. T., Wang, G., & Wereley, N. M. 2009 *Journal of Mechanical Design* **131** 091008
[5] Lin, S. C., & Wu, W. J. 2013 *Smart Materials and Structures* **22** 045016
[6] Yang, Z., Xia, G. G., Li, X. H., & Stevenson, J. W. 2007 *International Journal of Hydrogen Energy* **32** 3648
[7] Lee, B. S., Lin, S. C., Wu, W. J., Wang, X. Y., Chang, P. Z., & Lee, C. K. 2009 *Journal of Micromechanics and Microengineering* **19** 065014.

Acknowledgement

This work was supported in part by the Ministry of Science and Technology, National Taiwan University, MOST 105-2923-M-002-010

Photoelectric-Charging-Enhanced MEMS Electret Energy Harvester with Vacuum Packaging

Seonwoo Kim and Yuji Suzuki

Department of Mechanical Engineering, The University of Tokyo,
7-3-1 Hongo, Bunkyo-ku, Tokyo 113-8656, Japan
Email : skim@mesl.t.u-tokyo.ac.jp

Abstract. A novel MEMS electret energy harvester charged with UV light after sealed in a vacuum package is proposed. By 265 nm UV irradiation, electrons are generated inside the package through the photoelectric effect. Uniform surface potential on sidewalls of the comb drives has been obtained. With a MEMS electret generator in a 60 Pa package, 2.28 μW has been obtained at 1 g and 740 Hz, which is 10 times higher than the output power at the atmospheric pressure.

1. Introduction

Electret energy harvester attracts much attention as a promising power supply for low-power-consumption electronics, because of its potential for effective use of ambient low-frequency vibration [1]. Although the charge trapped in electret can be stabilized for tens of years [2], unwanted external influence such as moisture-induced discharge, structure jamming caused from dusts should be avoided. Therefore, to guarantee its stable operation in long term, the harvester needs to be packaged. In case of previous electret energy harvesters, electret is charged before packaging, so that the use of reliable thermal packaging methods were limited due to charge decay of electret under the thermal cycle. To address this issue, we previously proposed soft X-ray charging after packaging, in which vertical electret on comb drives can be charged through photo-ionization of gas in the package [3].

Improvement of effectiveness (the ratio of the electrical power output and VDRG limit [4]) also has been one of the most spotlighted issues for energy harvesters. We have shown that the comb-drive MEMS electret energy harvester can acquire effectiveness up to 57 % [5, 6]. By decreasing the viscous damping of the device, the output power at the resonant frequency and the effectiveness can be further improved. This should be realized with a vacuum package, but the soft X-ray charging cannot be used, because the soft X-ray charging is based on photoionization of gases; when a low-pressure package is used to reduce the viscous damping, even longer charging time should be required.

On the other hand, UV charging [7, 8] employs the photoelectric effect as well as photoionization to generate ions and electrons. In the present study, for the first time, we propose UV charging as the electret charging method for packaged devices in the vacuum/low-pressure environment.

2. Post-packaging UV Charging

Figure 1 shows the charging setup in a vacuum chamber using UV LED (VPC131, Nikkiso) as the light source. The intensity of UV-LED is 12 mW and the wavelength of light is 265 nm. The UV light penetrates the glass lid and absorbed to the metal film on the backside of the glass lid, generating electrons through the photoelectric effect. Photoemitted electrons are accelerated downward by the bias voltage. Then, more electrons and positive ions are generated in between the glass lid and the device electrodes through the Townsend avalanche, and collected to the electrodes.

As the photocathode, aluminum (Al) film is deposited on the backside of 1 mm-thick glass lid (Labo-USQ, Daico) by arc plasma gun (ARL-300, ULVAC). Work function of Al is 4.06-4.26 eV [8], which is lower than the energy of the UV light (4.67 eV). The transmission rate of the UV light is 91 % for the glass lid. The optical skin depth of Al is 11 nm [9], so that the Al film thickness is chosen in between 5-17 nm. The electrode for applying the bias voltage is covered with a 2 μm-thick nickel film (work function= 5.04-5.35 eV) to avoid unwanted photoelectric effect [8].

Figure 2a shows the ion current measured with different Al thicknesses at 2×10^{-2} Pa. The electrons purely generated from the photoelectric effect are collected with the bias voltage of 200 V. The peak value of 2.12 nA is observed with the lid with 8 nm-thick Al. Figure 2b shows the ion current versus the chamber pressure with 6N (99.99995%) N_2 gas. The ion current is increased with the pressure, and reaches its peak at 60 Pa. At this pressure, 12 folds amplification has been obtained by the amplification through the Townsend avalanche. With higher pressures, the ion current is decreased.

3. MEMS Electret Harvester with Comb Drives

Figure 3 displays fabrication process of the MEMS electret energy harvester [3, 5, 6]. Photolithography and DRIE processes are used to pattern pairs of comb drives, etching holes, and folded-beam suspensions (Fig. 3a). Then, the structure on the 8 μm-thick buried oxide layer is released with vapor HF process (Fig. 3b). After bonded to a package, 1.1 μm-thick parylene-C is deposited as the electret (Fig. 3cd). Finally, the lid is bonded to seal the device in vacuum condition (Fig. 3e) and UV light is irradiated for charging (Fig. 3f). Figure 4 shows overview of the MEMS electret energy harvester with comb drives. The seismic mass of 4.29 mg is supported with four folded-tilted springs, which enable traveling distance of 45 μm. 448 pairs of gap-closing comb drives are prepared as electrodes. The maximum/minimum gap of the electrodes is 47/2 μm, designating maximum variation of capacitance ΔC to be 23.6 pF.

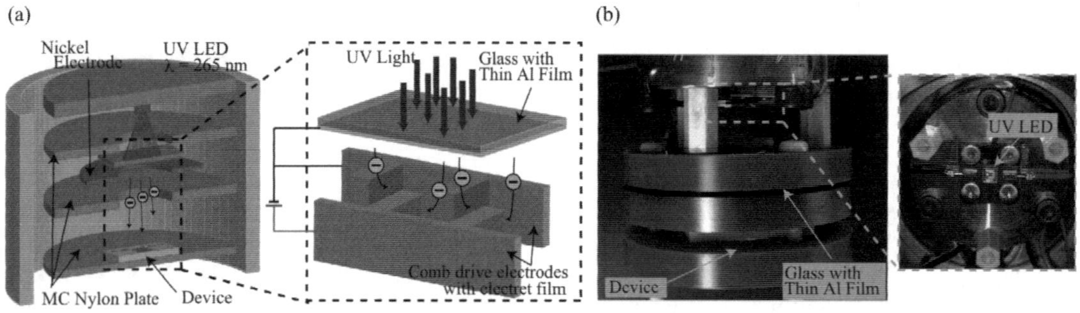

Figure 1. (a) Schematic of the post-packaging UV charging. (b) Picture of the charging setup.

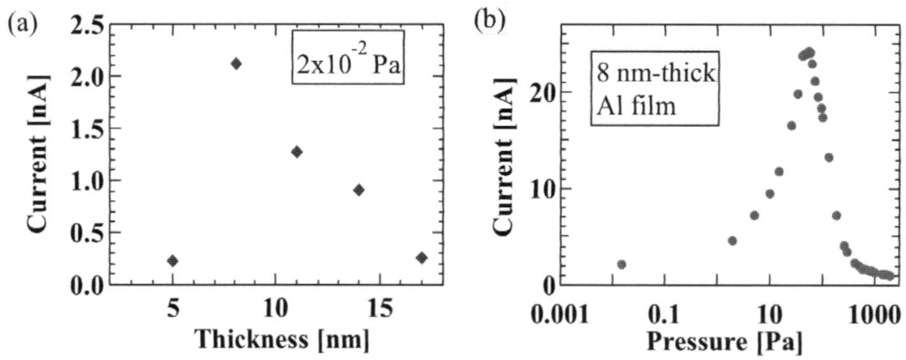

Figure 2. (a) Ion current versus Al film thickness at 2×10^{-2} Pa, (b) Ion current versus pressure for 8 nm thick Al.

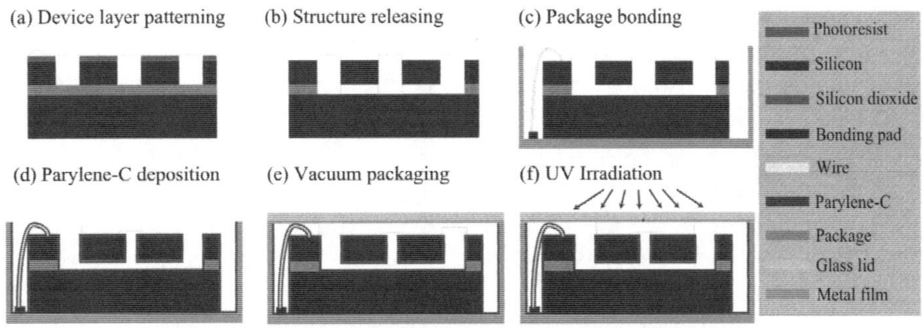

Figure 3. Fabrication process flow of the MEMS electret energy harvester.

Figure 4. Overview of the MEMS electret energy harvester with gap-closing comb drives [6].

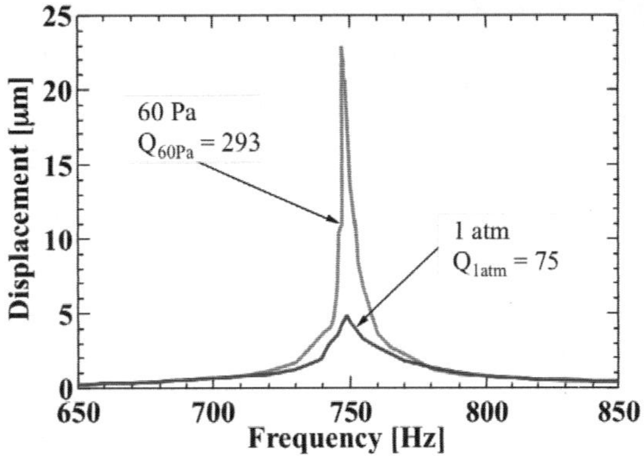

Figure 5. Frequency responses at 1 atm and 60 Pa.

Considering the shear flow and the squeezed film damping with the second-order slip flow model [11], viscous damping is found to be negligible below 500 Pa. As shown in Fig. 5, the Q factor is respectively 75 and 293 at 1 atm and 60 Pa, showing that the viscous damping has significant impact on the device performance.

4. Verification of Post-packaging UV Charging

The MEMS energy harvesters is exposed to 265 nm UV LED (B265SMDM1, Nikkiso, 4 mW) for 4 minutes with the bias voltage of 100V at the pressure of 70 Pa. Then, the sample is intentionally broken, and the surface potential on the sidewalls of the comb drives (Fig. 6a) is measured with Kelvin force microscopy (KFM, SPM-9600, Shimadzu). Figure 6b shows the surface potential distribution on the vertical/longitudinal direction. Although the charging condition is insufficient in this particular experiment, the surface potential is uniform at -22V in both directions, showing applicability of the present charging method for the comb drives.

Then, another device is exposed to UV LED (VPC131) for 10 min at the pressure of 60 Pa, and loaded into a vacuum container shown in Fig. 7a. Sinusoidal vibration is applied by an electromagnetic shaker (ET-126B-1, Labworks). The external load is 2.04 MΩ. Figure 7b shows the output power under 1 g acceleration with the requency sweeping rate of 0.25 Oct/min. At 60 Pa,

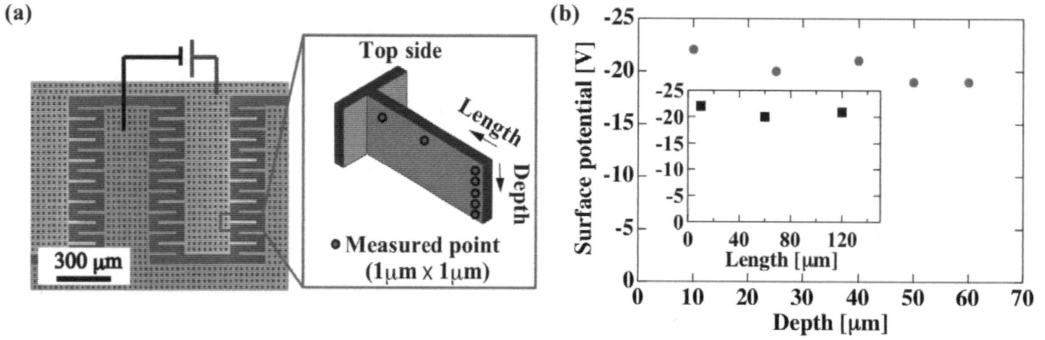

Figure 6. (a) Surface potential measurement positions. (b) Surface potential distribution of the vertical electrets on the comb drive.

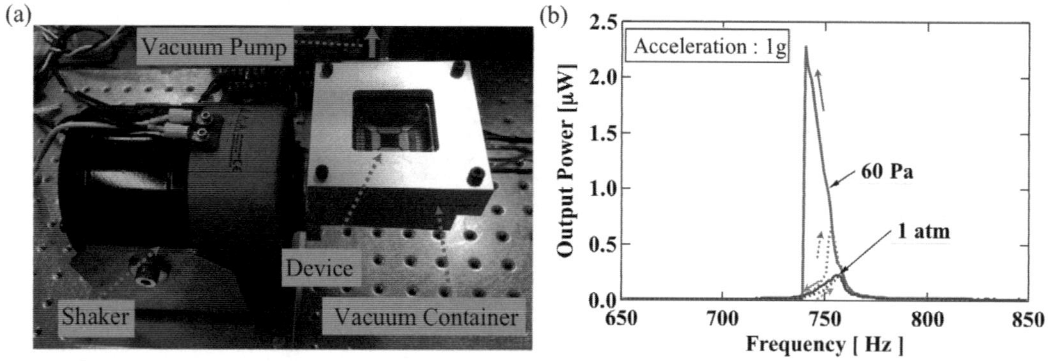

Figure 7. (a) Power generation setup, (b) Power output comparison at 1g acceleration

output power of 2.28 μW has been obtained at 740 Hz, which corresponds to the effectiveness as high as 52 %. On the other hand, only 0.23 μW is obtained at 1 atm.

5. Conclusion

We have developed a novel post-packaging UV charging method using the photoelectric effect on the backside of the transparent package lid. Uniform surface potential is obtained for the vertical electret on comb drives. A UV-charged MEMS electret energy harvester is developed, and up to 10 times higher output power has been realized at 60 Pa than the atmospheric pressure.

This work is partially supported by JST CREST program. SK is supported through the Leading Graduates School Program, "Global Leader Program for Social Design and Management" by MEXT.

References.

[1] Y. Suzuki, *IEEJ Trans. Electr. Electr. Eng.* **6** 101-111 (2011).
[2] G. M. Sessler, *Electrets 3rd edition.*, Laplacian Press (1998).
[3] S. Kim, et al., J. Phys.: Conf. Ser., **557**, 012063 (2014).
[4] P. D. Mitcheson, et al., *Proc. IEEE,* **96**, 1457-1486 (2008).
[5] Q. Fu, and Y. Suzuki., IEEE MEMS'14, San Francisco, 409-412 (2014).
[6] Q. Fu, and Y. Suzuki., Transducers '15, Anchorage, 1925-1928, (2015).
[7] M. Honzumi, et al., *Appl. Phys. Lett.*, **98**, 052901 (2011).
[8] K. Hagiwara, Ph.D Thesis, The University of Tokyo (2015)
[9] D. R. Lide, ed., *CRC Handbook of Chemistry and Physics*, CRC Press, 12-124 (2005).
[10] A. D. Rakic, et al., *Appl. Opt.*, **37**, 5271-5283 (1998).
[11] G. E. Karniadakis, et al., *Microflows and Nanoflows*, Springer, (2005).

COLLISION-FREE STRUCTURE USING THIN-FILM MAGNET FOR ELECTROSTATIC ENERGY HARVESTER

S Yoshii, K Yamaguchi, T Fujita, K Kanda and K Maenaka
Graduate School of Engineering, University of Hyogo
2167 Shosha, Himeji, Hyogo 671-2280, Japan

E-mail: ei16k019@steng.u-hyogo.ac.jp

Abstract. This paper proposes collision-free structure using NdFeB thin-film magnet for vibration energy harvesters. By using stripe shaped NdFeB magnet array on the Si MEMS structure, we finally obtained 3 mN of magnetic repulsive force on 8 × 8 mm^2 specimen with 40 μm air-gap.

1. Introduction

Recently, many researches of IoT (Internet of Things) using the wireless network for machinery, buildings and the human body monitoring [1] are actively studied. Energy harvesting techniques are also attracted attention for autonomous power source for IoT [2]. We study an electrostatic energy harvester among the vibration energy harvesters [3], which generates electric power by the changing in capacitance and the bias voltage between the electrodes when a mass vibrates. The high voltage and the narrow air-gap structure are required for high performance harvesters [4], however an electrostatic attractive force that causes pull-in phenomenon simultaneously is increased. In order to prevent the pull-in phenomenon, stopper mechanism [5], bipolar charging method [6], electrostatic repulsion [7] have been reported. Since the pull-in prevention measure with physical collision reduces the long-term reliability in durability, we propose a collision-free structure using a magnetic repulsive force with sputtering NdFeB/Ta multilayer thin-film magnet. NdFeB/Ta multilayer thin-film magnet have strong magnetic properties, however it difficult to fabricate in MEMS (Micro Electromechanical Systems) devices. In our previous studies, the deposition and microfabrication technology with RF magnetron sputtering of NdFeB/Ta multilayer was successfully developed [8][9].

In this study, the characteristics of magnetic repulsion force for vibration energy harvester are confirmed in simulations and measurements. Comparing with previous work of stripe shaped magnet [10], magnetic repulsive force (MRF) of striped magnet with optimized dimension shows 10 times larger than normal shape on 30 μm air-gap structure.

2. Demagnetizing field reduction by microfabrication magnet

2.1. Magnetic levitation

Figure 1 shows the preliminary experiment of magnetic levitation by MRF generated from 15 μm thick NdFeB/Ta film on a 8 × 8 × 0.5 mm^3 Si specimen guided by Teflon side-wall. One specimen floats 2 mm above another same specimen by the MRF.

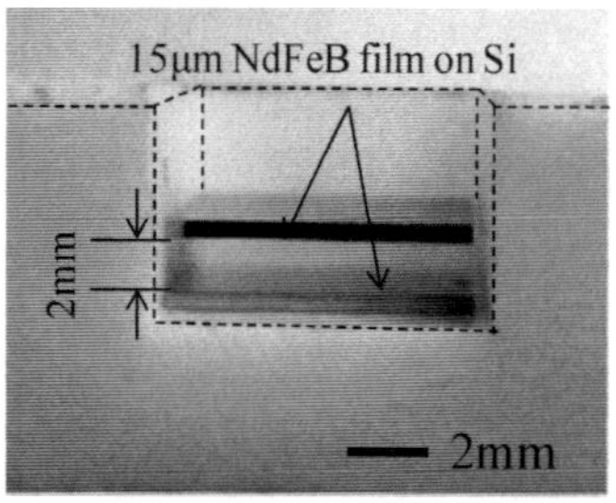

Figure 1. Preliminary result of magnetic levitation on Si specimen.

2.2. Electromagnetic field analysis

To realize an optimal dimension of magnet structure, an integral element method analysis (Qm ver 3.01, Shift lock Ltd., Japan) is performed in an effective volume of 8 mm × 8 mm × 15 μm. A magnetic flux density distribution at the position of 30 μm from the thin magnet surface is shown in Fig. 2 (a). Magnetic properties for simulation were from NdFeB/Ta laminated film produced in previous studies. The residual magnetic flux density of 1 T, the coercive force of 1 MA/m, the relative permeability 1.1 were adopted. Figure 3 shows a calculated magnetic flux density profile for the striped sample shown in Fig. 1. The magnetic flux density is accumulated on the edge and extremely low in the center part because of a demagnetization that is caused by a low structural aspect ratio. In order to increase the magnetic repulsive force, we reduced the demagnetizing field by increasing the aspect ratio by using a striped shape magnet. Figure 4 shows a calculated magnetic flux density profile for striped shape magnet with 50 μm width and 40 μm interval. Then the magnetic flux density is changed dramatically for transverse direction, however, uniformly spread along to a longitudinal direction.

The MRFs versus in-plane magnet positions for normal and striped shaped magnets are shown in Figs. 4 and 5. Figure 4 shows the MRF of 15 mN with 30 μm air-gap in a state that each magnets completely opposite (position = 0 μm). Because of the demagnetization, it can be confirmed that when one of the magnets is moved in the horizontal direction (X- or Y-direction), the MRF is greatly reduced and attractive force also appears in further movement. Figure 5 shows the MRF from 70 μm width and 70 μm interval striped magnet with 30 μm air-gap. There are two lines for transvers (X-direction) and longitudinal (Y-direction) movements. From the results, the MRF for transvers movement was changed according to stripe interval, however the MRF for longitudinal movement shows very low decrease. We confirmed that the MRF from striped shape magnet shows 10 times higher intensity than normal one and it has wide operation range along to longitudinal direction.

In our previous study, we investigate optimal dimensions i.e. magnet width, interval for several air-gaps by calculating the maximum MRF. For 30 μm air-gap, we obtained 70 μm width and 70 μm interval magnet shows the maximum MRF of 130 mN [10]. Each air-gap has optimum dimension and the MRF increases as narrowing gaps.

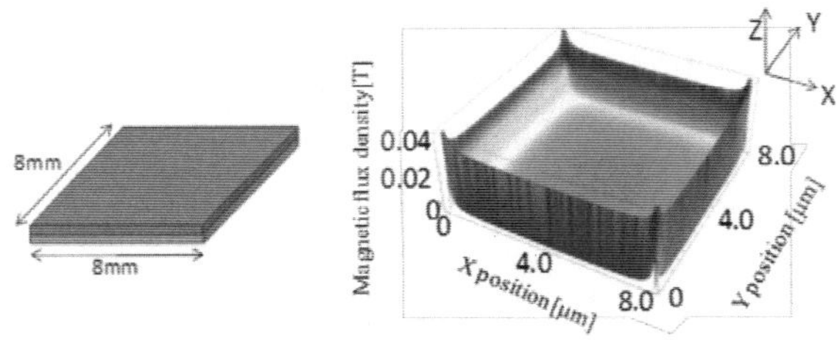

(a) Structure (b)Magnetic flux density
Figure 2. Analysis model and result of the thin-film normal shape magnet.

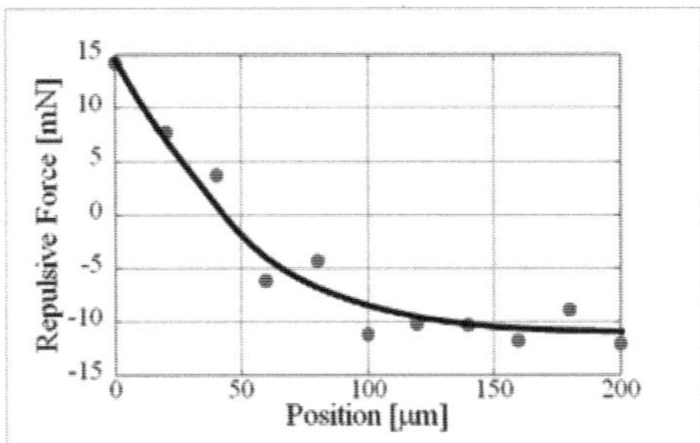

Figure 4. MRF vs. horizontal position shift for normal shape magnet.

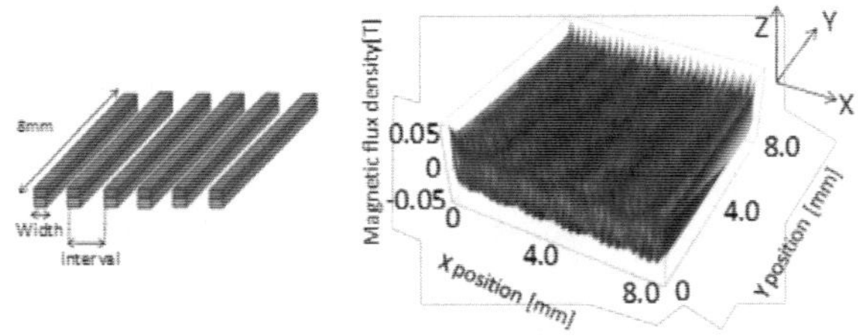

(a) Structure (b) Magnetic flux density
Figure 3. Analysis model and result of the striped magnet.

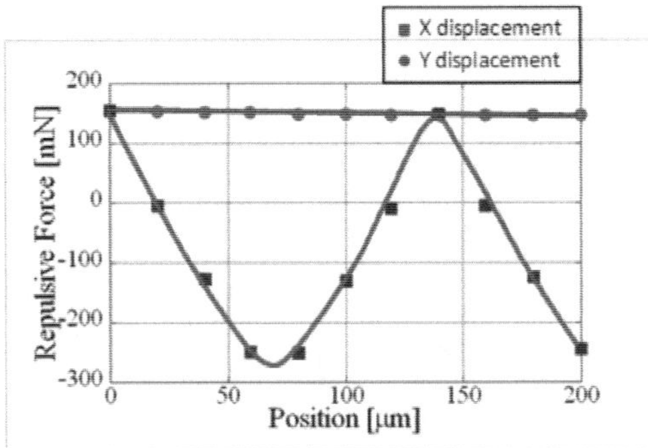

Figure 5. MRF vs. horizontal position shift for striped shape magnet.

Measurements of the MRF

To confirm the MRF from the fabricated device, we measured the MRF by using a precision electronic force balance. The specimen with magnetic film can be held in parallel and aligned by XYZ- θ stage (Fig. 6). Figure 7 shows the MRFs for transvers and longitudinal movement of 40 μm air-gap specimen. The magnet specimen was fabricated with dimension of 50 μm width and 40 μm interval in 8 x 8 mm^2 active area. The result shows similar tendency of the preliminary simulation result that shown in Fig. 5. However the MRF is extremely low compare to the simulation result with same dimensions (Fig. 8). The measured amplitude of the MRF was about 3 mN and 6% of the analysis one.

We assumed that the reduction of the MRF was caused by a degradation of the NdFeB material. We adopted the magnetic flux density of 1 mT from the previous result for simulation, however the measured result from the present NdFeB film shows 0.05 mT that is 4% of the analysis condition. It was confirmed that the MRF is reduced by the magnetic failure of the magnets. We are evaluating various sputtering conditions of NdFeB film, and trying to find a cause of the material degradation.

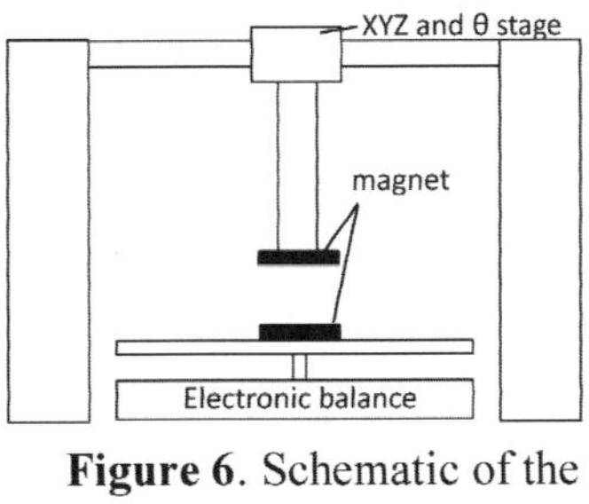

Figure 6. Schematic of the force measurement set-up

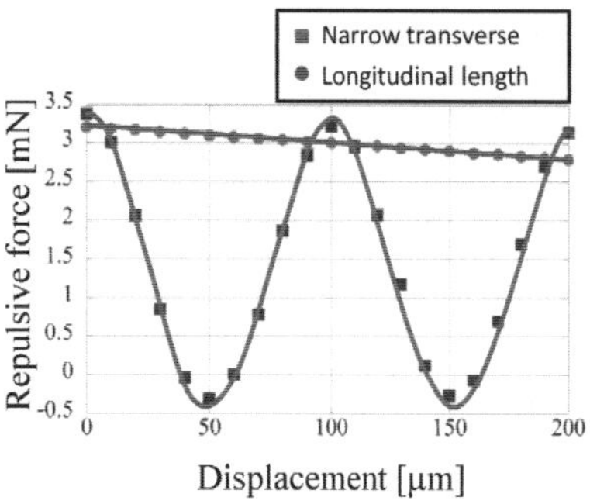

Figure 7. Experimental results of MRF of the striped magnets on positions shift for horizontal direction

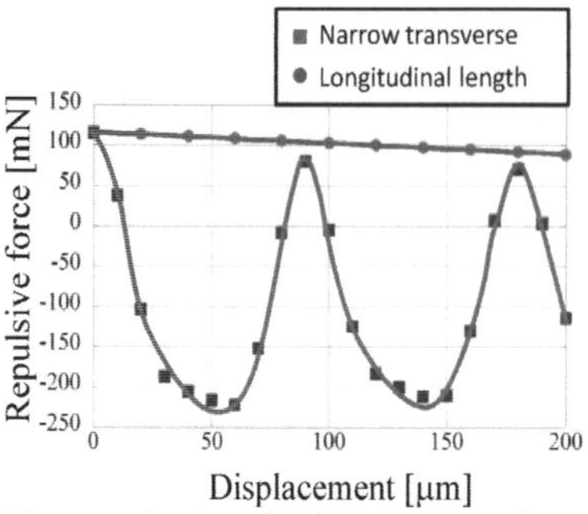

Figure 8. Analysis results of MRF of the striped magnets on positions shift for horizontal direction

3. Conclusion

In this study, we proposed collision-free structure using the NdFeB film for its repulsive force. The demagnetizing field in low aspect thin-film magnet can be reduced by using striped shape structure with optimum dimension. Although it showed low amplitude of the repulsive force because of material degradation, we confirmed that the striped magnet is useful mechanism for collision-free in wide travel range vibration harvesters.

Acknowledgement

This work was supported by JSPS(Japan Society for the Promotion of Science) KAKENHI Grant-in-Aid for Scientific Research(C) Number 16K06308.

References

[1] P. D. Mitcheson et al., "Energy Harvesting From Human and Machine Motion for Wireless Electronic Devices", Proceeding of the IEEE (vol.96, No.9), pp.1457-1486, 2008

[2] A. Darwish et al., "Wearable and Implantable Wireless Sensor Network Solutions for Healthcare Monitoring, " Journal of Sensors 2011, vol. 11, issue 6, pp. 5561-5595.

[3] K. Minami et al., "Electrostatic Energy Harvester High Density of Electrode for Higher Output Power", Proc. of PowerMEMS2015, Journal of Physics: Conference Series 660, 012107, 2015(12)

[4] J. Boland et al., "Micro electret power generator", The Sixteenth Annual International Conference on MEMS, Kyoto, Japan, pp. 538-541.

[5] K. Fujii et al., "Development of Electret Energy Harvester with Micro Spacer" Emerging Trends in Engineering and Technology (ICETET), 2012 Fifth International Conference on. IEEE, 2012. p. 48-51.

[6] Y. Tsurumi et al., "Non-Contact Electrostatic Micro-Bearing Using Polymer Electret" Proc. MEMS 2008, pp. 511-514.

[7] T. Fujita et al., "Bipolar Electret Charging Method for Energy Harvester." Proc. of PowerMEMS2012, pp. 436-439.

[8] K. Yamaguchi et al. "MEMS Narrow Gap Electromagnetic Harvester with Mitigation of Curvature Distortion." Journal of Physics: Conference Series. Vol. 660. No. 1. IOP Publishing, 2015.

[9] K. Yamaguchi et al., "Optimal Design of Electromagnetic Harvester with Sputtered Thin NdFeB/Ta Film and Considering Lorentz Force." Trans. IEEJ, Sensors and Micromachines 136.3 (2016): 72-76.

[10] S Yoshii et al., "Pull-in Prevention Method Using NdFeB Thin-Film Magnet for Electrostatic Energy Harvester" Proc. of APCOT 2016, Kanazawa, Japan

A Frequency-Independent Vibrational Energy Harvester using Symmetrically Charged Comb-Drive Electrodes with Heavily Doped Ion Electrets

H. Mitsuya[1], H. Ashizawa[1], K. Ishibashi[1], H. Homma[2], M. Ataka[2], G. Hashiguchi[3], H. Fujita[2], and H. Toshiyoshi[2]

[1]SAGINOMIYA SEISAKUSHO, Inc., 535 Sasai, Sayama, Saitama 350-1395, Japan
[2]Institute of Industrial Science, The university of Tokyo, 4-6-1 Komaba, Meguro-ku, Tokyo 153-8505, Japan
[3]Reserch Institute of Electronics, Shizuoka University, 3-5-1 Jyouhoku, Naka-ku, Hamamatsu, Shizuoka 432-8011, Japan

E-mail: hiro-mitsuya@saginomiya.co.jp

Abstract. An energy harvester has been developed to efficiently earn energy from both cyclic and impulse vibrations by using a symmetric pair of comb-electrodes that are heavily doped with potassium-ions to form electrets. By equalizing the electro-mechanical forces on the opposing comb-drives, energy conversion efficiency is enhanced for both impulses and broad-frequency harmonic vibrations.

1. Introduction

Energy harvesting is a vital component for Internet-of-Things (IoT) [1] sensing nodes that must operate without batteries. In such applications, batteries are not an appropriate power source because they must be frequently replaced. Typically, more than 1 mW is necessary as a power source for both sensors and wireless radio transmission devices. In order to generate large power, most existing devices are optimized in two ways; first type uses large electrostatic charges coupled with movable electrodes and the other type has a large resonant response, in harmonic terms a high Q-factor, at a selected frequency. However, these devices are ill-suited to capture energy from ambient vibrations such as human and vehicular motion which consist of random vibrational patterns and incoherent impulse rather than consistent harmonic vibrations. This deficiency arises because large electrical charges cause strong electrostatic attraction and tend to make the devices too stiff to respond to weak vibration. Furthermore, such devices are incapable of capturing wide spectrum vibrational energy from random and incoherent impulses due to a narrow resonant frequency peak represented by a high Q-factor. Therefore, these devices can only respond to either large amplitude impulse motions or consistent harmonic motions within a specific frequency range. We have developed a device that overcomes these limitations.

From amongst three major types of vibrational energy harvesters (i.e., piezoelectric, electro-magnetic, and electrostatic) for distributed sensor network nodes, we used the electrostatic type based on potassium-ion electrets because of the compatibility with impulse and broad-frequency harmonic vibrations. Compared to electrets made by corona discharge [2] or soft X-ray radiation [3], electrically

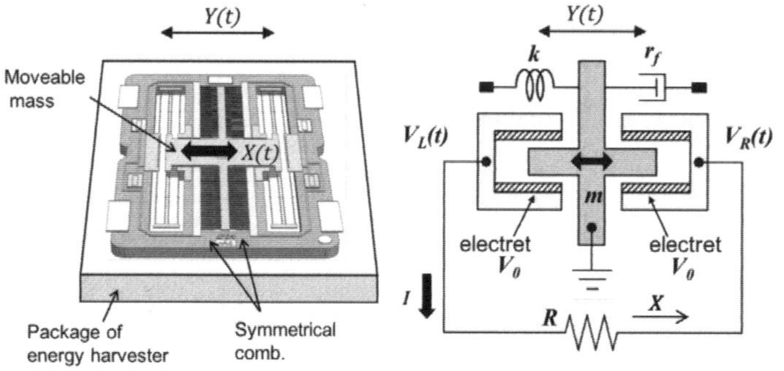

$$I = \frac{V_L - V_R}{R} \qquad (1.3)$$

$$Q = \frac{1}{\frac{1}{Q_m} + \frac{1}{Q_e}} \qquad (1.4)$$

$$m\ddot{X} = -m\ddot{Y} - kX - r_f\dot{X} - \frac{A}{2V_o}(V_L - V_o)^2 + \frac{A}{2V_o}(V_R - V_o)^2 \qquad (1.1)$$

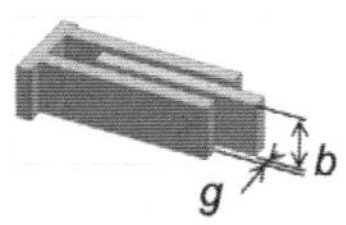

$$A = \frac{2n\varepsilon_o b V_o}{g} \ (force\ factor) \qquad (1.2)$$

$$Q_m = \frac{\sqrt{mk}}{r_f} \quad \left(\text{mechanical energy dissipation}\right)$$

$$Q_e = 1 + \frac{C_o k}{A^2} \quad \left(\text{electrical energy dissipation}\right)$$

Figure 1. Schematic of the device and an equation of the symmetrically comb-drive electrode

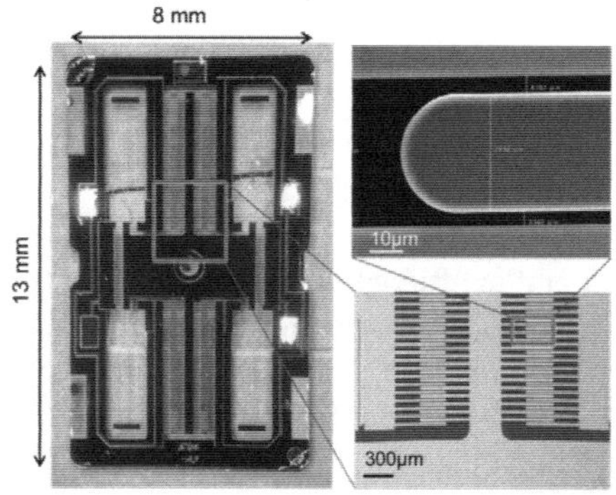

Figure 2. Photographs of the developed energy harvester chip (8 mm × 13mm): The chip consists of symmetrical pair of comb-drive-electrodes on both sides of the mass. These two opposing combs are 120 teeth of 30 μm in width. The overlap between the combs is 300 μm with a gap of 5 μm. The two electrets are equally biased at 150 V

activated solid ions in potassium-doped silicon dioxide can generate extremely strong electric fields even within the narrow gaps of a comb-drive [4]. In this work, we propose a novel device to achieve high efficiency energy conversion for both impulse and broad-frequency harmonic vibrations.

2. Concept and Device

2.1. Power generation method

Developed energy harvester shown in **Figure 1,** and the device uses symmetrically arranged comb-drives electrodes with charged electrets coupled with a moveable electrode mass. By using an electrically symmetrical configuration, the additional rigidity that usually arises from electrostatic attraction can be balanced and cancelled, thereby keeping the mechanical stiffness of the device independently from the electret bias voltage.

When the packaged device is mechanically shaken by an amount of Y from external vibration, the movable mass inside the energy harvester chip is displaced by X. This motion can be described by equation (1.1). As shown in **Figure 1** voltages V_L and V_R are generated at the left and right comb-drives, respectively, as a result of this motion. Consequently, the output current I (1.3) flows through the load resistor R. As can be seen in the expression for Q-factor (1.4), a large force factor A, achieved by applying a large electret voltage V_0 between the narrow gap g of the comb-drive (1.2), can reduce the Q-factor. The low Q-factor allows the device to freely capture low-amplitude impulse vibrations as well as wide-frequency harmonic motions by the efficient and fast conversion of the kinetic energy of the mass to electrical energy.

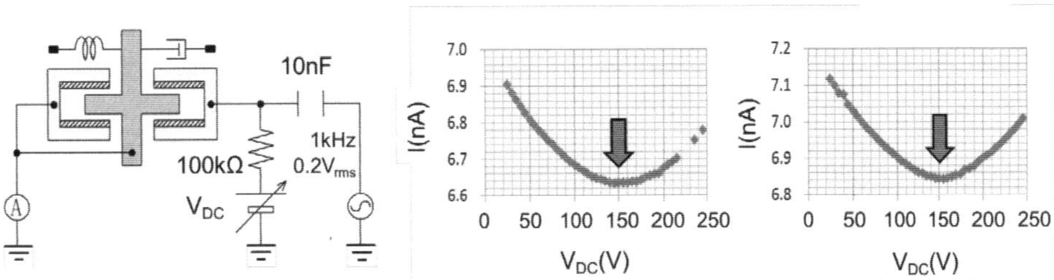

(a) Measurement setup (left electret) (b) Left electret (c) Right electret

Figure 3. Photographs of the developed energy harvester chip (8 mm x 13mm)

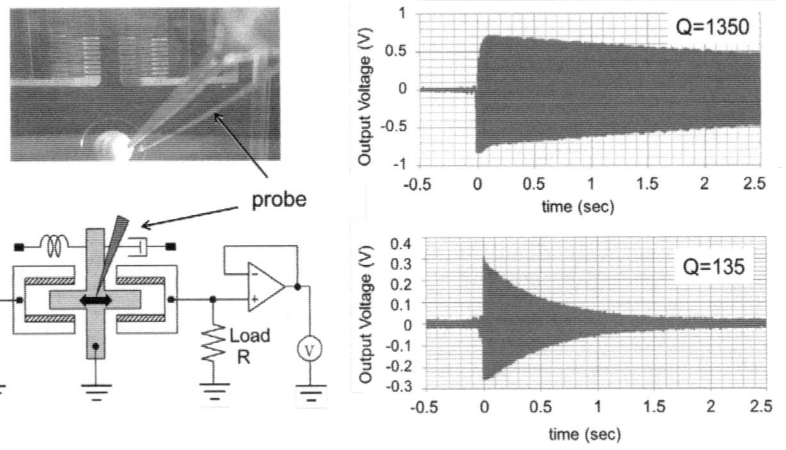

(a) Experimental setup (left electret) (b) Load R (top: ∞, bottom: 10M Ω)

Figure 4. Experimental setup and results: movable mass flicked by a probe

2.2. Fabrication and Device

The energy harvester device shown in **Figure 2** was fabricated by using double-sided Deep-RIE processes on an SOI wafer. The chip (8 mm x 13 mm) consisted of symmetrical pairs comb-drive-electrodes on both sides of the mass. The whole device was thermally oxidized at 980°*C* with an aqueous KOH bubbler to form the ion-rich silicon dioxide film. In order to prevent the oxidization of the electrical contacts, the area where contact pads were fabricated was covered with a 100-*nm*-thick silicon-nitride (Si_3N_4) film by low-pressure chemical vapor deposition (LPCVD) before the oxidation process. Finally, the potassium-doped-oxide was electrically polarized at approximately 500°*C* by applying a 150-V bias voltage for approximately 15 minutes to form a built-in electret with a large static charge.

2.3. Electret characteristics

In order to measure the electret charging voltage, we adopted the experimental setup shown in **Figure 3(a)**. We measured the current passing through the left electrode, by sensing the motion of the mass induced by AC excitation by 0.2 V at 1 kHz, while applying a voltage to the right side in order to gradually neutralize the static charge on that side only. We then repeated the experiment after exchanging the right and left side. As can be seen in **Figure 3(b)** and **(c)**, the both sides exhibited the same displacement dependence on DC bias, indicating that the two electrets are equally biased at 150 V.

3. Experimental & Simulation

3.1. Experimental setup and Results

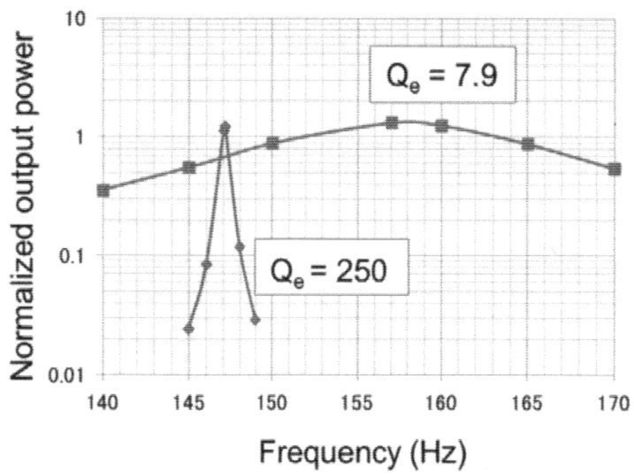

$r_f = 0$

$X_{max} = 200\ \mu m$

V_0	A	Q_e
300 V	13.5 μC/m	7.9
50 V	2.3 μC/m	250

Figure 5. Simulated frequency characteristics

Figure 4(a) shows the experimental setup. We performed a proof-of-concept experiment to demonstrate the electrical characteristics of the device by actuating the unpackaged harvester using a probe station. In **Figure 4(b),** the vertical axis represents the mechanical energy stored within the device after a manual impulse was applied to the device. In the top graph of **Figure 4(b)**, the device was not connected to a load resistance, and therefore did not lose a significant amount of energy over time. When a 10 MΩ load was connected in the bottom graph of **Figure 4(b)**, however, the full impulse energy was transmitted to the electrical load within less than two seconds. In other words, large electrical damping is introduced to the oscillator in a form of energy transmission to the load resistance.

3.2. Simulated Results and Discussion

Figure 5 shows the simulated output of two devices with electrets of different voltages. The output has been normalized to the applied acceleration required to displace the mass by 200 μm. As can be seen, the higher the electret voltage is, the lower the Q-factor, thereby allowing to efficiently harvest the energy across a wide frequency range. Moreover, this small Q-factor also enables the rapid decay of vibrations caused by energy conversion. It is ideal for capturing successive impulse vibrations.

4. Conclusion

In this study, we have proposed an energy harvester by using symmetrically charged comb-drive electrodes featuring very low Q-factor owing to the extremely high force factor *A* based on our potassium-ion-electret. The proposed technology allows us to achieve high efficiency energy conversion for both impulse and broad-frequency harmonic vibrations that are abundant around us.

5. Acknowledgments

This work is partly supported by New Energy and Industrial Technology Development Organization (NEDO).

References

[1] Luigi Atzori, Antonio Iera, Giacomo Morabito: "The Internet of Things: A survey", Computer Networks, Vol. 54, Issue 15 pp.2787-2805 (2010)

[2] R. A. C. Altafim, J. A. Giacometti, J. M. Janiszewski : "A novel method for electret production using impulse voltages" Proc. 7th International Symposium on Electrets (ISE 7), Berlin, Sep.1991, pp. 267-271 (IEEE Catalog No. 91CH3029-6).

[3] K. Hagiwara, M. Honzumi, M. Goto, T. Tajima, Y. Yasuno, H. Kodama, K. Kidokoro, K. Kashiwagi, Y. Suzuki, : "Novel Through-substrate Charging Method for Electret Generator Using Soft X-ray Irradiation" PowerMEMS 2009, Washington DC, Dec.2009

[4] T Sugiyama, M Aoyama, Y Shibata, M Suzuki, T Konno, M Ataka, H Fujita and G Hashiguchi: "SiO2 Electret Generated by Potassium Ions on a Comb-Drive Actuator" 2011 *Appl. Phys. Express.* **4** 114103-1-3.

Characterization of the capacitance variation of electrostatic vibration energy harvesters biased following rectangular charge-voltage diagrams

Armine Karami[1], Dimitri Galayko[1], Philippe Basset[2]
[1]Sorbonne Universités, UPMC Univ Paris 06, UMR 7606, LIP6, F-75005 Paris, France
[2]Université Paris-Est, ESYCOM, ESIEE-Paris, 93160 Noisy-Le-Grand, France

E-mail: armine.karami@lip6.fr

Abstract. This paper presents for the first time a method to measure the capacitance variation of electrostatic vibration energy harvesters (e-VEHs) that employ conditioning circuits implementing a biasing scheme that can be represented by a rectangular charge-voltage diagram. Given the increasing number of e-VEHs using such complex conditioning circuits and the complex dynamics that are induced from this type of biasing, a mean to assess this measurement is of primary importance for the analysis of e-VEHs. The proposed method is based on the inspection of the voltage evolution across two simple conditioning circuits implementing a rectangular charge-voltage diagrams biasing scheme. After the method is presented, it is carried out for the characterization of a state-of-the-art MEMS e-VEH.

1. Introduction

Recent research on micro-machined electrostatic vibration energy harvesters (e-VEHs) have shown that the electrical conditioning and load interfacing part of such systems is critical to the performances in terms of harvested power. In the light of this consideration, researchers have started to investigate new types of conditioning circuits [1, 2, 3]. These new conditioning circuits have in common that their biasing of the transducer, across one cycle of its capacitance variation, can be summarized by a rectangular charge-voltage characteristic diagram (QV diagram) [4].

However, the way this type of biasing affects the device's dynamics via the electromechanical coupling effect is still unclear: although semi-analytical approximations are possible for a limited range of input conditions [5], no measurement method allowed to measure the electromechanical coupling effect on the dynamics of a given e-VEH. In particular no method was reported for the measurement of the extremal values of capacitance variation for a device submitted to harmonic input excitation and biased following a rectangular QV diagram.

In this paper, a simple method is proposed to carry out such a measurement. This method only involves the use of characterization equipment that is typically found in MEMS e-VEHs characterization labs, i.e., generic electronics components, characterization gear, and a mechanical shaker. This method is based on the dynamics of two simple circuits implementing biasing schemes described by rectangular QV diagrams. After a brief presentation and analysis of these circuits, the method is presented and carried out as an example on a state-of-the-art micro-machined, symmetrical gap-closing geometry e-VEH.

2. Presentation of the circuits used for the characterization

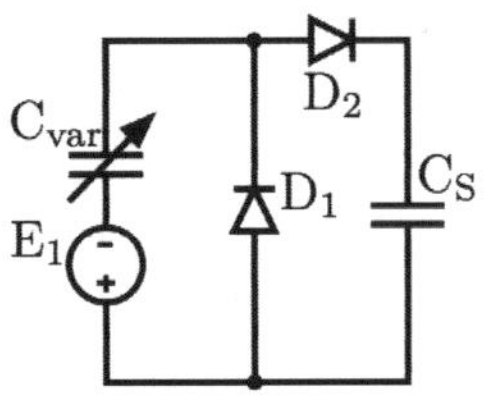

Figure 1. First conditioning circuit used for the characterization of C_{max} and C_{min}.

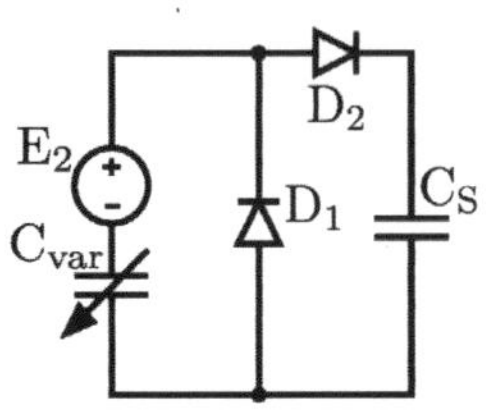

Figure 2. Second conditioning circuit used for the characterization of C_{max} and C_{min}.

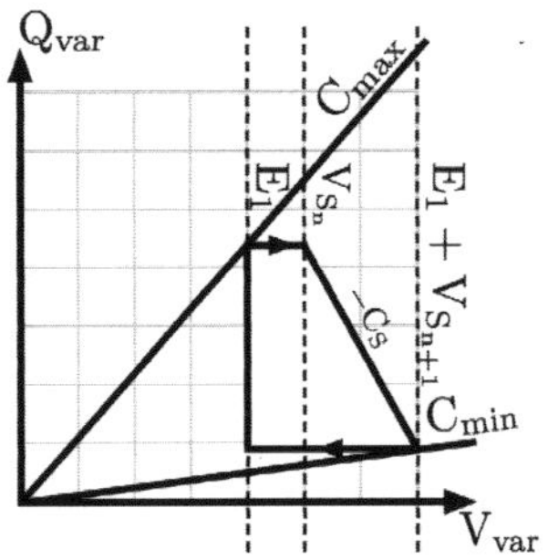

Figure 3. QV diagram for the circuit depicted in Fig. 1.

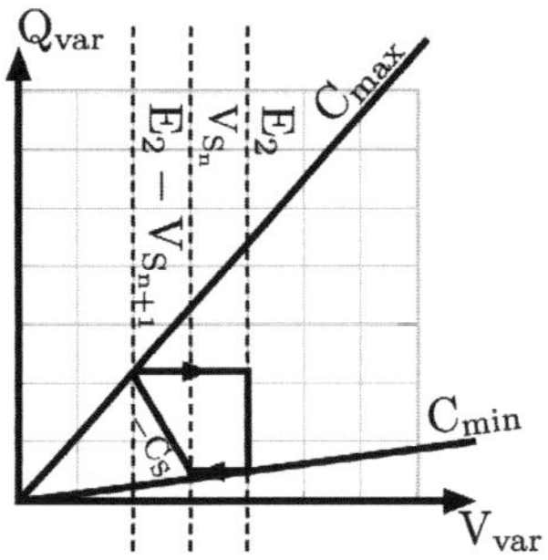

Figure 4. QV diagram for the circuit depicted in Fig. 2.

Consider the two conditioning circuits depicted in Fig. 1 and Fig. 2. The voltage source elements E_1 and E_2 model external voltage sources, eventually superposed with the effect of electret charging. The QV diagrams of the corresponding circuits are depicted in Fig. 3 and Fig. 4. The diodes are supposed to follow an ideal diode voltage-current law, with threshold voltage $V_T \geq 0$. Defining a variation cycle by the transducer's variable capacitance varying from a maximal value C_{max} to a minimal value C_{min} and then from C_{min} to C_{max}, it comes for the local evolution laws of each of the circuits that:

$$V_{n+1}^{\#1} = \frac{C_S V_n^{\#1} + (C_{max} - C_{min})E_1 - (C_{max} + C_{min})V_T}{C_{min} + C_S}, \tag{1}$$

$$V_{n+1}^{\#2} = \frac{C_S V_n^{\#HW2} + (C_{max} - C_{min})E_2 - (C_{max} + C_{min})V_T}{C_{max} + C_S}, \tag{2}$$

where $V_n^{\#1}$ and $V_n^{\#2}$ denote the value of the voltage across C_S at cycle index n for the circuits depicted in Fig. 1 and Fig. 2, respectively.

3. Presentation of the measurement method

From the preceding equations, denoting $\Delta_1 = V_{n+1}^{\#1} - V_n^{\#1}$, $\Delta_2 = V_{n+1}^{\#1} - V_n^{\#1}$, $V_1 = V_n^{\#1}$ and $V_2 = V_n^{\#2}$, it comes:

$$C_{min} = C_S \frac{\Delta_2(E_1 - V_T) - \Delta_1(E_2 - V_T - \Delta_2 - V_2)}{(E_1 + \Delta_1 + V_1 + V_T)(E_2 - \Delta_2 - V_2 - V_T) - (E_1 - V_T)(E_2 + V_T)}, \tag{3}$$

$$C_{max} = \frac{C_S \Delta_2 + C_{min}(E_2 + V_T)}{E_2 - V_2 - \Delta_2}. \tag{4}$$

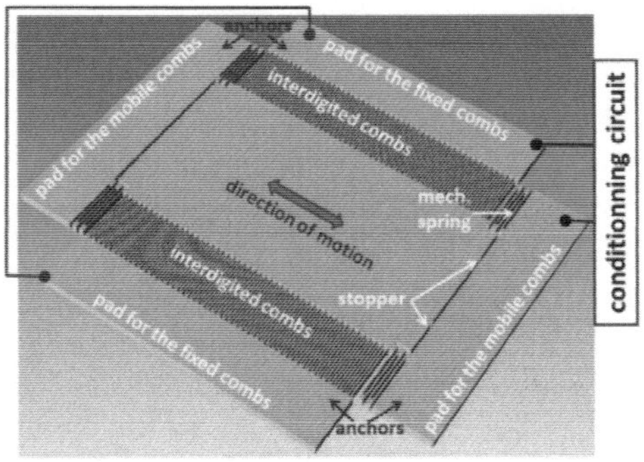

Figure 5. 3D schematic of the MEMS transducer that is characterized as an example of application of the presented measurement method.

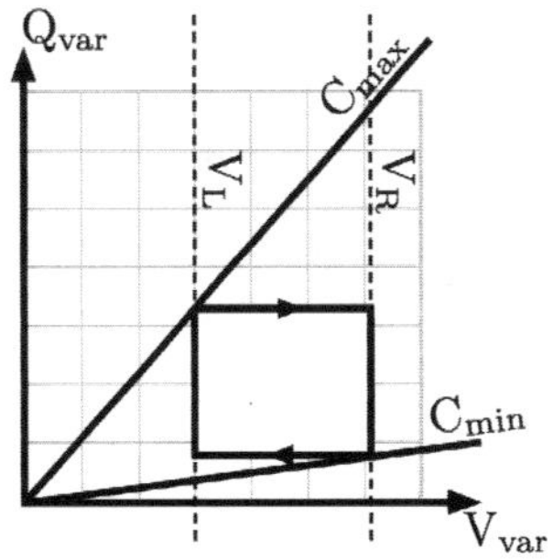

Figure 6. Generic rectangular QV diagram, with extreme voltages V_L and V_R.

Let's consider an e-VEH for which the value of maximal capacitance C_{max} is to be measured, when the device is submitted to an harmonic external excitation, and when its biasing across one cycle of variation is described by a rectangular charge-voltage diagram of extremal voltages V_L and V_R (see Fig. 6). Suppose the slope of the non-horizontal portions of the QV cycles of the two conditioning circuits is large compared to C_{max}, i.e., $C_S \gg C_{max}$ (see Fig. 3 and Fig. 4). By choosing the values of the external voltages sources and initial voltages across C_S such that:

$$E_1 = V_R, \tag{5}$$

$$E_2 = V_L, \tag{6}$$

$$V_1 = V_2 = V_R - V_L, \tag{7}$$

it comes that evaluating (3) and (4) with the values of Δ_1 and Δ_2 measured with these parameters yields the values of C_{max} and C_{min} as affected by the electromechanical coupling induced by the rectangular biasing depicted in Fig. 6.

Note that the values of C_{max} and C_{min} result from a dynamical process, and hence the measurement method is accurate only if the electromechanical coupling impact on the e-VEH's dynamics can be considered as a quasi-static process: otherwise, a systematic error is introduced in the measurement. This hypothesis is verified if C_S is chosen such that the values of Δ_1 and Δ_2 do not change abruptly from one cycle to the next, i.e., that C_S is chosen large enough. Also, this results in the QV diagrams in Fig. 3 and Fig. 4 to become more alike, which improves the measurement accuracy. However, a too large value of C_S results in Δ_1 and Δ_2 becoming indiscernible one from another because of a non-null noise floor due to random noise sources. Thus, prior to the measurement, a study should be done to chose the optimal value for C_S given those constraints.

4. Example of transducer characterization using the presented method

As an illustrating example, the following describes the application of the method to the measurement of the value of C_{max} for a practical transducer, that has been previously reported in [6]. The schematic of the transducer is depicted in Fig. 5. The measurement is carried out for the transducer biased following a rectangular QV diagram, with $V_L = 10$V and $V_R = 15$V, and mechanically forced by an external excitation of 1.5g amplitude, 150Hz frequency. Because of the symmetrical gap-closing geometry of the device, the value of C_{min} is fixed throughout

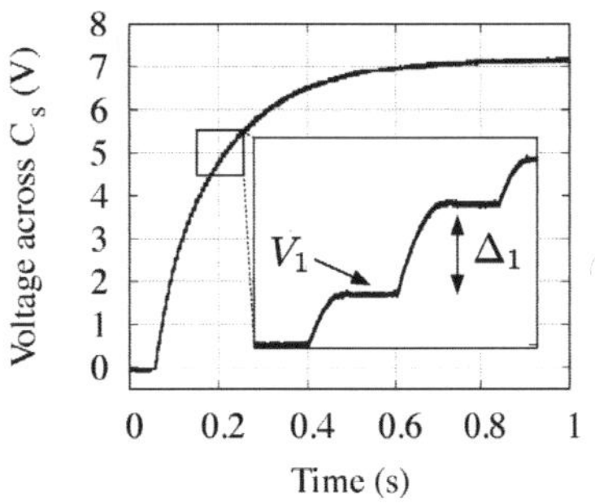

Figure 7. Complete time-evolution of the voltage across C_S for the circuit depicted in Fig. 1, with $E_1 = 10$V, with the transducer submitted to input acceleration of 1.5g amplitude at 150Hz frequency. The quantities V_1 and Δ_1 are highlighted.

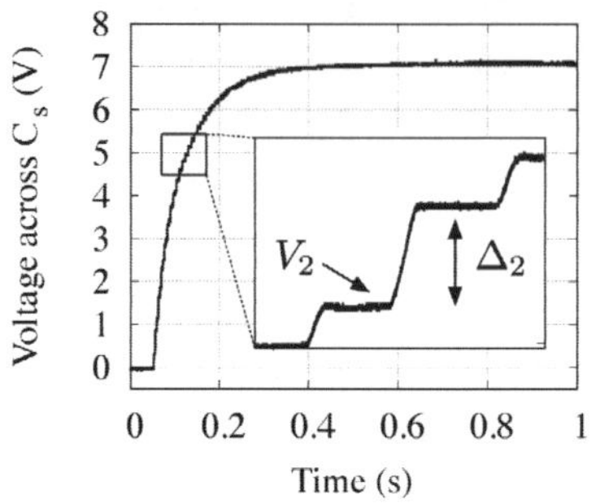

Figure 8. Complete time-evolution of the voltage across C_S for the circuit depicted in Fig. 2, with $E_2 = 15$V, with the transducer submitted to input acceleration of 1.5g amplitude at 150Hz frequency. The quantities V_2 and Δ_2 are highlighted.

the e-VEH's operation, and is measured by standard capacitance measurement means, as $C_{min} = 55$pF. In both circuits, the fixed capacitor is chosen such that $C_S = 500$pF.

Because C_{min} is fixed, only the circuit in Fig. 2 is necessary. The parameters for the measurements are $E_2 = 15$V, and $V_2 = 5$V (see (6) and (7)). Several measurements are carried out for the characterization, with approximatively equal values of V_2, to statistically estimate the uncertainty due to type A error sources on the measurements. With the obtained results for (Δ_2, V_2), applying (4) yields a value of $C_{max} = 94.7\text{pF} \pm 2.8\text{pF}$.

Equation (3) can be used to check the results as C_{min} is fixed, by carrying out the same measurement with the circuit in Fig. 1, choosing $E_1 = 10$V, and at $V_1 = 5$V (see (5) and (7)). If the geometry of the device is such that C_{min} is not unalterable, this latter measurement has to be carried out to measure the value of C_{min} as impacted by the electromechanical coupling resulting from the bias depicted in Fig. 6.

5. Conclusion

This work presented the first method to measure the capacitance variation of electrostatic MEMS transducers of an e-VEHs, when electrically conditioned following a rectangular charge-voltage diagram. This enables more thorough studies on the effect of the electromechanical coupling on the dynamics of e-VEH using electrical interfaces implementing these types of conditionings schemes, such as charge-pump conditioning circuits.

6. References

[1] De Queiroz A C M 2014 *2014 IEEE 5th Latin American Symposium on Circuits and Systems* 1–4

[2] Karami A, Galayko D and Basset P 2016 *IEEE Transactions on Circuits and Systems I: Regular Papers* **PP** 1–14 ISSN 1549-8328

[3] Wei J, Risquez S, Mathias H, Lefeuvre E and Costa F 2015 Simple and efficient interface circuit for vibration electrostatic energy harvesters *SENSORS, 2015 IEEE* pp 1–4

[4] Galayko D, Dudka A, Karami A, O'Riordan E, Blokhina E, Feely O and Basset P 2015 *IEEE Transactions on Circuits and Systems I: Regular Papers* **62** 2652–2663

[5] O'Riordan E, Dudka A, Galayko D, Basset P, Feely O and Blokhina E 2015 *IEEE Transactions on Circuits and Systems I: Regular Papers* **62** 2664–2673

[6] Lu Y, ORiordan E, Cottone F, Boisseau S, Galayko D, Blokhina E, Marty F and Basset P 2016 *Journal of Micromechanics and Microengineering* **26** 124004

A novel ethanol/oxygen microfluidic fuel cell with enzymes immobilized onto cantilevered porous electrodes

D Desmaële[1,2], T T Nguyen-Boisse[1], L Renaud[3] and S Tingry[1]

[1]Institut Européen des Membranes, UMR 5635, ENSCM-UMII-CNRS, place Eugène Bataillon, 34095 Montpellier, France

[2]Now at Istituto Italiano di Tecnologia (IIT), Center for Biomolecular Nanotechnologies, Via Barsanti, 73010, Arnesano, (Lecce), Italy

[3]Université de Lyon, Institut des Nanotechnologies de Lyon INL-UMR5270, CNRS, Université Lyon 1, Villeurbanne F-69622, France

E-mail: denis.desmaele@iit.it

Abstract. This paper introduces a novel design of membraneless microfluidic biofuel cell that incorporates three-dimensional porous electrodes containing immobilized enzymes to catalyze redox reactions occurring in the presence of ethanol/O_2 co-laminar flows. In order to maximize the penetration depth of the reactants inside the porous medium, we report on the preliminary evaluation of cantilevered bioelectrodes, namely the fibrous electrodes protrude along the internal walls of the miniature electrochemical chamber. As a first proof-of-concept, we demonstrate the integration of a bioanode and a biocathode into a lamination-based microfluidic cell fabricated via rapid prototyping. With enzymes deposited into the fibrous structure of 25 mm long, 1 mm wide and 0.11 mm thick carbon paper electrodes, the volumetric power density reached 1.25 mW cm^{-3} at 0.43 V under a flow rate of 50 μL min^{-1}. An advantage of the presented microfluidic biofuel cell is that it can be adapted to include a larger active electrode volume via the vertical stacking of multiple thin bioelectrodes. We therefore envision that our design would be amenable to reach the level of net power required to supply energy to a plurality of low-consumption electronic devices.

1. Introduction

Co-laminar enzymatic biofuel cells (CL-EFCs) are envisioned as a promising option for supplying power to low-consumption electronic devices [1]. It is indeed recognized that CL-EFCs can offer distinctive features: i) unlike conventional fuel cells, a proton exchange membrane is not required; ii) precious metal catalysts can be replaced by low-cost biocatalysts; iii) a variety of readily available fuels can be employed; iv) membraneless CL-EFCs are inherently well adapted to miniaturization. Despite such potential advantages, the practical deployment of CL-EFCs is still hindered by limited power yields. To alleviate this hindrance, three-dimensional porous electrodes (3D-PEs) show great potential. Indeed, 3D-PEs offer enlarged reactive surfaces compared to conventional two-dimensional electrodes. Notable performance improvements were therefore reported for co-laminar fuel cells exploiting 3D-PEs with chemical reactants [2]. Similarly, 3D-PEs would allow for higher enzyme loadings in CL-EFCs. To the best of the authors' knowledge, however, most designs of CL-EFCs that can be found in the literature still rely on soluble enzyme catalysts flowing over gold electrodes or on enzymes immobilized onto modified flat surfaces [3, 4]. Here, we introduce and evaluate a novel lamination-based CL-EFC that integrates cantilevered 3D-PEs with the aim to reach higher enzyme loadings and improved performance.

2. Experiments

Unless otherwise stated, all chemicals were obtained from Sigma-Aldrich and used as received.

2.1. Preparation of porous bioelectrodes

Toray carbon paper (TCP) was used as the electrode material. TCP sheets purchased were 110 μm thick (TGP-H-030, Toray). They were not coated for wet proofing. Electrode patterns were cut by xurography (i.e., razor writing) using a cutting plotter (CE 6000-40, Graphtec) [5]. Electrode areas which were not intended to be in contact with the reactants were coated with wax. The remaining hydrophilic zones devoted to the subsequent deposition of the enzymes (see dashed lines in figure 1B) were 25 mm long and 1 mm wide (i.e., the area and volume of the bioelectrodes were 0.25 cm^2 and $\approx$0.0028 cm^3, respectively).

For the bioanode, the enzyme alcohol dehydrogenase (ADH) and its cofactor β-nicotinamide adenine dinucleotide sodium salt (NAD^+) were employed for ethanol oxidation. Prior to enzyme immobilization, a first electropolymerized poly(methylene green) film was used as an electrocatalyst. A phosphate buffer solution containing the enzymes was subsequently pipetted onto the poly(methylene green) film and dried at room temperature by performing cyclic voltammetry under nitrogen atmosphere from -0.6 to 1.2 V/SCE (20 cycles, scan rate 50 mV s^{-1}) in a solution of methylene green 0.4 mM in 0.1 M borate buffer pH 9. The methylene green was used as electrocatalyst for NADH oxidation [6]. Then, a solution of 10 mg of ADH (150 U mg^{-1}), 67 μL of carbon nanoparticles powder (15 mg mL^{-1}, Super P, TIMCAL), 5.5 mg NAD^+, 5 μL polyethylenimine (5% wt), and 25 μL of 0.1 M phosphate buffer pH 7 was mixed in a vortex mixer and 5 μL of the final mixture was pipetted onto the poly(methylene green) film and dried at room temperature.

For the biocathode, Trametes versicolor laccase (13.6 U mg^{-1}) was entrapped along with 2,2-azinobis(3-ethylbenzothiazoline-6-sulfonate) diammonium salt (ABTS) in a polypyrrole film for O_2 electroreduction. First, 5 μL of a mixture containing 7.5 mg laccase, 100 μL of carbon nanoparticles powder (15 mg mL^{-1}, Super P, TIMCAL) in 0.1 M phosphate buffer pH 5 was drop-casted on the TCP. After drying at room temperature, pyrrole electropolymerization was performed at +0.6 V/SCE, by dipping the electrode in a solution containing pyrrole (200 mM), ABTS (20 mM) in 0.1 M $NaClO_4$ H_2O, until a charge of 0.5 C passed. The complete polymerization of any excess pyrrol that may have remained was achieved by cycling the electrode in phosphate buffer (pH 5) from 0 to 0.8 V (5 cycles, scan rate of 50 mV s^{-1}).

2.2. Fabrication of the CL-EFC

Both bioelectrodes were integrated in a T-shaped laminated CL-EFC composed of double-sided pressure adhesive (DSPA) films sandwiched between a glass slide and a Polyethylene naphthalate (PEN) capping layer. The DSPA layers (labeled $T_{j=1,3}$ in figure 1A) were 100 μm thick (Montex-DX 1, X-Film) whereas the PEN layer was 75 μm thick (ES361075, Goodfellow). DSPA and PEN layers were also prepared by xurography using a cutting plotter. For the assembly of the entire structure, we used a CNC-machined mechanical guide (alignment accuracy $\pm$100 μm). The bioelectrodes were placed in specific grooves patterned in T_2. The bioelectrodes were thereby centered in the middle of the microchannel (see cross section b-b' in figure 1). We intentionally favored such cantilevered (i.e., protruding) electrodes rather than electrodes conventionally placed at the bottom of the microchannel. Indeed, when fluids flow over bottom, reactants may solely interact with the top portion of the 3D-PEs [2]. For the delivery and removal of fluids, polytetrafluoroethylene tubing was firmly inserted into the holes of three access ports made of Plexiglas.

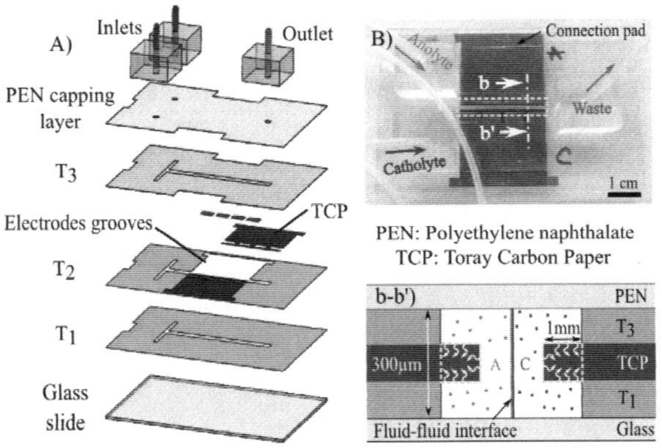

Figure 1. A) Expanded view detailing the design of the CL-EFC. B) Overview of the final CL-EFC after assembly. b-b') Cross-sectional view illustrating the central position of the cantilevered bioelectrodes.

After assembly, tests of the complete CL-EFC containing both bioelectrodes were performed at room temperature. The catholyte solution consisted of phosphate buffer solution (0.1 M, pH 3) saturated with O_2 (2-4 mM [7]). The anolyte solution contained ethanol (40 mM) and NAD^+ (5 mM) in phosphate buffer (0.1 M, pH 9). Solutions were then injected into the CL-EFC using a syringe pump (KDS200, KD Scientific). During experiments, the flow rate was set to 50 μL min^{-1} with the aim to avoid potential enzyme leaching. Steady-state polarization data were obtained under potentiostatic control (Autolab, Eco chemie).

3. Results

First, the enzymes immobilized onto the 3D-PEs were visually inspected via scanning electron microscopy (SEM). As can be viewed from figure 2A, the overall surface of the TCP is well covered by the enzyme layer. When observing the thickness of an enzyme-modified TCP, however, it appears that the enzymes do not fully penetrate into the fibrous electrode structure (see comparison to bare TCP in figure 2B). This suggests that our current immobilization protocols should be improved prior to be used with thicker TCP electrodes. Nonetheless, our CL-EFC could be adapted to increase the amount of enzymes exposed to the reactants by stacking multiple thin bioelectrodes such as the ones used here.

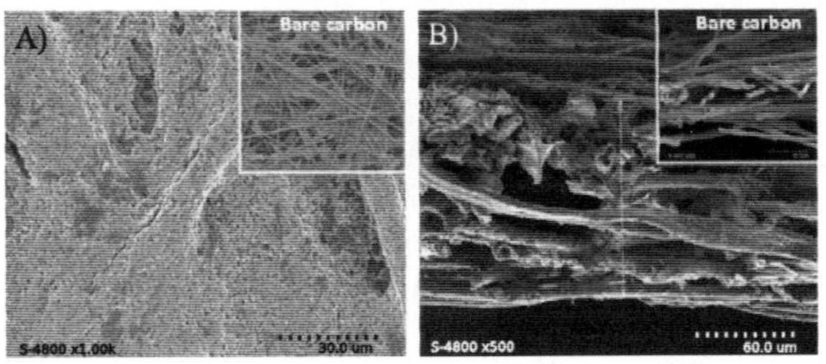

Figure 2. SEM micrographs of a 3D TCP porous bioanode before/after enzyme immobilization. A) Top view. B) Cross sectional view.

The performance of the complete CL-EFC was then evaluated. Figures 3A and 3B show the polarization curve and the power curve obtained, respectively. With no current flowing through an external load, the open circuit voltage is about 0.63 V, which is comparable with values reported elsewhere [8]. When considering the current and power densities normalized to the electrode projected surface area (PSA), our CL-EFC produces a maximum power density of $\approx$14 μW cm^{-2}. Although moderate, this power density still compares favorably with some previous works. In [9], a microchip-based bioanode paired with an external Pt cathode working from ethanol produced 5 μW cm^{-2} at 0.34 V. More recently, another ethanol/O_2 CL-EFC delivered 10 μW cm^{-2} when using gold bioelectrodes having a length-to-width aspect ratio (AR) of 20 (AR=25 in this work) [10]. A straightforward comparison between different configurations, however, remains delicate. Further, for 3D-PEs, the electrode PSA may be untrue [11]. As discussed elsewhere (e.g., see [2]), we believe that normalization based on the electrode volume provides a more accurate representation of our CL-EFC. In this case, the maximum volumetric power generated by our CL-EFC reaches ≈1250 μW cm^{-3} at 0.43 V. The corresponding volumetric current density is almost 3 mA cm^{-3}. We believe these volumetric values make more sense for our design, especially if considering that our CL-EFC could be scaled up with multiple stacked thin bioelectrodes offering higher enzyme loading capabilities.

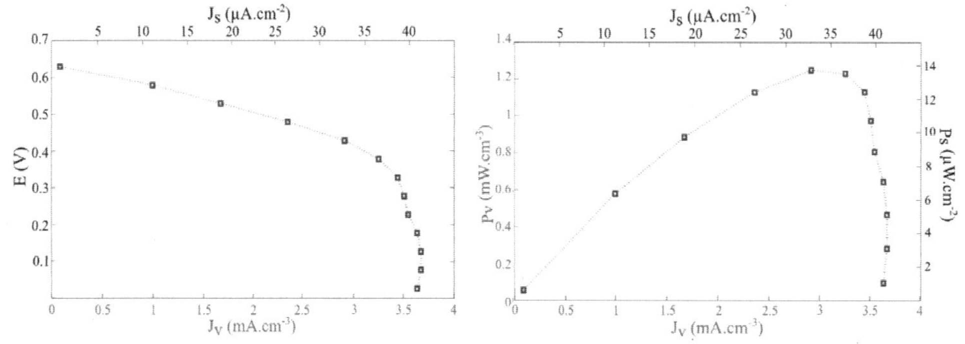

Figure 3. Performance measured for the ethanol/O_2 CL-EFC at 50 μL min^{-1}: (A) Polarization curve and (B) Power curve.

4. Conclusions

The possibility to reach a volumetric power density of ≈1.25 mW cm^{-3} at 0.43 V with the ethanol/O_2 CL-EFC presented here is encouraging. Indeed, even if considering the power density normalized by the projected surface area (≈14 μW cm^{-2}), the level of performance achieved can already be sufficient to drive a low-consumption electronic system, as demonstrated in [12]. In the near future, our efforts will be focused on the improvement of our immobilization protocols so that enzymes could be more efficiently attached to thicker carrier matrices. Additionally, we will try to scale up our design by stacking several bioelectrodes to increase the enzyme loading and the net power produced.

References

[1] Gellett W, Kesmez M, Schumacher J, Akers N and Minteer S D 2010 *Electroanalysis* **22** 727–31
[2] Kjeang E, Michel R, Harrington D A, Djilali N and Sinton D 2008 *J. Am. Chem. Soc.* **130** 4000–6
[3] Renaud L, Selloum D and Tingry S 2015 *Microfluidics and Nanofluidics* 1–10
[4] Beneyton T, Wijaya I P M, Salem C B, Griffiths A and Taly V 2013 *Chem. Commun.* **49** 1094–6
[5] Bartholomeusz D A, Boutté R W and Andrade J D 2005 *Microelectromechanical Systems, Journal of* **14** 1364–74
[6] Akers N L, Moore C M and Minteer S D 2005 *Electrochim. Acta* **50** 2521–5
[7] Jayashree R S, Gancs L, Choban E R, Primak A, Natarajan D, Markoski L J and Kenis P J 2005 *J. Am. Chem. Soc.* **127** 16758–9
[8] Neto S A, Forti J, Zucolotto V, Ciancaglini P and De Andrade A 2011 *Biosens. Bioelectron.* **26** 2922–6
[9] Moore C M, Minteer S D and Martin R S 2005 *Lab Chip* **5** 218–25
[10] Selloum D, Tingry S, Techer V, Renaud L, Innocent C and Zouaoui A 2014 *J. Power Sources* **269** 834–40
[11] Sharma M, Bajracharya S, Gildemyn S, Patil S A, Alvarez-Gallego Y, Pant D, Rabaey K and Dominguez-Benetton X 2014 *Electrochim. Acta* **140** 191–208
[12] Desmaële D, Renaud L and Tingry S 2015 *Sens. Actuators* B **220** 583–9

A monolithic integrated micro direct methanol fuel cell based on sulfo functionalized porous silicon

M. Wang[1, 2], Y. X. Lu[1, 3], L. T. Liu[1] and X. H. Wang[1, *]

[1]Insititute of Microelectronic, Tsinghua University, Beijing, 100084, P.R. China

[2]School of Computer and Communication Engineering, Zhengzhou University of Light Industry, Zhengzhou, 450000, P.R. China

[3]Université Paris-Est / ESCOM / ESIEE Paris, 93162, Noisy-le-Grand, France

Email: wxh-ime@tsinghua.edu.cn

Abstract. In this paper, we demonstrate a monolithic integrated micro direct methanol fuel cell (μDMFC) for the first time. The monolithic integrated μDMFC combines proton exchange membrane (PEM) and Pt nanocatalysts, in which PEM is achieved by the functionalized porous silicon membrane and 3D Pt nanoflowers being synthesized in situ on it as catalysts. Sulfo groups functionalized porous silicon membrane serves as a PEM and a catalyst support simultaneously. The μDMFC prototype achieves an open circuit voltage of 0.3 V, a maximum power density of 5.5 mW/cm^2. The monolithic integrated μDMFC offers several desirable features such as compatibility with micro fabrication techniques, an undeformable solid PEM and the convenience of assembly.

1. Introduction

With the increasing functions and shrinking sizes of the portable devices, for example, smart phones, the demand for efficient and integrated power sources increases radically. It becomes a huge challenge that how to improve the performance and integration of power sources, so that they can get more energy in the limited volume. Micro direct methanol fuel cells (μDMFCs), have drawn much attention, not only because of their high energy density, room temperature operation, simple and safe handling [1-2], but also because they can make full use of micro fabrication techniques developed for IC and MEMS to optimize the integration and sizes of fuel cells.

In recent years, varieties of μDMFCs have been developed [3-10]. Although a promising approach, such μDMFCs require a large volume due to the thickness of each stacked electrode and the inter-electrode space. The power density for such fuel cells is commonly reported in terms of electrode surface area instead of volume as is often the case for other power sources. In order to address the issue, a monolithic integrated μDMFC is a promising way to optimize the assembly and sizes, furthermore, is benefit to enhance the volumetric power density.

Here we present a monolithic μDMFC for the first time, integrated with silicon-based proton exchange membrane (PEM), nanocatalysts and current collector layer together. PEM is achieved by sulfo functionalized porous silicon membrane and 3D Pt nanoflowers are synthesized in situ onto it as nanocatalysts. Ag nanowires film serves as current collector layer. The monolithic μDMFC shows several key advancements in: 1) integrating all components on a chip and simplifying the structure and assembly of fuel cell; 2) solid PEM with little deformation and high proton conductivity; 3) in-situ synthesis of 3D Pt nanoflowers on a fuel cell body without the extra coating process; 4) compatible with micro fabrication techniques. As such, we believe this work is more attractive for system-on-chip.

2. Experimental

2.1. Design

The monolithic integrated μDMFC was designed as illustrated in Figure 1. The fuel cell combined PEM, nanocatalysts and current collector layer together on a chip, in which PEM was achieved by sulfo functionalized porous silicon membrane, Pt nanocatalysts were synthesized in situ on it and Ag nanowires were coated as current collector layer.

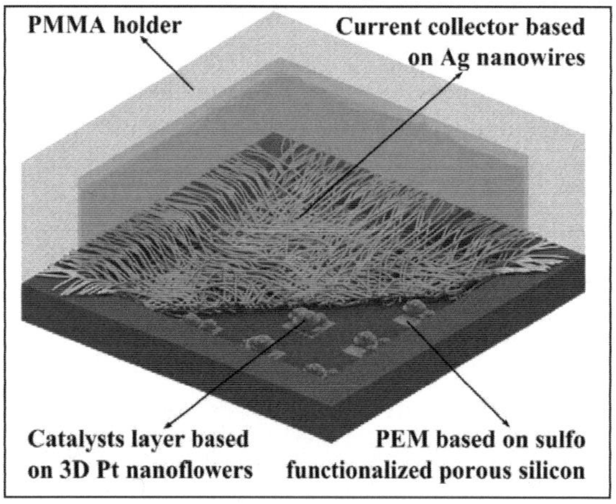

Figure 1. Schematic of a monolithic integrated μDMFC based on sulfo functionalized porous silicon membrane.

2.2. Fabrication

The fabrication process of the monolithic integrated μDMFC divided into four steps, shown as Figure 2(a). Porous silicon membranes were fabricated and 3D Pt nanflowers were synthesized in situ onto them using by the method proposed in our previous work [11-12]. After that, porous silicon membranes were functionalized with sulfonic acid groups via chemical grafting method. Then the collecting layer was formed by dropping Ag nanowires dispersion on the PEM and baking 30 min at 50 °C to dry. Finally, the monolithic integrated μDMFC prototype was obtained after assembly by PMMA holders.

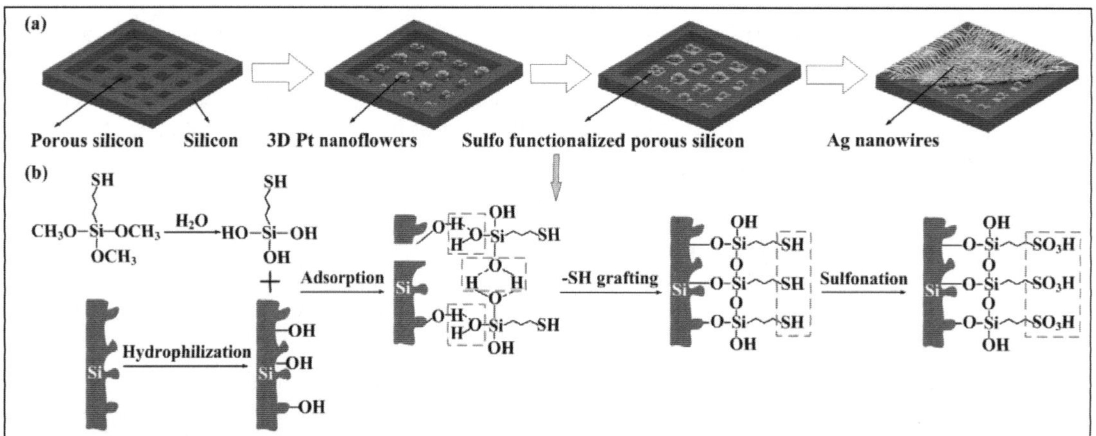

Figure 2. Fabrication of a monolithic integrated μDMFC (a); sulfo functionalization of the porous silicon membrane (b).

The porous silicon membranes were functionalized with sulfonic acid groups, according to the following steps, shown as Figure 2(b). Firstly, fabricated porous silicon membranes were immersed in a mixed solution of H_2SO_4 and H_2O_2 at the volume ratio of 3:1 for 30 minutes at 60 °C for hydrophilization. Simultaneously, Si-O-CH_3 groups in 3-mercaptopropyltrimethoxysilane (MPTMS) molecules were hydrolyzed to form Si-OH groups in mixed benzene solutions, with the MPTMS concentration of 30 wt. % and glacial acetic acid (GAA) concentration of 5 wt. %. Sencondly, Si-OH groups in hydrolyzed MPTMS were absorbed onto the walls of porous silicon nanoholes by immersing the hydrophilic porous silicon membrane into mixed benzene solutions. Thirdly, mercapto groups were grafted via dehydration condensation reaction with ultrasonically activating for 48 hours. Finally, mercapto groups were oxidize to sulfonic acid groups in an 30 wt. % HNO_3 aqueous solution for 3.5 hours with ultrasound at room temperature. Eventually, a functionalized porous silicon membranes were obtained, which were proton-conductive as a result of sulfonic acid terminal.

3. Results and discussion

Various tests had been done to describe the monolithic integrated μDMFC. The morphology characterizations were investigated by scanning electron microscopy (SEM), which were performed on a HITACHI S-5500 operating at an accelerating voltage of 5 kV. The performance of the monolithic integrated μDMFC were characterized using by a Solartron 1287 electrochemical interface coupled with a Solartron 1260B frequency response analyzer at room temperature.

SEM images of a monolithic integrated μDMFC based on sulfo functionalized porous silicon membrane are shown as Figure 3. As we can see, the monolithic integrated μDMFC is made up of PEM based on porous silicon, Pt nanoflowers and Ag nanowires layer by layer. The porous silicon membrane shows distinct nanoholes structure with the bore diameter of 50 nm, and there are many organic molecules attached onto the walls of porous silicon nanoholes, which are grafted -SO_3H groups. The sulfonic acid groups attached on the walls of porous silicon nanoholes will hydrolyze in water, thus the walls of porous silicon nanoholes will be negatively charged and the protons can transfer between neighboring oxygen atoms at the gradient of potential and concentration. 3D Pt nanoflowers are assembled on the PEM based on sulfo-functionalized porous silicon membrane, with the size of about 200-400 nm. Due to the unique morphology, 3D Pt nanoflowers exhibit excellent catalytic activity [12]. A uniform film formed by Ag nanowires with diameter of 40-60 nm serves as a collecting layer.

Figure 3. SEM images of a monolithic integrated μDMFC based on sulfo functionalized porous silicon: cross-sectional view (a); high resolution SEM image of sulfo functionalized porous silicon membrane (b), Pt nanoflowers (c) and Ag nanowires (d).

The novel fabricated monolithic integrated μDMFC prototype based on sulfo functionalized porous silicon membrane and its corresponding I-V curves were shown as Figure 4. When filled with 2M methanol, a constant voltage was applied and the output current was monitored for a period of 50 s until the final steady-state value was recorded, and the I-V curve was obtained. The results clearly show that the μDMFCs achieve an open circuit voltage of 0.3 V, a maximum power density of 5.5 mW/cm^2 and a maximum current density of 80 mA/cm^2.

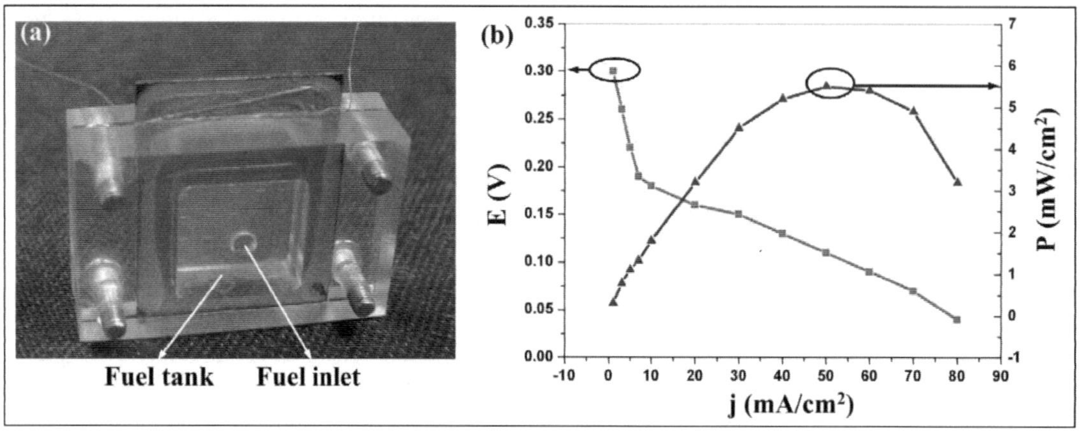

Figure 4. The monolithic integrated μDMFC prototype based on sulfo functionalized porous silicon membrane (a) and its corresponding I-V curves (b).

4. Conclusions

A monolithic μDMFC is demonstrated for the first time, integrated with silicon-based PEM, nanocatalysts and current collector layer together. The μDMFC prototype achieves an open circuit voltage of 0.3 V, a maximum power density of 5.5 mW/cm^2. These results demonstrate the proposed monolithic integrated μDMFC shows great potentials for optimizing the size and performance of μDMFCs, furthermore, are quite promising for integrated micro systems.

Acknowledgments
This work is supported by the National Natural Science Foundation of China (No. 61474071) and 973 program (No. 2015CB352100).

References

[1] N. Miyake, J. S. Wainright and R. F. Savinell, 2001, *J. Electrochem. Soc.*, 148, A905-9.
[2] H. Suzuki, Y. Yoshida, M. A. Mehta, M. Watanabe and T. Fujinami, 2002, *Fuel Cells*, 2, 46-51.
[3] Q. X. Wu, T. S. Zhao and R. Chen, L. An, 2013, *Appl. Energy*, 106, 301-6.
[4] X. H. Yan, T. S. Zhao, L. An, G. Zhao and L. Zeng, 2014, *Electrochimica Acta.*, 139, 7-12.
[5] H. C. Peng, P. H. Chen, H. W. Chen, C. C. Chieng, T. K. Yeh, C. Pan and F. G. Tseng, 2010, *J. Power Sources*, 195, 7349-58.
[6] Y. A. Zhou, X. H. Wang, X. Guo, X. P. Qiu and L. T. Liu, 2012, *Int. J. Hydrogen Energy*, 37, 967-76.
[7] Q. Zhang, X. H. Wang, Y. M. Zhu, Y. A. Zhou, X. P. Qiu and L. T. Liu, 2009, *J. Power Sources*, 192, 494-501.
[8] X. H. Wang, Y. A. Zhou, Q. Zhang, Y. M. Zhu and L. T. Liu, 2009, *J. Micromech. Microeng.*, 19, 094012.
[9] D. D. Meng and C. J. Kim, 2008, *Lab Chip*, 8, 958-68.
[10] N. Paust, S. Krumbholz, S. Munt, C. Miller, R. Zengerle, C. Ziegler and P. Koltay, *IEEE MEMS 2009*, Sorrento, Italy, 1091-4.
[11] M. Wang, X. H. Wang, S. Wu, X. Guo, Z. M. Tan and L. T. Liu, *Proc. NEMS 2011*, Kaohsiung, Taiwan, 968-71.
[12] M. Wang, X. H. Wang, L. N. Li and L. T. Liu, 2013, *J. Materials Chemistry A*, 1, 8127-33.

PERFORMANCE ENHANCEMENT OF MEMS-BASED MICROBIAL FUEL CELLS (μMFC) FOR MICROSCALE POWER GENERATION

Begüm Şen Doğan[1], Nilüfer Afşar Erkal[3], Ebru Özgür[3], Özge Zorlu[3] and Haluk Külah[1, 2, 4]

[1] Department of Micro and Nanotechnology, Middle East Technical University, 06800, Ankara, Turkey
[2] Department of Electrical and Electronics Engineering, Middle East Technical University, 06800, Ankara, Turkey
[3] Mikro Biyosistemler Inc., 06530, Ankara, Turkey
[4] METU-MEMS Research and Application Center, 06800, Ankara, Turkey

E-mail: kulah@metu.edu.tr

Abstract. This paper reports the design, fabrication, and testing of a microliter scale Microbial Fuel Cell (μMFC) based on silicon MEMS fabrication technology. μMFC systems are operated under different loads or open circuit to compare the effect of different acclimatization conditions on start-up time. *Shewanella oneidensis* MR-1 is preferred to be the biocatalyst. The internal resistance is calculated as 20 kΩ under these conditions. Acclimatization of μMFC under a finite load resulted in shorter start-up time (30 hours) when compared to the open load case. Power and current densities normalized to anode area are 2 μW/cm^2 and 12 μA/cm^2 respectively. When the load resistance value is closer to the internal resistance of the μMFC, higher power and current densities are achieved as expected, and it resulted in a shorter start-up time. Further studies focusing on the different acclimatization techniques for μMFC could pave the way to use μMFCs as fast and efficient portable power sources.

1. Introduction

Microbial fuel cells (MFC) are defined as bioreactors that convert the energy in the chemical bonds of organic compounds into electrical energy through catalytic activity of microorganisms under anaerobic conditions [1-3]. Adopting MFCs as portable power sources has been a popular research field for the last fifteen years [4-6]. However, the miniaturization of MFCs is necessary to be employed as portable power sources [7-8]. In this scope, MEMS technology is attractive for creating microscale microbial fuel cells (μMFC) due to the potential of miniaturization, economical mass production and large surface-area-to-volume ratio [9]. To be able to employ μMFCs as power sources, their performance parameters, namely power density, current density and start-up time, must be enhanced. The power and current densities depend on the biofilm (complex structure adhering to surfaces and consisting of colonies of bacteria) and the type of electrode surface. The biofilm varies from Gram-positive to Gram-negative bacteria and depends on the operational mode of the MFC, whether MFC is operated in closed circuit or open circuit system [10-11]. The growth of the microbial biofilm is found to decrease the anode polarization resistance and facilitate the kinetics of the electrochemical reactions [12]. Anode polarization resistance is one of terms forming the overall internal resistance, which limits the power and current generation, of µMFC. Thus, when the biofilm grows on the anode surface, the internal resistance decreases providing an enhancement on the performance [13]. In the light of these facts, to obtain higher performance µMFCs, the internal resistance of µMFC and the start-up time should decrease. Optimization of chamber and/or cell geometries, chamber or electrode materials, and electrode surface characteristics is crucial to increase µMFC performance. Furthermore, to enhance the biofilm formation, acclimatization of bacteria (giving time to bacteria to adjust to the conditions of growth) may be preferred. Thus, this study focused on a MEMS based µMFC with micro-liter volume to decrease the internal resistance and acclimatization of bacteria to decrease the start-up time.

2. Experimental methods and materials

2.1. Device fabrication and assembly

Since gold is biocompatible, conductive, and compatible with conventional microfabrication techniques, it was preferred as the electrode material. µMFC electrodeswere designed to have gold conductive areas and access holes inside the device. The masks were prepared with Cadence software, the flow inside the chambers modelled with COMSOL software (data not shown) and the microfabrication (figure 2) was performed in class 1000 clean room area.

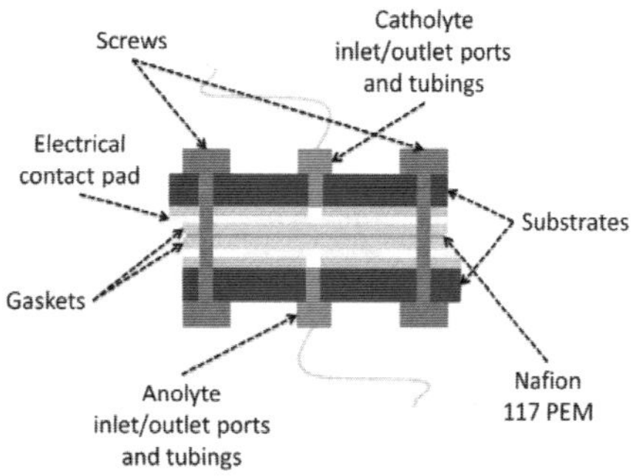

Figure 1. Schematic of µMFC.

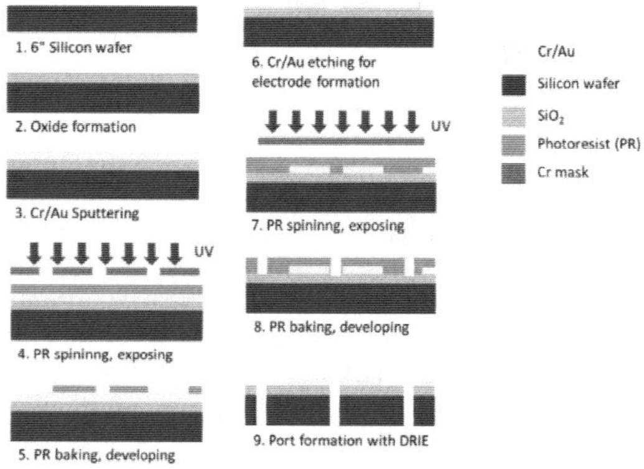

Figure 2. µMFC electrode fabrication steps.

Thermal oxide was formed on a polished 6″ polished silicon wafer by PECVD in order to provide passivation. Cr/Au layers (30 nm / 300 nm) were sputtered on the mentioned passivation layer and patterned by standard photolithography process. The access holes were drilled via DRIE. A picture of the microfabricated electrode is shown in figure 3.

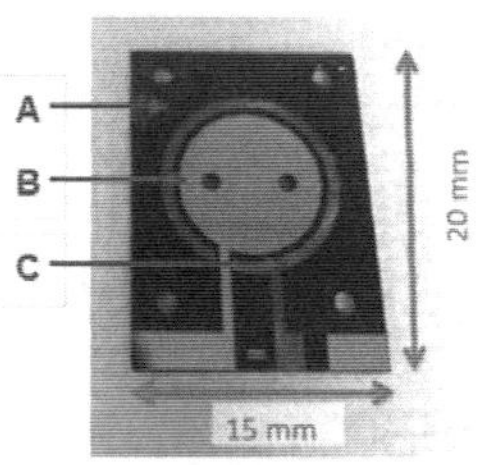

Figure 3. Microfabricated µMFC electrode. When two electrodes on substrates faced each other at 180°, the gold contact pads are exposed for electrical wiring. (A: Silicon substrate, B: Gold electrode, C: Silver electrode for further studies (the brownish thin circular electrode around the gold electrode is the oxidized silver layer to be processed as the reference electrode).

Two pieces of Gel-Pak WF 1.5-X4 gel films (170 µm x 15 mm x 17 mm) were sandwiched as gaskets between the two electrodes facing each other at 180°. The anode and cathode were separated by a proton exchange membrane (Nafion 117) placed between the gaskets. All layers were manually stacked and tightly kept together by screws and/or clamps. The assembled µMFC (figure 1) had four holes for fluidic inlet/outlet and eight holes for screws. It had two chambers (10.4 µL each) defined as anode and cathode chambers. The exposed conductive electrodes area per chamber was 0.61 cm^2. The inlet and outlet of the μMFC were accessed from the backside of the silicon substrates via nanoports (LabSmith) and transparent medical tubing (Ø 1 mm). Figure 1 depicts the schematic of the assembled μMFC.

2.2. Bacterial inoculum and μMFC operation
Shewanella oneidensis MR-1 (ATCC, USA), a facultative anaerobic electroactive bacterium, was grown on Triptic Soy Agar (Merck) at 30°C for 24 hours by streak plate method to obtain single colonies of bacteria. Then, the one grown single colony was cultured in Triptic Soy Broth (TSB) (Merck) medium on a shaker (150 rpm) at 30°C for 24 hours under aerobic conditions. To be fed as the anolyte, fresh TSB and bacteria inoculum was mixed (1:1). The anolyte (including inoculum) and catholyte solutions were continuously supplied using a syringe pump (KD Scientific) at rates of 3 μL/min and 5 μL/min respectively to anode and cathode chambers independently. The catholyte was 100 mM ferricyanide (Sigma-Aldrich) in a 100 mM phosphate buffer in which pH was adjusted at 7.5±0.1 with 0.1 M NaOH. The μMFC was operated at 25±1 °C. Three different μMFC assemblies were connected to either 10 kΩ or 25 kΩ external loads or operated under open circuit conditions.

2.3. Performance evaluation
The potential between the anode and the cathode was measured via a digital multimeter (Agilent 3441A) with a data acquisition system (National Instrument) and recorded the results every 1 min via Keysight IntuiLink interface. The start-up time of the biofilm formation was determined from the voltage versus time plot as the point when the voltage started to increase dramatically. The biofilm formation was accepted as terminated when the voltage reached a steady value. The polarization curve (V vs I) was obtained by changing the external resistors between 1 MΩ and 1 kΩ while recording the voltage. Linear fitting of the curve at the ohmic loss region resulted in the total internal resistance of the μMFC. The current through the resistors was calculated via Ohm's law, $I = V/R$, and the output power via Joule's law, $P = V \times I$. Current and power densities were normalized to the anode area (0.61 cm^2) and anode chamber volume (10.4 μL).

3. Results and discussion

3.1. Start-up performance
The start-up time of the biofilm formation was investigated by plotting voltage versus time plot (figure 4). Acclimatization of μMFC under a load resulted in shorter start-up time (table 1).

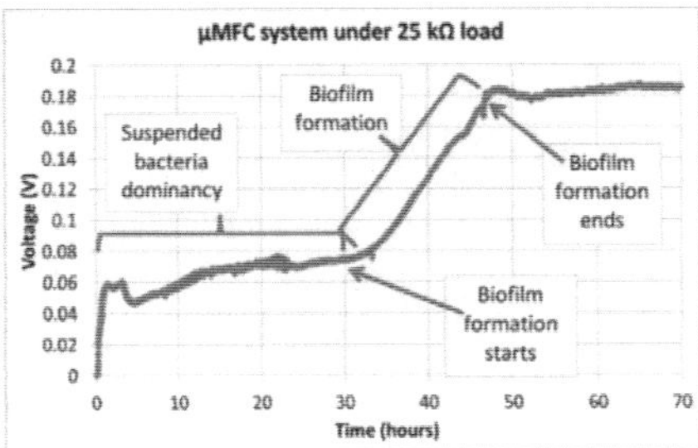

Figure 4. Effect of load on biofilm formation.

Table 1. Comparison of start-up values.

System	Start-up time of biofilm formation	Duration to reach maximum current
25 kΩ loaded μMFC system	~30 hours	~47 hours
10 kΩ loaded μMFC system	~31 hours	~84 hours
Open circuit μMFC system	~47 hours	~200 hours

3.2. Power generation performance
The results obtained showed the performance enhancement in terms of power and current densities and start-up time with respect to similar microliter scale microbial fuel cells with the same biocatalyst used given literature. Power density (133 $\mu W/cm^3$) and start-up time (30 hours) obtained with 25 kΩ loaded μMFC is better than to similar literature study by Qian et al. [7] and Li et al. [14] (table 2). The internal resistance was calculated as 20 kΩ under the mentioned conditions via polarization curve linear fitting (data not shown). It is observed that when the load is closer to internal resistance of the μMFC, higher power and current densities are achieved. However, power and current densities of this

study are smaller than the densities of MFC made of carbon paper by Vigolo et al. [15]. Bacteria prefer to adhere to carbon-based materials but they are difficult to integrate in MEMS processes when compared to gold as an electrode material.

Table 2. Comparison of performance values.

	25 kΩ loaded (this study)	10 kΩ loaded (this study)	Qian et al., 2009 [7]	Li et al., 2011 [14]	Vigolo et al., 2014 [15]
Bacteria	*S. oneidensis* MR-1	*S. oneidensis* MR-1	*S. oneidensis* MR-1	*S. oneidensis* MR-1	*S. oneidensis* MR-1
Fuel cell geometry	Double chamber with Nafion-117	Double chamber with Nafion-117	Double chamber with Nafion-117	Double chamber with laminar flow	Double chamber with Nafion-117
Anode area	0.61 cm^2	0.61 cm^2	0.15 cm^2	0.014 cm^2	0.5 cm^2
Anode volume	10.4 μL	10.4 μL	1.5 μL	0.3 μL	5 μL
Chamber depth	170 μm	170 μm	100 μm	50 μm	100 μm
Anode/cathode materials	Au/Au	Au/Au	Au/carbon cloth	Au / Au	Carbon paper / Carbon paper
Load	25 kΩ	10 kΩ	100 Ω	N/A	100 kΩ
Start-up time	30 hours	31 hours	47 hours	36 hours	15 hours
Internal resistance	20 kΩ	20 kΩ	30 kΩ	N/A	49 kΩ
Volumetric power density	133 μW/cm^3	26 μW/cm^3	15 μW/cm^3	N/A	900 μW/cm^3
Volumetric current density	716 μA/cm^3	500 μA/cm^3	1300 μA/cm^3	127 μA/cm^3	6180 μA/cm^3
Areal power density	2 μW/cm^2	0.4 μW/cm^2	0.15 μW/cm^2	N/A	9 μW/cm^2
Areal current density	12 μA/cm^2	9 μA/cm^2	13 μA/cm^2	2.542 μA/cm^2	61.8 μA/cm^2

4. Conclusions

A MEMS-based μMFC is demonstrated to operate with short start-up time in this study. With the designed 10.4 μL double-chamber μMFC, the internal resistance was calculated as 20 kΩ which is still an important bottleneck to overcome. Power and current densities obtained is comparable to similar literature study. In addition to higher power and current densities, when the bacteria is acclimatized under a load closer to the internal resistance of the μMFC, shorter start-up time is achieved. Further studies focusing on the different acclimatization techniques for μMFC could pave the way to use μMFCs as fast and efficient portable power sources.

Acknowledgments

This work was funded by TÜBİTAK in the scope of 113E195 Project.

References

[1] Du Z, Li H and Gu T 2007 *Biotechnol. Adv.* **25** 464–82
[2] Lovley D R 2006 *Nature reviews. Microbiology* **4** 7 497-508
[3] Rabaey K and Verstraete W 2005 *Trends in Biotechnology* **23** 6 291-8
[4] Bond D R and Lovley D R 2003 *Appl. Environ. Microbiol.* **69** 1548-55
[5] Dávila D, Esquivel, J P, Sabaté N and Mas J 2011 *Biosensors and Bioelectronics* **26** 5 2426-30
[6] Choi S 2015 *Biosensors and Bioelectronics* **69** 8–25
[7] Qian F, Baum M, Gu Q and Morse D E 2009 *Lab Chip* **9** 3076
[8] Choi S, Lee H S, Yang Y, Parameswaran P, Torres C I, Rittmann B E and Chae J 2011 *Lab on a chip* **11** 1110-7
[9] Lovley D R 2006 *Nature reviews. Microbiology* **4** 497-508
[10] Read S T, Dutta P, Bond P L, Keller J and Rabaey K 2010 *BMC Microbiol.* **10** 98
[11] Kumar R, Singh L, and Zularisam A W 2016 *Renew. Sustain. Energy Rev.* **56** 1322-36
[12] Ramasamy R P, Ren Z, Mench M M and Regan J M 2008 *Biotechnol. Bioeng.* **101** 1 101-8
[13] Reguera G, Nevin K P, Nicoll J S, Covalla S F, Woodard T L and Lovley D R 2006 *Appl. Environ. Microbiol.* **72** 11 7345-8
[14] Li Z, Zhang Y, LeDuc P R and Gregory K B 2011 *Biotechnol. Bioeng.* **108** 2061-9.
[15] Vigolo D, Al-Housseiny T, Shen Y, Akinlawon F O, Al- Housseiny S, Hobson R K, Sahu A, Bedkowski K I, DiChristina T J and Stone H A 2014 *Physical Chemistry Chemical Physics* **16** 24 12535-4

Water flow energy harvesters for autonomous flowmeters

Sebastien Boisseau[1,2], Alexandre-Benoit Duret[1,2], Matthias Perez[1,2], Emmanuel Jallas[3], Eric Jallas[3]
[1] Univ. Grenoble Alpes, F-38000 Grenoble, France
[2] CEA, Leti, Minatec Campus, 17 rue des Martyrs, F-38054 Grenoble, France
[3] ITK, Cap Alpha, F-34830 Clapiers, France

E-mail: sebastien.boisseau@cea.fr

Abstract. This paper reports on a water flow energy harvester exploiting a horizontal axis turbine with distributed magnets of alternate polarities at the rotor periphery and air coils outside the pipe. The energy harvester operates down to 1.2L/min with an inlet section of 20mm of diameter and up to 25.2mW are provided at 20L/min in a 2.4V NiMH battery through a BQ25504 power management circuit. The pressure loss induced by the insertion of the energy harvester in the hydraulic circuit and by the extraction of energy has been limited to 0.05bars at 30L/min, corresponding to a minor loss coefficient of K_{EH}=3.94.

1. Introduction – Water flow energy harvester concept

Flows and in particular water flows are a powerful source of energy to supply autonomous devices such as flowmeters, with typical applications in agriculture and urban water systems. This paper presents a water flow energy harvester based on a horizontal axis turbine with distributed magnets with alternate polarities at the rotor periphery and with coils outside the pipe (Figure 1). Contrary to most of water flow energy harvesters proposed in the state of the art [1-3], the choice has been made here to use a horizontal-axis turbine with a low electromechanical coupling to keep the pressure loss below 0.05bar@30L/min. As soon as the water flow enters in the energy harvester, the turbine starts rotating, leading to a rotational movement of the magnets. The movement of the magnets with alternate polarities (Figure 1b) induces a variation of magnetic flux in the air coils, which is finally turned into electricity (Lenz's law). An inlet cone has also been added at the turbine entry to increase the flow speed and then, the rotation speed of the turbine. Finally, the flow energy harvester is symmetric so that it could work in both directions.

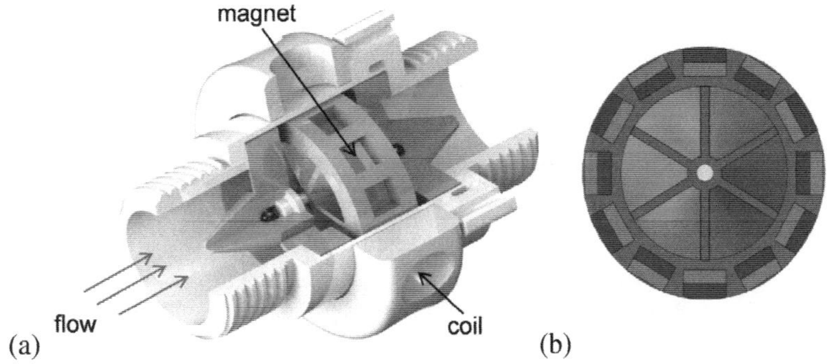

Figure 1. (a) Water flow energy harvester concept and (b) Magnets distribution at the rotor periphery

2. Models & equations – Water flow energy harvesters

2.1. Output powers and pressure losses

The output power of a turbine placed in a water channel is given by: $P = C_p \Delta p A U$, where Δp is the pressure loss induced by the turbine, A the turbine/channel surface, U the flow velocity and C_p the coefficient of power, which corresponds to the efficiency of the turbine, typically around C_p=2-5% for small-scale devices (some cm²). Water flows are then a powerful source of ambient energy with turbines' output powers reaching some milliwatts with a few L/min. Moreover, the pressure loss induced by the turbine Δp can be expressed by: $\Delta p = K_{EH} \rho U^2/2$, where K_{EH} is the minor loss coefficient of the energy harvester. K is typically comprised between 3 and 8 for turbine flowmeters. Minimizing the pressure loss Δp induced by the turbine and the extraction of energy is a strong constraint that has been solved by using horizontal axis turbines and a low electromechanical coupling (air core coils).

2.2. Turbine design considerations

The turbine has been designed by taking into account the following considerations: firstly, the inlet diameter has been set to 20mm to be equal to the diameter of the channel. Then, four other parameters have to be set: (i) the shape of the blades (NACA, flat plate, cambered plate…), (ii) the number of blades N, (iii) the evolution of the chord c(r), and (iv) the evolution of the blade angle β(r).

Horizontal axis turbines operate thanks to the lift force which provides motor torque, and are limited by the drag force which creates a resistant torque (Figure 2a). We must therefore select a profile which allows a high lift-to-drag ratio. We have decided to use simple flat plates which proved to be fairly efficient at low Reynolds number (Re≈10^4 in our case) compared to more complex profiles (Figure 2b), and easier to manufacture at small scale. The thickness of the flat blades has been fixed to 500µm in order to avoid potential cracks/breaks. The number of blades is also an important parameter. Overall, a large number of blades enables to decrease the depth of the turbine while maintaining the same motor toque. However, the depth of the turbine must not be too small to incorporate the permanent magnets (Figure 3b). The evolution of the chord enables to set the torque applied on the shaft and fixes the depth of the prototype at the same time. We restricted our study to the specific case in which the depth of the turbine H is constant, i.e. $H=c(r)\times sin(\beta(r))=cst$. Finally, the evolution of the blade angle enables to fix the rotational speed of the turbine (ωr in Figure 2a). Small blade angles lead to high rotational speeds but considerable pressure losses. On the contrary, high blade angles induce small pressure losses but low rotational speeds. In this paper, the blade angle follows a linear evolution from β(0)=10° to β (R) = θ, θ ∈[10°,35°].

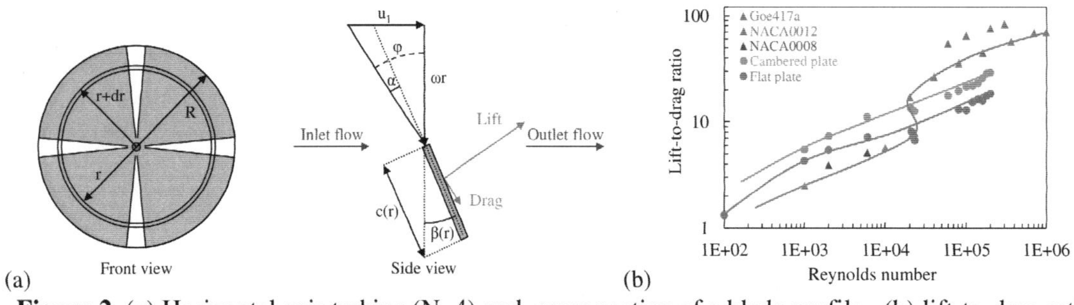

Figure 2. (a) Horizontal axis turbine (N=4) and cross section of a blade profile , (b) lift-to-drag ratio vs Reynolds number

3. Prototype

3.1. Turbines, coils and magnets

The turbine design has been experimentally optimized: more than 40 turbines with various blade angles, blade chords and number of blades have been manufactured and tested (Figure 3a) to maximize the output power of the energy harvester while keeping a low pressure loss. The optimal design was found to be a N=6-blades turbine with a blade angle of β(R)=30° and a depth of H=10mm (Figure 3b), able to generate power from 1.2L/min with an inlet diameter of 20mm. Twelve 5mmx5mmx3mm NdFeB magnets have been distributed at the rotor periphery with alternate polarities. Nine coils (7mmØ – 3000 turns – L_s=47mH – R_s=770Ω) have been placed at the outside of the pipe, in front of the magnets, to collect the variation of magnetic flux. The distance between the coils and the magnets is 2mm.

3.2. Energy harvester and output voltages

The energy harvester prototype is presented in Figure 3c; the active part sizes 60mmx45mmØ. The device has been connected to a standard domestic water line with 3 bars of static pressure and with water flow varying between U=0 and U=20L/min. An example of the output voltages of 3 adjacent coils is presented in Figure 3d for U=10L/min, showing that the output voltages of 2 adjacent coils are phase shifted of 120°.

The pressure loss has been measured in the worst condition with all the coils short-circuited. The pressure loss is equal to Δp=0.05bars at 30L/min in a 20mmØ pipe, corresponding to a minor loss coefficient of K_{EH}=3.94, and consistent with the standard minor loss coefficients of turbine flowmeters.

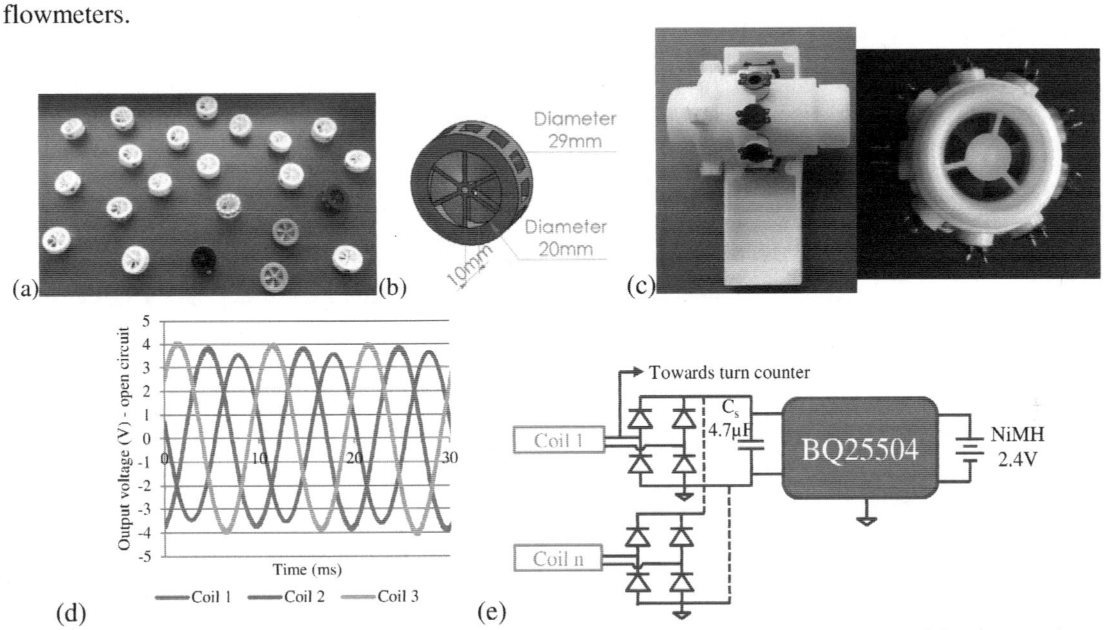

Figure 3. (a) examples of turbines characterized, (b) optimal design, (c) prototype – side view and top view, (d) output voltages at 10L/min in open circuit (3 adjacent coils) and (e) power management circuit managing the 9 coils and recharging a 2.4V NiMH battery

4. Power management circuit and delivered output powers

4.1. Power management circuit

The 9 coils have been connected to a storage capacitor C_s=4.7µF through diode bridges (BAS70). A BQ25504 circuit has been used to recharge a 2.4V NiMH battery and placed right after C_s (Figure 3e). The MPPT ratio of BQ25504 circuit has been set to 50% of the energy harvester's open circuit voltage. Then, every 16s, a research of the MPPT is performed by the power management circuit, maximizing the extraction of energy from the water flow energy scavenger.

4.2. Output power, rotation frequency and power supply of WSN

The energy harvester connected to its power management circuit starts to recharge the battery from 1.2L/min (4.8μW) and up to 25.2mW are harvested at 20L/min. The output powers as a function of the water flow are introduced in Figure 4a. The rotation frequency of the turbine is almost linear with the flow, as presented in Figure 4b, and comprised between 2.5Hz@2L/min and 25.8Hz@20L/min. To note that the rotation frequency of the turbine is impacted by the extraction of energy (-20% when energy is extracted), but remains linear. Then, the rotation frequency of the turbine can be directly used to measure the flow. This has been done by adding a turn counter on one coil (Figure 3e): the turn counter generates pulses each time the coil voltage reaches a maximum. Finally, the pressure drop induced by the turbine and the energy extraction is lower than 0.05bars on the whole operating range (1.2L/min-30L/min), as targeted; and, one of the lowest of the state of the art.

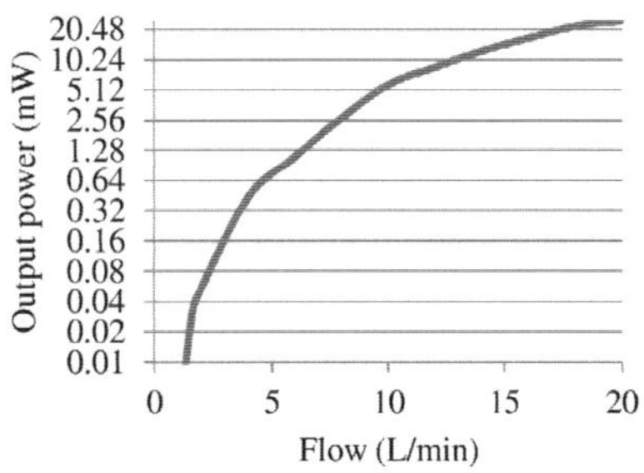

(a)

L/min	P(mW)
1.2	0.0048
1.6	0.03
2	0.055
4	0.4416
6	1.116
8	2.712
10	5.76
12	8.64
14	12.48
16	16.56
18	21.84
20	25.2

(b)

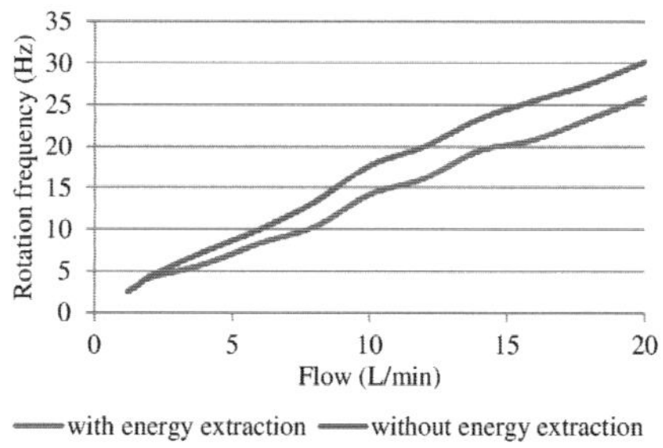

Figure 4. (a) output power vs water flow with the power management circuit – table & graph, (b) rotation frequency with and without energy extraction

This power has finally been used to supply a wireless sensor node (Bluetooth Low Energy) counting the number of rotations performed by the turbine (Figure 3e) and sending it by RF to a smartphone. The wireless sensor node is continuously connected to the 2.4V battery and exploits low-power modes to minimize its power consumption, waking up every 256 pulses thanks to a prescaler.

Conclusions & Perspectives

We have reported on a water flow energy harvester exploiting a horizontal axis turbine with distributed magnets at its periphery and coils at the outside of the pipe, with the aim of keeping a low pressure loss, typically under 0.1bar at 30L/min. The turbine has been experimentally optimized to maximize output powers and minimize head losses. The water flow energy harvester has been connected to a power management circuit to recharge a 2.4V NiMH battery from 1.2L/min (4.8μW); up to 25.2mW are generated at 20L/min. As targeted, the pressure loss is low, equal to 0.05bar@30L/min which corresponds to a minor loss coefficient of K_{EH}=3.94 for the energy harvester, and one of the lowest of the state of the art. Finally, the proportionality between the rotation speed of the turbine and the flow can be used to turn this energy harvester into an autonomous flowmeter. The next step is to test this water flow energy harvester in an irrigation system to measure and optimize flows and water consumption.

Acknowledgements

"This project has received funding from the European Union's Horizon 2020 research and innovation programme under grant agreement No 644856".

References

[1] Hao W.S. et al. 2014 Development of a Digital and Battery-Free Smart Flowmeter, *Energies*

[2] Becker P et al. 2013 Energy Autonomous Wireless Water Meter with Integrated Turbine Driven Energy Harvester, *Proc. PowerMEMS*

[3] Hoffmann D et al. 2013 Energy Harvesting from Fluid Flow in Water Pipelines for Smart Metering Applications, *Proc. PowerMEMS*

Journal of Physics: Conference Series **773** (2016) 012020 doi:10.1088/1742-6596/773/1/012020

Development of hydrodynamic micro-bearings

P. Wang[1], J. Zhang[1], H. A. Spikes[1], T. Reddyhoff[1] and A. S. Holmes[2]

1 Tribology Group, Department of Mechanical Engineering, Imperial College London, London SW7 2AZ, United Kingdom

2 Optical and Semiconductor Devices Group, Department of Electrical and Electronic Engineering, Imperial College London, London SW7 2AZ

peng.wang14@imperial.ac.uk

Abstract. This paper describes the modelling and testing of mm-scale hydrodynamic bearings which are being developed to improve the efficiency of a cm-scale turbine energy harvester, whose efficiency was previously limited by poorly lubricated commercial jewel-bearings. The bearings were fabricated using DRIE and their performance was assessed using a custom built MEMS tribometer. Results demonstrate that acceptably low friction is achieved when low viscosity liquid lubricants are used in combination with an appropriate choice of friction modifier additive. Further reduction in friction is demonstrated when the step height of bearing is adjusted in accordance with hydrodynamic theory. In parallel with the experiments, hydrodynamic lubricant modelling has been carried out to predict and further optimize film thickness and friction performance. Modelling results are presented and validated against experimental friction data.

1. Introduction

Despite significant advances in MEMS design and fabrication technology over the past few decades, problems of high frication coefficient and wear rates have severely limited the performance, and so far prevented the commercial exploitation, of devices that include sliding components [1-3]. These issues arise due to surface forces (viscous drag, van der Waals and capillary forces) that dominate at small length scales due to high surface to volume ratios. In addition to this, silicon – the most common material for MEMS components – is brittle and has high surface energy [4].

Many researchers have attempted to devise effective lubrication methods for MEMS devices. For instance, bonding self-assembled monolayers (SAMs) to surfaces has been shown to be an effective means of preventing stiction and reducing friction in MEMS [5]. Such an approach is partially effective since SAMs reduce surface energy and make the contacting surfaces hydrophobic – hence preventing water vapour adsorption [6,7]. Another lubrication method that has been applied to the silicon surfaces of MEMS is dry coatings, such as diamond-like carbon (DLC), which has shown to give friction coefficients of less than 0.01 [8]. The above two lubrication methods, however, are not suitable to be applied to contacting surfaces that involve large amount of sliding, such as the bearings within a rotating machine. This is because SAMs and DLC coatings cannot replenish themselves and so have a finite wear life [9]. Vapour phase lubrication is another way to lubricate MEMS, which overcomes this problem. However, the expense of this method and the requirement for specialized hermetic sealing preclude its use in many applications [1].

Liquid lubrication is widely applied in macro-scale applications but is generally considered to be unsuitable for micro-scale devices as it has been assumed to result in excessive viscous drag [3] – a claim supported by observations of over damping and the poor performance of certain liquid lubricated MEMS devices [10]. However, it should be noted that the viscosity of the liquids used in those studies was relatively high; whereas recent research at imperial college [2,3] has in fact shown that MEMS-type contacts can achieve low friction provided sufficiently low viscosity lubricants are used.

The aim of this research is to further demonstrate the effectiveness of liquid lubricating MEMS devices by designing and incorporating micro-hydrodynamic thrust pad bearings into an existing cm-scale energy harvester. The bearings were fabricated using deep reactive ion etching (DRIE) and tested on a custom designed tribometer. To optimize these components, the bearing step height was controlled and friction results are compared to results from a lubrication model based on a finite difference solution to Reynolds' equation. The test results illustrate that the friction coefficient of the thrust pad bearing is lower than 0.04, and can be further reduced by including amine additives.

2. Experimental methods

2.1. Bearing fabrication

A range of hydrodynamic micro-thrust step bearings were designed and fabricated on a bonded silicon on insulator (BSOI) wafer using photolithography and DRIE. The resulting bearings consist of pairs of circular pads with radial steps to pressurize the lubricant as they rotate against each other (Figure 1 shows an SEM image of the bearing geometry). The key dimensions of the bearings are a 1 mm radius, a recessed channel angle of 45°, an inner recessed region with a radius of 0.5 mm, an outer radius recessed region with a radius of 0.95 mm and a step height ranging from 1 to 50 μm.

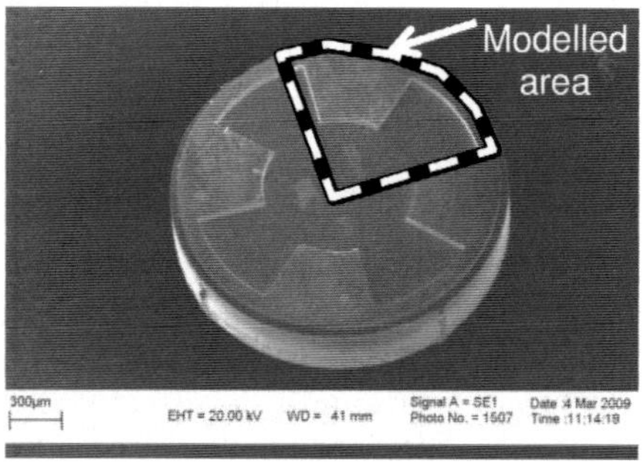

Figure 1. SEM image of a patterned MEMS thrust bearing [3].

2.2. Test rig

The bearings were evaluated using the custom built MEMS tribometer shown in Figure 2. This equipment is similar to the one used previously to study tribological behaviours of MEMS under high sliding conditions [11]. The setup uses a high-speed servo motor to rotate the bearings and measures the frictional torque from the rotation of the elastic suspension onto which the bearing is mounted by means of an optical lever technique.

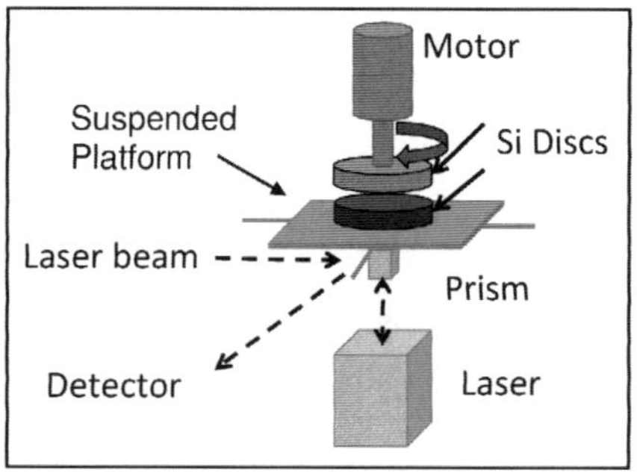

Figure 2. Schematic of tribometer used to characterize MEMS-scale bearings.

2.3. Pad bearing model

Modelling work was carried out in order to predict and further optimize micro-bearing film thickness and friction performance. As shown in figure 1, the bearing geometry can be divided into four identical parts, each containing a step, and the model is applied to each of the quarters. The quarter bearing geometry is then discretised into grid points and Reynolds' equation (in polar coordinates assuming only rotation and no squeeze film effects) is solved at each point using a finite difference approach:

$$p_{ij} = CN_{ij}\,p_{i,j+1} + CS_{ij}\,p_{i,j-1} + CE_{ij}\,p_{i+1,j} + CW_{ij}\,p_{i-1,j} + G_{ij} \qquad [1]$$

where $p_{i,j}$ is the hydrodynamic pressure at point *i,j* and coefficients *CN, CS, CE, CW* and *G* are derived from Reynolds' equation. By incorporating the stiffness of the bearing supports, this approach can predict bearing performance in terms of pressure distribution, friction coefficient and film thickness under hydrodynamic conditions.

3. Results and Discussion

Figure 3 shows friction vs. speed (Stribeck) curves for one particular design of thrust pad bearing lubricated with a very low viscosity liquid (toluene, v = 0.55 mPas); the step height of the bearing in this test is 50 μm. It can be seen that when speed is greater than 500 rpm, the contact operates in the full film lubrication regime (surfaces are fully separated by liquid) and the friction coefficient is around 0.1. At low speed, however, there is insufficient entrainment of fluid to separate the surface and friction is high due to solid-solid contact. This problem can be effectively solved by adding a surface active friction modifier, octadecylamine (ODA). When 0.1% weight of ODA is blended with the lubricant, the friction coefficient at low speeds is reduced from over 1 to less than 0.2.

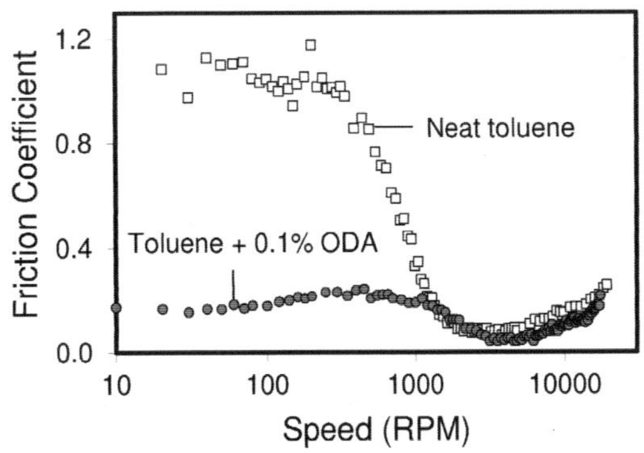

Fig. 3. Stribeck curve for MEMS thrust pad bearing showing effect of amine friction modifier.

Results from numerical simulations suggested that the performance of the mm-scale thrust pad could be further improved by adjusting the step height of the bearing. This is indeed the case, as shown experimentally in Figure 4, where it can be seen that the friction coefficient of the bearing reduces with decreasing step height. For instance, the friction coefficient of the bearing with a 1 μm-step at 10000 rpm is less than 0.02, whereas that for the 50 μm-step height bearing is approximately 0.3. This result is also in accordance with hydrodynamic theory. It should also be noted that the optimised step height of 1 μm provides low friction even when the higher viscosity liquid, hexadecane (v = 3.03 mPas), is used as the lubricant. This choice of lubricant is important in order to reduce evaporation.

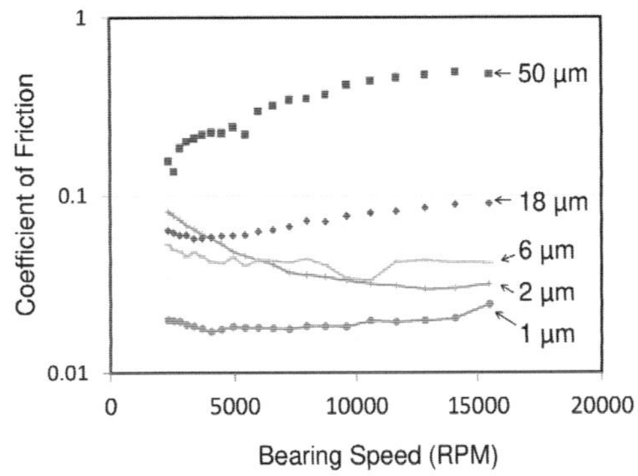

Fig. 4. Stribeck curve for MEMS thrust pad bearing showing effect of varying step height.

Bearing performance was also modelled using a finite difference solution to Reynolds' equation. The model was developed to predict the performance of the thrust pad bearing (*e.g.* pressure distribution, friction coefficient and film thickness) when operating in the full film hydrodynamic regime. The resulting hydrodynamic pressure distribution for one quarter of the 1 μm step height bearing is illustrated in Figure 5. The validation of the model is shown in Figure 6 where the friction coefficient vs. speed curves of a 1μm-step height bearing obtained from both simulation and experiment are compared. The film friction in the model is obtained by integrating the shear stress within the liquid, therefore, as expected, the simulation result does not match the experimental results when the speed is lower than 3000 rpm (*i.e.* when surface contact occurs and bearing operates in the boundary lubrication regime). However, when the speed is higher than 5000 rpm, and the bearing operates in the full film regime, the simulation result agrees well with the experimental result.

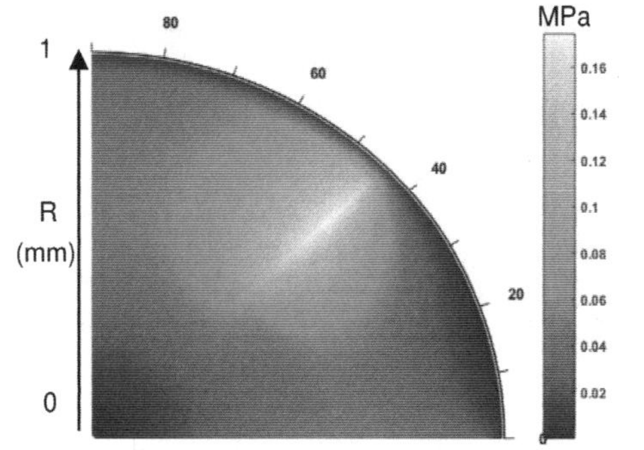

Figure 5. Hydrodynamic pressure in ¼ of pad bearing, predicted by finite difference model.

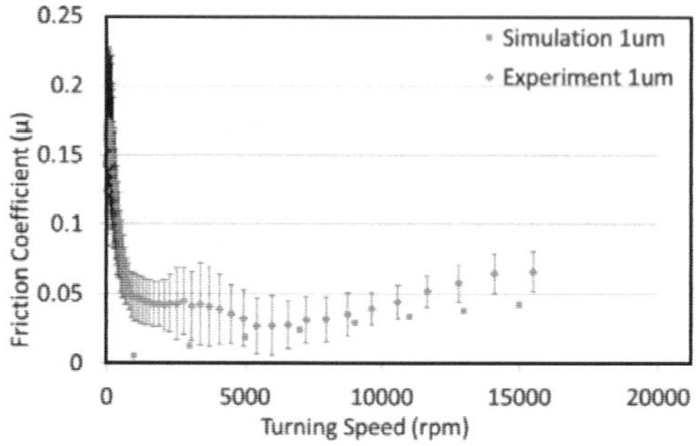

Figure 6. Comparison between of outputs from finite difference model and bearing test.

4. Conclusions

A series of mm-scale thrust bearings were successfully designed and fabricated with DRIE. With low viscosity lubricant, both simulation and experiment showed that the friction coefficient of an optimized bearing (1 μm step height) can be lower than 0.02 when operating in the high speed, full film regime. In addition, the performance of the bearing at low speeds can be significantly improved by including an amine friction additive friction modifier. The hydrodynamic thrust pad bearings will be integrated into the cm-scale turbine energy harvester shown in Figure 7, where it is expected they will offer improved performance compared to the existing jewel bearings.

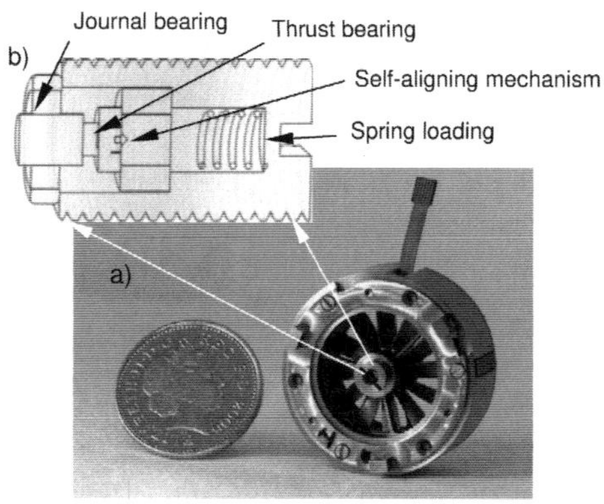

Figure 7. a) Photograph of turbine energy harvester [12], b) drawing of bearing and housing.

References

[1] Leong J and Satyanarayana N 2013 *Trib. Letts.* vol **50** pp 195-206.
[2] Reddyhoff T and Ku I 2010 *Trib. Letts.* vol **41** pp. 239-246.
[3] Ku I and Reddyhoff T 2012 *J. Trib.* vol **134** no. 1 p. 012002.
[4] Gavrilenko V and Novikov P 2009 *Proceedings of the SPIE.* 7405.
[5] Ashurst W and Yau C 2001 *J. Microelectromechanical Systems.* vol **10** no. 1 pp. 41-49.
[6] Srinivasan U and Houston M 1998 *J. Microelectromechanical Systems.* vol **7** no. 2 pp. 252-260.
[7] Teodorescu M and Theodossiades S 2009 *Trib. International.* vol **42** no. 2 pp. 197-205.
[8] Tagawa M and Ikemura M 2004 *Trib. Letts.* vol **17** no. 3 pp. 575-580.
[9] Smallwood S and Eapen K 2006 *Wear.* vol **260** no. 11-12 pp. 1179-1189.
[10] Mehregany M and Dhuler V 1992 *J. Micromech. Microeng.* vol **2** no. 1 pp. 1-3.
[11] Ku I and Reddyhoff T 2010 *Trib. International.* vol **43** no. 5-6 pp. 1087-1090.
[12] Deng K and Ramanathan G P 1994 *J. Micromech. Microeng.* vol 4 pp. 266–269.

Triboelectret-based aeroelastic flutter energy harvesters

Matthias Perez[1,2,3], Sebastien Boisseau[1,2], Matthias Geisler[1,2], Ghislain Despesse[1,2], and Jean Luc Reboud[1,3]
[1] Univ. Grenoble Alpes, F-38000 Grenoble, France
[2] CEA, Leti, Minatec Campus, 17 rue des Martyrs, F-38054 Grenoble, France
[3] Univ. Grenoble Alpes G2ELab, F-38000 Grenoble, France

E-mail: matthias.perez1@gmail.com

Abstract. This paper highlights some experimental results on several electrostatic membranes tested in a wind tunnel between 0 and $20 m.s^{-1}$ for airflow energy harvesting. The main idea is to use the aeroelastic behavior of thin flexible films to induce simultaneously the capacitance variations and the polarization required by the triboelectric/electrostatic conversion. This technology provides thin and flexible devices and avoids the issue of electrets discharge. Our prototypes ($<16 cm^2$) allowed a quick startup (from $3 ms^{-1}$), an electrical power-flux density from $0.1 \mu W.cm^{-2}$ to $60 \mu W.cm^{-2}$. In order to complete the energy harvesting chain, we have used a wireless sensor with temperature and acceleration measures coupled to a low power transmission (Bluetooth Low Energy) with reception on a smartphone.

Introduction

Airflow energy harvesting is an area of substantial and growing interest for the energy harvesting community. Indeed, airflows exist in various situations and offer a high density of kinetic power $((60\mu W/cm^2)/(m/s)^3)$. In order to convert the flow's kinetic energy into electrical energy, one common option is to go through an intermediary mechanical energy. This mechanical energy can take the form of a rotary or an oscillatory motion of a mechanical element, which can be rigid [1] or flexible [2]. In this paper, we use the fluttering phenomenon as a "kinetic-to-mechanical" energy converter, a triboelectret-based electrostatic transducer as a "mechanical-to-electrical" converter and a self-starting power management circuit to supply low power sensors with a RF communication.

1. Aeroelastic harvesters

A film at rest is more or less flexed by the gravity, according to its orientation and its intrinsic stiffness (Figure 1a). Once a low speed flow appears, the film is able to stretch in the direction of the flow thanks to viscous frictions (Figure 1b). Then, if we increase the velocity up to a critical speed U_c, the film starts to oscillate/flutter periodically (Figure 1c). As a consequence, a fraction of the kinetic energy can be converted into mechanical energy of oscillating motion, which can potentially be converted into electrical energy [3,4].

In the state of the art, many aeroelastic harvesters with different converters have been developed during the 10 past years. Overall, it may be noted that electromagnetic converters are not really suited, which explains the small number of publications [4-6]. However, the piezoelectric converters are much more suited to aeroelasticity and have been widely used [7-13]. Indeed, they provide flat and flexible harvesters and they do not require any shock/contact to operate.

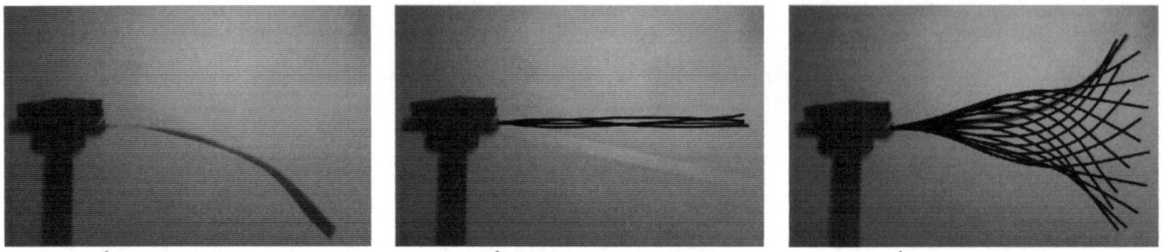

$U=0m.s^{-1}$: The film is flexed by the gravity | $U<6m.s^{-1}$: The film is stretched in the direction of the flow | $U_c=6m.s^{-1}$: The film starts to oscillate

Figure 1. Behavior of a PVDF film (L=3cm, H=1.5cm and h=25µm) as a function of the airflow speed.

2. Triboelectric conversion

The electrostatic conversion requires capacitance variations through relative motion between two elements to operate. In our case, this relative motion occurs thanks to the addition of two lateral walls (Figure 2). This will lead to mechanical contacts between the film and the lateral walls, simultaneously creating strong capacitance variations and the polarization of the structure by triboelectric effects.

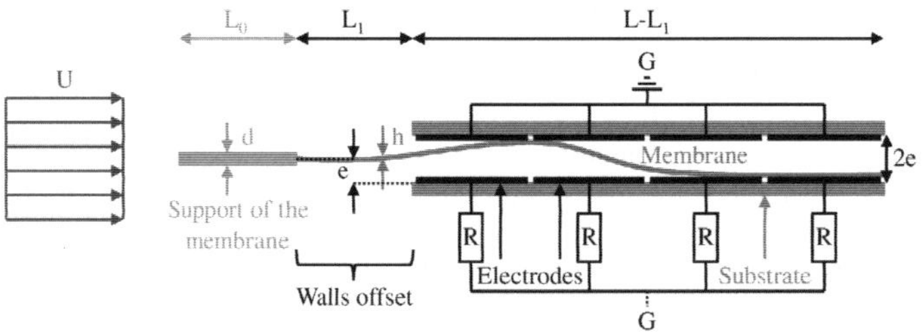

Figure 2. Functional schematic and parameters associated with the different prototypes.

In this paper, we used Teflon FEP uncharged membranes of various thicknesses, with the aim of using triboelectric effects to turn them into triboelectrets, and several electrodes located around (Figure 2). When the FEP membrane starts to oscillate and collides with the lateral electrodes,electric charges are exchanged by triboelectric effects ([14] and Figure 3). A portion of the negative charges are trapped ($<0.25mC/m^2$) in the dielectric membrane, turning the FEP membrane into a triboelectret (V=315V on a 25µm film; 410V on a 50µm film and 570V on a 125µm film at $15m.s^{-1}$; with V the surface voltage of the electret), and polarizing the variable capacitor.

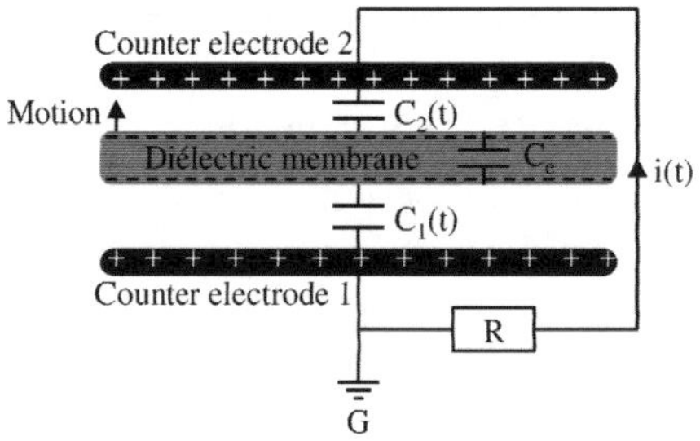

Figure 3. Polarization of the membrane by triboelectricity.

3. Experimental results

Various prototypes with various dimensions and parameters have been tested between 0 and $20m.s^{-1}$ (round markers). Whatever the configuration chosen (length, width, thickness, distance between the walls), the power-flux density exceeds $10\mu W.cm^{-2}$ from approximately $10m.s^{-1}$ (Figure 4a). At higher speeds (from $15m.s^{-1}$ to $20m.s^{-1}$), it is possible to reach larger power-flux densities (from $20\mu W.cm^{-2}$ to $40\mu W.cm^{-2}$). Overall, thin/flexible films should be favored at low speeds (the brown curve in Figure 4a has been achieved with a thin film of 25µm). In addition, the distance between the film and the walls must ideally be larger at low speeds and small at high speeds (the blue and green curves have been obtained with 2e=3mm and all others with 2e=11mm). Moreover, the high benefit of the multi-electrodes concept was also proven as it has enabled to multiply by three the output power of our devices compared to the two-electrode configuration (Figure 4b – E1, E2, E3 and E4 connected together).

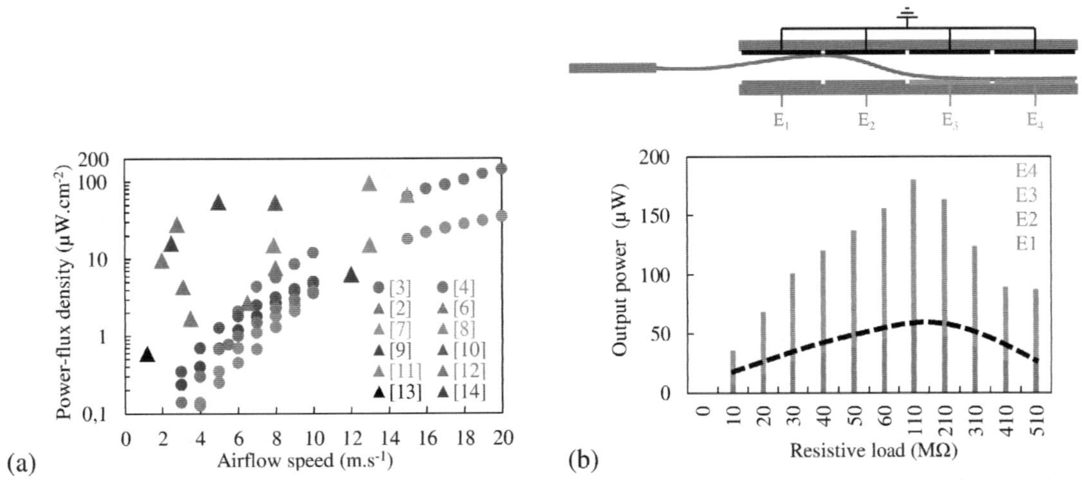

Figure 4. Evolution of the power-flux density of our prototypes (round tags) and comparison to the state of the art (triangular tags). (b) Impact of the multiplication of the electrodes on the output power (L=5cm, H=2cm, h=50μm, L_0=1cm, d=1mm, L_l=1cm and U=15m.s^{-1}), the black dashed line represents the case where the four electrodes (E1, E2, E3 and E4) are connected together

4. Self starting battery-free power management circuit

For this last reason, we decided to use a MPP (Maximum Power Point) high voltage circuit whose schematic is depicted in Figure 5. Several flutter converters (blue) are connected in parallel to diode bridges (red) and simultaneously recharge a capacitor C_0 (grey). C_0 is in the range of a few dozens of nanofarads in order to reach the optimum voltage V_{opt} (equal to $U_{open\text{-}circuit}/2$) in a few milliseconds. Then, C_0 is connected to a voltage divider to reduce the voltage across R_l to a few volts ($R_0/R_l \approx 100$), and used to measure V_{C0}, the voltage across C_0. The function of the "Flyback control" is to discharge a part of the energy stored in C_0 in the flyback (purple) as soon as the voltage at its terminal reaches V_{opt} in order to maintain V_{C0} close to V_{opt}. The electric energy is finally transferred into a buffer capacitance C_b (brown) to power the WSN (orange), as we did in [3] with a SECE circuit. A passive circuit (green) is added to supply the "Flyback control", through the charge of C_s (pink), at the start-up of the system, as presented in [15].

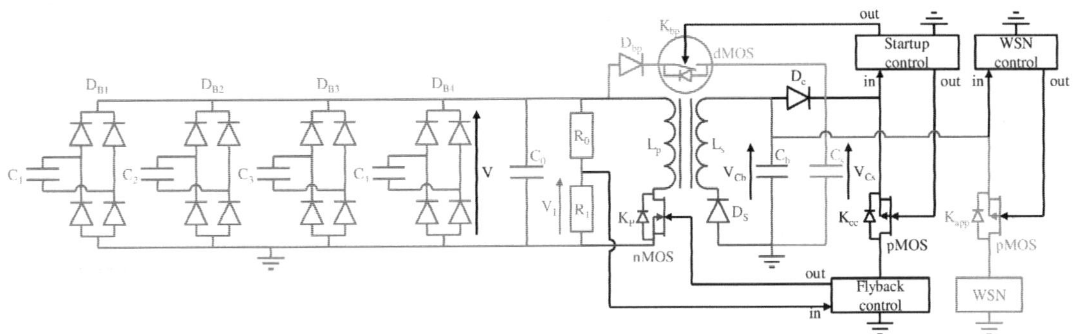

Figure 5. Schematic of the MPP circuit.

Thanks to the device illustrated in Figure 6b subjected to an airflow of 15m.s^{-1} and with C_b=100μF, the MPP active circuit turned out to be the most efficient with a first measurement of temperature and acceleration after 1 minute followed by other measurements every 9 seconds (green curve in Figure

6a). The SECE active circuit presented in [3] is less efficient, with a first measurement after 2 minutes followed by periods of 45 seconds (red curve in Figure 6a). However, it is still more efficient than a passive strategy with a simple diode bridge directly connected to the buffer capacitor C_b (blue curve in Figure 6a) In this case, the first measurement appears after 5 minutes and 35 seconds and every 1 minute and 41 seconds for the other measurements (Figure 6a).

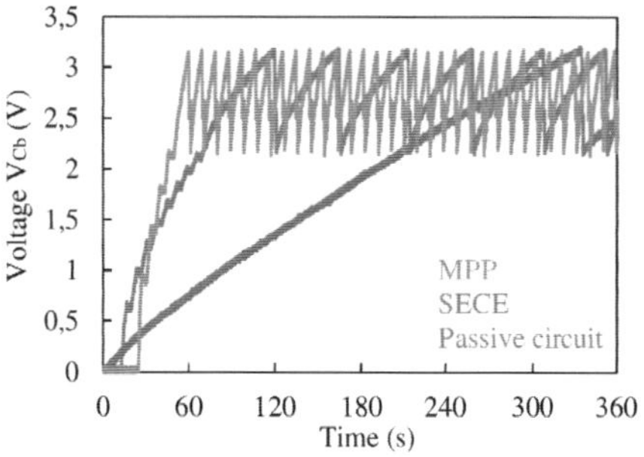

(a)

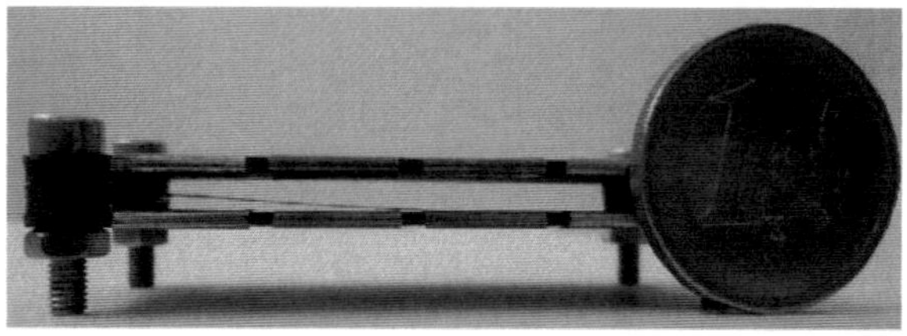

(b)

Figure 6. (a) Evolution of the voltage V_{Cb} during the measurements (L=5cm, H=2cm, h=50µm, L_0=1cm, d=1mm, L_1=1cm and U=15m.s^{-1}, C_b=100µF). (b) Picture of one of the harvesters.

References

[1] Perez M et al. 2016 A cm scale electret-based electrostatic wind turbine for low-speed energy harvesting applications *Smart materials and structures*

[2] Fei F and Li W.J. 2009 A fluttering-to-electrical transduction system for consumer electronics applications *Robotics and biomimetics*

[3] Perez M et al. 2015"An electret-based aeroelastic flutter energy harvester", Smart materials and structures (2015)

[4] Perez M et al. 2015 Electret-based aeroelastic harvester and its self-starting battery-free power management circuit *Proc. NEWCAS*

[5] Zhu D, Beeby S, Tudor J, White N and Harris N 2010 A novel miniature wind generator for wireless sensing applications *Sensors*

[6] Fei F, Mai J D and Li W J 2012 A wind-flutter energy converter for powering wireless sensors *Sensors and actuators A*

[7] Li S and Lipson H 2009 Vertical-stalk flapping-leaf generator for wind energy harvesting *Smart materials, adaptive structures and intelligent systems*

[8] Kwon S 2010 A T-shaped piezoelecric cantilever for fluid energy harvesting *Applied physics letters*

[9] St Clair D, Bibo A, Sennakesavababu V R, Daqaq M F and Li G 2010 A scalable concept for micropower generation using flow-induced self-excited oscillations *Applied physics letters*

[10] Li S, Yuan J and Lipson H 2011 Ambient wind energy harvesting using cross-flow fluttering *Journal of applied physics*

[11] Bryant M and Garcia E 2011 Modeling and testing of a novel aeroelastic flutter energy harvester *Journal of vibration and acoustics*

[12] Akaydin H D, Elvin N, and Andreopoulos Y 2012 The performance of a self-excited fluidic energy harvester *Smart materials and structures*

[13] Suwa E, Tsujiura Y, Kurokawa F, Hida H and Kanno I 2014 Airflow energy harvesters of metal-based PZT thin films by self-excited vibration *Journal of physics: Conference series*

[14] Yang Y et al. 2013 Triboelectric nanogenerator for harvesting wind energy and as a self powered wind vector sensor system *ACS Nano*

[15] Boisseau S et al. 2013 Self-starting power management circuits for piezoelectric and electret-based electrostatic mechanical energy harvesters *Journal of physics: Conference series*

Superhydrophobic surfaces' influence on streaming current based energy harvester

Florent Fouché, Thomas Dargent, Yannick Coffinier, Anthony Treizebré, Alexis Vlandas, Vincent Senez

Univ. Lille, CNRS, Centrale Lille, ISEN, Univ. Valenciennes, UMR 8520 - IEMN, F-59000 Lille, France
Thomas.Dargent@iemn.univ-lille1.fr

Abstract. The purpose of this paper is to report the design, fabrication and characterization of silicon-based microfluidic channels with superhydrophobic walls for energy harvesting. We present the fabrication step of silicon based streaming current energy harvester and the nanostructuration of the microchannel walls. We characterize the superhydrophobic properties of the surface in a closed system. Our preliminary results on the electrical characterization of the device show a 43% increase of power harvested with our superhydrophobic surface compared to a planar hydrophobic surface.

1. Introduction

Energy harvesting via streaming current has been investigated since the early 2000 with the growing field of microfluidics. Passing a fluid through a microchannel creates a net ionic current, which can be collected by an external electrical system. As predicted by physics, efficiency improvement has been reached by reducing the size of the microchannel, with efficiencies of 1.3% reached for micron-sized channels [1] and 3-5% for nanoslit [2]. In 2007, Pennathur *et al.*[3] have projected that 35% energy conversion efficiency could be obtained by using superhydrophobicity inside channels. Despite experiments showing liquid slip at the solid/liquid interface and the resulting decrease in pressure [3], [4], to the authors' knowledge, no experimental work has shown this promising predicated improvement in energy harvesting efficiency.

2. Design and fabrication

The first streaming current harvesting device built contains a "trench"-like microchannel. The accesses, reservoirs and channels are etched in a single process of deep-reactive ion etching (DRIE) on a silicon wafer (diameter 3", thickness 360 µm), and sealed by anodic bonding a 1 mm pyrex cover. The resulting microchannel is 1 mm long, 127 µm deep and 22 µm wide. In order to produce the superhydrophobic surface on vertical walls we adapted a version of the MACE process used to engineer silicon nanowires on orthogonal surfaces [5]. The structured surface is then made superhydrophobic by coating with a hydrophobic material (PFTS) in vapor phase. Water on a flat surface with these coated nanowires has a contact angle of 168° (6° hysteresis) and a stable Cassie-Baxter state.

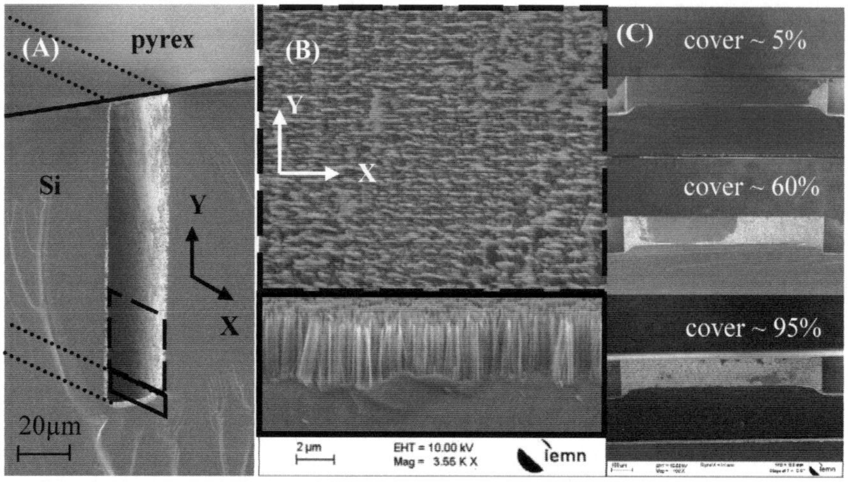

Figure 1 : SEM images of nanostructured microchannels, with (A) transverse cut of a finished microchannel, (B) close-up of a longitudinal cut, and (C) longitudinal cuts of full microchannels with varying nanowire covers (white area).

Our process generates a 2 μm-long nanowire structuration on both the bottom and the sides of the channel (Figure 1). Nanowire covering ranges from 5 to 100% of the surface (C). We used devices with more than 90% of the wall surface covered with nanowire structuration (3 out of 10 devices produced). However we could not verify the presence of the Cassie state in this geometry, which prompted the development of a second device.

The second device developed has a thin-slit microchannel and a pyrex cover allowing for optical observation of the surface. It decreases the percentage of structured surface of the microchannels to 50% but enables real-time optical monitoring of the wetting state of the surface independently from pressure measurements.

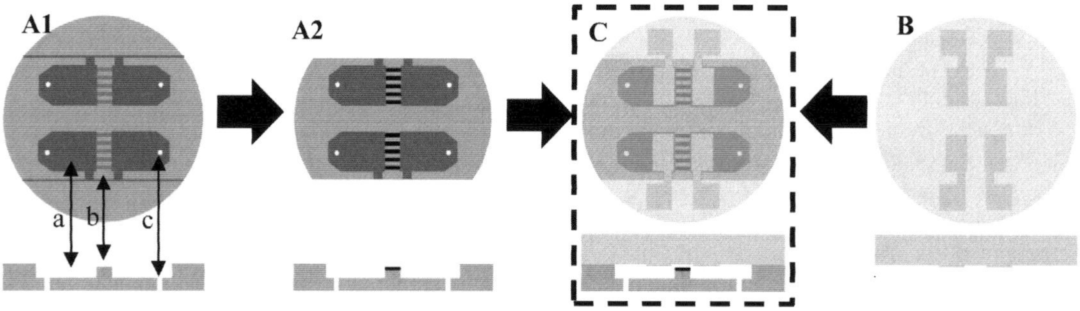

Figure 2 : Fabrication process of the second device. (A1) On a silicon wafer three levels of DRIE etching are followed by (A2) nanowire etching and PFTS deposition. (B) Gold microelectrodes are deposited on pyrex. (C) Both wafers are assembled by anodic bonding.

The microfabrication process is shown on Figure 2. Three DRIE steps (A1) are needed to etch (a) the reservoirs (100μm), (b) the channels and (c) the microfluidic accesses through the silicon. Each device contains five microchannels 10 μm deep, 1 mm wide and 5 mm long between reservoirs 100 μm deep. After etching the microchannels are structured (A2) by the same MACE process as previously used and coated with PFTS. The 100 nm thick gold microelectrodes are deposited on the pyrex wafer by sputtering and chemical etching (B), and silicon and pyrex are assembled by anodic bonding (C). The completed system is visible on Figure 3.

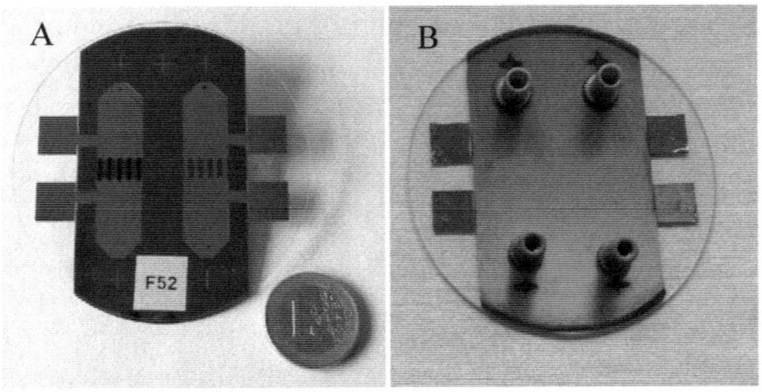

Figure 3: (A) Completed wafer with two devices, one with superhydrophobic microchannels (left) and one with flat hydrophobic surface (right). (B) backside of the device with microfluidic accesses.

3. Results

The electrical characterization of the hydrophilic silicon version of the first device is given at Figure 4.

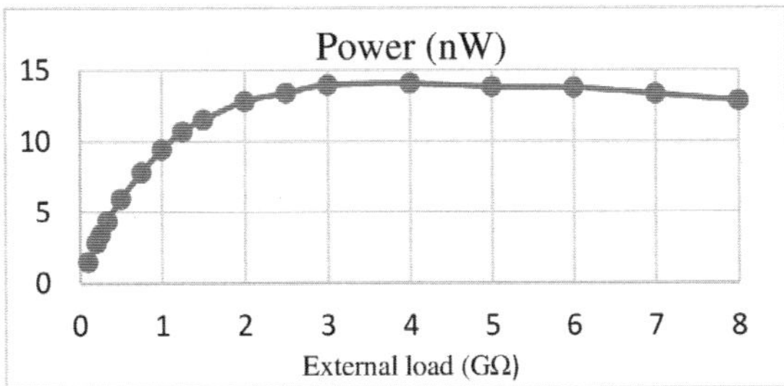

Electrical Characterization	
Streaming Potential	18 V
Streaming Current	5 nA
Efficiency	0.3%
Optimal Load	4 GΩ
Power Density	1.6 μW.cm^{-2}

Figure 4 : Electrical characterization of the hydrophilic trench-like structure.

No noticeable improvement could be observed in this device by using superhydrophobic surfaces. The high pressures needed to move the liquid in the channel may induce the impalement of the liquid on the nanotexturation. However, it is impossible to confirm this hypothesis.

The wetting state of the superhydrophobic surface of the second device has been monitored with an optical camera (Figure 5) showing either Cassie-Baxter (5.A) or Wenzel (5.B) states. The superhydrophobic surface can sustain overpressures superior to 170 kPa. After impalement, the superhydrophobic behaviour can be regenerated several times by drying.

Figure 5 : Optical picture from the glass side of the second device with 10 channels either in (A) a Cassie Baxter state (no wetting of the silicon nanowires) or (B) a Wenzel state (impalement of the liquid on the silicon-nanowires).

We did not observe noticeable pressure drop reduction in this device between the Cassie-Baxter state and the fully wetted Wenzel state. Two hypothesis can be formulated. First, since only 50% of the channel is covered by silicon nanowires (the remaining being the glass substrate), we have a strong heterogeneity in the slip conditions. It has been shown that for confined systems, the flow is strongly controlled by this heterogeneity [6]. Second, even if the optical characterization shows a Cassie Baxter state, we do not have any information about the degree of impalement of the liquid on the nanostructures and thus on the hysteresis (i.e. adhesion force). Additional characterization are mandatory.

Figure 6 (A) gives the variation of the streaming current as a function of pressure for the second device. At high pressures, streaming currents up to tens of nano-amperes and streaming potentials up to 3.5 V have been measured in the device. Preliminary results on the comparative power output of flat hydrophobic surfaces and superhydrophobic surfaces are reported on Figure 6 (B). One can observe an increase of up to 40% of the output power for superhydrophobic surfaces, depending on the load resistance of the external electrical circuit. Power measurement were made at a pressure of 20 kPa.

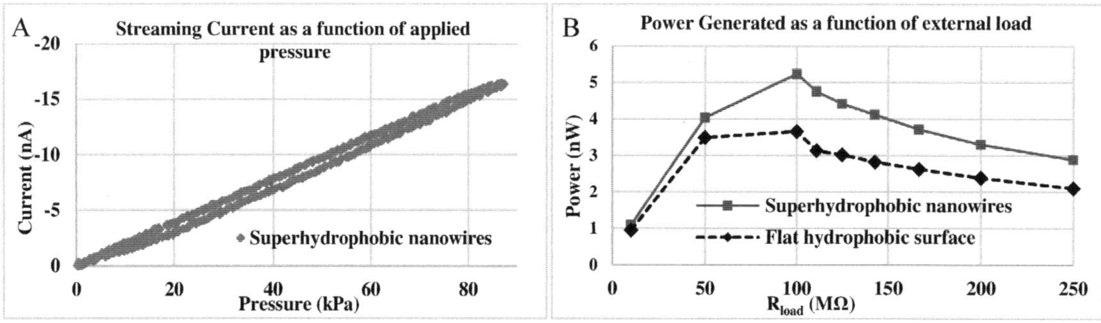

Figure 6 : Electrical characterizations of the second device consisting in a confined channel (10µm height) with the top surface made of flat glass and the bottom surface made of silicon nanowires.

4. Conclusion

Our preliminary results add experimental evidence that using superhydrophobic nanostructured surfaces in microchannels can increase the efficiency of energy generation through streaming current harvesting.

References

[1] Mansouri A and Kostiuk L W 2016 "Giant streaming currents measured in a gold sputtered glass microchannel array" *Chem. Phys. Lett.* **646** 81–86

[2] van der Heyden F H J, Bonthuis D J, Stein D, Meyer C and Dekker C 2007 "Power generation by pressure-driven transport of ions in nanofluidic channels" *Nano Lett.* **7** 1022–5

[3] Pennathur S, Eijkel J C T and van der Berg A 2007 "Energy conversion in microsystems: is there a role for micro/nanofluidics?" *Lab Chip* **7** 1234–7

[4] Choi C H, Ulmanella U, Kim J, Ho C M and Kim C J 2006 "Effective slip and friction reduction in nanograted superhydrophobic microchannels" *Phys. Fluids* **18**

[5] Piret G, Coffinier Y, Roux C, Melnyk O and Boukherroub R 2008 "Biomolecule and nanoparticle transfer on patterned and heterogeneously wetted superhydrophobic silicon nanowire surfaces.," *Langmuir* **24** 1670–2

[6] Yao Z, Lu Y W and Klandikar S G 2012 "Fabrication of nanowires on orthogonal surfaces of microchannels and their effect on pool boiling," *J. Micromechanics Microengineering* **22** 115005

Electret-based Unsteady Thermal Energy Harvester using Potassium Tantalate Niobate Crystal

Hong Xie[1], Kenichi Morimoto, Yuji Suzuki
[1]Dept. of Mechanical Engineering, The University of Tokyo
Email: hxie@mesl.t.u-tokyo.ac.jp

Abstract: An electret-based unsteady thermal energy harvester is proposed using potassium tantalate niobate ($KTa_{1-x}Nb_xO_3$, KTN) as a dielectric for the capacitor. By connecting in series the capacitor and an electret serving as a permanent voltage source, the capacitance change with temperature fluctuations alters the amount of induced charges thereby produces the external current. By using KTN having extremely-large temperature coefficient of permittivity together with the CYTOP electret, the output power of 572 nJ has been obtained from one heating cycle, which corresponds to 20 times higher output power than the previous result with $BaTiO_3$.

Keywords: Thermal energy harvester, Electret, Potassium tantalate niobate, Permittivity

1. Introduction

Thermal energy harvesting is a promising energy source for low-energy consumption devices like wireless sensor nodes for Internet of Things (IoT) [1]. Various thermoelectric generators based on the Seebeck effect are prototyped for power generation from the steady temperature gradient [2]. On the other hand, ambient temperature fluctuations can be another possible energy source for thermal energy harvesting. Thermoelectric harvesters with thermal capacitor are considered for transient thermal fields [3]. Pyroelectric materials such as PZT or PVDF, in which change of spontaneous polarizations within the crystal structure is used, has higher energy conversion efficiency [4, 5].

We previously proposed an electret-based electrostatic unsteady thermal energy harvester [6], and obtained the output power of 30 nJ by heating the $BaTiO_3$ from 20 to 100 °C in 300 s. The performance of the device was limited by the temperature sensitivity of the dielectric constant, although no heavy metal such as lead and cobalt is used for the electret and the capacitor.

In present study, an electret-based thermal energy harvester has been proposed with potassium tantalate niobate ($KTa_{1-x}Nb_xO_3$, KTN) crystal [6] with much higher temperature sensitivity.

2. Working principle

Figure.1 shows the working principle of the electret-based energy harvester [7]. The device consists of three parts; the electret serving as the permanent voltage source (capacitance: C_e, surface charge density: σ_e), the capacitor with temperature-sensitive dielectric (capacitance: C), and the external load. By connecting them in series, the permittivity of dielectric changes as ambient temperature varies, which leads to change of the amount of the induced charges. Thus, the external current is produced by the temperature fluctuations.

Based on the circuit model shown in figure 1, we employ the Kirchhoff's law, the Gauss's law and the conservation of charges:

$$V = IR = d_1E_1 - d_1E_1, I = -S_1\frac{d\sigma_1}{dt} = S_2\frac{d\sigma_2}{dt} \quad (1)$$

$$\sigma_1 = \varepsilon_0\varepsilon_1E_1, \quad \sigma_2 = \varepsilon_0\varepsilon_2(t)E_2 \quad (2)$$

$$S_1\sigma_e = -(S_1\sigma_1 + S_2\sigma_2) \quad (3)$$

where d_1, S_1, E_1, σ_1, ε_1 are the thickness, the area, the electrical field, the induced charge and the permittivity of the electret, while d_2, S_2, E_2, σ_2, $\varepsilon(t)$ are those of the dielectric. After the non-dimensionalization with $t^*=t/RC_e$, $\sigma^*= \sigma/\sigma_e$, $C^*(t)=C(t)/C_e$, we can get the differential equation:

$$I^* = V^* = \frac{d\sigma_1^*}{dt^*} = -\{(1+\frac{1}{C(t^*)})\sigma_1^* + \frac{1}{C(t^*)}\} \quad (4)$$

And the output power J can be written as,

$$J = V_{charge}{}^2C_eJ^* \quad (5)$$

where V_{charge}, C_e, and J^* are the surface voltage of electret, the capacitance of the electret, and the normalized output power derived from the non-dimensionalized differential equation, respectively. Therefore, the output power of the present unsteady energy harvester is mainly controlled by two main independent parameters, i.e., temperature sensitivity of the dielectric, and the surface potential of the electret.

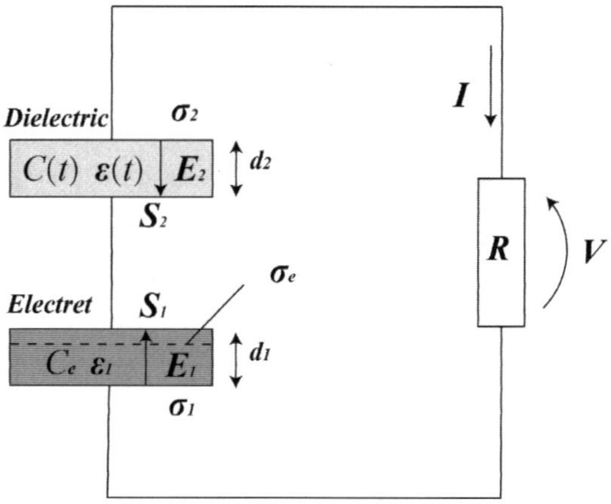

Figure 1: Circuit model of the electret-based energy harvester.

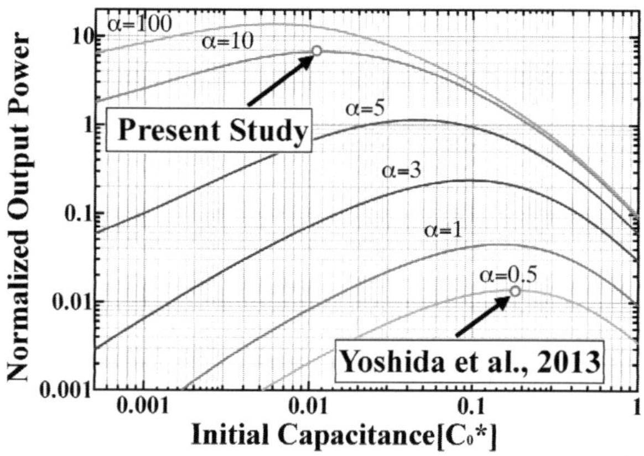

Figure 2: Normalized output power for different α.

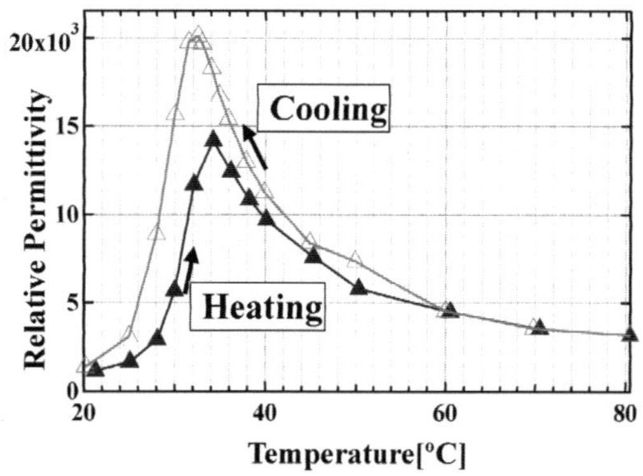

Figure 3: Permittivity change of the KTN crystal.

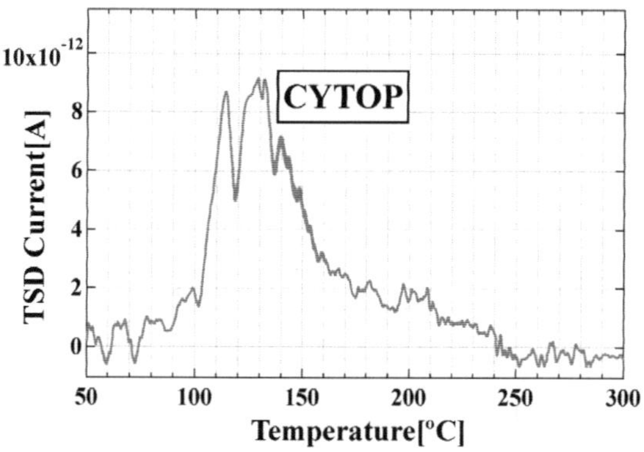

Figure 4: TSD spectra of the CYTOP electret.

Figure 2 shows the normalized output power versus the capacitance ratio $C_0^* = C_0/C_e$ (C_0: initial capacitance of the dielectric) for different permittivity increase ratios α, defined by the ratio of the maximum and minimum capacitances during one temperature change cycle. The normalized output power increases dramatically with α. In our previous study with a $BaTiO_3$ dielectric, α is as low as 0.5 [7]. Figure 3 shows the relative permittivity of the KTN crystal presently used in the temperature range between 20 °C and 80 °C. KTN shows very high permittivity change near its phase transition temperature [6], and its permittivity increase rate α is as high as 11 for heating and 13 for cooling.

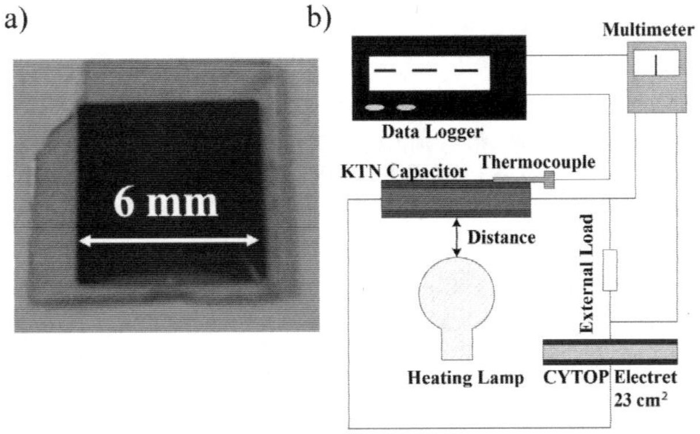

Figure 5: a) KTN crystal (size: 6 mm × 6 mm, thickness: 0.27 mm), b) Experimental setup for power generation.

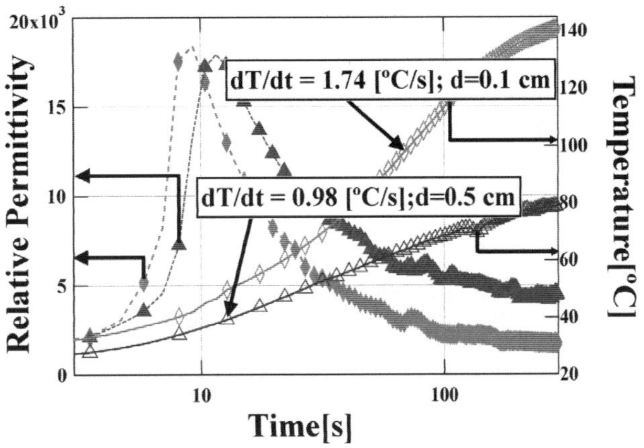

Figure 6: Temperature and permittivity for the KTN capacitor for different heating rates.

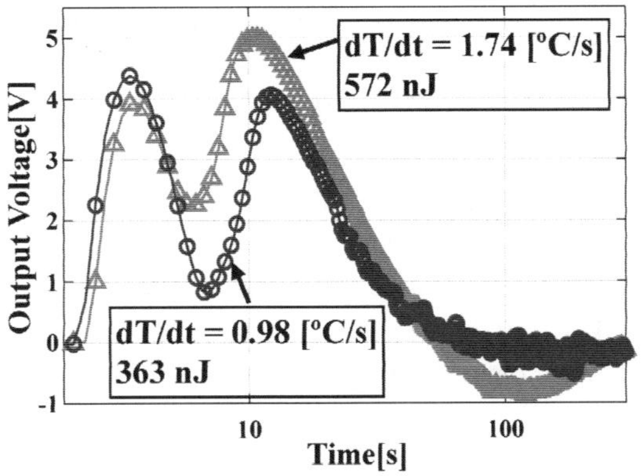

Figure 7: Voltage output during the single heating cycle.

3. Experimental set up and Results

Figure 5a shows the 0.27 mm-thick KTN crystal used in the present study. The size of the metalized area is 6 mm × 6 mm. 15 μm-thick CYTOP EGG [8] with double sides of electrodes on a silicon wafer is used as the electret, of which area is 23 cm^2. Thermal charging [9] at 120 °C with the bias voltage of 400 V is applied. To specify the charge properties of the electret, thermally-stimulated discharge (TSD) experiment [10] is used. The heating rate is 1 °C/min. Figure 4 shows the result of TSD experiment. By integrating the TSD current over the elapsed time, the stored charge density for the CYTOP electret is estimated to be 53 nC/m^2, which corresponds to the surface voltage of V_{charge} = 43 V.

Figure 5b shows the present experimental setup. A lamp bulb was used as the heating source of the KTN capacitor. Temperature of the KTN capacitor is monitored by a thin thermocouple attached to the surface. Permittivity is also monitored by measuring the capacitance.

Figure 6 shows the temperature and permittivity changes for the present KTN capacitor. With the distance between the KTN crystal and the lamp of 0.1 cm and 0.5 cm, different initial temperature gradients of 1.74 °C/s and 0.98 °C/s are achieved. Figure 7 shows the output voltage with the external load of 610 MΩ. With the heating rates of 1.74 °C/s and 0.98 °C/s, the output power of 572 nJ and 363 nJ have been obtained, respectively. With the heating rate of 1.74 °C/s, the output power is 20 times higher than that in our previous study [7]. In Fig. 6, at around t = 10 s, the time derivative of the relative permittivity is 1.5 times higher for the heating rate of 1.74 °C/s, which results in the higher output voltage and thus the higher output power as shown in Fig. 7. Because the surface voltage of the present CYTOP electret is much lower than its limit [8], much higher output power can be expected with improvement of the charging process.

4. Conclusion

An electret-based thermal energy harvester using KTN crystal has been proposed. The determining factors of the total output power are better illuminated by combining the simulation results and experimental data. KTN crystal is applied as temperature-sensitive dielectric with much higher permittivity change for the temperature variation. Surface voltage of 43 V is achieved by thermal charging of CYTOP. Up to 572 nJ output has been obtained in one heating cycle with the CYTOP electret, which is 20 times higher than our previous result with $BaTiO_3$.

The KTN crystal was provided by FDK Corp. This work is supported by JST CREST program.

References:

[1] Leonov V *et al* 2007 Thermoelectric converters of human warmth for self-powered wireless sensor nodes *IEEE Sensors J.* **7** 650
[2] Wang Z *et al* 2009 Realization of a wearable miniaturized thermoelectric generator for human body applications *Sensors Actuators, A* **156** 95
[3] Moser A *et al* 2012 Thermoelectric energy harvesting from transient ambient temperature gradients. *J. Electron. Mater.* **41** 1653
[4] Sebald G *et al* 2009 On thermoelectric and pyroelectric energy harvesting *Smart Mater. Struct.* **18** 125006
[5] Cuadras A *et al* 2010 Thermal energy harvesting through pyroelectricity *Sensors Actuators, A* **158** 132
[6] Imai T *et al* 2007 Crystal growth and electro-optic properties of $KTa_{1-x}Nb_xO_3$. *NTT Tech. Rev.* **5** 1
[7] Yoshida J *et al* 2013 Electrostatic thermal energy harvester using unsteady temperature change *J. Phys. Conf. Ser.* **476** 12079
[8] Sakane Y *et al* 2008 The development of a high-performance perfluorinated polymer electret and its application to micro power generation. *J. Micromech. Microeng.* **18** 104011
[9] Pretzsch G 1983 Charge State of PTFE Thermoelectrets. Phys. Stat. Sol. **79** 139
[10] Van Turnhout J 1971 Thermally stimulated discharge of polymer electrets *Polymer Journal* **2** 173

Coupled multiphysics finite element model and experimental testing of a thermo-magnetically triggered piezoelectric generator

Adrian Rendon-Hernandez[1,2] and Skandar Basrour[1,2]

[1] Université Grenoble Alpes, TIMA Laboratory, F38031 Grenoble, France
[2] CNRS, TIMA Laboratory, F-38031 Grenoble, France

E-mail: adrian.rendon@imag.fr

Abstract. This paper deals with the coupled multiphysics finite element modeling and the experimental testing of a thermo-magnetically triggered piezoelectric generator. The model presented here, which has been developed in ANSYS software and experimentally validated, promotes a better understanding of the dynamic behavior of proposed generator. Special attention was put into the coupled multiphysics interactions, for instance, the thermal-dependent demagnetization of soft magnetic material, the piezoelectric transduction and the output power. In order to characterize the power generator, many finite element simulations were conducted, included modal and transient analysis. To verify the effectiveness of the model, a prototype was built and tested. The findings thus obtained were compared with simulation results. Obtained results describe for the first time a fully coupled model of an innovative approach for thermomagnetic energy harvesting. Moreover, the total volume of our harvester (length x width x height: 20 x 4 x 2 mm) is 85 times lower than that of previous experimental harvester.

1. Introduction

Thermomagnetic generation of electricity is based on a change in magnetic properties of soft magnetic materials such as nickel-iron or iron-silicon alloys, which occurs when those materials, that are easily magnetised and demagnetised, are under temperature variations [1]. These temperature-dependent properties can be deployed for different purposes, for instance, power supply transformers, electromagnets, and quite recently, for energy harvesting applications [2]. Carlioz *et al.* [3] reported a thermomagnetic energy harvester able to harness the ambient thermal energy from a variation in environmental temperature and convert it into electrical energy to power wireless sensors. However, a key limitation of their power generator is that it is too bulky to be integrated into a wireless sensor node and studies on predictive modeling of such device are still lacking. With this goal, the present work seeks to offer a finite element model able not only to miniaturize the power generator but to predict its dynamic behavior, while maximizing its output energy. Furthermore, a test prototype is characterized to verify the validity of developed model. The remainder of the paper is organized as follows: the proposed power generator will be discussed in Section 2. The finite element model is presented in Section 3. Section 4 shows the experimental measurements of test prototype. Finally, Section 5 concludes with a summary.

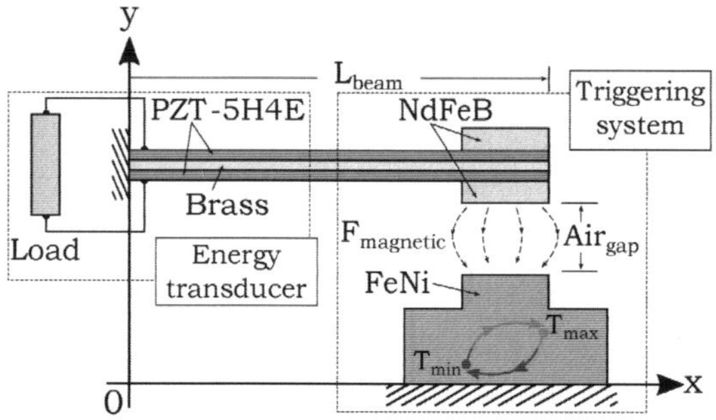

Figure 1. Side view of power generator. The variations in temperature on soft magnetic material triggers the piezoelectric transduction. Output power is dissipated on a load resistor, R_L.

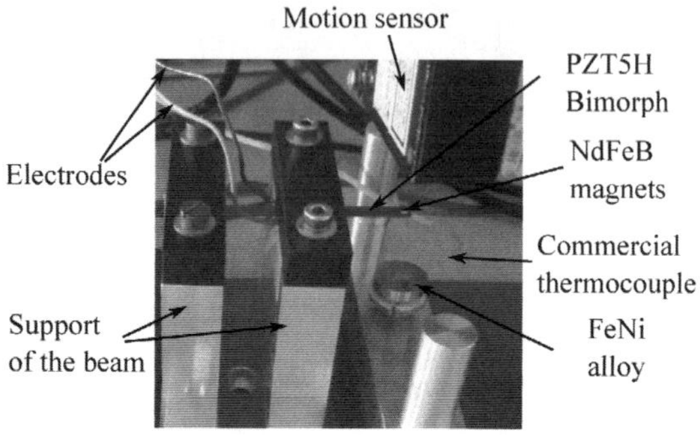

Figure 2. Test prototype. The in-house developed test bench is able to produce necessary conditions for the operation of the generator.

2. Design

As shown in Figure 1, the power generator has two main sections according to their respective functions: the energy transducer and the triggering system. The energy transducer is composed of a piezoelectric bimorph-type cantilever beam, namely PSI-5H4E from Piezo Systems, Inc with a series-type polarization. Concerning the triggering system, it consists of two permanent magnets (NdFeB), that are attached to the free end of the beam, and a soft magnetic material (FeNi), fixed under the free end of the beam, forming an air gap.

2.1. Working principle

Our power generator is based on a piezoelectric transduction, that is triggered by the interaction between a constant magnetic field and cycled variations in temperature acting along a permanent magnet and a soft magnetic material. The generator has two operation states: the closed position and the open one. In the initial state (open position), at a cold environmental temperature, the soft magnetic material is magnetized so the cantilever beam is pulled-down due to magnetic force. After, at closed position, when the environmental temperature increases, the magnetic properties of soft magnetic material change from ferromagnetic to paramagnetic, causing the magnetic force to dissapear. The beam is pulled back to its initial state because of the spring-back force of the beam. This process of pulling-down and pulling-back of the beam is periodically repeated as long as the energy harvester is placed in an environment with cycled variations in temperature. The basic generator parameters used in both simulation and test prototype have been listed in Table 1.

3. Modeling

Many finite element models have been reported to analyse the operation of energy harvesters. However, most of them do not take into account the prediction of the power output. We developed, in ANSYS software, a model able to predict the dynamic behavior and output power of proposed generator.

3.1. Configuration of finite element model

Our 2D model considers a sandwich cantilever beam with piezoelectric layers of opposite polarization, on top and bottom of a reinforcement brass layer, a pair of permanent magnets attached to its free end, a direct connection of a load resistor to piezoelectric layers, and a ferromagnetic block fixed under the free end of the beam. The adhesive thicknesses of the

Table 1. Basic parameters of proposed thermo-magnetically triggered piezoelectric generator.

Parameter	Symbol	Value	Units
Beam length	L_{beam}	20	mm
Beam width	W_{beam}	4	mm
Beam thickness	Th_{beam}	380	μm
PZT thickness	Th_{PZT}	130	μm
Brass thickness	Th_{brass}	120	μm
Magnet thickness	Th_{magnet}	1	mm
Air gap	Th_{air}	2	mm

beam are neglected because they are very thin compared to rest of the layers thicknesses. The element PLANE223 is used for modeling piezoelectric layers. Reinforcement layer is modeled by element PLANE183. For simulating permanent magnets, soft magnetic material and air gap, element PLANE13 is applied. Load resistor is simulated with CIRCU94 electric element. The pair magnetic contact, which occurs during the open-to-closed position switching, is modeled using CONTA172 and TARGE169 elements. Firstly, a modal analysis is carried out aiming at exploring the natural resonance frequency of the system. Special focus was given to modal damping by using QRDAMPED method. Secondly, a full transient analysis is realized to study the dynamic operation of the generator. The considered input, the variation in temperature as function of time, is as follows: the temperature of ferromagnetic alloy, is 45°C and kept constant for a short time. It is then decreased to 25°C and maintained there for a larger period of time before being increased to 45°C. The finite element model couples these time-dependent variations in temperature to temperature-dependent magnetic properties of the triggering system to compute the magnetic force and the mechanical displacement. Furthermore, the mechanical strain piezoelectric effect are coupled to calculate the electrical charge. Finally, the electrical charge is converted into voltage across the load resistor resulting in the dissipated power output of the model.

4. Experimental testing

Our test prototype consists in a platform for fixing both the cantilever beam and the ferromagnetic material. A DC current of 3A, flowing across the support base of the FeNi material, is periodically applied and cut to obtain cycled variations in temperature. To measure the temperature on the ferromagnetic material, a thermocouple type K is introduced, by a drilled hole, to the FeNi alloy and the output signal is then displayed in a multimeter in Celsius degrees. The electrodes of the energy transducer are connected to the terminals of a resistance decade box *CENTRAD DR07*. Through an oscilloscope *Tektronix DPO3014*, the differential voltage across the load resistor is measured. A laser-based motion sensor *ELITEC LAS-2010* is used to monitor the tip displacement of the beam. Figure 2 presents the developed test bench.

4.1. Results analysis

The dynamic operation of the power generator has been studied through many cycles of temperature variations. The values of temperature, cold and hot, were 25°C and 45°C, respectively. At cold temperatures, the generator is in closed position until a raise in temperature produces the closed-to-open position switching. The electrical power harvested was determined as a function of the load resistor. To determine the optimal load resistor value, the average of the dissipated power during a cycle of operation is calculated. Simulation and experimental results are illustrated in Figures 3, 4, and 5.

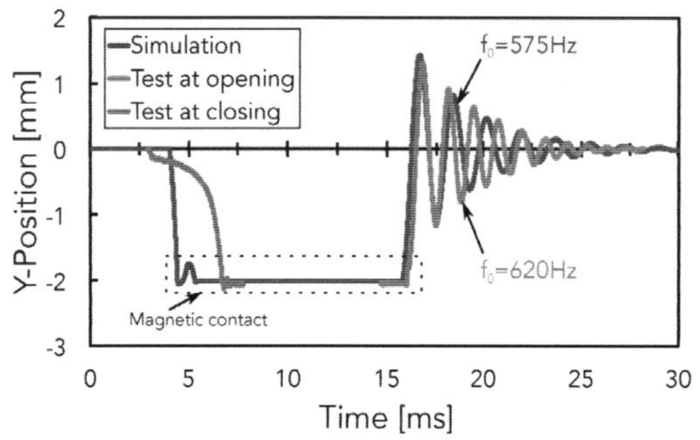

Figure 3. Tip displacement when $R_L = 10k\Omega$ is connected. Clearly, on simulation model, the rate of decreasing temperature is smaller than the experimental one. Magnetic contact is present at closed state.

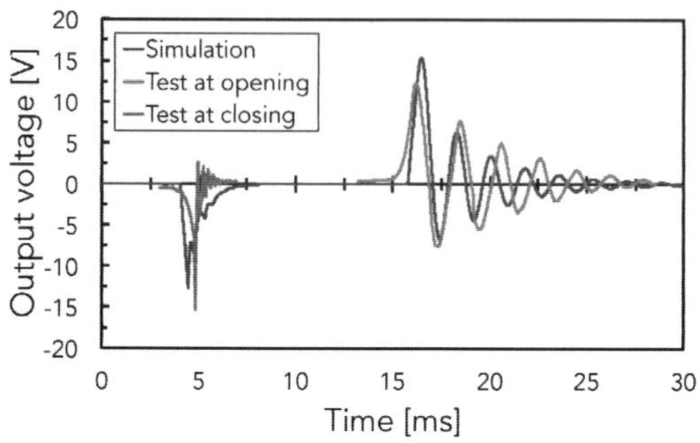

Figure 4. Output voltage when $R_L = 10k\Omega$ is connected. During the open-to-closed position switching, the output is a peak of voltage, whereas a sine wave output is exhibited during closed-to-open switching.

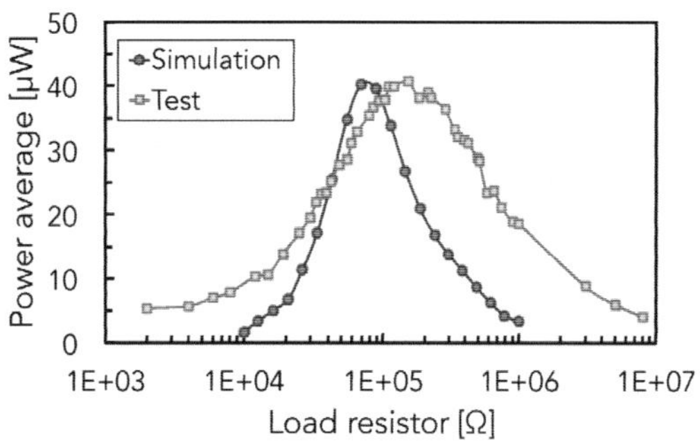

Figure 5. Harvested electrical power average as a function of load resistance. The power average during a complete cycle of operation is considered to find the optimal value of load resistor, $R_{Loptimal}$.

5. Conclusion

This paper has proposed a solution for thermomagnetic energy harvesters design. The maximum average power and average power density is 40μW and 250μW/cm^3, respectively. The data obtained show that smaller generators than that previously reported in [3] are feasible. Current research involves finding appropriate materials that favor the generator optimization.

Acknowledgments

Financial support from the National Science and Technology Council (CONACyT) of the Mexican government, Grant No. 315444 (PhD scholarship), is gratefully thanked.

References

[1] A. Post, C. Knight, and E. Kisi, "Thermomagnetic energy harvesting with first order phase change materials," *Journal of Applied Physics*, vol. 114, no. 3, 2013.

[2] K. B. Joshi and S. Priya, "Multi-physics model of a thermo-magnetic energy harvester," *Smart Materials and Structures*, vol. 22, no. 5, p. 055005, 2013.

[3] L. Carlioz, J. Delamare, and S. Basrour, "Temperature threshold tuning of a thermal harvesting switch," in *TRANSDUCERS 2009 - 2009 International Solid-State Sensors, Actuators and Microsystems Conference*, pp. 1385–1388, June 2009.

Thermoelectric Generator Design in Dynamic Thermoelectric Energy Harvesting

M E Kiziroglou[1,2,3], Th Becker[4], S W Wright[2], E M Yeatman[2], J W Evans[1] and P K Wright[1]
[1]University of California at Berkeley, CA, USA
[2]Imperial College London, U.K.
[3]ATEI Thessaloniki, Greece
[4]Airbus Group Innovations, Germany

Abstract. This paper reports an analysis of thermoelectric generator design for dynamic thermoelectric harvesting. In such devices, the available energy for a given temperature cycle is finite and determined by the heat storage unit capacity. It is shown by simulation and experimentally that specific thermoelectric generator designs can increase the energy output, by optimizing the balance between heat leakage and dynamic response delay. A 3D printed, double-wall heat storage unit is developed for the experiments. Output energy of 30 J from 7.5 gr of phase change material, from a temperature cycle between ± 22 °C is demonstrated, enough to supply typical duty-cycled wireless sensor platforms. These results may serve as guidelines for the design and fabrication of dynamic thermoelectric harvesters for applications involving environments with moderate temperature fluctuations.

1. Introduction

Dynamic thermoelectric energy harvesting allows electrical power generation from temperature fluctuations in an environment, such as a vehicle body or an industrial machine [1-2]. It employs a heat storage unit (HSU), with a phase change material (PCM) to increase heat storage, insulated from the environment and in thermal contact with a thermoelectric generator (TEG), as shown in Figure 1. The other side of the TEG is attached to the thermal mass in which the temperature variation occurs. The HSU follows this variation with a time delay, creating a temperature difference ΔT across the TEG. As heat flows in and out of the HSU, through the TEG, it is transduced into electrical energy with efficiency directly proportional to ΔT. The thermal response depends on the HSU heat capacity C, latent heat L, and thermal resistance R between the HSU and the environment, giving an exponential temperature profile with a time constant RC during non-phase-change operation. The operation of such devices has been studied analytically and numerically [3], and various implementations have been reported, including real flight testing [4-5]. In this paper, the role of TEG design on performance is studied by analysis and simulation, and a 3D printed HSU prototype is introduced.

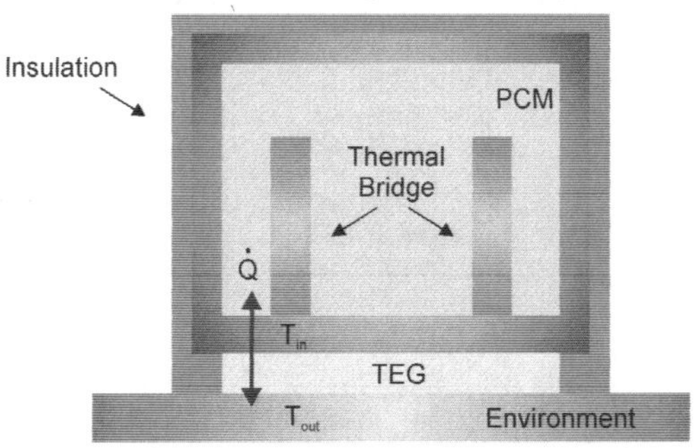

Figure 1. The concept of dynamic thermoelectric energy harvesting.

2. TEG design analysis and performance simulation

For the analysis of TEG performance in dynamic thermoelectric applications, a TEG of area A, thickness τ, N number of junctions and p and n materials with S_P and S_N Seebeck coefficients, and the same electrical and thermal resistivity ρ_E and ρ_{TH} are assumed. Assuming full area coverage with n and p pellets, the open circuit voltage, and the electrical and thermal resistances will respectively be:

$$V = N \cdot (S_p - S_n) \cdot \Delta T, \qquad R_E = \rho_E \cdot \frac{2 \cdot N \cdot \tau}{\frac{A}{2N}} = \rho_E \cdot \frac{\tau}{A} \cdot 4 \cdot N^2, \qquad R_{TH} = \rho_{TH} \cdot \frac{\tau}{A}$$

At a given ΔT and at matched electrical load conditions (maximum output power), the input thermal and output electrical power of the TEG will be:

$$P_{in} = \frac{\Delta T}{R_{TH}} \qquad P_{out} = \frac{V^2}{4R_E} = \frac{A}{16 \cdot \rho_E \cdot \tau} \left(S_p - S_n\right)^2 \cdot \Delta T^2$$

Defining the figure of merit ZT at device level:

$$ZT = N^2 \cdot \left(S_p - S_n\right)^2 \cdot \frac{R_{TH}}{R_E} \cdot T = \frac{1}{4} \cdot \frac{\rho_{TH}}{\rho_E} \cdot \left(S_p - S_n\right)^2$$

the device efficiency can be written as:

$$\eta = \frac{P_{out}}{P_{in}} = \frac{\rho_{TH}}{16\rho_E} \left(S_p - S_n\right)^2 \cdot \Delta T = \frac{ZT}{4} \cdot \frac{\Delta T}{T}$$

This formulation demonstrates the TEG efficiency independence from geometry. The efficiency depends only on the properties of the materials. In direct TEG applications, where thermal resistance matching is optimal, a design with suitable R_{TH} is typically selected. In dynamic thermoelectric harvesting, a high R_{TH} is desirable, to increase the $R_{TH}C$ time constant of the system and hence the ΔT that is achieved during a temperature sweep across the TEG [3]. On the other hand, a high R_{TH} results in increased leakage through the HSU insulation. Therefore, an optimum R_{TH} value must exist, that achieves a balance between $R_{TH}C$ and leakage, for maximum total energy from a given temperature cycle. This was studied by simulation, using the model presented in [3]. As an example case, a 30 mm a-side cubic HSU with 7.5 g of water as PCM and a TEG with properties corresponding to a Marlow TG12-2.5 TEG and parametrized dimensions was assumed. An insulation thickness of 4 mm was also assumed. The cumulative electrical energy output from a simulated 3 K/min temperature cycle between ± 20 °C was calculated, for different TEG area values. The results for insulation thermal conductivities of 0.1 W/mK and 0.01 W/mK are presented in Figure 2. They show that the overall maximum possible performance is achieved by thicker TEGs, and selecting a specific surface coverage area. However, if a the TEG area is bound by other limitations, such as cost and market availability, the optimum thickness varies. In practice, these results can serve as a guide for TEG selection. This method can be applied to other HSU geometries, TEG specifications and temperature profiles, by modifying the simulation parameters accordingly.

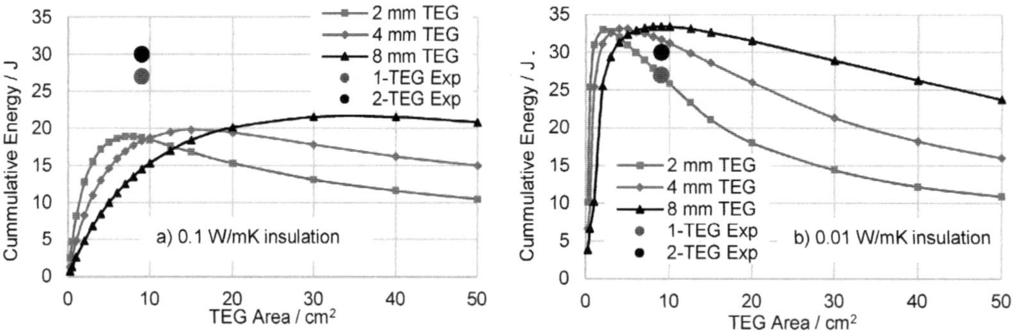

Figure 2. Cumulative energy from simulated 3 K/min cycles between ± 20 °C, for different TEG area and thickness. A 30 mm a-side cubic HSU with 7.5 g of water and 4 mm thick insulation was assumed. Figure 2a on the left assumes 0.1 W/mK whereas Figure 2b on the right assumes 0.01 W/mK insulation.

3. Fabrication and experimental results

A 32 × 32 × 26 mm HSU box was fabricated by Acrylonitrile-Butadiene-Styrene (ABS) 3D printing, with 3 mm walls. The side walls were designed to include a 1 mm internal airgap for enhanced thermal insulation. A 10 g Al internal heat bridge and 7.5 g of water as the PCM were used. The HSU was integrated with one and two (in thermal and electrical series) Marlow TG12-2.5 TEGs with dimensions 30 × 30 × 4 mm, in order to investigate the effect of TEG thickness to the device performance. Tests were performed using a 4 K/min temperature cycle between ±22 °C using a Tenney Engineering Inc. BTR mL environmental chamber. The TEG outside temperature and open circuit voltage were monitored using an Omega HH374 data logging thermometer and a VA18B data logging multimeter.

The measured temperature and voltage response of the 1-TEG and 2-TEG devices are shown in Figure 4, along with the corresponding power and cumulative energy, which were calculated by considering the 5 Ω electrical resistance of each TEG ($P_{OUT}=(V_{OC}/2)^2/R_E$). The results demonstrate 27J and 30J for the single and double TEG device respectively. In comparison with the simulation results of section 2 (red and black dots in Figure 2), the fact that the 2-TEG device gives a better performance indicates that these devices lie in the RC – limited range rather in a leakage-limited range. To confirm this, the devices were also tested using a 2 K/min cycle between 25 °C and 10 °C, where the slower temperature variation should lead to a clearer performance difference. The results are shown in Figure 5, demonstrating 22% better performance for the double TEG device.

Figure 3. (left) Double wall insulation heat storage unit, designed for 3D printing. (right) Assembled device with 3D printed HSU and two TEGs in serial thermal flow orientation.

4. Conclusions

In this paper, the design of TEG devices for dynamic thermoelectric applications was studied. An analytical formulation of performance was outlined, showing that the TEG geometry, though it does not affect the device transduction efficiency directly, it determines the thermal and electrical resistance and thereby the ΔT that is achieved across the device. In direct thermoelectric harvesting, an R_{TH} equal to the total series thermal resistance is optimal (for a given TEG material). In dynamic thermoelectric harvesting, the optimal R_{TH} depends on the balance between leakage and dynamic response delay ($R_{TH}C$), and hence on the insulation thermal resistance and the temperature variation speed.

A 3D printed HSU was fabricated, allowing the implementation of airgap walls for improved thermal insulation. An energy density of 3.6 J / ml of PCM and 4 J/ml of PCM was demonstrated from single and double TEG implementations of the device, from a 4 K/min temperature cycle between ±22 °C. This corresponds to 0.88 J/ml per device and 0.86 J/ml per device respectively. The demonstrated energy levels are enough to supply typical duty-cycled, wireless sensor platforms. As an example, for every temperature cycle, a 100 mW active mode sensor node can be powered for 5 minutes of continuous active operation, or for over 8 hours of 1% duty cycling. Further tests from a slower, 2 K/min temperature cycle show 22% better performance for the double TEG device, in accordance with the analysis and the simulation results. These results may serve as guidelines for the development of dynamic thermoelectric harvesting devices for applications involving moderate temperature rates.

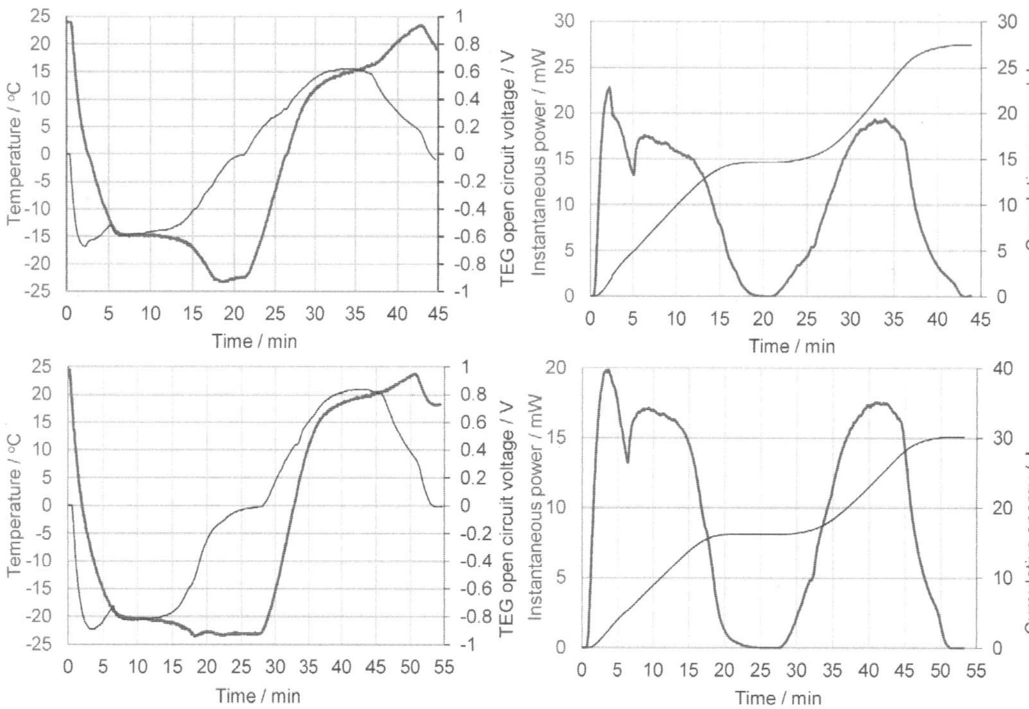

Figure 4. Measured temperature, voltage, power and energy response of the single (top) and double (bottom) TEG devices, demonstrating 27 J and 30 J of cumulative energy respectively from a 4 K/min temperature cycle between ±22 °C.

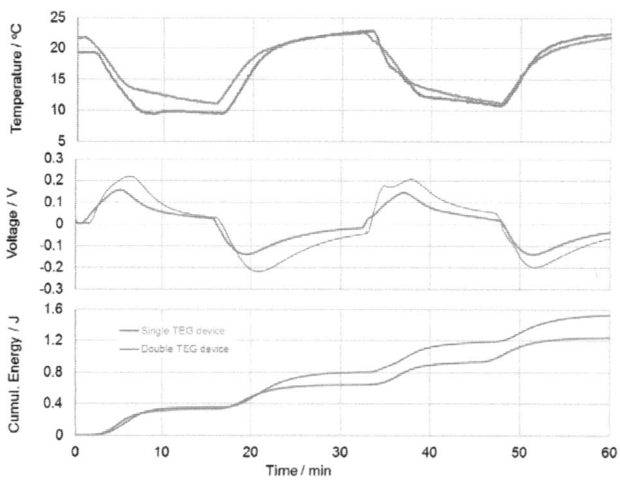

Figure 5. Response under 2 °C/min temperature variation between 22 °C and 10 °C. An energy increase of 22% is demonstrated for the double TEG device.

5. References

[1] Samson D, Otterpohl T, Kluge M, Schmid U and Becker Th. 2010 *J. Electr. Mater.* 39, 2095

[2] Bailly N, Dilhac J M, Escriba C, Vanhecke C, Mauran N and Bafleur M 2008 *PowerMEMS*, 205

[3] Kiziroglou M E, Wright S W, Toh T T, Mitcheson P D, Becker Th and Yeatman E M 2014 *IEEE Trans. Ind. Electr.*, 61, 302

[4] Elefsiniotis A, Samson D, Becker T and Schmid U 2013 *J. Electr. Mat.*, 42, 2301

[5] Kiziroglou M E, Elefsiniotis A, Kokorakis N, Wright S W, Toh T T, Mitcheson P D, Schmid U, Becker Th and Yeatman E M 2016 *Microsyst. Techn.* 22 1905

Optimization of power output in planar thermoelectric microgenerators based on Si nanowires

C. Calaza [1], I. Donmez [1], M. Salleras [1], G. Gadea [2], J. D. Santos [2], A. Morata [2], A. Tarancón [2] and L. Fonseca [1]

[1] IMB-CNM, CSIC, Campus UAB, Bellaterra, 08193, Barcelona, SPAIN

[2] Catalonia Institute for Energy Research (IREC), 08930, Barcelona, SPAIN

carlos.calaza@imb-cnm.csic.es

Abstract. This work reports the efforts to optimize the output power achieved with all-silicon thermoelectric microgenerators based on MEMS fabrication technology and silicon nanowires as active material. Recent improvements introduced in both the fabrication process and the device design are described, and a set of experimental results obtained with both device types (new vs. old design/process) are provided to assess the impact of the different modifications on thermal and electrical performance, as well as in the overall harvested power.

1. Introduction

The all-silicon thermoelectric microgenerators used in this work are based on an evolution of the silicon microplatforms reported [1], with low thermal conductivity supports for the suspended parts. The Si NW arrays used as active material have been integrated on these platforms using a bottom-up Si NW growth process formerly developed [2], after adapting process parameters to new structures. Modifications of device design and fabrication process sequence are projected to overcome the critical issues affecting performance that were identified in the 1st generation of Si platforms [3].

2. Device optimization

The fabrication process sequence has been rearranged to improve the production yield. After test of a number of feasible orders for the micromachining steps the sequence in Table 1 has been adopted. Starting from SOI wafers with a 15 μm thick doped device layer (0.1 Ω·cm) (<110> surface and flat aligned with vertical <111> planes) and a 500 μm thick handle wafer the microplatforms were created using a new photolithographic mask set with five levels.

A new zigzag arrangement for the low thermal conductivity metal supports has been implemented allowing arbitrarily long membranes without device parts misalignment (Figure 1). This new design provides lower internal resistances due to increased metal/window area ratio. Membranes are released with a 30 min. KOH etch step, while <111> aligned structures for Si NW arrays are preserved. The area covered by the Si_3N_4 film has been extended to prevent the KOH of coming into contact with the <110> Si surface during the membrane release. This Si_3N_4 film is patterned to perform a local B implantation on the Si-metal interface regions, which is intended to ensure the contact ohmic behavior after a thermal treatment.

Journal of Physics: Conference Series **773** (2016) 012026 doi:10.1088/1742-6596/773/1/012026

Table 1. Fabrication process sequence.

	Top side
•	LPCVD Si_3N_4 deposition and RIE etch. Patterning with mask #1.
•	B implantation, thermal annealing for activation and SiO_2 etch.
•	Metal deposition (Ti/W) and lift-off. Patterning with mask #2.
•	RTA for silicide formation in contact interface.
•	PECVD SiO_2 passivation deposition and partial SiO_2 RIE etch on pad areas. Patterning with mask #3.
•	PECVD SiO_2 and LPCVD Si_3N_4 RIE etch to expose top Si surface. Patterning with mask #4.
	Bottom side
•	Thick PECVD SiO_2 deposition and PECVD SiO_2 and LPCVD Si_3N_4 RIE etch to expose bottom Si surface. Patterning with mask #5.
•	DRIE of the 500 μm thick Si handle wafer (using protection photoresist on top side).
	After chip dicing
•	KOH etch to define top silicon structures and release the metal supporting membranes.
•	BOX removal with HF vapours.

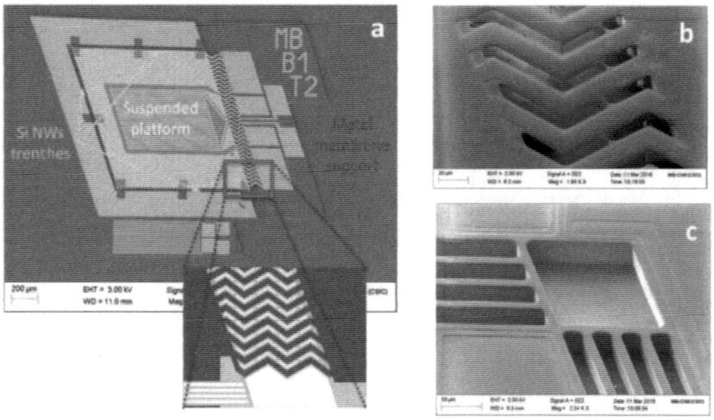

Figure 1. New arrangement for the low thermal conductivity membrane supports. a) Dielectric openings are aligned with Si substrate fast etching planes to allow Si under etch during KOH process. b) Number and width of metal lines is maximized to improve the internal resistance. c) Si platform and bars defining NW trenches are aligned with <111> planes to endure this etch process.

The microplatforms obtained with the fabrication process flow described in Table 1 and the new mask set have been characterized (before Si NWs integration) in order to validate the positive impact of the process and design changes in performance parameters.

2.1. Thermal isolation

The impact of the new dielectric membrane used as support for the metal interconnections on platform thermal isolation has been assessed by contrasting the total platform-to-rim thermal conductance of these new devices vs. a reference one using the previous support type (silicon beams). Figure 2 shows the power dissipated in a test element (heater) vs. temperature increase reached in the isolated platform (estimated from heater resistance change after calibration), for three devices with the same active area and different beam or membrane lengths. Despite reduced length (100 μm membrane vs. 200 μm beams), thermal isolation achieved with membranes outstrips that of previous silicon bridge supports.

Total platform-to-rim thermal conductance decreases from 1.99 to 1.26 mW/K, pointing out that bridge conductance in old designs accounts for more than one third of total thermal losses, turning into a significant limiting factor for thermoelectric performance. Negligible change in thermal conductance observed for devices with 100 and 200 μm membrane lengths (i.e. halved membrane contribution) confirms that in new designs the main contributions are linked to the surrounding air and the silicon bars that define the NWs trenches. Thus, longer membranes are not advised as no further benefit is expected in terms of thermal isolation and a major impact can be anticipated on internal resistance.

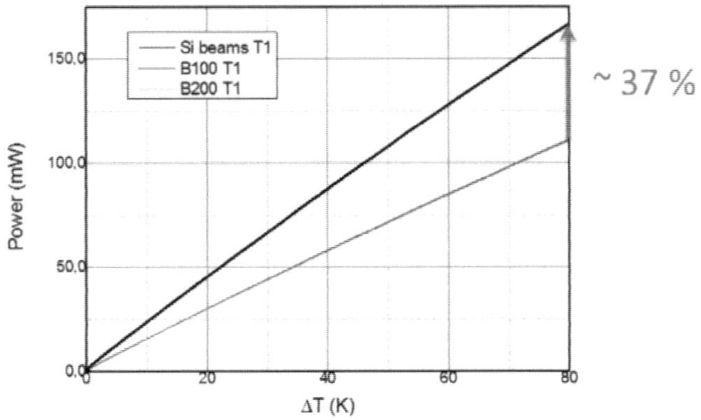

Figure 2. Comparison of platform-to-rim thermal conductance for three samples, a reference device with Si beams (200 μm length): 1.99 mW/K and two devices with Si_3N_4 membranes (100μm and 200μm length): 1.26 mW/K.

2.2. Internal resistance

The impact of process changes used to improve the Si-metal interface was assessed by measuring the contact resistance with a TLM (transmission line method) test structure. Two samples using a 20/200 nm Ti/W stack or a 230 nm W layer as metallization were considered. The TiW sample performed a previous B shallow implantation in contact regions (5e15 at/cm2 and 50keV) and both have undergone a rapid thermal annealing at 700°C for 30s to promote the silicide formation.

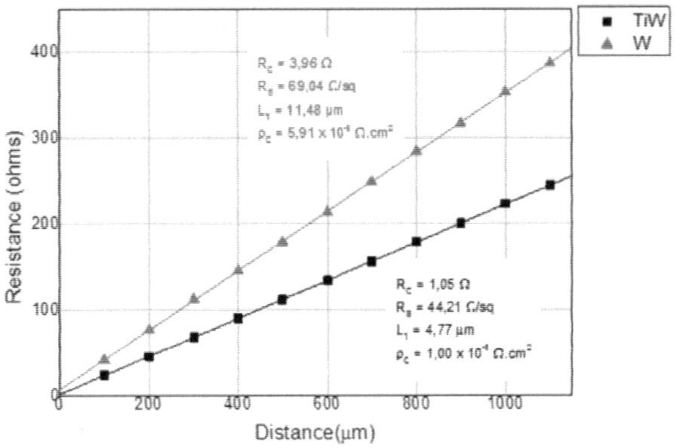

Figure 3. TLM measurement for two samples using 20/200 nm Ti/W and 230 nm W metal stacks, respectively. A rapid thermal annealing at 700°C for 30s was performed in both samples and a previous B implantation (5e15 at/cm^2 and 50keV) was executed in Ti/W sample.

Both metallizations provide ohmic behavior but TLM results in Figure 3 show that Ti/W improves the contact resistance by almost a factor 4 for pads with equal area. In addition, the adhesion problems observed for samples with pure W layers after passivation removal are prevented with the Ti addition.

3. Power generation

Performance enhancement attained with new microplatforms has been studied using a device with four structures (T1-T4) with a different number of 10 μm wide trenches between hot and cold parts, which allow the growth of silicon NWs with different effective lengths using a single batch process to create 10 μm long P-type silicon NWs (linking the 3 μm wide Si bars between trenches).

After NWs integration, the device under test was packaged to endure high temperatures and placed on a hotplate to evaluate its response in different operating conditions established with a temperature controller. The voltage generated by the thermocouple (Figure 4.a) surges with the increase of the effective NWs length due to a better thermal isolation between the hot and cold parts, which provides a higher temperature difference for each hotplate condition. Nevertheless, the low voltages measured (close to Seebeck coefficient of Si NWs) suggest that thermal gradient across the NWs is in the 1-5°C range. Therefore, progress is still required to integrate a heat exchanger element in device structure.

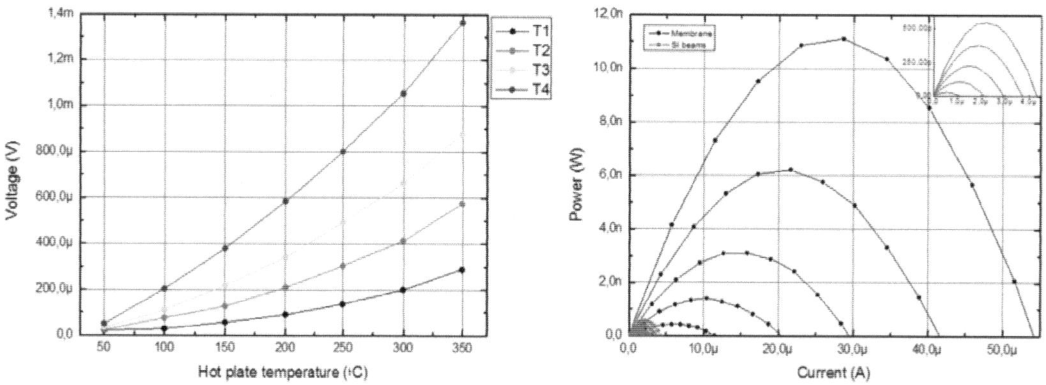

Figure 4. a) Thermoelectric voltage generated with devices using the new platform design. Impact of the effective NWs length (1 to 4 trenches). b) Power curves for T3 devices (three NWs trenches) at different temperatures (50 to 300°C, in 50°C steps) using the new optimized platform (dotted line) or the previous design with Si beams (inset).

Despite the limited temperature gradients the increase in generated voltage with the introduction of the membrane supports is higher than 60% when peer T3 devices (three NW trenches) are compared (540μV with beam supports vs. 880μV with new membrane supports, both @350°C). The advance in terms of power is even higher due to the additional enhancement of internal resistance, which has been reduced by a factor 10 (from ≈50 Ω to ≈5 Ω) through metal mask redesign and Si-W electrical contact enhancement. The comparison of power curves obtained with peer T3 devices (three NW trenches, Figure 4.b) shows an improvement of more than one order of magnitude in power levels, from 500 pW obtained with previous platforms [3] to 10 nW achieved with this new optimized version.

Acknowledgments

This work was supported by FP7-NMP-2013-SMALL-7, SiNERGY (Silicon Friendly Materials and Device Solutions for Microenergy Applications), Contract n. 604169.

References

[1] C. Calaza, L. Fonseca, M. Salleras, I. Donmez, A. Tarancón, A. Morata, J. D. Santos, G. Gadea, Journal of Electronic Materials 45 (2016), pp. 1689-1694.

[2] G. Gadea, A. Morata, J.D. Santos, D. Dávila, C. Calaza, M. Salleras, L. Fonseca, A. Tarancón Nanotechnology 26 (19):195302 (2015).

[3] D. Dávila, A. Tarancón, C. Calaza, M. Salleras, M. Fernández-Regúlez, A. San Paulo, L. Fonseca, Nanoenergy 1 (2012), pp. 812-819.

Triple Hybrid Energy Harvesting Interface Electronics

H Uluşan[1,*], S Chamanian[1], W M P R Pathirana[2], Ö Zorlu[3], A Muhtaroğlu[2], and H Külah[1]

[1] METU, Department of Electrical and Electronics Engineering, Ankara, Turkey
[2] METU Northern Cyprus Campus, Department of Electrical and Electronics Engineering, Güzelyurt, Mersin 10, Turkey
[3] Mikro Biyosistemler Inc., Ankara, Turkey

E-mail: hulusan@metu.edu.tr

Abstract. This study presents a novel triple hybrid system that combines simultaneously generated power from thermoelectric (TE), vibration-based electromagnetic (EM) and piezoelectric (PZT) harvesters for a relatively high power supply capability. In the proposed solution each harvesting source utilizes a distinct power management circuit that generates a DC voltage suitable for combining the three parallel supplies. The circuits are designed and implemented in 180 nm standard CMOS technology, and are terminated with a schottky diode to avoid reverse current flow. The harvested AC signal from the EM harvester is rectified with a self-powered AC-DC doubler, which utilizes active diode structures to minimize the forward-bias voltage drop. The PZT interface electronics utilizes a negative voltage converter as the first stage, followed by synchronous power extraction and DC-to-DC conversion through internal switches, and an external inductor. The ultra-low voltage DC power harvested by the TE generator is stepped up through a charge-pump driven by an LC oscillator with fully-integrated center-tapped differential inductors. Test results indicate that hybrid energy harvesting circuit provides more than 1 V output for load resistances higher than 100 kΩ (10 μW) where the stand-alone harvesting circuits are not able to reach 1 V output. This is the first hybrid harvester circuit that simultaneously extracts energy from three independent sources, and delivers a single DC output.

1. Introduction

Utilizing scavenged energy from environment becomes viable with the decreasing power demand of the new generation integrated circuits. Most common ambient sources are solar, thermal, vibration and RF energy where each micro-harvester by itself is typically limited in power generation capacity, and output voltage level. However, hybrid systems that effectively harvest multiple sources have potential to address this problem [1-3].

Several attempts have been reported in literature to build hybrid energy harvesters. Hybrid system presented in [4] combines RF and vibration based piezoelectric harvester structures to continuously power up a wireless sensor by switching between the two harvester outputs. Another hybrid structure, which utilizes TE and photovoltaic harvesters in order to generate individual power sources for multi-sensor wireless microsystem is described in [5]. The implemented system uses a microcontroller and arrange which power source to supply different sensors at the microsystem. These structures are

* Middle East Technical University, Department of Electrical and Electronics Engineering, Ankara, Turkey

useful when scavenged power from either harvester is sufficient to maintain the load power by itself, but do not couple multiple harvester outputs to increase the total output power.

In this paper, a triple hybrid structure is presented, which simultaneously combines scavenged power from TE, vibration based EM, and PZT harvesters to support higher loads at the output. The hybrid structure utilizes a power management circuit to generate a single DC power source from three different ambient sources. The AC power generated by EM and PZT harvesters are converted into DC while the low level DC output of the TE generator is boosted up to similar voltage level. The interface circuits have been designed, fabricated at UMC 180nm CMOS technology, and are connected in parallel with schottky diodes, before system validation. The organization of this paper is as follows: The proposed circuit topologies have been briefly described in Section II. Section III presents the test results from the triple hybrid structure. Finally, conclusions are summarized in Section IV.

2. Interface Circuits

Figure 1 shows the proposed triple hybrid energy harvesting system. Each interface circuit, designed and implemented in 180nm standard CMOS technology, contributes in parallel to power management to combine the harvested signals from EM, PZT and TE generators into a similar DC level. Each harvester sub-system is terminated with a schottky diode to avoid reverse current flow. The harvested AC signal from the EM harvester is rectified with a self-powered AC-DC doubler, which utilizes active diode structures to minimize the forward-bias voltage drop. The PZT interface electronics utilizes a negative voltage converter as the first stage to use full cycle of the input signal. The power extraction circuit also performs the DC-to-DC power conversion through a set of power switches and an external inductor. The ultra-low voltage DC power harvested by the TE generator is stepped-up by a charge-pump circuit driven by LC oscillator with fully-integrated centre-tapped differential inductors. The oscillator replaces digital oscillators with a low cost voltage-doubling LC tank implemented with on-chip inductors, eliminates large CMOS buffers, and thus provides high step-up ratio through reduced losses. Detailed description of TE interface electronics was given in [6] where details of EM and PZT interfaces were given below.

2.1. Rectifying Electronics for EM Harvester

EM harvesters are good candidates for low frequency ambient vibrations due to their low resonant frequency characteristics. The proposed system utilizes a self-powered AC/DC doubler as shown in Figure 2 which is the modified version of the rectifier presented in [7]. Rectification of the signal is obtained by active diodes, which is composed of a comparator and a PMOS switch. The active diodes enable low forward voltage drop and high current drive; however they require a DC supply to operate. In the proposed structure, positive and negative supplies of the comparators are provided internally by passive AC/DC positive quadrupler and passive AC/DC negative doubler, which are constructed with diode connected PMOS transistors. The input signal of these passive circuits are also maintained from the same harvesting source. Therefore the circuit operates without additional sources.

Operation principle of the circuit is as follows: When the input voltage is negative and V_x potential is lower than GND, the right sided comparator turns the transistor ON and the storage capacitor between the input and V_x node is charged up to positive peak voltage of the input. When V_x goes above ground potential, the switch is turned OFF, and a positive charge is stored on the capacitor. Similarly, when the input is positive and the V_x potential is above output voltage V_{rect}, left sided switch is turned ON, and output storage capacitor is charged. When V_{rect} falls below V_x, the switch is turned OFF, and a doubled voltage is stored on the output storage capacitor.

2.2. PZT Interface Electronics

The proposed hybrid system includes a PZT harvester subsystem, which effectively harvests high frequency vibrations. Therefore, both low and high frequency ambient vibration are captured by the system. The implemented electronic circuit for piezoelectric energy harvester is illustrated in Figure 3. In this circuit, the power extraction is realized through a set of power switches and an external inductor. Switching on the external inductor between piezoelectric clamped capacitance and the output buffer capacitance, based on Synchronous Electric Charge extraction (SECE) technique, enables boosting extracted power and transferring of the generated charge to the output capacitor [8].

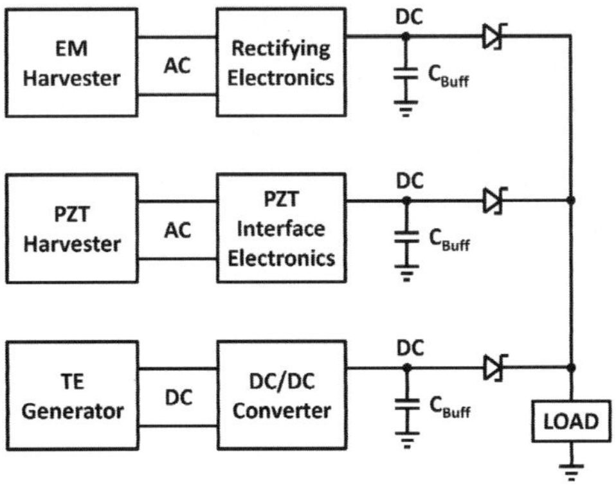

Figure 1. Triple hybrid energy harvester structure.

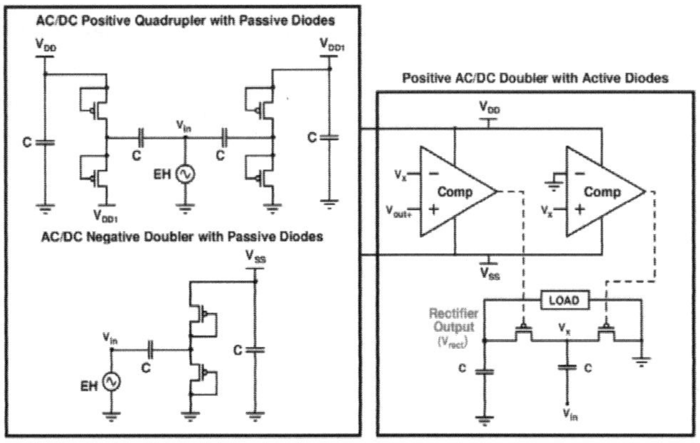

Figure 2. Self-powered rectifying electronics for low voltage EM harvester.

Initially the AC voltage generated on piezoelectric harvester is converted to positive signal through negative voltage converter (NVC). The power extraction is achieved in three phases by SECE circuit. In the first phase, all switches are turned OFF and the piezoelectric harvester vibrates in open circuit condition. The second phase is initiated by turning ON S_1 at maxima of the piezoelectric voltage. The established resonant circuit between piezoelectric clamped capacitor and the inductor transfers energy stored on the piezoelectric capacitor to the inductor. The third phase starts when piezoelectric capacitor voltage reaches to zero. At this moment, S_1 is turned OFF, S_2 and S_3 switches are ON, and the stored energy on the inductor is transferred to the storage buffer capacitance, C_S. This phase stays active until the inductor current reaches zero. After this point, a new energy transfer cycle is initiated by turning all switches OFF. High power efficiency is achieved for a wide range of output voltage by decreasing power losses in control switch circuitry and switch drivers.

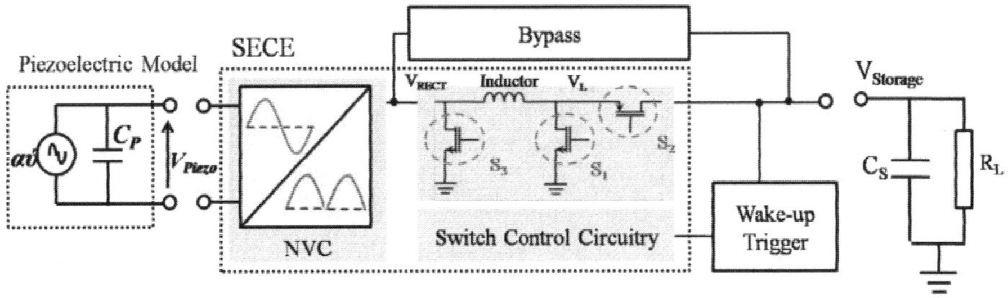

(a) (b) (c)

Figure 3. Schematic of self-powered SECE harvester: (a) piezoelectric model (b) implementation of SECE (c) storage capacitance and load.

3. Test Results and Discussion

Figure 4 depicts the die micrographs of the implemented CMOS harvester chips. EM rectifying electronics is excited with 5 Hz and 1.1 $V_{peak\text{-}to\text{-}peak}$ input during validation, PZT interface electronics has input signal with 69 Hz and 3 V open-circuit voltage where both sources can be generated with low level vibrations (<1g). TE interface electronics has 0.18 V input voltage which can be easily obtained with typical TE harvesters. Figure 5 shows variation of output voltage and power of the hybrid system and stand-alone interface electronics with respect to the load resistance. Although the stand-alone harvesting circuits are not able to reach 1 V output, hybrid system provides more than 1 V output for load resistances higher than 100 kΩ. Output power of the hybrid system is higher than stand-alone system outputs, which demonstrates the benefit of combining multiple sources at a single output. Although sum of power generated at the output of each interface is higher than total hybrid power the forward voltage drop and leakage currents at the schottky diodes leads to power losses.

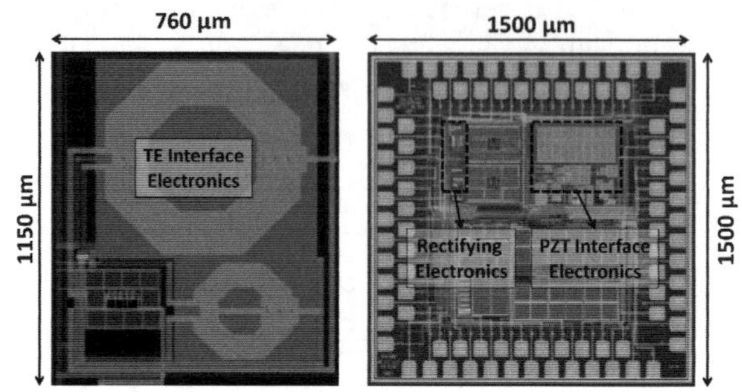

Figure 4. Die micrograph of the hybrid interface electronics in UMC 180nm CMOS technology.

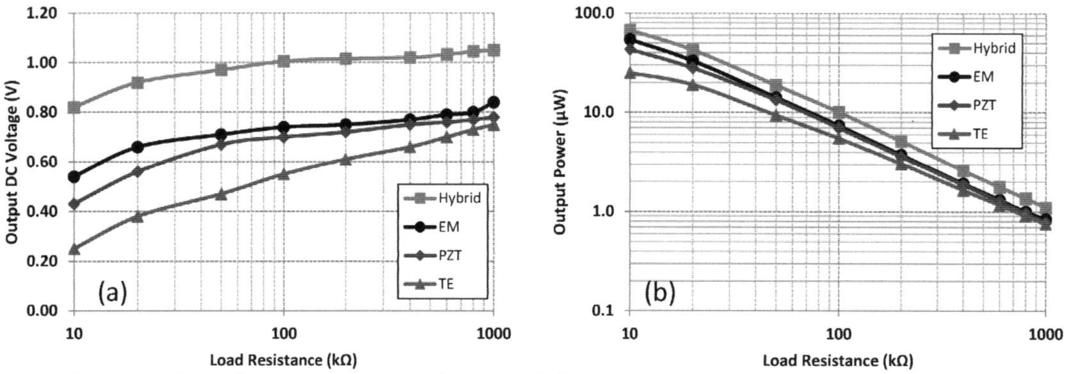

Figure 5. Variation of the (a) output voltage and (b) output power of the hybrid and stand-alone energy harvesting circuits with respect to load resistance.

4. Conclusion

A triple hybrid energy harvesting system that combines TE, vibration based EM and PZT harvester outputs at a single load is presented in this paper. The low frequency ambient vibrations are harvested by the EM harvester, and are rectified by a self-powered autonomous rectifier. The high frequency ambient vibrations are scavenged by PZT harvester, and converted into DC with a self-powered SECE interface. In addition to vibration sources, DC power scavenged from thermal energy is boosted and connected in parallel to other sources. Test results of the system showed that the hybrid structure was able to generate more than 1 V output for load resistances higher than 100 kΩ whereas stand-alone harvesting sources are not able to reach this voltage. The implemented structure is able to simultaneously combine harvested power from three different sources and generate enough voltage and power to drive low voltage sensors. It is expected that the efficiency of the hybrid system can be further increased in the future through the integration of the discrete components.

References

[1] Muhtaroğlu A, Yokochi A and Von Jouanne A 2008 *J. Power Sources* **178** 2–6.
[2] Sang Y, Huang X, Liu H, and Jin P 2012 *IEEE Transactions on Magnetics* **48** 4495–4498.
[3] Christman J F *et al.* 2011 *IEEE Design & Test of Computers* **28** 84-94.
[4] Huang C and Chakrabartty S 2011 *IEEE ISCAS* (Rio de Janeiro: Brasil) p 2473–2476.
[5] Christman J F *et al.* 2010 *IEEE Symp. on ASYNC* (Grenoble: France) p 62-71.
[6] Jayaweera H, Pathirana W, and Muhtaroğlu A 2015 *Journal of Physics: Conference Series* **660**.
[7] Ulusan H *et al.* 2015 *IEEE Transactions on Power Electronics* **30** 3712-3719.
[8] Lefeuvre E et al. 2006 *Sensors Actuators A Phys.* **126** 405–416.

A 12 mV start-up converter using piezoelectric transformer for energy harvesting applications

T Martinez[1], G Pillonnet[2] and F Costa[1,3]

[1]Laboratoire SATIE, ENS Paris-Saclay, France
[2]CEA-LETI, Université Grenoble Alpes, Grenoble, France
[3]Université Paris-Est, Créteil, France

E-mail: thomas.martinez@satie.ens-cachan.fr

Abstract. This paper presents a novel topology of start-up converter for sub 100 mV thermal energy harvesting based on an Armstrong oscillator topology using a piezoelectric transformer (PT) and a normally-on MOSFET. Based on a Rosen-type PT and off-the-shelf components, the proposed start-up topology begins to oscillate at 12 mV input voltage corresponding to a temperature gradient of 2°C and achieves 1 V output voltage with only 18 mV input voltage applied to the harvester.

1. Introduction

Energy harvesting devices are a key alternative to batteries to supply low-power devices. They allow those devices to be self-autonomous without need of maintenance and with longer lifetime. In particular, thermo-electric generators (TEG) provide a DC voltage by exploiting a gradient of temperature and constitute an interesting solution to supply devices consuming less than a few mW. However, in most applications, the TEG will only provide a voltage lower than the classical threshold voltage of semi-conductors (<100 mV) that cannot then be stepped-up by classical transistor-based converters. A cold-start circuit is required to start and charge a storage device at a sufficient voltage to allow another main converter to start and realize the optimized power extraction. The start-up converter power efficiency is therefore not the most important parameter due to its short working duration in the system and the focus in the design is put on optimization of minimum start-up voltage and voltage gain.

Resonant oscillator architectures and especially the Armstrong one have been used to start at voltages as low as 40 mV [1], [2] but they suffer from bulky magnetic transformers. Piezoelectric transformers (PT's) are a good alternative to magnetic ones as they present higher voltage gain and quality factor and limited electromagnetic radiation [3]. For exploiting PT properties in the cold-start context, the circuit proposed in [4] replaced the magnetic transformer by a piezoelectric one and used an architecture close to the half-bridge topology to obtain an inductor less converter that starts at 69 mV. Replacing a transistor with an inductance at the input allows them to decrease the start-up voltage to 32 mV.

This paper presents an alternate topology compared to [4] consisting in a self-start-up converter based on an Armstrong oscillator with a piezoelectric transformer. A biasing inductance is added at the input of the piezoelectric transformer to improve the performances. The converter has been realized experimentally and characterized with a TEG. This converter starts with an input voltage of 12 mV and the output reaches 1 V for an input voltage of 18 mV. The paper first describes the architecture of the converter before presenting the results of the measurements.

2. Start-up converter topology based on piezoelectric transformer

2.1. Converter Topology

The schematic of the proposed converter is presented in Figure 1. The ideal voltage source V_{in} and the resistance R_t in series respectively represent the TEG harvester's open-circuit voltage and internal impedance. The piezoelectric transformer consists in the two electrode capacitances C_1 and C_2, the RLC series branch representing the mechanical vibration and the ideal transformer characterizing the piezoelectric effect. The converter in itself is based on the Armstrong oscillator architecture where the magnetic transformer is replaced by a piezoelectric transformer. The active component is a depletion-mode MOSFET i.e. normally-on. The circuit works as a resonant oscillator. The inherent noise in the circuit is amplified and filtered by the PT. This signal is applied to the transistor and an AC current at the selected frequency appears at the primary branch of the circuit. The PT will further amplify the amplitude of the signal and, under particular conditions, an oscillating signal appears and its amplitude increases. This signal is rectified by the diode as soon as it reaches its threshold voltage. When loss and amplification are balanced, the amplitude stops increasing and the circuit is in steady-state.

At the beginning of operation, the transistor acts as a transconductance to amplify the feedback voltage applied at its gate. However, as contrary to a magnetic transformer, the two primary side branches of the piezoelectric transformer contain a capacitive part that prevent any DC current to flow through the transistor. Subsequently the transistor has no transconductance, cannot amplify the gate voltage and no oscillation appear in the circuit. In order to solve this issue, a biasing inductance L_{in} is added in parallel with C_1 in order to let a DC current flow through the FET during the start-up.

2.2. Start-up phase modelling

A small-signal analysis applied to the circuit in Figure 2 combined with the application of the Barkhausen criterion allows to find the free oscillation condition. The open-loop gain G_{OL} of the converter can be determined by opening the loop at the gate of the transistor. The Barkhausen criterion states that oscillation appears if the following conditions are respected:

$$|G_{OL}| > 1 \text{ and } \arg(G_{OL}) = 0\ [2\pi] \quad (1)$$

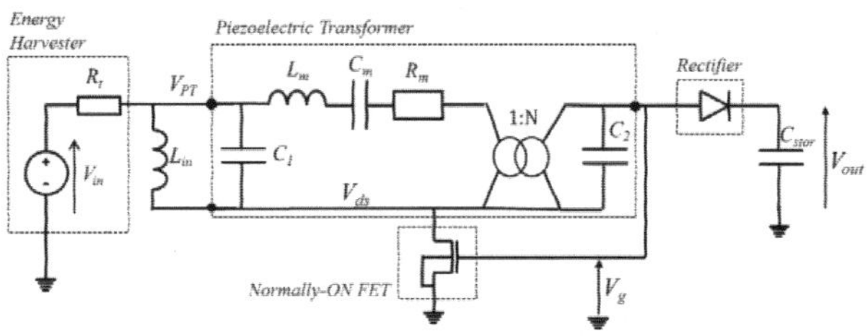

Figure 1. Circuit Schematic

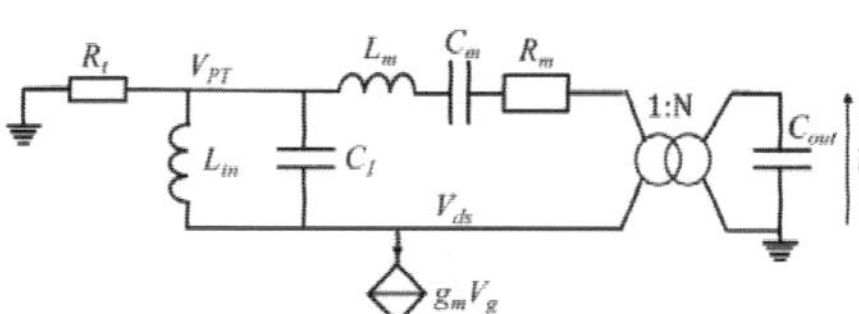

Figure 2. Small-signal equivalent circuit

The open-loop gain of the oscillator is characterized by the transconductance of the transistor, the input impedance of the piezoelectric transformer and its voltage gain. The current flowing through the FET is $i_{ds} = g_m V_g$. Thus the voltage at the input of the PT is expressed as:

$$V_{PT} - V_{ds} = g_m V_g Z_{in} \tag{2}$$

where $Z_{in} = \frac{Z_{PT}\, j\omega L_{in}}{Z_{PT} + j\omega L_{in}}$ and Z_{PT} is defined as $Z_{PT} = \frac{R_m + j.\left(L_m\omega - \frac{1}{C_{eq}\omega}\right)}{1 + jC_1\omega.\left(R_m + j.\left(L_m\omega - \frac{1}{C_{eq}\omega}\right)\right)}$ and $C_{eq} = \frac{C_m.N^2.C_{out}}{C_m + N^2.C_{out}}$

Finally, the open-loop gain is:

$$G_{OL} = G_{PT}\left(\frac{V_{PT} - V_{ds}}{V_g}\right) = \frac{-g_m}{Nj\omega C_{out}\left(1 + \left(\frac{1}{j\omega L_{in}} + jC_1\omega\right)\left(R_m + j\left(L_m\omega - \frac{1}{C_{eq}\omega}\right)\right)\right)} \tag{3}$$

Figure 3 shows the magnitude and phase of the input impedance Z_{in}, the voltage gain G_{PT} and the open-loop gain G_{OL}. Firstly, the addition of the inductance L_{in} modifies the resonance frequency of the input impedance. The oscillation frequency is defined as the one where the open-loop gain is equal to 0 according to (1). In this case, it happens close to the resonance frequency f_p of Z_{in} which is not a resonance for the PT voltage gain G_{PT}. As L_{in} increases, the oscillation frequency will be shifted away from the voltage gain resonance frequency thus reducing the voltage gain but the input impedance value Z_{in} will increase and so will do the input voltage $V_{PT} - V_{DS}$. A trade-off on the L_{in} value appears between the PT input voltage amplitude and the voltage gain of the PT.

3. Experimental Results

3.1. Setup description

The experimental setup is shown in Figure 4. The components were chosen to validate the model of the converter, and not in an approach of performances optimization. Their characteristics and references are summarized in table 1. The PT is a multi-layer Rosen-type PT with 55 kHz natural resonant frequency. Its output capacitance is small increasing the gain of the converter. The transistor consists in two identical depletion-mode MOSFET in parallel with a negative threshold voltage of -37 mV in order to let the current flowing at the beginning of operation. The rectifier is simply formed of a Schottky diode in series configuration. In order to test the circuit in real conditions, the measurements were realized with a TEG whose input resistance's value is 2.8 Ω. The TEG was combined to a controlled hot plate and a heat sink to create a temperature gradient and control the voltage provided by the TEG to the circuit.

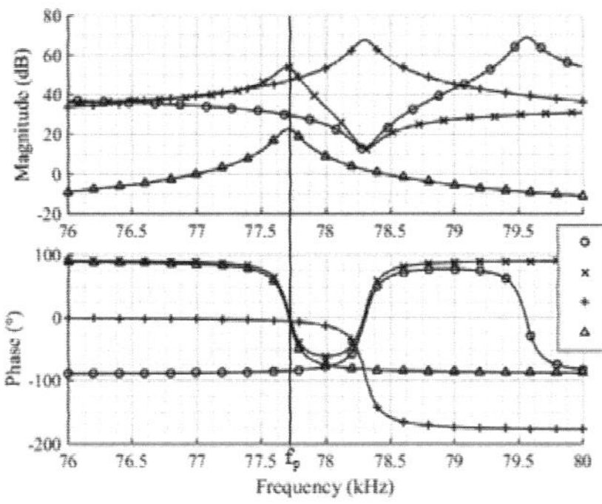

Figure 3. Magnitude and phase of the input impedance, the PT voltage gain and the total open-loop gain.

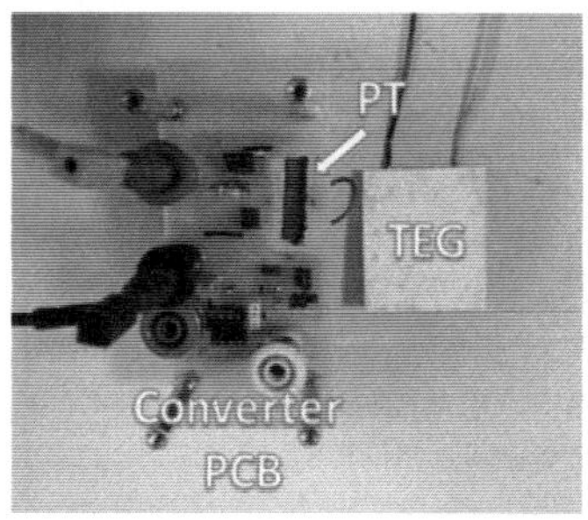

Figure 4. PCB of the proposed start-up converter.

Table 1. Values of the circuit parameters

PT SMMTF85P1S50	
C_1	90 nF
C_2	16 pF
R_m	0.9 Ω
L_m	1.61 mH
C_m	6 nF
N	45
Transistor ALD210800	
V_{th}	-37 mV
$\beta = \mu_0.C_{ox}.W/L$	22.4×10^{-3} A/V²
$C_{g,in}$	4.7 pF
Diode BAS70	
C_{diode}	2 pF
Forward Voltage V_F	0.41 V at I_F=1 mA

3.2. Measurements

Measurements are realized with different values of the input inductance L_{in}. The quality factor of these inductances is equal to 35. In Figure 5 are represented the output voltages V_{out} against V_{in} for different input inductance values. For higher inductances, the circuit starts at smaller voltages but as soon as the circuit starts, smaller inductances achieve higher output voltages. Furthermore, the estimated temperature gradient was extracted from the open-circuit voltage and Seebeck coefficient of the TEG. For an inductance L_{in} = 39 µH, the circuit starts for ΔT= 2°C which corresponds to an input voltage V_{in} of 12 mV. Furthermore, in this case, the output reaches 1 V for an input voltage slightly higher to 18 mV and an estimated ΔT = 3.3°C. Figure 6 represents the output voltage as a function of L_{in}. As expected, the higher the inductance is, the smaller is the output voltage and an optimum value exists for L_{in} that will lead to a maximum output voltage. This optimum value depends on the input voltage but in the studied range, it is between 15 and 20 µH.

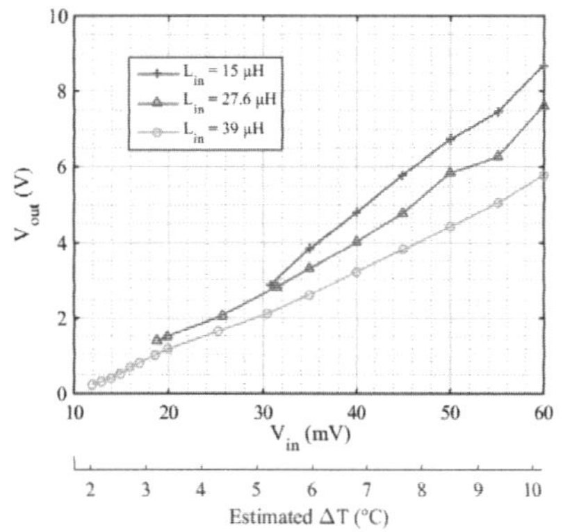

Figure 5. Voltage gain for different values of the biasing inductance as a function of the input voltage and ΔT

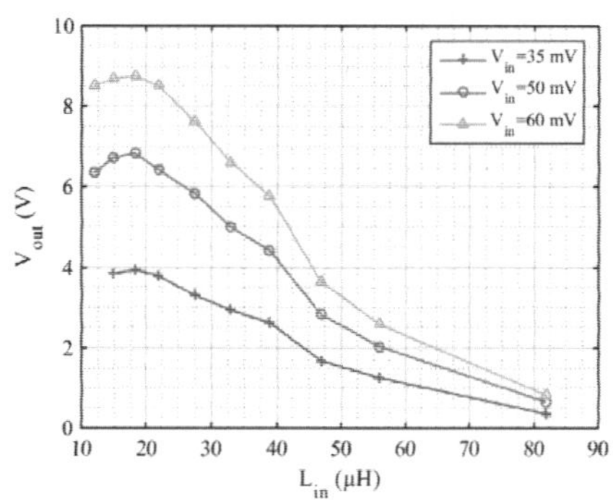

Figure 6. Evolution of the output voltage as a function of the biasing inductance.

Figure 7 represents the minimum input voltage to start-up the circuit as a function of the biasing inductance L_{in}. An optimum value of the biasing inductance exists also for the start-up voltage as expected from the modelling in 2.2. Its value is greater than the optimum value for maximising voltage gain. Secondly we can see in Figure 8 that when the voltage increases from 0 V, the circuit starts at a specific voltage but if the voltage then decreases, the circuit will continue to operate at voltages lower than the start-up voltage leading to an hysteretic behavior. Furthermore, this figure shows the trade-off that needs to be made during the design of the circuit. If the critical aspect is the start-up voltage, high value inductances will allow to start and work at smaller voltages. On the other hand, if the minimum input voltage provided by the harvester is higher, a smaller vallue inductance will increase the voltage gain.

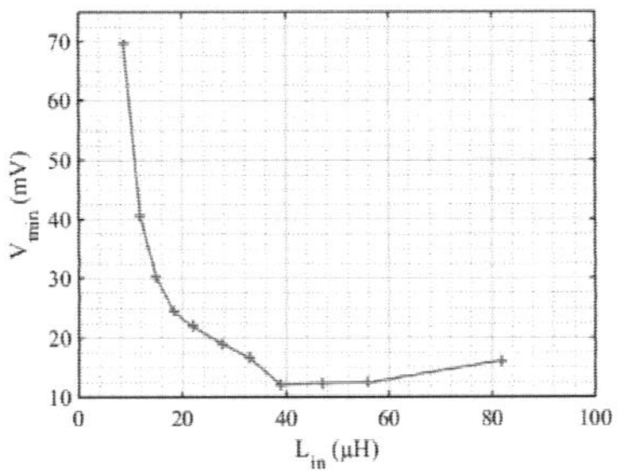

Figure 7. Minimum input start-up voltage of the circuit as a function of L_{in}

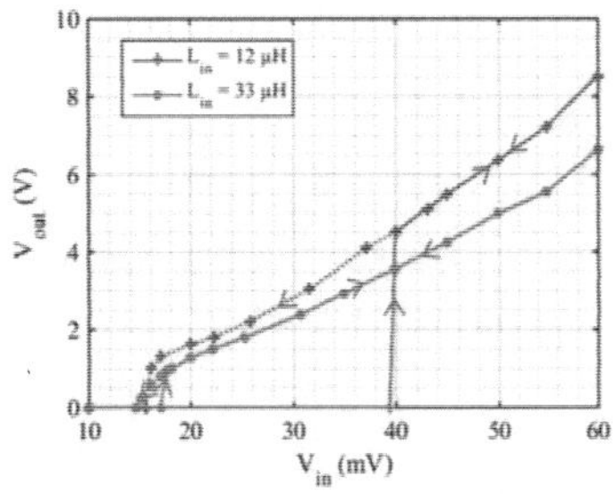

Figure 8. Hysteresis behaviour of the voltage gain

4. Conclusion

This paper presents a start-up converter with a new biasing scheme based on the Armstrong oscillator architecture with a piezoelectric transformer to achieve lower start-up input voltages. The circuit starts at voltages as low as 12 mV and can reach an output of 1 V for voltages as low as 18 mV once oscillation has started. The circuit was tested with a TEG and starts with a temperature gradient of only 2°C applied to the harvester. Compared to other start-up converters using piezoelectric transformers [4], this circuit achieves the lowest start-up voltage and highest voltage gain for similar piezoelectric transformer characteristics thanks to the proposed biasing scheme.

References

[1] J. P. Im, S. W. Wang, K. H. Lee, Y. J. Woo, Y. S. Yuk, T. H. Kong, S. W. Hong, S. T. Ryu, and G. H. Cho, "A 40mV transformer-reuse self-startup boost converter with MPPT control for thermoelectric energy harvesting," in *2012 IEEE International Solid-State Circuits Conference*, 2012, pp. 104–106.

[2] S. E. Adami, V. Marian, N. Degrenne, C. Vollaire, B. Allard, and F. Costa, "Self-powered ultra-low power DC-DC converter for RF energy harvesting," in *2012 IEEE Faible Tension Faible Consommation (FTFC)*, 2012, pp. 1–4.

[3] S. S. Bedair, J. S. Pulskamp, R. G. Polcawich, B. Morgan, J. L. Martin, and B. Power, "Thin-Film Piezoelectric-on-Silicon Resonant Transformers," *J. Microelectromechanical Syst.*, vol. 22, no. 6, pp. 1383–1394, Dec. 2013.

[4] A. Camarda, A. Romani, E. Macrelli, and M. Tartagni, "A 32 mV/69 mV input voltage booster based on a piezoelectric transformer for energy harvesting applications," *Sens. Actuators Phys.*, vol. 232, pp. 341–352, Aug. 2015.

An Integrated Circuit Design of High Efficiency Parallel-SSHI Rectifier for Piezoelectric Energy Harvesting

Y. C. Hsieh[1], J. J. Chen[1], H. S. Chen[2], W.J. Wu[1*]

[1] Department of Engineering Science and Ocean Engineering Science, National Taiwan University, Taipei, Taiwan

[2] Graduate Institute of Electronics Engineering, National Taiwan University, Taipei, Taiwan

*Email: wjwu@ntu.edu.tw

Abstract. This paper presents the design and implementation of a rectifier for piezoelectric energy harvesting based on the parallel-synchronized-switch harvesting-on-inductor (P-SSHI) technique, also known as bias flip circuit[1]. The circuit is implemented with 0.25 μm CMOS high voltage process with only 0.9648 mm^2 chip area. Post-layout simulation of the circuit shows the circuit extracts 336% more power compared with the full-bridge rectifier. The system's average control power loss is 26μW while operating with a self-made MEMS piezoelectric transducer with output current 25μA 120Hz and internal capacitance 6.45nF. The output power is 43.42 μW under optimal load of 1.5 MΩ.

1. Introduction

The concept of machine to machine(M2M) and internet of things(IOT) have drawn industry's much attention in recent years. In order to monitor different status of the environment, the requirement for wireless sensors and IOT devices is increasing. Due to the quick advancement of semiconductor fabrication technologies and ultra-low power integrated circuit design, the power consumption of wireless transceivers and microcontrollers were reduced to microwatts level. However, most wireless devices are still powered by batteries nowadays, and the reliability and performances of the device are limited by the use of batteries and battery lifetime. IoT devices and sensors in the future are better to be environmental friendly and maintenance free. By piezoelectric energy harvesting devices, we can transform vibration energy from the environment into electrical energy, thus solving the issue for powering IOT devices and sensors without using batteries.

Piezoelectric energy harvesting devices can potentially supply up to hundreds of μW of available power[2]. In order to use the harvested power, we need an interfacing circuit to transform the output energy from the transducer in AC (alternating current) form into stable DC (direct current) form required by the wireless transceivers and microcontrollers. Since the power consumed in the interface circuit reduces the amount of usable electrical power, the control of interface circuitry should be ultra-low power. In addition, the commonly used full-bridge rectifier limits the power extractable from piezoelectric devices. Rectifier structures other than full-bridge rectifier were researched and presented in order to extract more power. In this paper, we present a self-powered rectifier for piezoelectric energy harvesting based on the parallel-synchronized-switch harvesting-on-inductor (P-SSHI) technique, which was already proved to boost the output power of piezoelectric energy harvesters[3]. The performance of the full bridge rectifier and the P-SSHI interfacing circuit, the architecture of proposed integrated circuit implementation, and the experiment and SPICE simulation result are all detailed as follows.

2. Discussion of Full-Bridge Rectifier Interface Circuit and P-SSHI Interface Circuit

This section briefly discusses the performance of the full-bridge rectifier and P-SSHI interfacing circuit. Fig.1 shows a full-bridge rectifier interface circuit, also known as a standard interface circuit. Current source S_1 along with C_0 is piezoelectric device's equivalent circuit in current form. We assume all components are ideal and diode has zero voltage drop for simplicity. From the analysis of Wu et al.[4], the maximum output power of standard interface charging a capacitor can be calculated by:

$$P_{\max} = \frac{I_P^2}{2\pi\omega C_0} * \frac{C_1}{C_1 + 2C_0}$$

,where I_p is the amplitude of output current of piezoelectric device, and ω is the operating frequency.

Figure1. Full-bridge rectifier interface circuit

The P-SSHI interface circuit is shown in Fig.2. The only difference between P-SSHI and standard interface is that P-SSHI has an inversion phase while the other one does not. Inversion phase occurs at peak voltage of piezoelectric device's output voltage, which is generated by closing S_1 for half of the L_1C_0 oscillation period. If the voltage before inversion is V_{before}, the voltage after inversion V_{after} can be expressed as,

$$V_{after} = -\gamma \cdot V_{before}$$

,where γ is the inversion factor, a function of circuit's quality factor. By the same analysis method for full-bridge rectifier interface circuit based on Wu et al[4], the maximum output power of P-SSHI charging a capacitor can be expressed as,

$$P_{max} = \frac{I_P^2}{2\pi\omega C_0} * \frac{C_1}{C_1 + 2C_0} * \frac{2}{(1-\gamma)}$$

With a typical inversion factor of 0.7, the maximum output power of P-SSHI interface circuit can reach up to 667% of standard interface circuit without considering control and other power losses. Thus, we proposed a self-powered P-SSHI rectifier circuit for boosting output power of piezoelectric energy harvesters. The architecture and simulation results are explained in the following sections.

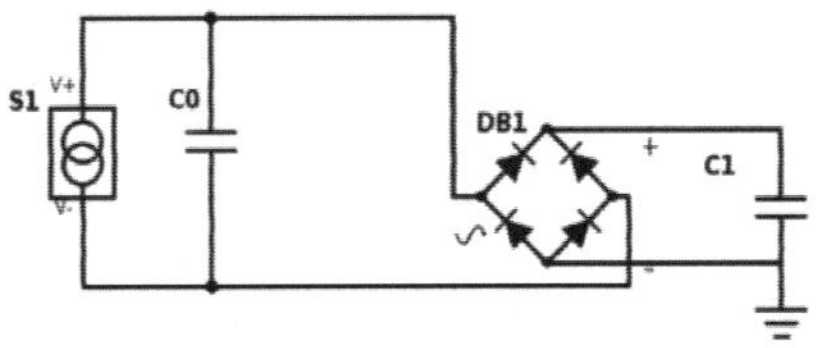

Figure1. Full-bridge rectifier interface circuit

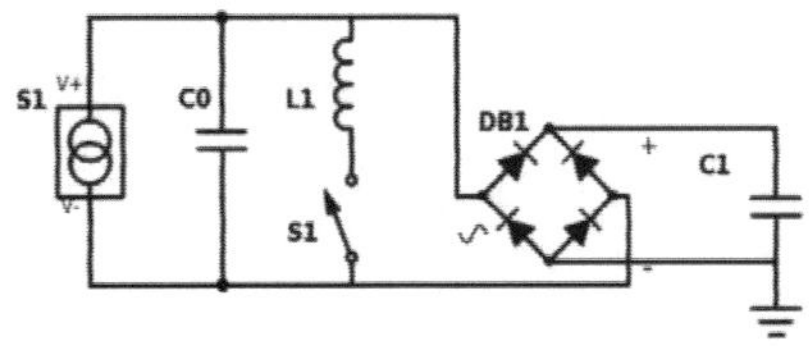

Figure2. P-SSHI rectifier interface circuit

3. Architecture of the Proposed SSHI Rectifier

Fig.3 shows a piezoelectric device's equivalent circuit along with the proposed circuit, which is consists of a peak detector, a comparator for SSHI switching instance control and a full bridge rectifier. The output power of this circuit is stored in the capacitor *Cr* for later use.

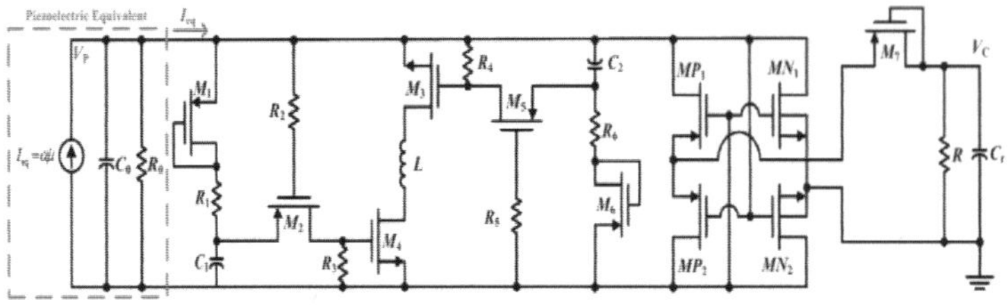

Figure3. Proposed SSHI rectifier circuit

The operation of the SSHI switching control and waveforms are shown in Fig.4. During the half cycle of the vibration, V_p will charge V_c up to $V_p - V_{th,M1}$, which is nearly the peak voltage of the piezoelectric transducer output. As V_p starts to fall, V_c will be held until V_p is smaller than $V_c - V_{th,M2}$. At this crossing point, transistor M_2 will be turn on, and thus turning on M_3 for a bias flip action. Resistor R_3 provides a current path for V_c to discharge. Diode D_1 is placed in series with the bias flip path to prevent over flipping, in the proposed design, the body diode of M_3 and M_4 in Fig.3 serves the function of the diode D_1. The proposed control circuit is modified and redesigned from previous work of Lallart et al.[5]. In this work, in order to implement the circuit with CMOS (complementary metal oxide transistors) integrated circuit design, the original BJT (bi-polar junction transistors) are replaced with all MOSFET (metal oxide field-effect transistors) design to lower the overall losses in the circuit and also the power consumption of control circuit.

The range of operating frequency of the proposed circuit depends on the value of R_1 and C_1. If the operating frequency is too high, the circuit will not be able to detect peak voltage of the piezoelectric waveform due to switching time delay caused by R_1C_1 time constant. In addition, the value of R_1 and C_1 has to be large enough to turn on the switch long enough for proper bias-flip action.

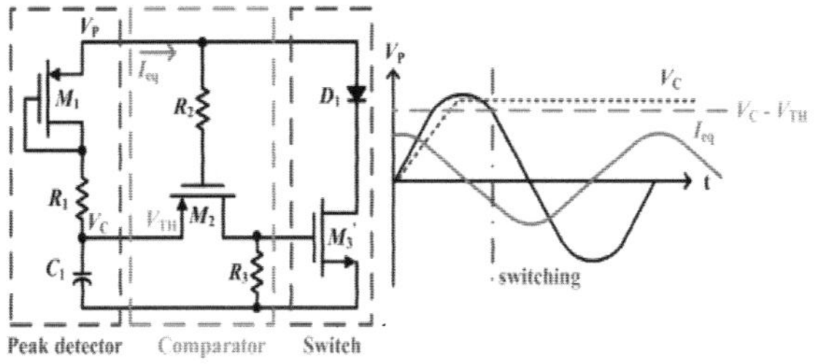

Figure4. Proposed SSHI rectifier circuit

4. Experiment and Simulation Results

A discrete components version of the proposed circuit was made, tested, and compared with standard interface and the circuit proposed by Lallart et al[5]. Since discrete components can handle larger power, tough the loss in the circuit is also higher, the circuit was tested with a self-made piezoelectric transducer with higher output power. The output current of the transducer is 70 μA 120Hz, and the internal capacitance is 7 nF .The tested output power result is shown in Fig.5, which verified the feasibility of the proposed design.

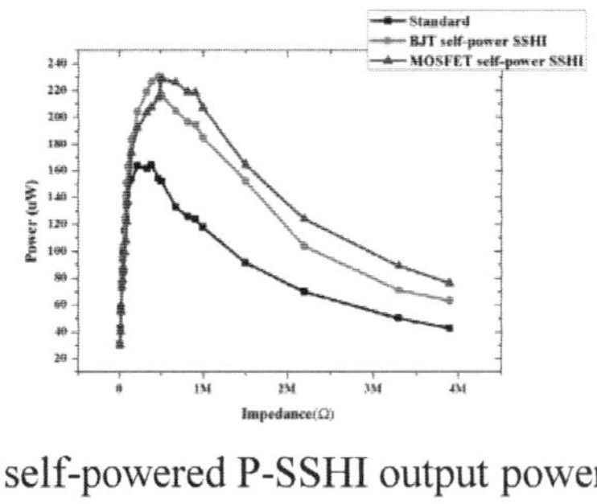

Figure5. Discrete version of self-powered P-SSHI output power verses output impedance

The circuit along with the model of our self-made MEMS piezoelectric transducer[2] is then simulated with Spice, the simulation result of output power verses output impedance is shown in Fig.6. The output current of our self-made transducer is 25 μA 120Hz, and the internal capacitance is 6.45 nF. The result shows that the proposed system is functional and provides 336% more power than standard interface. A table of the performance and working condition of this chip is presented in Table 1, and the layout diagram is shown in Fig.7. Simulated result of this chip compared to state-of-the-art circuits is presented in Table 2. The circuit has been taped out and now in fabrication process, and the test chips will be returned in November for further testing results.

Table 1. Performance and working condition of the proposed circuit

Parameter	
Operating Frequency (Hz)	<150Hz (This work 120Hz)
Amplitude of Current Ip	25μA
Internal Capacitance C_0	6.45 nF
Inversion factor	0.7
Output power	43.42μW
Performance Compared with Full Bridge Rectifier	3.36 x
Power Consumption	26μW
Chip Area	1.206*0.8 mm^2

Table 2. Comparison with state-of-the-art publications

Publication	**JSSC 10 [1]**	**TPE 11 [6]**	**A-SSCC 13 [7]**	**This work**
Process	0.35μm	Discrete	0.18μm	0.25μm HV
Operating Frequency (Hz)	225	185	200	120
Input Current Amplitude Ip	40μA	2mA	70μA	25μA
Piezoelectric Internal Capacitance C_0	12nF	330nF	25nF	6.45nF
Output power compared to standard	4x	2.3x	3.3x	3.36x

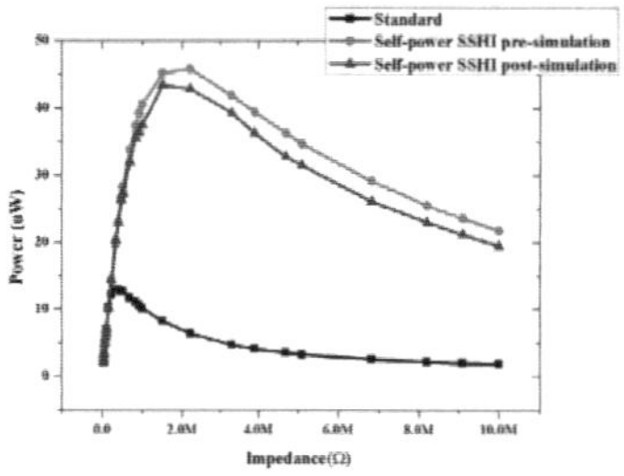
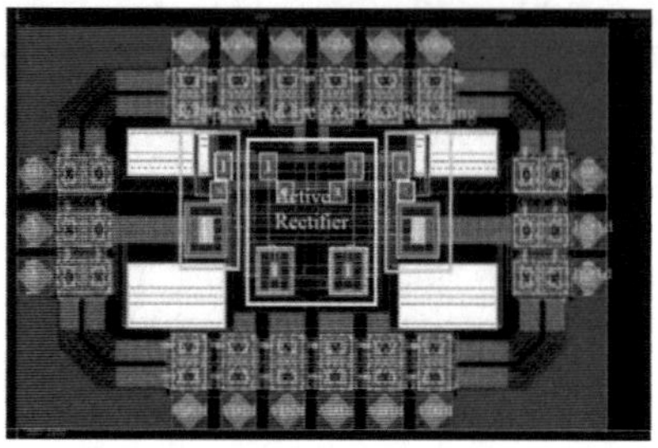

Figure6. Output power of the proposed P-SSHI circuit verses output impedance
Figure7. Layout diagram

5. Conclusion

This paper presented an integrated circuit design of a parallel-SSHI rectifier for piezoelectric energy harvesting, which can extract 336% more power than conventional full bridge rectifier with only 0.9648 mm^2 chip area. While operating with a self-made MEMS piezoelectric transducer with output current of 25μW 120Hz, internal capacitance 6.45nF, the output power can reach up to 43.42 μW. The maximum output power of conventional full bridge rectifier and parallel-SSHI is also studied and analyzed in this paper.

References

[1] Ramadass, Y.K. and A.P. Chandrakasan, *An Efficient Piezoelectric Energy Harvesting Interface Circuit Using a Bias-Flip Rectifier and Shared Inductor.* IEEE Journal of Solid-State Circuits, 2010. **45**(1): p. 189-204.

[2] Lin, S.-C. and W.-J. Wu, *Piezoelectric micro energy harvesters based on stainless-steel substrates.* Smart Materials and Structures, 2013. **22**(4): p. 045016.

[3] Guyomar, D. and M. Lallart, *Recent Progress in Piezoelectric Conversion and Energy Harvesting Using Nonlinear Electronic Interfaces and Issues in Small Scale Implementation.* Micromachines, 2011. **2**(4): p. 274-294.

[4] Wu, W.J., et al., *Modeling and experimental verification of synchronized discharging techniques for boosting power harvesting from piezoelectric transducers.* Smart Materials and Structures, 2009. **18**(5): p. 055012.

[5] Lallart, M. and D. Guyomar, *An optimized self-powered switching circuit for non-linear energy harvesting with low voltage output.* Smart Materials and Structures, 2008. **17**(3): p. 035030.

[6] Krihely, N. and S. Ben-Yaakov, *Self-Contained Resonant Rectifier for Piezoelectric Sources Under Variable Mechanical Excitation.* IEEE Transactions on Power Electronics, 2011. **26**(2): p. 612-621.

[7] Xuan-Dien Do, H.-H.N., Seok-Kyun Han, and Sang-Gug Lee *A Rectifier for Piezoelectric Energy HarvestingSystem with Series Synchronized Switch HarvestingInductor* Solid-State Circuits Conference (A-SSCC), IEEE Asian, 2013: p. 269-272.

Acknowledgments
This research was supported in part by the Ministry of Science and Technology of Taiwan MOST 105-2633-E-002-001, MOST 105-2923-E-002-006-MY3, National Taiwan University NTU-105R104045, Intel Corporation, and Delta Electronics.

Novel thick-foam ferroelectret with engineered voids for energy harvesting applications

Z Luo[1,2], J Shi[1] and S P Beeby[1]

[1]Electronics and Computer Science, University of Southampton, Southampton, SO171BJ, UK
[2]School of Water, Energy and Envinronment, Cranfield University, Cranfield, MK430AL, UK

E-mail: z.luo@soton.ac.uk

Abstract. This work reports a novel thick-foam ferroelectret which is designed and engineered for energy harvesting applications. We fabricated this ferroelectret foam by mixing a chemical blowing agent with a polymer solution, then used heat treatment to activate the agent and create voids in the polymer foam. The dimensions of the foam, the density and size of voids can be well controlled in the fabrication process. Therefore, this ferroelectret can be engineered into optimized structure for energy harvesting applications.

1. Introduction

Ferroelectret is a porous polymer film that can store charges in its internal voids after charging. It is able to convert compressive force into electric pulses, which can be used for both sensing and energy harvesting. Despite being developed into sensing material for more than a decade, ferroelectret has attracted research interest in energy harvesting only in the recent years [1-3]. In the past, we demonstrated that multilayer ferroelectret is capable of powering the transmission of a low-power wireless chipset [1], and fabricated prototypes of ferroelectret-powered wearable devices [4]. The ferroelectret we investigated was commercial polypropylene (PP), which was fabricated by stretching the original polyolefin material and expanded into a foam. Hence, it is a thin polymer film with thickness less than 70 μm. This is favourable for sensor and actuator applications, but not for energy harvesting, since one layer of PP ferroelectret is not sufficient to power any electronic chipset [1]. The multilayer structure of ferroelectret can significantly increase the energy output [1, 4], but also hugely increases the cost and complexity in manufacturing.

Recently we have developed a novel thick-foam ferroelectret that is specifically designed and engineered for energy harvesting. This material was fabricated by mixing a chemical blowing agent with a polymer solution. The polymer solution can be moulded and cured into different shape and thickness. Thus the ferroelectret foam can be fabricated with thickness from several hundred microns to several millimetres. A schematic diagram of the foam's fabrication comparing to the multilayer ferroelectret is shown in Figure 1. The blowing agent of the ferroelectret was activated at certain temperature and created cellular voids in the structure. The density of the voids was controlled by the concentration of the blowing agent, and the void size was controlled by the heating temperature and the heating time. Based on the model that we developed to predict the piezoelectricity of PDMS ferroelecctret [5], and the eletromechanical model to predict the energy output of ferroelectret [6], we were able to model and engineer the void dimensions of the ferroelectret. This optimized structure was engineered by adjusting the processing parameters during fabrication. Compared to the conventional thin ferroelectrets which are mostly fabricated by stretching or foaming, this moulded ferroelectret can be fabricated in desired dimensions and its void size/density can be engineered. This is useful for optimizing the structure of the ferroelectret for energy harvesting applications, and potentially can be used to produce large-area ferroelectret.

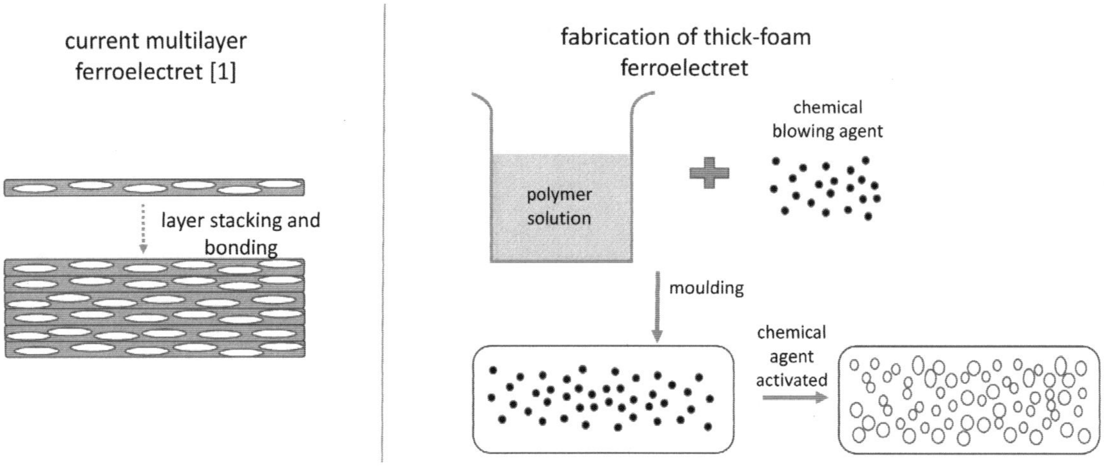

Figure 1. Schematic diagram of thick-foam ferroelectret's fabrication comparing to multilayer ferroelectret

2. Experimental

Low density Polyethylene (LDPE) was used as the polymer and Azodicarbonamide (ADZ) was used as the blowing agent. LDPE pellets were mixed with 0.2 wt% of ADZ power. They were dissolved in xylene by heat treatment and stirring, and then left to dry for a week so the solution can solidify. In order to fabricate ferroelectrets with dimensions of 45 mm × 45 mm × 0.6 mm in this work, 120 g of solidified polymer was used to fabricate one layer of ferroelectret. The weighted portion was hot pressed in a mould at 130 °C with 3 tons of pressing weight for 1 min. This process can shape the solution into the desired dimensions and remove its bubbles. To activate the blowing agent, the hot pressed samples were heat treated in an oven at 230 °C for different durations of 1, 2, 3 and 4 mins, respectively. In this process, internal voids were created in the samples.

Electrodes were deposited on the top and bottom surfaces of the voided samples. The deposition was achieved using a Leybold E-beam Evaporator Lab 600. The deposited layers were 10 nm of Cr then 1 μm of Al. After the deposition, the samples were charged in a self-built needle-to-grid corona charging apparatus, with an electric field of 35 kV for 1 min. The charged samples were then ready to be tested as ferroelectret.

3. Results and discussion

The piezoelectric coefficient d_{33} of the ferroeelctret was measured to be 200 pC/N using a Piezoetest PM300 d_{33} meter. Using a Wayne Kerr 4300 LCR meter to measure the electrical properties, the capacitance of the sample was 38 pF, resistance was 2.6 GΩ. Hence, the dielectric constant of the ferroelectret was calculated to be 1.27, indicating that a large number of voids are presented in the polymer.

During the heat activation, the ADZ blowing agents in the polymer decomposed and released gas. The expanded gas resulted in the void structure in the ferroelectret. A number of factors affects the density and size of the voids. Firstly, the concentration of the blowing agent obviously will determine the density of the void. Higher content of ADZ will create more voids in the sample. However, when the voids are too dense, the size of the void is more difficult to control since the neighboring voids tend to merge into a bigger one. Hence, 0.2 wt% of ADZ content was used in this work. This is not the optimized content of ADZ for energy harvesting, but will create voids that can be easily observed and measured. Secondly, the activation temperature of ADZ is about 200 °C. Heating the samples above this temperature will accelerate the decomposition of ADZ. Thus at the same heating time, higher temperature will result in more gas and larger void size. In this work, heating temperature of 230°C was used for all the samples so the void size was altered by heating time only. Lastly, increasing the heating time will allow more gas to be released, thus larger void size. This is demonstrated in Figure 2, where the samples were heated for different durations. Voids with diameter of 90 μm was observed in the samples heated for 1 min, 140 μm for 2 mins, 310 μm for 3 mins, and 340 μm for 4 mins, respectively. It also shows that the growing rate of void size decreased as the heating time increased, because the size of the void tended to saturate when the ADZ fully decomposed.

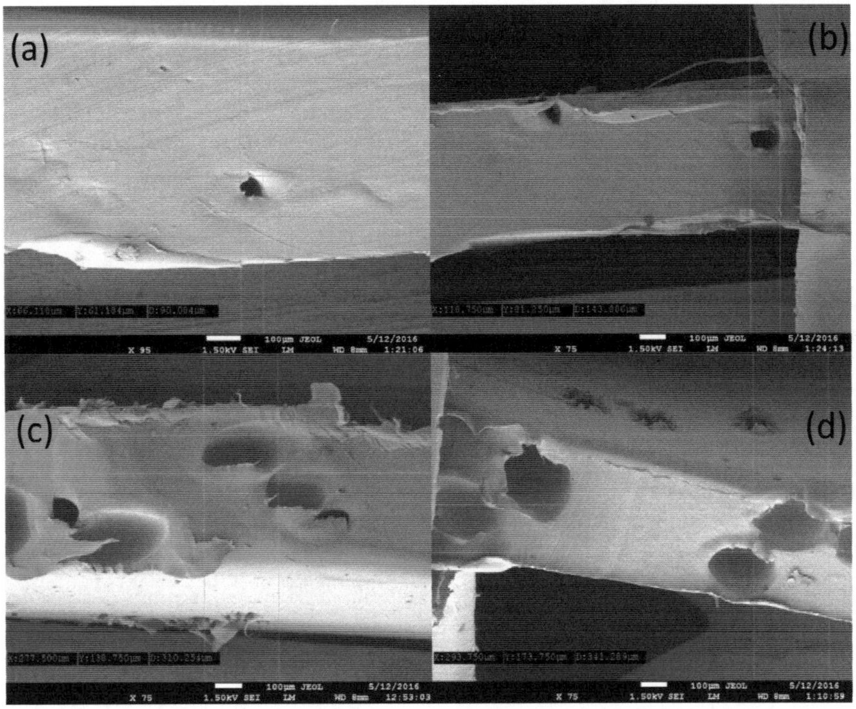

Figure 2. Fabricated ferroelectret with different void sizes of (a) 90 μm; (b) 140 μm; (c) 310 μm; (d) 340 μm by heating for durations of 1, 2, 3 and 4 mins respectively.

From our previous study in modelling the piezoelectric output in ferroelectret, the optimized void size for high piezoelectricity is in the range of 50 to 100 μm [5]. Therefore, it is anticipated that the sample heat treated for 1 min will have stronger piezoelectric response comparing to the one heat treated for 2 mins. This is supported by the result from the experiment where the samples were under trapezoidal function of compressive forces, which has maximum amplitude of 800 N. The function of the force is shown in Figure 3 (a). The output pulses of the ferroelectret samples measured by oscilloscope are shown in Figure 3 (b) & (c). It shows that the maximum output of the sample with void size of 90 μm is 0.46 V and -1 V, the one with void size of 140 μm is 0.26 V and -0.88 V. The voltage output of the former is stronger than the latter, indicating a stronger piezoelectric response.

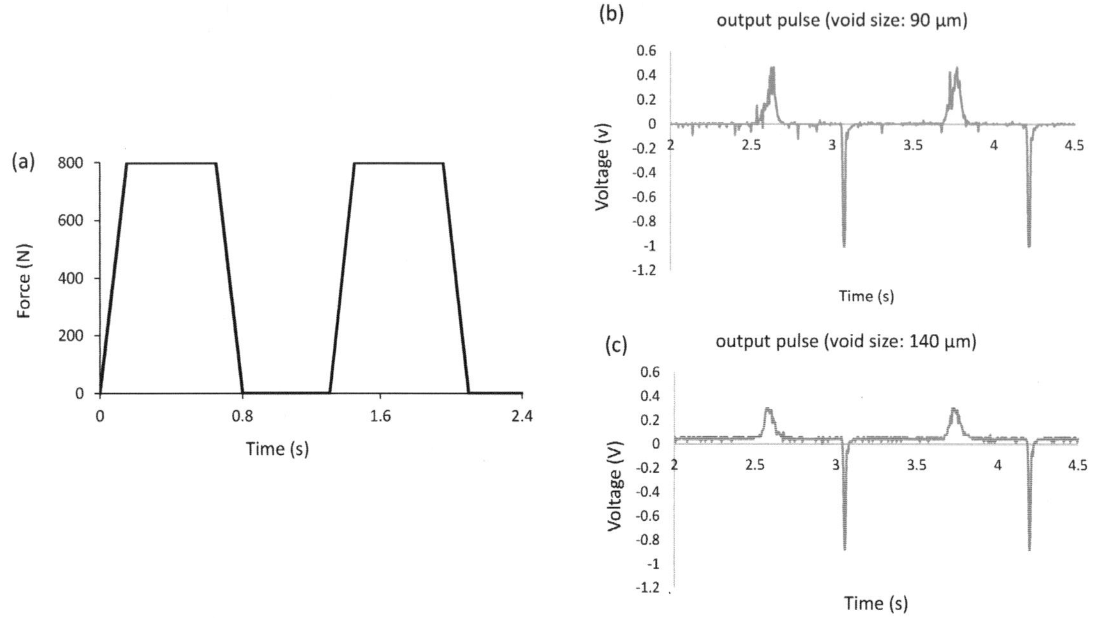

Figure 3. (a) Trapezoidal function of applied compressive forces; Output pulses of thick-foam ferroelectret with thickness of 600 μm and void size of: (b) 90 μm; (c) 140 μm.

4. Conclusion and future work

This paper reports the design and fabrication of a novel ferroelectret material. It is fabricated by mixing a polymer solution with a chemical blowing agent ADZ, then shaped into desired dimensions by casting. The blowing agent is activated by heat treatment and create voids in the polymer. By controlling the parameters in the fabrication process, such as heating temperature and time, the size of the voids can be engineered. Hence, numerical simulation can be used to design an optimal void size, and achieving this size by adjusting the processing parameters. The next step forward will be to improve the energy output of the ferroelectret by varying the ADZ's content and using different types of polymers as medium.

5. Reference

[1] Luo Z, Zhu D, Shi J, Beeby S, Zhang C, Proynov P and Stark B 2015 Energy harvesting study on single and multilayer ferroelectret foams under compressive force *IEEE Trans. Dielectr. Electr. Insul.* **22** 1360-1368

[2] Zhang X, Wu L and Sessler G M 2015 Energy harvesting from vibration with cross-linked polypropylene piezoelectrets *AIP Adv.* **5** 077185

[3] Anton S R, Farinholt K M and Erturk A 2014 An evaluation on low-level vibration energy harvesting using piezoelectret foam *J. Intel. Mat. Syst. Struct.* **25** 1681-1692

[4] Luo Z, Zhu D and Beeby S 2015 Multilayer ferroelectret-based energy harvesting insole *J. Phys. Conf.* **660** 012118

[5] Shi J, Zhu D, Cao Z and Beeby S 2015 Optimization of a PDMS structure for energy harvesting under compressive forces *J. Phys. Conf.* **660** 012041

[6] Luo Z, Zhu D and Beeby S 2016 An eletromechanical model of ferroelectret for energy harvesting *Smart Mat. Struct.* **25**, 045010

Acknowledgments

This work was performed under the SPHERE IRC funded by the UK Engineering and Physical Sciences Research Council (EPSRC), Grant EP/K031910/1.

Soft X-ray Charged Piezoelectret for Kinetic Energy Harvesting

Jia Lu[1], Hangjun Cho[1], and Yuji Suzuki[1]

[1]Dept. of Mechanical Engineering, The University of Tokyo, Tokyo, Japan

Email: jlu@mesl.t.u-tokyo.ac.jp

Abstract. Piezoelectret polymer attracts much attention for its high piezoelectric coefficient. Multilayered piezoelectret structures are often charged with corona discharge, but it is difficult to get high surface charge density. To address this issue, a multilayered piezoelectret structure with embedded electrode is proposed, which can be efficiently poled with soft X-ray charging. With the aid of embedded electrodes, the bias voltage is directly applied to each unit cell, rather than divided and distributed to multiple layers. With an early PTFE-based prototype, output power of 0.5 μJ has been obtained for 0.3 mm displacement in 0.2 s.

1. Introduction

Piezoelectrets are space-charged porous polymers, in which change of the macro dipole moment under structure deformation leads to piezoelectricity. They are gaining much attention in energy harvesting due to their softness and high piezoelectric coefficient [1]. Porous polypropylene is formed by gas expansion or stretching methods and then charged with contact or corona charging [2]. However, random pore dimensions and distributions would result in non-uniform polarization in those piezoelectrets and deteriorate the charge stability [3]. Alternatively, multilayered piezoelectret structures are often prototyped [4] as shown in figure 1a, where pores are charged with corona discharge.

Because the conventional charging method relies on the electrical breakdown in each void, the effective bias voltage per layer for the multilayered structure is much lower than the grid voltage. To address this issue, we propose a multilayered structure with embedded electrode that can be efficiently poled with the soft X-ray charging [5]. Figure 1b shows our concept for the improved piezoelectric structure. With the aid of the embedded electrode, the bias voltage is directly applied to each layer rather than divided and distributed to multiple layers. With the soft X-ray irradiation, positive and negative ions are generated in-situ through photo-ionization of gas molecules, and separated to the opposite sides by the bias voltage V_{bias}. With this approach, the surface charge density on pores could be much higher than that with the corona charging, therefore greatly increasing the output power.

2. Electromechanical Modelling

In order to study the output power of a multilayered piezoelectret under deformation, an electromechanical model shown in figure 2 is used. The electrostatic model [1] is employed with the assumption of 1-D electrostatic field, and the Euler-Bernoulli beam is used for estimating the deformation. Illustration of electric and mechanical models is shown in figures 2a and 2b.

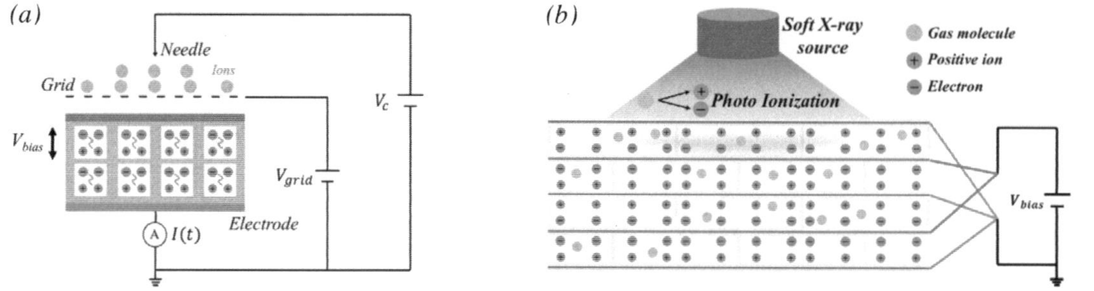

Figure 1. Charging methods for multilayered piezoelectrets. (a) Conventional corona charging. Bias voltage for each layer is much lower than the grid voltage V_{grid}, (b) Present soft X-ray charging of multilayered piezo-electrets with embedded electrodes.

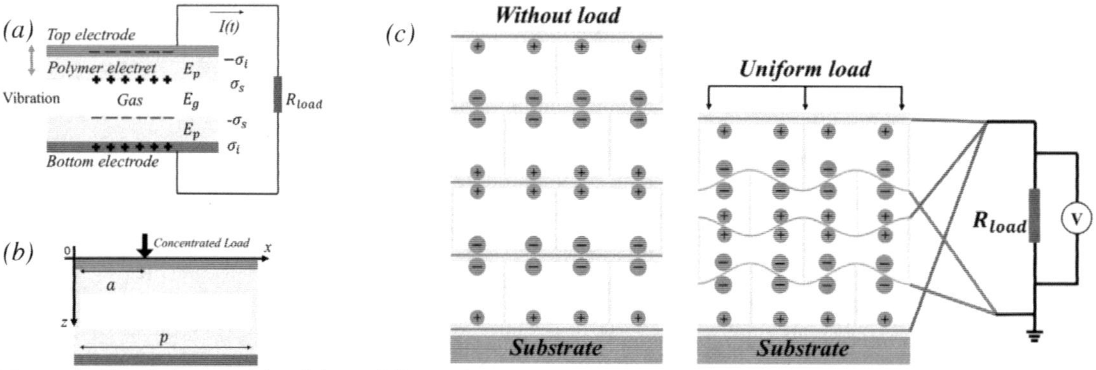

Figure 2. Simulation models of the multilayered structure. (a) Electrostatic model [1], (b) Mechanical model based on the beam theory, (c) Electromechanical model of the multi-layered piezoelectret.

In figure 2a, E is the electric field, with subscripts p and g standing for polymer electret and gas layers, respectively. σ_s and σ_i are the surface charge density on electret and induced charge density on electrode, respectively. According to the electrostatic model by Sessler [1], we have

$$\sigma_i = -\varepsilon_0\varepsilon_p E_p,\ \ \sigma_s = -\varepsilon_0\varepsilon_p E_p + \varepsilon_0\varepsilon_a E_g,\ \ E_p h + E_g g(t) + U(t) = 0 \tag{1}$$

where ε_0, ε_a (equal to 1) and ε_p represent the vacuum permittivity, the relative permittivity of air and polymer electret, respectively. h is the thickness of electret and $g(t)$ is the time-dependent gap. $U(t)$ is the voltage across the external load R_{load}. Using the Ohm's law and the conservation of charge, we get

$$I(t) = U(t)/R_{load},\ \ A d\sigma_i(t)/dt + I(t) = 0 \tag{2}$$

where A is effective surface area of electrode overlapped with electret coating. With an initial condition of $U(t)$ and prescribed temporal gap $g(t)$, the output voltage and thus power is obtained.

For the mechanical model, an Euler-Bernoulli beam with two fixed ends and concentrated load at any place of the beam is assumed. Beam deflection at x can be written as

$$\delta(x) = F_{load}x^2(p-a)^2\,[2a(p-x) + p(a-x)]/(6p^3EI),\ x \text{ in } [0, a] \tag{3}$$

where p is the pitch between two fixed ends, and EI is the flexural rigidity of beam. F_{load} is the concentrated force applied at $x_f = a$ from the origin, and consists of the mechanical force F_m applied at the top plate with the external force and the electrostatic force F_e between oppositely-charged electrets. The electrostatic force is obtained through integration of local electrostatic forces $\Delta F_e(x)$ as

$$Fe(t) = \int \Delta F_e(x,t)dx = \int (1/2)\varepsilon_0\varepsilon_a E_g(x,t)^2 w dx \tag{4}$$

where w is the depth of the beam (in y axis direction).

3. Fabrication Process, Mechanical and Charge Stability Test

Based on the present model, we design an early prototype using PTFE for proof of concept. The fabrication process is shown in figure 3. Firstly, an aluminum electrode is sputtered onto a 0.2 mm-thick PTFE film. Each film is bonded to a 0.4 mm-thick PTFE plate, in which pillar structures are formed by machining. This is followed by stacking and adhering five layers to form the multilayered structure with embedded electrodes. The as-fabricated prototype is shown in figure 3e, with dimensions of 30 mm × 30 mm × 2.8 mm.

Figure 4a shows the set-up for measuring the structure stiffness. The effective Young's modulus is as low as 0.26 MPa even with the PTFE plates, which is lower than the estimates as shown in figure 4b.

The surface potential of the PTFE film is measured after the soft X-ray charging. Initial voltage is 1 kV, which is the same as the bias voltage. It decreases with time, and reaches its stable value of 350 V after 10 days. The present result with the stable voltage at 35 % of the initial voltage is in accordance with the previous result [6].

4. Power Generation Experiment

Power generation experiment is conducted using a set-up shown in figure 6. The multilayered structure is poled using soft X-ray (peak photon energy: 58.9 keV) with the bias voltage of 3 kV for 3 min. By using a linear motor, prescribed total deformation of 0.1~0.3 mm in 0.2 s is applied to the multilayered structure, and the output voltage is recorded.

Figure 7 shows the temporal profiles of the output voltage for 1.1 GΩ load. The maximum voltage for the 0.3 mm deformation reaches 100 V, which corresponds to the output power of 0.52 μJ; 0.52 μJ

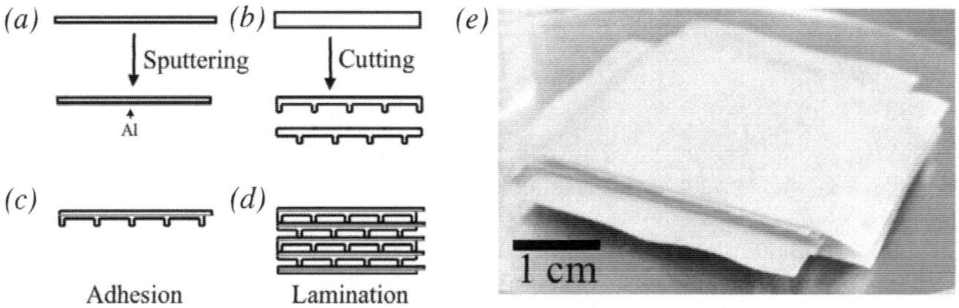

Figure 3. Fabrication process of the multilayered piezoelectret.

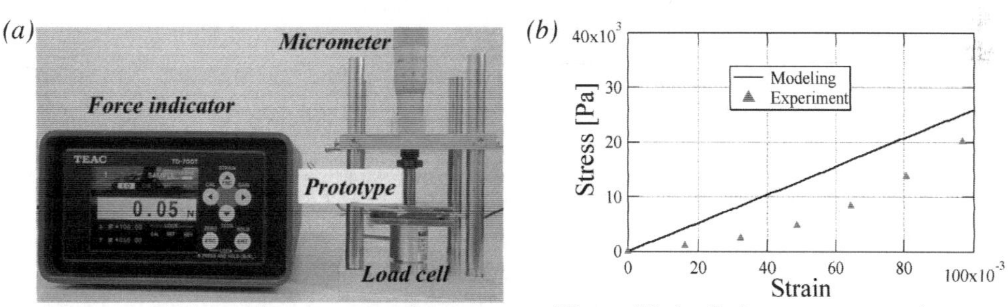

Figure 4. (a) Set-up for measuring the structure stiffness, (b) Applied stress versus strain.

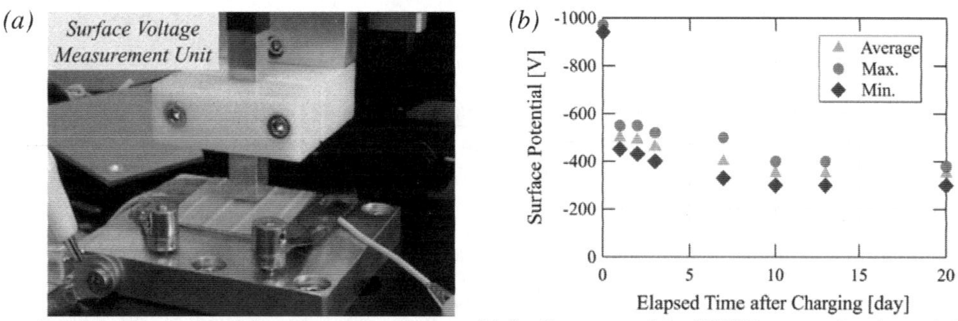

Figure 5. (a) Surface potential measurement set up. (b) Surface potential of PTFE versus the elapsed time.

Figure 6. Experimental set-up for the power generation with a linear motor, control interface, and the piezoelectret.

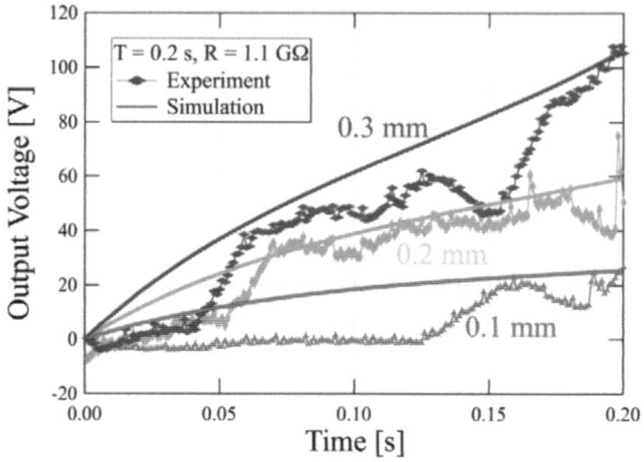

Figure 7. Output voltage waveforms with the prescribed deformation in comparison with simulation results. Up to 0.52 μJ has been obtained with the early prototype.

Table 1. Output power for different total deformation of the prototype.

Deformation amplitude	Experiment	Simulation
0.1 mm	**0.018 μJ**	**0.059 μJ**
0.2 mm	**0.22 μJ**	**0.28 μJ**
0.3 mm	**0.52 μJ**	**0.77 μJ**

corresponds to the piezoelectric coefficient of 155 pC/N. This is somewhat lower than the model prediction of 0.77 μJ, but the trend is in accordance with the simulation results. Output power for 0.1 mm and 0.2 mm deformation is also shown in table 1.

With multilayered structures using softer materials than PTFE, and with higher-performance electret material such as CYTOP EGG [7], piezoelectric coefficient d_{33} = 3000 pC/N could be realized, which corresponds to the output power of 200 μJ under the same condition.

5. Conclusions

We have proposed a soft-X-ray-charged multilayered piezoelectret energy harvester with embedded electrodes for the first time. An electromechanical model based on the electrostatic model [1] and the beam deflection theory is developed. With the aid of the embedded electrodes, the bias voltage is directly applied to each unit cell, which greatly enhances applied voltage per layer and results in much higher output power. Even with our early prototype using PTFE, where PTFE surface charge density is merely 0.05 mC/m^2, output power of 0.5 μJ has been obtained for 0.3 mm displacement in 0.2 s.

The authors would like to thank Dr. K. Hagiwara in Japan Broadcasting Corporation (NHK) for his cooperation and fruitful discussion in the course of this work.

References

[1] G.M. Sessler, and J. Hillenbrand 1999 *Appl. Phys. Lett.* **75** 3405-07
[2] M. Paajanen, H. Valimiiki and J. Lekkala 1999 *10th Int. Symp. on Electrets*
[3] X. Qiu, R. Gerhard and A. Mellinger 2011 *IEEE Trans. Dielectr. Electr. Insul.* **18** 34
[4] J.-J. Wang, et al. 2012 *J. Micromech. Microeng.* **22** 015013
[5] K. Hagiwara, et al. 2012 *IEEE Trans. Dielectr. Electr. Insul.* **19** 1291-98
[6] Z. Xia, et al. 1999 *J. Phys. D: Appl. Phys.* **32** L83–L85
[7] K. Kashiwagi, et al. 2011 *J. Micromech. Microeng.* **21** 125016

A paper-based electrostatic kinetic energy harvester with stacked multiple electret films made of electrospun polymer nanofibers

Y. Lu[1], D. Amroun[1], Y. Leprince-Wang[1] and P. Basset[1]

[1]Université Paris-Est / ESYCOM(EA2552) / UPEM – ESIEE Paris – CNAM, France,

Email: yingxian.lu@esiee.fr

Abstract. This paper reports the first flexible electrostatic kinetic energy harvester (e-KEH) with electret nanofibrous films obtained by electrospinning and paper-based electrodes. The nanofibrous electret outperforms plenary thin film Parylene in the storage stability of charge: the surface potential is stabilized within 1 day, without any obvious minishing during the following 9 days. The output power of the device is improved by implementing multiple electret layers, where the optimal number of electret layer is 3. With a finger tapping activation, this first prototype with the optimal configuration gives a maximum peak power of 45.6 μW with the optimal load of 16 MΩ. Working with a full-wave diode rectifier and a storage capacitor of 10nF, the voltage reaches 8.5 V with 450 strokes.

1. Introduction

Kinetic energy harvesters on soft substrates are applicable to wearable electronics. Previously reported works are mostly focused on either piezoelectric harvesters [1-3] or triboelectric nanogenerator [4]. Few others used thick materials as electret [5]. Both types can provide repetitive narrow peaks with pressing or bending external forces. In this work we present a flexible nanogenerator using multi-layered nanofibrous electret as the electrostatic source. The electret material is based on electrospun nanofibers with a thin layer of Parylene C, providing a charge storage stability even better than plenary Parylene films. The proposed device has a simple sandwich structure, and gives a maximum peak power of 45.6 μW. The design is light-weighted and is less than 1mm in thickness.

2. Device geometry

A schematic and a photograph of the proposed device are shown in Figure 1. The device has a sandwich structure with top and bottom electrodes of copper tapes sticking on paper. Between the two electrodes is a foam-like structure – a stack of negatively-charged nanofibrous films.

The working principle of the device is shown in Figure 1(c). By applying a mechanical force on the electrodes, the gaps between the nanofibrous films reduce and the capacitance of the device changes. Consequently, the distribution of the induced charges in the two electrodes vary, according to the principle of the electrostatic transduction [6].

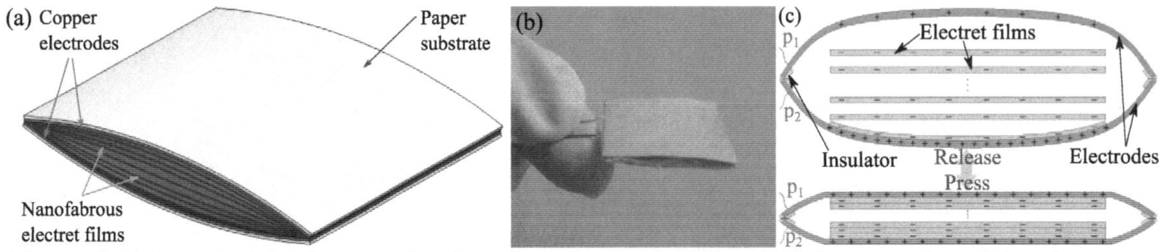

Figure 1. Schematic (a), photograph (b), and working principle (c) of the paper-based low-frequency e-KEH.

3. Material preparation

3.1. Preparation of nanofibrous electret

The nanofibrous films are fabricated from PVDF/PTFE solution through electrospinning. The component of the solvent is acetone and DMF with the volume ratio of 1:1 at room temperature (weight ratio 1:1.32), while the weight ratio of PVDF and PTFE is 5:1. The total weight proportion of the polymers in the solution is 18%. To prepare the solution, the 2 solvents are mixed by magneto stirring method at 60 °C for 30 min, and the powder of the 2 polymers are put into the solvent and stirred for 24 h at the constant temperature of 60 °C. The stirring speed is 50 rpm.

During the electrospinning process, the distance between the point and the receiver electrodes is 15 cm, where the receiver is grounded and the high voltage of 13 kV is applied at the point electrode. The supply rate of the solution is 0.7 mL/h, and the electrospinning process lasts for 2 h in total. The nanofibrous materials obtained are then processed by an additional deposition of Parylene C through Chemical Vapor Deposition process. Shown in Figure 2 are the microscopic photographs of the electrospun nanofibers before and after Parylene C deposition.

3.2. Precharging

The electret films are then charged by corona discharge at 12 kV/ 500 V/ 90 °C for 30 min. In Figure 3a, the comparison of the surface potential decay after corona charging are made between different materials: a) plenary Parylene C film, b) bare PVDF/PTFE nanofiber, c) PVDF/ PTFE nanofiber with Parylene C on top. Despite that the storage stability of charge in pure PVDF/PTFE nanofibrous material is quite poor (due to the property of the polymers themselves), the nanofibers covered by Parylene C provide even a higher storage stability than that of the plenary Parylene C. It is shown that the charge storage (surface potential) of the precharged PVDF/PTFE/Parylene material is stabilized within one day. With 2-μm-thick Parylene C, the nanofiber film provides both (i) a highly porous and flexible structure and (ii) a relatively more stable surface potential.

As shown in Figure 3b, the initial surface potential of a stack of electret films increases linearly with the number of electret layers. Thus, the density of the inducted charge on the electrodes is multiplied by increasing the number of electret layers, and increases the output capability of the device. However, this statement is true only for the case where the total thickness of the electret films is negligible compared to the size of the films, otherwise the adding effect of the surface potential minishes with the increase of the total film thickness: as for any e-KEH, energy conversion is obtained thanks to the device capacitance variation under external force.

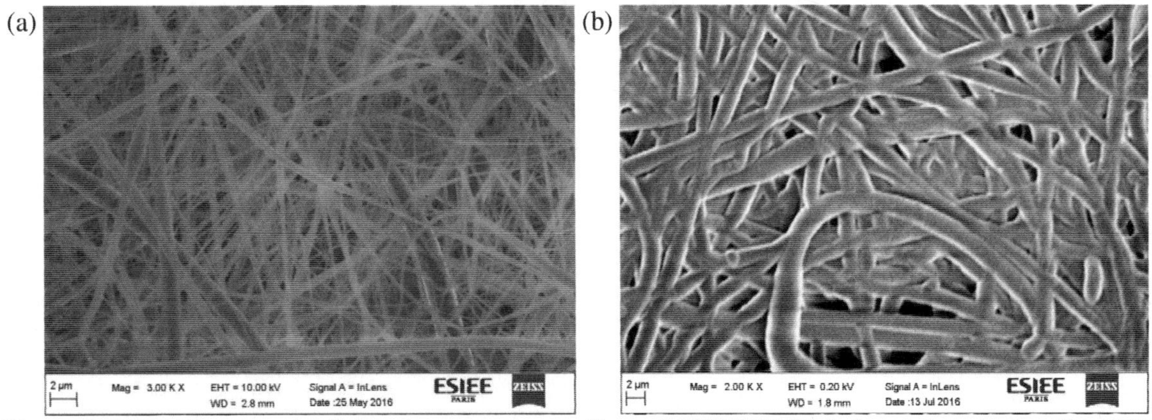

Figure 2. SEM Photograph of the electrospun nanofiber of PVDF/PTFE before (a) and after (b) CVD deposition of Parylene C

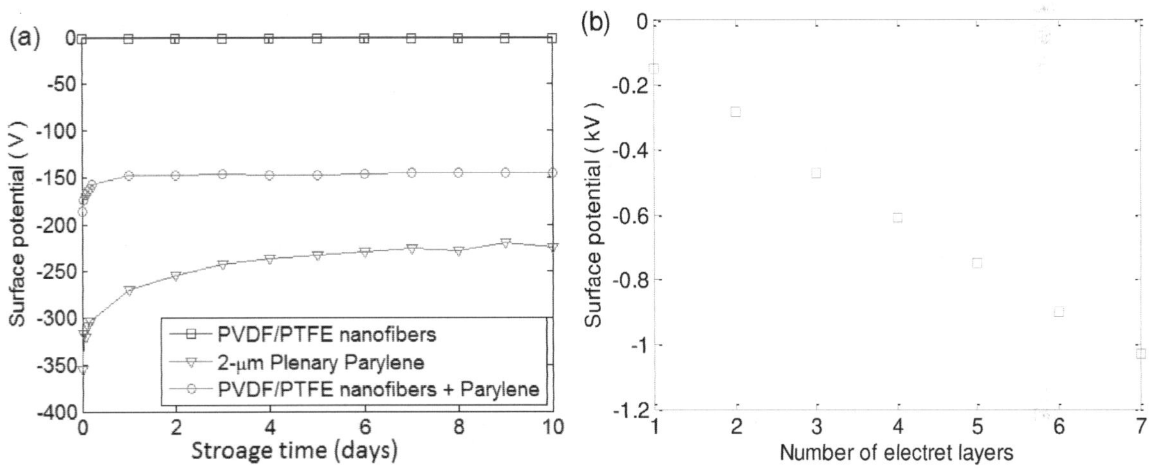

Figure 3. (a) Surface potential evolution of single-layer nanofabrous film after corona charging with and without parylene C. (b) Initial surface potential of stacked multi-layered nanofabrous electret.

4. Device performance

The device is tested with a varying number of electret layers and with a sweeping load. The output performance of the device is shown in Figure 4. The transient output signal during the test is plotted as the inset of Figure 4a. The average peak value of the output pulses (voltage and power) are calculated accordingly, and the variation of these values are plotted with varied load and varied number of electret layers in Figures 4a-b. As shown in this inset, sharp pulses with large amplitudes and small timespan are generated repetitively with the same frequency of the excitation force.

The increased number of electret layers results in an increased amount of induced charge on the electrodes. At the same time the maximum capacitance is reduced due to a larger gap between electrodes while the minimum capacitance is kept approximately the same. Therefore, there is an optimal number of electret layers. By pressing and releasing the device with finger, we get the average peak voltage of 27 V with the load of 16 MΩ and 3 layers of electret material (optimal condition), corresponding to the optimal instantaneous output power of 45.6 µW.

In an experiment charging a 10 nF capacitor with a full-wave diode rectifier, the voltage reaches 8.5 V with 450 strokes. In this experiment, the energy harvested by each cycle is far less than the case with the optimal resistive load. The reason for this difference comes from the voltage drop of the diodes and the impedance mismatch effect. The power transmitted through the rectifier can be further improved if a better matching is reached, or reduce the voltage drop from the rectifier.

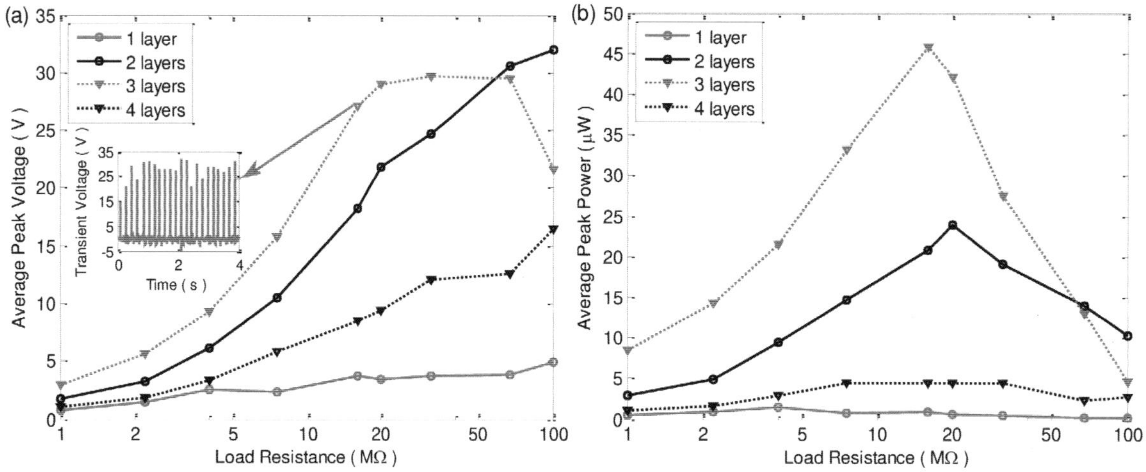

Figure 4. (a) Average peak voltage and (b) power with sweeping load. The inset plot in (a) is the transient output voltage when the device is working with the optimal configuration (3 layers of thin film electret, with the load of 16 MΩ), corresponding to the peak power of 45.6 μW.

5. Conclusion

In this article, we have proposed a non-resonant electrostatic KEH with soft substrate and stacked thin film nanofibrous electret. The charge storage stability of the proposed new type of electret is better than that of the plenary Parylene film electret. The power measurement shows the optimal resistive load of the device to be 16 MΩ, and the optimal number of the stacked electret films to be 3. The maximum peak power obtained is 45.6 μW. When the device is working with a full-wave rectifier and a storage capacitor of 10 nF, the voltage reaches 8.5 V with 450 strokes.

References

[1] Chen X, Xu S, Yao N, et al. Nano letters, 2010, 10(6): 2133-2137.
[2] Park K I, Xu S, Liu Y, et al. Nano letters, 2010, 10(12): 4939-4943.
[3] Park K I, Son J H, Hwang G T, et al. Advanced materials, 2014, 26(16): 2514-2520.
[4] Fan F-R, Tian Z Q and Wang Z L, NanoEnergy, 2012.01.004
[5] McKay T G, Rosset S, Anderson I A and Shea H, Smart Mater. Struct. 24 (2015) 015014
[6] Basset P, Blokhina E and Galayko D, Electrostatic Kinetic Energy Harvesters (2016), Wiley

Wireless sensor node demonstrating indoor-light energy harvesting and voltage-triggered duty cycling

M. A. Cowell[1], B. P. Lechene[1], P. Raffone[2], J. W. Evans[1], A. C. Arias[1], and P. K. Wright[1]

[1] Advanced Manufacturing for Energy, University of California, Berkeley CA 94720, USA
[2] Politecnico di Milano, Piazza Leonardo Da Vinci, Italy

E-mail: martin.cowell@berkeley.edu

Abstract. We demonstrate a wireless sensor node (WSN) that operates purely on harvested ambient indoor light energy and regulates its duty cycle via voltage-triggered sensing and transmission. The extremely low-power light found in indoor environments can be considered at first sight as unsuitable for high-power load demands characteristic of radios [1], but by leveraging spectrum-tailored solar cells, trickle charging a high-power energy reservoir, and implementing triggered duty cycling, we show that these power demands can be consistently met. All energy harvesting and storage components are fabricated in our labs.

1. Introduction

Indoor ambient light is a readily available source of energy for microelectronic devices, but its low power density has typically meant it is not considered for wireless snesor node (WSN) applications [1]. We extend prior literature on indoor photovoltaic energy harvesting by fabricating and demonstrating a wireless sensor node in which a spectrum-tailored organic solar cell is used to charge a supercapacitor, which in turn intelligently powers a radio payload using a state-based control strategy.

1.1. Application Environment

Because solar cells are generally optimized for outdoor use, their performance is traditionally measured under 1-sun standard irradiance of 100 mW/cm^2 with AM 1.5G spectrum. However, indoor conditions differ considerably from this conventional scenario. First, indoor light intensity is characterized in illuminance, as seen by the human eye. Standard indoor illumination at 300 lux corresponds to 100-500 $\mu W/cm^2$ irradiance depending on the light source. Second, the light spectrum is very different than the sun. The absorption spectrum of organic photo-active materials is well-suited to absorb indoor light spectra, so they are a natural solution for indoor light harvesting. Figure 1 illustrates that the solar cell's absorption spectra overlaps well with high fractions of LED and CFL emission spectra. By tailoring the energy harvester to the intended environment, we are able to operate in very low power-availability environments.

Indoor lighting intensity in workspaces and industrial settings is internationally regulated and must be above a minimum threshold [2, 3]. This lighting level is adequate to power our WSN with a periodicity below one minute, a frequency applicable for intermittent sensing applications like environmental monitoring.

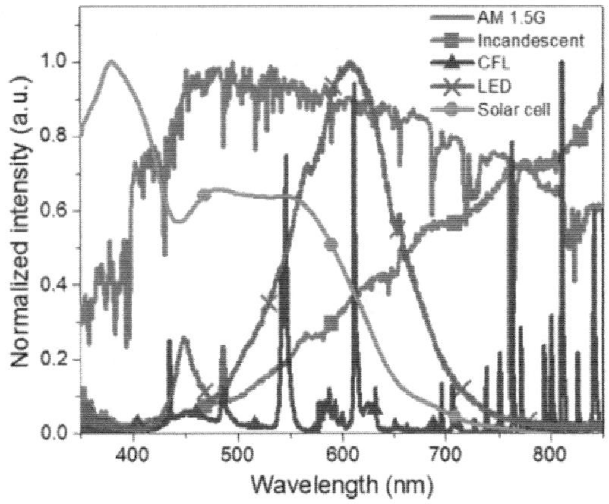

Figure 1. Absorption spectrum of the solar cell (pink), and normalized emission spectra of AM1.5G, incandescent, compact fluorescent, and LED sources.

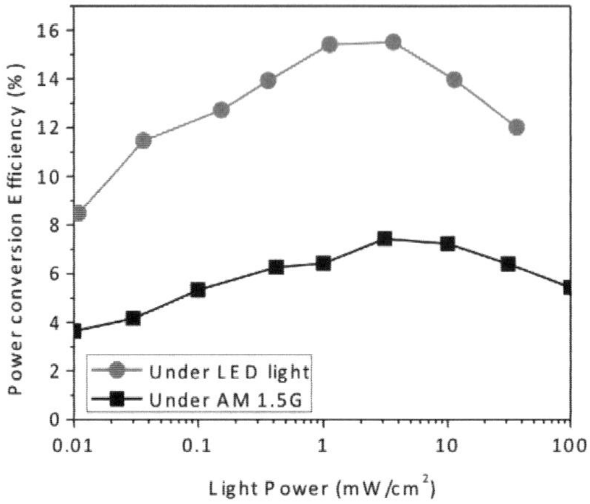

Figure 2. Evolution of PCE for the solar cell under LED and AM 1.5G sources.

2. Our System

2.1. Organic Solar Cells

Organic solar cells (OSC) currently achieve more than 10% power conversion efficiency (PCE) under 1-sun conditions [1, 2] and can be stable over several years [3, 4]. The materials involved in their fabrication can be completely printed on flexible substrates, allowing the use of large- area, high-throughput, low-cost manufacturing processes such as roll-to-roll printing. Here, OSCs based on the blend of poly[N-9'-heptadecanyl-2,7-carbazole-alt-5,5-(4,7-di-2-thienyl-2',1',3'-benzothiadiazole] (PCDTBT) and [6,6]-phenyl-C71-butyric acid methyl ester (PC71BM) (1:3.7 in ortho-dichlorobenzene with 5% dimethyl sulfoxide) are chosen. The architecture of the cell is shown in figure 3 and the fabrication followed the procedure described in [4]. We leverage spectrum-tailored OSCs to make the best use of indoor ambient energy, and are able to continuously harvest 20 $\mu W/cm^2$ [4].

For a given light source, the light power (P_{min}) at which a solar cell will stop performing well can be linked to its short circuit current (J_{sc}) under $P_{100} = 100\ mW/cm^2$ by: $J_{dark}/J_{sc} \approx P_{min}/P_{100}$. In order to optimize OSCs for indoor light, it is critical to maintain their dark current as low as possible. Control of the dark current can be achieved by varying the thickness of the top polyethylenimine ethoxylated (PEIE) layer. Figure 2 shows the evolution of the PCE of PCDTBT:PC71BM solar cell under an AM1.5G light source and a standard LED light source for varying light intensities. The cells have a dark current of 2 $\mu A/cm^2$ at a bias of -0.5 *volts*, while their J_{sc} under 1-sun is 12 mA/cm^2. As a result, they achieve more than 4% of power conversion efficiency under 10 $\mu W/cm^2$ of AM 1.5G light, from 5.4% under 100 $\mu W/cm^2$. Moreover, because of the adequacy of the absorption spectrum of PCDTBT:PC71BM, those solar cells under an LED light source demonstrate PCEs between 7% and 15%. Figure 4 shows the typical J-V curve of the PCDTBT based solar cells under typical indoor conditions: 340 *lux* of either AM1.5G light (corresponding to 310 $\mu W/cm^2$) and LED light (corresponding to 124 $\mu W/cm^2$ of power). The maximum power produced by this cell is 15.75 $\mu W/cm^2$ under LED light and 19.6 $\mu W/cm^2$ under AM 1.5G. Despite the almost three-fold difference in light power, the solar cell provides similar amounts of power under them.

2.2. Supercapacitor

To meet these power demands, energy harvested from the OSC accumulates in the high-power energy reservoir of a printed carbon supercapacitor [5, 6] for use during radio transmission. Figure 5 shows our reservoir is able to supply 9.8 mW/cm^2 of reservoir footprint, and therefore

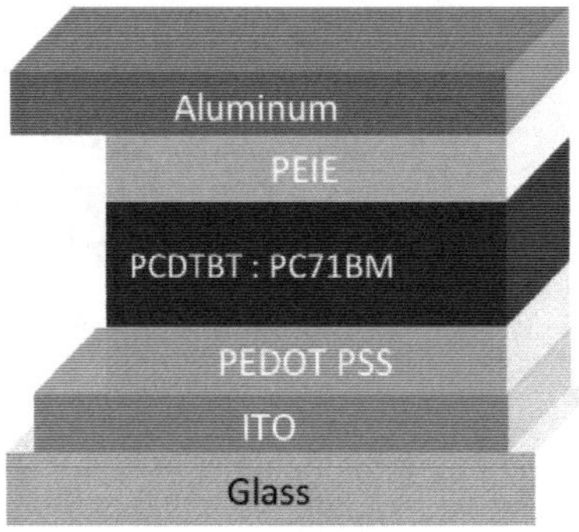

Figure 3. Architecture of the solar cell considered.

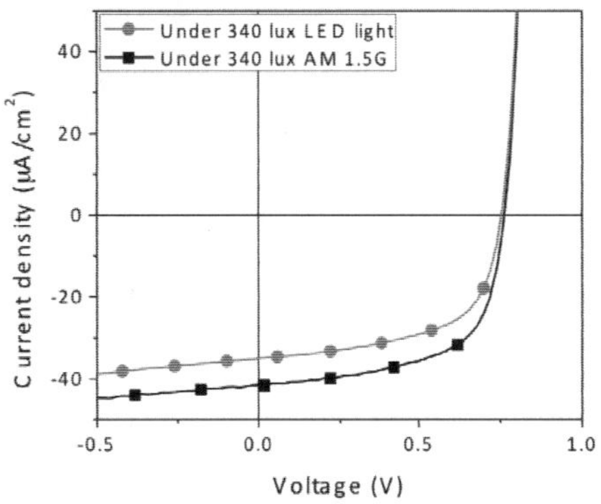

Figure 4. J-V curve of a solar cell under 340 *lux* from LED and AM1.5G source.

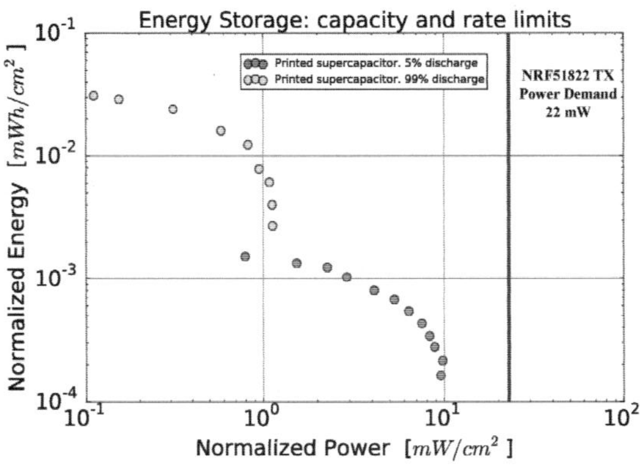

Figure 5. Ragone plot: power capabilities of printed supercapacitor. NRF51822 max power requirement is blue vertical line.

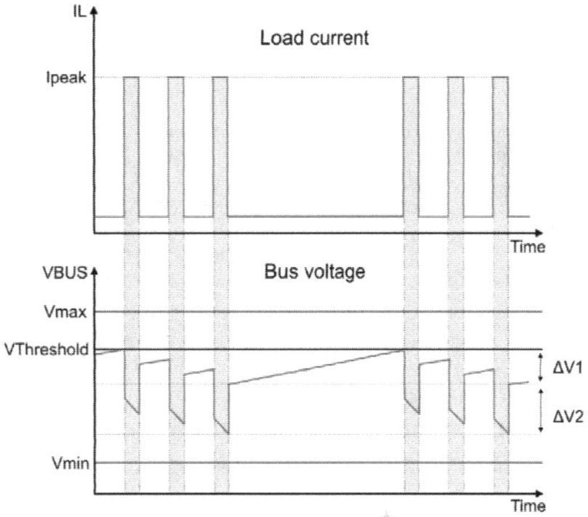

Figure 6. Acquisition and transmission is triggered once the threshold voltage is reached

a minimum supercapacitor footprint of 2.2 cm^2 is necessary to meet the radio power demands. It is important to avoid oversizing the supercapacitor in this low-power harvesting environment to mitigate leakage currents. An excessively leaky reservoir can quickly make the WSN inoperable in low-power harvesting environments. We leverage printed supercapacitor manufacturing because it enables the energy storage parameters to be tailored by controlling footprint geometry, layer thickness, and parallel or series inter-cell connections.

2.3. Radio and Sensor

We selected Nordic Semiconductors Bluetooth low energy radio NRF51822 for its low peak-power demands. This 2.4 gigahertz transceiver is seen transmitting 56 packets in figure 7 while drawing 6.9 mA peak current at 3.2 *volts*. The peak power of these transmissions is therefore 22 mW, nearly 1000 times the available power from the OSC. The radios sleep current is 3 μA. Additionally, we avoid energy-costly handshaking protocols by transmitting data using advertising packets.

NRF51822 can operate at voltages from 1.8 to 3.6 *volts*. To measure accelerations, we used the low power 10 bit accelerometer ADXL345 from Analog Devices sensing at 200 Hz with 145 μA current consumption. By leveraging these low operating and sleep currents, were able to best mitigate the costs of performing our sensing and transmission protocols.

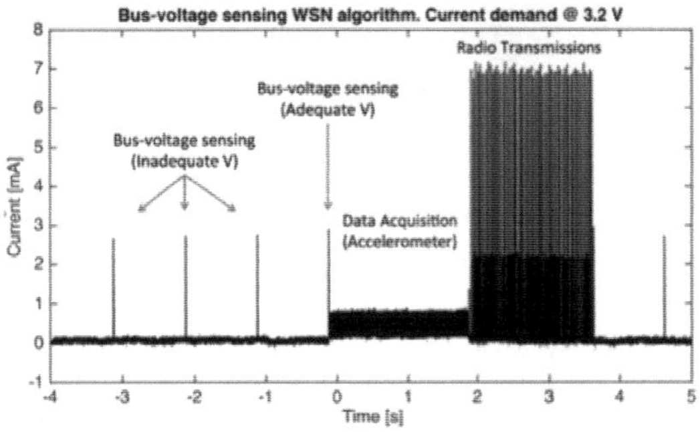

Figure 7. Voltage-dependent triggering protocol. Sensing and transmission only occurs when bus-voltage threshold is reached.

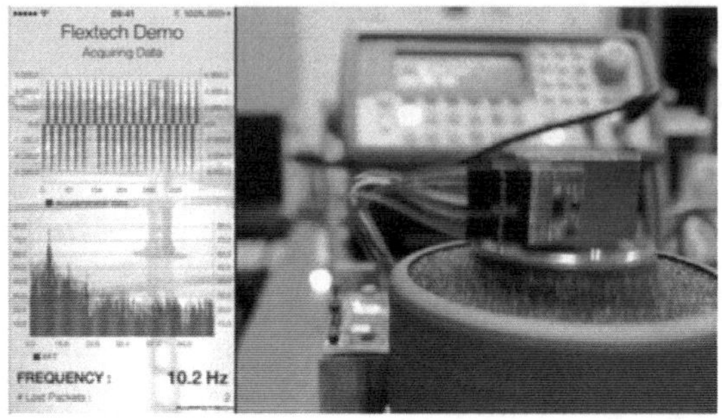

Figure 8. Frequency processing of data acquired from sensor mounted on voice coil oscillating at 10.2 Hz.

2.4. Triggering Algorithm

We implement a triggering algorithm wherein the WSN lies dormant until the energy stored in the supercapacitor is adequate to power sensing and transmission protocols. The WSN cold-starts when the radio's low voltage bound is reached at 1.8 *volts* and immediately enters a sleep state. The supply voltage continues to scale linearly as the supercapacitor accumulates charge. The WSN wakes periodically to measure the reservoir's bus voltage (V_{bus}). Once (V_{bus}) exceeds a threshold voltage, the WSN senses from the accelerometer and transmits via advertising packet. The triggering voltage must be above the cold start voltage plus the worst case voltage drop $\Delta V1 + \Delta V2$ seen in figure 6.

In our tests, the coulombic cost of acquisition and transmission are calculated to be 2.2 mC. This provides a $\Delta V1$ voltage drop of 55 mV for a supercapacitance of 40 mF. Our peak current of 6.9 mA produces a $\Delta V2$ ohmic drop of 280 mV. These voltage drops sum to 335 mV. To provide an adequate safety factor for our demonstration, we set the triggering voltage to 2.3 *volts*, 500 mV above our cold-start voltage.

3. Demonstration and Conclusion

We show a functional WSN in figure 8 that integrates a printable organic solar cell and printed supercapacitor to power the WSN in indoor light. Power is managed with a simple voltage-based triggering algorithm that only enables data acquisitions and transmission once the supercapacitor is adequately charged. This demonstration shows time-series acceleration data being transmitted to a base station where it is processed to view the frequency response.

3.1. Conclusion

With this work we demonstrate a wireless sensor node capable of running exclusively on indoor light, illustrating feasibility for nodes that can operate autonomously in low-light and indoor environments.

References

[1] Tuukkanen S, Välimäki M, Lehtimäki S, Vuorinen T and Lupo D 2016 *Nature Publishing Group* 1–9
[2] DiLaura D *et al.* 2011 *Illuminating Engineering Society of North America* **120**
[3] EN12464-1 *European Standard*
[4] Lechêne B P *et al.* 2016 *Nano Energy* **26** 631–640
[5] Cowell M *et al.* 2014 *Journal of Physics: Conference Series* **557** 012061–5
[6] Kötz R and Carlen M 2000 *Electrochimica Acta* **45** 2483–2498

Vibration Powered RF-Transponder for Sensing Low Frequency Motion Events

S K Gupta, V Pinrod, S Nadig, B Davaji and A Lal

SonicMEMS Laboratory, School of Electrical & Computer Engineering, Cornell University, Ithaca, NY, USA

E-mail: skg73@cornell.edu

Abstract. Vibration energy harvesting offers a pathway to developing battery-less sensing solutions to be deployed in wireless sensor network nodes. The integration of the energy harvesters require regulation by power conditioning and control circuitry that consume some of the energy generated, reducing the effective energy available for node function. By designing a unique 3D-printed plastic structure for low frequency sensitivity and mechanical switching, and a lateral PZT bimorph for capturing energy from environmental vibrations, we report a zero-power consumption RF-transponder capable of detecting and reporting motion events without a battery. We have successfully picked up wireless transmissions on an external receiver placed ~25cm away from the transponder, shaken at 0.75 g and 20 Hz. We have additionally demonstrated the ability to harvest energy from 5 Hz vibrations up to just under 150 Hz. When placed on an oil-based electric generator, which vibrates when operating, the RF-transponder has successfully picked up the differing harmonics to identify the mode of operation as the economy or regular power setting.

1. Introduction

There is a need for reliable, battery-less, motion event detectors. Such devices are useful for detecting early vibrations on pipes that can lead to sudden implosions, fall events from hospital patients, excessive vibration of operating machinery, etc. [1]. While the power available to a vibration energy harvester is inversely proportional to frequency [2], these harvesters must be large to remain effective at low frequencies [3]. Frequency up-conversion is one remedy [4]. Because the power harvested is typically small, the power consumption of associated circuitry for node integration and transmitters can be detrimental to device utility. Our group has demonstrated up-conversion energy harvesting from a radioisotope source with a self-reciprocating actuator and switch [5]. He *et al.* [6] reported a sensor architecture using a mechanical switching mechanism that eliminated associated circuitry. The electromagnetic harvester directly drove the antenna, primed only by the voltage output of the sensor. Jiang *et al.* [7] removed the need for a priming voltage with a 75x27x25 mm^3 design. Additionally, 3D-printing has been shown useful in housing energy harvesters [8]. We report a novel apparatus of a compact 3D-printed structure that exploits a mainly lateral serpentine design to excite the impulse response of a piezoelectric lateral bimorph, providing multiple resonant peaks at frequencies ranging from 5 Hz to under 150 Hz. Lateral structures presented in this work have the potential of providing a thin overall package needed for convenient attachment to two-dimensional surfaces. The mechanical switching mechanism eliminates the need for CMOS ICs needed in more conventional approaches.

 doi:10.1088/1742-6596/773/1/012034

2. Device Concept

A piezoelectric lateral bimorph [9] is anchored on top of a 3D-printed plastic structure (figure 1). The structure consists of a proof mass connected to a fixed frame through a serpentine spring providing it with substantial motion at low-frequencies from near zero to under 150 Hz. Figure 2 illustrates the sequence of events leading to RF transmission. When vibrating, the plastic proof mass and mounted bimorph collide with the 3D-printed structure's frame, inducing the impulse response of the piezoelectric energy harvester. The induced electrical charge is rectified and stored onto a capacitor. An integrated contact switch closes after the spring induces the impulse, triggering the LC tank oscillations and RF pulses that can be detected using a tuned radio. Since the contact switch only closes upon sufficient amplitude of motion, the switch gap can be adjusted so that it only switches upon a desired threshold of motion, or event. By using a sufficiently small capacitor, the time it takes for the contact to close will be significantly longer than the charge time of the capacitor. This will ensure the capacitor is fully charged before transmission.

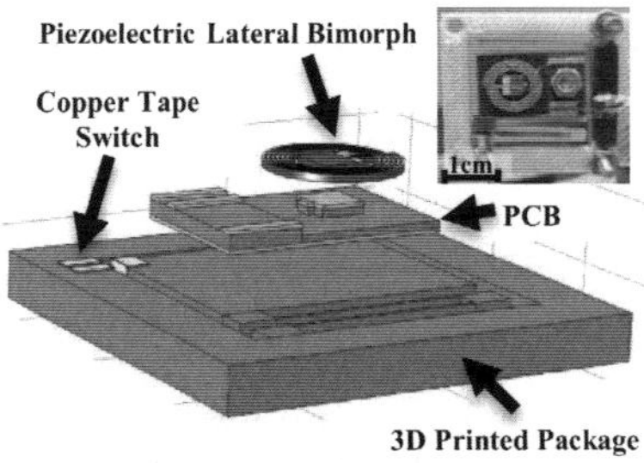

Figure 1. Assembly of transponder. Inset: Photograph of fabricated

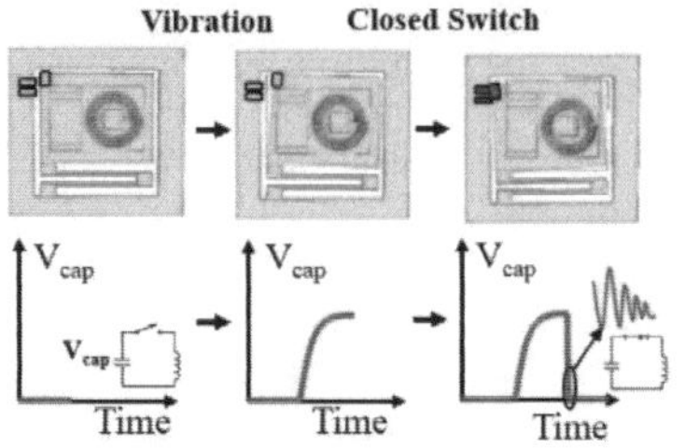

Figure 2. The capacitor charges as the proof-mass hits one side of the frame and discharges as it hits the other.

3. Device Model

Both electrical and mechanical processes define the system. The RF-transponder can be represented as the combination of the equivalent circuit and mechanical system model shown in figure 3.

3.1. Equivalent Circuit

The electrical model for a piezoelectric energy harvester (*EH*) has already been shown to be a current source with a shunt capacitor and resistor [10]. The current source, *i(x)*, is dependent on the displacement of the bimorph, *x*. The charge from the piezoelectric is then rectified using the configuration presented in [11]. The rectified charge is stored onto a capacitor (C_s) which connects to an inductor (*L*) through the contact switch (*SW*).

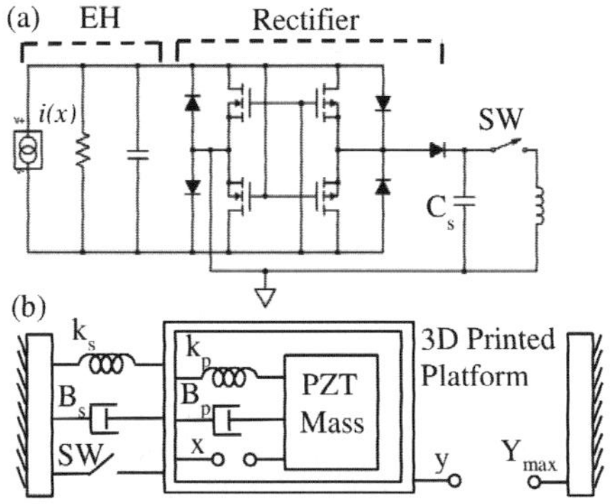

Figure 3. Electrical and mechanical model.

3.2. Mechanical Model

The RF-transponder consists of two spring-mass systems. The 3D-printed structure is anchored on one side with a spring constant (k_s), damping force (B_s) for the stage, electro-mechanical switch (*SW*) and mass (m_{total}). On top of the proof mass, the second spring-mass system is defined by the piezoelectric bimorph. It too has an associated spring constant (k_p), a damping force (B_p) and mass (m_p). The displacement of the piezoelectric and 3D-printed platform is represented by x and y respectively. The damping of the piezoelectric is the mechanism of energy extraction and so defines the upper boundary of power generation [12]. The 3D-printed platform is a driven harmonic oscillator with displacement amplitude ($|Y_0|$) in response to an acceleration (a_{in}). For maximum power transfer, we consider the case where the 3D-printed package is excited at its resonance (ω_0):

$$|Y_0| = a_{in}\left((\omega_0^2 - \omega^2)^2 + \frac{B_s^2}{m_{total}^2}\omega^2\right)^{-\frac{1}{2}} = \frac{a_{in}m_{total}}{B_s\omega_0} \quad (1)$$

The model has two cases defined by whether the proof mass hits the surrounding frame (gap of Y_{max}).

3.2.1. Case one: $|Y_0| < Y_{max}$. The maximum power achievable is calculated in [12] where the stage and drive resonance frequencies (ω_0) are much less than the bimorph structure resonant frequency (ω_n):

$$P_{max} = \frac{Y_0^2 \omega_0^5 m_p}{4\omega_n^2} \left| \frac{\omega_0^2}{\omega_n^2} - 1 \right|^{-1} \approx \frac{a_{in}^2 \omega_0^3 m_{total}^2 m_p}{4\omega_n^2 B_s^2}, \qquad a_{in} < \frac{Y_{max} B_s \omega_0}{m_{total}} \tag{2}$$

Here we have assumed that the damping of the PZT is much less than the damping of the stage.

3.2.2. Case two: $|Y_0| = Y_{max}$. In this case, the proof mass will hit the sidewall and induce the bimorph impulse response. The bimorph is a damped harmonic oscillator that experiences a jump in velocity due to its momentum before collision. We assume a perfectly elastic collision where all the momentum from the bimorph and plastic platform remain in their respective systems at time t_i where $y(t_i) = Y_{max}$:

$$\dot{x}(t_i) = \frac{a_{in} m_{total}}{B_s} \cos\left(\sin^{-1}\left(\frac{Y_{max} B_s \omega_0}{a_{in} m_{total}}\right)\right) \tag{3}$$

Because the piezoelectric is driven well below its resonant frequency before impact, $x(t_i) \approx 0$ relative to the displacement caused after collision. Thus, the solution to the damped harmonic oscillator is:

$$x(t) = \frac{\dot{x}(t_i)}{\omega_1} \exp\left(-\frac{B_p(t-t_i)}{2m_p}\right) \sin(\omega_1(t-t_i)), \qquad \omega_1 = \left(\omega_n^2 - \frac{B_p^2}{4m_p^2}\right)^{\frac{1}{2}}, \qquad t > t_i \tag{4}$$

The maximum mechanical power harvestable by the piezoelectric at time t after the impulse is then:

$$P(t) = B_p \dot{x}^2(t), \qquad P(t_i) = B_p \frac{a_{in}^2 m_{total}^2}{B_s^2} \cos^2\left(\sin^{-1}\left(\frac{Y_{max} B_s \omega_0}{a_{in} m_{total}}\right)\right) \tag{5}$$

Figure 4 shows the maximum mechanical power of the bimorph for a device with the net effective mass 1 gram for different accelerations. The power is determined by the ratio of the bimorph damping factor over the 3D-printed structure's damping factor squared. We need an additional derivation to convert this mechanical power into electrical power. The energy harvested can be optimized by minimizing the damping of the 3D-printed structure relative to the lateral bimorph.

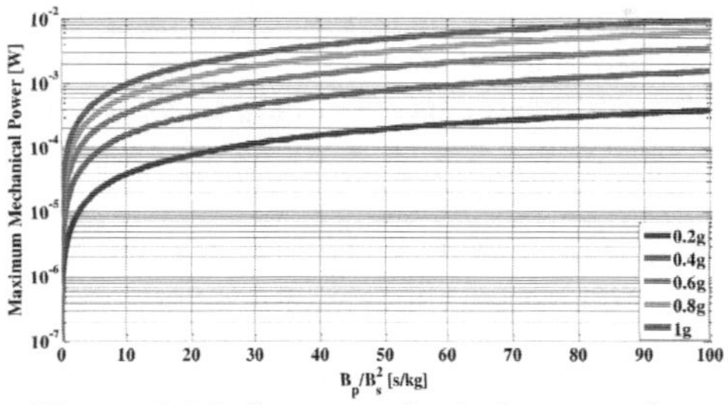

Figure 4. Maximum mechanical power of bimorph over different accelerations as a function of damping ratio. m_{total} = 1 gram.

4. Device Fabrication and Testing

The PZT lateral bimorph [9] with two-axis sensitivity uses a compact spiral design shown in figure 5 to decrease the required area and increase in-plane symmetry. An additional groove cut is added along the middle of the spiral bimorph to increase out-of-plane sensitivity, providing full three-axis sensitivity for all angle energy harvesting. The bimorph is fabricated by selective scanning of the LPFK Protolaser U, allowing a simple, single step process for achieving a smooth circular geometry that is otherwise challenging with other processes. The plastic spring-mass of thickness 3 mm is 3D-printed from a MakerBot Desktop. The serpentine spring segments have lateral

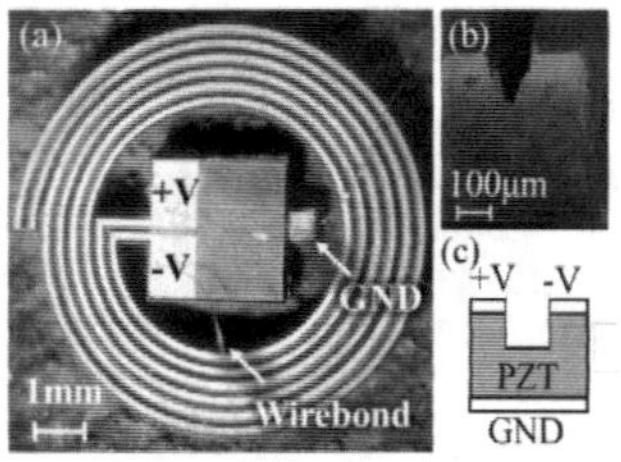

Figure 5. Photograph, SEM and schematic of bimorph and cross section

dimension 15.5x0.3 mm^2 and 2x2.5 mm^2. The proof mass has an area of 18x14 mm^2 and the inertial frame has inner dimensions of 20x20 mm^2 with a 4 mm width. Holes with a 2 mm diameter are included for mounting purposes.

The 3D-printed spring-mass structure has a resonance frequency of 24 Hz, a Q factor of 17 and spring constant of ~66 N/m. The PZT bimorph has a resonance frequency of 416 Hz, a Q factor of 173.5, an output capacitance of ~0.2nF and spring constant of ~653 N/m. This gives a maximum achievable power of ~1.9 nW just before collision and ~11 mW/g^2 after collision when excited at resonance.

The RF-transponder was mounted on a tilt to a shaker table to measure the voltage induced on the bimorph for accelerations. A small magnetic ball mass was added to the proof mass to further increase the output of the system. The results are shown in figure 6a. It should be noted that the 1 MΩ load was not optimized for maximum power transfer. Figure 6b shows a wireless transmission in the 20-100 MHZ range picked up by an LC tank 9 inches from the transponder. Figure 7 shows the bimorph captured different motion signatures when placed on a generator.

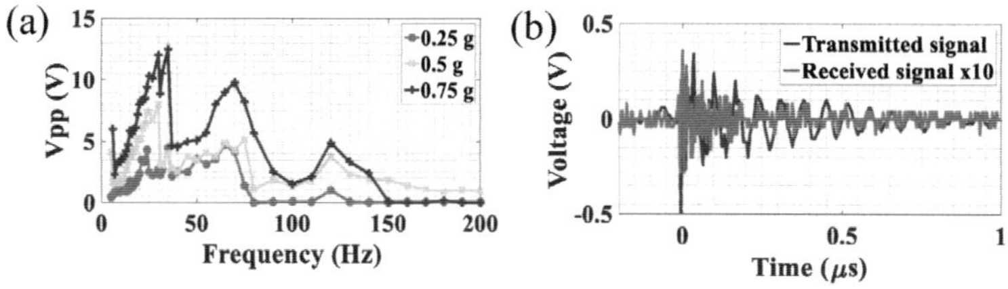

Figure 6. Frequency response of bimorph and transmission of transponder excited at 20 Hz and 0.75g by a shaker table.

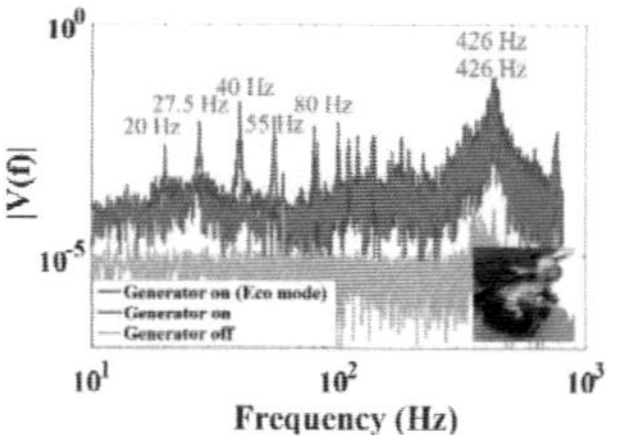

Figure 7. FFT of the bimorph output on the generator with modes of operation.

5. Conclusions

A RF-transponder has been built that transmits to an external RF receiver upon the onset of a motion event which is greater than a set motion threshold, without the use of any power conditioning circuitry, battery, control circuitry or external sensors. While a PZT lateral bimorph is used for capturing energy from vibrations, it can be swapped out with any type of energy harvesting device capable of converting the energy from impact into electrical energy. This device is applicable in wearables, industrial machinery, medical patient monitoring, intruder detection and automotive sensing.

6. Acknowledgements

This work was performed in part at the Cornell NanoScale Facility, a member of the National Nanotechnology Infrastructure Network, which is supported by the National Science Foundation (Grant ECCS-0335765). The SEM in figure 4b was taken by Leanna Pancoast. Funding for this project was provided by the DARPA NZERO and PASCAL programs.

References

[1] Akyildiz I F, Su W, Sankarasubramaniam Y and Cayirci E 2002 **38** *Computer Networks* 393-422
[2] Roundy S, Wright P K and Rabaey J M 2004 *Energy Scavenging for Wireless Sensor Networks* (Boston: Kluwer Academic Publishers) pp 31-39
[3] Roundy S 2005 **16** *JIMSS* 809-23
[4] Ashraf K, Khir M H, Dennis J O and Baharuddin Z 2012 *Proc. PowerMEMS* (Atlanta, USA) pp 416-9
[5] Lal A, Duggirala R and Li H 2005 **4(1)** *IEEE Pervasive Computing* 53-61
Tin S and Lal A 2009 *Proc. IEEE IEDM* (Baltimore, USA) pp 539-42
[6] He C, Kiziroglou M E, Yates D C and Yeatman E M 2011 **11** *IEEE Sensors* 3437-45
[7] Jiang H, Kiziroglou M E, Yates D C and Yeatman E M 2015 **2** *IEEE IoT* 5-13
[8] Constantinou P and Roy S 2015 *Proc. PowerMEMS* (Boston, USA)
[9] Nadig S, Ardanuc S and Lal A 2015 *Proc. IEEE MEMS* (Estiril, Portugal) pp 1129-32
[10] Kiran S, Selvakumar D, Mervin J and Pasupuleti H 2015 *Proc. ICCEREC* (Bandung, Indonesia) pp 63-9
[11] Romani A, Filippi M, Dini M and Tartagni M 2015 **12** *ACM JETC* 7:1-17
[12] Mitcheson P D, Green T C, Yeatman E M and Holmes A S 2004 **13** *J MEMS* 429-40

Tunable Ultrasonic Energy Harvesting for Implantable Biosensors and Medical Devices

M Pellegrino[1,2,3], B E Eovino[1], L Beker[1], T Bourouina[2] and L Lin[1]

[1]University of California, Berkeley, California, USA,
[2]Univeristé Paris-Est, Marne-la-Vallée, France
[3]Politecnico di Milano, Milan, Italy

E-mail: massimopellegrino@berkeley.edu

Abstract. This work reports a tunable ultrasonic energy harvesting (UEH) device capable of high power output and/or large bandwidth based on concentric piezoelectric ring-shaped structures. Two different designs are presented: (1) the single ring-shaped UEH (r-UEH), and (2) concentric r-UEHs. Concentric r-UEHs can save space and therefore can provide benefits in powering low-power implantable biosensors and medical devices. This paper presents results of simulation studies and initial experiments of a single r-UEH.

1. Introduction

With the recent development of the Internet of Things (IOT), wireless communication technology, and wireless sensors node (WSN) architectures, the lack of long-term batteries has become a practical problem for microsystems operating in hazardous and inaccessible areas such as the human body. In this regard, finding a replacement for the battery is one of the main issues of modern biosensors and implantable medical devices (IMDs) like implantable cardiac defibrillators, deep brain stimulators or pacemakers. Depending on how active the devices are, patients with IMDs currently need to undergo a surgical procedure for battery replacement every few years, increasing the medical risk and costs. As such, recent research efforts on wireless power transfer technology have received great attention and have potential to mitigate this problem.

Researchers have focused on RF and magnetic based wireless power transfer technologies due to their ease of application. However, because of their short transfer distances and low efficiencies caused by heavy attenuation in water (e.g., the human body), these approaches are limited when the human body is considered as the transfer medium. Furthermore, electromagnetic coupling with other devices is another concern that needs attention [1]. For deep IMDs, the ultrasonic frequency range is an advantageous alternative. On the other hand, the main challenge to using ultrasound energy harvesters (UEHs) is the selection of the frequency, which should carefully guide the transducer design [2]. This work reports a tunable acoustic energy harvesting (AEH) device capable of high power output and/or large bandwidth based on concentric piezoelectric ring-shaped structures as shown in Fig. 1. The silicon substrate acts as a mechanical anchor so that the topographic geometry of each ring-shaped structure can be represented by a mean radius r_0 and width w.

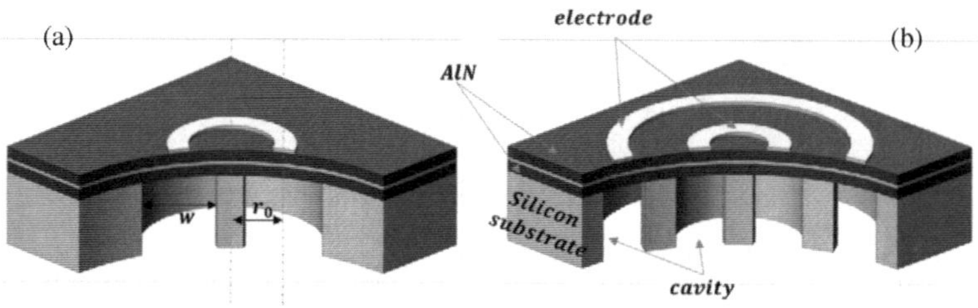

Figure 1: Model representation of: (a) single r-UEH, (b) concentric r-UEHs, where r_0 is the mean radius and w the width of the ring.

2. Ring-shape AEH Designs

Two different designs are presented: (1) the single ring-shaped UEH (r-UEH), and (2) concentric r-UEHs, see Fig. 1. The former shows the working principle of the r-UEH, while the latter is an extension that aims to improve the output power or operation bandwidth with multiple individual r-UEHs that can harvest either the same or different targeted frequencies. Contrary to conventional circular UEHs, concentric r-UEHs can save space and target several frequencies, thereby providing critical benefits in powering low-power implantable biosensors and devices.
Taking advantage of the fabrication process used in our group for other projects [3-4], the fabricated devices are shown in Fig. 2

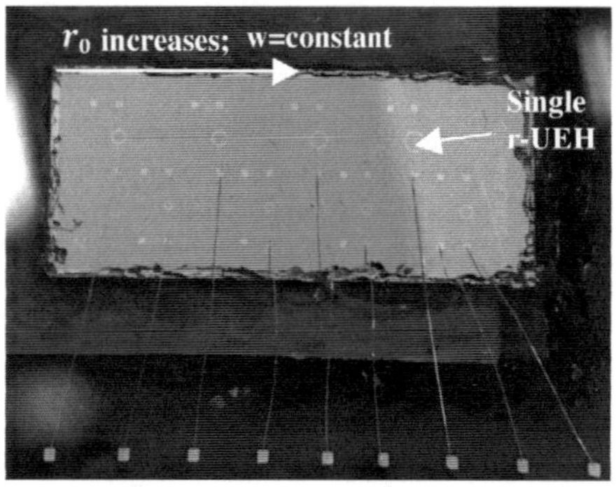

Figure 2: eight fabricated r-UEHs in the same chip mounted in a PCB, with wire bonding. In the same chip, the ring width (w) of the r-UEHs is kept constant and the mean radius (r_0) increases from left to right, bottom to top.

2.1. Key parameters and simulations of r-UEHs

One advantage of the r-UEHs is that the resonance frequency is not significantly dependent on the mean radius r_0 and rather only depends on the width of the ring w, as shown by simulation in Fig. 3. Fig. 3a shows the I_{sc} for fixed w as we change r_0, resulting in a mere 9% change in resonance frequency as r_0 is increased by 3x; Fig. 3b presents the effect of changing w at constant r_0=250μm, showing the dependence of the resonance frequency on the ring width. This allows for the design of concentric r-UEHs with the same resonance frequencies in order to achieve high power output, or with different resonance frequencies (by changing w) to achieve a large bandwidth. Moreover, it is trivial to target lower frequencies by increasing the width of the rings, however, other parameters such as the thickness or material composition of the diaphragm could also be changed. In this work, we assumed the latter parameters, among others, as constraints in order to match the devices fabricated from previous works in our group [4].

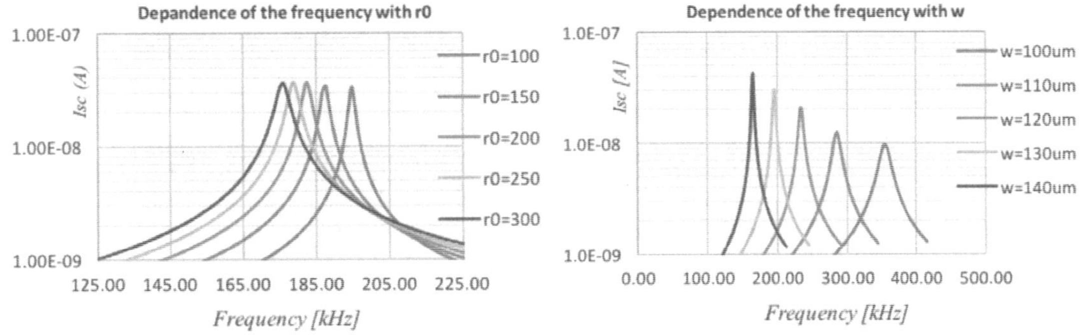

Figure 3: Short circuit current (Isc) vs. frequency for water-coupled devices. (a) effect of changing r_0 at constant w=135μm, showing less than a 9% change in resonance frequency as r_0 is increased by 3x; (b) effect of changing w at constant r_0=250μm, showing the dependence of the resonance frequency on the ring width.

2.1.1. Power output simulation. As a representative case, COMSOL simulations in Fig. 4 indicate that the normalized power output (P_o) is P_o = 2.4nW/(Pa*cm^2), which shows promising results and guidelines for the reported UEHs. As an example, in the case of pacemakers, it is expected to have ~3-4 cm^2 available for the power systems and several Pascal of applied pressure, suggesting that the power output can be on the order of μW. This is taking into account the concentric r-UEHs with the assumption that all of the 3-4 cm^2 of available space is used.

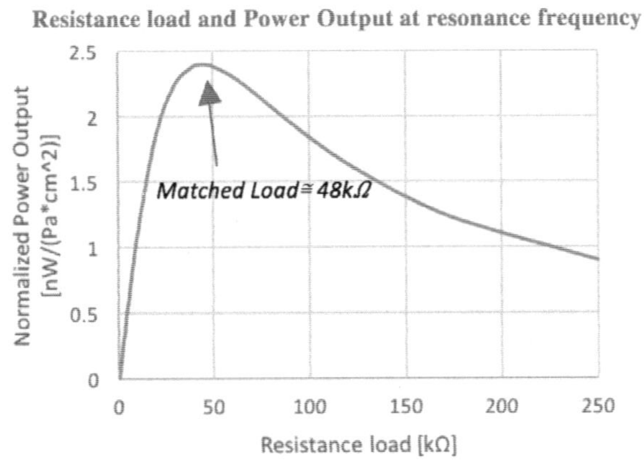

Figure 4: Power output for a single r-UEH with r_0=100μm and w=150μm operating at its resonance frequency with a variable external resistance load.

3. Experimental setup and single r-UEHs results

Fig. 5 shows the experimental setup and the first promising experimental results of a single r-UEH with r_0=200μm and w=135μm, operating in "reverse-mode" by driving the r-UEH with an applied 500mV AC voltage peak to peak and reading V_{pp}=220mV from a commercial ultrasonic transducer (Noliac UZ250). It is believed that due to acoustic reciprocity, these results should be indicative of the ability of r-UEH to receive ultrasonic energy as well; however, an optimized readout circuit is needed to acquire a reliable signal from the r-UEHs.

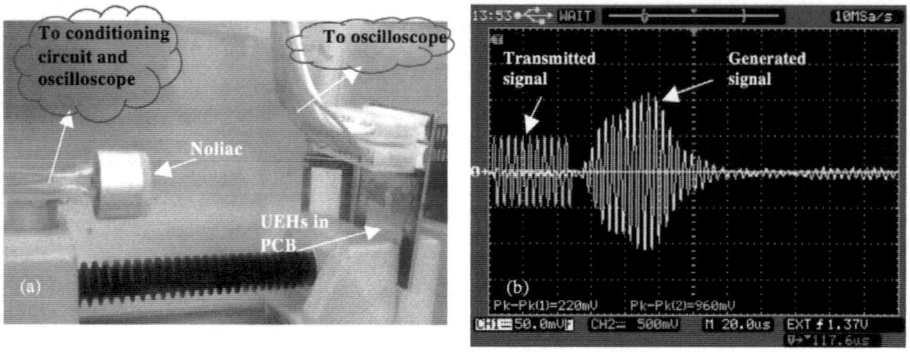

Figure 5: (a) Experimental setup and (b) promising experimental results of single r-UEH at r_0=250μm and w=135μm at 240kHz, in "reverse-mode" operation if applied 500mV AC voltage and reading the voltage, which is proportional to the pressure generated due to the reciprocity in the transduction of the energy

4. Discussions

In this work we present the design, simulations, and experimental results of a ring-shaped acoustic energy harvester (r-UEH), which is an advantageous design as the mean radius of the rings can be changed as needed without significantly affecting the resonance frequency. As a consequence, it is possible to design concentric r-UEHs in order to save space and therefore provide benefits in powering low-power implantable biosensors and medical devices. Depending on the applications, it is also possible to harvest different frequencies by using concentric r-UEHs with different widths in order to create a large bandwidth array. In the case of implantable medical devices, because ultrasonic waves are safe for the human body, one could conceivably think to reduce the dimensions of the battery - and therefore the battery life - and charge the devices more often (e.g., once a year) using r-UEHs for wireless power transfer.

References

[1] Y. Zhu, S. O. R. Moheimani, and M. R. Yuce, "*Ultrasonic Energy Transmission and Conversion Using 2-D MEMS Resonator*", IEEE Electron device letters, Vol. 31, No. 4, April 2010

[2] A. Denisov and E. Yeatman, "*Ultrasonic vs. inductive power delivery for miniature biomedical implants. In Body Sensor Networks (BSN)*", 2010 International Conference on, pages 84 –89, June 2010

[3] S. Akhbari, F. Sammoura, C. Yang, M. Mahmoud, N. Aqab and L. Lin, "*Bimorph pMUT with dual electrodes*", 28yh IEEE International Conference on Micro Electro Mechanical Systems (MEMS), pp.928-931, Estroil, Portugal 2015

[4] B.E. Eovino, S. Akhbari, and L. Lin, "*Ring-Shaped piezoelectric micromachined ultrasonic transducers (PMUT) with increased pressure generation*", in Hilton Head 2016

ZnO photoluminescent quantum dots with down-shifting effect applied in solar cells.

A Zazueta-Raynaud[1,2], J E Pelayo-Ceja[1,3], R Lopez-Delgado[1,2] and A Ayon[1]

[a] University of Texas at San Antonio, Dept. of Physics and Astronomy, MEMS Research Lab, One UTSA Circle, San Antonio, TX 78249,USA.
[b] Universidad de Sonora, Departamento de Física, Blvd. Luis Encinas y Rosales, Col. Centro, Hermosillo, Sonora, México. CP 83000
[c] Universidad de Guadalajara, Centro de Ciencias Exactas e ingenierías, Blvd. Gral. Marcelino García Barragán 1421, Olímpica, 44430 Guadalajara, Jal., México.

aldo.zazuetaa@gmail.com

Abstract. We report the synthesis of Zinc Oxide (ZnO) quantum dots (QDs) and their influence on the power conversion efficiency of photovoltaic devices. With an excitation wavelength of 340 nm the down-shifted emission peaks were observed to be located at 510 and 540 nm for colloidal solutions with pH values of 10 and 12, respectively. The largest PCE variation was observed to increase from 14.60% to 15.49% when dispersing the QDs extracted from a 4 mL colloidal solution that were subsequently dispersed in PMMA. This represents an improvement of ~6.1%.

1. Introduction

The solar spectrum comprises wavelengths extending form the infrared to ultraviolet, however, in silicon electron-hole pairs are only produced by photons that fall within the visible range [1-2], obviating the need to extend the range of photons producing electron-hole pairs. Specifically, Zinc Oxide (ZnO) quantum dots have the capability to absorb photons in the UV region of the solar spectrum and subsequently emitting a photon within the region of interest for silicon photovoltaic structures. The colloidal ZnO quantum dots were synthesized in ethanol-based solutions obtaining different sized nanoparticles, i.e., emitting different wavelengths. The emission wavelength of the synthesized ZnO quantum dots can, in fact, be tuned by adjusting the pH value of the corresponding solution. Subsequently, the quantum dots were collected by adding hexane to the ethanol solution to promote their sedimentation [3]. Different concentrations of zinc oxide quantum dots were collected and dispersed in Poly methyl methacrylate (PMMA) to be deployed on the window side of solar cells using a standard spin coating approach.

2. Experimental details.

2.1. Synthesis of ZnO quantum dots.

Colloidal Zinc Oxide Quantum Dots (ZnO QDs) were synthesized employing a controlled-precipitation method [4]. In a typical synthesis, 0.02M zinc acetate solution was made by dissolving 0.256 gr of zinc acetate (99.99%, Sigma-Aldrich) in 70 mL of pure ethanol, and 0.01M lithium hydroxide solution was prepared separately by dissolving 0.125 gr of LiOH ($\geq$ 98.0% Sigma-Aldrich) in 30 mL of pure ethanol. The reaction was carried out at room temperature by dropwise addition of LiOH solution to zinc acetate solution under constant stirring until the pH values of 10 and 12 were reached. Once the sought pH values were obtained, the solution was placed in an ultrasonic bath for 3 hours and it was observed to be transparent. In order to elucidate the process of reaction in the synthesis, a chemical reaction model was proposed where the interaction between zinc acetate and ethanol carries an esterification reaction that promotes the formation of water, ester and also the growth of nanoparticles of ZnO, according to the following reaction [5].

$$Zn(CH_3COO)_2 + 2R\text{-}OH \rightarrow ZnO + 2CH_3\text{-}CO\text{-}R + H_2O$$

The size of the produced quantum dots is determined by the pH of the solution during the reaction synthesis, as previously discussed.

2.2. Collection of ZnO quantum dots.

Once the reaction was brought to completion, the synthesized quantum dots were purified by extracting the unreacted products and collecting the nanoparticles from the ZnO solution. Several groups have reported the utilization of organic "nonsolvents" such as heptane and hexane to promote the precipitation of the particles in the solvent [3, 6]. However, in this exercise hexane was added to the ZnO QD solution in a volume ratio of 3:1 and allowed to stand for 24 hours. Subsequently, the supernatant was removed and the precipitated ZnO QDs were washed three times in ethanol and redispersed also in ethanol for storage.

2.3. Application of ZnO quantum dots over solar cells.

After collecting the quantum dots, several QD samples with different concentrations of the purified ZnO QDs were prepared. The purified samples were centrifuged in order to promote the precipitation of the ZnO particles suspended in ethanol, after that, the solvent was removed leaving only ZnO quantum dots precipitated at the bottom of the vial. In order to incorporate the synthesized QDs to solar cells, they needed to be dispersed in a polymer matrix. Polymethylmethacrylate (495 PMMA A2 from Microchem) was selected for its properties of transparency, resistance to ultraviolet light, and its reported low brittleness. Thus, 1 mL of PMMA was added to the vials with the precipitated QDs. Polycrystalline silicon, commercial solar cells, with dimensions of 52mm x 38mm and a thickness of 200μm were employed to quantify the influence of the synthesized ZnO QDs. The PMMA+QD films were spin cast on the window side of the aforementioned solar cell. The spin coating was carried out starting with 300 rpm for 10 secs, followed by 4,000 rpm for 45 seconds [7]. After spin coating the solar cells with PMMA + QDs, they were placed on a hot plate at 200 °C for 60 secs to promote the evaporation of the solvents within the PMMA. Device characterization was carried out before and after the deployment of the aforementioned PMMA+QD film.

3. Results and discussions.

During ZnO quantum dot synthesis, the pH of the solution was adjusted to different values before submitting to reaction process in order to obtain different sizes as well as different luminescent and absorption effects. The pH values selected during synthesis were approximately 10 and 12.

3.1. Photoluminescence characterization.

Employing an excitation wavelength of 340 nm, the produced ZnO QDs exhibited photoluminescence spectra extending from approximately 420 nm to 650 nm and with luminescence peaks at 510 and 540 nm for pH values of 12 and 10, respectively. Figure 1-a shows the normalized spectra –obtained by dividing the collected spectra obtained by the maximum of their excitonic emission- obtained in luminescence measurements for different characterized samples. It can be observed that the maximum luminescent peaks in the samples have a wavelength difference between them of 30 nm, centred around 525 nm, evincing that the luminescent characteristics of the produced particles, as well as their size, are directly linked to the reaction conditions related with the rate of pH in the solution during the synthesis and the luminescence effects can be tunable. The yellow-greenish and blue-cyan photon emission for the pH values of 10 and 12, respectively (see figure. 1-b) [8], were observed using a handheld UV lamp light over the synthesized samples.

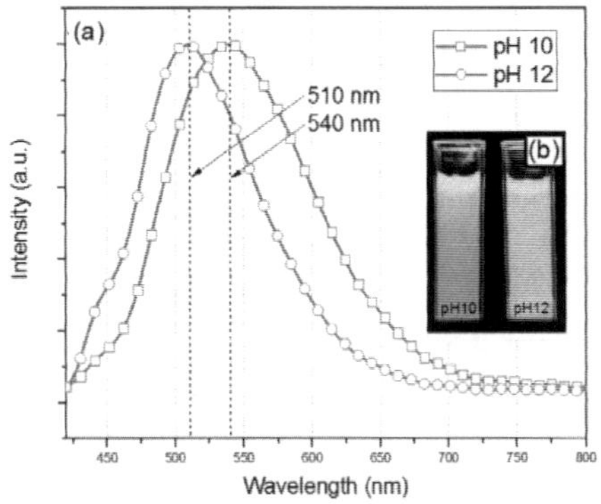

Figure 1. (a) Synthesized ZnO QDs photoluminescence spectra with an excitation wavelength of 340 nm.
(b) Visible light emission of samples with pH values of 10 and 12 when the samples are exposed to UV light.

3.2. Absorption characterization.

The UV/Vis absorption spectra were characterized with a UV-VIS-NIR Cary 5000 spectrometer. Absorption measurements of ZnO quantum dots in ethanol exhibit a spectrum starting around ~250 nm and extending up to ~400 nm, except for the sample made with a pH value of 12, which shows a gradual downward curve to ~800 nm (see figure 2). This behaviour may be attributed to the turbidity of the solution at the end of the synthesis process in contrast to the transparency observed on the samples with pH values of 8 and 10. The solutions of ZnO quantum dots show their spectra centred around ~ 300 nm and an approximate width of ±50 nm consistent with values reported elsewhere [9].

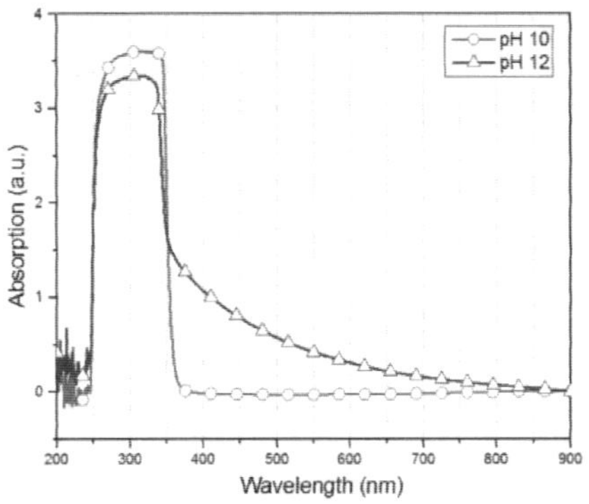

Figure 2. Absorption spectra for ZnO quantum dots synthesized with different pH values

3.3. Solar cell characterization.

To measure the performance of the solar cells, we used a solar simulator (Oriel Sol2A) to evaluate the major values of the photovoltaic device related with open circuit voltage (V_{OC}), short circuit current (J_{SC}), Fill Factor (FF%) and Power Conversion Efficiency (PCE%). The collected measurements are shown in Table 1. [10-12]. All measurements were made before and after deploying the synthesized quantum dots on the window side of the solar cells with a film of PMMA + ZnO QDs obtained by spin casting.

Table 1. Summary of measured performance of photovoltaic devices before and after deploying PMMA layers with different ZnO QD concentrations.

	2 mL of ZnO QDs		4 mL of ZnO QDs		6 mL of ZnO QDs	
	Before	After	Before	After	Before	After
Voc (V)	0.62	0.62	0.63	0.62	0.62	0.63
Jsc (mA/cm^2)	36.90	36.59	37.03	37.91	36.75	37.19
FF (%)	67.06	67.58	63.10	65.79	65.43	64.27
PCE (%)	15.40	15.42	14.60	15.49	14.87	14.94

4. Conclusions

The size and, therefore, the emission wavelength of Zinc Oxide quantum dots can be tuned by varying the pH value of the solution during synthesis. The largest PCE variation was observed to increase from 14.60% to 15.49% when dispersing the QDs extracted from a 4 mL colloidal solution that were subsequently dispersed in PMMA. This represents an improvement of ~6.1%.

Acknowledgments

The authors would like to acknowledge the U.S. Army Research Office (Grant W911NF-13-1-0110), CONACYT, the Physics and Astronomy Department of the University of Texas at San Antonio and the Physics Department of University of Sonora, for the financial support provided for this research project.

References

[1] Carlson D E and Wronski C R, 1976 *Applied Physics Letters*. **28**(11): p. 671-673.
[2] Semonin O E, Luther J M, Choi S, Chen H-Y, Gao J, Nozik A J and Beard M C, 2011 *Science*. **334**(6062): p. 1530-1533.
[3] Meulenkamp E A, 1998 *The Journal of Physical Chemistry B*. **102**(29): p. 5566-5572.
[4] Rodrıguez-Paez J, Caballero A, Villegas M, Moure C, Duran P and Fernandez J, 2001 *Journal of the European Ceramic Society*. **21**(7): p. 925-930.
[5] Tonto P, Mekasuwandumrong O, Phatanasri S, Pavarajarn V and Praserthdam P, 2008 *Ceramics International*. **34**(1): p. 57-62.
[6] Sun D, Wong M, Sun L, Li Y, Miyatake N and Sue H-J, 2007 *Journal of sol-gel science and technology*. **43**(2): p. 237-243.
[7] Bajuri S N M, Halim N H A, Nor M N M and Hashim U. *PMMA characterization and optimization for nano structure formation*. in *Proc. of 1st National Conference on Electronic Design*. 2005.
[8] Zhou X, Xiao X, Xu J, Cai G, Ren F and Jiang C, 2011 *EPL (Europhysics Letters)*. **93**(5): p. 57009.
[9] Rodríguez-Gattorno G, Santiago-Jacinto P, Rendon-Vazquez L, Németh J, Dékány I and Díaz D, 2003 *The Journal of Physical Chemistry B*. **107**(46): p. 12597-12604.
[10] You J, Dou L, Yoshimura K, Kato T, Ohya K, Moriarty T, Emery K, Chen C-C, Gao J and Li G, 2013 *Nature communications*. **4**: p. 1446.
[11] He Z, Zhong C, Su S, Xu M, Wu H and Cao Y, 2012 *Nature Photonics*. **6**(9): p. 591-595.
[12] Sun Y, Seo J H, Takacs C J, Seifter J and Heeger A J, 2011 *Advanced Materials*. **23**(14): p. 1679-1683.

Fabrication of an Omnidirectional 2D Photonic Crystal Emitter for Thermophotovoltaics

Veronika Stelmakh[1,2], Walker R. Chan[1], Michael Ghebrebrhan[3], Marin Soljacic[4], John D. Joannopoulos[1,4] and Ivan Celanovic[1].

[1] Institute for Soldier Nanotechnologies, Massachusetts Institute of Technology, 77 Massachusetts Ave., Cambridge, MA 02139, USA

[2] Department of Electrical Engineering, Massachusetts Institute of Technology, 77 Massachusetts Ave., Cambridge, Massachusetts 02139, USA

[3] U.S. Army Natick Soldier Research, Development, and Engineering Center, 1 4th Ave., Natick, Massachusetts 01760, USA

[4] Department of Physics, Massachusetts Institute of Technology, 77 Massachusetts Ave., Cambridge, Massachusetts 02139, USA

E-mail: stelmakh@mit.edu

Abstract. In a thermophotovoltaic (TPV) system, a heat source brings an emitter to incandescence and the spectrally confined thermal radiation is converted to electricity by a low-bandgap photovoltaic (PV) cell. Efficiency is dominated by the emitter's ratio of in-band emissivity (convertible by the PV cell) to out-of-band emissivity (inconvertible). Two-dimensional photonic crystals (PhCs) offer high in-band emissivity and low out-of-band emissivity at normal incidence, but have reduced in-band emissivity off-normal. According to Lambert's law, most thermal radiation occurs off-normal. An omnidirectional PhC capable of high in-band emissivity at all angles would increase total in-band power by 55% at 1200°C. In this work, we present the first experimental demonstration an omnidirectional hafnia-filled 2D tantalum PhC emitter suitable for TPV applications such as combustion, radioisotope, and solar TPV. Dielectric filling improved the hemispherical performance without sacrificing stability or ease of fabrication. The numerical simulations, fabrication processes, and optical and thermal characterizations of the PhC are presented in this paper.

The increasing power demand of communication equipment, sensors, micro-robotic platforms, and portable electronics has driven recent interest in micro- and millimeter-scale thermophotovoltaic (TPV) generators. Thermophotovoltaic energy conversion is a solid-state high-temperature heat-to-electricity conversion scheme with no moving parts that allows for scalable energy production with high specific energy from a variety of heat sources. In TPV systems, thermal radiation from a heat source at high temperature drives a suitable low-bandgap photovoltaic cell (PV). The heat can be produced by hydrocarbon combustion, ideal for lightweight, high specific energy portable power sources [1]; by radioisotope decay, ideal for space missions and remote missions requiring power sources with long lifetimes and low maintenance [2]; or by concentrated solar radiation [3]. A selective emitter between the source and the cell enables high heat-to-electricity conversion efficiency by enhancing in-band (convertible) radiation and suppressing out-of-band (nonconvertible) radiation that would otherwise be wasted. TPV has only recently been enabled by advances in two areas: low-bandgap III-V semiconductors [4] and high-temperature photonic crystal selective emitters [5]. A viable TPV selective emitter requires: high temperature stability for a long operational lifetime, good optical performance, and a simple fabrication process capable of producing large area samples. Most of the selective emitters currently available (fabricated as 1D, 2D, and 3D photonic crystals, metamaterials, as well as from natural materials) only meet one or two of

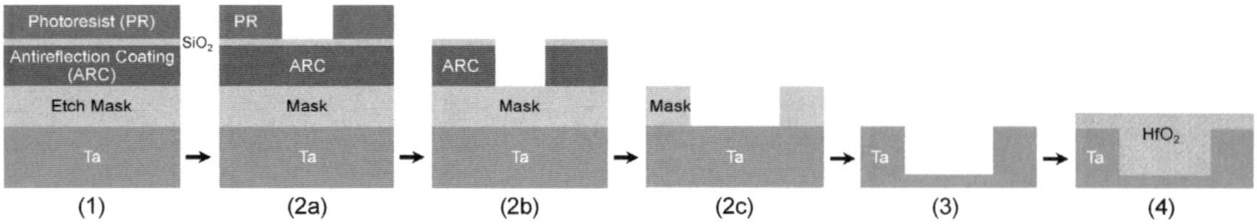

Figure 1: A general process flow for the fabrication of our 2D PhC involves (1) pattern generation, (2) pattern transfer into a hard mask by etching, (3) pattern transfer into the metallic substrate by etching, and (4) filling of the cavities with HfO_2 by ALD.

these criteria. Indeed, while multilayer stacks [6], cermets [7] emitters are easy to fabricate, heterogeneous platforms are subject to thermo-mechanical stresses and chemical reactions at material interfaces that are initiated at elevated temperatures. Even homogeneous material platforms, while not being subjected to stresses, can degrade at high temperature: radius of curvature driven surface diffusion shortens the lifetime of complex structures such as 3D photonic crystals [8]. Our 2D metallic photonic crystal meets all of the requirements of a selective emitter for high-temperature energy conversion.

Our phonic crystal consists of a square periodic array of holes etched into the low-emissivity polished metallic substrate, polycrystalline tantalum. A refractory metal is used as the substrate for the photonic crystal due to its high melting point, low vapor pressure, advantageous low emissivity in the infrared, and ability to be etched. The emissivity of the substrate material is selectively enhanced to near that of a blackbody emitter by the resonant cavity modes resulting from the etched holes. The spectral range of enhancement can therefore be easily tailored to the specific system needs by tuning cavity geometry to match the photonic crystal to different low-bandgap PV cells. A conformal passivation layer of a dielectric material, hafnium dioxide (HfO_2) is used [9] to further achieve high-temperature stability and prevent structural and optical degradation after long operational lifetimes.

The performance of our conformally coated photonic crystal has been consistently good at normal incidence [9, 10], however most of the thermal radiation actually occurs off-normal according to Lambert's law. While metamaterial designs based on metal-dielectric stacks [11] and 2D metallic pyramid arrays [12] have shown promise, they are difficult to fabricate and have not been experimentally demonstrated at high temperatures under extended operation. A simple approach of filling a 2D metallic photonic crystal with a dielectric was introduced [13] to obtain omnidirectional, polarization insensitive, wavelength selective thermal emission. In this paper, we present the first *experimental* demonstration of a HfO_2-filled 2D tantalum photonic crystal selective emitter that further increases TPV system efficiencies by increasing total in-band power (at all polar angles and polarizations) without sacrificing ease of fabrication.

In our 2D photonic crystal selective emitter, the high emittance at short wavelengths is achieved by matching the radiative rate to the absorptive rate for each resonance in the square periodic array of holes. This is assisted by the fact that both the radiative and absorptive rates increase with mode number and hence frequency. At long wavelengths the emittance is due to a surface area weighted effective impedance created by the holes [14]. The result is near-blackbody emittance at short wavelengths as well as emittance almost as low as that of polished tantalum at long wavelengths, with a sharp cutoff between the two regimes. For off-normal incidence however, this performance significantly drops [13]. The spectral emissivity of the photonic at normal incidence and the hemispherically-averaged emissivity for the photonic crystal structure was simulated by the Fourier Modal Method using a freely available software package [15]. The material properties of the substrates were taken into account using a Lorentz-Drude model fitted to the elevated temperature emissivity to capture the optical dispersion of the substrate at high temperature. Nonlinear optimization determined the period, cavity radius, and height with bounds based on the desired cutoff wavelength and other fabrication constraints.

Figure 2: SEM micrographs of the photonic crystal (a) before and (b) after filling of the cavities with HfO_2 by ALD. (c) Cross section after filling.

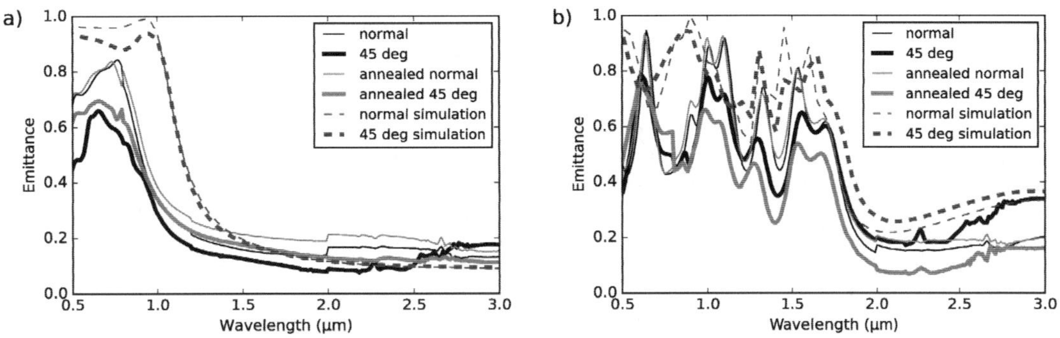

Figure 3: Normal and 45° incidence measurements (before and after anneal at 1000°C for 24 hours) of the (a) unfilled and (b) HfO_2-filled photonic crystals compared with their normal incidence and hemispherically averaged simulations.

The photonic crystal was fabricated in the same way as in our previous work [16]. The fabrication process began with by defining a periodic pattern in photoresist (PR) by interference lithography. Two orthogonal exposures in a Mach-Zehnder setup generated the pattern. A numerically optimized anti-reflection coating (ARC)/SiO_2/PR stack was used where the SiO_2 served as a protection layer for the ARC while etching the PR. Using reactive ion etching (RIE), the pattern was transferred through the tri-layer stack into the SiO_2 mask deposited by plasma-enhanced chemical vapor deposition (PECVD). The tantalum was etched by deep reactive ion etching (DRIE) using a Bosch process. After the final pattern transfer, the residual passivation layer was removed by oxygen plasma and the residual SiO_2 mask was removed by hydrofluoric acid. The fabrication steps are described in Figure 1. The final step - the filling of the photonic crystal cavities by atomic layer deposition (ALD) - was done in the same way as when depositing the conformal layer, but simply depositing for a longer period of time until the cavities are filled.

The fabricated photonic crystal was measured by scanning electron microscopy (SEM) and found to have a period $a \sim 0.5$ μm, radius $r \sim 0.2$ μm, and cavity depth $d \sim 1.5$ μm before filling, as shown in Fig. 2 (a). The surface of the photonic crystal after filling is shown in Fig. 2 (b). The filling of the holes was found to be relatively uniform, as show in Fig. 2 (c) in cross-section. The spectral emissivity at room temperature of the photonic crystal before and after filling was obtained by measuring near-normal incidence reflectance and reflectance at 45° incidence using a Cary 5000 spectrophotometer. The same reflectance measurement was also performed after annealing the photonic crystal for 24 hours at 1000°C in a quartz tube furnace in vacuum (5×10^{-6} Torr, heating and cooling rate of 10°C/minute). A 20 nm conformal coating was deposited on the non-filled photonic crystal to prevent degradation of the optical properties during anneal. In simulations, the unfilled photonic crystal was found to have effective (least squares fit of sigmoid with 1200°C blackbody weighting) hemispherical in-band emissivity of 0.594 and out-of-band emissivity of 0.168 whereas a photonic crystal with HfO_2 filled cavities would have an in-band emissivity of 0.922 and out-of-band emissivity of 0.154, corresponding to a 55% increase in in-band emissivity without a change in out-of-band emissivity [13]. Thus, in an unfilled photonic crystal, the emissivity at 45° (approximately the hemispherically averaged emissivity) is about half of that at normal incidence, but in a filled photonic crystal the two emissivities are about equal. As shown in Fig. 3 (b), the peak normal in-band emittance of our fabricated filled photonic crystal was 0.9, and the 45° peak in-band emittance was 0.8, while the out-of-band emittance approached the emissivity of flat Ta. A sharp cutoff was observed in the desired wavelength range. The optical properties remained remarkably stable even after annealing. The measured emittance of the photonic crystal matched the simulated emittance relatively well, as shown in Fig. 3. This work demonstrates that dielectric filling improved the hemispherical performance of our 2D photonic crystal selective emitter without sacrificing stability or ease of fabrication allowing for further increase in TPV system efficiencies.

Acknowledgments

The authors thank J. Daley and T. Savas at NSL (MIT), and R. Geil at the University of North Carolina at Chapel Hill for assistance with fabrication. Fabrication was done in part at NSL at MIT and at CNS at Harvard University, a member of the National Nanotechnology Infrastructure Network (NNIN), supported by the National Science Foundation (NSF) under NSF Award No. ECS-0335765. Fabrication was supported by the Solid-State Solar-Thermal Energy Conversion Center (S3TEC), an Energy Frontier Research Center funded by the U.S. Department of Energy (DOE), Office of Science, Basic Energy Sciences (BES), under Award # DE-SC0001299 / DE-FG02-09ER46577. Research was also supported by the U.S. Army Research Laboratory and the U.S. Army Research Office through the Institute for Soldier Nanotechnologies, under Contract # W911NF-13-D-0001.

References

[1] Chan W R, Wilhite B A, Senkevich J J, Soljacic M, Joannopoulos J D and Celanovic I 2013 *J. Phys. Conf. Ser.* vol 476 p 012017 ISSN 1742-6588
[2] Crowley C J, Elkouh N a, Murray S and Chubb D L 2005 *AIP Conf. Proc.* **746** 601–614 ISSN 0094243X
[3] Lenert A, Bierman D M, Nam Y, Chan W R, Celanovic I, Soljačić M and Wang E N 2014 *Nat. Nanotechnol.* **9** 1–5 ISSN 1748-3387
[4] Wang C A, Choi H K, Ransom S L, Charache G W, Danielson L R and DePoy D M 1999 *Appl. Phys. Lett.* **75** 1305–1307
[5] Rinnerbauer V, Ndao S, Yeng Y X, Chan W R, Senkevich J J, Joannopoulos J D, Soljačić M and Celanovic I 2012 *Energy Environ. Sci.* **5** 8815 ISSN 1754-5692
[6] Narayanaswamy A, Cybulski J and Chen G 2004 *AIP Conf. Proc.* **738** 215–220 ISSN 0094243X
[7] Kennedy C E 2002 *NREL Tech. Rep.* 1–58
[8] Fleming J G, Lin S Y, El-Kady I, Biswas R and Ho K M 2002 *Nature* **417** 52–55
[9] Stelmakh V, Chan W R, Joannopoulos J D, Soljacic M, Celanovic I and Sablon K 2016 *Proc. ASME 2016 5th Int. Conf. Micro/Nanoscale Heat Mass Transf.* pp 1–5
[10] Stelmakh V, Rinnerbauer V, Geil R D, Aimone P R, Senkevich J J, Joannopoulos J D, Soljacic M, Celanovic I, Soljačić M and Celanovic I 2013 *Appl. Phys. Lett.* **103** 0–4 ISSN 00036951
[11] Arpin K, Losego M D, Cloud A N, Ning H, Mallek J, Sergeant N P, Zhu L, Yu Z, Kalanyan B, Parsons G N, Girolami G S, Abelson J R, Fan S and Braun P V 2013 *Nat. Commun.* **4** 2630 ISSN 2041-1723
[12] Rephaeli E and Fan S 2009 *Opt. Express* **17** 15145–15159 ISSN 1094-4087
[13] Yeng Y X, Chou J B, Rinnerbauer V, Shen Y, Kim S G, Joannopoulos J D, Soljacic M and Celanovic I 2014 *Opt. Express* **22** 21711 ISSN 1094-4087
[14] Ghebrebrhan M, Bermel P, Yeng Y X, Celanovic I, Soljacic M and Joannopoulos J D 2011 *Phys. Rev. A - At. Mol. Opt. Phys.* **83** 1–6 ISSN 10502947
[15] Liu V and Fan S 2012 *Comput. Phys. Commun.* **183** 2233–2244 ISSN 00104655
[16] Stelmakh V, Chan W R, Senkevich J, Ghebrebrhan M, Joannopoulos J D, Soljačić M and Celanovic I 2016 *IEEE Trans. Nanotechnol.* vol 15 pp 303–309 ISBN 9781467381550

Design and test of a Bennet's doubler device with mechanical switches for vibrational energy harvesting

M. A. Ben Ouanes[1,2], Y. Lu[2], H. Samaali[1,2], P. Basset[2] and F. Najar[1]

[1]Applied Mechanics and Systems Research Laboratory, Tunisia Polytechnic School, BP 743, La Marsa 2078, University of Carthage, Tunisia.

[2]Université Paris-Est, ESYCOM, ESIEE Paris, Noisy-le-Grand, France.

E-mail: benouanesaymen@gmail.com

Abstract. In this work, we demonstrate that the use of self-synchronized mechanical switches in replacement of diodes into electrostatic vibration energy harvesters (e-VEH) can lead to better power generation. Indeed, mechanical switches have the advantage of no leakage current and no threshold voltage. As a proof of concept, we use the Bennet's doubler electrostatic generator. The proposed e-VEH is composed of two variable capacitors triggered by a central electrode taken as an inertial mass. Ambient vibrations induce inertial forces on the central electrode, as a result a voltage doubling is obtained at each operating cycle. The mechanical switches are directly fixed to the moving electrode. In addition, no dedicated pre-charge is required: the system starts with ambient electrical charges. The device is fabricated and tested under harmonic motion. A comparison between the proposed design and those using diodes under the same operating conditions shows an experimental direct increase of the harvested electrical power of around 28%.

1. Introduction

Vibration energy harvesting (VEH) from mechanical motions is becoming a promising technology that allows a better spreading of smart and micro-systems in many new applications. This energy source offers extraordinary capability to new technological paradigms, such as the Internet of Things and Smart Cities.

Many VEH devices have been developed during the last decade. Several transduction mechanisms were tested to transform the kinetic energy of the environment into useful electrical energy. Three types of transduction dominate the major part of the published designs: (i) Electromagnetic transduction [1] described by a high power generation, but require complex design and large dimensions, (ii) Piezoelectric and smart materials [2, 3, 4] characterized by high power generation density, but require special and costly fabrications steps, and (iii) Electrostatic transduction [5] known for its high integration, but generally generates low power density.

In 1787, Bennet proposed the principle of an electrostatic generator by continuously doubling the charges on a conductor after repeating a given sequence of contacts between three plates [6]. Recently, Queiroz [7] proposed to harvest kinetic energy based on the Bennet doubler principle where the contacts are replaced by diodes. This new conditioning circuit for electrostatic Vibration Energy Harvesters (e-VEH) was used for the first time with a capacitive MEMS transducer dedicated to harvest kinetic energy from vibrations by Dorzhiev et al. [8].

In this work we aim at designing and testing the first Bennet's electrostatic generator dedicated to vibration energy harvesting where the diodes introduce by Queiroz et al [7] are removed. Indeed, by replacing the diodes with self-synchronized mechanical switches, we get the advantage of no leakage current and no threshold voltage. An expected increase of the power generation is sought along with an elimination of the device control during the operating cycle.

2. Bennet doubler for e-Vibration Energy Harvesting with self-synchronized mechanical switches

The functional representation of the proposed Bennet's doubler for e-VEH is shown in Figure 1. It is composed by two variables capacitors C_{ab} and C_{bc}, and a restoring capacitor C_{res}. They are made of 3 aluminum electrodes A, B and C covered with a Parylen layer (Figure 1). Three mechanical switches S_1, S_2 and S_3 are also implemented to play the role of classically used diodes D_1, D_2 and D_3, respectively. The equivalent electrical circuit with diodes or mechanical switches is shown in Figure 2.

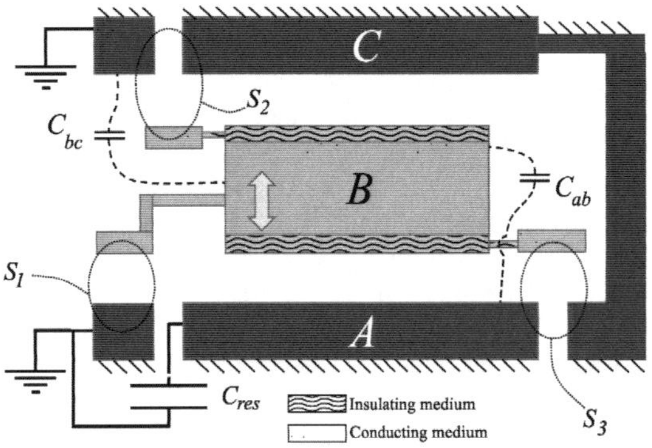

Figure 1. Functional representation of the e-VEH Bennet's doubler device with mechanical switches.

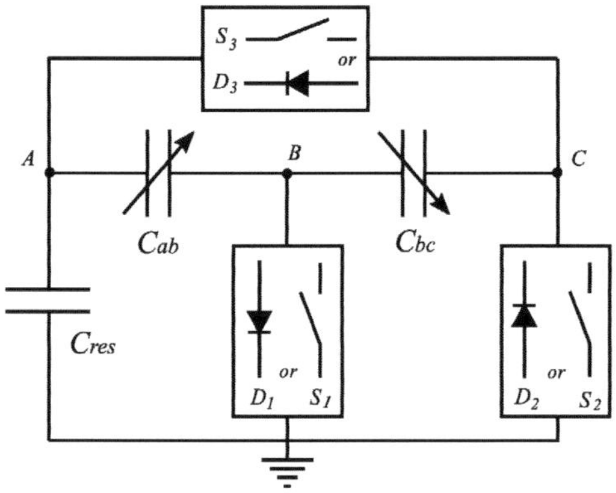

Figure 2. Electrical circuit of the e-VEH Bennet's doubler device with diodes or mechanical switches.

The basic principle is the following: the central electrode B moves between electrodes A and C thanks to inertial forces due to external excitations. As shown in Figure 3, the induced motion generates out of-phase variations of C_{ab} (between A and B) and C_{bc} (between B and C). For our demonstrator, the value of C_{ab} varies between 20pF and 80pF, while C_{bc} varies between 1020pF and 1100pF.

The stoppers S_1 and S_2 are used as mechanical switches, which relate respectively the electrodes B and C to the ground. The mechanical switch S_3 relates electrode C to A when B is at its lowest position with respect to A.

3. Operating cycle of the e-VEH Bennets doubler device

Initially, there are some charge Q on the electrode A due the ambient electromagnetic noise. The operating cycle starts when the central electrode B is at its lowest position with respect to A, which activates the switch S_1, connecting B to the ground (Figure 3-a). This brings an induced charge $-Q$ on B and turns the capacitor C_{ab} to its maximum value C_{ab}^{Max}, and the capacitor C_{bc} to its minimum value C_{bc}^{Min}. At this configuration, all the capacitances are in parallel. Then the electrode B moves towards C under inertial forces. When B reaches C at its highest position (Figure 3-b), the electrode C is grounded through the activation of the switch S_2, and a charge $+Q$ is induced on C. All the capacitances are now connected in series, and C_{ab} and C_{bc} are set to C_{ab}^{Min} and C_{bc}^{Max}, respectively.

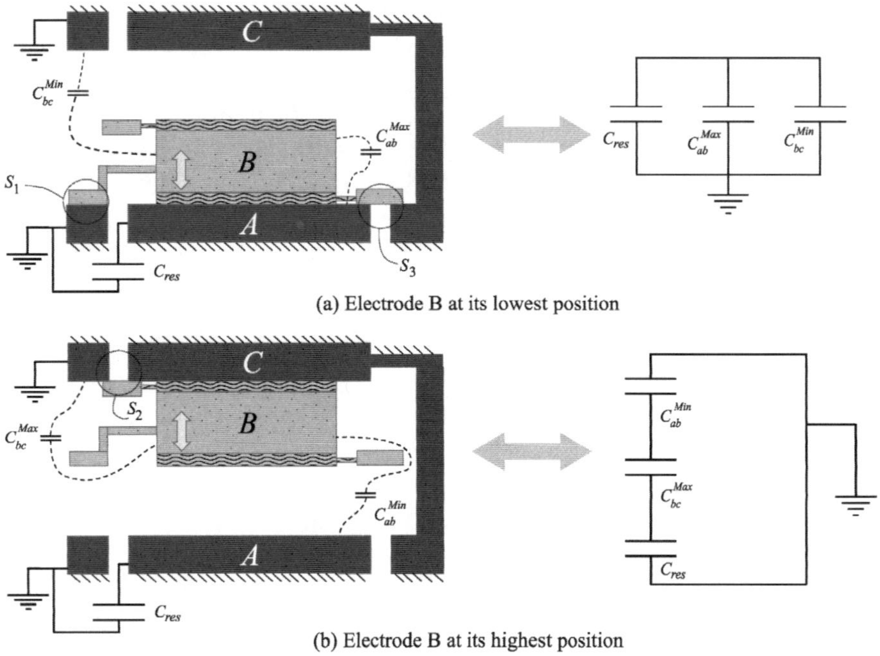

(a) Electrode B at its lowest position

(b) Electrode B at its highest position

Figure 3. Operating cycles of the e-VEH Bennet's doubler device with the corresponding equivalent circuit.

The last step of this cycle is initiated when the electrode B moves back to its lowest position. There, B is grounded again through S_1 and A is connected to C through S_3 (Figure 3-a) and the new B electrode's charge is equal to $-2Q$. All the capacitances are again in parallel. The operating cycle is repeated again and again, to double the charge at each period.

4. Experimental validation of the e-VEH Bennets doubler with mechanical switches

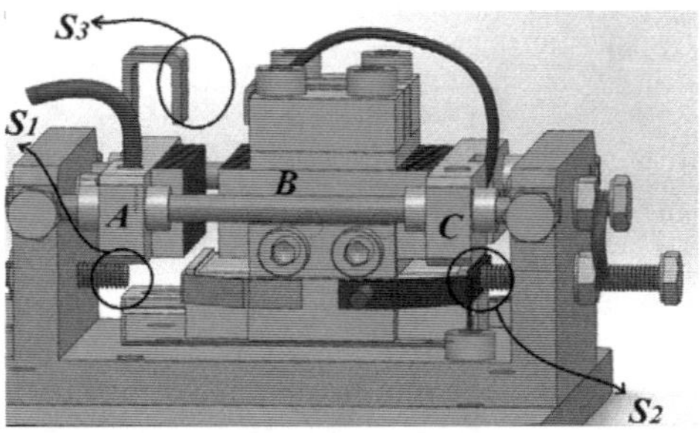

Figure 4. 3D model of the fabricated e-VEH Bennet's doubler device.

Figure 5. Fabricated e-VEH Bennet's doubler with aluminium and polyamide materials mounted with a shaker.

The e-VEH Bennet's doubler device is modeled (Figure 4) and fabricated (Figure 5) according to the described operating cycle in the previous section. The general dimensions of the device are 5cm×3cm×2cm. Interdigited electrodes with triangular shape is adopted on the two faces of electrode B and the active faces of electrodes A and C, so that an increase of the extrema values of C_{ab} and C_{bc} is obtained. The device is mounted to a shaker that is excited with harmonic motion.

The objective of these experiments is to validate the proposed proof of concept by comparing the designed device with a similar device using a set of diodes replacing the mechanical switches. Using the same external excitation conditions (12Hz at $1g_{rms}$) for both designs, we measure the voltage across C_{res}. In Figure 6-a, we plot the voltage variations as the number of cycles is increasing. As expected, a voltage saturation is observed for both designs (with mechanical switches and with diodes). However, for the mechanical switches design a voltage gain of almost 80% is observed (25V against 14V). The harvested power is measured by deriving the instantaneous electrical energy in C_{res}. The results, shown in Figure 6-b, depict an increase of around 28% in favor of the mechanical switch design.

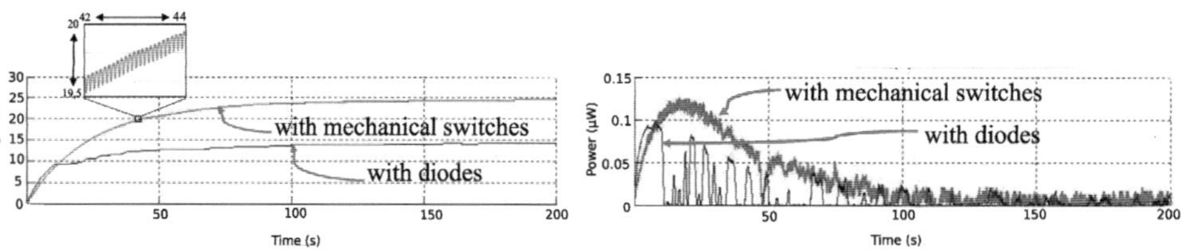

(a) Harvested voltage (b) Harvested power

Figure 6. Measurements of voltage and harvested power across C_{res} for both diode and mechanical switch designs.

5. Conclusion

Using an electrostatic vibration energy harvester (e-VEH) and adopting a Bennet's doubler design as conditioning circuit, we showed that replacing the classically used diodes by self-actuated and self-synchronized mechanical switches leads to a performance increase of the harvester. We designed and fabricated a new e-VEH and conduct a series of experiments to compare diode and mechanical switch based designs. The experimental results showed that a net increase of the saturation voltage was observed around 80%. This corresponds to a harvested power improvement of 24%. The power generation profile was not only higher for mechanical switches design, but also showed a smoother behavior as function of time leading to further increase in terms of total harvested energy. More analysis are needed to estimate and reduce the mechanical losses and to reduce the device size.

6. References

[1] Naifar S, Bradai S, Viehweger C and Kanoun O 2016 *Measurement* In Press.
[2] Marzencki M, Defosseux M and Basrour S 2009 *J. MEMS* **18** 1444-1453
[3] Li H, Tian C, and Deng Z D 2014 *Applied Physics Reviews* **1** 041301
[4] Jemai A, Najar F, Chafra M and Ounaies Z 2016 *Composite Structures* **135** 176190
[5] Amri M, Basset P, Cottone F, Galayko D, Najar F and Bourouina T 2011 *Proc. PowerMEMS2011* Seoul 15-18 Nov.
[6] Bennet A and Kaye R 1787 *Philosophical Trans. of the Royal Society of London* **77** 288-296
[7] de Queiroz A C M 2010 *53rd IEEE Int. Midwest Sym. on Circuits and Systems* Seattle WA 1-4 Aug.
[8] Dorzhiev V, Karami A, Basset P, Marty F, Dragunov V and Galayko D 2015 *IEEE Elec. Device Letter* **36** 183135

First experimental demonstration of a Self-Oscillating Fluidic Heat Engine (SOFHE) with piezoelectric power generation

T. Monin[1,2,3], A. Tessier-Poirier[3], E. Léveillé[3], A. Juneau-Fecteau[3], T. Skotnicki[1], F. Formosa[2], S. Monfray[1], L.G. Fréchette[3]

[1]STMicroelectronics, Crolles, France, [2]Université Savoie Mont-Blanc, Annecy-le-Vieux, France, [3]Laboratoire Nanotechnologies et Nanosystèmes, CNRS UMI-3463, Université de Sherbrooke, Canada.

thomas.monin@univ-smb.fr

Abstract. *In this paper, we present the working principle and first experimental demonstration of an innovative approach to harvest low-quality heat sources, the Self-Oscillating Fluidic Heat Engine (SOFHE). Thermal energy is first converted into pressure pulsations by a self-excited thermo-fluidic oscillator driven by periodic phase change of a fluid in an enclosed channel. A piezoelectric membrane then converts this mechanical energy into an electrical power. After describing the working principle, an experimental demonstration is presented. The P-V diagram of this new thermodynamic cycle is measured, showing a mechanical power of 3.3mW. Combined with a piezoelectric spiral membrane, the converted electrical power generation achieved is close to 1µW in a 1MΩ load. This work sets the basis for future development of this new type of heat engine for waste heat recovery and to power wireless sensors.*

1. Introduction

In this paper, we present for the first time a new principle to convert heat into electricity using a Self-Oscillating Fluidic Heat Engine (SOFHE) [1]. From a constant heat source and heat sink, a thermo-fluidic engine develops cyclic pressure variations that deform a piezoelectric membrane to generate electricity. This approach is suited to take advantage of low temperature waste heat (T<200°C), which is one of the most pervasive energy sources in our environment, and is implemented with low cost component.

In the following, we will first explain the working principle of the SOFHE and its two-stage conversion mechanism. We will then describe in more details the components of the SOFHE as well as the experimental conditions. Finally, we present the performances of the SOFHE both as a heat engine and a thermo-electrical converter.

2. Working principle

The SOFHE piezoelectric demonstrator, presented in Figure 1, is based on an innovative, two-stage, simple conversion mechanism. First, a constant heat source supplies a passive, auto-oscillating mechanism through an evaporator (component 1 in Figure 1) generating mechanical energy via the capillary tube (component 2 in Figure 1). This mechanical energy is then converted into electricity using a piezoelectric membrane (component 3 in Figure 1) during the second step of the conversion.

2.1. *Thermo-fluidic oscillator: from thermal to mechanical energy*

The proof of concept of the thermo-fluidic oscillator is formed by a capillary tube, closed at one end and initially filled with liquid water. A part of the glass tube is inserted in a larger glass tube filled with constant temperature glycerin while the remaining section is cooled by ambient air convection. A vapor bubble is formed in the hot zone. Upon heating, the vapor volume expands until it reaches its thermal equilibrium position. Depending on the temperature gradient at the interface of the vapor and liquid phases, the meniscus begins to oscillate as presented in Figure 2. As a phenomenological approach, this first stage of conversion can be considered as a damped mechanical oscillator excited by the working fluid phase change, as displayed in Figure 3. Evaporation induced when the meniscus enters the hot zone builds pressure in the vapor bubble, pushing the liquid column out until it reaches the cold zone. There, vapor condensates, lowering the pressure in the bubble and pulling the liquid column back into the hot zone. This results in self-sustained periodic oscillations generating pressure fluctuations (Fig. 2) that can be used for mechanical work, driven from a constant source of heat. These dynamics are further analyzed in a complementary paper [2].

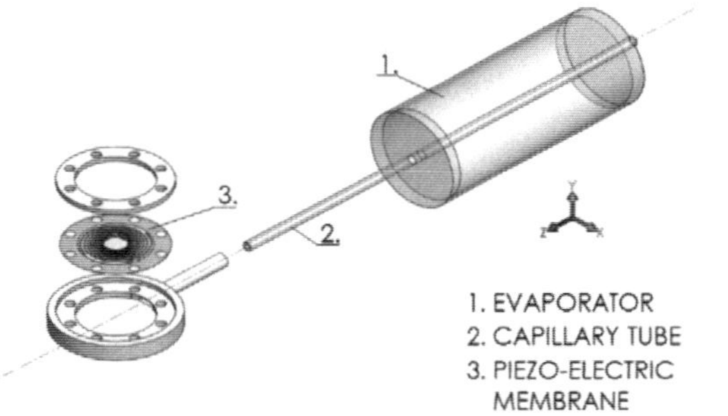

Figure 1. Self-Oscillating Fluidic Heat Engine demonstrator.

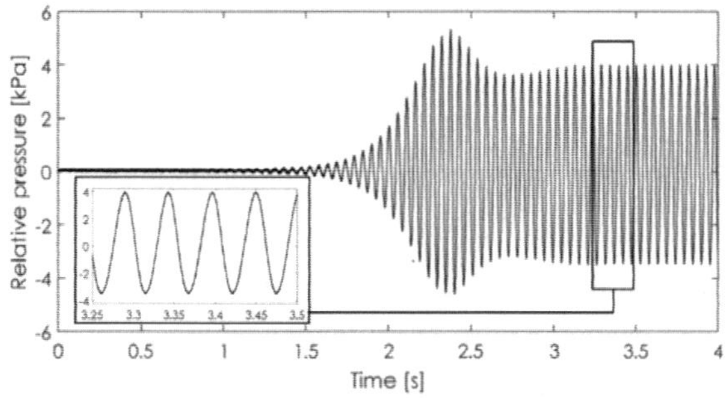

Figure 2. Oscillation start-up regime. At first, the meniscus stays still (t<1s). After a transient regime (1s<t<3s), the amplitude of meniscus oscillations stabilizes.

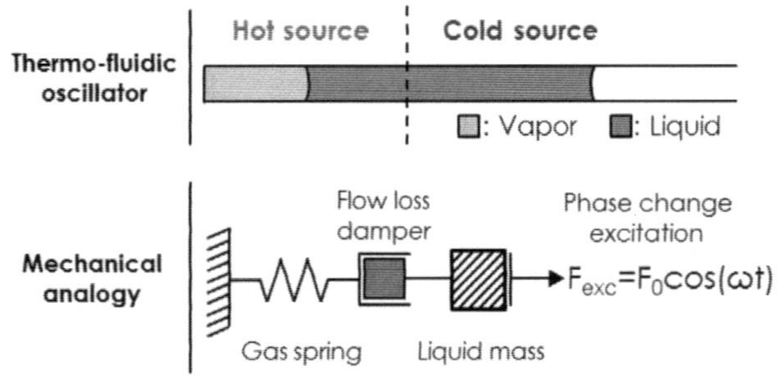

Figure 3. Mechanical analogy of the thermo-fluidic oscillator. The vapor phase volume acts as a gas spring while the liquid column is represented as an inertial mass. The viscosity of the fluid limits the displacement of the liquid column. The phase change, which occurs at the meniscus, can be seen as the excitation force F_{exc}.

2.2. *Piezoelectric conversion: from mechanical to electrical energy*

A dedicated-piezoelectric membrane (Figure 4) is used for electromechanical transduction. A silicone layer allows the sealing for a piezoelectric bimorph planar spiral to be deformed by the oscillating pressure. In order to improve the conversion efficiency, the mechanical impedances of the thermo-fluidic engine and the transducer should be matched. The planar spiral shape, allows control of the mechanical stiffness to adapt its impedance.

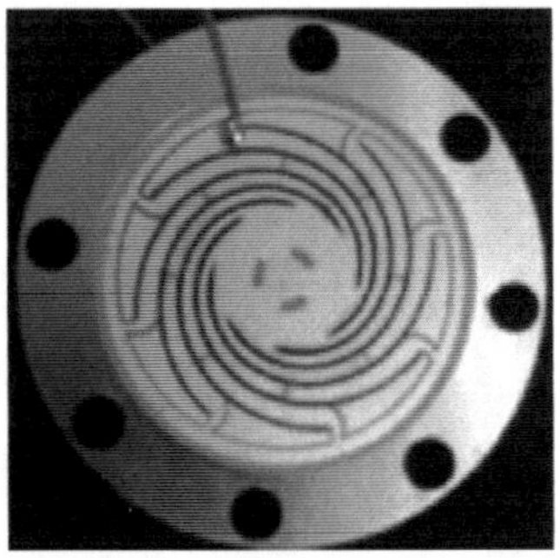

Figure 4. Piezoelectric bimorph spiral electromechanical transducer.

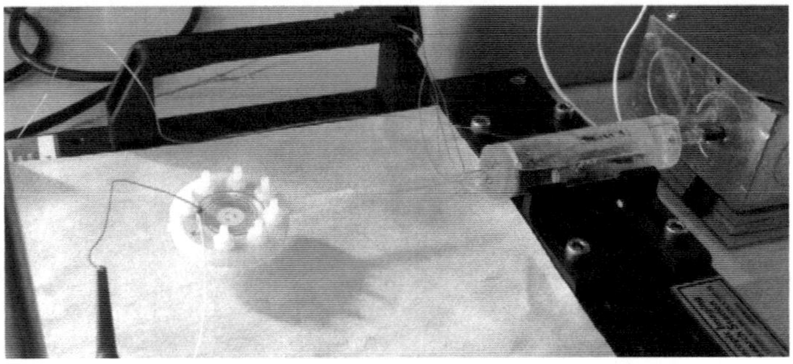

Figure 5. SOFHE generator and associated test-bench.

3. Experimental set-up

The experimental set-up is shown in Figure 5 and detailed hereafter.

3.1. SOFHE piezoelectric generator components and instrumentation

The evaporator is a 30mm diameter glass tube filled with glycerin. A heater located inside the evaporator and connected to a PID controller allows the control of the temperature. A 2mm/4mm inner/outer diameters capillary tube is filled with deionized water and inserted through the glycerin evaporator. One end of the tube is connected to a PX26 pressure sensor (OMEGA Engineering) while the other end is connected to the piezoelectric membrane. The membrane is made by laser micromachining a spiral shape into a commercially available piezoelectric buzzer (PUI Audio Inc. AB4113B). The membrane is mounted to the end of the capillary tube with a 3D printed support. During operation, the meniscus position is captured with a PCO high speed camera, the pressure in the tube is measured, and the membrane output voltage is collected with an Agilent DSO6000 oscilloscope. A data acquisition board (KPCI-3110) enables the acquisition and synchronization of the data.

3.2. Experiments protocol

The initial setup of the SOFHE starts with the cleaning of the capillary tube with a flush of acetone, followed by IPA and rinsed heavily in deionized water. After its insertion through the evaporator, the capillary tube is filled with deionized, degassed water. The pressure sensor is then connected to one of its end and the membrane to the remaining one.
The evaporator temperature is gradually set to a value above 100°C. Once the oscillations appear, the vapor pressure variations are registered and monitored. For a constant temperature, the generator is considered as stable whenever the pressure variation amplitude is constant for 10 minutes at least. The PV-diagram is then registered and the electrical power output is determined.

4. Results and discussion

4.1. Performances of the SOFHE

4.1.1. First stage of conversion – Thermal to Mechanical. At first, the thermo-fluidic oscillator is studied alone, as a heat engine would be (i.e. no load attached to it). For a hot source temperature set to 115°C, the resulting P-V diagram is obtained (Fig. 6). The calculated mechanical power of the thermodynamic cycle is 3.3mW. Although this power is consumed through viscous energy dissipation of the liquid column at the capillary walls, it gives an order of magnitude of the thermodynamic power that can be delivered by this new cycle. Considering the dimensions of the thermo-fluidic oscillator (length=20cm, section=3.14mm^2), the calculated power density is 5mW.cm^{-3}.

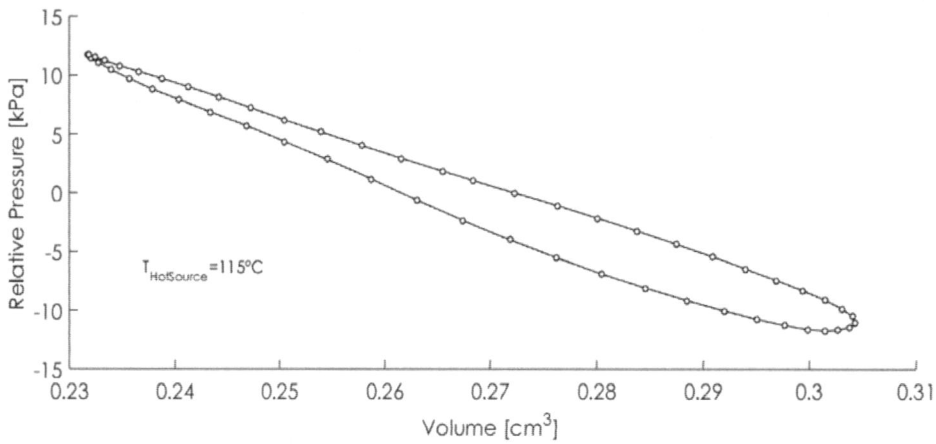

Figure 6. P-V diagram for a free-loaded SOFHE thermo-fluidic oscillator.

4.1.2. Second stage of conversion – Mechanical to Electrical. The piezoelectric membrane is attached to the thermo-fluidic oscillator and the temperature is raised to 140°C. The output voltage is measured with respect to time with an oscilloscope having an input impedance of 1MΩ. The corresponding graph is presented in Figure 7.

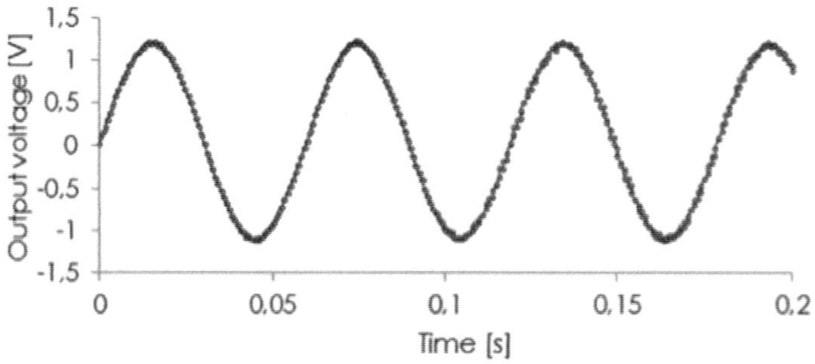

Figure 7. SOFHE voltage output in a 1MΩ load.

The calculated power dissipated through the electrical load is 0.69μW. From this initial demonstration, an overall conversion efficiency from the cycle's mechanical power to electrical power of less than 0.1% was shown.

5. Conclusion

We report a 3.3mW thermodynamic mechanical power and a converted electrical power generation close to 1μW in a 1MΩ load from a low quality heat source. This figure of power generation can be greatly increased by finding the optimal electrical load for the generator as well as further optimization of the piezoelectric membrane. Nonetheless, compared to Seebeck-effect generators, this innovative approach is not limited by material properties and significant engineering improvements can be achieved such as for controlled phase change and better impedance matching between the two stages of conversion.

References

[1] "Dispositif de récupération d'énergie thermique", S. Monfray, T. Skotnicki, L.G. Freéchette, G. Misheraki, E. Léveillé, patent pending.

[2] A. Tessier-Poirier, T. Monin, L.G. Fréchette, "Influence of the nonlinearities on the steady-state regime of the pulsating heat pipe", submitted at PowerMEMS2016, Paris, December 06-09, 2016.

A water-powered Energy Harvesting system with Bluetooth Low Energy interface

M. Kroener, K. Allinger, M. Berger, E. Grether, F. Wieland, S. Heller, P. Woias

Laboratory for Design of Microsystems, Department of Microsystems Engineering, Georges-Koehler-Allee 102, 79110 Freiburg, Germany

kroener@imtek.de

Abstract. This paper reports the design, and testing of a water turbine generator system for typical flow rates in domestic applications, with an integrated power management and a Bluetooth low energy (BLE) based RF data transmission interface. It is based on a commercially available low cost hydro generator. The generator is built into a housing with optimized reduced fluidic resistance to enable operation with flow rates as low as 6 l/min. The power management combines rectification, buffering, defined start-up, and circuit protection. An MSP430FR5949 microcontroller is used for data acquisition and processing. The data are transmitted via RF, using a Bluegiga BLE112 module in advertisement mode, to a PC where the measured flow rate is stored and displayed. The transmission rate of the wireless sensor node (WSN) is set to 1 Hz if enough power is available, which is the case for flow rates above 5.5 l/min. The electronics power demand is calculated to be 340 μW in average, while the generator is capable of delivering more than 200 mW for flow rates above 15 l/min.

1. Introduction

The first hydroelectric power plant, converting water flow to electrical energy, began operation in Appleton (Wisconsin) in 1882. While on the larger scale this method has been used for a long period, water-powered energy harvesting systems in the mW-range are pretty rare. Recent research in this field covers smart-water meters [1], and some special generation principles, such as piezoelectric harvesting from vortices [2], and graphene thin film based harvesting in the μW-range [3].

An early version of a water-powered energy-autonomous sensor system has been realized in the lab with a price-awarded student project [4]. The research shown in this paper aims at an expansion of the idea of water-powered WSNs for smart home applications. For reasons of simplicity an off-the-shelf "micro hydro power generator" is used here. The WSN is intended to measure water temperature and flow as a standalone system in the water conduit of a smart home, to determine the total water usage and energy consumption. For this purpose, the recorded data are transmitted to a receiver, processed and displayed continuously. The presented system is limited to water flow measurements only, as the characterization and power management stood in the focus of this demonstrator, as it turned out to be the most complex part.

2. System design and measurements

This system consists of an optimized turbine housing, the generator itself, an electronic power management, the microcontroller-based data acquisition, and a Bluetooth low energy (BLE) transmission unit. The component description is given in detail below.

2.1. Generator and turbine housing

The micro hydro generator used in this project is commercially available (Manufacturer: Yoyo Zeng Shenzhen Slinya Electronic Co., Ltd). The turbine is made from a pot-shaped rotor with turbine blades at the outside, a belt-shaped magnet with six poles at its inner side and a central axis fitting into a brass gliding bearing in the turbine housing. The generator part consists of a three phase design with 9 coils that are mounted liquid-tight in the stator and protrude into the pot-shaped rotor, thus facing the ring magnet at a short distance given by the turbine housing in-between. If water flows through the turbine housing it starts rotating the magnet around the stator, thus inducing a voltage in the stator coils due to the dynamic change of the magnetic field. The data sheet of the original device gives a maximum output power of 10 W, with 80 V output and a maximum water pressure of 1.2 MPa (12 bar). The minimum pressure for rotation of the turbine is specified to 0.06 MPa, i.e. 0.6 bar.

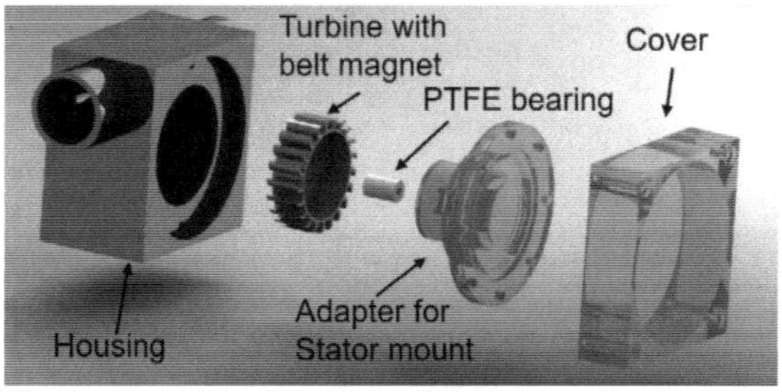

Figure 1. *CAD drawing of the specially designed housing for the turbine with reduced fluidic resistance. The PTFE bearing reduces friction losses an uneven rotation compared to the original brass bearing. The transparent PMMA adapter and cover allows for observation of the turbine during operation.*

For optimization the turbine housing was replaced by a custom-made brass housing with optimized fluid flow and reduced fluidic resistance (see Figure **1**). Instead of the original brass bearing, a PTFE bearing was used to reduce friction losses and uneven rotation. A custom adapter and cover were fabricated from PMMA to embed the stator and to have an optical access to the rotating turbine.

2.2. Power management, data processing and transmission

Frequently, power management in energy harvesting means to optimally distribute a small amount of energy available. In this work particular attention had to be payed to overpower and overvoltage protection. Rectification from the generator's AC-signal is achieved with Schottky diodes (BAT54SW) on the three-phases. A buffer capacitor was chosen with a capacitance large enough to supply one complete measurement and wireless transmission cycle, an ultra-low-power hysteretic switch enables defined electronics start-up [5], providing under voltage protection. Over voltage protection of the circuit is achieved with a voltage divider and an NMOS transistor (TI CSD17313Q2), which converts abundant power to heat via opening a path to GND for voltages on the buffer capacitor above 5 V. The PCB is shown in Figure 2.

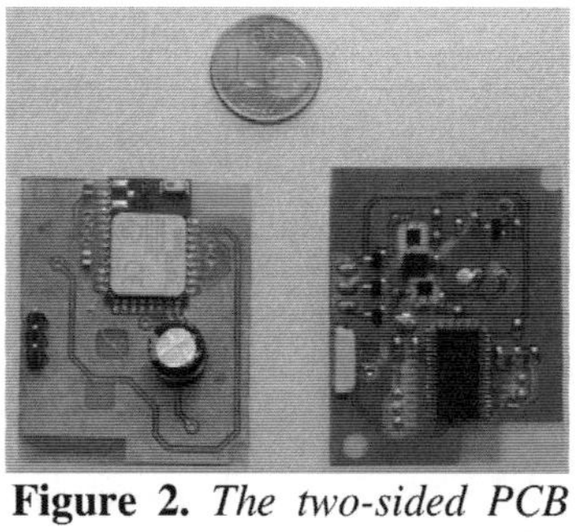
Figure 2. *The two-sided PCB combines, a BLE module for wireless transmission (left), and an μC for data acquisition and processing (right). On top: 1 €-cent for size comparison.*

A 16-Bit MSP430FR5949 (TI) microcontroller (μC) is used for data acquisition and processing. It measures the rotational speed of the turbine by measuring the frequency of the AC voltage delivered by the generator. This data is then correlated to external flow rate measurements to get calibrated values. During measurement only the μC is active as a load. Once the data are acquired they are transmitted to a PC via RF, using a Bluegiga BLE112 Bluetooth low energy module in advertisement mode. The advertisement mode transmits the measurement data in the header, elsewhere used to establish a linked connection, which is not done here. Thereby the power consumption of the whole system is lowered remarkably, because it is only on for 4 ms during a full measurement and transmission cycle. The average power consumption of every part of the electronic is derived from the datasheet values and given in Table 1. The average power consumption with a transmission rate of 1 Hz is 340 μW. We are currently working on a method to assess the power consumption of such a system with high time and low current resolution [6].

Table 1. Power consumption of the electronic components.

Component	U in V	I in mA	t in ms	Avg. Power P=U*I*t/1s
MSP430 – Active (1 MHz)	3.3	0.1	10	3.3 μW
MSP430 – Sleep (LPM3.5)	3.3	0.00025	990	0.82 μW
BLE Modul – Sleep (PM2)	3.3	0.0009	996	2.96 μW
BLE Modul – Transmit	3.3	25	4	330 μW
Sum total				**~340 μW**

2.3. Measurements

The system can be operated from water flow rates between 5.5 l/min and 15 l/min, which lies in the range of household applications. Below flow rates of 5.5 l/min operation is limited, because the voltage generated by the turbine is too low (<4 V), while still approximately 10 mW of power is available. To characterize the system several measurements have been conducted, all in the optimized housing mentioned before:

1. Output power and rotational frequency vs. load resistance for two different flow rates
2. Output power at optimal load resistance vs. rotational frequency of the turbine
3. Pressure drop between inlet and outlet vs. water flow rate

3. Results

The measurement of the generator's output power as a function of the load resistance, together with the rotational frequency of the turbine is shown in *Figure 3*. The curve was measured for two different flow rates. The higher flow rate equals to a maximum flow rate of 12.5 l/min, and results in a shifted curve for output power and rotational frequency. Water flow is taken from the building supply. Pressures oscillations at the supply line are reflected in the curves. For these measurements the turbines were uncalibrated with respect to flow rate. As can be seen, the rotational frequency of the turbine decreases from high to low load resistances by about 25 % (e.g. 200 Hz to 150 Hz), i.e., there is a strong feedback of the generator to the fluid flow. A successive measurement calibrates the flow rate versus rotational frequency including the system electronics connected, as is shown in Figure 4. During these measurements only the timer and the comparator module of the microcontroller is active for frequency measurement, to simulate no-load conditions. Only by this the correct flow rate can be determined, which then goes linearly with the rotational frequency of the turbine. The lower limit of 6 l/min was chosen, because from this flow rate on the system is capable of transmitting data once per second. The pressure drop between inlet and outlet, as a function of flow rate were also measured, without electronics connected (see Table 2). The pressure loss for 15 l/min is 360 mbar, while for 5 l/min it was measured to be 30 mbar. This shows the gain in performance achieved by the optimized turbine housing, as the original system, according to the datasheet requires a pressure drop of 0.6 bar for start-up.

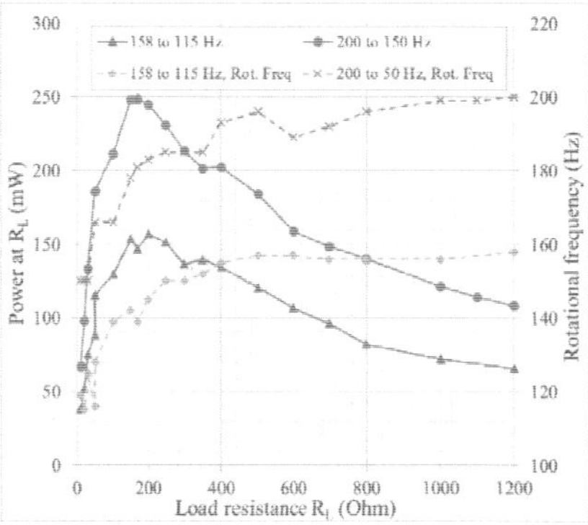

Figure 3. *Measured output power (solid) and rotational frequency of the turbine (dashes) vs. load resistance for two flow rates. The optimum load is in the range of the generators internal resistance (185 Ohm). The rotational frequency of the turbine decreases with smaller load resistances.*

Table 2. Measured pressure drop between inlet and outlet, with disconnected electronic.

Flow rate in l/min	15	14	13	12	11	10	9	8	7	6	5
Pressure drop in bar	0.36	0.31	0.29	0.22	0.19	0.15	0.10	0.07	0.06	0.04	0.03

4. Conclusion and Outlook

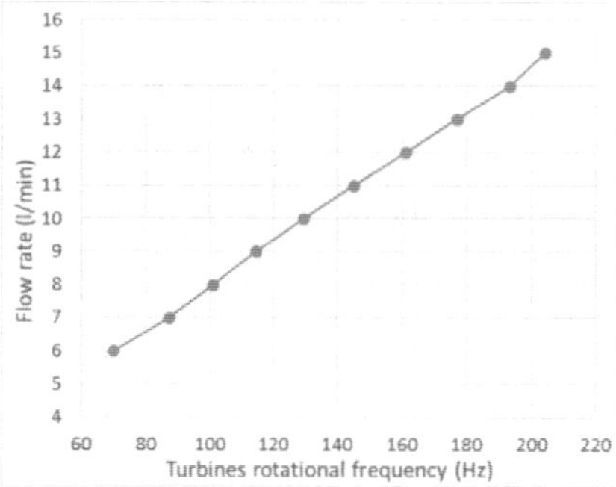

Figure 4. *Measurement of the flow rate through the prototype with connected electronics. During the frequency measurement only the timer and comparator module of the microcontroller is active to achieve a no-load measurement.*

A system capable of powering itself from the fluid flow with typical flow rates in domestic application (5-15 l/min) was successfully demonstrated. A great deal of development was put into the power management, with a hysteresis switch cutting off the system electronics from a buffer capacitor for voltages below 3.3 V and enabling operation from ~4 V. The capacitor was laid out such, that one full cycle of measurement and transmission was guaranteed. This ensures, that the BLE module does not undergo any undervoltage situation, which otherwise may be permanently damaged. Since for high flow rates the generator can easily deliver 100 mW and voltages above 20 V, the system electronics was protected by an active shunt resistor, if buffer voltages exceeded 5 V. The turbine shows strong feedback onto the flowrate depending on the electrical load, therefore measurements are conducted with only very light load, i.e. only the microcontroller's timer and comparator module is used. The power measurements show, that for flow rates of only 4 l/min enough power is available. However, at this rotational speed the generator only produces voltages of ~3.7 V (the voltage drop over the diodes is around 0.3 V), which is too small to charge the capacitor to the needed 4.0 V.

To conserve the energy below and above the supply voltage, in future a small battery will be integrated into the system, in combination, with a Delon voltage-doubler or a buck-boost converter. Further this would allow for more functionality, like a full, synchronized and encrypted BLE connection, unlike the advertisement mode which is currently used to save power. In future several other signal acquisition lines, such as for water temperature and pressure drop will be implemented into the system, which will enable the user to be provided with more data for saving hot water and giving feedback via Bluetooth Apps on the smart phone, or into a home automation system. Perspective even an automatic control of valves could be implemented, for a fast flow regulation in shower rooms (e.g. in a gym), so that the water temperature can remain more constant for all users.

References

[1] D. Hoffmann, A. Willmann, R. Göpfert, P. Becker, B. Folkmer, und Y. Manoli, „Energy Harvesting from Fluid Flow in Water Pipelines for Smart Metering Applications“, *J. Phys. Conf. Ser.*, Bd. 476, Nr. 1, S. 12104, 2013.

[2] R. Song, X. Shan, F. Lv, und T. Xie, „A study of vortex-induced energy harvesting from water using PZT piezoelectric cantilever with cylindrical extension“, *Ceram. Int.*, Bd. 41, Supplement 1, S. S768–S773, Juli 2015.

[3] P. Dhiman, F. Yavari, X. Mi, H. Gullapalli, Y. Shi, P. M. Ajayan, und N. Koratkar, „Harvesting Energy from Water Flow over Graphene“, *Nano Lett.*, Bd. 11, Nr. 8, S. 3123–3127, Aug. 2011.

[4] http://partner.vde.com/cosima-mems/cosima2015/Seiten/InternationalerWettbewerbiCan2015.aspx, accessed on 29/09/2016

[5] M. Wischke, M. Masur, M. Kröner, und P. Woias, „Vibration harvesting in traffic tunnels to power wireless sensor nodes“, *Smart Mater. Struct.*, Bd. 20, Nr. 8, S. 85014, 2011.

[6] S. Heller et al, "Electrical current measurement system for energy harvesting applications", Proceedings PowerMEMS 2016, IOP Conference Series.

Energy-autonomous wireless sensor nodes for automotive applications, powered by thermoelectric energy harvesting

P. Mehne, F. Lickert, E. Bäumker, M. Kroener, P. Woias
Laboratory for Design of Microsystems, IMTEK, University of Freiburg, Georges-Koehler-Allee 102, 79110 Freiburg, GERMANY

E-mail: philipp.mehne@imtek.uni-freiburg.de

Abstract. In this paper we will first present the measurement of temperatures on different positions at a diesel-powered car. As a result, several locations are identified as suitable to implement a wireless sensor node powered by thermal energy harvesting. Based on the data gained a thermoelectric generator (TEG) has been selected, and measurements of energy generation have been performed. Further, a complete energy-autonomous wireless sensor node was designed, including the TEG with its mounting bracket, an electronic power management, and a Bluetooth Low Energy (BLE) sensor node. Based on temperature differences from -10 K up to 75.3 K occurring in test drives, a low power set up was chosen to achieve a system start-up time below 10 minutes and to ensure service even under difficult ambient conditions, like high ambient temperatures or a slow movement of the car in stocking traffic. 2 minutes after starting the engine a power about of 10 mW is available from the chosen TEG, and in peak the power exceeds 1 W. In a 50 minute test drive it was possible to generate 650 J of energy. This information was used to develop the complete system, demonstrating the opportunity to deploy energy-autonomous wireless sensor nodes in a car, e.g. for exhaust gas monitoring. The system is used to gather sensor data, like temperature and humidity, and transmits data successfully via BLE to a prepared main node based on a Raspberry Pi.

1. Introduction

Owing to a growing number of sensors and electronic control units (ECUs) [1], there is a motivation for the application of wireless sensor nodes (WSN) in cars to replace wired sensors with all their drawbacks. A modern mid-size passenger car carries 20 to 40 ECUs with, in total, approximately 3 km of wires for the supply of electric components and communication [2]. This requires complicated and error-prone power/data grids, fabricated and installed by hand, not easily maintained, and providing a substantial cost factor.

Even highly efficient combustion engines are exhausting more than 60 % of fuel energy by heat dissipation. This leads to the opportunity to save wires, therefore space and weight in cars, and to add some flexibility through the integration of waste heat-powered WSNs at different parts of the car, which is shown in this paper. This work differs from other experiments, as all systems presented are deployed on-site in a car, and not tested in a laboratory setup [3]. Thermal measurement data gathered from test drives show several positions, where thermal energy can be harvested. Based on these data, a thermoelectric generator has been selected and power measurements were performed. In the most promising application targeted here, a peak power of 1 W is harvested from heat losses. The development of the WSN is ongoing, and will demonstrate the potential of retrofitting a car with, e.g., sensors for exhaust gas monitoring.

2. Temperature measurement set up

Figure 1 shows the measurement setup in the motor compartment of a Citroen C3 Picasso BlueHDi 100: There are five PT1000 1/3 DIN sensors at the (A) oil-level dip stick, (B) manifold, (C) radiator support panel top and bottom (D) as well as a sensor for ambient temperature (E) in the motor compartment next to the battery. In (F) a data logger is placed (LETE LE-LOG_S0), prepared with a four wire sensing interface. A second setup is used to record data at the bottom back of the car, with sensors mounted at the exhaust pipe [G] and the crossbar [H]. The test drive track (see Fig. 2 [6]) contains sections of highway, country road, and city traffic to determine differences in the thermal conditions at the selected positions. The positions were selected based on their basic thermal behaviour and expected temperature difference. All track sections are logged with an external GPS Logger (ASUS K013), giving precise information about velocity and position.

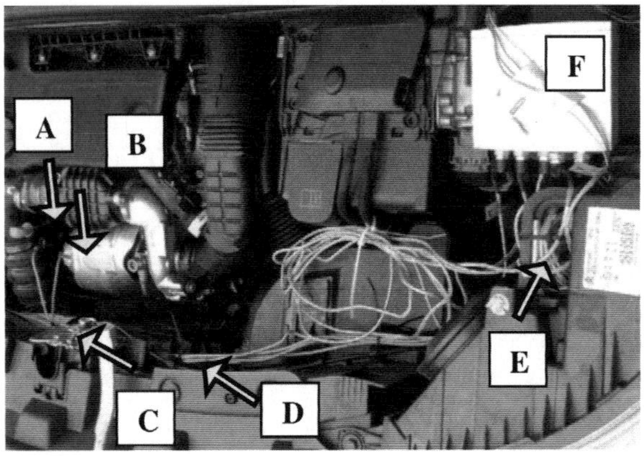

Figure 1. Motor compartment of a Citroen C3 Picasso Blue HDI 100 with five installed temperature sensors [A]-[E], and the data logger [F]. Sensors [B] and [D] are not directly visible, as hidden behind engine parts.

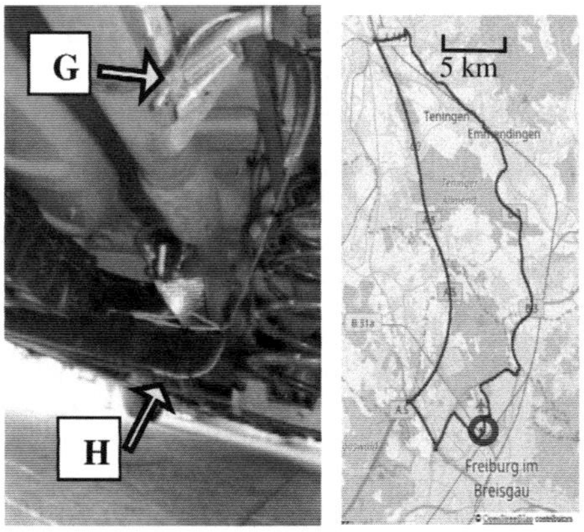

Figure 2. Temperature sensors at the end of the exhaust pipe [G] and crossbar [H]. The right picture shows the drive test track on a map [6], with the start and stop position indicated by the mark O.

3. Temperature measurements

Three test drives are realized to measure and compare temperature behavior, one is exemplary shown in Figure 3. The measured results are depending of the ambient temperature, especially with respect to the maximum temperature difference between exhaust pipe and crossbar temperature, the latter representing ambient temperature the closest (no direct sunlight and forced convection). With an ambient temperature around 22 °C, an average ΔT of 18.7 K can be reached at the exhaust pipe, while at approximately 11 °C ambient temperature an average ΔT of 41.4 K is found. In particular when the car is on the highway, the influence of the vehicle speed is clearly visible. This is related to higher forced convection at the mounted cooler, as well as the higher heat generation that comes with higher exhaust gas flow.

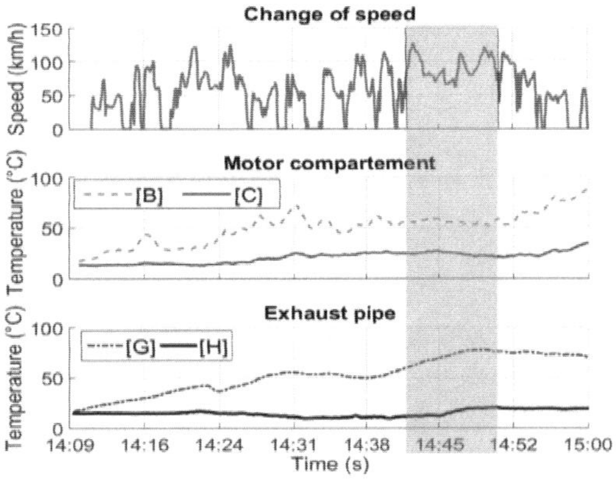

Figure 3. Measurements on May, 20th. 2016. The top plot shows the car's speed. The middle plot shows the temperatures at manifold [B] and radiator support panel [C] as the most promising positions for heat extraction in the motor compartment. The bottom plot shows the temperature at the end of the exhaust pipe [G] at the back of the car and at the crossbar [H]. The grey field is highlighting the highway section of the route.

4. TEG system design and power management

A suitable TEG (TEG-071-150-22, Thermalforce [4]) has been chosen and mounted at the exhaust pipe with a custom-made retainer (Figure 5, Top). Thermal coupling of the generator to the air is realized with a passive heat sink (Fischer Elektronik ICK S, 50x50x40 mm, [5]). Figure 4 shows measurements performed on the same test track as used for the temperature measurements. At the beginning of a drive the TEG delivers only a power output around 10 mW. Figure 5 bottom shows the increase of power dependent from the temperature difference between exhaust pipe and ambient temperature and in Figure 6 the system with main parts is displayed. The middle of the test drive shows the highway section, where a maximum power above 1 W is available. Nevertheless a low voltage DC/DC converter has been selected (enOcean ECT310) to enable a short power-up time after engine start.

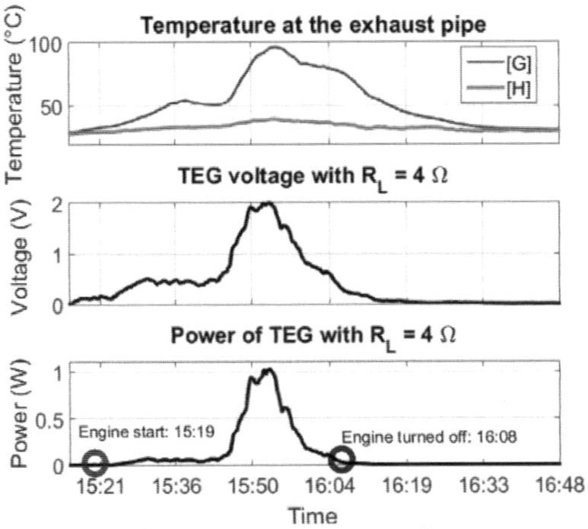

Figure 4. Measurement of temperature difference, output voltage, and output power of the mounted TEG [4]. Test drive from June, 28th 2016.

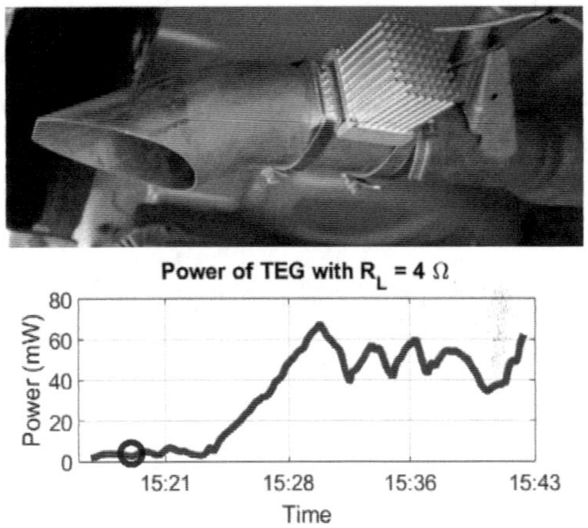

Figure 5. Top: TEG mounted at the exhaust pipe. Bottom: Starting behaviour of the TEG, and electric output power measured with a 4 Ω load resistor.

4.1. *Power Management*

Harvested energy is stored on a large 4 mF tantalum capacitor to ensure the start-up of a microcontroller (MSP430F5529) and the connected Bluetooth Low Energy (BLE) module (Bluegiga BLE112). A start-up switch is designed based on a comparator MIC833 (Microchip, [7]) with a supply current of 1 μA. It controls the activation of both modules, microcontroller and BLE, depending on the voltage level present at the capacitor. The chosen voltage level for start-up is 4 V. This relatively high value is required to ensure a complete start-up of the BLE module from stored charge during the first initialization, with an associated high power drain of appr. 8 mA over 500 ms. After startup, the microcontroller and the initialized BLE module continue operation in a timed sleep mode, with short wake-up periods for sensing and data transmission. The start-up circuit does also ensure deactivation of the microcontroller and the BLE module as soon as the capacitor voltage drops to a threshold of 1.8 V, thus resulting in a voltage hysteresis for operation in a predefined range between 4V and 1.8 V.

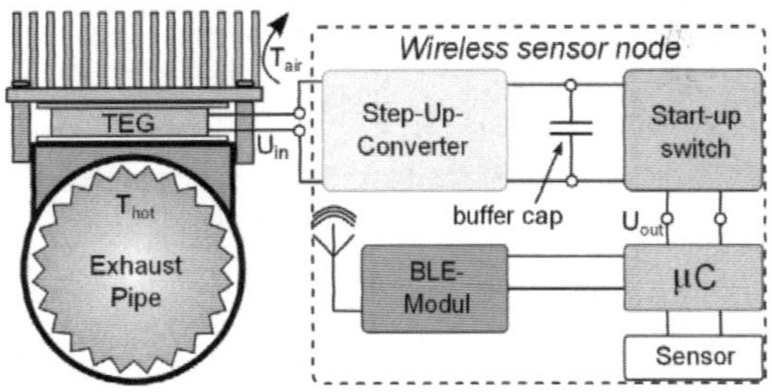

Figure 6. System overview with all main parts

The complete set up was tested with the engine running in idle state. The power management circuit and the radio module are connected to the TEG at the exhaust pipe. The harvested energy is sufficient to power the system. As seen in Figure 4 the output power during driving is much higher, thus giving the potential to perform tasks like data transmission with short sending intervals and powering of various sensors, e.g. a CO_2 or NOx sensor.

4.2. Wireless communication

BLE provides a broad availability in modern systems combined with a frequently advertised low power functionality. However, this argument has to be discussed critically in an application that requires frequent reconnection of a synchronized wireless BLE link. The current consumption of the BLE module used here (Bluegiga BLE112) is about 8,4 mA base current and 25 mA peak current per telegram over a period of 1ms to 10 ms during the transmission of 21 data packets with a payload of 4 byte each. As measured, approximately 1.5 mJ are needed for the transmission of such a single data packet with a payload of 4 byte in bi-directional communication, plus 3 mJ for the establishment of the synchronized wireless connection. Average power consumption is therefore 4.5 mW for 1 telegram per second. Using advertisement mode with a reduced data overhead would decrease the power consumption a lot. However, in automotive applications a bidirectional communication is desirable. From that reason the more power-hungry synchronized communication mode was kept as a realistic scenario. In comparison to that the microcontroller draws negligible power in the range of several µW.
In a laboratory setup using the designed system consisting of a low voltage step-up converter, the developed start up switch, the microcontroller and the BLE module, successful transmission of collected temperature data to a receiver station (Raspberry Pi) was achieved. In its current implementation the capacitor voltage is read by the microcontroller in a predefined interval, simultaneously with the aggregation of measurement data.

5. Conclusions and Outlook

The applicability of a thermally powered wireless sensor node at the exhaust pipe of a passenger car has been demonstrated successfully. The TEG applied for this purpose is working stable since its installation, and has shown stable output over several 1000 km of driving distance, proving again the robustness of the thermoelectric energy harvesting principle. Its functionality will be tested further on with respect to long-term stability and mechanical defects. The WSN will typically turn on within 15 minutes after the start of the car engine. This relatively long period is solely caused by the high energy demand of the BLE module in use, and could be significantly reduced by more sophisticated and low-power wireless communication.
This project shows further possibilities to investigate different places in a car to deploy TEGs, for example in the motor compartment. As well the power management has to be improved, to store excessive energy in times of high power generation, e.g. with a super capacitor. This will allow a faster start up time of the sensor node in the next driving cycle.

6. Acknowledgement

The authors would like to thank the German Science Foundation DFG for funding this project under the graduate research training program – GRK 1322 – Micro energy harvesting.

References

[1] J. Leohold, Communication Requirements for Automotive Systems, keynote talk at the 5th IEEE Workshop on Factory Communication Systems, Vienna, Sept. 22-24, 2004, *retrieved from http://www.ict.tuwien.ac.at/wfcs2004/downloads/*

[2] N. Navet, Y Song, F Simonot-Lion, C Wilwert, Trends in Automotive Communication Systems, *Proc. IEEE* 93 (6), pp. 1204-1223, June 2005.

[3] A. Costa, et. Al., Autonomous wireless sensor with low cost TEG for application in automobile vehicles, Eurosensors 2014, Procedia Engineering 87, 1226 – 1229.

[4] Thermalforce, TEG-071-150-22, data sheet: http://thermalforce.de/de/product/thermogenerator/TG71-150-22h.pdf

[5] Fischer Elektronik, ICK S, 50 x 50 x 40 mm: http://www.fischerelektronik.de/

[6] openstreetmap.com

[7] Microchip, MIC833, data sheet, http://ww1.microchip.com/downloads/en/DeviceDoc/MIC833.pdf

Evaluating Micro-Power Management of Solar Energy Harvesting using a Novel Modular Platform

J. Kokert, T. Beckedahl and L.M. Reindl
Department of Microsystem Engineering - IMTEK, Freiburg, Germany
E-mail: jan.kokert@imtek.uni-freiburg.de

Abstract. Micro-Power Management (µPM) is essential to supply power to autarkic sensor nodes from energy harvesting sources. As there are numerous ways to realize a µPM, the question arises of how to benchmark different managements under reproducible boundary conditions. In this paper we present these conditions for solar harvesting.Further, we propose a system efficiency definition, which is applicable to all self-powered systems. For verification, we use our modular construction kit, which is used to set up four different µPM configurations. We examined the interplay of state-of-the-art power converters with a supercapacitor array. As one result, the improvement of using a buck converter compared to an LDO was quantified by an increase of 10 percentage points in the system efficiency. The experiments show that the modular setup and the boundary conditions are suitable for such investigations.

1. Introduction

A Micro-Power Management (µPM) is essential to supply power to wireless sensor nodes (WSN) from energy harvesting sources. There are various ways of realizing a micro power management, either with discrete electronic components or with commercial power management integrated circuits (PMICs). It is a challenging task to find the optimal solution: although the efficiency of individual components can be determined at certain operation points, it is not clear which operating point is most dominant in interplay with other components and in a realistic scenario. Moreover, additional losses may originate from component interaction. Furthermore, the question arises of how to measure the energy balance and benchmark different systems under realistic and reproducible boundary conditions. In this paper we answer these questions.

1.1. Related work

Several publications ([1], [2]) document that µPMs are realized in various ways. In these publications the system evaluation is usually limited to a single test run of a few days in a non-repeatable real-word scenario. Different approaches to address this issue and to estimate the system performance in an adequate way are presented in the following:

In battery-powered wireless sensor networks (WSNs) the lifetime can be used as a performance indicator as done in [3]. The authors show that using a dc-dc converter prolongs the lifetime to 30%. In [4] the authors implement an energy-aware duty cycle and take the number of active time slots as a figure of comparison. The boundary conditions are vague, using a real solar panel and a fluorescent lamp. The authors of [5] present a energy harvesting platform for evaluation of double-layer capacitor (EDLC, short: supercaps) and thin-film batteries in conjunction with power converters. A flaw is that the energy harvester (a solar cell) is oversimplified to a constant voltage power supply.

1.2. General structure of micro-power managements

Figure 1 shows a general structure of a µPM. The two blocks shown in green (*energy extraction* and *voltage supply*) are power converters. The *energy extraction* block is required to extract as much energy from a harvester as possible. Hence, the output impedance of the harvester and the input impedance of the extraction block need to match. The *voltage supply* block is required to supply a constant voltage (e.g. 3.3 V) to a load.

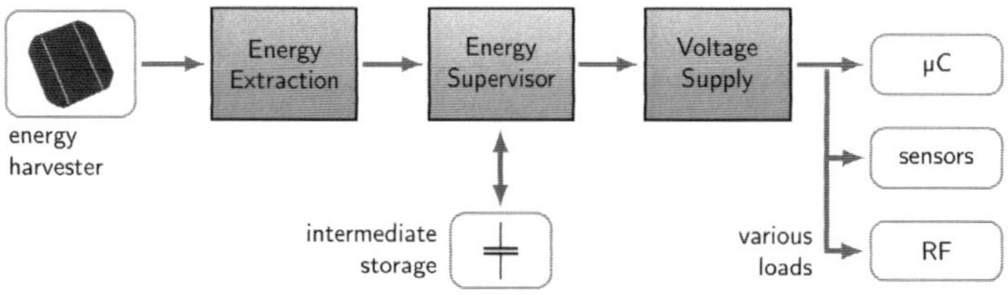

Figure 1: The micro-power management supplies regulated power to loads from energy harvesting sources

2. Methods and Materials

First, we present the harvesting scenario and propose the system efficiency as a performance indicator. Afterwards, we present the experimental setup comprising the modular construction kit, the solar cell simulator, the intelligent dummy load and the energy measurement. Then we illustrate the realization of the four different µPM configurations with the construction kit.

2.1. Harvesting scenario and system efficiency

Our design goal was to set up a harvesting scenario for all experiments, where the maximum amount of 200 J can be harvested and 100 J are consumed per day (E_{harv} and E_{cons}). The energies equate to an average power of 2.31 mW and 1.16 mW, respectively. An energy storage is used, which is large enough to buffer the theoretic maximum of excess energy of 100 J ($\Delta\hat{\text{E}} = E_{\text{harv}} - E_{\text{cons}}$). However, in reality there are several losses in the system mainly due to power conversion and only a certain amount of energy is stored (E_{stor}). To combine all quantities, we define the *system efficiency* η in Eq. (1). An $\eta > 50\%$ means that excess energy can be stored.

$$\eta = \frac{\text{result}}{\text{efford}} = \frac{E_{\text{cons}} + E_{\text{stor}}}{E_{\text{harv}}} \tag{1}$$

Table 1: Overview of the energy amounts used to determine the system efficiency

E_{harv}	E_{cons}	$\Delta\hat{\text{E}}_{(\eta=100\%)}$	$E_{\text{stor}}\vert_{\eta=90\%}$	$E_{\text{stor}}\vert_{\eta=75\%}$	$E_{\text{stor}}\vert_{\eta=50\%}$	$E_{\text{stor}}\vert_{\eta=25\%}$
200 J	100 J	100 J	80 J	50 J	0 J	−50 J

Journal of Physics: Conference Series **773** (2016) 012042 doi:10.1088/1742-6596/773/1/012042

2.2. Modular construction kit

To systematically evaluate different µPMs we developed a modular construction kit, which is shown in Fig. 2. It consists of standardized modules and a base board. Each module incorporates a dedicated power management function like a power converter, current sensor, comparator, power switch, energy storage or a load. Banana plugs are installed to connect harvesters, loads and laboratory equipment. Further details are available in our previous work [6].

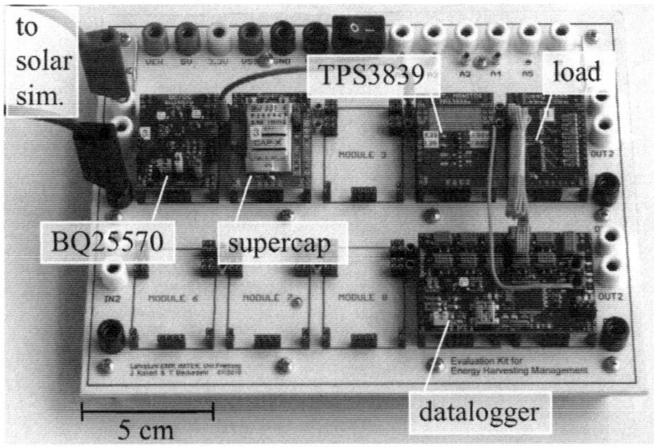

Figure 2: Construction kit which is used to set up the experiments in an example configuration

2.3. Performance of the solar cell simulator

The design goal was to represent a solar harvester which can deliver up to a total amount of energy of 200 J per day. To accomplish that, we assumed that a solar cell had an area of $A = 0.67\,\mathrm{cm}^2$ and took the standard global radiation of $G = 100\,\mathrm{mW\,cm}^{-2}$. The relation between the irradiance intensity and the short-circuit current I_{SC} is linear. Laboratory devices have measured I_{SC} of over $42\,\mathrm{mA/cm}^2$ and commercial solar cells are between $28\,\mathrm{mA/cm}^2$ and $35\,\mathrm{mA/cm}^2$ [7]. We took $30\,\mathrm{mA/cm}^2$ as a representative figure, resulting in a total I_{SC} of 20 mA of the cell.

The solar cell simulator is based on the one-diode solar model utilizing a real diode and an adjustable current source. This effort is needed to simulate the non-linear harvester impedance and to create the appropriate conditions for MPPT. We chose the GBJ1506 as diode and a Keithley 2400 source meter as adjustable current source. The mock-up harvester was characterized by an impedance sweep, resulting in an I-V curve, shown in Fig. 3. The fraction between the voltage at maximum power point (V_{mpp}) and the open-circuit voltage (V_{oc}) is 81.4%. The peak cell efficiency at an intensity of one sun ($\eta_{\mathrm{cell,p}}$) is 19.1%.

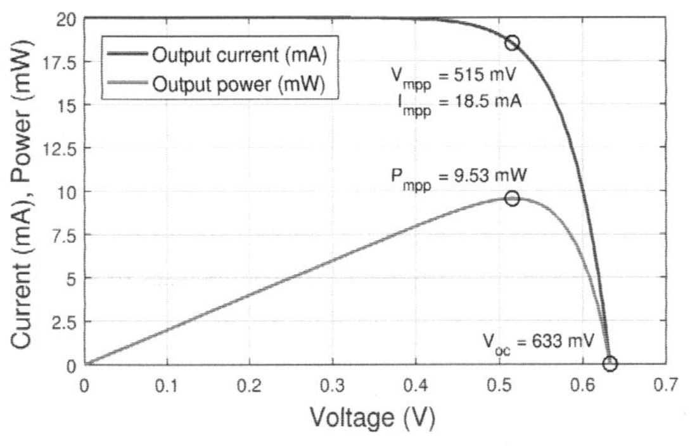

Figure 3: I-V curve of the solar cell simulator. The short-circuit current I_{SC} is 20 mA

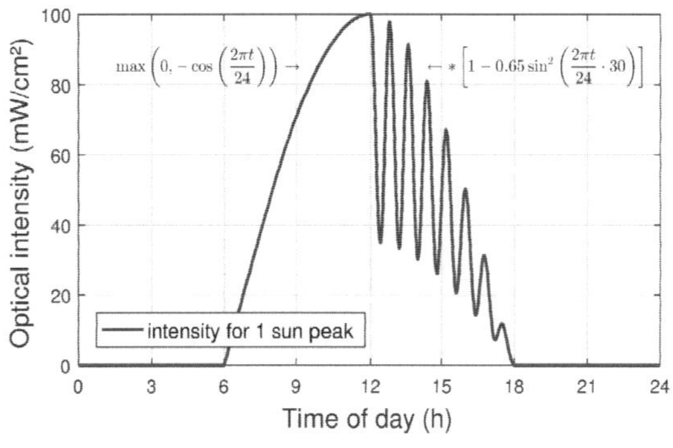

Figure 4: Standard solar day defined by trigonometric functions

2.4. Standard solar day scenario

The standard solar day in our scenario consists of a sunny morning and a cloudy afternoon and is well-defined by trigonometric functions, as given in Fig. 4 where the time t is given in hours. As the cell efficiency varies with the irradiance intensity, the electrical output power was evaluated at every point of the day curve. The integration over 24 h of the electrical output power yields a total amount of energy $E_{PV,day}$ of 212.6 J. This is slightly higher than the desired $E_{\mathrm{harv}} = 200\,\mathrm{J}$ and accounts for losses due to non-perfect impedance matching in reality.

2.5. Intelligent dummy load

An intelligent load like a microcontroller and a sensor is represented in our test setup by two precision resistors (tolerance = 0.1%) and two nMOS transistors. The design goal was to represent a system which has an energy consumption of 100 J per day, which equates to an average power consumption of 1.16 mW. For a simple but realistic scenario we chose the period to 3 s and the on-time to 30 ms, resulting in a duty cycle d of 1%. Furthermore, we chose 100 Ω for the on-resistance R_{on} and 100 kΩ for the off-resistance R_{off}. The average power is then calculated by Eq. (2) and yields 1.20 mW for $V_{\mathrm{cc}} = 3.3\,\mathrm{V}$:

$$P_{\mathrm{avg}} = d \cdot P_{\mathrm{active}} + (1-d) \cdot P_{\mathrm{sleep}} = V_{\mathrm{cc}} \cdot \left(d \cdot \frac{V_{\mathrm{cc}}}{R_{\mathrm{on}}} + (1-d) \cdot \frac{V_{\mathrm{cc}}}{R_{\mathrm{off}}} \right) \tag{2}$$

2.6. Current and voltage measurement

To determine actual values for E_{harv}, E_{stor} and E_{cons} of the µPM, currents and voltages are measured at three nodes. Namely, before the energy extraction, to and from the storage and after the voltage supply. For the measurements, modules with an INA226 current-shunt monitor IC are used. The shunt resistors used have a resistivity of 1 Ω with a precision of 0.2%. For readout, a datalogger module was developed, which further controls the dummy load and makes all data accessible via USB. All sensor modules and the datalogger are energy independent (powered over USB) to not falsify the energy balance of the system under test.

2.7. Power management setups

Table 2 shows the four chosen µPM configurations. We varied the energy extractor with state-of-the-art ICs, namely the BQ25570 and the ADP5090, and furthermore, we varied the voltage supply block to compared buck converter vs. LDO, both by simply replacing the modules.

In the experiments the excess energy is buffered in two supercaps connected in series (WPL2R72561626 from YEC). One capacitor has a nominal capacity of 25 F and tolerates a voltage of 2.7 V. The energy storage capability from $V_{\mathrm{c,min}} = 3.3\,\mathrm{V}$ to $V_{\mathrm{c,max}} = 5.4\,\mathrm{V}$ can be estimated to 114 J, which is larger than $\Delta\hat{E}$ and thus is sufficient for the experiments.

Figure 5 shows the setup of configuration #3 with the modular construction kit in full detail. The thick green wires are power connections and the thin orange wires are control signals (I^2C and digital I/O). In the experiments we cut the first 6 h of the standard solar day (Fig. 4) and started the experiment at t = 6 h with a pre-charged supercap of 3.4 V. We conducted the experiments in a room with a controlled temperature of 22 °C.

Table 2: Configurations of power managements used for the examination in this paper

config	energy extractor	voltage supply
#1	BQ25570	internal buck of BQ
#2	BQ25570	LDO MCP1700
#3	ADP5090	TPS62736
#4	ADP5090	LDO MCP1700

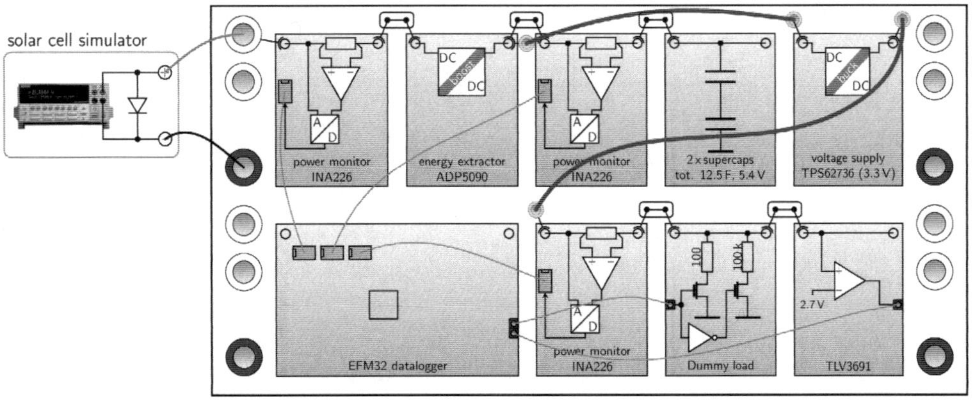

Figure 5: Schematic of the construction kit with the complete µPM setup of configuration #3

3. Results

In the following we compare the results of the four system configurations. In Fig. 6 the top series of curves shows the trend of the harvested energy E_{harv} over one day. The final values at $t = 30\,\mathrm{h}$ are summarized in Table 3 and range from 184.7 J and 192.4 J. We observed a mismatch (% mism.) of 11.3% in average with respect to the theoretic maximum of 212.6 J. In Fig. 6 the bottom series of curves shows the trend of the consumed energy over one day. The consumed energies are slightly (max. 8%) above the desired value of 100 J. The standard deviation $\sigma_{\mathrm{cons}} = 0.94\,\mathrm{J}$ underlines the reproducibility of the experiments.

In Fig. 7 the trend of the stored energy (E_{stor}) over one day is shown. The final values differ significantly, ranging from 11.2 J to 38.8 J. As comparison the theoretic maximum of storable energy ($= E_{\mathrm{harv}} - E_{\mathrm{cons}}$) is shown in dashed lines, which represents a system efficiency of 100%. Based on the values E_{cons}, E_{stor} and $E_{\mathrm{harv}} = 212.6\,\mathrm{J}$, the system efficiency at $t = 30\,\mathrm{h}$ is calculated using Eq. (1) as listed in Table 3.

The trend of the stored energy can be separated into three sections: from $6\,\mathrm{h} < t < 6.7\,\mathrm{h}$ the harvested energy is too low to fully supply the load and thus the storage is discharged. From $6.7\,\mathrm{h} < t < 16.5\,\mathrm{h}$ the power-converters generate excess energy which is stored in the supercap. At $16.5\,\mathrm{h} < t$ the energy balance is negative and thus the storage is discharged again.

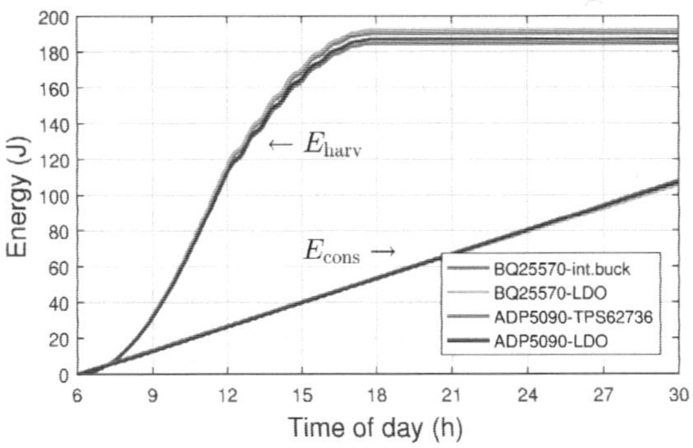

Figure 6: Trend of harvested energies (top series of curves) and consumed energy amounts (bottom series of curves) over one day

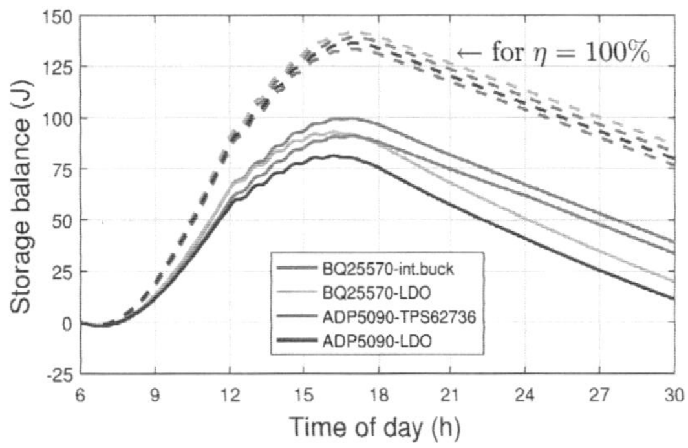

Figure 7: Trend of the theoretic maximum of storable energy (top series) and the actual stored energies (bottom series) over one day

Table 3: Final values at $t = 30\,\mathrm{h}$ for the internal energies and system efficiencies

config	$E_{\mathrm{harv.}}$	% mism.	E_{cons}	E_{stor}	η
#1	190.3 J	10.5%	107.1 J	38.8 J	68.6%
#2	192.4 J	9.5%	105.4 J	19.8 J	58.9%
#3	184.7 J	13.1%	108.0 J	33.5 J	66.6%
#4	186.9 J	12.1%	107.0 J	11.2 J	55.6%

4. Discussion

Although the stored energy E_{stor} differs significant between the four experiments, it is not that obvious in the calculated system efficiency. The efficiency can represent quite extreme cases: an $\eta = 0\%$ represents a system where no load is supplied and no storage is charged, whereas $\eta = 100\%$ represents a perfect system, where all excess energy can be stored. Moreover, both input power converters operate in their supposed operating region. An under- or overdimensioned converter will reduces the system efficiency significantly. Furthermore, the storage was large enough. A too small storage ($< 100\,\mathrm{J}$) reduces the capability of storing excess energy, which then reduces the system efficiency further.

Nevertheless, a clear difference of 10 percentage points of higher system efficiency is notable between buck converter and LDO as voltage supply. Especially during noon, where the storage voltage is high, the LDO conversion efficiency is low which is visible in a steeper slope for $t > 18\,\mathrm{h}$ in Fig. 7. The mismatch of E_{harv} to the 212.6 J is assumed to arise from the FOCV method. A system efficiency of max. 69% seems to be low but it is due to permitting all possible system losses and not due to the testbed.

5. Conclusion and outlook

Four different configurations of µPM where tested. The fixed boundary condition generate reproducible results and the proposed system efficiency makes it possible compare different systems. The approach is adaptable to other types of harvesters, such as thermoelectric and vibration harvesters. Further research will focus on varying the input power and storage capacity.

References

[1] J. Varley, M. Martino, S. Poshtkouhi, and O. Trescases, "Battery and ultra-capacitor hybrid energy storage system and power management scheme for solar-powered wireless sensor nodes," *IECON 2012 - 38th Annual Conference on IEEE Industrial Electronics Society*, pp. 4806–4811, 2012.

[2] C. Park and P. H. Chou, "Ambimax: Autonomous energy harvesting platform for multi-supply wireless sensor nodes," *2006 3rd Annual IEEE Communications Society on Sensor and Adhoc Communications and Networks, Secon 2006*, vol. 1, no. c, pp. 168–177, 2007.

[3] D. Oletic, T. Razov, and V. Bilas, "Extending lifetime of battery operated wireless sensor node with dc-dc switching converter," in *Instrumentation and Measurement Technology Conference (I2MTC), 2011 IEEE*. IEEE, 2011, pp. 1–5.

[4] P. Lee, Z. A. Eu, M. Han, and H.-P. Tan, "Empirical modeling of a solar-powered energy harvesting wireless sensor node for time-slotted operation," in *2011 IEEE Wireless Communications and Networking Conference*. IEEE, 2011, pp. 179–184.

[5] S. Bader and B. Oelmann, "Short-term energy storage for wireless sensor networks using solar energy harvesting," in *Networking, Sensing and Control (ICNSC), 2013 10th IEEE International Conference on*, April 2013, pp. 71–76.

[6] J. Kokert, T. Beckedahl, and L. M. Reindl, "Development and evaluation of a modular energy management construction kit," in *18. GMA/ITG-Fachtagung Sensoren und Messsysteme 2016*, 2016, pp. 84–91.

[7] C. Honsberg and S. Bowden. (2014) A collection of resources for the photovoltaic educator. [Online]. Available: http://www.pveducation.org/pvcdrom/short-circuit-current

WSN Nodes: Design Considerations and Energy Management

V.Boitier[1,2], **P.Tajan**[1], and **J.M.Dilhac**[1,3]
[1] CNRS, LAAS, 7 Ave du colonel Roche, F-31400 Toulouse, France
[2] Univ de Toulouse, UPS, LAAS, [3] Univ de Toulouse, INSA, LAAS
E-mail: vboitier@laas.fr, dilhac@laas.fr

Abstract. This paper details some practical considerations about the implementation of low power WSN, with a focus on energy consumption.

1. Introduction

Wireless Sensors Networks (WSN) are necessarily autonomous in energy, particularity that involves the knowledge and control of the energy consumption.

Some aspects related to energy are largely developed in scientific literature, such as reconfigurable interconnections between nodes or on line modifications within the node to adapt functionalities depending on the available energy [1, 2]. Another point is related to Harvesting or Scavenging Systems devoted to power battery-free WSN nodes [3, 4, 5]. However, few papers [6] are dedicated to minimizing the consumption of the measurement sub-part, and on the other hand the complete energetic budget is too rarely described. In this paper, we choose to develop these points based on a practical implementation of an outdoor WSN.

The first part of this paper details the structure of the deployed WSN. The second part explains the End Device structure and its operations. The last part presents some energy considerations (unexpected start-up, wake-up time, energy leakage, autonomy …). Those aspects are critical for a good design. In this article, they are experimentally illustrated.

2. WSN solution

Each node of the network is based on Jennic 5148 module which has been specifically developed for low power WSN. We have chosen this module because the user has not to interfere with the network layer. The user very simply will have to use a template in order to implement the protocol of its embedded system. We chose a tree topology (fig. **1**). The network is made of a supervisor, routers and end devices. The **coordinator** is always ON, aggregating data from sensors and sending them to a computer through a USB line. **Routers** are also always ON, relaying data between end devices and supervisor. Maximum distance between supervisor and end devices ranges between 45 meters (open space) and 10 meters (indoor). The **end devices** perform sensing and transmission of readings.

3. End Device Node (ED)

3.1. Structure of the ED

To offer modularity, each node is made of three parts (fig.**2** and fig.**3**): one part is composed of the Jennic 5148 module (microcontroller and radio), one stack is devoted to energy (supercapacitors storage in [7], secondary battery in this paper with or without a harvesting stage -BQ25504-) and the last part is focused on the measurement circuit and depends of the required specifications.

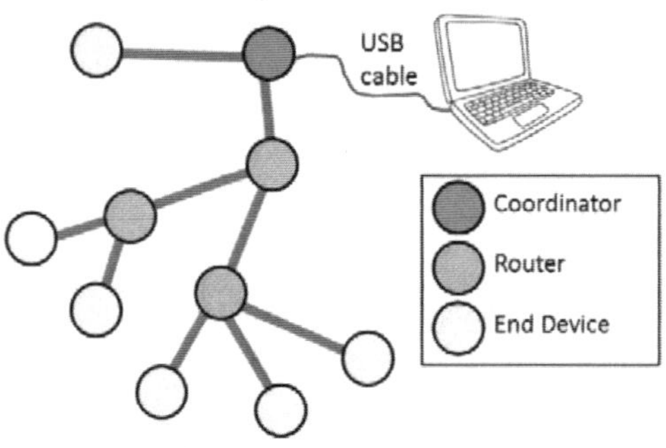

Figure 1. *Topology tree.*

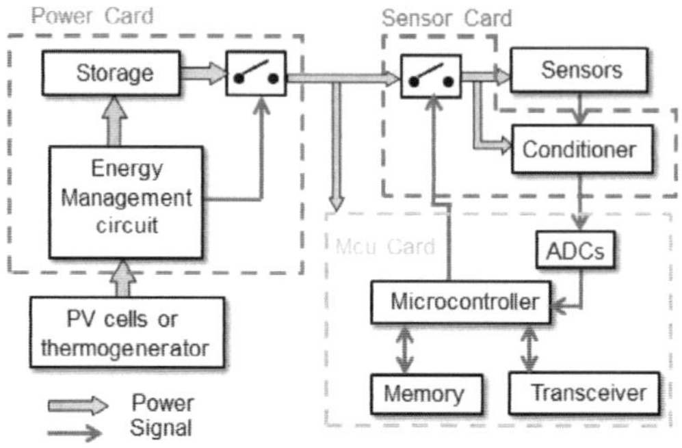

Figure 2. *Generic structure of an ED node.*

The **Mcu Card** is composed of a JN5148 module which is powered through a LDO Regulator TPS78227 (V_{out}=2.7V), which efficiency is greatly better than the TPS73633 previously used. The **Power Card** includes a secondary Li-Ion Coin Cell Battery - LIR2450 3.6V 120mAh and may incorporate a BQ25504 CI which harvests and manages energy from PV cells or from a thermogenerator (STPV1040 circuit has also been tested as a photovoltaic harvester circuit). A load switch (TPS2101) offers the ability to disconnect the battery and the load in order to protect the battery against deep discharge.

The **Sensor Card** adapts the sensor outputs to the voltage range required by the ADCs of the Mcu. A load switch controlled by the Mcu will stop sending power to the board in order to minimize the consumption as the Mcu enters in sleep mode. Several cards with different conditioning stages have been designed accordingly with sensors. Only analogic sensors are presented in this paper.

3.2. Working principle

Medium Access Control: to initiate connection (fig.**4**), a large power level is necessary. Once connection established, to maintain it, a beacon is used (10ms emission with a period up to 6s). The average power required is then of 13mA under 3.6V.

Steady state (fig. **5**): after a sleep time phase (depending upon module type, Isleep ≈ 4 to 10μA @ 3.6V), with periodic beacon activity, the wake-up is triggered during 100ms every 24s; the microcontroller activates sensing and sends data to the coordinator. Then after turning off the sensing devices, it goes back into the sleep mode. This sequence is active till network is active and node is powered.

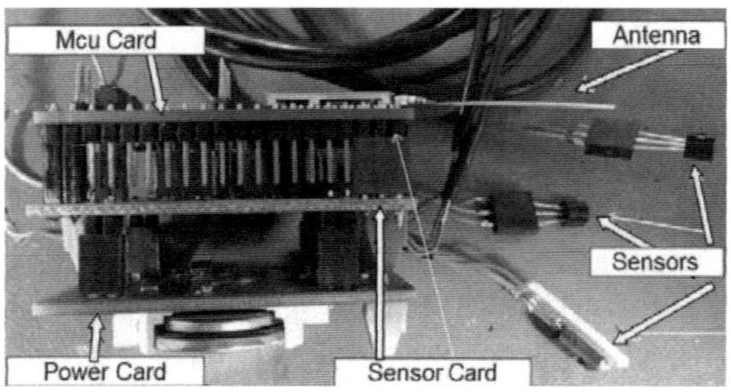

Figure 3. *Assembled End Device.*

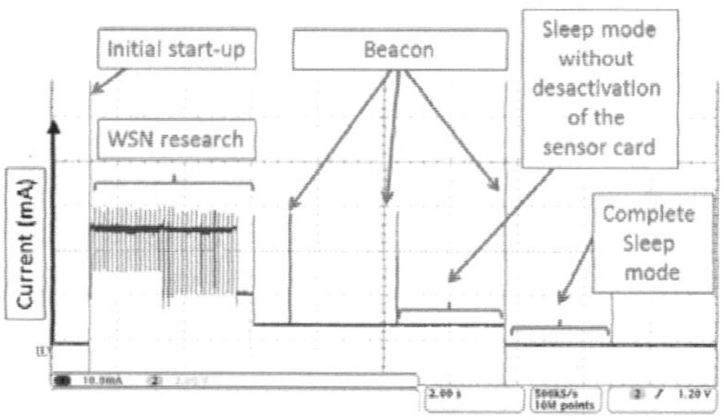

Figure 4. *Current furnished to ED during start-up.*

4. Energy issues

4.1. Setting of End Devices (ED)

A reduced sampling frequency (for instance one measurement or message each 15 min) is required to minimize energy consumption, but is not enough in the deployment / initialization phase of the network, particularly if the end device is positioned close to the maximum connecting range.

Consequently, we incorporated an initial phase, 5 min in duration, during which each 6s a counting data is transmitted. This allows the operator to properly adjust position of end device together with aerial orientation, therefore helping for a proper start-up of network. Of course, this implies an added, small but unavoidable, energy budget.

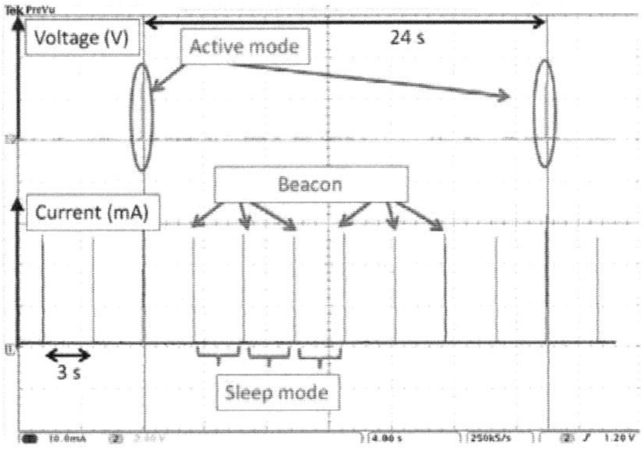

Figure 5. *(up) Sensor card supply voltage, (down) Battery current.*

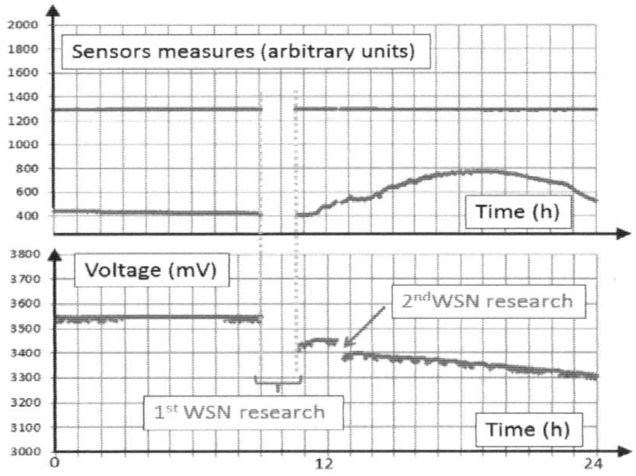

Figure 6. *(up) Measurements performed by ED, (down) Battery voltage.*

4.2. Sleep mode.
A proper setting of the sleep mode is mandatory. Experimentally, we noticed that an inadequate programming of microcontroller leads to an extra consumption in the sleep mode (400 μA instead of 5 μA). Care must be taken to actually switch off important functions such as UART, Flash Memory, ADC, together with setting unused inputs and outputs to high impedance mode.

4.3. Re-establishing radio link.
If the connectivity is lost, the end devices concerned enter a high consuming phase ($<i_{ED}> = 4$mA) while trying to re-establish the connection with routers. In fig.**6**, router has been inadvertently switched off for *100* minutes, and after being turned on again, network connectivity is re-established but the voltage level at battery terminals of the ED dropped from *3.55*V to *3.45*V.

4.4. Active mode duration.
When the microcontroller is turned on, data sampling through the ADC must be delayed until sensors output transient phase is terminated (1ms in our case). In active mode, the end device consumption is of the order of about 20mA for Jennic module, and between 40 and 100mA for Sensor Card. Therefore the active mode duration will be adjusted to a minimum.

4.5. Using an Energy Harvester.
The energy harvester must be properly designed because, if not, the energy captured would ne insufficient to power the devices. This is illustrated by fig.**7**: the voltage between the battery terminals at end of day is there less than at the beginning, showing that the energy balance is negative and that the energy reserve is decreasing. On the other hand, when the photovoltaic cell connected to BQ 25504 is not illuminated, the integrated circuit consumes a rather large amount of energy (I_{BQ} = 14μA), to be compared to I_{JENNIC} = 4μA when the Jennic module is in sleep mode. In a similar situation, with STPV1040 device, the consumed current is even higher: 100μA.

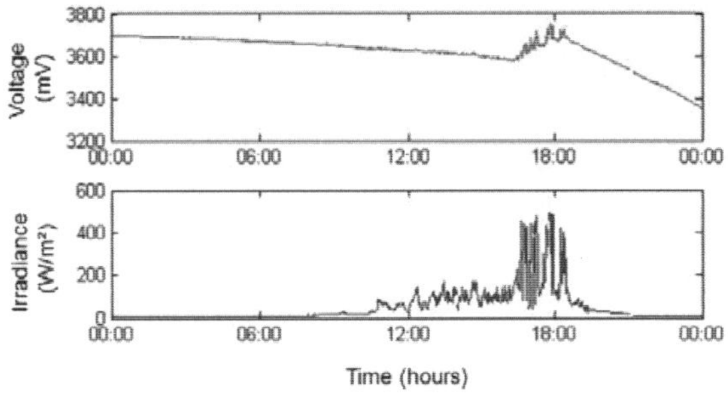

Figure 7. *(up) Coin cell voltage, (down) solar irradiance.*

mode :	init	beacon (7 beacons then 1 measure)	measure and communication	sleep	total
time	4 s	12 ms each 3.3s	110ms each 26.2s		
current	25mA	12 mA	13.6 mA	4.4μA	
voltage max	4V	4V	4V	4V	
energy	0,4J	4mJ for one beacon and 28 mJ /26.2s	5,8mJ /26,2s	0.46 mJ / 26.2 s	10.3mJ /26,2s

Tab.1. *Detailed consumption of a node (according to fig 5).*

4.6. Autonomy vs temperature

If during a lab test, the autonomy (@T ≈23°C) more or less fitted with calculations (within a 5% error margin), when performing similar tests outdoor in Summer (Tab.1) with a cell coin LIR 2450 submitted to daily temperature variations (15°C to 47°C), the lifetime was then much lower than expected (≈20 days instead of 45 days).

4.7. Router consumption.

In the above context, the router consumption is constant and quite large, 45 mA, and implies either to have a very large energy storage and harvesting system, or alternatively to have the router plugged to the mains.

5. Conclusion

We have presented the practical implementation of a low power wireless network, stressing the issues associated with the reduction of energy consumption and network initial setting. With respect to this last point, the connecting range of nodes is limited and environment sensitive. Initial deployment is therefore facilitated by a preliminary phase (even if energy demanding) in favor of telecommunication setting and optimization. With respect to the environment, high temperatures and humidity jeopardize energy autonomy. Harvesting ambient energy may balance these effects, but the harvester consumption when no ambient energy is accessible must be taken into account.

At the software level, active phases must be limited to the minimum required, and at the hardware level sensors must be chosen as energy efficient and exhibiting short transients when switching from sleep to active modes. Provided that these general guidelines are considered, the concept of autonomous wireless sensor network makes sense.

6. References

[1] C. Alippi, G. Anastasi et al., Energy Management in Wireless Sensor Networks with Energy-hungry Sensors, *IEEE Instrumentation and Measurement Magazine*, Vol. 12, N. 2, April 2009, pp. 16-23

[2] Hana Besbes et al., Analytic Conditions for Energy Neutrality in Uniformly-Formed Wireless Sensor Networks , *IEEE Transactions on Wireless Communications*, 2013, Vol.12, Issue: 10 Pp. 4916 - 4931,

[3] A. Liberale; E. Dallago; A.Lazzarini Barnabei, Energy harvesting system for wireless body sensor nodes, *IEEE Biomedical Circuits and Systems Conference (BioCAS)*, 2014 pp 416 - 419,

[4] J. Pedro Amaro et al., Energy harvesting for Zigbee compliant Wireless Sensor Network nodes , *IECON, 2012*, pp :2583 - 2588,

[5] R. Joshi, J. Zhu, Ambient energy harvester design for a wireless sensor network, *IEEE International Conference on Electro/Information Technology (EIT)*, 2015 pp 246 - 250

[6] A. Giuliano, Enhanced piezoelectric energy harvesting powered wireless sensor nodes using passive interfaces and power management approach, *Thesis of Cranfield University*, 2014.

[7] V. Boitier et al., Under Voltage Lock-Out Design Rules for Proper Start-Up of Energy Autonomous Systems Powered by Supercapacitors, *PowerMEMS 2013, IOP J. of Physics: Conference Series 476* (2013) 012121

Scaling effects in a non-linear electromagnetic energy harvester for wearable sensors

M. Geisler[1,2], S. Boisseau[1,2], M. Perez[1,2], I. Ait-Ali[4] and S. Perraud[3]

[1] Univ. Grenoble Alpes, F-38000 Grenoble, France
[2] CEA, Leti, Minatec Campus, 17 rue des Martyrs - 38054 Grenoble Cedex 9, France
[3] CEA, Liten, 17 rue des Martyrs - 38054 Grenoble Cedex 9, France
[4] CityzenSciences, 208 rue Garibaldi – 69003 Lyon – France

matthias.geisler@cea.fr

Abstract. In the field of inertial energy harvesters targeting human mechanical energy, the ergonomics of the solutions impose to find the best compromise between dimensions reduction and electrical performance. In this paper, we study the properties of a non-linear electromagnetic generator at different scales, by performing simulations based on an experimentally validated model and real human acceleration recordings. The results display that the output power of the structure is roughly proportional to its scaling factor raised to the power of five, which indicates that this system is more relevant at lengths over a few centimetres.

1. Introduction

The energetic self-sufficiency of body-worn sensors is an important matter to enhance usage convenience and durability. In certain applications such as sportswear and smartwatches, one relevant approach consists in converting a small fraction of the user's mechanical energy into the required electrical energy. In a previous study [1], we proposed an original optimization method for an inertial generator whose structure is well-known and studied in several works [2-4]. A prototype roughly the size of an AA battery (15mm Ø x 52mm length) carried in an armband by a subject running on a treadmill produced high power densities, up to 550μW/cm³ in a resistive load.

This type of inertial harvester complies well with several wearable systems requirements, such as low cost and easy hermetic packaging. However, the comfort of the solutions remains a challenge, because of their inherent rigidity and volume, and pushes towards the continuous reduction of their dimensions. Therefore, we examine in this paper the impact of the down-sizing of the previously studied structure.

2. Generator structure and model

The generator is composed of a magnetic mass moving between two repulsive magnets inside a straight-lined tube, around which are wrapped several identical and independent coils to enable Faraday's induction (*Figure 1a*). The moving mass includes several magnets with alternate polarities, separated by ferromagnetic spacers. This configuration aims at providing high electromagnetic damping levels [4].

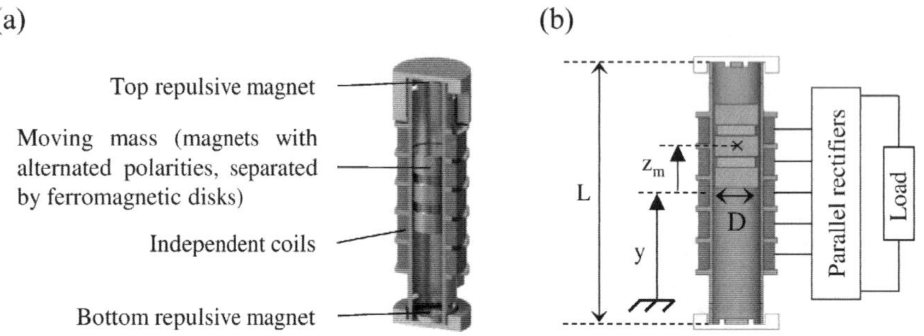

Figure 1 – (a) Structure of the non-linear harvester. (b) Model parameters

Model parameters for this structure are illustrated in *Figure 1b*. The evolution of the position of the moving mass m in the reference frame of the generator when the latter is subject to an exterior stimulation $\ddot{y}$, and the instantaneous dissipated electrical power P_e are expressed in equations (1) and (2), respectively. c_m is the constant mechanical damping coefficient (fixed at 0.01 N.s/m for all this study). c_e is the electrical damping coefficient, approximated by a periodic function of the mass position z_m, with an amplitude A_{Ce} and a spatial period p equal to the magnetic period of the multipolar mass (3). F_{rep} is the magnetic repulsion force estimated numerically (for both top and bottom repulsive magnets). The complete model for this system and its experimental validation are detailed in [1].

$$m\ddot{z}_m + c_e(z_m)\dot{z}_m + c_m\,\dot{z}_m - F_{rep}(z_m) = -m\ddot{y} \tag{1}$$

$$P_e = c_e(z_m)\,\dot{z}_m^2 \tag{2}$$

$$c_e(z_m) = A_{Ce}\cos^2(\frac{2\pi}{p}z_m + \beta) \tag{3}$$

3. Method

Using Matlab Simulink, we simulated the performances of the previous modelled system for various generator sizes, with a recorded acceleration sample as the mechanical input $\ddot{y}$ (subject running at 4.8 km/h on a treadmill, with the system fixed at his arm). The ratio between the guiding tube's length L and inner diameter D was kept constant and equal to $L/D=6.25$. The scaling factor k is defined as $k=D/D_0$, where $D_0=8mm$ is the inner diameter of the AA battery-sized reference energy harvester presented in [1]. For each value of k, all other parameters were optimized numerically to yield the maximum electrical power, while preventing shocks between the moving mass and the repulsive magnets. Output voltage was also considered to control the application-readiness. From the value of the optimal electrical damping determined in the power optimization step, the parameters of the multipolar magnetic mass, and the choice of some coils properties (60µm ø copper wire, 0.5 copper fill factor), we evaluated an optimal number of turns of the coils and the associated electrical resistance R_c. Given that all coils are identical, and considering a unique resistive load R_l at the output of the parallel rectifying circuits, the portion of the electrical power dissipated at the load $\frac{P_l}{P_e} = \frac{R_l}{R_l+R_c}$ is identical for each coil. This configuration allows to calculate the RMS value of the output voltage V_{RMS} from the average dissipated electrical power $P_{e,ave}$, with $V_{RMS} = \sqrt{P_{e,ave} \cdot \frac{R_l^2}{R_c+R_l}}$.

4. Simulation results and discussion

The dimensions of the optimized systems at several different scaling factors k are displayed in *Table 1*. It is worth noticing that the shapes of the optimal repulsive magnets are rather 'flat', leaving a maximum length available for the inertial mass displacement. The simulated electrical performances are plotted in *Figure 2*.

Scaling factor k	L (mm)	D (mm)	Multipolar magnetic mass			Repulsive magnets	
			Magnet unit (Ø mm x h mm)	Number of poles	Mass (g)	Bottom magnet (Ø mm x h mm)	Top magnet (Ø mm x h mm)
0.38	18.75	3	3 x 3	2	0.34	2.6 x 0.1	0.6 x 0.2
0.5	25	4	4 x 3	2	0.64	4 x 0.6	1 x 0.2
0.63	31,25	5	5 x 4	3	2.2	2 x 1	1 x 1
0.75	37.5	6	6 x 4	3	3.2	6 x 0.5	4 x 0.1
1	50	8	8 x 4	4	7.8	8 x 1.5	4 x 0.5
1.25	62.5	10	10 x 4	4	9.2	10 x 3	10 x 0.5

Table 1 – Parameters of the optimized systems for the considered values of the scaling factor

4.1. Load power

The average electrical power dissipated by the load ($P_{l,ave}$) is plotted in *Figure 2b*. It varies from 38 μW (k=0.375) to 12.42 mW (k=1.25). An adjusted exponential trend αk^{β} with α=4.7 and β=4.8 is plotted alongside the simulated results, and indicates that the down-scaling possibilities of this structure are quickly limited. Several factors can explain this observation. Smaller moving masses obviously provide lower mechanical energy inputs. Besides, it was shown in [1] that limiting the amplitude of motion of the moving mass can also reduce the performances of inertial harvesters, and especially with human-type inputs. Finally, size reduction limits the levels of electromagnetic coupling achievable (*Figure 2c*), as shown in another study [5]. This effect can be observed here with all cases $k \leq 0.75$, for which the optimal electrical damping corresponds to the maximum value that can be provided. This "saturation" lessens the power furthermore at the affected scales.

4.2. Output voltages

The calculated RMS voltages at the output load are indicated in *Figure 2d.* A particularity of the "periodic" electromagnetic damping provided by this generator structure is that a given value of its amplitude may correspond to different possible sets of coil parameters (length, number of turns). In particular, coils configurations with greater numbers of turns yield higher voltages. This effect especially visible for the greatest scales (for k=1.25, output voltages range from 0.7 V to 25 V). While the size reduces, the optimum coupling factor becomes equal to its highest value, thus reducing the number of optimum coils configurations and the relative output voltage range. At the lowest dimensions ($k \leq 0.63$), levels of output voltages produced by the coils in the optimum configurations become too low to be directly rectified with diodes (V_{RMS}<200 mV, which is the forward voltage drop of a BAS70 diode). Since the coils must be kept independent in this structure to produce a sufficient level of electromagnetic damping, this issue is not easily solvable and requires voltage multipliers circuits [6].

4.3. Power density

Power density is a key parameter to describe the evolution of the performance while down-sizing the structure. To each coils configuration corresponds a given coil thickness. From its value and the length of the system, an estimation of the volume of the harvester and hence of the power density can be determined (Power density = Average load power/generator volume). For comparison, this power density was calculated for the coils configurations corresponding to the extremum values of the output voltages studied previously. The results (*Figure 2e*) show that its value decreases exponentially along with the scaling factor, from 1.586 mW.cm^{-3} (k=1.25) to 44 μW.cm^{-3} (k=0.38). This evolution confirms the limitation of this structure in terms of down-scaling possibilities. Besides, it is worth noting that higher-voltage coils configurations are associated with lower power densities, which may call for a compromise when designing the system, depending on the application.

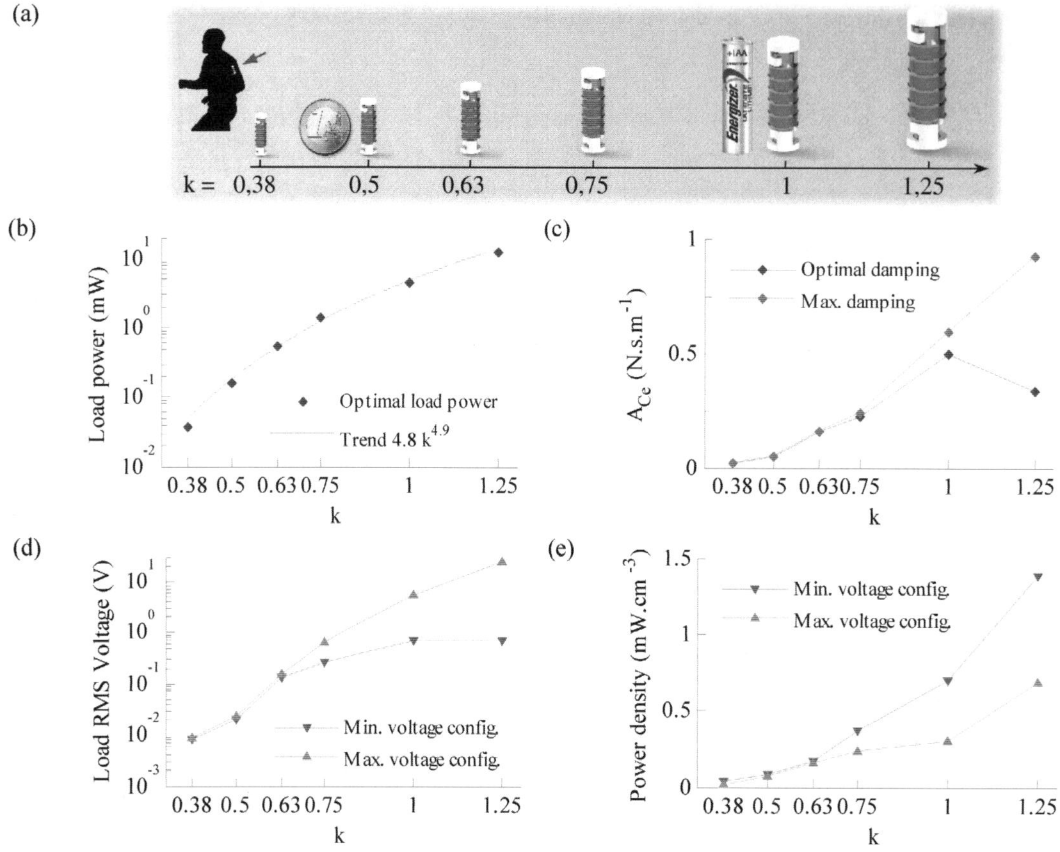

Figure 2 – Simulation results ("4.8 km/h running" input) (a) Computed scales (b) Optimum load power in a matching resistance (c) Electrical damping amplitude A_{Ce} (maximum and optimum). (d) Load RMS voltage and (e) power density, for both the min. and max. voltage coil configurations associated with the optimum load power value.

5. Conclusions

The performances of this type of energy harvester for human motion are strongly impacted by the down-scaling process. Output voltages are the first limiting factor if the device is used with rectifying diodes. In this matter, the critical size appears to be around k = 0.63 (L= 31.25mm D = 5mm), which corresponds to an average load power of 530 μW, but for an RMS output voltage of 150 mV only. More complex power management circuits are thus required to handle the output effectively at these smallest sizes.

References

[1] Geisler M, Boisseau S et al. 2016 Human-motion energy harvester for autonomous body area sensors, *submitted.*

[2] Ylli K, Hoffmann D et al. 2014 Human Motion Energy Harvesting for AAL Applications *J. Phys. Conf. Ser.* **557** 012024

[3] Nico V, Boco E et al. 2015 Dynamics of a C-battery Scale Energy Harvester *ICT- Energy Lett.* 7

[4] Cheng S and Arnold D P 2009 A study of a multi-pole magnetic generator for low-frequency vibrational energy harvesting* *J. Micromechanics Microengineering* **20** 025015

[5] O'Donnell T, Saha C, Beeby S and Tudor J 2007 Scaling effects for electromagnetic vibrational power generators *Microsyst. Technol.* **13** 1637–45

[6] Arroyo E and Badel A 2011 Electromagnetic vibration energy harvesting device optimization by synchronous energy extraction *Sens. Actuators Phys.* **171** 266–73

PowerMEMS 2016 IOP Publishing
Journal of Physics: Conference Series **773** (2016) 012045 doi:10.1088/1742-6596/773/1/012045

Electrostatic Energy Harvesting Circuit with DC-DC Convertor for Vibration Power Generation System

J Wei[1a], E Lefeuvre[a], H Mathias[a], F Costa[b]

[a]C2N – CNRS, Univ. Paris Sud/Univ. Paris Saclay, Orsay, France
[b]SATIE – CNRS, Univ. Paris Est Créteil, France

E-mail: jie.wei@u-psud.fr

Abstract. This paper presents an interface circuit with power control features for electrostatic vibration energy harvesting. A DC-DC convertor is used to control the output voltage of a diode-based charge pump circuit. Therefore, the maximum and minimum voltage across the variable capacitor of the energy harvester may be adjusted to track the maximum power point of the system. The power conversion function of the DC-DC convertor depends on the switches configuration. An example of Maximum Power Point Tracking (MPPT) for different conversion function is presented in this paper. Simulation results show that at least 10 μW is generated.

1. Introduction

Replacement or recharging of batteries is a major problem for wide development of microscale devices such as wearable electronics, medical implants and wireless sensor networks. Energy harvesting provides a solution. One of the possible power sources in this case is ambient vibration. Electrostatic transduction is a promising way to convert ambient vibrations into electricity. In functional systems, energy harvesting device generally require circuits interface between the harvester and the energy storage element to provide power management function. Since energy harvesting devices produce relatively low output power (often only microwatts), specialized power electronic circuit are necessary.

In this paper, an interface circuit with power control features is presented. It includes a N-cells diode-based interface circuit implementing rectangular charge-voltage cycles [1, 2]. The simplicity and the high efficiency of this interface circuit enabled to implement ultra-low-power electrostatic energy harvesters [3]. However, this diode-based circuit cannot optimize the performance of the system by itself. Due to mechanical damping induced by energy conversion, previous works showed that there exists an optimal operating point at which the output power is maximum [4]. In practice, the operating point can be controlled by adjusting the output voltage value of the N-cells circuit. The association of this circuit with a DC-DC converter enables to achieve this control.

[1] To whom any correspondence should be addressed.

2. Theoretical analysis of the proposed circuit

The proposed circuit is presented in Figure 1. It consists of an N-cells diode-based interface circuit and a DC-DC convertor. The N-cells diode-based interface circuit is used to provide initial charge to the variable capacitor and transfer the generated energy to the storage capacitor (or bulk) C_{res}. The converter's input is connected to the C_{res} and the output is here connected to a rechargeable battery. The start switch is used to provide initial charge to the variable capacitor. The DC-DC converter consists of an inductor and four switches. Its power conversion function depends on the switches configuration and drive.

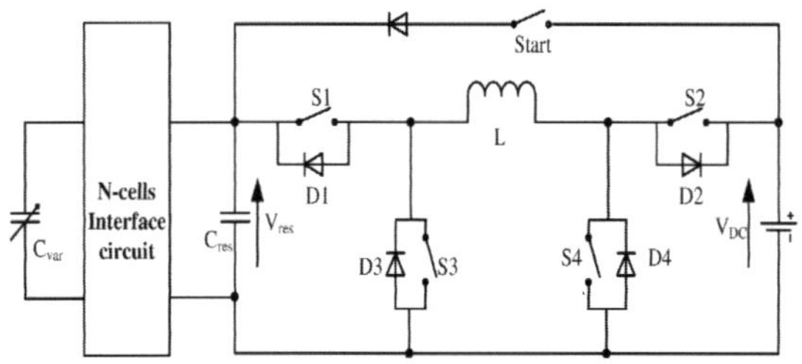

Figure 1. Schematic diagram of the proposed circuit

		Controlled switches	State of the rest switches: Open(O) or Closed(C)
Unidirectional	buck	S1	S2(C),S3(O), S4(O)
	boost	S4	S1(C),S2(O), S3(O)
	Buck-boost	S1, S4	S2(O),S3(O)
Reversible	Buck-boost	S1,S2, S3,S4	

Table 1. States of switches for different type of converter

Table 1 presents the control of the switches for different possible features. the converter is used to transfer the energy from the storage capacitor to the battery. Hence, the converter function to be used is determined by the optimal voltage range [V_{CresL}, V_{CresH}] across Cres and the voltage V_{DC} of the battery. If $V_{CresL}>V_{DC}$, a buck converter is required and a boost converter is necessary when $V_{CresH}<V_{DC}$. Indeed, they can all be replaced by a buck-boost converter, but with more switches to control, which requires more power to drive the switches. A reversible converter can be also achieved using this generic topology. The use of a reversible converter enables dynamic control of the voltage V_{Cres} by controlling energy flows between the battery and the capacitor C_{res} in both directions.

3. Spice Simulation

To study the performance of the proposed circuit, an out-of-plan gap-closing MEMS energy harvester [5] was modelled using LTspice software. In the simulations, to get to a high level power range, the gap is modified from 40.5 μm to 81 μm and the mass is changed from 0.066 g to 0.198 g. Figure 2 shows the model schematic of the harvester with a two-cells interface circuit. In this model, the excitation mechanical force is represented as a voltage. The linear mechanical resonator is modeled by a L-C-R circuit, these components representing respectively the inertial mass, the spring compliance and the mechanical damping. The electromechanical coupling is modelled by two behavioral voltage sources. The theoretical expressions of the MEMS capacitance and the electrostatic force as a function of the the displacement x from the rest position are given respectively by equations (1) and (2).

$$C_{var} = \varepsilon * S(\frac{1}{d-x} + \frac{1}{d+x}) \quad (1)$$

$$F_{elec\_mech} = \frac{1}{2} * V(C_{var})^2 * (\frac{1}{(d-x)^2} - \frac{1}{(d+x)^2}) \quad (2)$$

In the DC-DC convertor, the ON and OFF states of the electronic switches were modelled by 1 Ω and 10^{10} Ω resistors respectively. The switches were operated periodically at 100 kHz with a duty cycle of 50%. The inductor value is 1 mH.

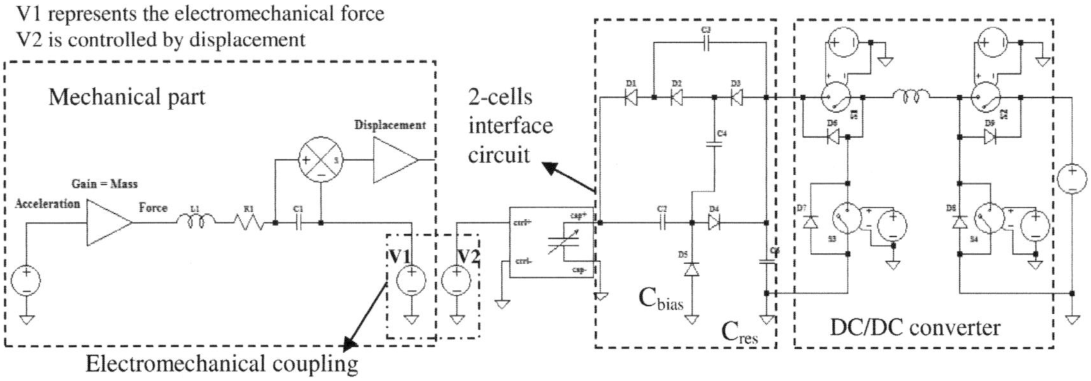

Figure 2. LTspice model of the MEMS e-VEH and the proposed circuit

In the simulation, the acceleration was progressively increased at first and then kept constant when it reached 16 m/s^2. The waveform of acceleration and the simulation results of power of C_{res} and V_{Cres}, when all the switches are open, are presented in Figure 3. One can clearly see that there exists an optimal voltage value of V_{Cres} for which the harvested power is maximum. This is due to the electrostatic force, which induces strong damping of the resonator at higher voltages. One can clearly see that the maximum power is 11 μW when V_{Cres} is about 22V. Choosing a voltage interval of V_{Cres} from 18V to 27V enables to get at least 92% of the maximum power.

Simulated waveforms of power of C_{res} and V_{Cres} with different topologies of convertor for MPPT are presented in figure 4. A buck DC-DC convertor was connected to a battery of 10V. Hence, the capacitor C_{res} was initially charged to 10V. The convertor didn't work with the switches S1, S3, S4 open and S2 closed. After 45s, V_{Cres} reached 27V and the switch S1 was periodically turned on and off with 50% duty cycle at 100 kHz. After several cycles, V_{Cres} went down to 18V and the convertor stopped. V_{Cres} came back to 27V after 18s and the convertor started to work with the switch S1 periodically turned on and off. A boost DC-DC convertor was connected to a battery of 40V. This voltage was too high to start up the system with a weak vibration at the beginning. Hence, at first, to discharge the capacitor C_{res}, the boost convertor worked with S1 closed, S2, S3 open and S4 periodically turned on and off. A reversible convertor was also connected to a battery of 40V. Compared to the boost and buck ones, it had not to use the start switch and the capacitor C_{res} was able to be charged to any voltage at the beginning. In this simulation, V_{Cres} was initially charged to 14V with the switches S2, S3 open and the switches S1, S4 periodically turned on and off with 50% duty cycle at 100kHz. After 45s, V_{Cres} was discharged when the switches S1, S4 open and the switches S2, S3 periodically turned on and off. For the 3 different topologies of convertor, one can clearly see that at least 10 μW is generated.

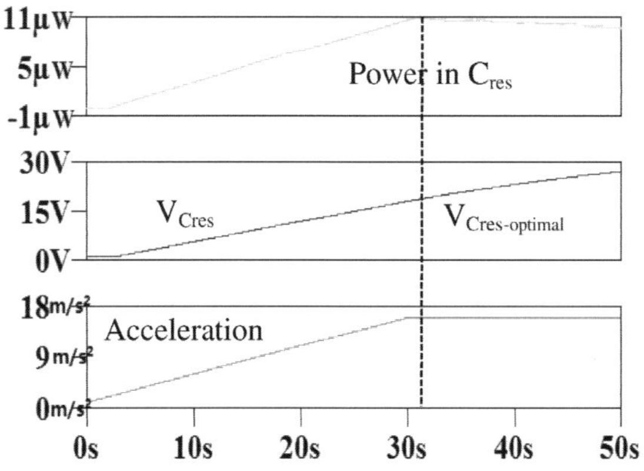

Figure 3. Simulated power in C_{res}, V_{Cbias}, and V_{Cres} vs time

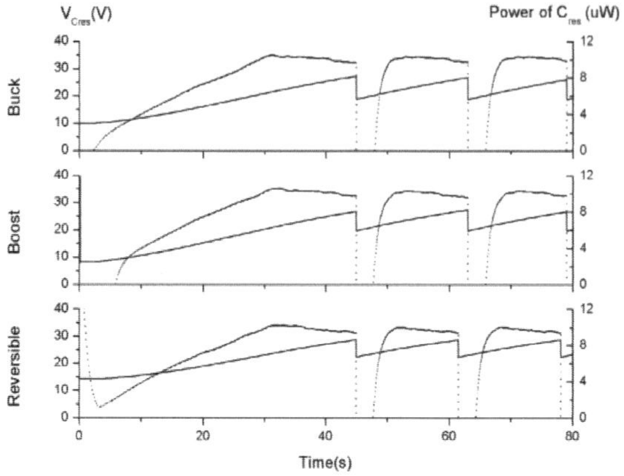

Figure 4. Simulated waveforms of power in C_{res} (dotted lines)and V_{Cres} (solid lines)with different topologies of convertor for MPPT

4. Conclusion

This interface circuit for electrostatic energy harvesters enables to adjust the voltage across the variable capacitor to optimize the performance of the system. Simulation results show that Tracking of maximum power point is achievable for different functions of DC-DC convertor. Our future work will focus on the strategy of switch control and on the design of the integrated control circuit.

5. Reference

[1] De Queiroz A C M, Domingues M 2011 the doubler of electricity used as battery charger IEEE Trans. Circ. Syst. II: Express Briefs 58(12) pp 797-801

[2] Lefeuvre E, Risquez S, Wei J, Woytasik M, Parrain F 2014 Self-Biased Inductor-less Interface Circuit for Electret-Free Electrostatic Energy Harvesters J. Phys.: Conf. Series 557 012052.

[3] Wei J, Risquez S, Mathias H, Lefeuvre E, Costa F 2015 Simple and Efficient Interface Circuit for Vibration Electrostatic Energy Harvesters Proc. IEEE Sensors Conf. 2015

[4] Dorzhiev V, Karami A, Basset P et al. MEMS electrostatic vibration energy harvester without switches and inductive elements. Journal of Physics: Conference Series 557 (2014) 012126.

[5] Basset P, Galayko D, Cottone F, Marty F, Dudka A, Bourouina T 2014 Electrostatic vibration energy harvester with combined effect of electrical nonlinearities and mechanical impact J. Micromech. Microeng. 24 035001

Acknowledgments

This work is supported by the "IDI 2014" project funded by the IDEX Paris-Saclay, ANR-11-IDEX-0003-02

MEMS ion-sorption high vacuum pump

T Grzebyk, P Knapkiewicz, P Szyszka, A Górecka-Drzazga and J A Dziuban

Faculty of Microsystem Electronics and Photonics, Wroclaw University of Science and Technology, 11/17 Janiszewskiego Str., 50-372 Wroclaw, Poland

tomasz.grzebyk@pwr.edu.pl

Abstract. In the article a miniature MEMS-type ion-sorption vacuum pump has been presented. The influence of electric and magnetic field, as well as horizontal and vertical dimensions of the micropump and type of material used for electrodes on the pump properties has been investigated. It has been found that the micropump works efficiently as long as the magnetic field is higher than 0.3 T, and pumping cell is larger than $1\times1\times1$ mm^3. The pump allows generating vacuum at the level of 10^{-7}–10^{-9} hPa in 100 mm^3 volume.

1. Introduction

Different miniature sensors and actuators take advantage of working in vacuum [1]. Moreover, new vacuum micro/nanoelectronics instruments like array X-ray sources, THz sources, and other devices basing on field emission mechanism have been proposed in the literature [2].

Generation, stabilization and measurement of high vacuum inside macro instruments is not a problem up to 10^{-11} hPa or better. In opposite, in microscaled/microengineered MEMS devices the technical limit of long term stabilization of vacuum inside tightly sealed microchamber is still near 10^{-3}–10^{-4} hPa. The most common methods here are vacuum sealing of MEMS followed by gettering of inner gaseous atmosphere. This approach is insufficient because several new possible MEMS instruments should work in at least 10^{-5} hPa kept inside microchamber by at least several hours/days.

Our team has focused onto high vacuum MEMS for at least 5 years. In our previous publications we demonstrated possibility of on-chip pumping of MEMS structures by integrated MEMS ion-sorption micropumps [3, 4]. Here we report a micropump which allows reaching 10^{-7}–10^{-9} hPa in about 100 mm^3 microchamber of a MEMS device.

2. Results and discussion

The micropump reported here is a silicon-glass sandwich structure, consisting of two planar cathodes and one anode with a square via-hole made in its central part (fig 1a, b). Its work is based on the ignition of an electric discharge between the electrodes. The discharge is initiated either by spontaneous or field-emitted electrons cruising inside the micropump cavity, due to the presence of a strong magnetic field generated by two permanent magnets (fig. 1c). Since their paths are very long, they finally hit residual gas molecules and ionize them. The ionized particles are attracted towards the cathodes where they sputter surface layer. In the same time secondary electrons are also generated and a discharge is sustained. Molecules are removed from the microchamber due to chemical reactions with sputtered material and to physical trapping under re-deposited layer.

The construction and working principle of the micropump is similar to the Penning pumps [5], and in part to other partially miniaturized pump presented by Green *et al.*[6]. However, our device is the only one, which is technologically compatible with other MEMS. It is fabricated of silicon and glass by the use of MEMS techniques. It is hermetically sealed by an anodic bonding process and does not require any additional housing. It starts working at about 1 hPa, which can be ensured during encapsulation process performed in the vacuum chamber.

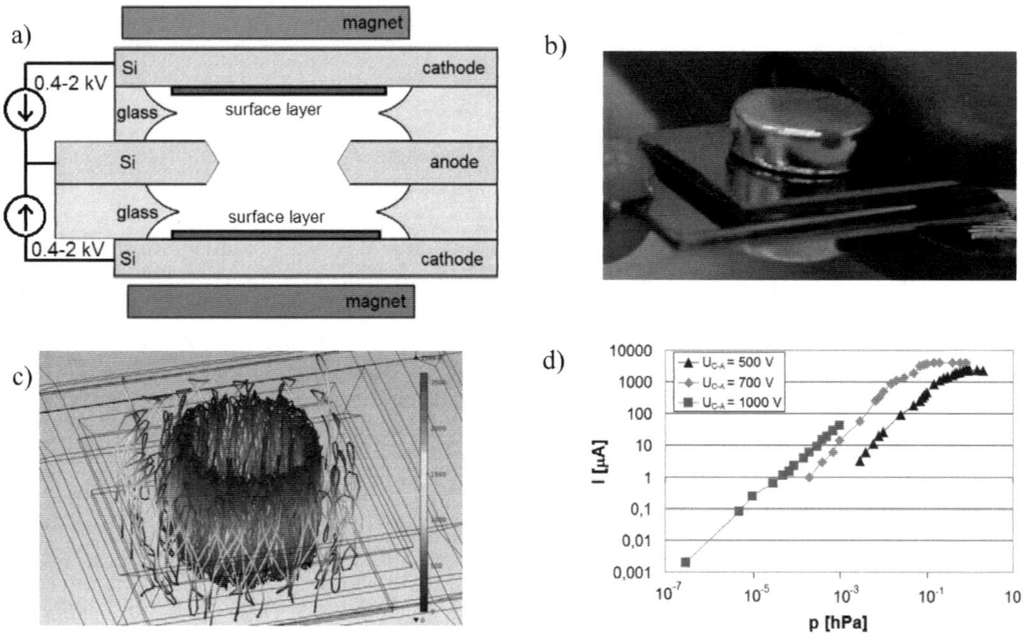

Fig. 1. Ion-sorption vacuum micropump: a) schematic cross-section, b) realization, c) results of simulation of electron trajectory inside a micropump cavity (COMSOL 4.3), d) discharge current vs pressure (calibration curves of the micropump).

The first version of the micropump was 20×12×3.4 mm^3 large (pumping chamber was equal to 6×6×3.2 mm^3). Cathodes of the pump were covered with titanium layer. This devices allowed obtaining 10^{-7} hPa pressure within about 30 min (fig. 1d) [4]. The micropump is still developed, structures with different geometrical and material parameters have been prepared and their properties investigated in different conditions (electric and magnetic field, external pressure level).

Strength of electric and magnetic fields are the most important parameters influencing the micropump operation. Operating voltage varies depending on the pressure, but in any case it should not be lower than 400 V. Without a magnetic field any discharge inside a micropump cavity can be ignited below 10^{-1} hPa (fig. 2a). The larger the magnetic flux density, the smaller the electron orbit and longer the trajectory. Its critical value is equal to about 0.3 T; in this case the diameter of the orbit is smaller than the diameter of the pumping cavity and electron path is almost infinite.

Geometrical parameters (spacers thickness and size of the hole in the anode – equivalents of height and diameter of Penning pump cells) also influence the properties of the micropump (fig. 2b, c). Optimal values are 0.7 mm for the spacer and 2×2 mm^2 for the hole in the anode – they provide the highest discharge current in wide pressure range. The minimal spacer thickness is 0.3 mm and minimal hole in the anode – 1×1 mm^2 – it is still possible to maintain a gas discharge, but the operating voltage needs to be increased to about 1500 V. In smaller microchambers gas ionization is no longer possible.

Journal of Physics: Conference Series **773** (2016) 012047 doi:10.1088/1742-6596/773/1/012047

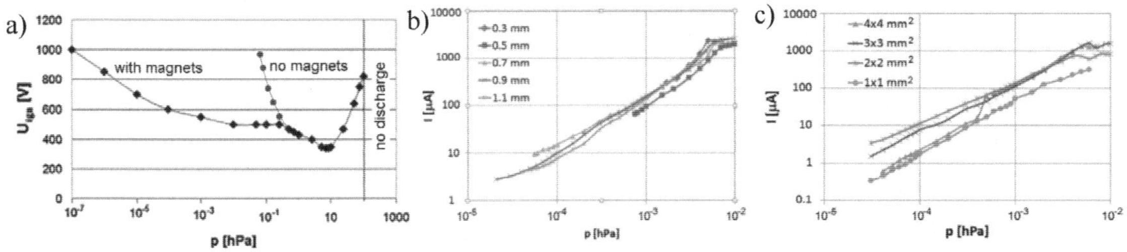

Fig. 2. Results of characterization of the micropump: a) influence of a pressure level on a discharge ignition voltage for structures with and without magnets; b) discharge current vs pressure for different spacer thickness, c) discharge current vs pressure for different size of square hole in the anode.

Apart from the impact of geometrical parameters on micropump performance, also material properties have been examined. Different materials form chemical compounds with different gases and have different sputtering and secondary emission coefficients. Measurements showed, that the *I-p* characteristics obtained for various cathode materials did not differ significantly (fig. 3a), but the highest values of the current were obtained for copper and aluminum layers, while the lowest for tungsten.

Moreover materials exhibiting some "special" properties (ensuring field emission or very high secondary emission) have also been tested (fig. 3b). Cathodes covered with carbon nanotubes (CNT), magnesium oxide and made of porous silicon exhibited superior performance – discharge current increased 10-1000 times in high vacuum range compared to flat silicon [7]. Measurements showed that, it is very likely that the application of MgO allows reaching ultrahigh vacuum inside hermetically sealed MEMS structures (fig. 4), probably even at the level of 10^{-9} hPa (according to the calibration curve). However, this result should be further verified.

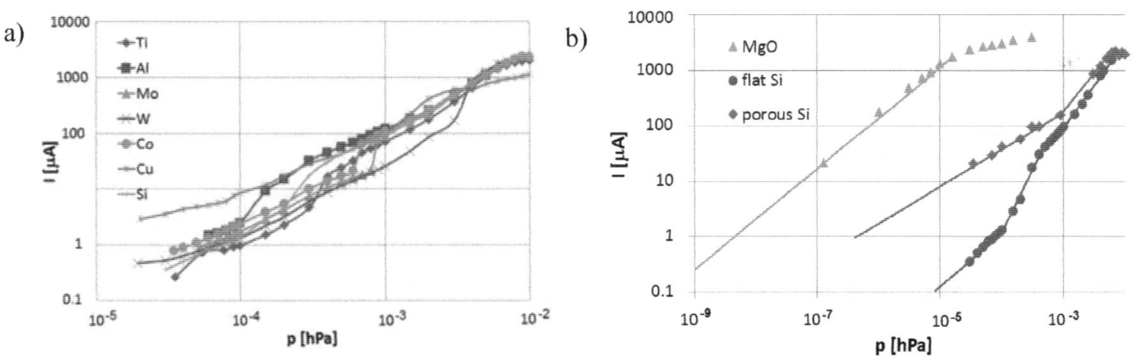

Fig. 3. Characteristics of discharge current vs pressure for: a) silicon cathodes covered with various thin film metals, b) comparison of a flat and porous silicon cathodes and a cathode with MgO layer.

The obtained result – generation of high vacuum (10^{-7}– 10^{-9} hPa) by a MEMS micropump – opens a way towards invention of a new class of high-vacuum microdevices and integrated MEMS instruments [8]. Our recent works are focused onto elaboration of Rb/Cs optical MEMS cells equipped with on-chip micropumps in which at least 10^{-7} hPa will be stably maintained. An example of high-vacuum optical cell is shown in the fig. 4.

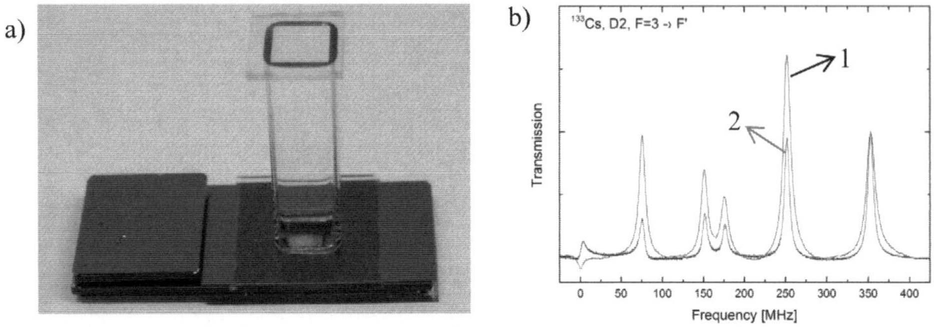

Fig. 4. Cesium optical MEMS cell integrated with a micropump: a) image of the structure, b) Doppler-free absorption spectroscopy (black line 1 – MEMS cell, blue line 2 – reference cell). The cell sealed anodically, pumped by an on-chip micropump and Cs laser-dispensed, final vacuum ~10 $^{-6}$ hPa.

3. Conclusion

In the article we have demonstrated the first miniature ion-sorption pump which is able to generate high/ultrahigh vacuum in MEMS devices. The influence of magnetic and electric field, as well as geometrical and material parameters of the micropump on the ionization process have been carefully investigated. Micropump works well until magnetic field is larger than 0.3 T, voltage is properly adjusted to the pressure level and pump microchamber is at least $1\times1\times1$ mm^3. Application of magnesium oxide on the cathode allowed significant improvement of micropump performance and obtaining 10^{-9} hPa. The micropump can be self calibrated and can work also as a vacuum gauge.

Realization of the micropump which enables ensuring stable high vacuum in microvolume opens a way for fabrication of a new class of microsystems which works in vacuum range well below 10^{-5} hPa.

Acknowledgment

Authors would like to thank the group of prof. Wojciech Gawlik from Jagiellonian University in Kraków for spectral measurements. This work was supported by the statutory grants of Wroclaw University of Science and Technology (0401/0154/16 and 0402/0139/16) and Polish National Science Center, project no UMO2013/09/B/ST7/01602.

References

[1] Esashi M 1994 Encapsulated micro mechanical sensors *Microsystem Technologies* **1**, Springer-Verlag.

[2] van Someren B, van Bruggen M J, Zhang Y, Hagen C W, Kruit P 2006 Multibeam electron source using MEMS electron optical components *Journal of Physics: Conference Series* **34** 1092–1097.

[3] Grzebyk T, Górecka-Drzazga A, Dziuban, J A, Zawada A, Konarski P 2014 Micropump for generation and control of vacuum inside miniature device *Journal of Microelectromechanical Systems* **23** 50-5

[4] Grzebyk T, Górecka-Drzazga A 2014 Glow-discharge ion-sorption micropump for vacuum MEMS *Sensors and Actuators A* **208** 113-119.

[5] Penning F M 1936 Die Glimmentladung bei niedrigem Druck zwischen koaxialen Zylindern in einem axialen Magnetfeld (sputtering apparatus paper) *Physica* **3** (9) 873-894

[6] Green S R, Malhotra R, Gianchandani Y B 2013 Sub-Torr chip-scale sputter-ion pump based on penning cell array architecture *Journal of Microelectromechanical Systems* **22** 309-317.

[7] Grzebyk T, Górecka-Drzazga A, Dziuban J 2016 Improved ionization efficiency in MEMS-type ion-sorption micropump *Technical Digest of 29th IVNC*, pp. 1-2.

[8] Grzebyk T, Górecka-Drzazga A 2016 Lateral MEMS-type field emission electron source *IEEE Transactions on Electron Devices* **63** (2) 809-813.

MEMS electrostatic influence machines

Cuong Phu Le and Einar Halvorsen
University College of Southeast Norway, Norway
E-mail: Einar.Halvorsen@hbv.no

Abstract. This paper analyses the possibility of MEMS electrostatic influence machines using electromechanical switches like the historical predecessors did two centuries ago. We find that a generator design relying entirely on standard silicon-on-insulator(SOI) micromachining is conceivable and analyze its performance by simulations. The concept appears preferable over comparable diode circuits due to its higher maximum energy, faster charging and low precharging voltage. A full electromechanical lumped-model including parasitic capacitances of the switches is built to capture the dynamic of the generator. Simulation results show that the output voltage can be exponentially bootstrapped from a very low precharging voltage so that otherwise inadequately small voltage differences or charge imbalances can be made useful.

1. Introduction

A vibration energy harvester typically consists of a spring-mass system and an electromechanical transducer. The electrostatic transducer is particularly suitable for realization in microfabrication processes [1, 2, 3]. A capacitive structure with electrodes whose relative motion is driven by ambient mechanical oscillations constitutes the transducer. The electrostatic energy harvester can be operated in either continuous or switched mode [4].

In the continuous mode, the harvester is biased by external sources in the form of a provided voltage or charge, or by internal sources such as work function differences between electrode materials [5], precharged floating electrodes [6] or electrets [7]. While using an external bias delegates the problem of charge or energy retention to other parts of the system such as the power electronics, or is used solely for test purposes, an internal bias provides a more complete and self-sufficient integration that contains the solution to the problem on chip. As far as the MEMS device is concerned, the internal bias poses challenges regarding more sophisticated materials technology, additional demanding process steps and limitations on operating temperature.

In the switched mode, a control circuit is utilized to operate the harvester in a power conversion cycle that can be synchronous or asynchronous with respect to the vibration. Several circuit topologies [2, 8, 9] show the ability of making the harvester system self-sustaining. However, they employ switched inductor circuits that possibly face problems of system integration and power loss due to switching.

Recently, influence machines have inspired or been the subject of several works based on the Bennet or Nicholson [10] doubler principle which is shown in figure 1a. By a sequence of grounding or connecting capacitor plates when one plate C is moved so that it alternates between being aligned with the fixed plates A or B, a doubling of charge is achieved. For repeated cycles, the charge then grows exponentially. The control was originally realized by parts making or breaking contact as the plate moves. By similar means, an electrostatic harvesters should be

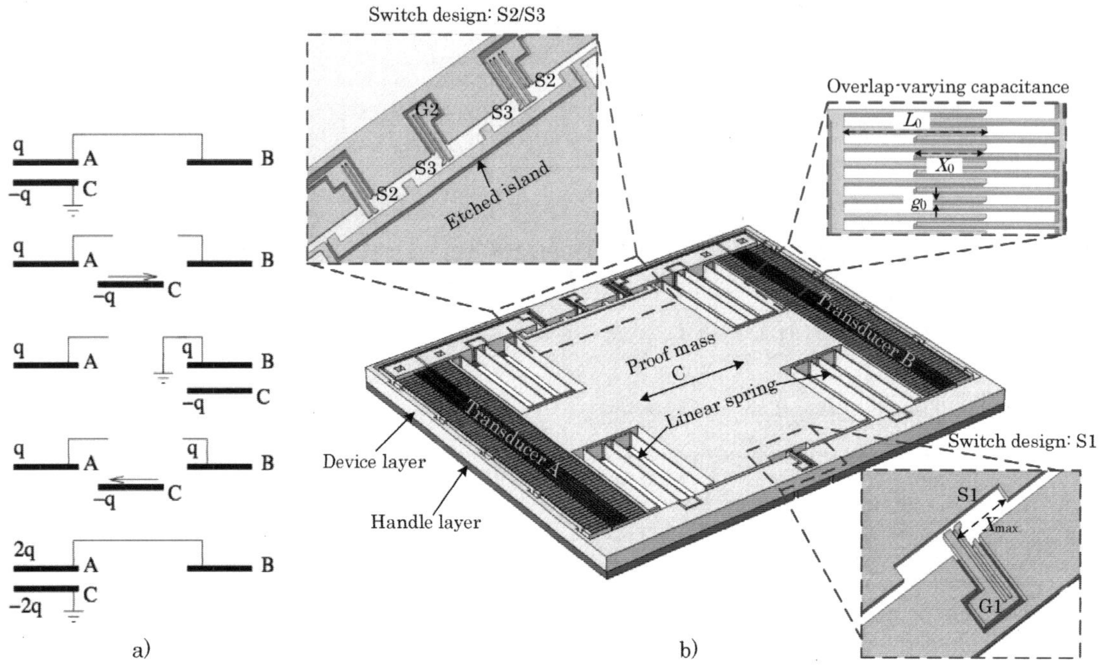

Figure 1. a) Principle of Nicholson influence machine and b) Schematic design for MEMS-implementation of influence machine using standard silicon-on-insulator(SOI) process.

possible to operate without an electric control circuit. A macroscale rotating doubler [11] was capable of increasing output voltage to hundreds volt from an initial voltage of 9.0 V. Several designs for small scale generators have shown doublers of electricity with one single capacitor or two anti-phase capacitors [12, 13, 14]. However, their circuits rely on diodes that result in reduced charging current and a large required minimum precharge voltage.

In this work, a MEMS generator concept proposed overcomes these challenges by reverting to electromechanical switches. By design, the leakage current due to the diode switches can be avoided so that the charging current is improved over the conventional topology. The electromechanical switches enables operation of the bootstrapping generator from a low precharging level.

2. Design description

Details of the microscale design are shown in figure 1b. The necessary functionality can be realized with standard microfabrication structuring of an SOI wafer. Both handle and device layers need to be patterned. The three capacitor plates A-B-C in Nicholson's device are made of two ordinary overlap-varying transducers in the device layer. For a design constrained by minimum feature size, it is advantageous to choose the device layer thick to increase the nominal capacitance and its variation. We choose the device layer thicker than the handle layer.

The proof mass C suspended by linear folded-springs is excited laterally by vibrations. Three electromechanical switches S1-S2-S3 can close at contact between flat surfaces and cantilevers with bumps when the proof mass moves to either of two extremes. A small, electrically isolated island on the proof mass is etched through the device layer for the switches S2/S3. This island is then supported by the handle layer which must be present on the proof mass. The switching arrangement allows three terminals differently connected or grounded to G1/G2 at the maximum displacement amplitude $X_0 = 500$ μm. The charge will increase by a factor $r(\alpha) \in (1, 2)$ on

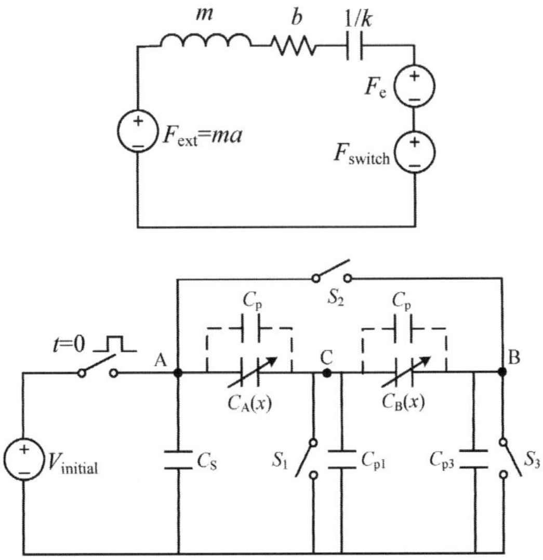

Figure 2. Equivalent circuit lumped-model of the bootstrapping generator.

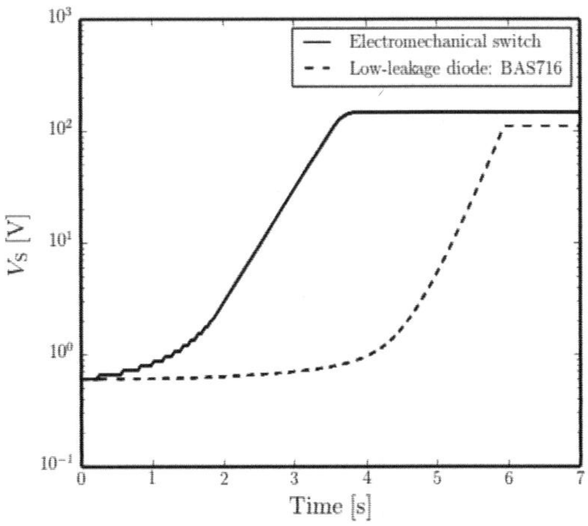

Figure 3. Comparison between two doubler-circuit configurations: i) electromechanical switch and ii) low-leakage diode BAS716 for $A = 1.0g$ and $V_{\text{initial}} = 0.6$ V.

each cycle, where $\alpha = \frac{C_{\text{max}}}{C_{\text{min}}}$ including parasitic capacitances C_{p}. The mathematical form $r(\alpha)$ can be found in the previous analysis by [11]. In the following analysis, we consider a device with total active area 1.2 cm^2, device layer thickness 400 μm, handle layer thickness 50 μm, proof mass $m = 60.6$ mg, stiffness $k = 215.3$ N/m, mechanical damping $b = 2.28 \times 10^{-4}$ Ns/m, nominal capacitance $C_0 = 8.4$ pF and $\alpha = 4.35$ with $C_{\text{p}} = 5.0$ pF. The resonant frequency is then $f_0 = 300$ Hz. Each cantilever of the switches has a stiffness $k_{\text{s}} = 891.2$ N/m.

3. Lumped-model

The bootstrapping generator is represented by equivalent circuits for mechanical and electrical domains in figure 2. The generator is driven by an external force $F_{\text{ext}} = ma$ and the electrostatic force acting on the proof mass is $F_{\text{e}} = \frac{1}{2}\left(\frac{1}{C_{\text{A}}(x)+C_{\text{p}}}\right)' Q_{\text{A}}^2 + \frac{1}{2}\left(\frac{1}{C_{\text{B}}(x)+C_{\text{p}}}\right)' Q_{\text{B}}^2$, where Q_{A} and Q_{B} are the total charges on the transducers A and B respectively, and $C_{\text{A,B}}(x) = \frac{C_0}{1 \pm \frac{x}{X_0}}$. F_{switch} is the force acting on the proof mass when the switch cantilevers contact the proof mass at two extreme positions.

The doubler circuit with the electromechanical switches is shown in the bottom of figure 2. The circuit includes stray capacitances for the switches $C_{\text{p1}} = C_{\text{p2}} = 2.0$ pF. The voltage source V_{initial} is a way to include a precharging function in the simulation in order to initiate operation of the bootstrapping generator. The output is connected to a capacitor $C_{\text{S}} = 1.0$ nF as an energy reservoir.

4. Simulation results

Figure 3 shows increase of the voltage on the reservoir capacitor for $V_{\text{initial}} = 0.6$ V and an acceleration amplitude $A = 1.0g$. For comparison to the diode doubler, a circuit with the low-leakage diode BAS716 is simulated with the same transducer design. The diode has a junction capacitance of the same magnitude as the parasitic capacitance of the switches. The comparison shows that the mechanically switched device bootstraps to the voltage $V_{\text{s}} = 145$ V in $t = 3.8$ s, while these values are $V_{\text{s}} = 110$ V in $t = 5.9$ s for the diode scheme. The initial voltage of 0.6 V

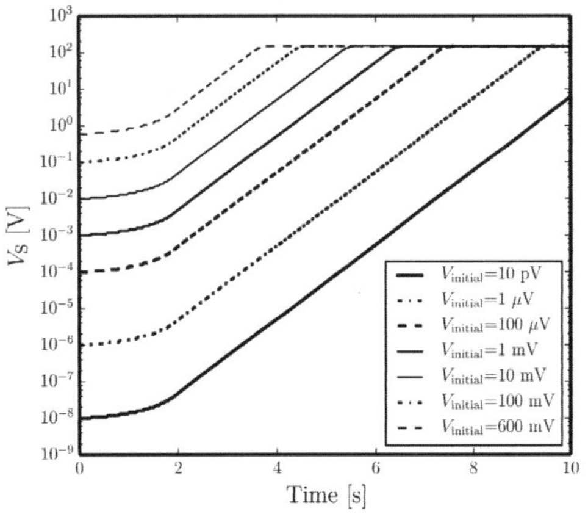

Figure 4. Voltage on reservoir capacitor for various initial voltages and $A = 1.0g$.

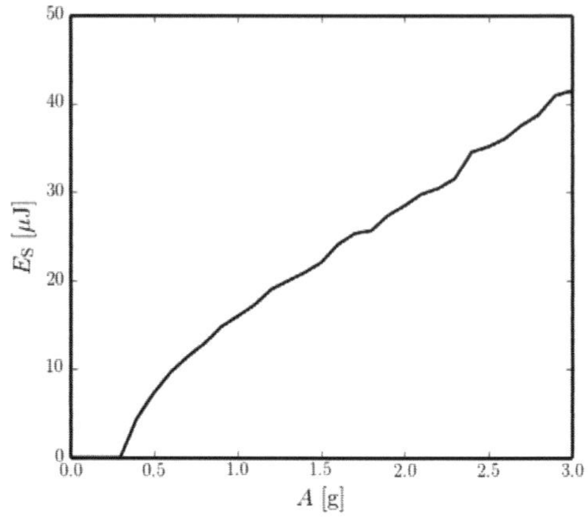

Figure 5. Energy storage on C_s vs. acceleration amplitudes for $t = 10$ s and $V_{\text{initial}} = 1.0$ mV.

used in the comparison is approximately the minimum threshold for the diode doubler to allow the operation.

Figure 4 shows an advantage that the design of the electromechanical switches can bootstrap to high voltages even when very small initial voltages are applied. The precharging voltage in range of mV, μV or even pV can enable the bootstrapping operation, but the smaller V_{initial} needs longer time to reach the maximum level. With respect to realistic devices, it shows that a miniscule voltage difference or charge imbalance is sufficient to make the output voltage increase based on the doubler mechanism. In practice the lower limit will possibly be limited by leakage. However, with the mechanical switches, such a leakage will be orders of magnitude less than for a diode.

Figure 5 shows energy obtained in $t = 10$ s under increase of acceleration amplitudes for $V_{\text{initial}} = 1.0$ mV. For low acceleration amplitudes, the proof mass displacement is insufficient to achieve a capacitance variation α beyond the required lower limit for the operation. It takes an acceleration amplitude beyond a certain critical value A_c to initiate the bootstrapping. We found $A_c = 0.3g$ for this design. The obtained energy increases with the acceleration amplitude, e.g. $E = 41.8$ μJ for $A = 3.0g$.

5. Concluding remarks

A first analysis of a MEMS electrostatic influence machine, or doubler of electricity, based on a standard silicon-on-insulator(SOI) process was made. The doubler circuit relies fully on electromechanical switches to improve the charging current and to minimize the precharging voltage. The advantages overcome the challenges of a diode-based doubler configuration. The output can be bootstrapped to the maximum voltage from extremely small voltages. This opens up for the possibility to use priming techniques for electrostatic energy harvesters that in themselves provide an insufficient bias. The use of work-function differences for example, will provide a best case priming voltage of only a few volts [5] if the device is optimized for it. However, with the doubler configuration, this or other techniques that can provide a charge imbalance would be useful even if only a mV or lower bias voltage can be achieved.

Acknowledgments
This work was supported by the Research Council of Norway through the Grant no. 229716/E20.

References

[1] Sterken T, Altena G, Fiorini P and Puers R 2007 *Proc. MEMS/MOEMS-DTIP2007 (Stresa, April)* 297–300
[2] Yen B C and Lang J H 2006 *IEEE Trans. Circuits and Systems-I: Regular Papers* **53** 288–295
[3] Mitcheson P D, Sterken T, et al. 2008 *Measurement and Control* **41** 114-119
[4] Mitcheson P D and Green T C 2012 *IEEE Trans. Circuits and Systems-I: Regular Papers* **12** 3098--3111
[5] Kuehne I, Frey A, et al. 2008 *Sensors and Actuators A: Physical* **142** 263–269
[6] Ma W, Zhu R, et al. 2007 *J. Microelectromechanical Systems* **16** 29–37
[7] Suzuki Y 2011 *IEEJ Trans. Elec. Electron Eng.* **6** 101--111
[8] Galayko D, Pizarro R, et al. 2007 *Proc. IEEE BMAS Conf. (San Jose, September)* 115–120
[9] Torres E O and Rincon-Mora G A 2009 *IEEE Trans. Circuits and Systems I: Regular Papers* **56** 1938–1948
[10] Nicholson W 1788 *Phil. Trans. R. Soc.* **78** 403–407
[11] Queiroz A C M and Domingues M 2011 *IEEE Trans. Circuit and Systems-II* **58** 797–801
[12] Dorzhiev V, Karami A, et al. 2014 *Proc. PowerMEMS2014 (Awaji, November)* **557** 012126
[13] Queiroz A C M and Filho L C M 2015 *Proc. IEEE* 6th *LASCAS, (Montevideo, February)* 15439332
[14] Galayko D, Dudka A, et al. 2015 *IEEE Trans. Circuits and Systems-I: Regular Papers* **11** 2652–2663

Crystallographic investigation of aluminium nitride thin films on stainless steel foil for highly efficient piezoelectric vibration energy harvesters

N. Moriwaki[1,2], L.V. Minh[1], R. Ohigashi[2], O. Shimada[2], H. Kitayoshi[1] and H. Kuwano[1]

[1] Department of Robotics, Tohoku University, 6-6-01, Aramaki-Aoba , Aoba-ku Sendai, Miyagi, 980-8579, Japan
[2] Dai Nippon Printing Co., Ltd, 250-1 Wakashiba, Kashiwa, Chiba, 277-0871, Japan

moriwaki@nanosys.mech.tohoku.ac.jp

Abstract. This study reports piezoelectric properties and crystallographic microstructures of aluminium nitride (AlN, wurtzite structure) thin films on 50 μm thick stainless steel foil. The transverse piezoelectric coefficient $d_{31,f}$ and $e_{31,f}$ of 10 μm thick AlN films were estimated as -1.42 ± 0.08 pm/V and -0.48 ± 0.03 C/m^2 from a tip displacement of the piezoelectric cantilevers. Dielectric constant ε_{33} was measured as 10.5 ± 1.0. An electron beam diffraction pattern by a high-resolution transmission electron microscope and x-ray diffraction pattern showed that abundance ratio of the orientation such as <101>, <102> and <103> of AlN crystal on stainless steel foils increased with increasing thickness.

1. Introduction

Recently, there have been growing research and development activities of a power source of MEMS-based wireless sensor modules. Aluminium nitride (AlN) films are expected to be the most attractive piezoelectric material due to relatively high electromechanical coupling coefficient and no environmental pollution in comparison with lead zirconate titanate [1]. A structural layer constitution of AlN on stainless steel foil (SUS304) which features high fracture toughness in contrast to Si-based MEMS is useful for mm-scale MEMS applications. And it also has the advantage of process compatibility based on a traditional mass-production technology. Therefore, we have been developing the piezoelectric vibration energy harvesters using AlN/ SUS304 foil [2].

In this study, piezoelectric properties of AlN thin films on SUS304 foil deposited by reactive RF-magnetron sputtering were measured from a tip displacement of the cantilevers. And the crystalline structure was investigated using a x-ray diffractometer, a scanning electron microscope (SEM) and HR-TEM in order to discuss the relationship between it and piezoelectric properties.

2. Experimental procedure

Two micrometer thick AlN thin films on 50 μm thick mirror polished SUS304 foil substrate were deposited by reactive RF-magnetron sputtering. The AlN crystal growth process on SUS304 and Si (as a reference sample) substrates in same batch is shown in figure 1 as SEM images. X-ray diffraction (XRD) patterns and x-ray rocking curves (XRC) of AlN films including a 10 μm thick sample for the cantilevers were measured by x-ray diffractometer (Bruker D8 ADVANCE) as indicated in figure 2 and 3. The electron beam diffraction patterns with bright field and dark field TEM images of 2 μm thick AlN thin films on SUS304 foil obtained by HR-TEM (JEM ARM 200FB) are shown in figure 4.

The fabrication process flow of a piezoelectric cantilever for the measurement of the piezoelectric properties of AlN films is shown in figure 5. First, ten micrometer thick AlN films were deposited at the same condition as the above samples on both-sides of 4 inch mirror polished SUS wafers (a). The AlN film on the back side of the SUS substrate is the layer for suppressing the warpages of wafers. Next, AlN films were etched by a heated TMAH solution (b). Thirdly, upper electrode of 0.2 μm Au/ 2 μm Ni/ 3.0 μm Cu was patterned on the entire wafer by electroplating using thin Cu/Ti seed layer as a power feed electrode and adhesive layer (c). Then, a patterning of the SUS substrate into the cantilever shape was carried out by a ferric chloride solution (d). Finally, the SUS wafers were diced (e) and the piezoelectric cantilever chips were adhered on printed circuit boards by a epoxy resin (f). The dimensions of a cantilever were 6.5 mm length and 3.0 mm width.

Piezoelectric vibration was generated by applying sine wave voltage with the half DC bias voltage between upper and bottom electrodes, and its frequency was 500 Hz which was much less than the mechanical resonance frequency 1650 Hz. The tip displacement as a function of applied voltage in figure 6 was measured by the setup using laser vibrometer (Onosokki LV-1710) [3]. The electrical properties were evaluated by semiconductor characterization system (KEITHLEY 4200-SCS).

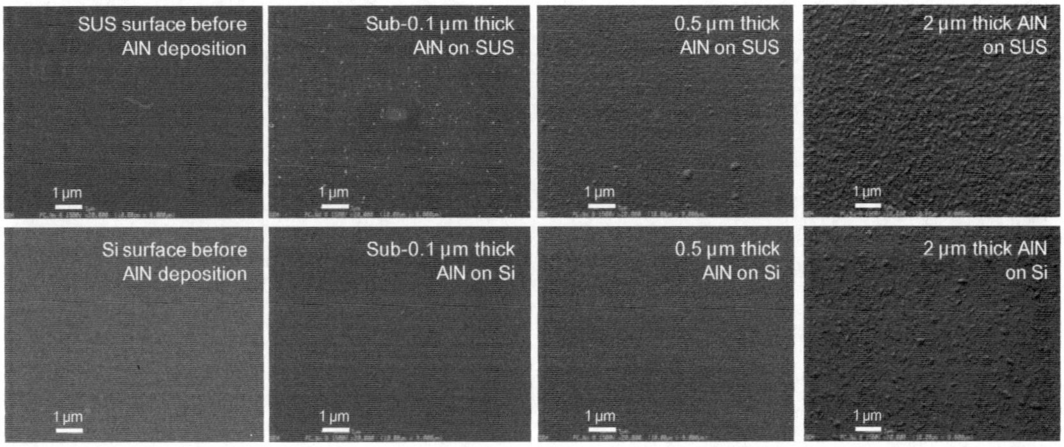

Figure 1. SEM images of the AlN crystal growth process on SUS304 and Si (as a reference sample) substrates in same batch.

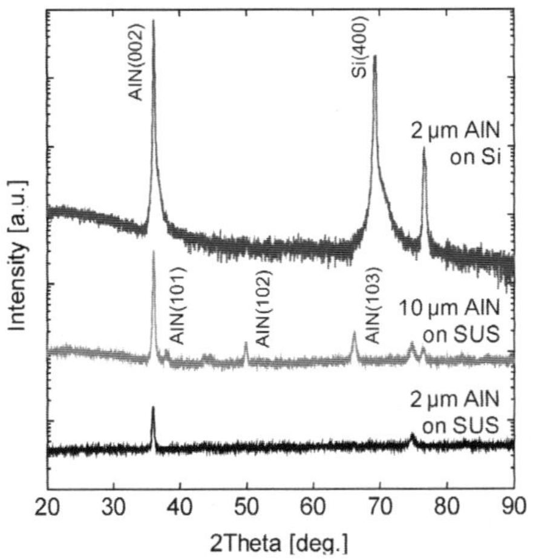

Figure 2. XRD pattern of the AlN films on SUS and Si substrates.

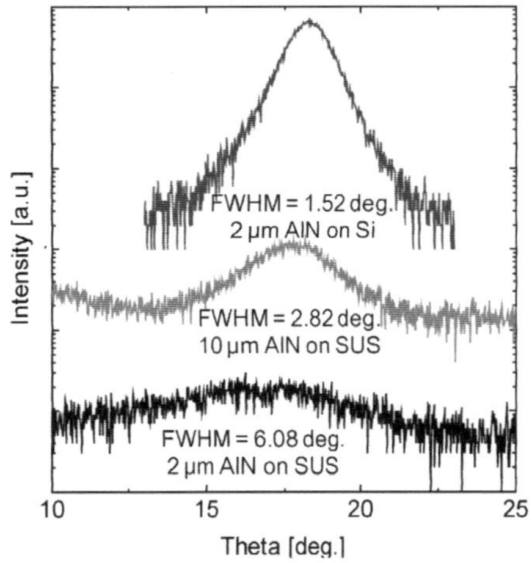

Figure 3. X-ray rocking curves of AlN(002) of the AlN films on SUS and Si substrates.

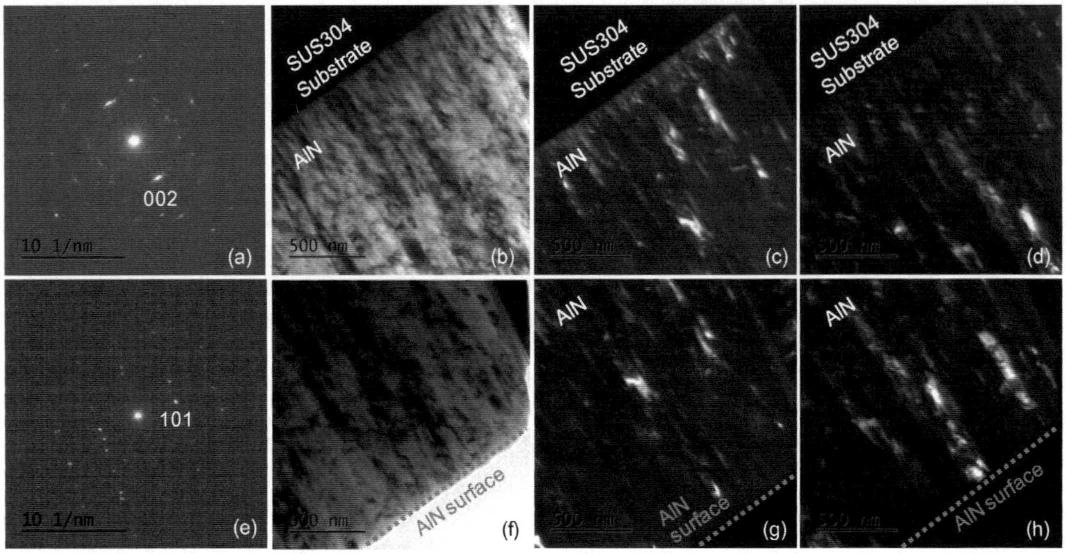

Figure 4. The electron beam diffraction patterns in (a) and (e), bright field TEM images in (b) and (f). Dark field TEM images using (002) reflections in (c) and (g), and (101) reflections in (d) and (h). The results (a), (b), (c) and (d) were taken near the SUS304 substrate, and others were taken near the AlN surface.

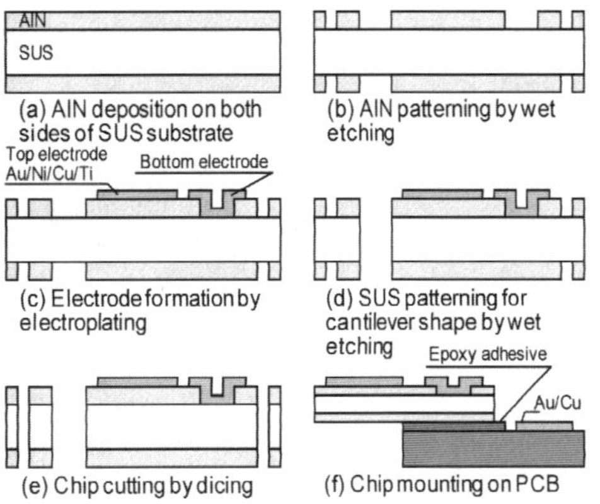

Figure 5. Schematics of the fabrication process for the piezoelectric cantilevers.

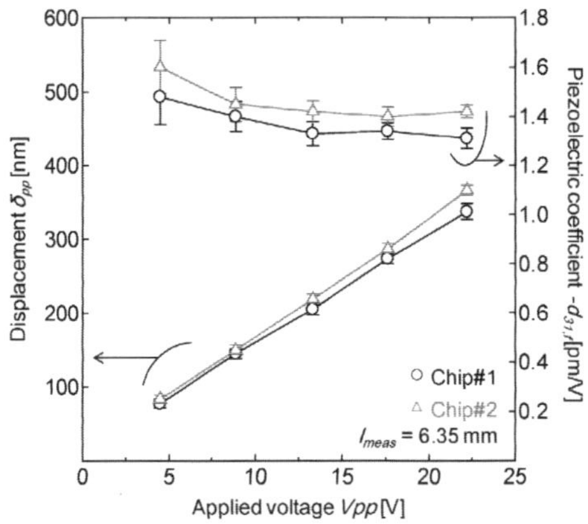

Figure 6. Tip displacement and estimated $-d_{31}$ of the two cantilevers as a function of the applied voltage. The length between the laser spot and fixed end was 6.35 mm.

3. Results and discussion

The x-ray crystal structural analysis in figure 2 and 3 indicates the difficulty of the AlN(002) orientation growth on the SUS304 substrate focusing on its intensity and the value of full width at half maximum (FWHM) of XRC. As shown in figure 1, Stranski-Krastanov growth of AlN was occurred on the Si substrate. On the other hand, Volmer-Weber growth was observed in case of the SUS304 substrate. Because the roughness values of the Si and SUS304 substrate were almost same ~Ra = 0.3 – 0.6 nm, it is thought that one of the reasons is difference of chemical potential of the substrate surface. SUS304 is the alloy which consists of about 10 % Ni, 20 % Cr and Fe. It is estimated that the distribution of adatom interactions depending on the difference of the element at substrate surface caused Volmer-Weber growth on the SUS304 substrate. There is possibility that the ± 10 deg. tilting and the low abundance ratio of the AlN(002) orientation shown in figure 4 were occurred on the same mechanism. The lateral size of AlN columns was in the range of less than 50 – 300, and the random orientation including AlN(101), (102) and (103) was seen in a crystalline columns. In addition, there are no single-oriented crystalline columns. This is striking difference between the AlN film on Metal/Si and on SUS substrate [4], [5]. And the results shows the piezoelectric properties of AlN on the SUS304 substrate could not be estimated from the crystallographic film quality such as the value of FWHM of AlN(002) peak based on the relationship in case of Si substrate [6].

The transverse piezoelectric coefficient $d_{31,f}$ and $e_{31,f}$ of the AlN films were estimated from a tip displacement of the piezoelectric cantilevers using the constituent equation for multiple layer cantilever [7]. The structural layer constitution was defined as 0.2 μm Au/ 2 μm Ni/ 3.0 μm Cu/ 15 nm Ti/ 10 μm AlN/ 50 μm SUS/ 10 μm AlN, and the mechanical properties of bulk were used for the calculation [2], [8]. The results as a function of the applied voltage were shown in figure 6. The comparison of the electrical and piezoelectric properties of various AlN films on Si substrates [9], [10] and on the SUS304 substrate was summarized in table 1. The piezoelectric coefficient of the AlN film on the SUS304 substrate was significantly less than the values of AlN films on Si substrates. It is thought that the complicated crystalline structure as above caused the low piezoelectric properties in this experiment.

Table 1. Comparison of the electrical and piezoelectric properties of various AlN films on Si and SUS304 substrates. Symbol ※ means that the value was calculated by the author. E_{Bulk} = 340 GPa.

Ref.	Growth method	Substrate	Dielectric constant	Piezoelectric coefficient $d_{31,f}$ [pm/V]	Piezoelectric coefficient $e_{31,f}$ [C/m^2]
M.-A. Dubois *et al.* [6]	Pulsed DC sputtering	$Pt/Si_3N_4/SiO_2/Si$	10.4 ± 0.4	※-$d_{33,f}$ /2 = -1.7 ± 0.12	- 1.02 ± 0.04
A. Ababneh *et al.* [7]	DC magnetron sputtering	(100) Si	Unclear	-1.0 to -1.8	※$d_{31,f}$ × E_{Bulk} = -0.34 to – 0.61
In this study	RF magnetron sputtering	Stainless steel SUS304	10.5 ± 1.0	-1.42 ± 0.08	-0.48 ± 0.03

4. Conclusion

Piezoelectric properties and crystallographic microstructures of AlN thin films on 50 μm thick SUS304 were investigated. The transverse piezoelectric coefficient $d_{31,f}$ and $e_{31,f}$ of 10 μm thick AlN films were estimated as -1.42 ± 0.08 pm/V and -0.48 ± 0.03 C/m^2 , and its dielectric constant was 10.5 ± 1.0. It is thought that these low piezoelectric properties were due to the complicated crystalline structure including the fine grain and the random orientation of AlN<101>, <102> and <103>.

References

[1] J. Zhang, Z. Cao and H. Kuwano, *Jpn. J. Appl. Phys.* **50** (2011), 09ND18
[2] Z. Cao, J. Zhang and H. Kuwano, *Sens. Actuators A* **179** (2012), pp. 178-184
[3] I. Kanno, H. Kotera and K. Wasa, *Sens. Actuators A* **107** (2003), pp. 68-74
[4] A. Artieda, M. Barbieri, C. S. sandu and P. Muralt, *J. Appl. Phys.* **105** (2009), 024504
[5] A. T. Tran, G. Pandraud, F. D. Tichelaar, M. D. Nguyen, H. Schellevis and P. M. Sarro, *Appl. Phys. Lett.* **103** (2013), 221909
[6] K. Tonisch, V. Cimalla, Ch. Foerster, H. Romanus, O. Ambacher and D. Dontsov, *Sens. Actuators A* **132** (2003), pp. 658-663
[7] D. L. DeVoe and A. P. Pisano, *IEEE J. Microelectromech. Syst.* **6** (1997), 3
[8] Chronological Scientific Tables, Maruzen (2015) [In Japanese]
[9] M.-A. Dubois and P. Muralt, *Appl. Phys. Lett.* **74** (1999), 20
[10] A. Ababneh, U. Schmid, J. Hernando, J. L. Sánchez-Rojas and H. Seidel, *Mater. Sci. Eng. B* **172** (2010) , pp. 253-258

Energy Autonomous Wireless Sensing System Enabled by Energy Generated during Human Walking

Yang Kuang, Tingwen Ruan, Zheng Jun Chew and Meiling Zhu*

College of Engineering, Mathematics and Physical Sciences, University of Exeter, Exeter, EX4 4QJ, UK

(*corresponding author: m.zhu@exeter.ac.uk)

Abstract. Recently, there has been a huge amount of work devoted to wearable energy harvesting (WEH) in a bid to establish energy autonomous wireless sensing systems for a range of health monitoring applications. However, limited work has been performed to implement and test such systems in real-world settings. This paper reports the development and real-world characterisation of a magnetically plucked wearable knee-joint energy harvester (Mag-WKEH) powered wireless sensing system, which integrates our latest research progresses in WEH, power conditioning and wireless sensing to achieve high energy efficiency. Experimental results demonstrate that with walking speeds of 3~7 km/h, the Mag-WKEH generates average power of 1.9~4.5 mW with unnoticeable impact on the wearer and is able to power the wireless sensor node (WSN) with three sensors to work at duty cycles of 6.6%~13%. In each active period of 2 s, the WSN is able to measure and transmit 482 readings to the base station.

1. Introduction

Wearable energy harvesting (WEH), which scavenges energy around human body and converts it to electric energy, has been the subject of scientific and technological effort worldwide in the past two decades [1]. This has been driven by the demand to prolong the life time and reduce the maintenance of body sensors, which are increasingly used for human health condition monitoring to support diagnosis and improve treatment from a medical point of view [2]. Those sensors are currently powered by batteries with limited capacity and thus limited life time but are usually required to operate in long-term. WEH has been considered as a potential technology to provide sustainable electric power supply for those sensors to extend the life time of or fully replace the batteries to establish energy autonomous wireless sensing systems.

Although there are variable energy sources such as solar illumination, radio-frequency energy and thermal energy available around human body, WEH from kinetic energy has attracted the most attention due to the large amount of energy available during human motion and has been implemented by using piezoelectric, electromagnetic and electrostatic mechanisms [3]. To date, a large amount of wearable energy harvesters have been proposed. However, most of these harvesters were tested under simulated input excitations, and limited research has been devoted to integrating them with wireless sensors and testing them in system-level under real world settings so as to advance the technology towards application.

In previous work [4], the present authors developed a magnetically plucked knee-joint energy harvester (Mag-KEH), which is able to generate an average power of 5.8 mW when actuated by simulated knee-joint motion at 0.9 Hz. In this work, the Mag-KEH was modified to a wearable prototype and integrated with a wireless sensor node to establish a magnetically plucked wearable energy harvester (Mag-WKEH) powered wireless system. The performance of the system, including the power generation, the capability of the wireless sensing and transmission, has been characterized as a human subject carrying the system walks at different speeds.

2. Description of the Mag-WKEH powered wireless sensing system

The Mag-WKEH powered wireless sensing system is shown in Figure 1. The whole system is fixed to the leg through an upper brace and a lower brace. Two circuits are fixed on the top of the Mag-WKEH: one is a rectifying circuit, and the other is a circuit comprising a power management module (PMM), an energy aware interface (EAI) and a wireless sensor node (WSN).

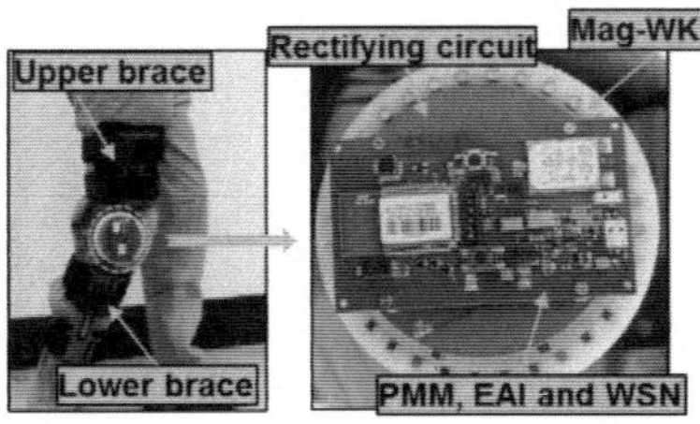

Figure 1 The magnetically plucked wearable knee-joint energy harvester powered wireless sensing system

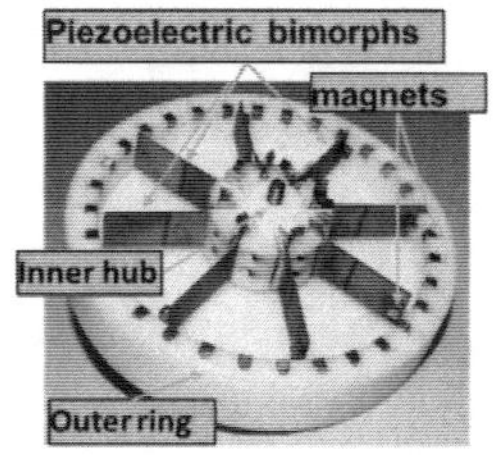

Figure 2 The magnetically plucked wearable knee-joint energy harvester

The Mag-WKEH, shown in Figure 2, used in this study is a wearable version of the Mag-KEH developed in [4]. Compared with the Mag-KEH, which was designed to test on a stepping motor, the Mag-WKEH was designed with adapters to connect to the lower and upper braces. The Mag-WKEH comprises eight piezoelectric bimorphs fixed on an inner hub. Each bimorph tip was glued with a permanent magnet. Thirty two permanent magnets were equally positioned along the inner edge of the outer ring. The inner hub was connected to the lower brace, while the outer ring was connected to the upper brace. As the knee-joint motion actuates the relative rotation between the inner hub and the outer ring through the two braces, the piezoelectric bimorphs are plucked by the intermittent magnetic forces, i.e. they are first deflected and then released to free vibration at their resonance frequency. The plucking mechanism converts the low frequency knee joint motion to the high resonant vibration of the piezoelectric bimorphs, thus achieving high power generation. More details of the dynamic responses of the magnetically plucked piezoelectric bimorphs can be found elsewhere [4]. The Mag-WKEH has a diameter of 90 mm and height of 25 mm. The total weigh is 300 g excluding the braces.

A schematic of the system is presented in Figure 3. The energy generated by the Mag-KEH is rectified and then processed by the PMM. The rectifier circuit comprises eight full-bridge rectifiers. Each bimorph was connected to a rectifier individually and the outputs of the eight rectifiers were connected in parallel and then to the PMM. The output of the PMM charges a 10 mF storage capacitor C_{CS} and the energy stored in C_{CS} is used to power the WSN under the supervision of the EAI.

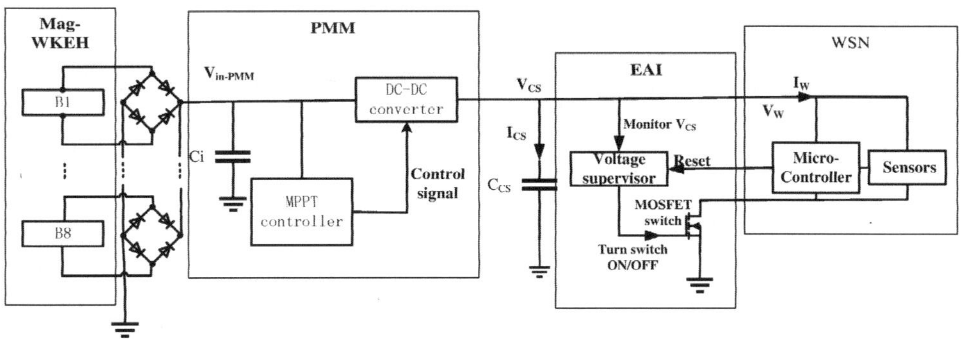

Figure 3 A schematic of the wearable knee-joint energy harvester powered wireless sensing system

The PMM was designed with maximum power point tracking (MPPT) function to address the low power transfer efficiency resulted from the electric impedance mismatch between the Mag-WKEH and the succeeding circuits. Initially, the DC-DC converter is disabled and the smoothing capacitor C_i is charged up by the Mag-WKEH. As the voltage across C_i, which is $V_{in\text{-}PMM}$, increases to half of the open circuit voltage V_{OC} of the Mag-WKEH, the DC-DC converter is enabled by the MPPT controller and the energy stored in C_i is transferred to C_{CS}. As $V_{in\text{-}PMM}$ drops to below $V_{OC}/2$, the MPPT controller disables the DC-DC converter. C_i is charged up again and the cycle is repeated. More details of the design and implementation of the MPPT controller can be found in [5].

The energy aware interface (EAI) uses a voltage supervisor to monitor the voltage V_{CS} across the storage capacitor C_{CS}. When V_{CS} increases to be higher than V_1 (3.15 V), the voltage supervisor turns on the MOSFET switch and the energy stored in C_{CS} is discharged to the WSN, which is then turned to its active mode to measure and transmit data. When V_{CS} decreases to be lower than V_0 (determined by the WSN) due to the discharge, the WSN finishes its tasks and then reset the MOSFET switch. As a result, C_{CS} is disconnected to the WSN and the WSN is turned to its sleeping mode.

The WSN uses a JN5148 microcontroller unit (MCU) to coordinate the wireless transmission and data sampling from three sensors: an ADXL 335 3-axis accelerometer, a MCP9700 temperature sensor and a humidity sensor. In each active phase, the MCU takes one reading from the temperature sensor and the humidity sensor one by one. Then it takes 3 readings from the 3-axis accelerometer every 10 ms. Each time after 48 readings have been taken, the MCU measures V_{CS} to judge whether the energy in the C_{CS} is enough to take another 48 readings and transmit all the data before V_{CS} drops to too low (<2.45 V) for the operation of the WSN. If so, the MCU will take another 48 readings and the cycles repeat. If not, the MCU will turn on the transceiver, transmit all the data to the base station, and then reset the voltage supervisor in the EAI to turn off the MOSFET switch.

3. Experimental Methods

Experiments were performed with a human subject wearing the Mag-WKEH powered wireless sensing system and walking on a treadmill at speeds ranging from 3 to 7 km/h. A base station was placed at a distance of 4 m and connected with a laptop, which was used to monitor the data received by the base station. The voltages and currents input to the PMM, the storage capacitor C_{CS} and the WSN (denoted in Figure 3) were recorded to calculate the electric energy and power generated or consumed.

4. Results and Discussions

Figure 4 shows the average power output of the Mag-WEKH at different walking speeds, which is the sum of the power input to the PMM and the power lost in the rectifying circuit. As the walking speed increases from 3 to 7 km/h, the power output grows from 1.9 to 4.5 mW. The expected increase of the power with the walking speed is attributed to the more frequently and more quickly plucking of the piezoelectric bimorphs at higher walking speeds.

Figure 5 shows the voltage across and the current through C_{CS} and the WSN at a walking speed of 4 km/h. In phase (1), C_{CS} is charged up by the Mag-WKEH and its voltage V_{CS} increases steadily from zero to 3.15 V. In this phase, V_W and I_W are zero because the EAI disconnects the WSN to C_{CS} and the WSN is at its sleeping mode. The time period of phase (1) is referred to as cold start time t_c, which is 62.2 s in this case. In phase (2), the EAI turns on the switch. As a result, the energy stored in C_{CS} is discharged to the WSN. The WSN is switched to its active mode to measure data from the three sensors at a sampling frequency of 100 Hz. Because the power consumption of the WSN is higher than the charging power provided by the Mag-WKEH, V_{CS} decreases during phase (2). Before V_{CS} decreases to be too low (<2.45 V) to operate the WSN, the WSN transmits all the measured data to the base station and then resets the EAI to turn off the switch and consequently disconnect the WSN to the C_{CS}, after which phase (3) starts. In phase (3), the WSN is turned to its sleeping mode again and C_{CS} is charged up by the Mag-WKEH. As soon as V_{CS} increases to 3.15 V, the EAI turns on the switch and another phase (2) starts. The cycles are then repeated. The time period of phase (3) is referred to as warm start time t_w. Each active mode of the WSN (phase (2)) lasted 2 s (referred to as active time t_a), during which 482 reading were sampled from the sensors (1 from the temperature sensor, 1 from the humidity sensor and 480 from the 3-axis accelerometer) and successfully received by the base station.

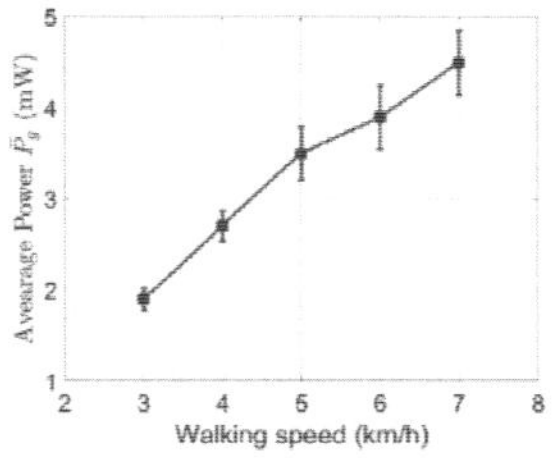

Figure 4 Average power output of the Mag-WKEH at different walking speeds

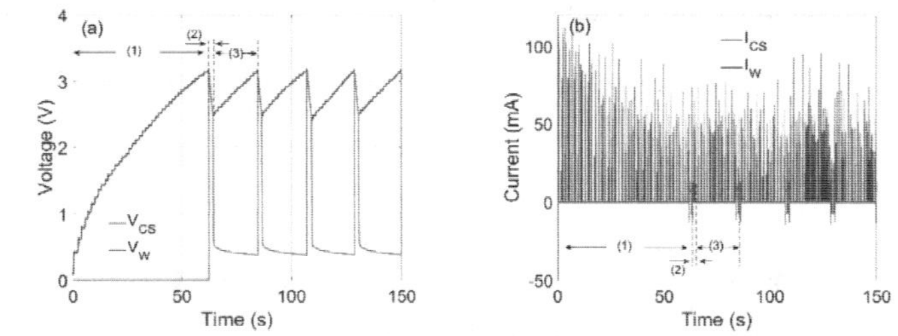

Figure 5 Voltage (a) across and current (b) through the storage capacitor and the wireless sensor node at walking speed of 4 km/h

The cold start time t_c, the warm start time t_w and the duty cycle of the system at different walking speeds are presented in Figure 6(a), (b) and (c), respectively, where the duty cycle is calculated as $t_a/(t_a + t_w)$. As the walking speed increases, both t_c and t_w decreases. This is because the power output of the Mag-WKEH increases with the walking speed. As a result, both the time required to charge the C_{CS} from zero to 3.15 V, i.e. the cold start time, and the time required to recharge the C_{CS} to 3.15 V in phase (3), i.e. the warm start time, are reduced. It should be noted that the active time t_a of the WSN keeps almost constant at different speeds because it is mainly determined by the capacitance of the C_{CS}. Because of the decrease of t_w, the duty cycle increases with the walking speeds.

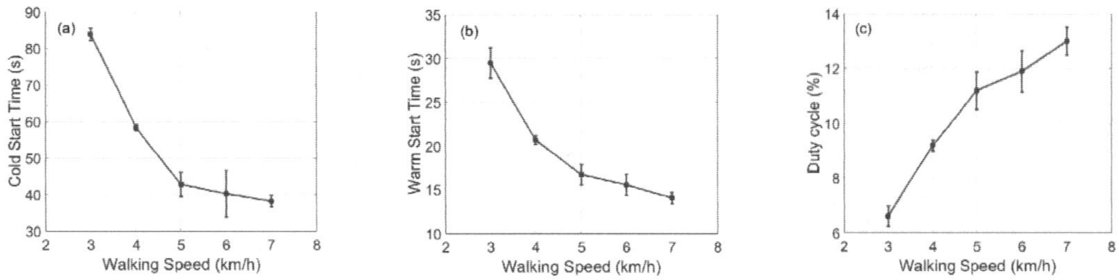

Figure 6 Cold start time (a), warm start time (b) and the duty cycle of the system at different walking speeds

5. Conclusions

This paper reported the implementation and characterisation of the Mag-WKEH powered wireless sensing system enabled energy generated during human walking. The power output of the Mag-WKEH was measured at different walking speeds and its capability to power the wireless sensor node through the power management circuit and energy aware interface was tested. At all the walking speeds tested (3~7 km/h), the Mag-WKEH is able to produce enough energy to power the wireless sensor node to measure and transmit data with a working duty cycle up to 13% while imposing unnoticeable reactive impact on the wearer. With proper sensors deployed in the WSN, the developed system can be potentially used for long-term body/health condition monitoring.

Acknowledgment

The authors gratefully acknowledge the financial support from the EPSRC through project EP/K017950/2.

References

[1] Hannan M A, Mutashar S, Samad S A and Hussain A 2014 *Biomedical Engineering Online* **13**
[2] Jurik A D and Weaver A C 2009 *Software, IEEE* **26** 71-3
[3] Mitcheson P D, Yeatman E M, Rao G K, Holmes A S and Green T C 2008 *Proceedings of the IEEE* **96** 1457-86
[4] Yang K, Zhihao Y and Meiling Z 2016 *Smart Materials and Structures* **25** 085029
[5] Chew Z J and Zhu M 2015 In: *Journal of Physics: Conference Series*: IOP Publishing pp 012022

PDMS/PVA composite ferroelectret for improved energy harvesting performance

J Shi[1], Z Luo[2], D Zhu[3] and S P Beeby[1]

[1]Electronics and Electrical Engineering Group, Electronics and Computer Science University of Southampton, Southampton, SO17 1BJ, UK
[2]Cranfield University Cranfield, MK43 0AL, UK
[3] School of Computing, Electronics and Maths, Faculty of Engineering, Environment and Computing, Coventry University, Coventry, CV1 5FB

E-mail:js4g11@soton.ac.uk

Abstract. This paper address the PDMS ferroelectret discharge issue for improved long- term energy harvesting performance. The PDMS/PVA ferroelectret is fabricated using a 3D-printed plastic mould technology and a functional PVA composite layer is introduced. The PDMS/PVA composite ferroelectret achieved 80% piezoelectric coefficient d_{33} remaining, compared with 40% without the proposed layer over 72 hours. Further, the retained percentage of output voltage is about 73% over 72 hours.

1. Introduction

Ferroelectret are thin films of polymer foams which are can store electric charges in its internal voids, presenting strong piezoelectric-like properties after electric charging. The resulting macroscopic behavior of ferroelectret is very similar with the well-known piezoelectric material; but its charge generation mechanism is entirely different from the latter. For ferroelectret, the interior voids not only reduce the overall mass and stiffness, but also form great dipoles when the separated positive and negative charges are trapped on the top and bottom gas-polymer interfaces in the voids. When the ferroelectret is compressed by an external mechanical load, it can generate electric charges because the majority of dipole moment change take place in the voids [1]. Due to its high piezoelectric properties, ferroelectret are utilized as functional materials in electromechanical sensors and actuators also as electrically charged layers in electrostatic transducers [2, 3].

PDMS (Polydimethylsiloxane) is highly compliant and therefore suitable for integrating into wearable devices for harvesting kinetic energy from human motion. The PDMS gives conformal contact over large areas, even for substrates with uneven surfaces, and is easy to fabricate inexpensively. In addition, the PDMS is sufficiently elastic that it can be separated from the microstructure without destruction or distortion. However, the PDMS ferroelectret discharge significantly and only a small surface charge density which implies a very short lifespan of harvester and lower piezoelectric activity.

It is essential to solve the lower charge density and instability charge issue for PDMS ferroelectret. In this paper, a method of adding a functional PVA (Polyvinyl alcohol) composite film is presented to improve the lifespan of the harvester and enhance its piezoelectric effect due to PVA composite's high dielectric permittivity and thermal stability.

2. Concept

As is shown in Figure 1, a PVA composite layer is introduced at the gas-polymer interfaces which is used as electrets to store charges. Our previous work has shown the relationship between the level of piezoelectric activity and the geometry of the composite structures and its material properties [4]. Therefore, for a certain structure, the level of piezoelectric activity is proportional to the charge density of material. Compared with PDMS, the PVA composite has higher charge density and better thermal

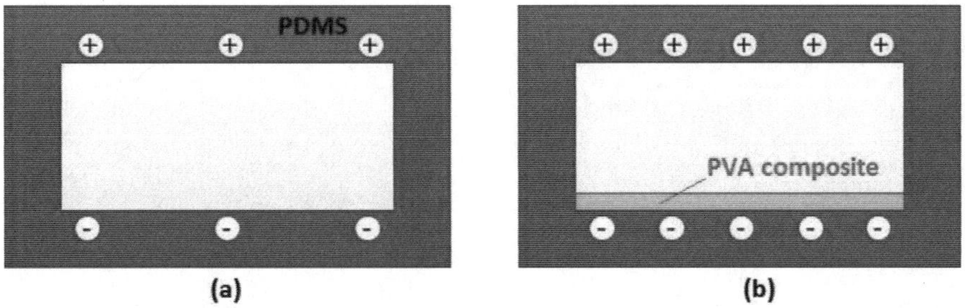

Figure 1. (a) concept of PDMS ferroelectret (b) concept of PDMS/PVA ferroelectret

stability [5]. The combined PDMS/PVA ferroelectret harvester can further improve the energy harvesting performance and stability.

3. Experiment

3.1. Fabrication

The proposed PDMS/PVA composite ferreoelectret is fabricated using a moulding process as illustrated in Figure 2. The plastic moulds were 3D-printed by a Connex350 TM 3D printing system (Stratasys, MN,USA). VerClear^{TM} was selected as the mould material because it is transparent, which makes it easy to observe the entire fabrication process. Because the 3D printable material inhibits PDMS polymerization and the printed structure tends to warp to some extent when it is removed from the printer, the resulting 3D-printed mould is not immediately suitable for PDMS casting. Therefore the moulds are baked in oven at 80 °C for 24 hours. Next, the moulds were glued using a drop of degassed non-polymerized PDMS to a piece of clear glass. After this step, the glass-backed moulds were exposed to a silane vapour for 1 hour to be coated with a thin layer of trichloro(1H,1H,2H,2H-perfluorooctyl) silane (Sigma Aldrich, MO, USA). The final, baked moulds were flat and no longer inhibited the PDMS.

Once the plastic moulds are fabricated, liquid PDMS and curing agent (Sylgard 184 from Dow Corning, MI, USA) were mixed at a 10:1 weight ratio and the mixture degassed in a vacuum desiccator. The degassed PDMS was poured into the glass-backed mould which were again degassed and then baked at 80 °C for 1 hour. When the degassed PDMS was poured into the mould, the fluid level in the mould was monitored to ensure the degassed PDMS would not overflow. This step is used to control the thickness of the foam. Next, the PVA solution was prepared by dissolving 10g in 200 ml of distilled water in a glass beaker and stirred for 30 mins while being heated at 40 °C due to the higher solubility PVA at this temperature. 2 g of ammonium salt was then dissolved in 100 ml of distilled water in another beaker and stirred for 30 mins at room temperature. The PVA solution and ammonium salt solution were then mixed together by electromagnetic stirring for 10 mins at 40 °C. The PVA composite solution is deposited by doctor blading to the flat PDMS layer. The composite films was baked in oven for 5 mins at 60 °C.

There are two possible methods to bond the two PDMS parts together: applying an adhesive or utilizing oxygen plasma machine. For the first method, degassed PDMS can be used as the. After detaching the polymerized PDMS from the moulds, the liquid degassed PDMS was painted on the part 1 with smooth surface and then the other PDMS layer was placed on to it, with its patterned surface in contact with the smooth surface (see figure 2). This bonding method was easy to implement but its was found to be difficult to control the amount of adhesive and overflow of the PDMS into the void would change the geometry of the whole cellular structure. In the second method, after detaching the polymerized PDMS from the moulds, an oxygen plasma treatment (Femto Asher, Diener, Germany, 30 S at 35-40W) was applied to both patterned surface of the PDMS part 2 and the smooth surface of PDMS part 1 (see figure 2) which are then bonded together. The bonded PDMS/PVA composite ferroelectret was baked in oven at 80 °C for 1 hour. A 10 N force was applied to the sample during the bake to improve the bond quality, and also recover the original hydrophobic PDMS surface chemistry. The image of a fabricated PDMS/PVA composite ferreoelectret is shown in Figure 3.

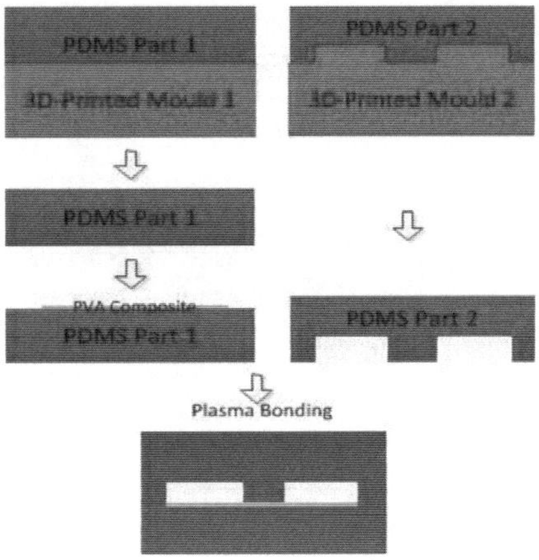

Figure 2. Schematic of fabrication processes

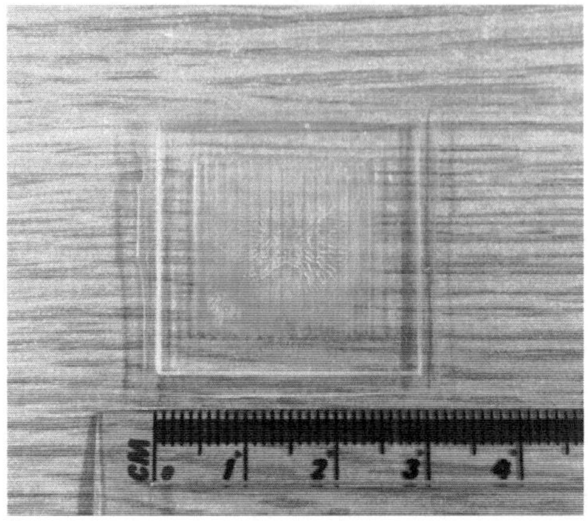

Figure 3. The image of PDMS/PVA composite ferroelectret

3.2. *Testing*

To investigate how the properties of the material are affected by the PVA composites, another pure PDMS ferroelectret was fabricated which is used for comparison. Both of these samples were polarized by a corona charging. A corona-tip voltage of -25 kV and a charging time of 2 mins were employed. The piezoelectric charge constant d_{33} was measured using a PiezoMeter (PM300, Piezotest Ltd). To evaluate its energy harvesting performance, the output voltage of these sample were recorded using an oscilloscope. To quantify the applied forces on the test samples, an Instron electrodynamic instrument (ElectroPuls E1000, Instron Ltd) was used. The maximum dynamic force this equipment can produce is 1000 N and it can provide several types of compressive force including square, sinusoidal and triangular wave forces. The frequency of the compressive forces applied on the test samples was varied from 0 to 100 Hz. These parameters can be modified in the control software.

4. Results and discussion

Figure 4 shows a comparison of piezoelectric performance between PDMS ferroelectret and PDMS/PVA ferroelectret. The initial piezoelectricity of PDMS/PVA ferroelectret was about 110 pC/N which is 1.5 times greater than that of the PDMS ferroelecrret. The PDMS/PVA ferroelectret retained

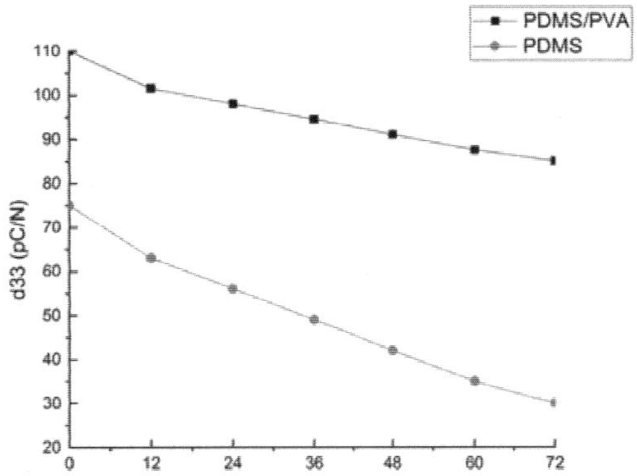

Figure 4. Piezoelectricity d_{33} decay cures for 72 hours

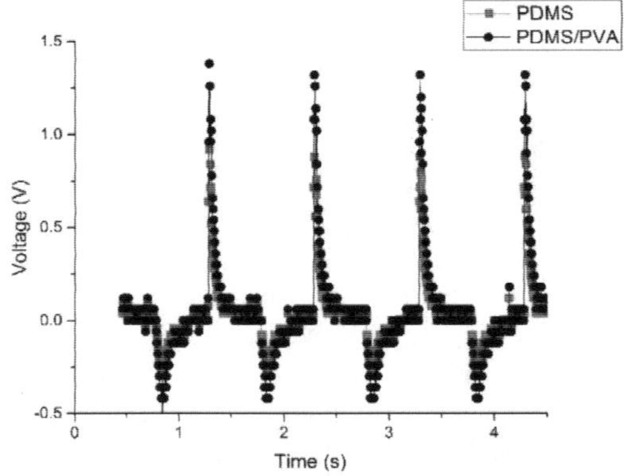

Figure 5. Measured voltage output for PDMS and PDMS/PVA ferroelectret at 1 hour after corona charging

around 80% of its initial piezoelectric coefficients d_{33} over 72h comparing with 40% for the pure PDMS material over 72h.

The output voltage of the PDMS ferroelectret and PDMS/PVA composite ferroelectret under compressive force takes the form of a voltage pulse as shown in figures 5 and 6. From figure 5, the maximal peak voltages measured 1 hour after corona charging for PDMS ferroelectret and PDMS/PVA composite ferroelectret were about 1 V and 1.5 V, respectively. In figure 6, the maximal peak voltage at 72 hours after corona charging for PDMS ferroelectret and PDMS/PVA composite ferroelectret were about 0.5V and 1.1 V, respectively. Thus the PDMS/PVA retains about 73% of its output voltage after 72 hours.

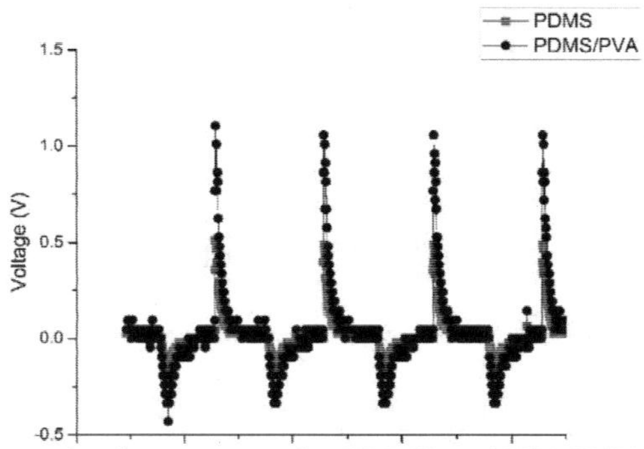

Figure 6. Measured voltage output for PDMS and PDMS/PVA ferroelectret at 72 hour after corona charging

5. Conclusion

This paper presents PVA composite layer to address the PDMS ferroelectret discharge issue for improved energy harvesting performance. The PDMS/PVA ferroelectret achieves 80% piezoelectric coefficients retain rate over 72 hours, compared with 40% without the proposed layer over 72 hours. Device voltage retain rate is about 73% over 72 hours. These results prove that PVA composite is effective in extending the charge lifespan and enhancing the piezoelectric activity for PDMS ferroelectret. Future work will focus on further improving the charge density, stability for using another electret material.

Acknowledgements

This work was performed under the SPHERE IRC funded by the UK engineering and Physical Sciences Research Council (EPSRC), Grant EP/K031910/1.

References

[1] M. Paajanen, The cellular polypropylene electret material-Electromechanical properties, Dissertation, VTT Publications Vol. 436, Technical Research Centre of Finland, Tampere University of Technology, 2001.

[2] J. Backman. Audio applications of electrothermomechanical film (ETMF), *J. Aud. Eng. Soc.*, vol. 38 (5), pp. 364-371, 1990.

[3] M. Paajanen, et al. Electromechanical film (EMFi) - a new multipurpose electret material, *Sens. Actu. A: Phys.*, vol. 84 (1-2), pp. 95-102, 2000.

[4] J. Shi, et al. An investigation of PDMS structures for optimized ferroelectret performance. *Journal of Physics: Conference Series* , 557, 012104. 2014.

[5] M. Uddin, et al. Ammonium Dihydrogen Phosphate/PVA Composite Films with High Dielectric Permittivity and Enhanced Thermal Stability. Advanced Science, Engineering and Medicine 5(2): 126-132. 2013.

A step towards on-chip biochemical energy cascade of microorganisms: carbon dioxide generation induced by ethanol fermentation in 3D printed modular lab-on-a-chip

A Podwin, W Kubicki, K Adamski, R Walczak and J A Dziuban.

Faculty of Microsystem Electronics and Photonics, Wrocław University of Technology, 11/17 Janiszewski St., 50-372 Wrocław, Poland

agnieszka.podwin@pwr.edu.pl

Abstract. The concept of biochemical energy cascade of microorganisms towards oxygen generation in 3D printed lab-on-a-chip has been presented. In this work, carbon dioxide – a product of ethanol fermentation of yeasts has been utilized to enable light-initialized photosynthesis of euglenas and as a result of their metabolic transitions produce pure oxygen.

1. Introduction

Culturing of microscopic creatures in lab-on-a-chip environment is one of the most investigated subjects standing on the borderline of microfluidics and microbiology [1-4]. For instance, several works concerning the growth and study of euglena and yeast biological potential have been demonstrated recently [5-8]. Euglenas are commonly known as distinctive microorganisms that easily generate oxygen – a bypass product of their metabolism, based on CO_2 photosynthesis process. On the other hand, generation of CO_2 is an obvious effect of organic compounds' fermentation, especially ethanol fermentation of yeasts.

In this paper, a lab-on-a-chip, in which both of the aforementioned processes are used to obtain an energy cascade, has been presented. In the cascade, a colony of yeasts generates gaseous carbon dioxide (CO_2) which next, diffuses through a semi-permeable membrane and nourishes a colony of euglenas. Specified metabolism of euglenas (photosynthesis) results in production of gaseous oxygen (O_2) that penetrates through the next diaphragm and reaches the final microchamber. Eventually, gaseous oxygen can be used to supply a new generation of bio-based fuel cells.

Among different fabrication methods of the lab-on-a-chip devices, 3D print technique has been selected as the most appropriate in this solution. The lab-chip construction and principle of working, as well as results of preliminary experiments are shown in the next section.

2. Experiment

As mentioned above, two colonies of microorganisms have been cultured in the lab-on-a-chip device (Fig. 1).

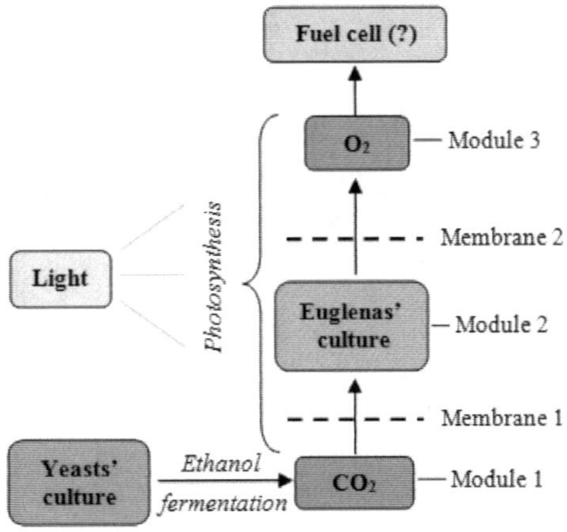

Figure 1. Scheme of the microorganisms' biochemical energy cascade.

In the module 1, solution of baker's yeasts (*Saccharomyces cerevisiae*), glucose and H_2O mixed in the relation of 1:4:10 (w/w) is used to start the ethanol fermentation. The photosynthesizing microorganism, *Euglena gracilis* (Blades Biological Ltd) is cultured in a buffer prepared according to the guidelines (Blades Biological Ltd, Protozoa and algae culture instructions) in the module 2. As a result of euglenas' photosynthesis (process initialized by CO_2 and light) gaseous oxygen is generated and transferred through the membrane to the module 3.

2.1. Lab-on-a-chip construction

All of the processes leading to the microorganisms' energy conversion have been conducted inside the 3D printed modular lab-on-a-chip. The 3D printer (model: Projet 3510, 3D Systems, USA) utilizing ink-jet UV photo curable materials (Visijet M3 Crystal and support – Visijet S300, 3D Systems, USA) has been exploited for fabrication of the device. The use of 3D printing method and modularity of the chip structure enables for a reconfigurable and rapid extension of the system. As mentioned, the chip contains three modules. The modules are arranged horizontally, creating a sandwich structure of the chip (Fig. 2). Every module has a cavity for o-ring sealing, which provides tightness and protects samples from the external environment. Volume of each module chamber is 132 µl. Semi-permeable polydimethylsiloxane (PDMS) membranes are indispensable elements of the chip, since they separate both the second and the third modules, enabling the diffusion of gases from one chamber to another. The membranes have been fabricated utilizing moulding technique. Glass slide (Borofloat 3.3, Schott) have been covered with polymer mask (Avery Graph) with laser-cut pattern of the membrane and isotropically etched in the solution of 50% HF:69% HNO_3. Next, polydimethylosiloxane (Sylgard 184, Dow Corning) mixed with its curing agent at the ratio of 10:1 (w/w) and degassed in ultrasonic washer, has been casted onto master mould. Ultimately, after 24 hours of polymerization, circle-shaped membranes with a diameter of 12 mm and thickness of 120 µm have been obtained. Every diaphragm is combined with the module by a pressing ring, which prevents it from damage while screwing. On a top of the chip, a cap with a glass window is attached for microscopic observations and illumination of euglenas' colony.

3. Results and discussion

Yeasts have been cultured in the module 1, according to the conditions mentioned earlier. After 24 hours, generation of carbon dioxide has been observed in the module 2 lasting for 5 days. The detection of CO_2 has been obtained by the hydrogencarbonate indicator [9], solution that changes its colour from red to orange, if pH value in the environment increases. The indicator has been introduced to the module 2 and full colour pictures have been taken by CMOS minicamera to visualize the solution colour before and after the start of ethanol fermentation. Acquired pictures have been then

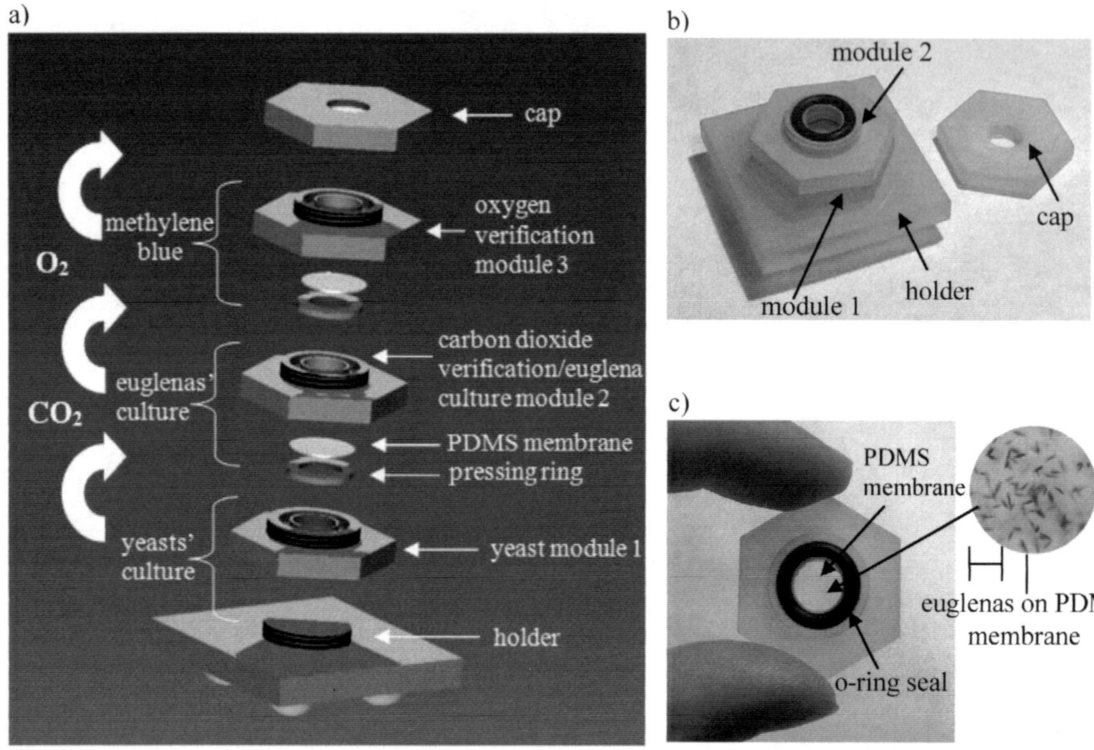

Figure 2. Construction and functioning of the 3D printed lab-chip: a) exploded schematic view, b) lab-chip at a glance, c) euglenas' colony in the module 2, scale bar – 100 μm.

digitized to obtain basic RGB colour histograms (Fig. 3). In the next experimental step, colonies of yeasts and euglenas have been cultured in the module 1 and 2, respectively. CO_2 generated by yeasts in the module 1 permeates through the first membrane and supplies euglenas' colony, cultured in the module 2 and lightened by daylight. The colony of euglenas absorbs CO_2 and generates O_2. Production of O_2 is in turn detected by solution of reduced methylene blue [10], introduced into the module 3. On the contrary to the long-lasting colony of yeasts (5 days easily), in our first experiment the culture of euglenas has survived solely 3 hours. It seems that concentration of generated in the system CO_2 (>0.04%, according to the colour transition of hydrogencarbonate indicator) might be too high and causes evaporation of the buffer with euglena colony. The culturing conditions of this microorganism have to be optimized in the future works to enable the implementation of the energy cascade.

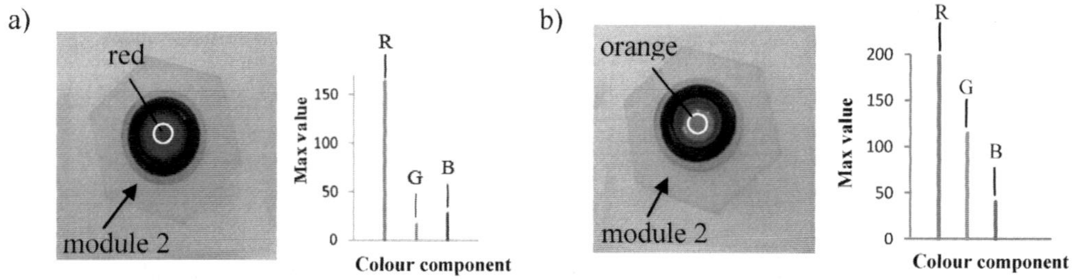

Figure 3. The change of colour of hydrogencarbonate (pH) indicator in the lab-on-a-chip chamber after 24 hours of yeasts' culturing: a) on the left: initial red colour of solution in the chamber, on the right: colour histogram of the representative area, b) on the left: change of the colour to orange hue in the chamber, on the right: colour histogram of the representative area.

4. Conclusions

3D printed lab-on-a-chip consisting of the three modules has been presented. The chip has provided diffusion of gases between the modules and enabled for simultaneous yeasts' and euglenas' culturing. In the first module, generation of CO_2 has been obtained and next utilized as nourishment for euglenas. Euglenas cultured in the second module, has absorbed CO_2 in the photosynthesis process and produced O_2 that has been then gathered in the third module of the chip. The culture of yeasts has lasted for 5 days, while euglenas' stimulated by based on ethanol fermentation CO_2 – 3 hours.

It may be contended that the construction of the chip and application of 3D printing technique is suitable to fit the requirements of the microorganisms' energy cascade, however further works have to be conducted to improve the functioning of the biological system and allow for its future application in fuel cells.

Acknowledgments
The works were financed by NCN OPUS 10; project no 2015/19/BST8/01110.

References

[1] Boehm A D, Gottlieb A P and Hua S Z 2007 On-chip microfluidic biosensor for bacterial detection and identification *Sensors and Actuators B: Chemical* **126** pp 508–14
[2] Tang Y, Mingzhe G, Xie Y, Li X and Chen L 2014 Fast screening of bacterial suspension culture conditions on chips *Lab Chip* **14** pp 1162-67
[3] Bouguelia S, Roupioz Y, Slimani S, Mondani L, Casabona M G, Durmort C, Vernet T, Calemczuk R and Livache T 2013 On-chip microbial culture for the specific detection of very low levels of bacteria *Lab Chip* **13** pp 4024-32
[4] Au S H, Shih S C C and Wheeler A R 2011 Integrated microbioreactor for culture and analysis of bacteria, algae and yeast *Biomedical Microdevices* **13** pp 41-50
[5] Itoh A, Houri S and Tamura W 2008 Micro parts assembly by formation controlled euglena group using their phototaxis Proceedings of IMECE2008 2008 ASME International Mechanical Engineering Congress and Exposition October 31-November 6 2008 Boston Massachusetts USA pp 2207-08
[6] Li M, Gao X and Wu B 2014 Microalga euglena as a bioindicator for testing genotoxic potentials of organic pollutants in Taihu Lake China *Ecotoxicology* **4** pp 633-40
[7] Euglena corporation website: http://euglena.jp/en/, visited on 03.01.2015
[8] Hilbert W and Jakoby B 2015 Ethanol fermentation as the basis for autonomous, long-term and high-pressure fluid transport in microfluidics *Procedia Eng.* **150** pp 100-05
[9] Preparation of hydrogencarbonate solution: http://www.saps.org.uk, visited on 05.05.2016
[10] Akhtar M, Brandhoff L and van den Driesche S 2015 Air-droplets as gas reservoir to provide O_2 to the stored-aqueous droplets in micro-channels *Procedia Eng.* **120** pp 92-5

Red-shift of the photoluminescent emission peaks of CdTe quantum dots due to the synergistic interaction with carbon quantum dot mixtures

E Pelayo[1,2], A Zazueta[1,3], R López-Delgado[1,3], E Saucedo[2], R Ruelas[2] and A Ayón[1]

[1]University of Texas at San Antonio, Dept. of Physics and Astronomy, MEMS Research Lab, One UTSA Circle, San Antonio, TX 78249, U SA.

[2]Universidad de Guadalajara, Centro de Ciencias Exactas e Ingenierías, Blvd. Gral. Marcelino García Barragán 1421, 44430 Guadalajara, Jal, México.

[3]Universidad de Sonora, Departamento de Física, Luis Encinas y Rosales S/N, Hermosillo, Son, 83000, México.

Email: aayon@utsa.edu.

Abstract. We report the relatively large red-shift effect observed in down-shifting carbon quantum dots (CQDs) that is anticipated to have a positive impact on the power conversion efficiency of solar cells. Specifically, with an excitation wavelength of 390 nm, CQDs of different sizes, exhibited down-shifted emission peaks centered around 425 nm. However, a solution comprised of a mixture of CQDs of different sizes, was observed to have an emission peak red-shifted to 515 nm. The effect could arise when larger carbon quantum dots capture the photons emitted by their smaller counterparts followed by the subsequent re-emission at longer wavelengths. Furthermore, the red-shift effect was also observed in CdTe QDs when added to a solution with the aforementioned mixture of Carbon QDs. Thus, whereas a solution solely comprised of a collection of CdTe QDs of different sizes, exhibited a down-shifted photoluminescence centered around 555 nm, the peak was observed to be further red-shifted to 580 nm when combined with the solution of CQDs of different sizes. The quantum dot characterization included crystal structure analysis as well as photon absorption and photoluminescence wavelengths. Subsequently, the synthesized QDs were dispersed in a polymeric layer of poly-methyl-methacrylate (PMMA) and incorporated on functional and previously characterized solar cells, to quantify their influence in the electrical performance of the photovoltaic structures. We discuss the synthesis and characterization of the produced Carbon and CdTe QDs, as well as the observed improvement in the power conversion efficiency of the fabricated photovoltaic devices.

1. Introduction

Carbon, CdTe, Si, CdSe and ZnO quantum dots (QDs), among others, are frequently explored nanostructures that exhibit the property of capturing high energy photons and subsequently emitting lower energy ones that are more suitable for producing electron-hole pairs in silicon solar cells. The wavelength of the emitted photons is directly related to QD size, specifically, the larger the synthesized QD the longer the wavelength of said photons. The size of the produced Carbon quantum dots (CQDs) is controlled by the applied current during synthesis, while that of CdTe QDs is determined by the refluxing time. In addition to the aforementioned down-shifting effect, the colloidal mixtures of C and CdTe QDs of different sizes, exhibit significant differences. Specifically, the photoluminescent peak of a colloidal mixture of CdTe QDs (CdTeMS) is observed to be centered around the collection of the individual peaks, thus, there is no discernible red-shift or otherwise a deviation from what could be anticipated a priori. However, the CQD colloidal mixture (CMS) behaves differently since the peak is observed to be red-shifted respect to the position of the collection of the individual photoluminescent peaks. This additional red-shift should prove useful in boosting the power conversion efficiency of solar cells.

2. Experimental details

2.1. Synthesis of Carbon QDs

The Carbon nanostructures with a distribution of sizes (see Figure 1) were synthesized employing an alkali-assisted electrochemical fabrication method utilizing graphite rods for both the anode and the cathode [1] [2], while varying the applied current between 10 and 60 mA in increments of 10 mA. The graphite rods employed had a diameter of 5 mm, a separation of anode to cathode of 25.4 mm, and were submerged 30 mm in an 100 ml electrolyte solution composed of ethanol and water with a volume ratio of 99.5/0.05 to which 0.3g of NaOH were added. The current was applied for one hour immediately upon the submersion of the graphite rods within the specified current range. Subsequently, the samples were stored for 48 hours at room temperature to stabilize them, and the produced solutions were allowed to evaporate until obtaining 5 ml for every 100 ml of quantum dot solution. Upon the completion of the evaporation step the samples were separated employing a silica-gel chromatography column with an 100 ml mixture of petroleum ether and diethyl ether with a volume ratio of 30/70. The final step was to evaporate the solvents thoroughly in each vial to increase the CQD concentration.

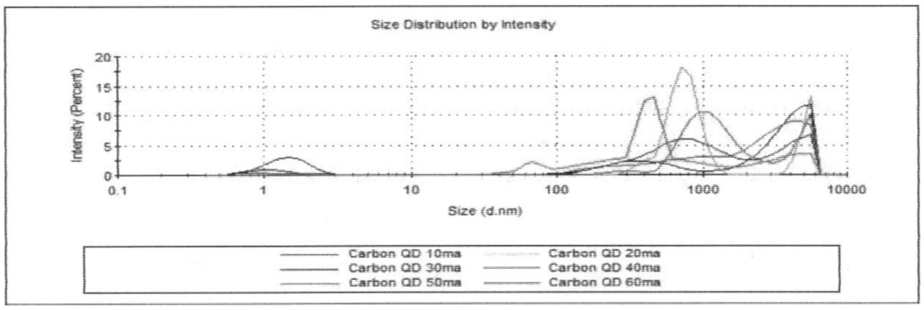

Figure 1.- Carbon quantum dots size distribution.

2.2. Synthesis of CdTe QDs

CdTe QDs of measured sizes between 2 to 18 nm (see Figure 2) were obtained employing a chemical synthesis scheme. Specifically, 0.0533 g of cadmium acetate dihydrate ($Cd(CH_3COO)_2 \cdot 2H_2O$, 99.5%) were dissolved in 50 ml of deionized (DI) water in an 125 ml Erlenmeyer flask, subsequently 18 µl of thioglycolic acid (TGA, 90%) were added, and the pH was adjusted with 1 M sodium hydroxide (NaOH) set solution until reaching a value between 10.5 to 11 in the pH scale, and stirred for 5 minutes. Separately, 0.0101 g of potassium tellurite (K_2TeO_3, 95%) were dissolved in 50 ml of DI water in an 125 ml Erlenmeyer flask, stirred for 5 minutes and 0.0101 g K_2TeO_3 were added to this second solution. Subsequently, the previously prepared solutions were mixed, 0.08 g of sodium borohydride ($NaBH_4$, 99.99%) were added to the mixture, and the reaction was allowed to proceed for 5 minutes. The mixed solution was then transferred to a single-neck, round-bottom flask that was attached to a Liebig condenser, which was stirred at 500 rpm while being refluxed. During the refluxing times of 15 min, 30 min, 1 h, 3 h, 5 h, 6 h, 8h and 12h, the flask remained submerged in laboratory oil whose temperature was maintained at 100°C. QD size and photoluminescent emission wavelengths were determined by the refluxing time [3] [4].

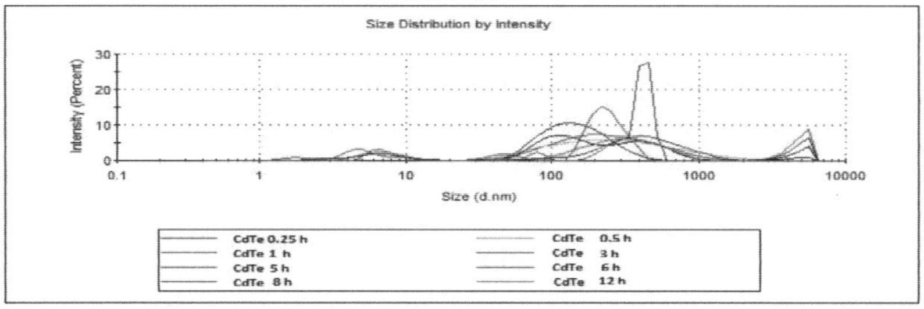

Figure 2.- CdTe quantum dots size distribution

3. Results and discussion

3.1. QD Colloidal Solution Characterization and Solar Cell Deployment

The colloidal mixture of Carbon QDs of different sizes (CMS) was observed to produce a spectrum with a red-shifted peak located at ~515 nm (see Figure 3 part (a.)) while the spectra of the previously synthesized CQDs with the selected current values, were observed to be centred around ~425 nm. As far as the CdTe QDs are concerned, their ultimate size can be adjusted by the refluxing time, and size is reflected in the emission wavelength of the synthesized nanostructures (see Figure 3 part (b.)). Specifically, with an excitation wavelength of 390 nm, CdTe QDs with a refluxing time of 1 h have a characteristic emission peak at ~548 nm, while the characteristic emission peak is located at ~570 nm for a refluxing time of 8 hours. When any single-sized colloidal CdTe solution is combined with the aforementioned CMS, the respective spectra is red-shifted (see Figure 4). In order to incorporate the synthesized QDs on functional solar cells, they had to be dispersed in a matrix layer [5]. For this purpose, we selected Polymethylmethacrylate (495 PMMA A2 from Microchem) as the matrix in which to disperse the CdTeMS + CMS. Upon nanostructure dispersion, the PMMA + QDs solutions were spin cast at 4,000 rpm on the window side of 52mm x 38mm, commercially available polysilicon solar cells (Eco-worthy Company) with a nominal thickness of 200μm. The resultant thin films had an average thickness of ~1050 nm. Solar cell performance was quantified using an Oriel Sol2A solar simulator under standard testing conditions. Specifically, measurements were collected before and after the deployment of the PMMA+QDs. The external quantum efficiency (EQE) was measured with an Oriel Quantum Efficiency Measurement kit (QE-PV-SI) using a spot size of approximately 2 mm^2. Solar cells performance improvement was corroborated in the measured values of the overall efficiency of the cells (see Figure 5) which improved from 12.43% to 13.03% upon QD deployment.

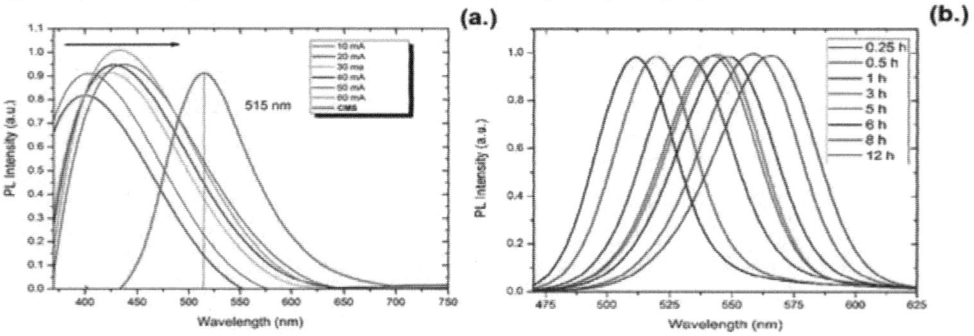

Figure 3.- (a.) Carbon QDs have size-dependent, down-shifted photoluminescent emission peaks centered around 425 nm, however, the solution comprising all the carbon QDs previously synthesized, exhibits a red-shifted photoluminescent emission peak centered at ~515 nm (b.) Emission wavelength of CdTe QDs with different refluxing times under 390 nm excitation.

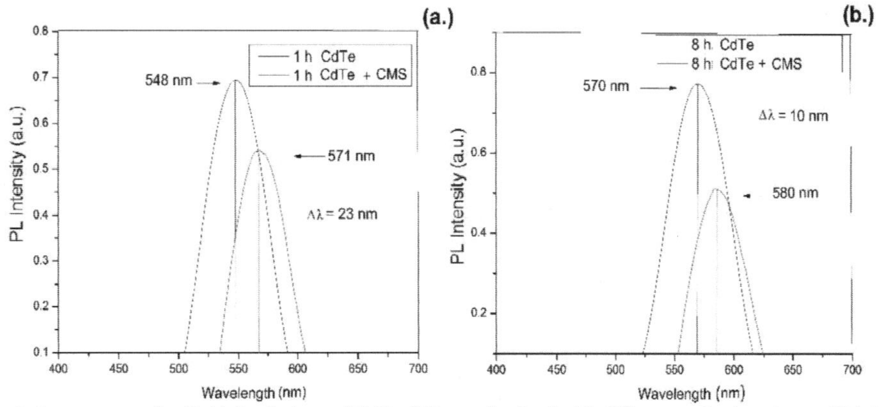

Figure 4.- Emission spectra of colloidal solutions of CdTe QDs synthesized with different refluxing times of 1 h in (a) and 8 h in (b) and their shifted spectra when combined with CMS.

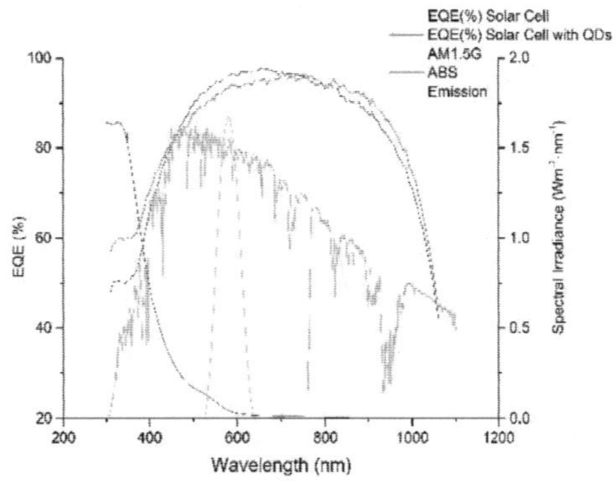

Figure 5.- *A Comparison of the EQE before (Black line) and after the addition (Red line) of CMS + CdTeMS. The absorption and emission spectra are overlain.*

4. Conclusions

The experimental observations of the interaction of the incorporation of PMMA thin films with downshifting CMS, CdTeMS and CdTeMS + CMS on the window side of solar cells indicate a relatively small but not negligible increase in the power conversion efficiency (PCE) of solar cells.

Acknowledgments

The authors would like to acknowledge the U.S. Army Research Office (Grant W911NF-13-1-0110) and CONACYT for the financial support for this project.

References

[1] Haitao L, Xiaodie H, Hui H, Yang l, Suoyuan L 2010 Water-Soluble fluorescent carbon quantum dots and photocatalyit desingn *Angew Chem Int Edit.* **49** 4430

[2] Zheng X T, Ananthanarayanan A, Luo K Q, and Chen P 2015 Glowing graphene quantum dots and carbon dots: properties, syntheses, and biological applications *Small.* **11** 1620

[3] Suli W, Jun D, Jie Z and Shufen Z 2012 A simple and economical one-pot method to synthesize high-quality water soluble CdTe QDs *J Mater Chem.* **22** 14573

[4] Yuan Z, Yang P and Cao Y 2012 Time-resolved photoluminescence spectroscopy evaluation of CdTe and CdTe/CdS quantum dots *SRN Spectroscopy.* 2012

[5] McIntosh K R, Lau G, Cotsell J N, Hanton K, Bätzner D L, Bettiol F and Richards B S 2009 Increase in external quantum efficiency of encapsulated silicon solar cells from a luminescent down-shifting layer *Research and Applications.* **17** 191

Journal of Physics: Conference Series **773** (2016) 012054 doi:10.1088/1742-6596/773/1/012054

ON-CHIP SUPERCAPACITOR ELECTRODE BASED ON POLYPYRROLE DEPOSITED INTO NANOPOROUS AU SCAFFOLD

P Lu, P Ohlckers, X Y Chen
Department of Micro- and Nanosystem Technology, University College of Southeast Norway, Campus Vestfold, Raveien 215, 3184 Borre, Norway

E-mail: Xuyuan.Chen@hbv.no

Abstract. On-chip supercapacitors hold the potential promise for serving as the energy storage units in integrated circuit system, due to their much higher energy density in comparison with conventional dielectric capacitors, high power density and long-term cycling stability. In this study, nanoporous Au (NP-Au) film on-chip was employed as the electrode scaffold to help increase the electrolyte-accessible area for active material. Pseudo-capacitive polypyrrole (PPY) with high theoretical capacitance was deposited into the NP-Au scaffold, to construct the tailored NP-Au/PPY hybrid on-chip electrode with improved areal capacitance. Half cell test in three-electrode system revealed the improved capacitor performance of nanoporous Au supported PPY electrode, compared to the densely packed PPY nanowire film electrode on planer Au substrate (Au/PPY). The areal capacitance of 37 mF/cm^2~10 mV/s, 32 mF/cm^2~50 mV/s, 28 mF/cm^2~100 mV/s, 16 mF/cm^2~500 mV/s, were offered by NP-Au/PPY. Also, the cycling performance was enhanced via using NP-Au scaffold. The developed NP-Au/PPY on-chip electrode demonstrated herein paves a feasible pathway to employ dealloying derived porous metal as the scaffold for improving both the energy density and cycling performance for supercapacitor electrodes.

1. Introduction

supercapacitor has been attracting tremendous interest owing to its advantage in high power density and long cycling stability [1]. Recently, the promising application of supercapacitor in micro-electro-mechanical-system (MEMS) has been posed and acknowledged as a feasible route to replace the widely used micro-battery that is suffering poor cycling life and low charge/discharge rate [2]. On-chip supercapacitor device therefore becomes the focal point in energy storage field.

At the current stage, one main obstacle in the further development of on-chip supercapacitors is their insufficient energy density [3]. The mainly employed carbon-based active electrodes for on-chip supercapacitors store energy based on double layer mechanism, of which the areal capacitance is still low ranging from several hundred μF to several mF [4]. To use pseudo-capacitive active materials with high theoretical capacitance for on-chip supercapacitors has been regarded as one effective route to energy enhancement. PPY, one typical conductive polymer has been well studied as pseudo-capacitive material and successfully used in on-chip supercapacitor device [5]. In this study, we demonstrated nanoporous Au film on-chip was used as the electrode scaffold to help increase the electrolyte-accessible area for PPY, and thus realize a high areal capacitance on-chip supercapacitor electrode.

2. Experimental

Alloy thin films (~$Au_{0.3}Cu_{0.7}$) with the given thickness of 750 nm were deposited on the silicon wafers via co-sputtering of Au (DC-model, 35 W) and Cu (RF-model, 200 W) using a AJA sputter. To selectively etch Cu component, the alloy films coated silicon wafers were dipped in concentrated nitric acid (70 vol%) at room temperature for 3 hours. Afterwards, the thoroughly cleaned porous Au film coated Si wafer was used as the substrate for deposition of active material PPY. The electrochemical deposition of PPY was performed in three-electrode setup, at a constant current density of 1 mA/cm^2 for 10 min, in the plating bath: 0.2 M phosphate buffer (pH = 6.86) containing 0.2 M pyrrole and 0.02 M p-toluenesulfonate acid.

Conventional three-electrode system was employed to study the supercapacitor performance. Pt counter electrode, and Ag/AgCl (in saturated KCl) reference electrode were used for test. A Zahner IM6 electrochemical workstation was used for electrochemical measurements. Cyclic voltammograms (CV) were conducted in 1 M Na_2SO_4 electrolyte. Electrochemical impedance spectra (EIS) were measured over the frequency range from 100 kHz to 100 mHz with an AC perturbation of 10 mV at the open circuit potential. The areal specific capacitance of the on-chip electrodes was calculated by integrating the CVs.

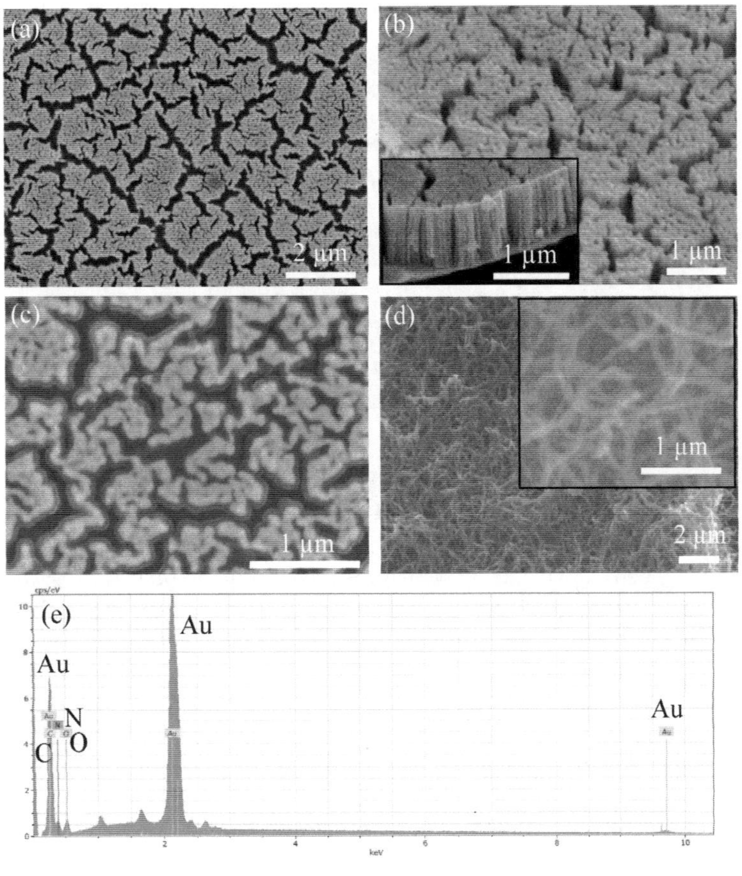

Figure 1. SEM images of (a-b) NP-Au, (c) NP-Au/PPY, (d) Au/PPY. (e) EDX result of NP-Au/PPY

3. Results and discussion

Figure 1(a-b) are SEM images of the obtained NP-Au film after selective etching of Cu component in the alloy film. It can be found that hierarchical porous Au scaffold with both macropores (> 50 nm) and mesopores (2~50 nm) was obtained. From the inset SEM image in Figure 1b, the real thickness was estimated to be 700~800 nm, in accordance with the given thickness (~750 nm) in the sputtering operation. When employed as electrode scaffold, the nanoporous structure could endow large surface areas and favourable pathways for electrolyte ion diffusion. After electro-deposition, the PPY coating with the thickness less than 10 nm was grown into the nanopores of NP-Au scaffold, as shown in Figure 1c. The successful growth of PPY can also be confirmed by the EDX (energy dispersive X-ray) analysis, shown in Fig. 1e. Besides the Au signal, C, N, O elements can also be detected. C, N signals are the main element components of PPY, while the O element is from the doped anions in PPY matrix. Without the NP-Au scaffold, densely aggregated PPY nanowires are formed on the flat Au film, as shown in Figure 1d, with the diameter of 50~100 nm (see inset).

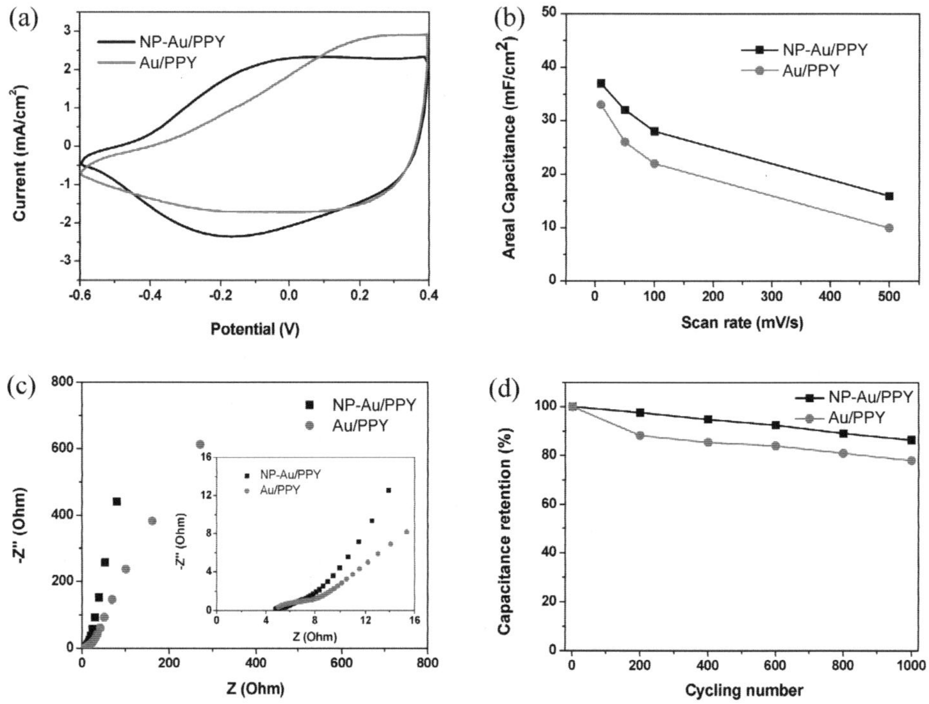

Figure 2. Supercapacitor performance of NP-Au/PPY and Au/PPY: (a) CV at a scan rate of 50 mV/s, (b) Capacitance at varying scan rates, (c) EIS study, and (d) cycling performance at 100 mV/s

Half cell test in three-electrode system revealed the improved supercapacitor performance of NP-Au/PPY compared to the densely stacked PPY nanowire electrode on planer Au substrate. The pseudo-capacitive character of PPY was displayed by Figure 2a, characterized by the existence of prominent redox peaks that is distinct from the carbon-based electric double layer electrode. The larger enclosed CV area of NP-Au/PPY than that of Au/PPY also indicated the larger capacitance can be realized by NP-Au/PPY electrode. As for on-chip supercapacitors, areal capacitance and volumetric capacitance should be the most valuable indicator to evaluate their capacitance, rather than gravimetric capacitance. Therefore, only the capacitance data per unit area and volume are presented herein. The calculated areal capacitance at varying scan rates is presented in Figure 2b. The areal capacitance of 37 mF/cm^2~10 mV/s, 32 mF/cm^2~50 mV/s, 28 mF/cm^2~100 mV/s, 16 mF/cm^2~500 mV/s, were offered by NP-Au/PPY. While Au/PPY electrode delivered the areal capacitance of 33 mF/cm^2~10 mV/s, 26 mF/cm^2~50 mV/s, 22 mF/cm^2~100 mV/s, 10 mF/cm^2~500 mV/s. Considering the thickness of Au scaffold part (~750 nm), very high volumetric capacitance can be attained by NP-Au/PPY (493 F/cm^3~10 mV/s, 213 F/cm^3~500 mV/s). The employed NP-Au scaffold can help increase the electrolyte-accessible area for active materials, and thus led to the capacitance enhancement. EIS study (Figure 2c) was employed to illustrate the improved supercapacitor performance of NP-Au/PPY compared to Au/PPY. The quasi semi-circle part in the high-frequency range of Nyquist plot characterizes the charge transfer resistance at the electrode/electrolyte interface. The smaller charge transfer resistance was achieved by NP-Au/PPY. This is benefiting from the hierarchical porous structure with both macropores and mesopores, which can help improve the ionic conductivity [6]. Also the more favourable ion-diffusion realized by porous conductive scaffold can also be revealed by larger slope of Nyquist plot in the low-frequency range. It is well known that the volume of PPY dramatically varies during the doping and de-doping of electrolyte ions when applied as supercapacitor electrode, which leads to the poor cycling performance. This problem can be alleviated by using NP-Au scaffold that can provide the buffer zone for volume expansion and help avoid the pulverization of PPY. As shown in Figure 2d, after charging/discharging for 1000 cycles at 100 mV/s, the NP-Au/PPY exhibited a 87% capacitance retention, better than the retention ratio of 78% for Au/PPY.

4. Conclusions

We demonstrated the nanoporous Au fabricated by dealloying of sputtered Au-Cu alloy films can be employed as an effective scaffold to develop high performance on-chip supercapacitor electrode. Pseudo-capacitive PPY as the active materials was deposited into the NP-Au scaffold by electrochemical plating method. The hierarchical porous structure of NP-Au with both macropores and mesopores is beneficial for shortening the ion diffusion path and decreasing the ionic resistance, thus led to improved performance of NP-Au/PPY compared with Au/PPY. A high areal capacitance of 37 mF/cm^2 and volumetric capacitance of 493 F/cm^3 can be realized. More significantly, the use of metal scaffold on-chip would readily facilitate the integration of supercapacitor with other MEMS energy harvesting systems through metal-metal bonding technique.

Acknowledgments

Significant discussions with Professor Einar Halvorsen are acknowledged. Financial support from the Research Council of Norway under Grant No. 229716 is acknowledged.

References

[1] Huang P, Lethien C, Pinaud S, Brousse K, Laloo R, Turq V, Respaud M, Demortiere A, Daffos B, Taberna P L, Chaudret B and Gogotsi Y 2016 *Science* **351** 691
[2] Lu P, Ohlckers P, Müller L, Leopold S, Hoffmann M, Grigoras K, Ahopelto J, Prunnila M and Chen X Y 2016 *Electrochem. Commun.* **70** 51
[3] Lin T Q, Chen I W, Liu F X, Yang C Y, Bi H, Xu F F and Huang F Q 2015 *Science* **350** 1508
[4] Wu Z S, Feng X L and Cheng H M 2014 *Nat. Sci. Rev.* **1** 277
[5] Sun W, Zheng R L and Chen X Y 2010 *J. Power Sources* **195** 7120
[6] Xu J, Tan Z Q, Zeng W C, Chen G X, Wu S L, Zhao Y, Ni K, Tao Z C, Ikram M and Ji H X 2016 *Adv. Mater.* **28** 5222

Power density improvement of the power conditioning circuit for combined piezoelectric and electrodynamic generators

H Zessin, P Spies and L Mateu

Fraunhofer Institute for Integrated Circuits IIS, Nordostpark 84, 90411 Nuremberg

E-mail: henrik.zessin@iis.fraunhofer.de, loreto.mateu@iis.fraunhofer.de, peter.spies@iis.fraunhofer.de

Abstract. In this study, we report a power management circuit for a combined piezoelectric-electrodynamic generator. A piezoelectric element is bonded to a spring steel cantilever beam and a magnet, used as tip mass, oscillates through a coil. This principle creates the combined generator. A test setup has been created to automate the characterization of the piezoelectric generator and its power management circuit. Three different power management circuits for the piezoelectric part of the combined generator have been analysed: a bridge rectifier, an SSHI circuit with an external inductance and an SSHI circuit which utilizes the coil of the electrodynamic generator as circuit element. The three circuits are compared in terms of their output power, efficiency and power density. The SSHI circuit with an external inductance has the highest output power and efficiency, followed by the SSHI circuit with the electrodynamic generator coil. The power density of the bridge rectifier is the highest but for higher efficiency the power density of the SSHI circuit with the coil of the electromagnetic generator reaches the best results.

1. Introduction

Ambient vibrations are an energy harvesting source available in several application scenarios. An inertial generator converts the environmental vibrations into electrical energy to supply low power electronic circuits like micro-controllers, sensors and RF communication modules. The electrical output power of an energy harvesting power supply can be improved on the generator side as well as on the power management circuit side. Moreover, power density is usually employed for a comparison between inertial generators since output power scales directly with the volume. However, power density has not been a figure of merit for the power management circuit until now.

This paper presents a coupled piezoelectric-electromagnetic energy harvester that converts ambient vibrations into electrical vibrations increasing the power output density compared to a standalone piezoelectric (PEG) or electrodynamic generator (EDG) [1].

Piezoelectric generators provide AC power that has to be rectified. Nonlinear circuits can be more efficient than full-wave rectifiers when the inertial generators are excited at their resonant frequencies [2]. Thus, the paper compares the power harvested by the piezoelectric part of the coupled generator employing a diode full-wave rectifier and a synchronized switch harvesting on inductor (SSHI) circuit [3]. By using the inductor of the electrodynamic generator in the SSHI circuit, the output power density of the power management circuit is increased since no extra inductor is required.

The structure of the paper is as follows. Section 2 describes the coupled piezoelectric-electromagnetic generator and the setup of the measurement system. Section 3 describes the standard power management circuit. Section 4 shows the advanced power management circuitry employing an external coil and employing the coil of the electrodynamic part of the generator, respectively. Finally, Section 5 compares the different power management circuits and draws the main conclusions achieved in the paper.

2. Measurement setup

A schematic of the piezoelectric-electrodynamic generator (PEDGen) presented in this paper is shown in **Figure 1**. The piezoelectric generator part consists of a unimorph cantilever beam with tip mass. A permanent magnet has been employed as tip mass and together with a coil around it forms the electrodynamic generator. Therefore, an inertial generator based on the piezoelectric and electrodynamic effect has been presented.

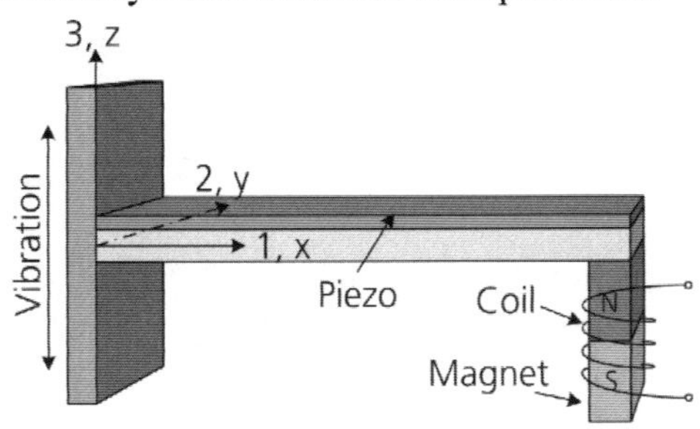

Figure 1. Schematic of the piezoelectric-electromagnetic energy harvesting generator (PEDGen).

Figure 2. Picture of the piezoelectric-electromagnetic generator prototype.

Figure 2 displays the prototype of the piezoelectric-electromagnetic prototype used for testing in the paper. A P-876.A15 DuraAct transducer by PI Ceramic with the dimensions 61x35x0.4 mm^3 is glued to a 97x30x0.3 mm^3 spring steel beam. A neodymium magnet (NdFeB) that weighs 5 g and has a diameter of 12 mm and a height of 6 mm is placed at the free end of the cantilever beam. The coil has 1500 windings with 0.15 mm enameled copper wire resulting in an electrical resistance of 187 Ω and an inductance of 47 mH.

Figure 3 shows the combined output power at the resonance frequency of 42.44 Hz of the piezoelectric-electromagnetic generator without rectification. The maximum output power is 282 μW at the load resistances R_{EDG} = 3.9 $k\Omega$ and R_{PEG} = 48.9 $k\Omega$. This and all following measurements were performed with an amplitude of 0.1 g since it can be observed that many real ambient vibrations have an amplitude around this value [2].

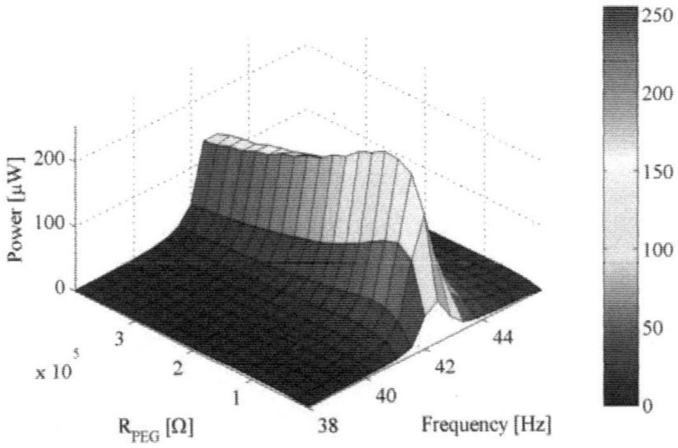

Figure 3. AC power output of the piezoelectric generator.

3. Bridge rectifier

The commercially available power management circuits for piezoelectric generators utilize a bridge rectifier for the rectification stage, e.g. [4]. For this reason, the first measurements are done with a bridge rectifier.

Compared to the AC power with no losses in power management, the bridge rectifier reaches an efficiency of $\eta = 80\ \%$ at the resonance frequency and optimum load resistance without impedance matching. Note that this number is not the efficiency of the circuit but the ratio of available AC power compared to the rectified DC power. This is the reference measurement which means that any advanced or improved power management circuit should deliver more power or rather have a higher efficiency than the bridge rectifier. Otherwise, there is no reason to use a more complicated circuit that utilizes more components and needs to create switching signals for active parts and power them. The effect of the nonlinear circuits on the output power depends on the kind of piezo and on the mechanical setup [5].

4. SSHI circuit

A more advanced power management for piezoelectric generators is the SSHI circuit. It allows a higher efficiency for the rectification stage. The circuit depicted in **Figure 4** is a modified parallel SSHI circuit presented by [3].

The SSHI circuit uses an inductance, which is by far the biggest component of the power management circuit. This is a big disadvantage for miniaturizing the circuit and also because inductances of such high values are relatively expensive.

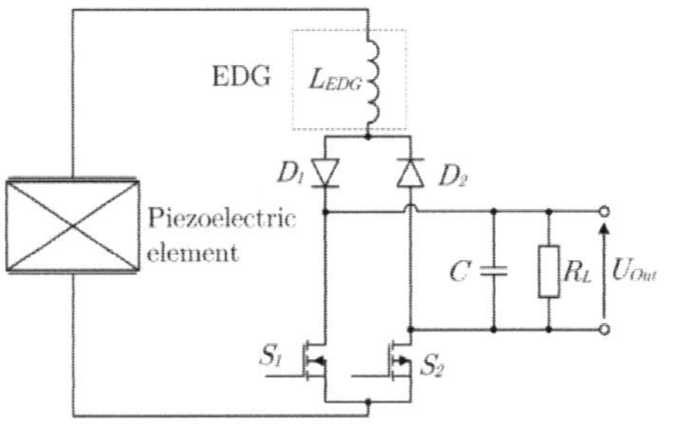

Figure 4. SSHI circuit using the EDG coil.

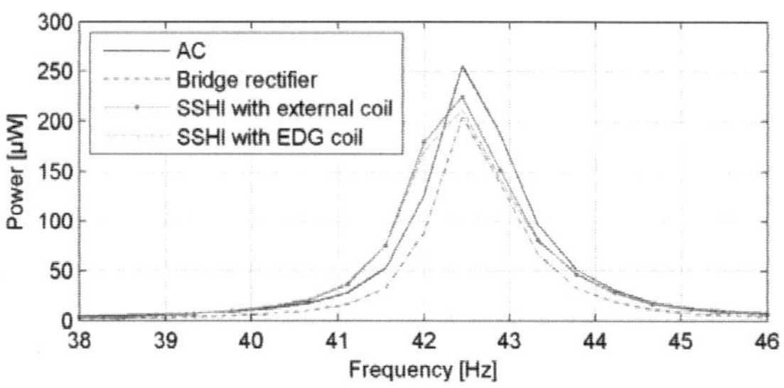

Figure 5. Power output comparison.

The proposed circuit in **Figure 4** utilizes the coil of the electrodynamic generator as the inductance. This is possible because the voltage on the piezo and the induced voltage in the coil have a 90° phase shift. The piezoelectric generator has its maximum voltage at the turning point of the vibration where the mechanical stress is maximum. The electrodynamic generator has its maximum voltage at the highest speed of the magnet, which happens exactly between the two turning points. The results are compared against an SSHI circuit using an external inductance ($L = 48\ mH$ and $R = 190\ \Omega$). The values of the inductance of the electrodynamic generator are $L_{EDG} = 40\ mH$ and $R_{EDG} = 187\ \Omega$.

5. Conclusions and future work

Table 1 and **Figure 5** show the comparison of the different measurements. The AC measurement can be considered as the reference measurement that delivers the most power. If a circuit matched this power, it would have an efficiency of 100 %.

It is clear from **Table 1** that the SSHI circuit has advantages over the standard bridge rectifier topology. With an external inductance, an improvement of 10 % and with the EDG coil an improvement of 3.4 % over the bridge rectifier is possible for this generator.

Table 1. Comparison of measurements

	AC	Bridge rectifier	SSHI external inductance	SSHI EDG coil
f_{Res}	42.44 Hz	42.44 Hz	42.44 Hz	42.44 Hz
P_{Out}	255 μW	204 μW	224 μW	211 μW
R_{Opt}	92.8 $k\Omega$	137 $k\Omega$	100 $k\Omega$	110 $k\Omega$
η		80 %	88 %	83 %
L_{SSHI}			48 mH	40 mH
R_{SSHI}			190 Ω	187 Ω
Power density		1.17 $\mu W/mm^3$	0.42 $\mu W/mm^3$	0.56 $\mu W/mm^3$

The power density of the bridge rectifier is the highest for the three power management solutions, as it requires only minimal board space. Since there is no need for a control circuit or inductance, it has the smallest footprint. If an advanced rectification circuit is employed, the utilization of the EDG coil for the power management significantly increases the circuit's power density as the inductance is by far the biggest circuit component.

A 33 % increment in power density, compared to the SSHI circuit with an external inductance, can be achieved when using the EDG coil.

A power management circuit for the piezoelectric part of a hybrid piezoelectrodynamic generator has been presented. The improvement over the usual SSHI circuit is the utilization of the electrodynamic generator as a circuit component instead of using an external inductance. That way the power management can be realized in a smaller footprint.

A comparison of the different power management circuits has been given with respect to their efficiencies and improvements over the other solutions.

Future work will be developing a power management circuit for the whole generator, which will extract the power of the electrodynamic generator and store the power of both generators in the same storage element. The power density of a combined power management can then be compared to state-of-the-art techniques.

References

[1] Challa V R, Prasad M G and Fisher F T 2009 A coupled piezoelectric-electromagnetic energy harvesting technique for achieving increased power output through damping matching *Smart Mater. and Struct.* **18**

[2] Spies P, Pollak M, and Mateu L 2015: Handbook of Energy Harvesting Power Supplies and Applications, *CRC Press*

[3] Mateu L, Lühmann L, Zessin H, and Spies P 2011 Modified Parallel SSHI AC-DC Converter for Piezoelectric Energy Harvesting Power Supplies, *Telecommunications Energy Conference (INTELEC)*

[4] Linear Technology (2015): "LTC3588-1 - Nanopower Energy Harvesting Power Supply," URL *http://www.linear.com/product/LTC3588-1*

[5] Priya S and Inman D J 2008 Energy Harvesting Technologies *Springer Publishing Company Incorporated*

Design Optimization of a Magnetically Levitated Electromagnetic Vibration Energy Harvester for Body Motion

K Pancharoen[1], D Zhu[2], and S P Beeby[1]

[1] School of Electronics and Computer Science, University of Southampton, SO17 1BJ, UK
[2] College of Engineering, Mathematics and Physical Sciences, University of Exeter, EX4 4QF, UK

E-mail: kp6e12@ecs.soton.ac.uk

Abstract. This paper presents a magnetically levitated electromagnetic vibration energy harvester based on magnet arrays. It has a nonlinear response that extends the operating bandwidth and enhances the power output of the harvesting device. The harvester is designed to be embedded in a hip prosthesis and harvest energy from low frequency movements (< 5 Hz) associated with human motion. The design optimization is performed using Comsol simulation considering the constraints on size of the harvester and low operating frequency. The output voltage across the optimal load 3.5kΩ generated from hip movement is 0.137 Volts during walking and 0.38 Volts during running. The power output harvested from hip movement during walking and running is 5.35 μW and 41.36 μW respectively.

1. Introduction

In-vivo monitoring of joint replacement has been proposed to address the problem of joint replacement failure leading to difficult surgery and rising costs of healthcare [1,2]. To power instrumented hip implants, batteries are not an option because of limited lifetime and replacement requires surgery. Therefore, an electromagnetic vibration energy harvester based on magnetic levitation is presented as an alternative power supply that is compatible with low frequency, high amplitude excitation such as that associated with human motion. The constraints on the size of the harvester due to the volume of the hip prosthesis makes designing an effective energy harvester operating at a frequency below 5 Hz a significant challenge.

The magnetically levitated configuration has been developed previously, with some variations in design and arrangement of the moving-mass components. One of the approaches used to improve the power generated is adding arrays of levitated magnets or coils [3–5]. However, this has to be optimized properly with consideration of the space available for the displacement of the moving magnet in particular applications such as the hip prosthesis. If the moving mass is too large its displacement and velocity will be limited resulting in lower output power [6,7].

In this paper, we present a magnetically levitated electromagnetic vibration energy harvester based on magnet arrays. It has a nonlinear response that extends the operating bandwidth and enhances the power output of the harvesting device. The design investigation using Comsol simulation is described along with experimental results that demonstrate an improvement in performance of the harvester based on the magnet arrays.

2. Structure of the energy harvester

Two harvester configurations were investigated as shown in Figure 1. They both consist of a fixed magnet at the bottom of the glass tube and a moving mass inside the tube, with some variations in the design and arrangement of the moving-mass components as shown.

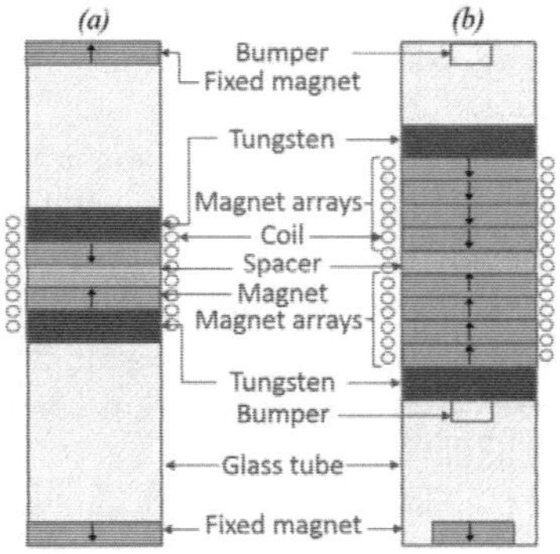

Figure 1. Schematic of electromagnetic vibration energy harvesters: (a) two-magnet and (b) magnet array

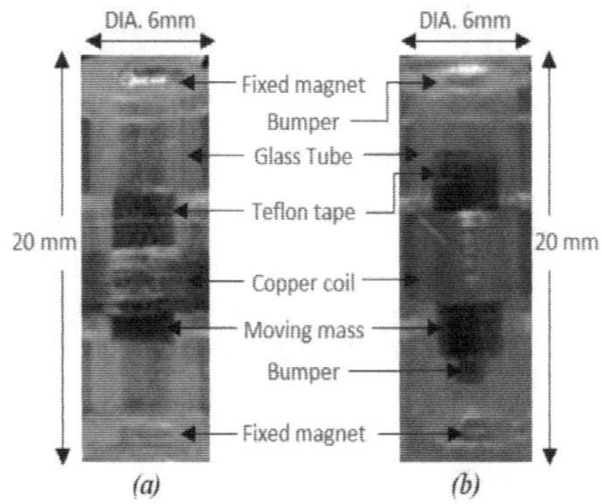

Figure 2. Fabricated harvester (a) two-magnet and (b) magnet array

Both harvester designs were assembled and are shown in Figure 2. The glass tube is 20 mm in length with inner and outer diameters of 5 mm and 6 mm respectively. The 4 mm-diameter NdFeB cylindrical magnets, two cylindrical Tungsten pieces and a ferrite spacer (1.4 mm thickness) have been glued together to form the moving masses. The thickness of each NdFeB magnet and Tungsten piece is 1 mm. The mass of each inertial mass configuration is 1.1 grams and 1.49 grams for the two-magnet harvester and the magnet array harvester respectively. To levitate the moving masses inside the glass tubes, the fixed bottom magnets were attached at bottom of each tube. The coil is wrapped around the outside of the tube and is wound from 30μm-diameter enamelled Copper wire. The inner and outer diameters of the coil are 6 mm and 7 mm respectively with fill factor of 0.54. The coil length was optimized by Comsol (see section 3) for each design and is 3.4 mm for the magnet pair and 5 mm for the array. Teflon tape of 0.1 mm thickness was wrapped around the moving masses to reduce the friction between moving mass and inner surface of the tube. Other variations in the designs include the addition of bumpers made from crepe rubber for the magnet array harvester and the use of a fixed top magnet at top of the tube for the two-magnet harvester.

3. Design investigation

The component dimensions, the number of magnet pairs within the array of levitated magnets and the length of the coil have been investigated given the size constraint of the harvester. This was done using Comsol finite element simulation, which was used to predict the induced voltage given variations in the number of magnet arrays and the coil length.

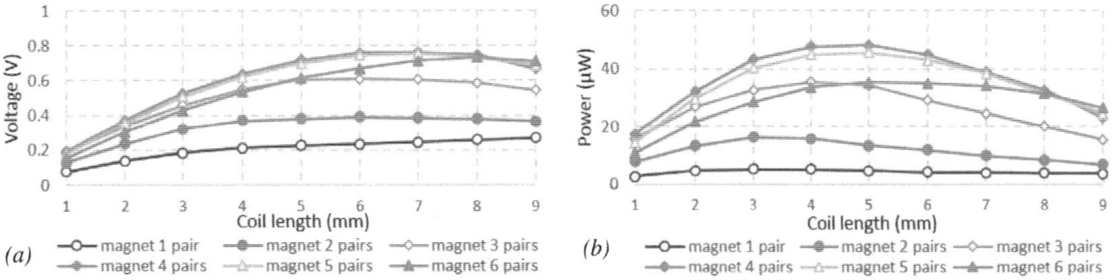

Figure 3. The simulation results of (a) voltage induced and (b) power output with variations in the number of magnets arrays and coil length

The voltage was induced by the varying flux gradients intersecting the coil when the levitated magnet moves from the top to bottom of the tube. The variation in the number of magnets arrays and length of a stationary copper coil lead to the change in the voltage induced as shown in Figure 3. The power output as shown in Figure 3(b) was calculated from the simulation results using a load resistance equal to the coil resistance. It shows that the highest power output can be generated from a 5 mm long coil and a magnet array consisting of 4 pairs of magnets to form the moving mass.

4. Experimental procedures

The experiments have been separated into two sections. In first section, the electromagnetic energy harvesters shown in Figure 2 were tested on an electrodynamic shaker (ET-126B) under sinusoidal acceleration with different levels (0.3g, 0.5g, and 0.8g). The induced voltage was recorded across the frequency range to determine the resonant frequency of the harvester. In the second section, the harvesters were mounted at the hip to record the output voltage during walking and slow running on a treadmill. The walking and running velocities were 2.6 km/h and 7 km/h respectively.

5. Results and discussion

5.1. Experimental results on an electromagnetic shaker

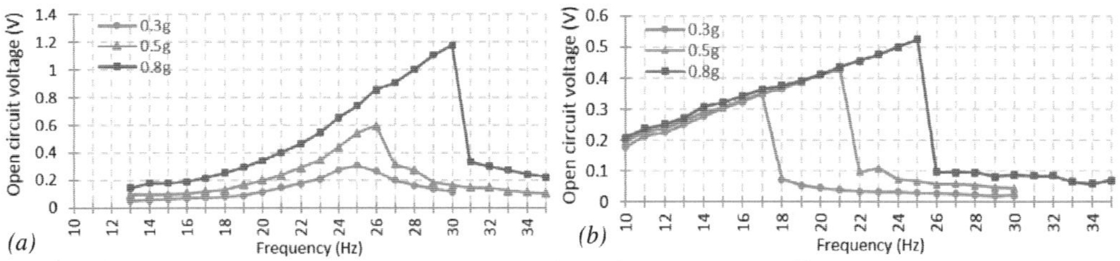

Figure 4. The open-circuit voltage versus exciting frequency at different levels of acceleration of the harvester for (a) two-magnet and (b) magnet arrays.

In Figure 4, the shape of the plot of open circuit voltage versus exciting frequency of the two harvesters at different levels of acceleration tested on the shaker indicates a hard nonlinearity that will influence on the bandwidth and output power of such harvesters. Their resonant frequencies change with increasing acceleration amplitudes going from 25 Hz to 30 Hz and 17 Hz to 25 Hz for the two-magnet magnet array harvesters respectively. The operating frequency of the magnet array harvester is lower than that of the two-magnet harvester. This is due to the removal of the top fixed magnet, the size reduction of the bottom fixed magnet, and the increased mass of moving mass in the tube. The lower resonant frequency implies that the magnet array harvester is more suitable for harvesting energy from human movement, which is characterized by low frequency high amplitude displacements.

5.2. Experimental results on treadmill

The output voltage and the power output of the two-magnet harvester and the magnet array harvester were recorded during walking and running on treadmill. In contrast to the sinusoidal excitation, in this case the output voltage of the two-magnet arrangement is less than that of the magnet array. Results for voltage and power output are given in Table 1 with optimal load resistance of 3.5 kΩ for the magnet array and 3 kΩ for two-magnet harvesters. The experimental results demonstrate that the voltage and power generated by the magnet array harvester is higher than those generated by the two-magnet harvester.

Table 1. Closed circuit voltage and output power harvested during walking and running.

	Closed circuit voltage (V)		Output power (μW)	
	Walking	Running	Walking	Running
Two-magnet harvester	0.012	0.146	0.05	7.08
Magnet array harvester	0.137	0.380	5.35	41.36

The influence of the bumpers on the performance of the magnet array harvester was also investigated and the closed circuit voltage recorded during running on the treadmill with and without bumpers is presented in Figure 5. The voltage signal of the magnet array harvester without bumpers displays higher decay rate compared with that of the magnet array harvester with bumpers. The damping ratio, determined from the decay in the voltage output after stopping the driving vibrations on the shaker is 0.026 and 0.019 with and without bumpers respectively. This implies lower energy loss during the collision between moving mass and top/bottom of the tube when bumpers are present.

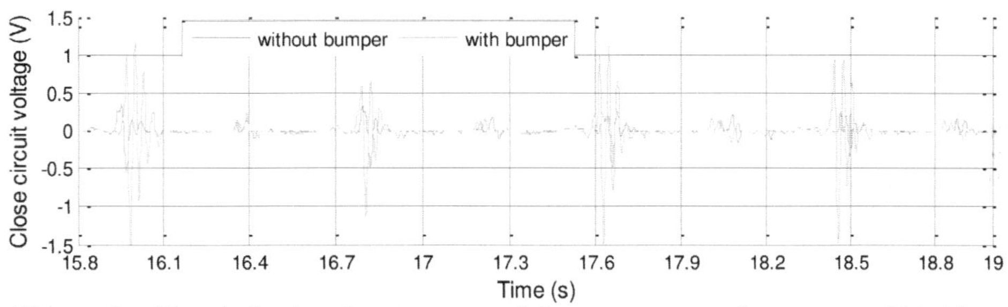

Figure 5. Closed circuit voltage output of the magnet array harvesters with/without bumpers when mounted at the hip and with the user running at 7 km/h.

6. Conclusion

These experimental results have demonstrated a significant improvement in performance of the harvester based on magnet arrays. The operating frequency of the magnet array harvester shifts towards the frequency range of human movements due to increasing mass and weak magnet forces. The integrated bumpers conserve the kinetic energy, covering unavoidable collisions due to the form factor constraints. This optimum design promises sufficient voltage and power output for powering an instrumented hip implant according to experimental results.

Acknowledgement
This work was performed under the SPHERE IRC funded by the UK Engineering and Physical Sciences Research Council (EPSRC), Grant EP/K031910/1. All data supporting this study are openly available from the University of Southampton repository at http://dx.doi.org/10.5258/SOTON/381021.

References

[1] Pynsent P B, Carter S R and Bulstrode C J 1996 *J. Public Health Med.* **18** 157–68
[2] Kandala N-B, Connock M, Pulikottil-Jacob R, Sutcliffe P, Crowther M J, Grove a., Mistry H and Clarke a. 2015 *Bmj* **350** h756–h756
[3] Zhang Q, Wang Y and Kim E S 2014 *J. Appl. Phys.* **115**
[4] Munaz A and Chung G-S 2015 *Microsyst. Technol.*
[5] Abed I, Kacem N, Bouhaddi N and Bouazizi M L 2016 *Smart Mater. Struct.* **25** 25018
[6] Ylli K, Hoffmann D, Becker P, Willmann A, Folkmer B and Manoli Y 2014 *J. Phys. Conf. Ser.* **557** 12024
[7] Munaz A, Lee B and Chung G 2013 *Sensors Actuators A. Phys.* **201** 134–40

Graphene synthesized on porous silicon for active electrode material of supercapacitors

B B Su, X Y Chen and E Halvorsen

Department of Micro- and Nanosystem Technology, University College of Southeast Norway, Campus Vestfold, Raveien 215, 3184 Borre, Norway

E-mail: Xuyuan.Chen@hbv.no

Abstract

We present graphene synthesized by chemical vapour deposition under atmospheric pressure on both porous nanostructures and flat wafers as electrode scaffolds for supercapacitors. A 3nm thin gold layer was deposited on samples of both porous and flat silicon for exploring the catalytic influence during graphene synthesis. Micro-four-point probe resistivity measurements revealed that the resistivity of porous silicon samples was nearly 53 times smaller than of the flat silicon ones when all the samples were covered by a thin gold layer after the graphene growth. From cyclic voltammetry, the average specific capacitance of porous silicon coated with gold was estimated to 267$\mu F/cm^2$ while that without catalyst layer was 145$\mu F/cm^2$. We demonstrated that porous silicon based on nanorods can play an important role in graphene synthesis and enable silicon as promising electrodes for supercapacitors.

1. Introduction

Supercapacitors have attracted great interest for their high power density compared to batteries and high energy density compared to traditional capacitors [1-2]. Generally supercapacitors have an energy density 10 to 100 times larger than electrolytic capacitors, and less charging time than batteries, as well as longer charge- and discharge-cycle life [3-4]. Therefore, supercapacitors are broadly used in computer systems, grid power buffers, street lights, automobiles, and other power supply systems. Graphene is very popular for building carbon-based supercapacitors. It is an atomically thin two-dimensional sheet of sp2 bonded carbon atoms which gives rise to exceptional electronic, magnetic, and optical properties. Each atom has four bonds, one σ bond with each of its three neighbours and one π-bond that is oriented out of plane [5]. Chemical vapour deposition (CVD) has been extensively explored to grow high quality graphene layers of large area and allows growth of single or multiple layers of graphene directly on a substrate (usually metal) using hydrocarbon precursors. However, the graphene layer must usually be transferred to other preferred substrates for subsequent use and is easy to damage during the segregation from the growth substrate. Recently it has been reported that graphene can be grown on silicon templates thus transforming the substrates into stable electrodes for electrochemical devices while avoiding the exfoliation from the templates [6]. A major restriction of direct graphene growth on silicon-based substrates is that the pristine silicon is chemically inert due to the weak reaction between silicon and hydrocarbons. Hence application of surface engineering techniques for the CVD growth of graphene becomes an enabling process when silicon is used as a substrate.

2. Process for graphene growth

In this work, we propose to synthesize graphene on porous silicon through a CVD process with a catalysis mechanism. We apply a layer of gold as the catalyst. Gold has a high melting point of 1337.33K and a relatively low thermal expansion coefficient of 13. Therefore it can stay solid during the CVD process and introduce ignorable thermal stress within the graphene layer. In addition, gold doesn't react with oxygen so there is no worry about oxidation inside the reaction chamber even if the process gas is not entirely oxygen-free [7]. Yet a thick layer of metal catalyst will compromise the main advantage of a high surface area of porous silicon by filling voids and thereby reducing area. In our case, a gold layer catalyst thickness of 3 nm was chosen based on the dimension of the nanostructures. The layer was deposited with an AJA sputtering system.

Subsequently the samples were loaded into a CVD system and placed at the center of the tube furnace. The system provided an inert environment under temperatures ranging from 35°C to 850°C. Argon and Hydrogen were mixed and introduced in the tube to maintain atmospheric pressure with a flow rate of 1 SLM and 100 SCCM respectively. Then 10 SCCM of acetylene was added after reaching 650°C. The furnace was then heated to 750°C and 850°C internally with a holding time of 15 minutes in both temperature stages. Then acetylene was shut off and the samples were cooled off to room temperature in the presence of Argon and Hydrogen.

3. Results and discussion

Scanning electron microscopy (SEM) was carried out using Scanning Electron Microscope SU3500 with an acceleration voltage of 5 to 15kV to identify graphene layers on porous silicon. The SEM images show the nanoscale features on porous silicon. It is clearly illustrated in figure 1(a) and figure 1(b) that there is a thin layer (bright field) on the nanorods (grey field) on porous silicon after graphene growth process. More clearly, the cross-sectional images of porous silicon of figure 2(a) and figure 2(b) show obvious difference before and after direct-graphene coated process and indicate a continuous layer covered on the nanorods although it was not as smooth as might have been expected.

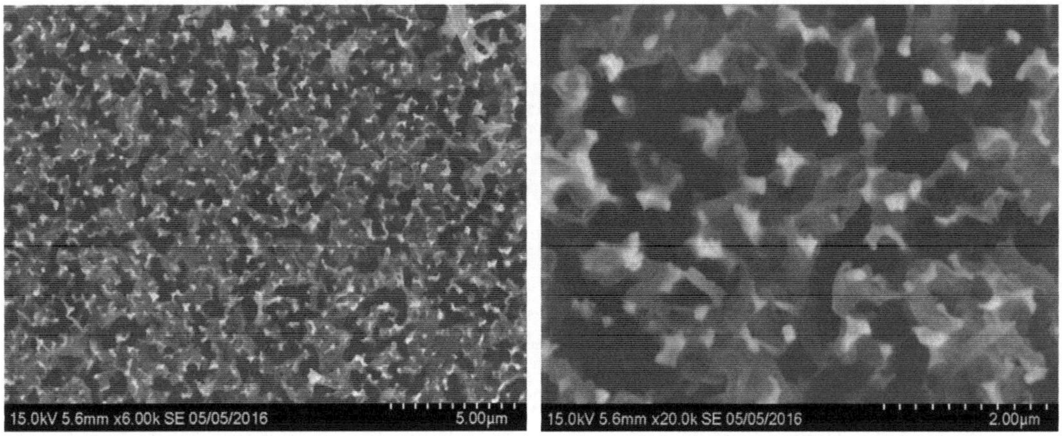

(a) (b)

Figure 1 SEM images of porous silicon showing the graphene layer coated on the nanorods, (a) in a small magnification; (b) in a large magnification.

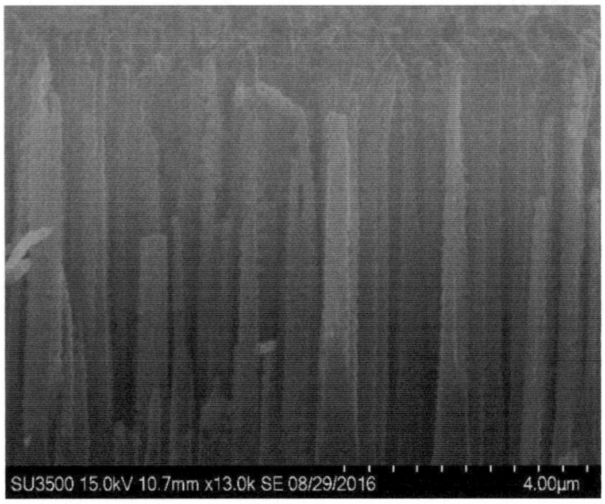

(a)

(b)

Figure 2 SEM images of (a) cross-section of porous silicon before direct-graphene coated process, (b) cross-section of porous silicon after direct-graphene coated process indicating layers coated between the nanorods.

Four-point-probe measurements were conducted to provide qualitative analysis on the conductivity of direct-graphene coated and catalyst-graphene coated flat silicon and porous silicon respectively. Measurements of the resistance of the prepared samples were performed in a collinear configuration. The ratio of voltage and current, the slope in the figure 3, presents the resistance of the tested samples. Among the P-doped flat silicon substrates, the resistance is almost 0.15Ω after catalyst-graphene coating and 1.56Ω after direct-graphene coating. The results are consistent with the expectation that pure silicon is not a good substrate for graphene growth due to the low carbon solubility in silicon while the gold is a good substrate due to the adsorption of C atoms on its surface. This adsorbing behaviour further accelerates the graphene growth when acetylene was introduced [8].

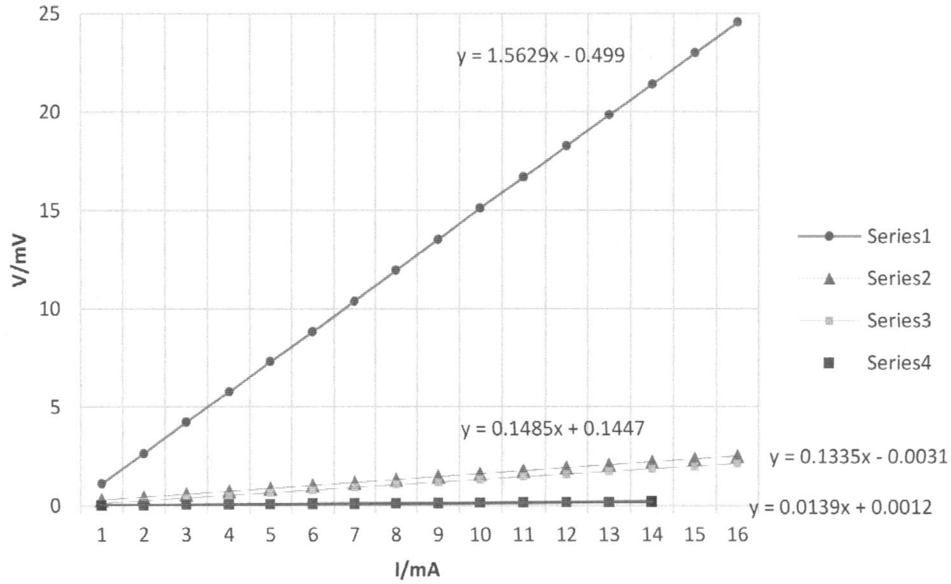

Figure 3 Four-point-measurement results for tested samples of direct-graphene coated silicon (series 1), catalyst-graphene coated silicon (series 2), direct-graphene coated porous silicon (series 3) and catalyst-graphene coated porous silicon (series 4).

For the porous silicon, figure 3 shows that the direct-graphene coated porous silicon has the resistance reduced to approximately 0.13Ω compared to 1.56Ω of the direct-graphene coated flat silicon. The improvement on the conductivity largely benefited from the nanoporous structure on the surface of the porous samples. Because the nanoporous structure enlarges the surface area, carbon atoms have greater opportunity to adsorb on the surface when the reactive gases flow over the samples in the furnace. Therefore, the nanoporous structure can be seen as a pool of carbon atoms, which provided an effective path of supplying carbon source for graphene growth. In addition, for the porous silicon samples, figure 3 also shows that the catalyst-graphene coated porous silicon has resistance greatly reduced to 0.0028Ω. This significantly improved conductivity can be attributed to both the gold catalyst layer and the nanoporous structure, which allowed more C atoms to be incorporated for coating on the sample surface, thus resulting in continuous graphene film. The results reveal that porous silicon with the gold catalyst layer can be considered as an excellent functional structure base for graphene growth and can be a candidate for active electrodes of supercapacitors [9].

As the active electrodes for supercapacitors, the electrochemical performance, e.g., the electrochemical double layer capacitance (EDLC) of the samples was investigated by cyclic voltammetry (CV). The catalyst-graphene coated porous silicon exhibits almost double the specific capacitance of the direct-graphene coated porous silicon (figure 4). The average specific capacitance of the catalyst-graphene coated porous silicon was estimated to be 267μF/cm^2 while that of the direct-graphene coated porous silicon was 145μF/cm^2. Xia al. have conducted an experimental determination of EDLC capacitance of graphene to be 21 μF / cm^2 [10]. The results attained in this experiment indicate a substantial improvement on the electrochemical performance of both the catalyst-graphene coated and direct-graphene coated porous silicon samples, which suggests the porous silicon yields a reliable functional structure to graphene formation. An ideal CV curve of EDLC capacitor should be approximately rectangular in shape. The samples of both catalyst-graphene coated and direct-graphene coated porous silicon exhibited resistive behaviour, which makes the CV plots differ from the shape of an ideal EDLC capacitor.

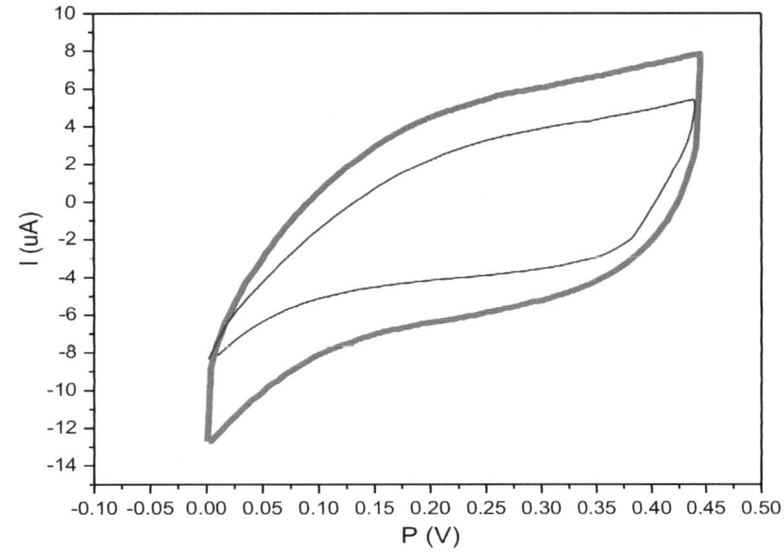

Figure 4 Cyclic voltammetry plots for catalyst-graphene coated porous silicon (thicker line) and direct-graphene coated porous silicon (thinner line).

4. Conclusions

In summary, we have successfully managed to grow graphene on nano structured silicon samples for making active electrodes for EDLC supercapacitors. The conductivity of the direct-graphene coated porous silicon was 10 times better than that of the direct-graphene coated flat silicon, which can be explained by a high quality graphene layer because the nano structure provides large surface area, thus resulting in a large amount of carbon atoms segregating on the surface of the samples. A gold catalyst layer can further effectively promote the formation of graphene layer on the nano structure surface, which is supported by the observation of 100 times lower resistance in catalyst-graphene coated porous silicon than the one in direct-graphene coated flat silicon. Subsequently, cyclic voltammetry proved the superiority of graphene as active electrode materials for electrochemical double layer capacitors, indicating an EDL capacitance of 145μF/cm^2.

Acknowledgement

Financial support from the Research Council of Norway under Grant No. 229716 is acknowledged.

Reference

[1] Stoller M D, Park S J, Zhu Y W, An J H and Ruoff R S 2008 *Nano Lett.* 8 3498
[2] Novoselov K S, Geim A K, Morozov S V, Jiang D, Zhang Y, Dubonos S V, Grigorieva I V and Firsov A A 2004 *Science* 306 666
[3] Yoo E, Okata T, Akita T, Kohyama M, Nakamura J and Honma I 2009 *Nano Lett.* 9 2255
[4] Li Y F, Zhou Z, Yu G T, Chen W and Chen Z F 2010 *J. Phys. Chem. C* 114 6250
[5] Loginova E, Bartelt N C, Feibelman P J and McCarty K F 2008 *New J. Phys.* 10 093026
[6] Oakes L, Westover A, Mares J W, Chatterjee S, Erwin W R, Bardhan R, 2013 *Scientific Reports* 3
[7] Yazyev O Y and Katsnelson M I 2009 *Phys. Rev. Lett.* 100 047209
[8] Zhang Y H, Chen Y B, Zhou K G, Liu C H, Zeng J, Zhang H L and Peng Y 2009 *Nanotechnology* 20 185504
[9] Kovács Gy J, Bertóti I and Radnóczi G 2008 Thin Solid Films 516 7942
[10] Xia J L, Chen F, Li J H, Tao N J 2009 *Nanotechnology* 4 101038

A low frequency rotational energy harvesting system

M Febbo[1], S P Machado[2], J M Ramirez[2] and C D Gatti[2]

[1]Instituto de Física del Sur, CONICET and Departamento de Física. Universidad Nacional del Sur (UNS). Av Alem 1253, 8000 Bahía Blanca, Argentina

[2]Grupo de Investigación en Multifísica aplicada, Universidad Tecnológica Nacional FRBB and CONICET. 11 de Abril 461, 8000 Bahía Blanca, Argentina.

mfebbo@uns.edu.ar

Abstract. This paper presents a rotary power scavenging unit comprised of two systems of flexible beams connected by two masses which are joined by means of a spring, considering a PZT (QP16N, Midé Corporation) piezoelectric sheet mounted on one of the beams. The energy harvesting (EH) system is mounted rigidly on a rotating hub. The gravitational force on the masses causes sustained oscillatory motion in the flexible beams as long as there is rotary motion. The intention is to use the EH system in the wireless autonomous monitoring of wind turbines under different wind conditions. Specifically, the development is oriented to monitor the dynamic state of the blades of a wind generator of 30 KW which rotates between 50 and 150 rpm. The paper shows a complete set of experimental results on three devices, modifying the amount of beams in the frame supporting the system. The results show an acceptable sustained voltage generation for the expected range, in the three proposed cases. Therefore, it is possible to use this system for generating energy in a low-frequency rotating environment. As an alternative, the system can be easily adapted to include an array of piezoelectric sheets to each of the beams, to provide more power generation.

1. Introduction

It is still a remained challenge to design energy harvesting (EH) systems for real applications. This means a scavenging mechanism capable of working at low frequencies (<100 Hz) and low accelerations (<1g). Most of previously large amount of work on EH systems were developed for harvesting energy in vibrating environments [1, 2]. However, a major shortcoming is the high selectivity in the frequency range where the generation is significant, since most of them work at resonance [3]. Energy harvesting from rotational motion is one of the alternatives to this problem when rotational kinetic energy is available (rotating machines, shafts, wheels, tires). It is demonstrated that under a proper design, the resonant frequency of the harvester can match the frequency of the rotation over a wide frequency range [4]. Khameneifar et al. [5] modelled and analyzed an EH system comprising a cantilever beam with a tip mass, on which a PZT is mounted. All the system is fixed on a rotating hub. Then, mechanical vibrations are generated by the gravitational force while the hub is in rotary motion. They demonstrated that the device could be used as a power generator to provide enough energy for charging batteries of wireless sensors in rotating mechanisms. Hsu et al [6] presented a self-frequency tuning EH system for rotational motion, consisting on a cantilever beam mounted on a rotating axis similar to the configuration of ref. [5]. However, a more complete FEM formulation is considered in this case, since it takes into account shear deformation, piezoelectric plate, and the centrifugal force in the beam harvester. Their results show a relatively good power generation but fails to predict a precise voltage peak generation. More recent, Guan and Liao [7] reported a generation about 80-800 μW on a rotating prototype cantilever piezoelectric beam. The system is aimed at introducing a rotational EH system, which can generate a high output voltage at low speeds.

This paper presents an innovative design of a piezoelectric EH system which represents an alternative to the most common cantilevered beam systems to generate energy on a rotating environment. The system can be adapted to add/eliminate beams in the device to explore its dynamics with the aim to maximize voltage generation. Numerical modelling and experimental data sustain the validity of the projected mechanism.

2. Design and Modelling

The proposed EH system comprises two systems of flexible aluminum beams connected by two masses jointed by a spring with a PZT (QP16N, Mide Corporation) piezoelectric sheet mounted on one of the beams. The device is placed rigidly on a rotating hub, at some distance of the axis of rotation, which is parallel to the span of the system of beams (see figure 1).

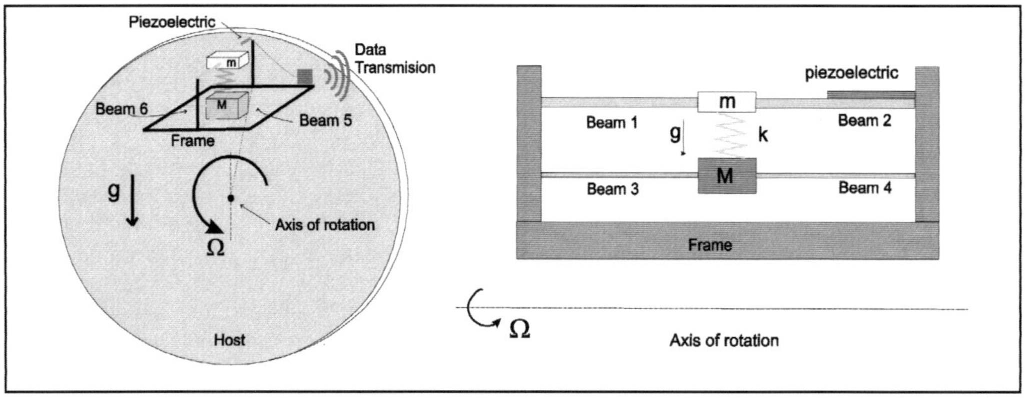

Figure 1: Schematic picture of the system comprising six beams and a piezoelectric sheet mounted over one of them. Front view (left) and lateral view (right).

Table 1. Parameters of beam structure and PZT element.

	Length (mm)	Width (mm)	Thickness (mm)	Density (kg/m^3)	Young`s modulus (GPa)	Coupling d_{31}(pm/V)	Capacitance (nF)	Permittivity
Beams 1-2	75	25	0.5	2700	70	-	-	-
Beams 3-6	75	18	0.5	2700	70	-	-	-
QP16N	45.9	20.57	0.25	2700	67	-190	90.78	1500 ε_0

The mathematical formulation of the problem is based on a finite element (FE) unidimensional description. In contrast with analytical models, the FE method allows to build complex geometrical models such as the proposed here. The main aspects of the present piezoelectric rotating beam formulation are as follows. The kinematics is based on Timoshenko's beam theory, the electrical potential is a linear interpolation through the piezoelectric's thickness and a linear constitutive model for the piezoelectric material is adopted in this primary approach to the problem. However, the authors recognize the failures of the linear model as has been suggested by other researchers [8] for similar levels of excitation. In this sense, future work will be conducted to consider this point.

The system is designed to test different arrange of beams in the device: two beams in different planes, four beams in different planes and six beams, which allow different possibilities of voltage generation (see fig 2). All three cases possesses one piezoelectric sheet attached to one of the beams. However, the prototype is thought to include an array of piezo elements to provide more power generation.

2.1 Mathematical formulation

As it has been stated before, a FE formulation is proposed to model the EH system of figure 1. This method provides a convenient approach for the discretization of a continuum in case of complex geometry. Regarding the FE construction, we build the elements in such a way that each beam has 2 nodes per element with 12 degrees of freedom (6 at each node=3 displacements + 3 rotations). The discrete matrix form of the dynamics governing equations can be written as:

$$\boldsymbol{M}\,\ddot{\boldsymbol{U}} + \boldsymbol{D}\,\dot{\boldsymbol{U}} + (\boldsymbol{K}_M + \boldsymbol{K}_G + \boldsymbol{K}_R)\boldsymbol{U} + \boldsymbol{C}_c\dot{\boldsymbol{U}} - \boldsymbol{\Theta}\,\boldsymbol{V} = -\boldsymbol{F}_E \tag{1}$$

$$\boldsymbol{\Theta}^T\dot{\boldsymbol{U}} + \boldsymbol{C}_P\,\dot{\boldsymbol{V}} + \frac{\boldsymbol{V}}{R_E} = \boldsymbol{0} \tag{2}$$

where is $\boldsymbol{U}$ the nodal displacements and $\boldsymbol{M}, \boldsymbol{K}_R, \boldsymbol{D}, \boldsymbol{C}_C,$ are the mass, rotation stiffness, damping and Coriolis force matrices respectively. Additionally $\boldsymbol{K}_M, \boldsymbol{K}_G, \boldsymbol{\Theta}, \boldsymbol{C}_P$ are the material, geometric stiffness, electromechanical and capacitance matrices respectively with R_E being the resistive load and $\boldsymbol{F}_E$ a vector representing the external forces which includes the acceleration due to gravity [10].

2.2 Numerical validation of FE with ABAQUS

A commercial software ABAQUS is used to validate the three dimensional piezoelectric rotating finite beam element. The results are shown in figure 2, for the material constants of table 1 and the other parameters are: k = 568 N/m, m=31.6 gr and M=76 gr. From the figure it is possible to observe an excellent agreement of both methods for the first frequency at a speed of rotation of 120 rpm (2 Hz). Superior natural frequencies also show a similar agreement, but they are not shown here for a matter of space availability.

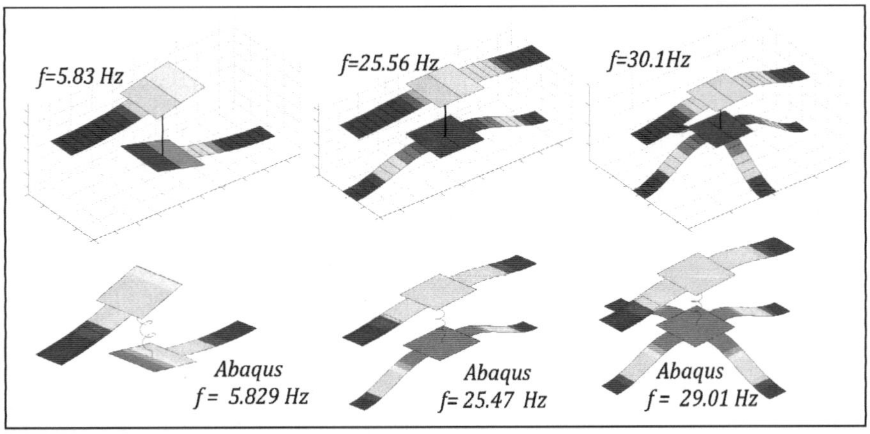

Figure 2. Comparison between the FE method developed here and the commercial software ABAQUS. First natural frequency for a speed of rotation of 120 rpm (2 Hz).

3. Experiments and Results

Experimental tests for the three proposed cases were performed in the following way. An electric motor with a variable speed controller provides the rotational motion of the system which was rigidly mounted to the hub as observed in figure 3. Instead of using a more common slip-ring system which is a major shortcoming to properly acquire the voltage signal, an Arduino board with a Bluetooth connection was used instead, that sends the collected data to a pc at a rate of 100 samples/sec. A load resistor of 56 and 10Kohm was used in order to capture the voltage generation for different loads. However, the data acquisition system presents a maximum possible read of 5 volts (Arduino setting) which restricts the load selection to limited values.

The experimental voltage generation is presented in figure 4 for the desired frequency range (50-150 rpm). From the analysis of the results, a common feature of all three cases can be pointed out. There is a region between 0.8-3 Hz where the generation increases and then the voltage drops up for increasing values of frequency.

Figure 3. Photo of the experimental setup used in the experiments, the EH system is mounted over an electrical motor which induces the rotation of the system

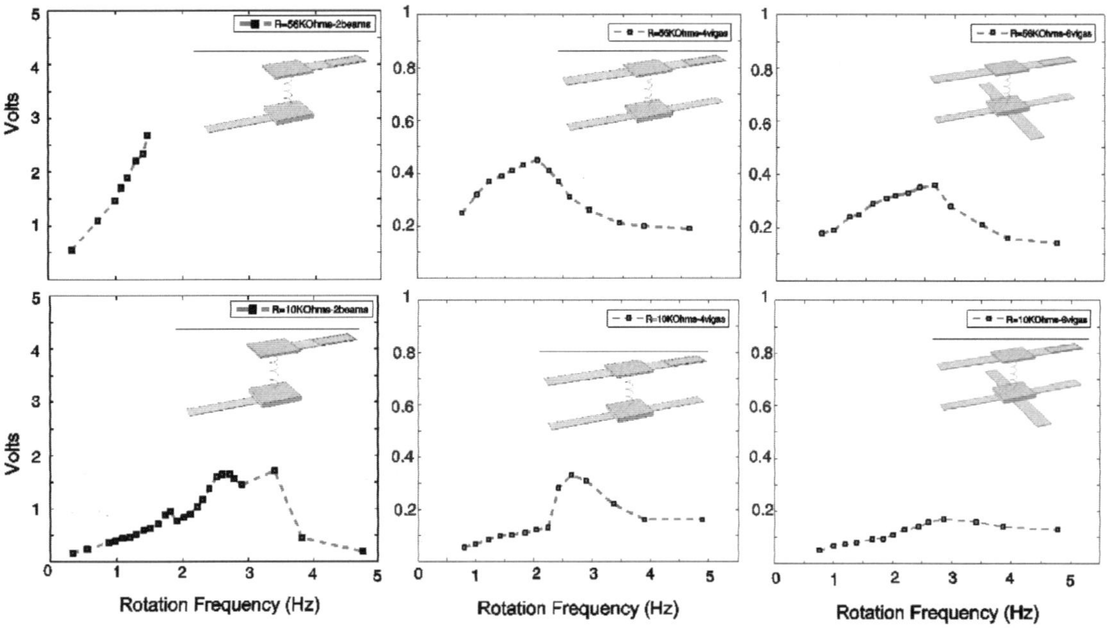

Figure 4. Influence of rotation speed on the voltage generation for the different proposed designs of EHs.

From the proposed cases, the one with two beams has the largest voltage generation. It is observed that for R=56 Kohm the generation exceeds the maximum limit of the data acquisition system (5 Volts) so the readings are cut in this point. FE predictions for this case are summarized in figure 5. There, it is possible to find the voltage generation as a function of speed of rotation, the natural frequency of beam system and the speed of rotation. An inspection of this figure reveals the important fact of a matching between the natural frequency of the system and rotation velocity for a certain frequency range due to the softening effect of the centrifugal force. It is expected that the voltage generation becomes maximum in this range. Compared with the experimental data, there exists a discrepancy in the frequency of maximum generation. This may be due to several omitted effects in the FE formulation: nonlinearities in the piezoelectric material [8], imperfect clamped boundary conditions of the beams, consideration of point masses and perfect bond of the spring, between others. Due to this limitation, FE predictions are presented only for the case of two beams. For the other two cases, no evidence of self-tuning is found in the desired driving frequency range. However, there exist an acceptable voltage generation despite of the fact that the system is in an off- resonance condition. Nevertheless, experimental data of a narrow frequency band where the generation raises, is found.

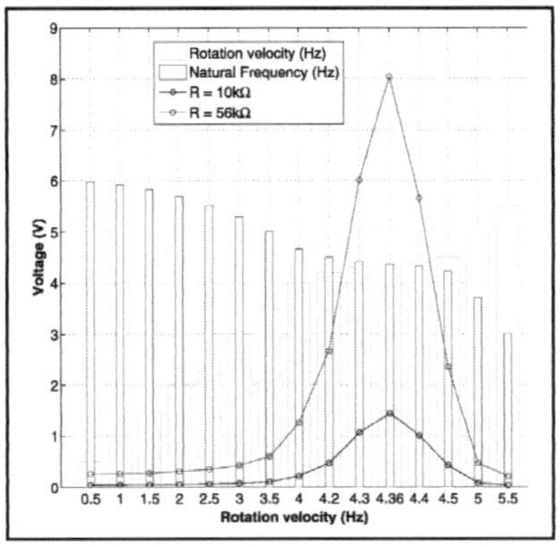

Figure 5. FE predictions of the generated voltage for the case of two beams. Variation of the natural frequency of the system as a function of speed of rotation.

4. Conclusions

This paper presents an alternative of a power scavenging system for a rotating environment comprising a set of beams, masses, spring and a piezoelectric sheet with different geometric configurations. The gravitational force causes sustained oscillatory motion in the flexible beams provided by the rotary motion. A FE beam model is proposed to describe the complex geometry of the system which was validated with available commercial FE software (ABAQUS). The experiments show an acceptable sustained voltage generation for the desired frequency range, in the three proposed cases. However, the case with two masses has a noticeable better voltage generation compared with the other two, principally due to a self-tuning effect, which was numerically confirmed. Future works will be conducted to improve the numerical model including the effect of the nonlinearity in the piezoelectric material, imperfect boundary conditions and large masses. The intention is also to study the generation when an array of piezoelectric sheets is added.

Acknowledgments

The authors wish to thank CONICET (PIP N°:11220120100346), UNS, UTN FRBB and ANPCyT under grant PICT 2013-2065 for their financial support.

References

[1] Beeby S P *et al* 2006 *Meas. Sci. Technol.* **17** 175–195
[2] Mitcheson P D *et al* 2008 Proc IEEE **96** 1457–86
[3] Roundy S *et al* 2005 *IEEE Pervas. Comput.* **4** 28-36
[4] Gu L and Livermore C 2012 *Smart Mater. Struct.* **21** 015002
[5] Khameneifar F *et al* 2011 *J. Vib. Acoust.* **133** 011005
[6] Hsu J-C *et al* 2014 *Smart Mater. Struct.* **23** 075013
[7] Guan M and Liao W-H 2016 *Energy Convers. Manage.* **11** 239-244
[8] Leadenham S and Erturk A 2014 *Nonlinear Dyn.* **79** 1727-1743
[9] Yang Y *et al* 2015 *Smart Mater. Struct.* **24** 125011
[10] Ramirez, JM 2016 *PhD Dissertation* (internal report) in Spanish.

Active AC/DC control for wideband piezoelectric energy harvesting

A Morel[1,2], R Grézaud[1], G Pillonnet[1], P Gasnier[1], G Despesse[1] and A Badel[2]

[1] Univ. Grenoble Alpes, CEA, LETI, MINATEC, F-38000 Grenoble, France
[2] Univ. Savoie Mont Blanc, SYMME, 74000 Annecy, France

adrien.morel@cea.fr

Abstract. This paper proposes a simple interface circuit enabling resonant frequency tuning of highly coupled piezoelectric harvesters. This work relies on an active AC/DC architecture that introduces a tunable short-circuit sequence in order to control the phase between the piezoelectric current and voltage, allowing the emulation of a capacitive load. It is notably shown that this short-circuit time increases the harvested power when the piezoelectric operates outside of resonance. Measurements on a piezoelectric harvester exhibiting a large global coupling coefficient (k^2=15.3%) have been realized and have proven the efficiency and potential of this technique.

1. Introduction

The last decade has seen a growing interest in new sustainable energy sources that could replace batteries. Mechanical energy harvesting is a good alternative to solar or thermal generators, since vibrations can be found in closed confined environments. Piezoelectric elements are of particular interest because of their high energy densities and integration potential [1,2].

Piezoelectric Energy Harvesters (PEHs) are relatively efficient when the vibration frequency matches the harvester's resonance frequency. However, environmental excitations are subject to variations that can lead the vibration frequency to shift away from the generator resonance frequency, reducing considerably the efficiency of the PEH and thus the extracted energy [3].

In order to extend the frequency range where a large amount of energy can be harvested, it is possible to dynamically adjust the interface circuit, which has an influence on the harvester due to the electromechanical coupling [4]. In addition to the well-known resistive tuning, it has been recently shown that adding capacitances in parallel of a highly coupled piezoelectric material allows the tuning of the stiffness of the harvester, leading to an adaptation of its resonance frequency. However, this strategy requires the use of a large off-chip capacitive bank that needs to be tuned step by step [5,6].

In this paper, we propose a new solution to emulate a capacitive behavior by short-circuiting the PEH during a tunable time, as shown in figure 1. This enables the control of the phase between the piezoelectric voltage and current, allowing the emulation of a complex load that has a direct influence on the resonance frequency of the harvester. Using this strategy, a continuous tuning of the resistive (R_{in}) and capacitive (φ_2) load is achievable with components that can easily be integrated in a small chip.

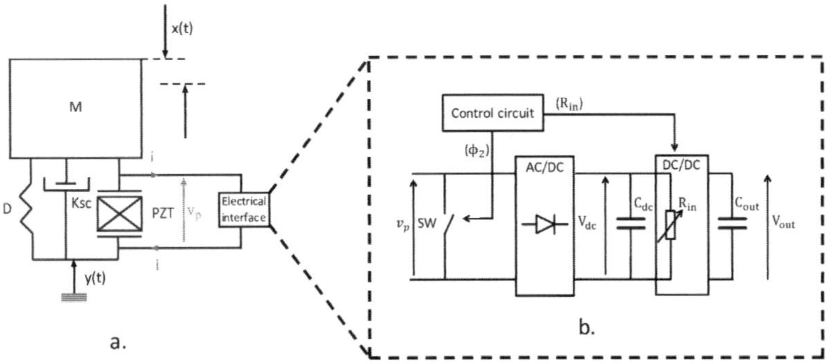

Figure 1. (a) Generic mechanical to electrical conversion model and (b) proposed interface circuit

2. Theoretical analysis

The study of this strategy is derived through a theoretical demonstration that is introduced in the following. The standard mass-spring-damper type PEH model with a single degree of freedom is shown in figure 1. Assuming that the strain of the piezoelectric material is purely sinusoidal, with a constant amplitude ($x(\theta) = X_m \cdot \cos(\theta) = X_m \cdot \cos(\omega t)$), its governing equations are given by (1).

$$\begin{cases} F = M \cdot \gamma = M \cdot \ddot{x}(\theta) + D \cdot \dot{x}(\theta) + K_{sc} \cdot x(\theta) + \alpha \cdot v_p(\theta) \\ i(\theta) = \alpha \cdot \dot{x}(\theta) - C_p \cdot \dot{v}_p(\theta) \end{cases} \quad (1)$$

where F, M, D, K_{sc}, α refer to the driving force, the dynamic mass of the system, the mechanical damping, the global equivalent stiffness and the piezoelectric coupling coefficient respectively. C_p is the dielectric capacitance of the harvester, v_p and i are the piezoelectric voltage and the load current respectively. The expression of the piezoelectric voltage during a half-period is given by (2).

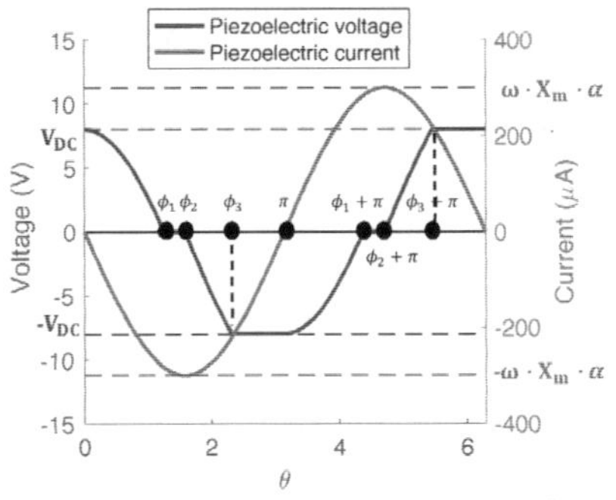

Figure 2. Typical waveforms using the active AC/DC strategy

$$v_p(\theta) = \begin{cases} V_{dc} + \dfrac{1}{C_p}\displaystyle\int_0^\theta \alpha \cdot \dot{x}(\theta)d\theta, & \forall\theta \in [0, \varphi_1] \\ 0, & \forall\theta \in [\varphi_1, \varphi_2] \\ \dfrac{1}{C_p}\displaystyle\int_{\varphi_2}^\theta \alpha \cdot \dot{x}(\theta)d\theta, & \forall\theta \in [\varphi_2, \varphi_3] \\ -V_{dc}, & \forall\theta \in [\varphi_3, \pi] \end{cases} \quad (2)$$

Where φ_1 is the angle when the piezoelectric voltage reaches 0 and when the generator should be short-circuited, φ_2 the angle when the short-circuit is opened, φ_3 the angle when the piezoelectric voltage reaches $-V_{dc}$, and V_{dc} the output voltage of the AC/DC rectifier, directly controlled by the input impedance R_{in} of the DC/DC converter. We can express the parameters φ_1, φ_3 and V_{dc} as in (3).

$$\begin{cases} \varphi_1 = \arccos\left(1 - \dfrac{V_{dc} \cdot C_p}{X_m \cdot \alpha}\right) \\ \varphi_3 = \arccos\left(\cos(\varphi_2) - \dfrac{V_{dc} \cdot C_p}{X_m \cdot \alpha}\right) \\ V_{dc} = \dfrac{X_m \cdot \alpha \cdot R_{in} \cdot \omega}{\pi + R_{in} \cdot C_p \cdot \omega} \cdot (1 + \cos(\varphi_2)) \end{cases} \quad (3)$$

From (1), (2) and (3), we can express the Fourier series representation of v_p as a function of φ_2 and R_{in}. The fundamental of this series is given by (4).

$$\begin{cases} \widehat{v_p}(\theta,\varphi_2,R_{in}) = \dfrac{2\cdot X_m}{\pi}\left(a(\varphi_2,R_{in})\cdot\cos(\theta)+b(\varphi_2,R_{in})\cdot\sin(\theta)\right) \\ a(\varphi_2,R_{in}) = \dfrac{\alpha\cdot R_{in}\cdot\omega}{\pi+R_{in}\cdot C_p\cdot\omega}\cdot[1+\cos(\varphi_2)]\cdot[\sin(\varphi_1)+\sin(\varphi_3)]+\dfrac{\alpha}{C_p}\cdot[\dfrac{\varphi_3+\varphi_1-\varphi_2}{2}+ \\ \dfrac{\sin(2\varphi_3)+\sin(2\varphi_1)-\sin(2\varphi_2)}{4}-\sin(\varphi_1)+\cos(\varphi_2)\cdot\sin(\varphi_2)-\cos(\varphi_2)\cdot\sin(\varphi_3)] \\ b(\varphi_2,R_{in}) = \dfrac{\alpha\cdot R_{in}\cdot\omega}{\pi+R_{in}\cdot C_p\cdot\omega}\cdot[1+\cos(\varphi_2)]\cdot[-\cos(\varphi_1)-\cos(\varphi_3)]+\dfrac{\alpha}{C_p}\cdot[\cos(\varphi_1)-1+ \\ \dfrac{\sin^2(\varphi_3)+\sin^2(\varphi_1)-\sin^2(\varphi_2)}{2}+\cos(\varphi_2)\cos(\varphi_3)-\cos^2(\varphi_2)] \end{cases} \quad (4)$$

$a(\varphi_2,R_{in})$ and $b(\varphi_2,R_{in})$ are the Fourier coefficients of the piezoelectric voltage's first harmonic. Assuming that only the fundamental of the voltage affects the dynamic mass motion, the differential equation of the piezoelectric system given by (1) can be rewritten in the Fourier domain as shown by (5).

$$F = M\cdot\underline{\ddot{x}}(\theta)+D\cdot\underline{\dot{x}}(\theta)+K_{sc}\cdot\underline{x}(\theta)+\alpha\cdot\underline{\widehat{v_p}}(\theta,\varphi_2,R_{in}) \quad (5)$$

Applying the Laplace transform on (5), isolating $\underline{x}(\theta)$, and getting its amplitude X_m, we obtain the expression of the strain amplitude given by (6).

$$X_m = F\cdot[(K_{sc}-M\cdot\omega^2+\alpha\cdot 2\pi^{-1}\cdot a(\varphi_2,R_{in}))^2+(\omega\cdot D-\alpha\cdot 2\pi^{-1}\cdot b(\varphi_2,R_{in}))^2]^{-\frac{1}{2}} \quad (6)$$

The expression of the harvested power transmitted in the DC/DC's input resistance is given by equation (7).

$$P = {V_{dc}}^2\cdot R_{in}^{-1} \quad (7)$$

Combining (3), (6) and (7), we can determine for any parameter couple $(\boldsymbol{\varphi_2},\mathbf{R_{in}})$ the harvested power with this strategy.

3. Experimental results

A PEH having a high electromechanical coupling has been used for the experimental validation of the proposed strategy (figure 3). The characteristics of the PEH are given in Table 1. The experimental setup consists in an electromagnetic shaker that can simulate an input vibration. The cantilever displacement and acceleration are sensed by a laser placed a few decimeters upon the shaker. The piezoelectric device is placed on the shaker and is directly connected to the electrical interface.

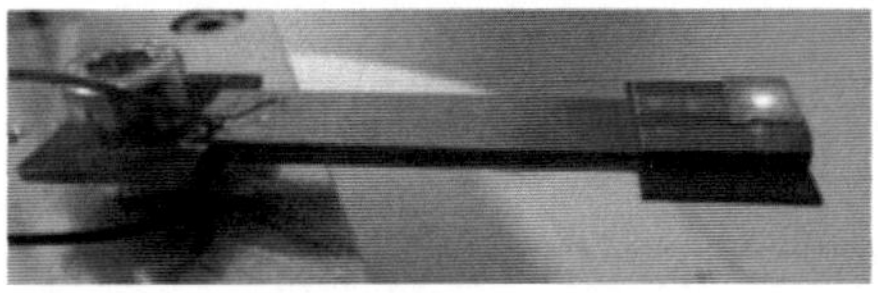

Figure 3. Highly coupled piezoelectric harvester

Table 1. PEH parameters

Parameters names	Parameters values	Units
D	0.18	$N.s.m^{-1}$
α	0.0053	$m.s^{-2}$
k^2	15.3%	-
M	5.6	g
K_{sc}	14151	$N.m^{-1}$
R_p	420000	Ω
C_p	11	nF
V_{diode}	0.25	Volts

The experimental results have been compared with the theoretical results, obtained thanks to the computation of the equations (6) and (7), as shown on figure 4. The piezoelectric voltage, the short-circuit control and the piezoelectric system acceleration, respectively in yellow, pink, and blue, can be observed on figure 5.

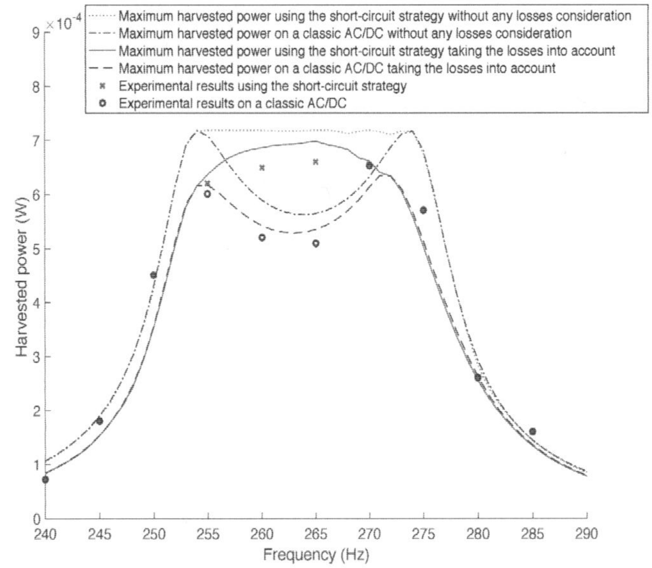

Figure 4. Comparison between theoretical and experimental harvested power

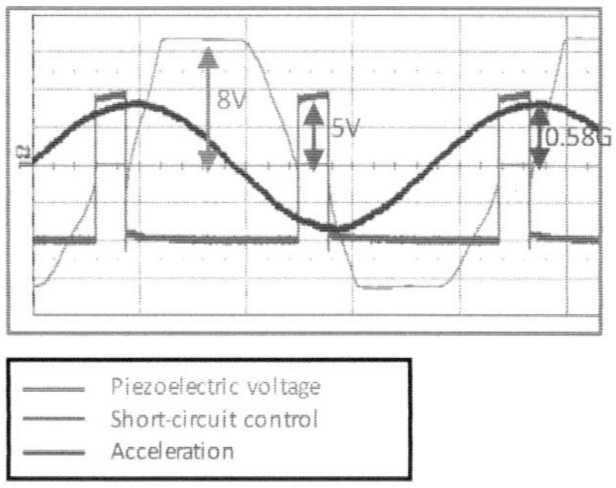

Figure 5. Experimental acceleration and voltage waveforms

4. Discussions
We can observe that the short-circuit technique improves the performances of the PEH between its resonance (253Hz) and anti-resonance (275Hz) frequencies, due to its capacitive effect. Experimentally, under a sinusoidal ambient acceleration γ of 0.58G, we have been able to harvest more than 600µW over a 20Hz bandwidth. In order to take into account the losses in the piezoelectric material ($R_p = 420k\Omega$) and the voltage drops across the diodes ($V_{diode} = 0.25V$), we added these parameters in our theoretical models, as shown on figure 3 and figure 4. In order to reduce these losses, an interface circuit implementing this strategy based on an active AC/DC that includes a short-circuit sequence control should be designed.

5. Conclusion
This paper presents a new interface circuit that can dynamically tune the resonance frequency of a highly coupled PEH. Using an active AC/DC that introduces a short-circuit sequence, this interface is able to synthesize a capacitive load. Through a theoretical analysis and an experimental validation, we proved that this strategy enhances up to 35% the off-resonance performances of the PEH.

References
[1] Calio R, Rongala U B, Camboni D, Milazzo M, Stefanini C, de Petris G and Oddo C M, 2014, "Piezoelectric Energy Harvesting Solutions". *Sensors*, **14**(3), 4755-4790.
[2] Roundy S, Wright P K and Rabaey J, 2003, "A study of low level vibrations as a power source for wireless sensor nodes" *Comput. Commun.* **26**, 1131-1144.
[3] Zhu D, Beeby S, Tudor J, Grabham N, White N and Harris N, 2011, "Performance of a Piezoelectric Energy Harvester under Vibrations Taken from a Helicopter". Proc. PowerMEMS 2011.
[4] Badel A and Lefeuvre E, 2014, "Wideband Piezoelectric Energy Harvester Tuned Through its Electronic Interface Circuit", *In Journal of Physics: Conference Series* 557 012115.
[5] Ahmed-Seddik B, Despesse G, Boisseau S and Defay E, 2013, "Self-powered resonant frequency tuning for Piezoelectric Vibration Energy Harvesters", *In Journal of Physics: Conference Series* 476 012069.
[6] Ibrahim S W and Ali W G, 2012, "A review on frequency tuning methods for piezoelectric energy harvesting systems", *Journal of Renewable and Sustainable Energy*, vol. **4**, 062703.

PowerMEMS 2016 IOP Publishing
Journal of Physics: Conference Series **773** (2016) 012060 doi:10.1088/1742-6596/773/1/012060

Energy Analysis of *n*-Dodecane Combustion in a Hetero/Homogeneous Heat-Recirculating Microreactor for Portable Power Applications

C M Waits[1], E D Tolmachoff[2], W R Allmon[1], and N E Zecher-Freeman[1]

[1]US Army Research Laboratory, 2800 Powder Mill Road, 20783, USA
[2]Naval Air Warfare Center Weapons Division, 1 Administration Circle, China Lake, CA 93555, USA

Abstract. An energy analysis is presented for *n*-dodecane/air combustion in a heat recirculating Inconel microreactor under vacuum conditions. Microreactor channels are partially coated with platinum enabling operating with coupled heterogeneous and homogeneous reactions. The radiant efficiency, important for thermophotovoltaic energy conversion, was found to decrease from 57% to 52% over 5 different runs covering 377 min of operation. A similar decrease in combustion efficiency was observed with 6%-8% energy lost to incomplete combustion and 5%-6% lost through sensible heat in the exhaust. The remaining thermal loss is from unusable radiation and conduction through inlet and outlet tubing. Changes in the Inconel microreactor geometry and emissivity properties were observed.

1. Introduction

Hybridized thermophotovoltaic (TPV) power sources with batteries can provide lighter and longer running power systems than battery-only systems. Efficient microreactors are required to convert the high specific energy contained in tactical fuels such as JP-8 into radiated heat for portable power applications. Microreactors have challenges with efficiency at the high operating temperatures (>1000°C) required for TPV conversion and stability owing to short residence times and the high radiated heat.

Hybrid hetero/homogeneous catalytic microreactors show promise as stable heat sources for portable power devices due to their ability to resist both extinction under high heat loss and blowout under high throughput conditions. Confinement and fuel decomposition were previously shown to play a critical role in determining *n*-dodecane combustion performance for hybrid microreactors without heat recirculation and under ambient conditions [1]. In this work we experimentally analyze the performance of a hybrid microreactor in a vacuum environment with heat recirculation to improve efficiency [2].

2. Microreactor

Planar Inconel 600 microreactors with 20 mm x 20 mm radiating surfaces designed on two sides were machined having two outer channels and one inner channel with 0.5-mm and 1.0-mm gaps respectively (figure 1). All walls are 0.5 mm thick making the total microreactor thickness 4 mm. Previous work predicted improved chemical-to-radiation efficiency for partially filled channels [3]. Thus, platinum was deposited into only 10 mm of the channels by pooling chloroplatinic acid in the microreactor for 24 hr.

The microreactors are operated in <10^{-5} Torr environment with a 100-W equivalent *n*-dodecane flow to mimic expected conditions in ultra-compact thermophotovoltaic sources for portable power applications. n-dodecane, used as a surrogate for JP-8, is injected approximately 100 mm upstream of the microreactor inlet into a dry air stream pre-heated between 423 K and 433 K. The n-dodecane/air mixture enters through the inner channel and undergoes a series of surface reactions and gas-phase decompositions once reaching the platinum coated region [1]. Heat exchanged from recirculated exhaust and conduction along the inner walls heats incoming reactants upstream and improves efficiency.

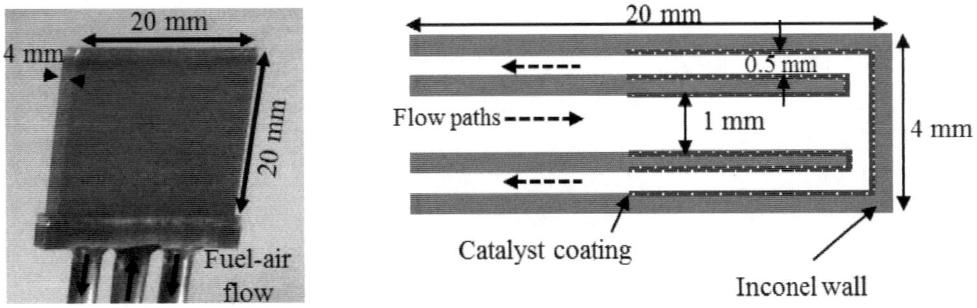

(a) (b)

Figure 1. Microreactor (a) image and (b) schematic showing microreactor dimensions, fuel/air flow paths, and approximate catalyst coating location.

3. Energy Analysis Measurements

The microreactor energy balance in a vacuum environment includes radiation, conduction along the inlet and outlet tubes, sensible heat contained in the exhaust, and unburned products from incomplete combustion. Maximum radiant energy from the microreactor surface to a photovoltaic cell is desired for TPV conversion. The radiant efficiency, η_{rad}, is defined as the power radiated, Q_{rad}, from the 20 mm x 20 mm microreactor faces divided by the fuel input power, Q_{chem}, as determined by the lower heating value. A pyrometer (Raytek MI3-2M) mounted on an X-Y stage was used to measure the surface temperature, $T_{surface}$, through a sapphire window. The surrounding vacuum chamber wall temperature, $T_{surrounding}$, is assumed to be significantly less than the microreactor temperature. The radiant efficiency is calculated by:

$$\eta_{rad} = \frac{Q_{rad}}{Q_{chem}} = \frac{\varepsilon\sigma \iint \left(T^4_{surface} - T^4_{surroundings}\right) dA}{Q_{chem}}. \tag{1}$$

Combustion efficiency, η_{comb}, is estimated by carbon counting of the gaseous exhaust products. The carbon-content of the liquid is assumed to be negligible. The gaseous products are analyzed using micro gas chromatography (Agilent 490) capable of measuring up to C3. The chemical-to-thermal conversion, or heat release, efficiency is estimated by

$$\eta_{heatrelease} = \left[\sum_i \frac{X_i}{X} n\Delta H_f(i) + n_{H_2O}\Delta H_f(H_2O) - n_{C_{12}H_{26}}\Delta H_f(C_{12}H_{26})\right] / LHV_{C_{12}H_{26}}, \tag{2}$$

where X_i/X is the measured molar ratio of species i, n is the total molar flow rate, ΔH_f is the heat of formation, and LHV_{C12H26} is the lower heating value for n-dodecane.

The sensible heat lost in the exhaust is estimated by the temperature difference between the outlet temperature during combustion and the inlet temperature during pre-heating. K-type thermocouples were inserted into the inlet and outlet tubes to approximately 20 mm from the microreactor to measure the n-dodecane/air inlet and exhaust temperatures.

4. Results and Discussion

Figure 2 shows an image of the microreactor during and after sustained combustion. A bulge in the outer channel on both sides of the microreactor was observed. The bulge is caused by the stress induced by >1 atm pressure difference caused by operating in the vacuum environment and weakening of Inconel at high temperatures [4]. A peak deflection of 1.0 mm was measured in the central region of the microreactor suggesting the outer channel heights on both sides each increase from 0.5 mm to 1.0 mm. This change can affect the heat exchanged from the walls to and from the gas and may in turn affect the coupling between the heterogeneous and homogeneous reactions if the reactions are not completed in the inner channel [5]. Approaches to strengthen the outer microreactor walls are required to maintain the geometry integrity and may include adding in a rib structure or thickening the outer Inconel wall.

Figure 2. Optical image of the microreactor (a) during combustion operation and (b) post combustion. The change in surface optical properties can be observed in (b).

A large error exists in measuring the absolute temperature of the microreactor surface due to the unknown emissivity of Inconel 600 in our vacuum environment. The surface visually varies from an oxidized surface to a polished un-oxidized surface (figure 2b). The total emissivity can be as high as 0.82 at 1253 K [4], but data is inconclusive at high temperatures in vacuum where surface oxides can be reduced. Oxidized Inconel 718 total hemispherical emissivity measurements ranged between 0.35 and 0.5 for different oxide formation conditions at conditions up to 1200 K in vacuum [6]. Although, Inconel 718 has a lower nickel content than Inconel 600, similar oxides may form. The analysis performed in this work used an emissivity of 0.65. The actual average surface temperature can range from 1339 K to 1094 K for an emissivity of 0.4 to 0.9 respectively. The impact, however, to radiant efficiency measurements using pyrometer data is <1% for the range of emissivity since the pyrometer measures the amount of radiation absorbed by its detector.

Figure 3 shows the variation of $T_{surface}$ for an average emissivity of 0.65. A temperature drop of 103 K from the microreactor tip to the base is measured by averaging the temperature across the microreactor width. The temperature drop along the microreactor could be more significant since the surface emissivity is expected to increase from right to left causing the actual temperature to be lower towards the inlet and outlets and higher at the microreactor tip. Further studies may optimize the surface temperature and uniformity through a combination of catalyst coating location [3] within the inner and outer channels, as well as the level of confinement as determined by the channel height [1,5].

The radiant efficiency calculated from the surface temperatures in figure 3 is 57% assuming $T_{surface}$ is the same on both sides. TPV radiative to electrical conversion efficiencies of 20% and higher have been demonstrated for InGaAs and InGaAsSb photovoltaic cells [6,7] suggesting greater than 10% total conversion efficiency is feasible if combined with this microreactor. Other possible losses include cavity loss, photovoltaic cell cooling, as well as energy required to pump the fuel and air.

The radiant efficiency change over 5 different combustion runs is shown in figure 4. Each run lasted between 43 min and 100 min with the total time of 377 min. Prior to Run #1, the microreactor underwent various combustion events including propane combustion outside the vacuum chamber and ignition experiments with *n*-dodecane inside the vacuum chamber. A decrease in the combustion efficiency and heat release efficiency was observed (figure 4) suggesting a decrease in catalytic activity. Although the combustion efficiency ranged from 85% to 90% the heat release efficiency ranged from 92% to 94%. A complete evaluation of the performance versus different input power and fuel-to-air ratios is necessary to determine conditions for peak performance, expected occur with more lean conditions. Furthermore, studies comparing with a fresh microreactor operation and inspection of the catalyzed surfaces overtime is required to determine if the efficiency drop is due to catalyst deactivation, catalyst surface area reduction, or other effects.

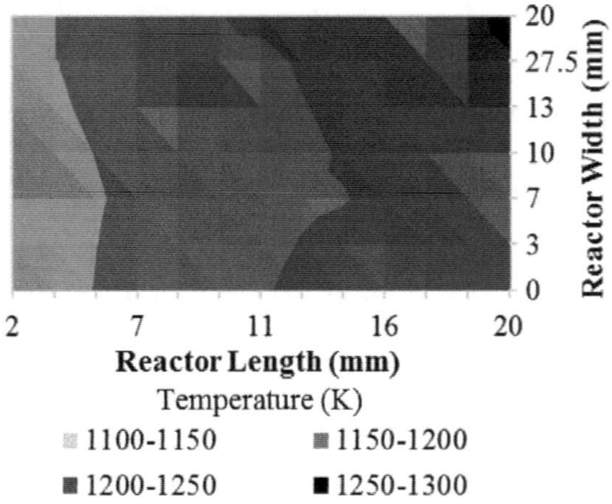

Figure 3. Temperature map for run #3 generated by pyrometer measurements taken across the 20 mm x 20 mm Inconel using an emissivity of 0.65

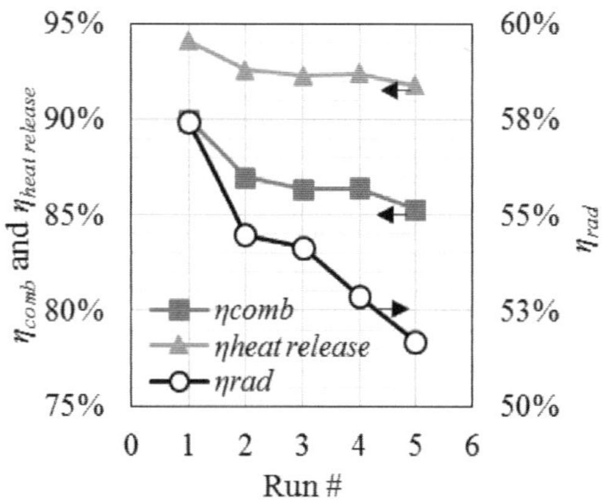

Figure 4. Experimental combustion, heat release, and chemical-to-radiation efficiencies versus the run # data was collected.

The exhaust temperature decreased from 963 K to 923 K over the 5 runs, averaged between the two exhausts, resulting in 5.3% to 5.6% sensible heat lost in the exhaust. The observed inlet temperature was higher than the outlet temperatures for all runs. This is believed to be due to a small amount of catalyst depositing in the inlet tube during the drying process. In further experiments, not shown here, the inlet temperature reduced below the outlet suggesting the surface reactions were eliminated. The radiant efficiency did not change noticeably as a result of the possible pre-combustion in the inlet tube.

Radiation from the top and sides of the microreactor were estimated to be 15% based on the measured edge temperatures measured in figure 2. The remaining 22% is assumed to be lost by conduction along the tubes and radiation from the tubes and plenum surfaces.

5. Conclusion

The performance of a microreactor combining heat recirculation and hybrid hetero/homogeneous reactions via channels partially coated with platinum was evaluated for *n*-dodecane fuel and stoichiometric air flow. The Inconel microreactor requires strengthening to maintain geometrical integrity for vacuum operation. Radiant efficiency up to 57% with a 94% heat release efficiency has been demonstrated. Both the radiant efficiency and the combustion efficiency were observed to reduce over time suggesting reduced catalytic activity, but further experiments are necessary to determine the cause of the degradation. The sensible heat lost in the exhaust was measured to be 5%, and the remaining 32% was lost to unusable radiation and through conduction along the inlet and outlet tubing.

6. References

[1] Tolmachoff E D, Booth A D, Lee I C, Allmon W R, and Waits C M 2015 *Combustion and Flame* **162** 3674-80
[2] F. Weinberg 1971, *Nature* 233 **239**–41.
[3] Tolmachoff E D and Waits C M 2015 *Proc. Western States Section of the Comb. Inst. Fall Meeting*
[4] Inconel 600 Datasheet, *Special Metals Corporation*, 2008
[5] Tolmachoff E D, Allmon W R, and Waits C M 2014 *Applied Energy* **128** 111-18
[6] BP Keller, SE Nelson, KL Walton, TK Ghosh, RV Tompson, and SK Loyalka 2015 *Nucl Eng Des* **287** 11-18
[7] Wernsman *et al.* 2004 *IEEE Transactions on Electron Devices* **51** 512-15
[8] Dashiell *et al.* 2006 *IEEE Transactions on Electron Devices* **53** 2879-91

Optimal component sizing in a two-reservoir passive energy harvesting system

E. Munsing[1], M. Cowell[2], S. Moura[1] and P. Wright[2]

[1] Energy Controls and Applications Laboratory, University of California, Berkeley CA 94720
[2] Advanced Manufacturing for Energy, University of California, Berkeley CA 94720, USA

E-mail: e.munsing@berkeley.edu

Abstract. We utilize particle swarm optimization to reduce the size of the energy management components in an energy harvesting system, allowing us to eliminate the need for voltage regulators or DC-DC converters without affecting system performance. Prior literature on optimal power management in microelectronics [1, 2] has relied on engineering estimates or exhaustive parameter searches to optimize system design. No prior literature has considered the optimal design of a device with only passive components [3]. By using particle swarm optimization, we demonstrate a 55% reduction in device size relative to conventional engineering calculations of an optimal device design.

1. Introduction

Energy management systems for wireless sensor nodes must deliver reliable power and voltage while being small and low-cost. In prior literature, power reliability is maintained by using active energy management components (voltage regulators, DC/DC converters), but these impose efficiency losses [4] while increasing device cost and size [1, 2]. While the use of an active energy management system allows for a convex formulation of the optimal sizing problem in a two-reservoir battery/capacitor system [5], the passive energy management problem is nonconvex and has not been explored in prior literature.

Building on previous work developing flexible printed batteries [6], capacitors [7], and generators [8, 9] we propose a design which eliminates the need for active power management, and instead passively maintains device performance through optimal sizing of generators, batteries, and supercapacitors. To overcome the nonconvexity in this problem, we employ a Particle Swarm Optimization (PSO) algorithm to identify an optimal design [10], and are able to demonstrate a significant reduction in system size while meeting all payload power requirements.

1.1. Novel Contributions

This work extends prior literature on energy management in energy harvesting systems in the following ways:

- Demonstrates a method for designing a passive energy management system without the use of voltage regulators or DC/DC converters, and
- Applies particle swarm optimization to microelectronic energy management system design.

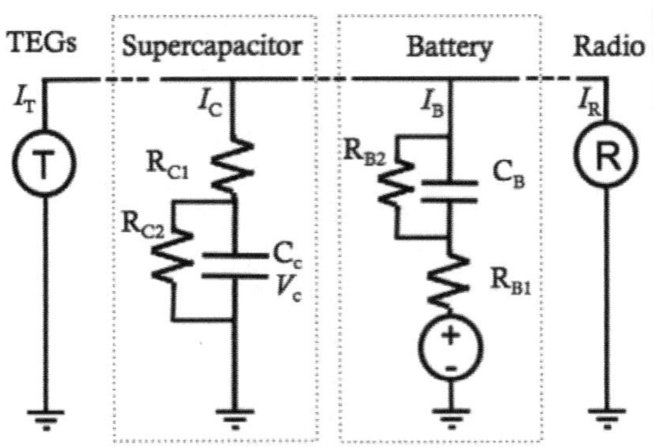

Figure 1. Equivalent Circuit Model; the design can be adjusted by changing the number of TEG elements in series in each string, or adding parallel TEG strings, supercapacitors, or batteries in parallel.

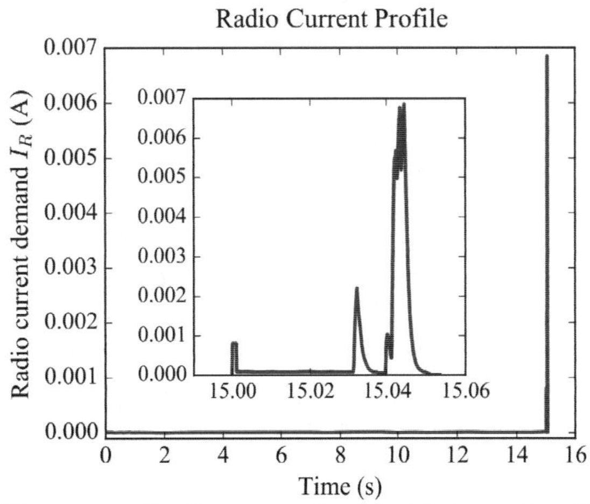

Figure 2. Radio power draw, with inset detail of the data acquisition/transmission period.

2. System Model

We consider a circuit model shown in figure 1 in which power is harvested through strings of thermoelectric generator cells (TEGs) in series; the approach we consider can also accommodate photovoltaics or other energy harvesting modes [9]. Charge is stored in parallel banks of supercapacitors and batteries; the components are linked by a main bus at voltage V. A wireless sensor node (WSN) is attached to the bus, and is modeled as a current sink with 15-second cycles with a deterministic current profile. The radio current draw I_R is shown in figure 2: a 15-second inactive period is followed by a 50-millisecond data acquisition and transmission load. We seek to design an energy management system which can repeat this load cycle indefinitely.

The system designer chooses the number of TEG elements in series s, number of TEG strings in parallel p, number of batteries b, and number of capacitors c. As the batteries have a low current capacity and supercapacitors have low energy capacity, we expect that an optimal design may include both storage reservoirs. We assume that each component has a constant per-unit cost, captured in the TEG unit cost α, battery unit cost β, and supercapacitor unit cost γ; the net system cost is thus $\alpha ps + \beta b + \gamma c$.

The system dynamics can be derived through the application of Kirchoff's Current and Voltage Laws (KCL and KVL) and a simple battery model where open circuit voltage OCV is linearized within the acceptable ranges of State Of Charge $\mathrm{SOC_{min}}$ and $\mathrm{SOC_{max}}$. These governing equations and limits on variables are captured below, where italic fonts are used to denote optimization variables:

$$\min \quad \alpha ps + \beta b + \gamma c \tag{1}$$

Subject to:

KCL for circuit: $$pI_T + cI_C + bI_B = \mathrm{I_R} \tag{2}$$

TEG dynamics: $$I_T + \frac{1}{s}V\zeta = \mathrm{T_0}\Delta\mathrm{K} \tag{3}$$

KVL in Capacitor: $$V + I_C\mathrm{R_{C1}} - V_C = 0 \tag{4}$$

KCL in Capacitor: $$\dot{V_c} + \frac{1}{\mathrm{C_c}}I_C + \frac{1}{\mathrm{C_cR_{C2}}}V_C = 0 \tag{5}$$

KVL in Battery: $$V + I_B\mathrm{R_{B1}} - V_B - OCV = 0 \tag{6}$$

KCL in Battery: $$\dot{V_B} + \frac{1}{\mathrm{C_B}} I_B + \frac{1}{\mathrm{C_B R_{B2}}} V_B = 0 \quad (7)$$

Battery OCV linearization: $$OCV - \nu SOC = \mathrm{V_0} \quad (8)$$

Battery Charging/Discharging: $$I_B - I_{BC} - I_{DC} = 0 \quad (9)$$

Battery State of Charge Conservation: $$\dot{SOC} + \frac{\eta}{\mathrm{Q}} I_{BC} + \frac{1}{\eta \mathrm{Q}} I_{DC} = 0 \quad (10)$$

Inequality Constraints: $$b, c, s, p \geq 0 \quad , \quad \mathrm{V_{cutoff}} \leq V \leq \mathrm{V_{max}} \quad (11)$$

$$\mathrm{I_{Cmin}} \leq I_C \leq \mathrm{I_{Cmax}}, \quad \mathrm{I_{Bmin}} \leq I_B \leq \mathrm{I_{Bmax}}, \quad \mathrm{I_{Bmin}} \leq I_{BC} \leq 0 \quad , \quad 0 \leq I_{BD} \leq \mathrm{I_{Bmax}} \quad (12)$$

3. System Design

3.1. Conventional Engineering Calculations

We expand on the methods in [1, 2, 3, 11] to provide a baseline estimate of system design using conventional engineering calculations.

To calculate the number of TEG elements in series within each string, and number of TEG strings in parallel, we use the linear TEG dynamics found in equation (3) and consider a constant temperature difference $\Delta\mathrm{K} = 20$. We determine the number of TEGs in series by dividing the nominal bus voltage by the midpoint voltage of the TEG output curve found in [8]. Similarly, we calculate the necessary number of parallel TEG strings by dividing the average WSN current demand by the midpoint current of a single TEG module.

We calculate the number of supercapacitors required to ensure that maximum payload current $\mathrm{I_{max}}$ can be satisfied without violating $V \geq \mathrm{V_{cutoff}}$. We estimate the minimum number of parallel supercapacitors by dividing $\mathrm{I_{max}}$ by the maximum current a single supercapacitor is able to deliver, $\mathrm{I_{Cmax}}$. We then verify that the combination of coulombic voltage drop and ohmic voltage drop at $\mathrm{I_{max}}$ will not depress the bus voltage below $\mathrm{V_{cutoff}}$.

Because the current limit of the batteries is 2.5% that of the supercapacitors, the battery's contribution to the radio's current demand during peak discharge is assumed to be negligible. Instead, the battery is sized such that it will be able to provide a nominal baseload power, here considered as 10% of $\mathrm{I_{max}}$, and has sufficient capacity to trickle charge the supercapacitor over the cycle should the TEG power drop.

3.2. Particle Swarm Optimization

Particle Swarm Optimization is a nonlinear optimization method in which a large number of particles explore the parameter space to identify a low-cost solution [10]. We consider an optimization space in the state variables $SOC(0)$, $V_C(0)$, and $V_B(0)$ in addition to the design variables variables p, s, b, and c. We provide a brief description of our implementation below; additional details can be found in [10] and associated references.

We assign a set of n particles to random locations and velocities in the optimization space, and use the equations above to simulate the system dynamics at each particle's coordinates. If constraints are violated a high penalty is assigned, and the total cost (penalty plus the value of the objective function) is saved for that point. After simulating the system at each particle's location, the location of the lowest-cost point is broadcast to all particles in the swarm. The velocities of all particles are then updated as the weighted sum of the particle's own velocity and the vector to its own best location and the swarm's overall best location. The process is repeated until the swarm converges on a low-cost point, though this may not be the global optimum.

4. Results

The results for both design methods are shown in table 1 and the resulting bus voltage during the data transmission period is shown in figure 3. We assume that a constant per-unit cost for each component, $\alpha = \beta = \gamma = 1$. We see that both designs are feasible, though the nonlinear

	Engineering Calculations	Nonlinear Optimization
Battery Cells	3	0
Supercapacitors	2	3
TEGs in series	50	22
TEG strings	1	1
Total units	55	25

Table 1. System designs resulting from conventional engineering and optimization-based approaches; a 55% decrease in cost is shown.

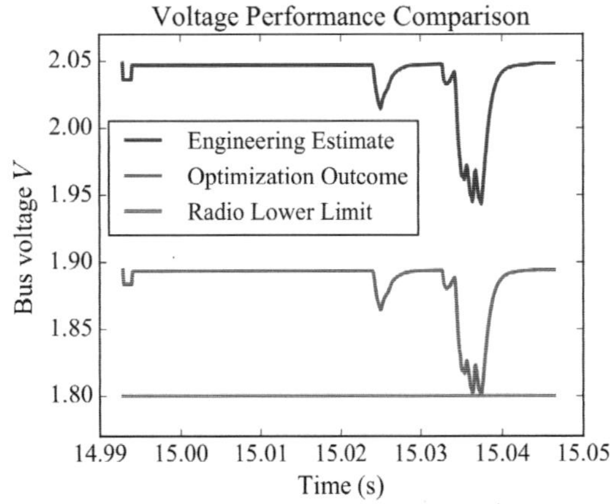

Figure 3. Bus voltage during radio dispatch, showing how the nonlinear optimization reduces costs by operating closer to system bounds.

optimization method converges on a solution which does not use batteries and more closely tracks the system constraints. This contributes to a 55% decrease in the number of components required to power the payload.

A number of other demand profiles were considered, including those in which the TEGs were not able to produce power for a portion of the load cycle. In these scenarios, the nonlinear optimization resulted in a design in which batteries are added to maintain bus voltage through the drop in generation.

We note that the convergence rate of the nonlinear optimization is dependent on the penalty function, and was found to slow significantly for longer study periods.

5. Conclusions

We demonstrate a method for energy management in a wireless sensor node which does not require the use of voltage regulators or DC-DC converters, instead satisfying payload requirements through optimal component sizing. Using particle swarm optimization, we find a design with significant reductions in system cost relative to conventional engineering calculations. While this nonlinear optimization technique does not guarantee a globally optimal design, it can be readily applied to other demand profiles and generation/storage technologies and we expect that similar results may be found in other energy management problems.

References

[1] Vullers R J M, van Schaijk R, Doms I, Van Hoof C and Mertens R 2009 *Solid-State Electronics* **53** 684–693 URL `http://dx.doi.org/10.1016/j.sse.2008.12.011`
[2] Harb A 2011 *Renewable Energy* **36** 2641–2654 URL `http://dx.doi.org/10.1016/j.renene.2010.06.014`
[3] Lee D, Dulai G and Karanassios V 2013 *SPIE Defense* 87280S–87280S–9
[4] Maxim Integrated 2012 *Application Note* number 751
[5] Hu X, Murgovski N, Johannesson L M and Egardt B 2014 *IEEE Transactions on Intelligent Transportation Systems* **15** 1193–1205
[6] Kim B, Winslow R, Lin I, Gururangan K, Evans J and Wright P 2015 *JP:CS* **660** 012009
[7] Cowell M, Winslow R, Zhang Q, Ju J, Evans J and Wright P 2014 *JP:CS* **557** 012061–5
[8] Chen A, Madan D, Wright P K and Evans J W 2011 *J. of Micromechanics and Microengineering* **21** 104006–9
[9] Lechêne B P, Cowell M, Pierre A, Evans J W, Wright P K and Arias A C 2016 *Nano Energy* **26** 631–640
[10] Poli R, Kennedy J and Blackwell T 2007 *Swarm Intelligence* **1** 33–57
[11] Srinivasan V and Weidner J 1999 *Journal of The Electrochemical Society* **146** 1650–1658

Shoe-mounted vibration energy harvester of PZT piezoelectric thin films on metal foils

T Nishi, T Ito, H Hida and I Kanno

Department of Mechanical Engineering, Kobe University, 1-1 Rokkodai-cho, Nada-ku, Kobe, Japan

E-mail: kanno@mech.kobe-u.ac.jp

Abstract. This paper describes shoe-mounted piezoelectric vibration energy harvesters (PVEHs). The PVEHs were fabricated from $Pb(ZrTi)O_3$ (PZT) thin films which were directly deposited onto Pt/Ti-coated stainless steel foil by rf-magnetron sputtering. We experimentally and theoretically evaluated impulse responses of the PVEHs by applying a simple impulse input on the energy harvesters, typical damped free vibration behaviour was clearly observed, and the output signal was in good agreement with the theoretical value. We measured the output power by applying the impulse input with an optimal load resistance of 33.9 kΩ. The maximum output power was approximately 20 μW, which correspond with the calculated value based on theoretical equation. From these results, the theoretical equation we derived might be helpful for design purposes of the shoe-mounted PVEHs.

1. Introduction

Piezoelectric vibration energy harvesters (PVEHs) have been of great interest as promising power sources for networks of wireless sensor nodes. Among a variety of target applications, shoe-mounted PVEHs have been thought to be promising products as battery-free wearable system because of their high power generation efficiency [1]. To mount the PVEHs on shoes, high flexibility and toughness are strongly required. We have previously fabricated PVEHs of $Pb(Zr,Ti)O_3$ (PZT) thin films on stainless steel foil and confirmed their excellent performance such as flexibility, toughness, and high-efficiency of power generation [2], which would be suitable characteristics for the shoe-mounted PVEHs. However, several technical issues still remain, for example, establishment of design guideline and prediction of power generation on the shoe-mounted PVEHs. The practical design of the PVEHs is based on the experimental results and theoretical calculation of electro-mechanical models. Conventionally PVEHs have been evaluated under continuous sinusoidal excitation. However, the shoe-mounted PVEHs are placed under the periodic impulsive excitation environment and the theoretical evaluation should be considered according to the specific input waveforms including impulse input.

In this study, we fabricated PVEHs composed of PZT thin films on stainless steel foils deposited by rf-magnetron sputtering. We evaluated both of harmonic and impulse response of the PVEHs and compared with the calculation results of impulse response of piezoelectric unimorph cantilevers. Finally, we discussed applicability of PZT thin-film PVEHs for shoe-mounted power generators.

2. Experiments

2.1. Device fabrication
Figure 1 shows a schematic illustration of the PZT thin-film PVEHs, which are mounted in an empty space at heel part. The PZT thin film was deposited on the Pt/Ti-coated stainless steel foil by rf-magnetron sputtering. The thickness of the PZT and stainless steel are 2.5 μm and 40 μm, respectively. After the PZT deposition, post annealing was conducted to ensure the crystallization of perovskite phase. Subsequently, the Pt top electrode was prepared through a shadow mask. To adjust the resonant frequency, we attached the seismic mass of 0.93 g at the tip of cantilever as shown in Fig. 1. We confirmed that the pyrochlore-free polycrystalline PZT thin films were prepared on the stainless steel foils by X-ray diffraction measurement. The measurement values of the relative dielectric constant and dielectric loss were 285 and approximately 5 %.

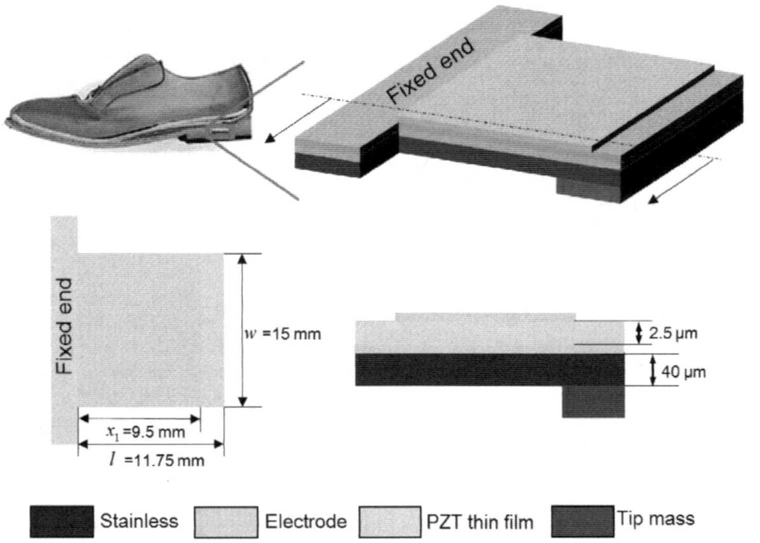

Figure 1. Schematic illustration of PVEH. 2.5-μm-thick PZT films were deposited on 40 μm-stainless steel foils by rf-magnetron sputtering.

2.2. Frequency response
We measured the frequency response of the tip displacement of the PZT thin films PVEHs to determine the resonance frequency. The PVEH was attached to a vibration exciter and frequency of periodic vibration was swept under the acceleration of 0.1, 0.2 and 0.3 m/s^2. We also characterized the load resistance dependence of the output power with a variable external load resistance at the resonance frequency.

2.3. Power generation performance
The measurement setup of the power generation performance is shown in Fig. 2. The impulsive force was inputted at the fix end with a hammer. The impulse response of output voltage and tip displacement was measured using an oscilloscope and laser Doppler vibrometer, respectively.

The optimal load resistance is assumed by the following equation,

$$R_{opt} = \frac{1}{\omega C} \tag{1}$$

where ω and C denote natural angular frequency and capacitance respectively.

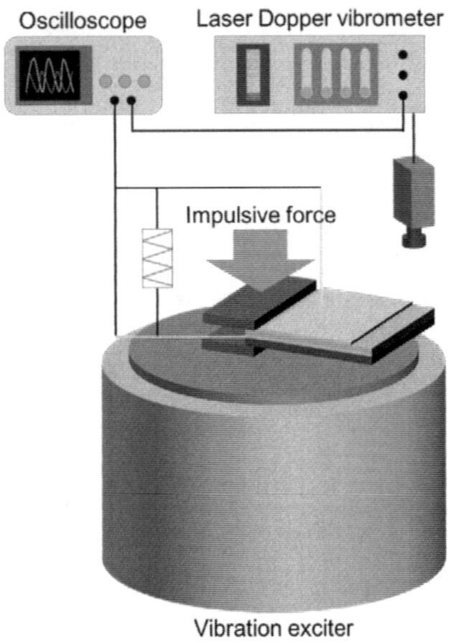

Figure 2. Schematic illustration of measurement setup (left) and photo (right).

3. Results and discussion

Figure 3 shows the tip displacement as a function of the frequency. We clearly observed non-linear resonances, which is due to large deformation of the PVEHs, in the frequency range from 20 Hz to 30 Hz. Furthermore, the resonance frequency was decreased with increasing the acceleration because of softening spring effect. Before the measurement of impulse response, we applied a negative unipolar electric voltage (1 kHz, $20V_{p\text{-}p}$) on the top of electrode for one minute as a polling treatment of the PZT thin films. After that, we measured the impulse response of the output voltage and results are shown in Fig. 4. Typical damped free vibration curve was clearly observed.

We derived the theoretical equation of the impulse response of the piezoelectric cantilever as the following equation,

$$V = e_{31,f} \frac{3wh_s x_1 (1-\nu_s)(2l-x_1)}{4l^3 C} \frac{\Delta v}{\omega_d} e^{-\zeta\omega_n t} \sin\omega_d t \tag{2}$$

where w, h_s, x_1, ν_s, l, C, Δv, ω_n, ω_d, ζ and t denote the cantilever width, substrate thickness, distance from fix end to the edge of upper electrode, Poisson's ratio, cantilever length, capacitance, velocity change before and after impulse, natural angular frequency, damping natural angular frequency, damping and time, respectively. The velocity change after impulse was derived from the curve fitting with the experiments. The values for calculation are listed in Table 1. As shown in Fig. 4, calculated curves are in good agreement with the experimental results. We connected the optimal load resistance of 33.9 kΩ, and measured output power after impulse, and the results are shown in Fig. 5. These results imply that theoretical equation based on the impact model would be applied for the design of the shoe-mounted PVEHs and be helpful to improve the performance of the PVEHs.

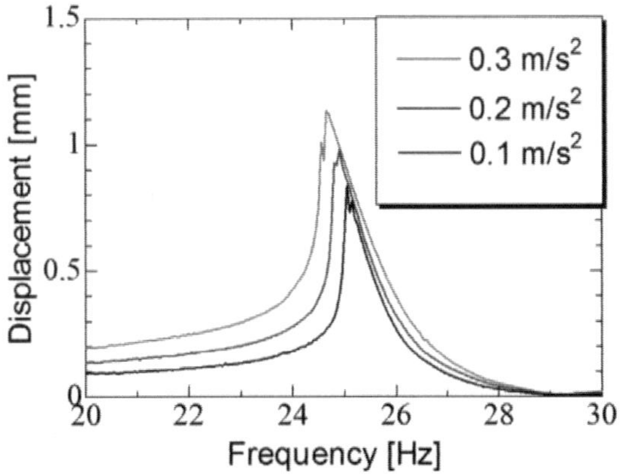

Figure 3. Displacement of the PVEH in an open-circuit state as a function of frequency. The nonlinear resonance was clearly observed due to the large deflection of the PVEH.

Table 1. The input values for calculation.

Cantilever width: w	15 mm
Substrate thickness: h_s	40 μm
Distance from fix end to the edge of upper electrode: x_1	11.8 mm
Poisson's ratio: v_s	0.3
Cantilever length: l	15 mm
Capacitance: C	193.3 nF
Velocity change before and after impulse: Δv	Open 0.2 m/s^2 Short 3.0 m/s^2
Damping natural angular frequency: ω_d	157 rad/s
Natural angular frequency: ω_n	157 rad/s
Damping: ζ	0.008
Piezoelectric coefficient: $e_{31,f}$	1.5 C/m^2

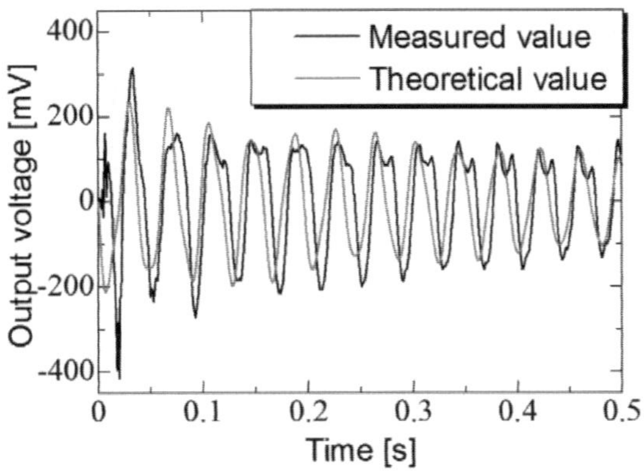

Figure 4. Impulse response of output voltages at the open-circuit state. Theoretical calculation were in good agreement with the experimental result.

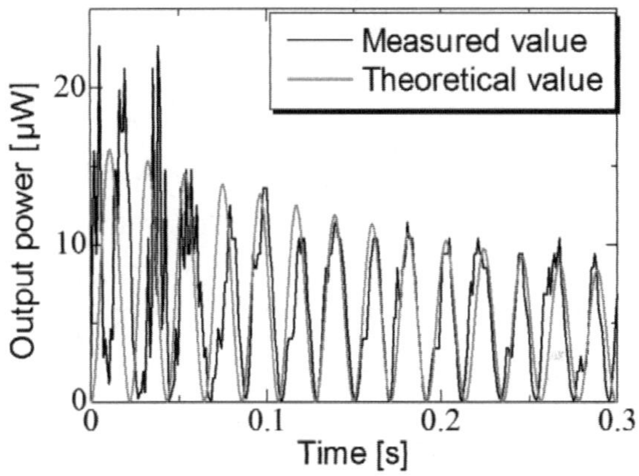

Figure 5. Impulse response of output powers connected with the optimal load resistance.

4. Conclusions

In this study, we developed PZT thin films PVEHs, and evaluated impulse response aiming for shoe-mounted PVEHs. We derived the theoretical equation of output power under the impulsive excitation and compared with the experimental results of impulse response of output power. The calculated curves showed good agreement with measured values. The PVEHs of this study is expected to generate power levels enough for the operation of sensor-nodes by heel strikes.

References

[1] Moro L and Benasciutti D 2010 *Smart Mater. Struct.* 19, 115011(12pp)

[2] Tsujiura Y, Suwa E, Kurokawa F, Hida H and Kanno I 2014 *Proc. PowerMEMS 2014*, pp 482-486

Connection Configurations to Increase Operational Range and Output Power of Piezoelectric MEMS Vibration Energy Harvesters

Sijun Du, Shao-Tuan Chen, Yu Jia, Emmanuelle Arroyo and Ashwin Seshia

Department of Engineering, University of Cambridge, Trumpington Street, Cambridge CB2 1PZ, UK

E-mail: sd672@cam.ac.uk; stc33@cam.ac.uk; yu.jia.gb@ieee.org; emmanuelle.arroyo@gmail.com; aas41@cam.ac.uk

Abstract. Among the various methods of extracting energy harvested by a piezoelectric vibration energy harvester, full-bridge rectifiers (FBR) are widely employed due to its simplicity and stability. However, its efficiency and operational range are limited due to a threshold voltage that the open-circuit voltage generated from the piezoelectric transducer (PT) must attain prior to any energy extraction. This voltage linearly depends on the output voltage of the FBR and the forward voltage drop of diodes and the nature of the interface can significantly limit the amount of extracted energy under low excitation levels. In this paper, a passive scheme is proposed to split the electrode of a micromachined PT into multiple (n) equal regions, which are electrically connected in series. The power output from such a series connected MEMS PT allows for the generated voltage to readily overcome the threshold set by the FBR. Theoretical calculations have been performed in this paper to assess the performance for different series stages (n values) and the theory has been experimentally validated. The results show that a PT with more series stages (high n values) improves the efficiency of energy extraction relative to the case with fewer series-connected stages under weak excitation levels.

1. Introduction

Along with increasing research interests on piezoelectric vibration energy harvesting (PVEH), efficient interface rectification circuits have drawn much attention and interests with respect to extracting as much energy as possible from miniaturized piezoelectric transducers (PT) [1,2]. A typical PT can generate a raw output power of around 10 - 500 $\mu W\,cm^{-2}$, and hence interface circuits should be designed to be power-efficient and stable [3]. Full-bridge rectifiers (FBR) are widely employed in commercial energy harvesting systems due to its stability and simplicity; however, they set high threshold voltages for the generated energy in the PTs to be extracted by the rectifiers [4]. This limitation prevents the system from operating if the environmental excitation is not high enough to attain the required operational threshold voltage, particularly considering miniaturized formats. Hence, the vibrational energy under such conditions is not transferred to the energy storage device [5,6]. Furthermore, for excitation resulting in harvester output slightly greater than the threshold voltage, a very significant amount of energy is wasted as a result.

A full-bridge rectifier (FBR) usually consists of four diodes connecting, whose input is connected to the piezoelectric harvester and output is connected to a capacitor C_S in most implementations. A PT vibrating at or close to its resonance frequency can be modeled as a current source I_P in parallel with a capacitor C_P and a resistor R_P [7]. In order to transfer energy

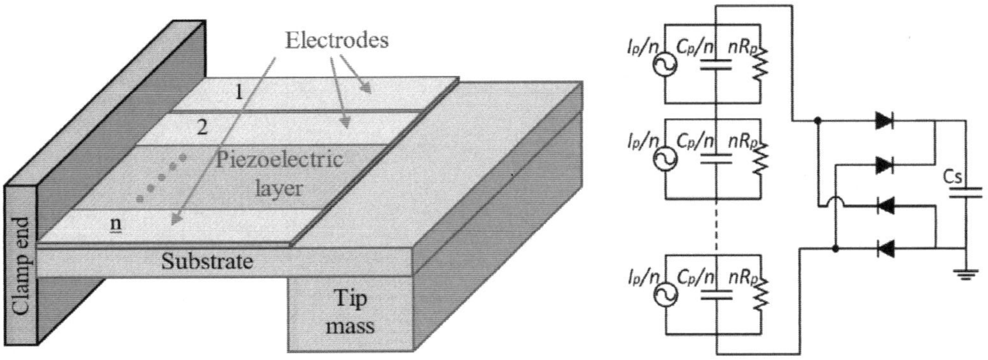

Figure 1: Segmentation of a monolithic cantilevered PT into multiple regions (left) and series connection with a FBR (right).

through the FBR, the voltage across the PT (V_{piezo}) needs to attain $V_S + 2V_D$ (or $-(V_S + 2V_D)$) to overcome the threshold voltage set by the FBR, where V_S is voltage across C_S and V_D is the voltage drop of the diodes. In order to charge the internal capacitor C_P from $V_S + 2V_D$ to $-(V_S + 2V_D)$ (or vice-versa), some charge is wasted, which can be expressed as:

$$Q_{wasted} = 2C_P(V_S + 2V_D) \tag{1}$$

Assuming $V_{pp(open)}$ as the peak-to-peak open-circuit voltage generated from the PT, the condition $V_{pp(open)} > 2(V_S + 2V_D)$ should be satisfied to transfer energy from the PT to C_S. Otherwise, all of the generated energy by the PT is wasted for discharging and charging C_P. In this paper, a passive approach is proposed to segment the top and bottom electrode layers of a monolithic piezoelectric transducer (PT) into multiple regions and electrically connect these regions in series before the PT is connected to a FBR. Figure 1 shows the electrode segmentation of the proposed scheme. In the following sections, the power performance for different number of series stages is theoretically studied and experimentally validated.

2. Theoretical derivation

In this section, theoretical models are developed to establish the effect of series connected PTs on the output power of a full-bridge rectifier. Assuming that the input excitation is sinusoidal, the current source generated by the monolithic model (before splitting the electrode) can be written as $I_P = I_0 \sin \omega t$, where $\omega = 2\pi f_P$. The total charge generated by the PT in a half cycle ($\frac{T}{2}$) should first be calculated, which can be written as:

$$Q_{total} = \int_0^{\frac{T}{2}} I_0 \sin(\omega t)\mathrm{d}t = \frac{2I_0}{\omega} \tag{2}$$

After the electrode is split into n equal regions, the piezoelectric layer area for each region is $1/n$ times of the monolithic model; hence, values of the corresponding current source, inherent capacitor and resistor for each region are $\frac{1}{n}$, $\frac{1}{n}$ and n times of the monolithic model respectively, where the number n is the number of regions that the electrode is split into.

Before the FBR becomes conductive,a certain amount of generated charge is used to charge the internal capacitor C_P from V_S+2V_D to $-(V_S+2V_D)$, or vice-versa. For the monolithic model, this amount of charge is specified by equation (1). In the split electrode case,, the wasted charge from each individual region can be expressed as $Q_{wasted1} = C_{p1}\frac{2}{n}(V_S + 2V_D) = \frac{2C_p}{n^2}(V_S + 2V_D)$. After V_{piezo} attain $V_S + 2V_D$ (or $-(V_S + 2V_D)$), the FBR becomes conductive transferring the charge from C_P to C_S. As only one region is considered, the current sources of all other $n-1$ regions can be turned off using superposition theory. Therefore, the charge transferred into C_S from one single source can be calculated and the total charge flowing into C_S from all of the n sources is n times of the charge from one source:

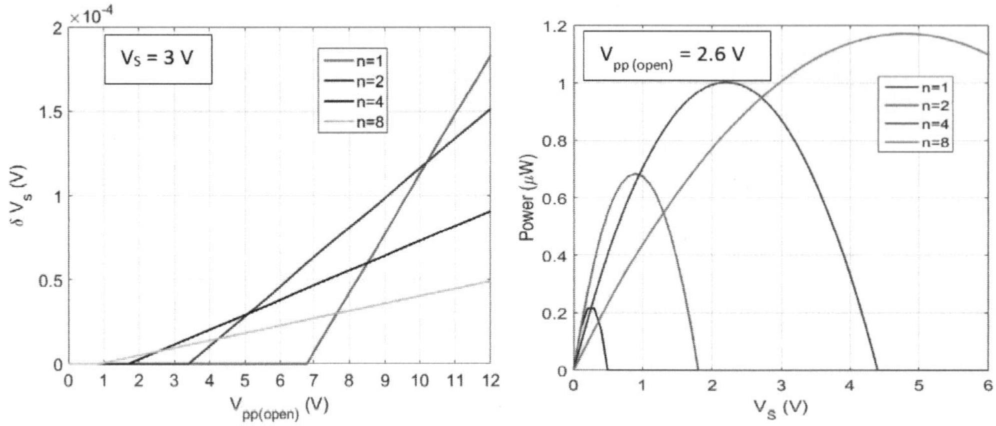

Figure 2: Theoretical output power for different numbers of series stages (n = 1, 2, 4 and 8)

$$Q_{S(n)} = \sum_n Q_{S1} = \frac{2C_P}{n}\left(\frac{I_0 R_P}{\sqrt{1+\omega^2 R_P^2 C_P^2}} - \frac{V_S + 2V_D}{n}\right) \tag{3}$$

The voltage increase of C_S in half an oscillation period for series-connected n sources is:

$$\Delta V_{S+(n)} = \frac{2C_P}{C_S}\left(\frac{V_{pp(open)}}{2n} - \frac{(V_S + 2V_D)}{n^2}\right) \tag{4}$$

where the subscript "+" represents series and "n" is the number of series stages. The open-circuit voltage is calculated as $V_{pp(open)} = \frac{2I_0 R_P}{\sqrt{1+\omega^2 R_P^2 C_P^2}}$. As the energy extracted and stored in C_S is calculated as $\Delta E_{\frac{T}{2}} = \frac{1}{2}C_S((V_S + \Delta V_S)^2 - V_S^2)$. The corresponding reactive output power by a FBR for a n-region series PT can be expressed as:

$$P = \frac{\Delta E_{\frac{T}{2}}}{T/2} = 2f_P \Delta E_{\frac{T}{2}} = f_P C_S((V_S + \Delta V_S)^2 - V_S^2) \tag{5}$$

where f_P is the excitation frequency and ΔV_S is given in equation (4). While n is set to 1, 2, 4 and 8, the output power is plotted in figure 2 under different excitation levels and different V_S values.

3. Experiment

In this section, experiments are performed to validate the theoretical results and demonstrate the performance improvement of the proposed approach. Figure 3 shows the optical micrograph of the test MEMS device and the measured output power for different series stages. The device consists of 8 cantilevers, which are strongly coupled with a monolithic proof mass at their free ends. Hence, the eight cantilevers vibrate in the same amplitude, phase and frequency and their electrodes can be perfectly connected in series or in parallel. The cantilever is excited on a shaker at its natural frequency of 211 Hz under an excitation level of 2.0 g. The open-circuit peak-to-peak voltage generated by the cantilever while all the 8 regions are connected in parallel is 2.6 V. In the experiment, the storage capacitor connected at the output of full-bridge rectifier is a super capacitor of value C_S = 1 mF. A full-bridge circuit is built using four diodes with a measured forward voltage drop of around 0.5 V each.

For each cantilever, there are 2 electrode pads shown in the figure, summing up to 16 electrode pads for 8 cantilevers. The experiments have been performed by connecting the electrodes in

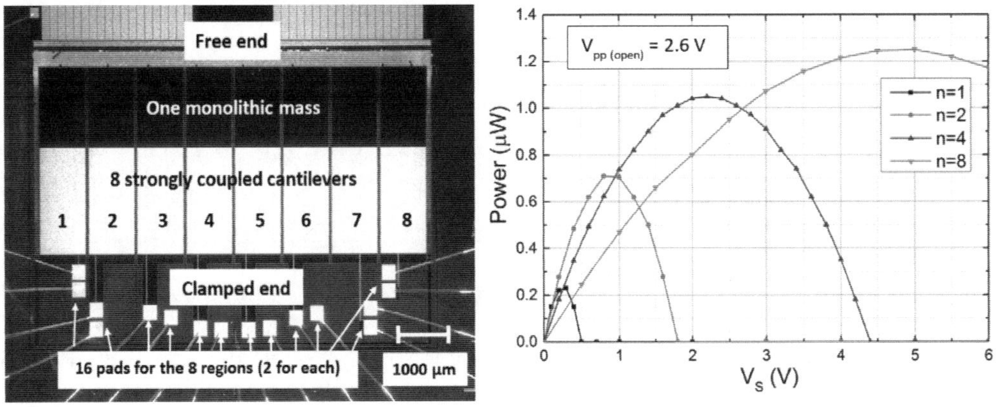

Figure 3: Optical micrograph of the test MEMS device and measured electrical output power

parallel and in series for n = 1, 2, 4 and 8 series stages. The results show that splitting the electrode into 8 regions in series can increase the output power by up to 5.4× compared to the monolithic model when all the 8 regions are connected in parallel. The slight deviation on power for theoretical and measured results is due to the non-ideal diodes used in the experiment.

4. Conclusion

This paper addresses the issue of the high threshold voltage that exists in a full-bridge rectifier (FBR), which should be overcome prior to any energy extraction from the piezoelectric transducer (PT). In order to flip the voltage generated by the PT between the positive and negative threshold voltages, a significant amount of energy is wasted. A passive scheme of segmenting the electrode layers of a PT into n equal regions connected in series is proposed to lower threshold voltages and decrease the energy wasted to overcome the thresholds, hence, increasing the power efficiency. Compared to active interface circuits, this scheme significantly decreases system volume without an inductor and increases stability without consuming any quiescent power by simply employing a single FBR. Furthermore, the performance of a 8-region PT is improved by 5.4× compared to a monolithic PT and this approach is particularly preferred in MEMS piezoelectric harvesters due to their relatively low open-circuit voltage compared to macroscopic harvesters. A MEMS cantilever with segmented electrodes has been integrated with a FBR circuit and experimentally tested to validate the theory.

References

[1] M. Belleville, H. Fanet, P. Fiorini, P. Nicole, M. J. M. Pelgrom, C. Piguet, R. Hahn, C. Van Hoof, R. Vullers, M. Tartagni, and E. Cantatore, "Energy autonomous sensor systems: Towards a ubiquitous sensor technology," *Microelectronics Journal*, vol. 41, no. 11, pp. 740–745, 2010.

[2] G. D. Szarka, B. H. Stark, and S. G. Burrow, "Review of power conditioning for kinetic energy harvesting systems," *Power Electronics, IEEE Transactions on*, vol. 27, no. 2, pp. 803–815, 2012.

[3] S. Du, Y. Jia, C. D. Do, and A. A. Seshia, "An efficient sshi interface with increased input range for piezoelectric energy harvesting under variable conditions," *IEEE Journal of Solid-State Circuits*, vol. PP, no. 99, pp. 1–14, 2016, DOI: 10.1109/JSSC.2016.2594943.

[4] Y. Sun, N.-H. Hieu, C.-J. Jeong, and S.-G. Lee, "An integrated high-performance active rectifier for piezoelectric vibration energy harvesting systems," *Power Electronics, IEEE Transactions on*, vol. 27, no. 2, pp. 623–627, 2012.

[5] H. S. Kim, J.-H. Kim, and J. Kim, "A review of piezoelectric energy harvesting based on vibration," *International journal of precision engineering and manufacturing*, vol. 12, no. 6, pp. 1129–1141, 2011.

[6] S. Du, Y. Jia, and A. Seshia, "An efficient inductor-less dynamically configured interface circuit for piezoelectric vibration energy harvesting," *IEEE Transactions on Power Electronics*, vol. PP, no. 99, pp. 1–1, 2016, DOI: 10.1109/TPEL.2016.2587757.

[7] G. K. Ottman, H. F. Hofmann, A. C. Bhatt, and G. A. Lesieutre, "Adaptive piezoelectric energy harvesting circuit for wireless remote power supply," *Power Electronics, IEEE Transactions on*, vol. 17, no. 5, pp. 669–676, 2002.

INNOVATIVE ENERGY HARVESTER DESIGN USING BISTABLE MECHANISM WITH COMPENSATIONAL SPRINGS IN GRAVITY FIELD

Bogdan Vysotskyi[1,2], Fabien Parrain[1], Denis Aubry[2], Philippe Gaucher[2] and Elie Lefeuvre[1]
[1]Centre for Nanoscience and Nanotechnology (C2N) – Site d'Orsay, Univ. Paris-Sud, Université Paris-Saclay, Bât. 220, Rue Ampère, 91405 Orsay, France

[2]Laboratoire Mécanique des Sols, Structures et Matériaux, CentraleSupélec, Université Paris-Saclay, Châtenay-Malabry F-92295, France

E-mail: bogdan.vysotskyi@c2n.universite-paris-saclay.fr

Abstract. The purpose of the presented work is to introduce the novel design of electrostatic energy harvester using bistable mechanism with compensational springs in gravity field capable of providing the output of several μW under the excitation of extremely small amplitude (up to 0.2g) and low frequency (10-100Hz). Presented energy harvester uses the bistable hysteresis modification to achieve low-frequency low-amplitude sensibility. It was demonstrated with finite element modelling (FEM) that hysteresis width produced by bistability is changing with a constant linear coefficient as a function of a compensational spring stiffness and thus a device sensitivity could be adjusted to the minimum point for the amplitude of external excitation. Further, highly non-linear bistable double curved beam mechanism assures the high sensitivity in frequencial domain due to the non-defined bandwidth. The equivalent circuit technique is used for simulating the device performance.

1. Introduction

Today energy harvesting aims towards offering the solutions for supplying the variety of standalone devices with the electricity. The possible applications for compact self-recharging sources of energy are immense: from structures health monitors and tire pressure sensors to biomedical devices.

Creation of the energy harvester that will be able to replace a conventional battery in task of powering the pacemaker had been considered as one of the major challenges in energy harvesting domain during the last several years. Typical requirements for such a device is relatively small size (volume <1cm^3), long lifetime (more than 20 years) and the power output delivering 1-10μW. In this paper we propose a concept, design and simulations of the device that is expected to meet the required characteristics.

The bistable mechanism had been chosen for such a device in order to achieve high bandwidth in low frequency domain. It is explained by the fact that the switching between stable positions depends only on the value of the force applied, and frequency dependence is significantly suppressed due to the presence of strong non-linearity.

Another innovation of the offered mechanism is the use of a gravitational effect. Generally, in MEMS domain gravitational offset is either neglected due to the low masses involved [1], either is meant to be perpendicular to the operational plane of the device [2]. In this case, when the device is aligned with gravitational force it is not expected to be functional. However, proposed energy harvester needs to be placed into gravity field in order to achieve its full sensibility.

Movement of a seismic mass is achieved by switching between stable positions of bistable hysteresis modified by the compensational springs, in contrast to previously reported hysteresis shift [3].

2. Design of the energy harvester

Presented energy harvester consists of bistable double curved beam mechanism [4], compensational springs, seismic mass and transducer part. The schematic representation of the device is shown on the Fig. 1. With the external excitation, the inertial force is acting upon the seismic mass, which leads to the switching between stable positions of bistable mechanism. However, due to the presence of the compensational springs and accurate choice of the mass, the required switching force is significantly lowered in comparison to non-compensated bistable system. It happens due to bistable hysteresis modification with linear spring. For the variable capacitance, an overlap geometry had been chosen in order to make use of all displacement amplitude of the seismic mass.

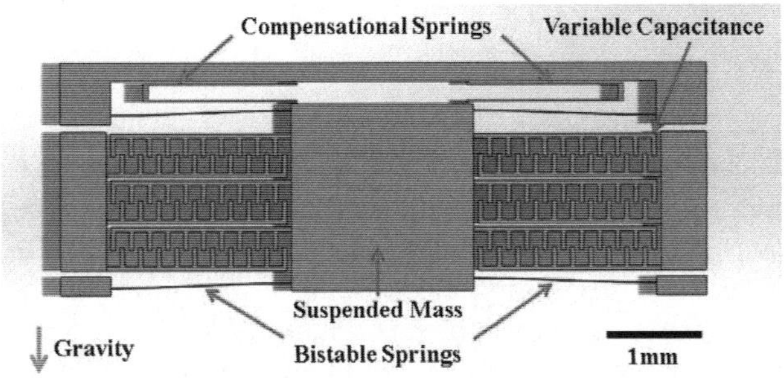

Figure 1. Schematic representation of the device using bistable mechanism and compensational springs in gravity field for energy harvesting.

Parameter	Value
Curved beam length, mm	2
Capacitive finger length, μm	120
Number of fingers	2688
Seismic mass, g	0.5
Compensational spring stiffness, N/m	60
Inter-finger distance, μm	5
Cmax, pF	35
Cmin, pF	239

Table 1. Summary of designed parameters.

2.1. Bistable hysteresis modification

A finite element modelling (FEM) with ANSYS© had been performed on the double curved beam bistable mechanism with linear compensational springs attached. It had revealed that a usual bistable hysteresis is significantly modified in the presence of the linear spring.

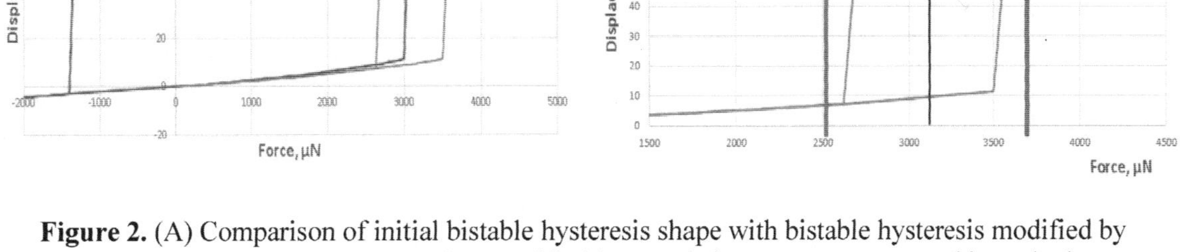

Figure 2. (A) Comparison of initial bistable hysteresis shape with bistable hysteresis modified by compensational spring with k=50N/m. Modified hysteresis becomes narrower and is pushed to a higher forces of switching; (B) Image showing the movement available with the sinusoidal excitation at 0.2g amplitude. With the chosen mass system is placed in the center of modified bistable hysteresis, making the seismic mass unstable and, thus, sensible to the small excitations.

As it could be seen on the Fig. 2 (A), compensational spring makes it harder to push the bistable system to another stable position, but the return into initial state is easier due to the pulling from the linear spring.

2.2. Use of gravitational offset
Hysteresis modification phenomenon could be used along with gravitational offset to achieve the high device sensibility in terms of acceleration. If the mass is chosen carefully, it will be possible to put the system in a highly unstable state by applying a gravity field (typical example is given on Fig. 2 (B)). Due to the positioning in the centre of hysteresis, even a small external acceleration (considerably less than 1g) could produce a switching of states in bistable mechanism.

3. Spice model

3.1. Mechanical part
For the simulation of the device performance a classical approach of building an equivalent circuit is used. Linear part of the harvester is simulated with RLC oscillator, whereas all other forces are included into the voltage behavioral source. Thus, additional external source takes into account the non-linearity that comes from the bistable system, force of gravity, electrostatic attraction inside the variable capacitance and the restoring force of the linear compensational springs. A sinusoidal source is used to simulate the external acceleration of a variable frequency and amplitude. The schematics of the equivalent circuit is shown on the Fig. 3. The resulting displacement over time is calculated. It is shown on the Fig. 4: switching is followed by near stable position vibration. With the lower amplitude of excitation the system behaves mainly in chaotic way, whereas when the amplitude is increased the movement reaches higher periodicity, up to the point of total following the excitation near the cut-off frequency.

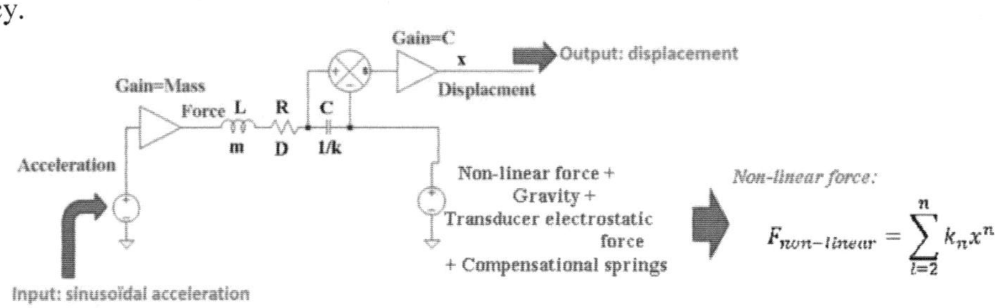

Figure 3. Equivalent circuit schematics of the energy harvester using bistable mechanism and compensational springs in gravity field.

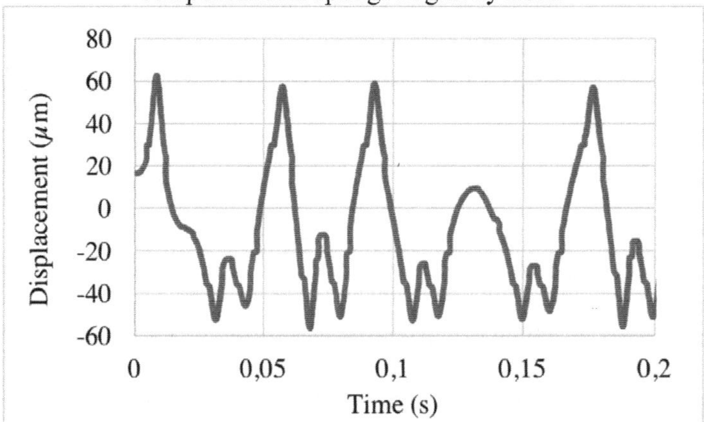

Figure 4. Spice simulation revealing the displacement of the seismic mass over time under 20Hz, 0.2g sinusoidal excitation.

3.2. *Power output*

To simulate a power output of the energy harvester, an interface circuit have to be chosen. For this work, we had taken the simplest one proposed by Lefeuvre et al. [5] in order to check the device performance.

Several simulations under different conditions had been performed. The excitation frequency of sinusoidal excitation and its amplitude had been varied in order to obtain the power output dependence (Fig. 5). It could be seen that harvested power is proportional to the excitation frequency. However, input acceleration affects only the bandwidth, keeping only the small impact on the amount of energy harvested. It could be easily explained by the fact that the movement is achieved by the switching of the bistable mechanism, whose design imposes almost constant displacement magnitude. Thus, the input excitation could either produce the switching which will lead to the power generation or not. Dependence of the bandwidth on the input acceleration comes from the strong non-linearity present in the system.

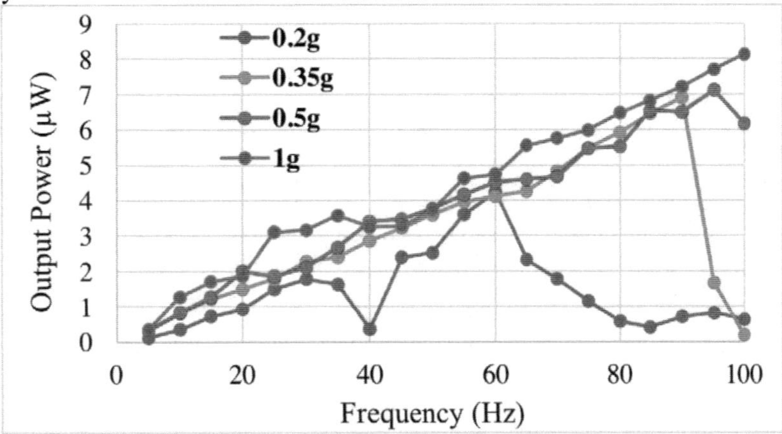

Figure 5. Simulated power output as a function of sinusoidal excitation amplitude and frequency. Note the changes in the bandwidth with the excitation amplitude whereas the power output is maintained.

4. Conclusions and future perspectives

In this work a novel technique of using the bistable hysteresis modification by compensational spring is employed in order to achieve extremely low values of harvestable accelerations (>0.2g) and frequencies (10-100Hz). Simulated power output lies in the range of several µW depending on the given excitation.

In the nearest future the device is going to be fabricated in Silicon with classical MEMS technologies and then characterized.

Acknowledgements

This work is supported by the "IDI 2015" project funded by the IDEX Paris-Saclay, ANR-11-IDEX-0003-02.

References

[1] D. S. Nguyen and E. Halvorsen, Proc. PowerMEMS 2009, pp. 411-414;
[2] Y. Lu, F. Cottone, S. Boisseau, F. Marty, D. Galayko, P. Basset, Proc. MEMS 2016, pp. 33-36;
[3] K. Yamamoto, T. Fujita, A. Badel, F. Formosa, K. Kanda and K. Maenaka, Proc. PowerMEMS 2015, pp. 1-4;
[4] J. Qiu, J. H. Lang, and A. H. Slocum, J. Microelectromech. Syst., 13 (2004), pp. 137-146;
[5] E. Lefeuvre, S. Risquez, J. Wei, M. Woytasik and F. Parrain, Proc. PowerMEMS 2014, 557.

Flexible energy harvesting from hard piezoelectric beams

Aidin Delnavaz and Jérémie Voix

École de technologie supérieure, Montreal, Quebec, Canada

E-mail: jeremie.voix@etsmtl.ca

Abstract. This paper presents design, multiphysics finite element modeling and experimental validation of a new miniaturized PZT generator that integrates a bulk piezoelectric ceramic onto a flexible platform for energy harvesting from the human body pressing force. In spite of its flexibility, the mechanical structure of the proposed device is simple to fabricate and efficient for the energy conversion. The finite element model involves both mechanical and piezoelectric parts of the device coupled with the electrical circuit model. The energy harvester prototype was fabricated and tested under the low frequency periodic pressing force during 10 seconds. The experimental results show that several nano joules of electrical energy is stored in a capacitor that is quite significant given the size of the device. The finite element model is validated by observing a good agreement between experimental and simulation results. the validated model could be used for optimizing the device for energy harvesting from earcanal deformations.

1. Introduction

Energy harvesting from human body power can make wearable technologies pervasive by providing alternative, clean, free, and life-long energy sources. Mechanical energy of human body can be transfomed to electrical current running the electronic circuits of wearable devices, more efficiently among others, by piezoelectric devices. For example, piezoelectric transducers can be incorporated in backpacks [1], gloves [2], helmet chinstraps [3], shoes [4] and garments [5] for energy harvesting from different types of human motions. Integration of flexibility and stretchability to wearable piezoelectric generators is the key to make them convenient to wear. Polymer-based piezoelectric materials such as PVDF are intrinsically flexible, but they have a low electromechanical coupling and hence barely efficient for energy harvesting. Ceramic-based piezoelectric materials like PZT are very efficient in terms of energy conversion, but too rigid to conform to the curves of the human body. To tackle this defect, several composite piezoelectric structures have been developed in recent years by integrating piezoelectric ceramics in form of microfibers [6], nanowires [7] or ribbons [8] onto a flexible substrate. However, they present complex mechanical structure, need fabrication of interdigitated electrodes and suffer from the small total volume of the piezoelectric elements contributing to the power output.

This paper reports a new architecture that integrates a bulk piezoelectric ceramic onto a flexible platform for energy harvesting from the human body pressing force. The key advantage of the proposed energy harvesting device is its simplicity of manufacturing while using a bulk PZT beam for the maximum efficiency. Also, the flexible silicon body of the energy harvesting device makes it safe for biomedical applications.

2. Design and finite element modeling

The proposed energy harvesting device consists of a piezoelectric beam cut in form of "I" from a PZT-5A sheet (Piezo Systems Inc, USA) with the square section of 1 mm at the beam stem. Two tiny aluminum washers are placed at two extremes of the beam stem and the whole space between them is molded by a flexible medical grade liquid silicon rubber (MED4910, NuSil Technology Inc., USA) as shown in Fig. 1. Any pressing force applied to the silicon cylinder stretches the piezoelectric beam through the washer caps, and generates an electric charge in response to the applied mechanical stress.

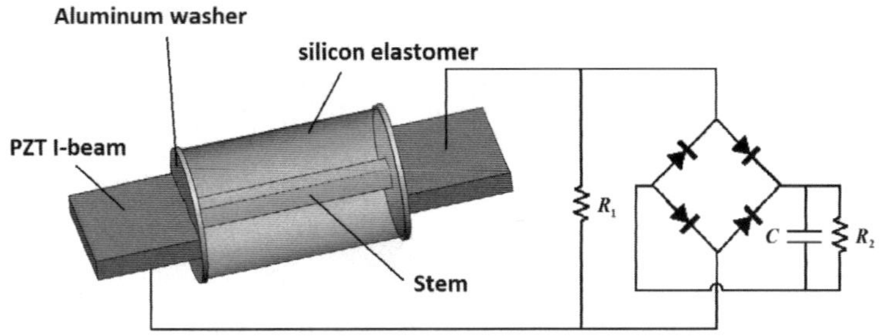

Figure 1. Energy harvesting device

A multiphysic finite element 1/4 symmetric model of the device has been developed in COMSOL and simulated for 10 seconds under the sinusoidal mechanical load with the very low frequency of $f = 1$ Hz that represents the average frequency of normal human motions. The finite element model includes the piezoelectric model of the PZT beam and the linear elastic model of the non-piezoelectric parts. The coupled piezoelectric and mechanical structure models are linked to the electrical circuit model that consists of a diode bridge, a $C = 22\mu F$ capacitor and two resistive loads representing the impedance of the measuring probe (R_1) and the leakage current (R_2) of the capacitor as illustrated in Fig. 1.

3. Prototype and testing

A prototype of the energy harvester was fabricated as illustrated in Fig. 2 and tested by applying controlled deformations using a mechanical wave driver (shaker) as shown in Fig. 3.

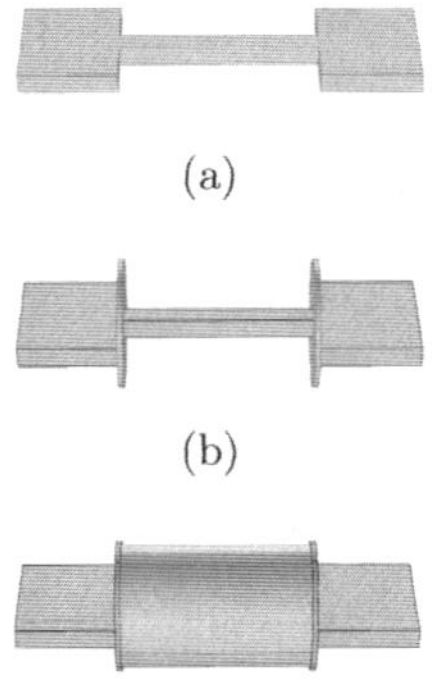

(a)

(b)

(c)

Figure 2. Fabrication steps: (a) PZT cut (b) washer installation (c) silicon molding

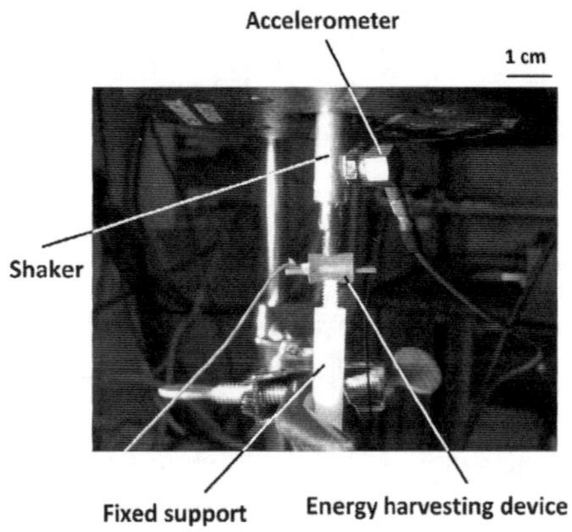

Figure 3. Test setup

The energy harvesting device is clamped between the shaker arm and a fixed support. An accelerometer is used to measure the displacement of the shaker arm that implies the deformations applied to the energy harvesting prototype.

4. Results and conclusion

The device is designed for the maximum silicon deformation of 0.5 mm. Figure 4(a) shows the maximum deformation of the silicon body under the transversal pressing force. The von Mises yield criterion is investigated using finite element analysis to make sure that the PZT beam tolerates the applied stresses. The simulation results demonstrated in Fig. 4(b) confirm that the maximum von Mises stresses in the beam is about 1 MPa which is substantially below the tensile strength of PZT-5A (80 MPa). The maximum beam strain and voltage output are also computed and the results are illustrated in Figs. 4(c) and 4(d)

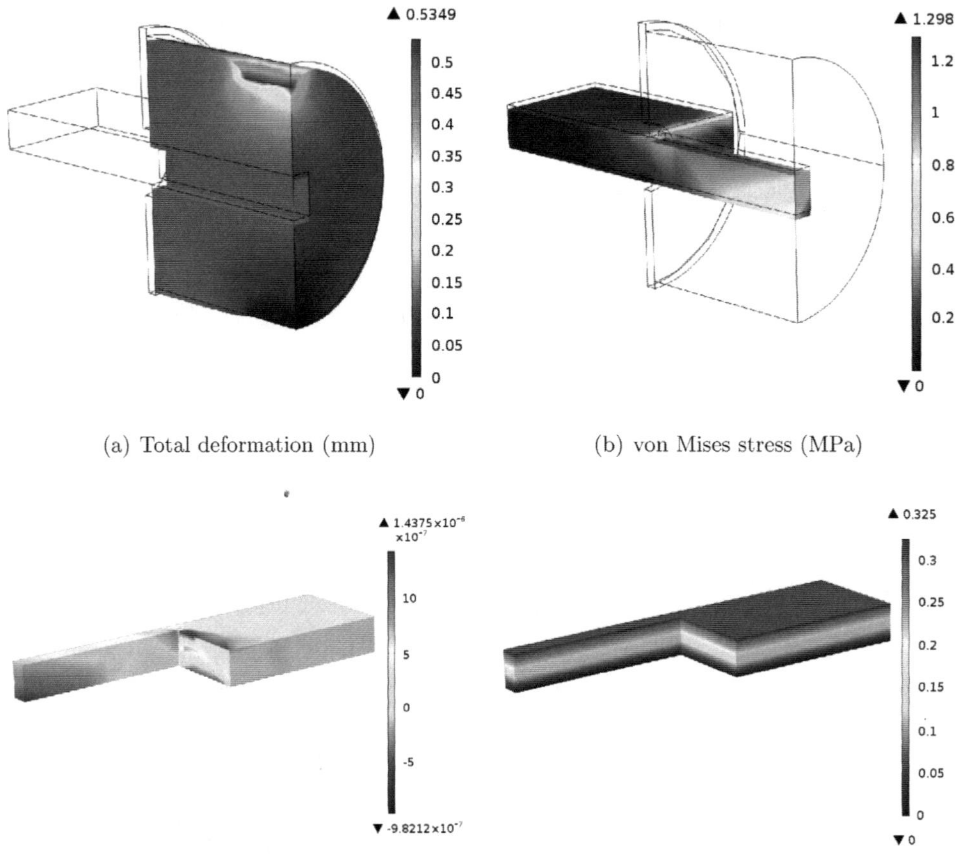

(a) Total deformation (mm)

(b) von Mises stress (MPa)

(c) Axial strain

(d) Voltage output (V)

Figure 4. Finite element modeling of the flexible energy harvesting device

The finite element model is then verified by comparing the experimental and simulation results shown in Fig. 5 and is validated with reasonably little differences between the numerical and measured results. The relatively large amplitude of the voltage output show that the proposed energy harvesting device is efficient in terms of transferring the applied deformation from the flexible medium to the rigid part. Also, the voltage peaks in Fig. 5 are not symmetric in positive

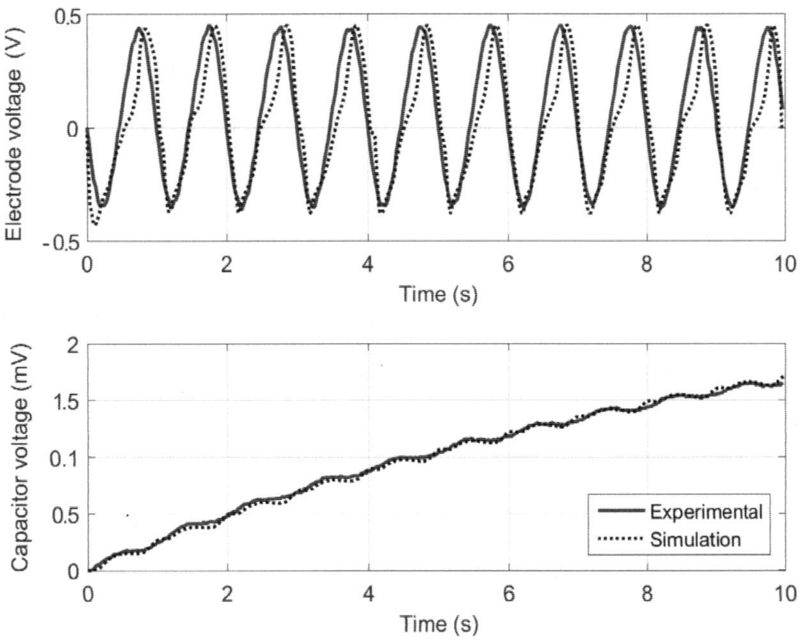

Figure 5. Experimental and simulation results

and negative regions that can be explained by the elastic hysteresis associated with the flexible silicon body.

Moreover, the energy stored in the capacitor is calculated to be $Q = \frac{1}{2}CV^2 \simeq 4\,\mathrm{nJ}$. The developed finite element model could be used to investigate the effect of geometric parameters and electric components on the energy output and enables optimizing the device dimensions and design parameters. The proposed device can be primarily used for energy harvesting from the earcanal deformations, but could also be used for energy harvesting from other human power sources such as heel strikes and finger presses.

References

[1] Jonathan Granstrom, Joel Feenstra, Henry A Sodano, and Kevin Farinholt. Energy harvesting from a backpack instrumented with piezoelectric shoulder straps. *Smart Materials and Structures*, 16(5):1810–1820, oct 2007.

[2] Youngsu Cha, Jin Hong, Jaemin Lee, Jung-Min Park, and Keehoon Kim. Flexible Piezoelectric Energy Harvesting from Mouse Click Motions. *Sensors*, 16(7):1045, jul 2016.

[3] Aidin Delnavaz and Jérémie Voix. Flexible piezoelectric energy harvesting from jaw movements. *Smart Materials and Structures*, 23(10):105020, oct 2014.

[4] Jingjing Zhao and Zheng You. A Shoe-Embedded Piezoelectric Energy Harvester for Wearable Sensors. *Sensors*, 14(7):12497–12510, jul 2014.

[5] Jin-Hee Yang, Hyun-Seung Cho, Seon-Hyung Park, Seung-Hwan Song, Kwang-Seok Yun, and Joo Hyeon Lee. Effect of garment design on piezoelectricity harvesting from joint movement. *Smart Materials and Structures*, 25(3):035012, mar 2016.

[6] L M Swallow, J K Luo, E Siores, I Patel, and D Dodds. A piezoelectric fibre composite based energy harvesting device for potential wearable applications. *Smart Materials and Structures*, 17(2):025017, apr 2008.

[7] Xi Chen, Shiyou Xu, Nan Yao, and Yong Shi. 1.6 V nanogenerator for mechanical energy harvesting using PZT nanofibers. *Nano letters*, 10(6):2133–7, jun 2010.

[8] Yi Qi, Noah T Jafferis, Kenneth Lyons, Christine M Lee, Habib Ahmad, and Michael C McAlpine. Piezoelectric ribbons printed onto rubber for flexible energy conversion. *Nano letters*, 10(2):524–8, feb 2010.

Hotspot Liquid Microfluidic Cooling: Comparing The Efficiency between Horizontal Flow and Vertical Flow

Yuki Okamoto[a], Hiroyuki Ryoson[b], Koji Fujimoto[b], Keiji Honjo[b], Takayuki Ohba[b] and Yoshio Mita[a]

[a] Dept. EEIS, the Univ. of Tokyo, 7-3-1 Hongo, Bunkyo-ku, Tokyo (113-8656), Japan
[b] Tokyo Institute of Technology, Bldg. J3, 4259 Nagatsuta-cho, Midori-ku, Yokohama, Kanagawa (226-8501), Japan

E-mail: mems@if.t.u-tokyo.ac.jp

Abstract. This paper reports a novel cooling method for a local high-temperature block in an integrated circuit, which is called a "hotspot". The method is to cool the chip in out-of-plane (3-D) direction to overcome efficiency limit of traditional horizontal (2-D) cooling. Our result indicates that high-temperature (over 180 ℃) circuit block such as a phase-locked-loop (PLL), which is a performance limiting block in a modern CPU, can more efficiently be cooled by the vertical (3-D) cooling scheme.

1. Introduction

The more transistors are integrated to an LSI, the larger the power dissipation becomes. Due to the large power dissipation, especially of high-performance chips, chip-level cooling has become important today [1]. For effective cooling of a chip, several works focused on liquid cooling with integrated microfluidics since increasing the volume and the surface area of solid heat-sink is not always possible. While some studies such as Ref.[2] and Ref.[3] propose using microfluidics to replace the solid heat-sink, these methods need many external equipment like a pump, which leads to expand the size of the whole system.

To avoid increase of the size, some studies [4], [5] use both integrated microfluidics and solid heat-sink. In such a usage, the major role of an integrated microfluidics cooling system is as a heat spreader that reduce thermal non-uniformity of a LSI chip, which reduces in a performance of a microprocessor [6], caused by heavy loaded circuit block like a PLL.

To date, most of the previous research on microfluidics hotspot cooling systems has focused on improving the micro pump performance and characterization of in-plane micro-channels. A horizontal

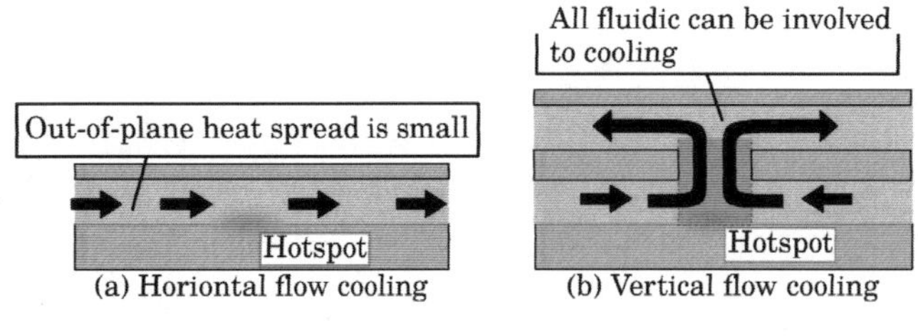

Figure 1. (a) Cooling a hotspot by horizontal flow. Out-of-plane heat spread is small due to laminar flow. (b) Cooling a hotspot by vertical flow. All flow can contribute cooling a hotpot over the hotspot.

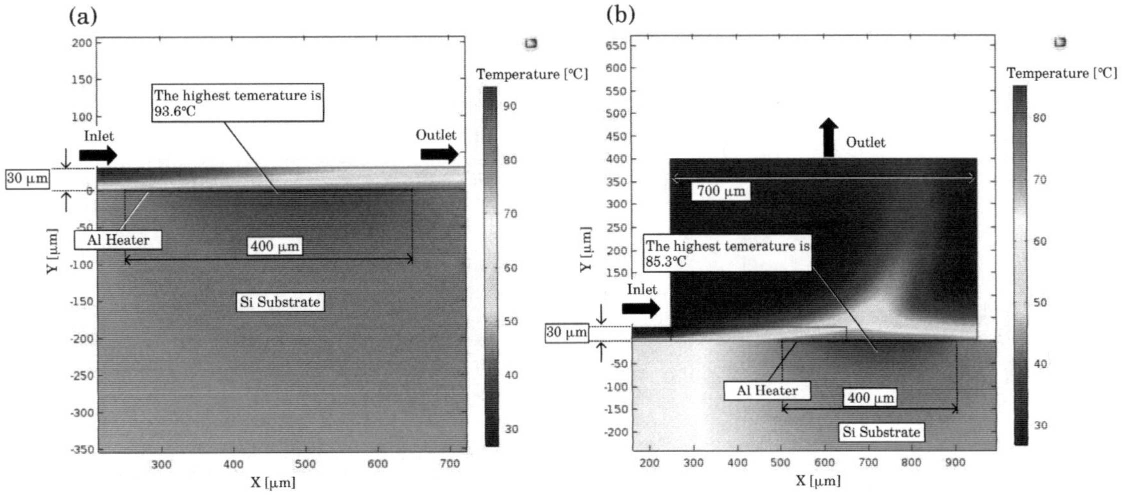

Figure 2. FEM simulation of (a) horizontal in-plane cooling (b) vertical out-of-plane cooling

in-plane micro-channel, however, can involve only the surface of the micro-flow as shown in Figure 1-(a). To involve all the liquid over the hotspot, the authors propose to make use of vertical flow (Figure 1-(b)). In this paper, we report the efficiency of vertical flow in order to cool down a hotspot with a micro aluminum a heater and a thermal sensor.

2. Experimental results

2.1. FEM simulation

Figure 2-(a) and (b) show the simulated temperature of a hotspot cooled with horizontal in-plane flow and vertical out-of-plane flow. In the simulation, 400 μm × 400 μm square Al heater is put on Si substrate as a hotspot and microchannel is placed on the Al heater. In the in-plane cooling simulation, a height of the channel is 30 μm, and the flow is made from left to right (Figure 2-(a)). In the case of the out-of-plane channel, the inlet is the 30 μm channel and the 750 μm outlet is above the hotspot (Figure 2-(b)). The input power is supposed to be 2 W and input liquid is supposed to be DI-water at a flow-rate of 500 μm/min. While the highest temperature with horizontal in-plane cooling is 93.6 ℃, the highest temperature with vertical out-of-plane coolingis 85.3 ℃. It means that the vertical flow can take out the heat from the hotspot thanks to the flow direction and the larger volume of cooling involved liquid.

2.2. Temperature measurement of in-plane cooling and out-of-plane cooling

To measure the temperature and to heat locally, we fabricated micro resistors as thermal sensors and an artificial hotspot as shown in Figure 3. The design of the device follows the simulated layout.

One micro heater was placed under the horizontal flow and the other was placed under the vertical flow. For fabricating a bottom plate, SiO_2 was sputtered by 500 nm as an insulator (Figure 4-(a-1)), and then Al was sputtered by 500 nm and patterned (Figure 4-(a-2)). After the patterning, 50 nm SiO_2 was sputtered again to avoid an electrolysis when liquid flowing (Figure 4-(a-3)). As a top plate with micro-channels, we patterned PDMS using a SU-8 mold (Figure 4-(b-1)). After peeling off the PDMS from the mold, inlet and outlet holes were made (Figure 4-(b-1)). Finally both of them were bonded after O_2 plasma activation (Figure 4-(c)).

The micro-channel was connected to an external pump, whose flow rate was 450 μL/min. The micro-heater as a hotspot was connected to a power source and the thermal sensor was connected to an LCR meter (HP 4284A) as shown in Figure 5. After the correspondence of the change of the resistance against

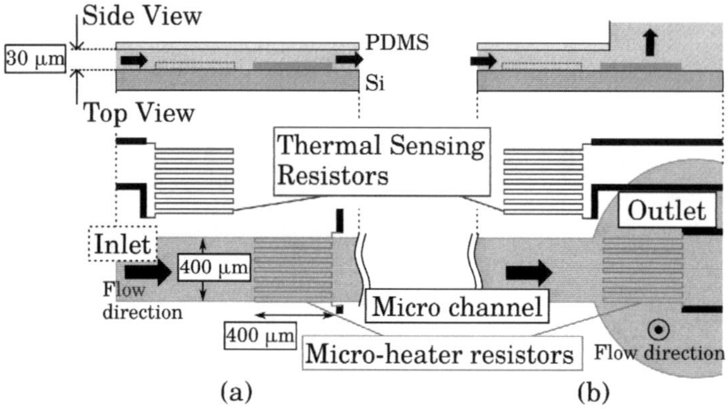

Figure 3. Schematic of the fabricated device to measure the impact of (a) horizontal flow and (b) vertical flow to cool the hotspot. A resistor is placed next to a micro-heater to measure the temperature .

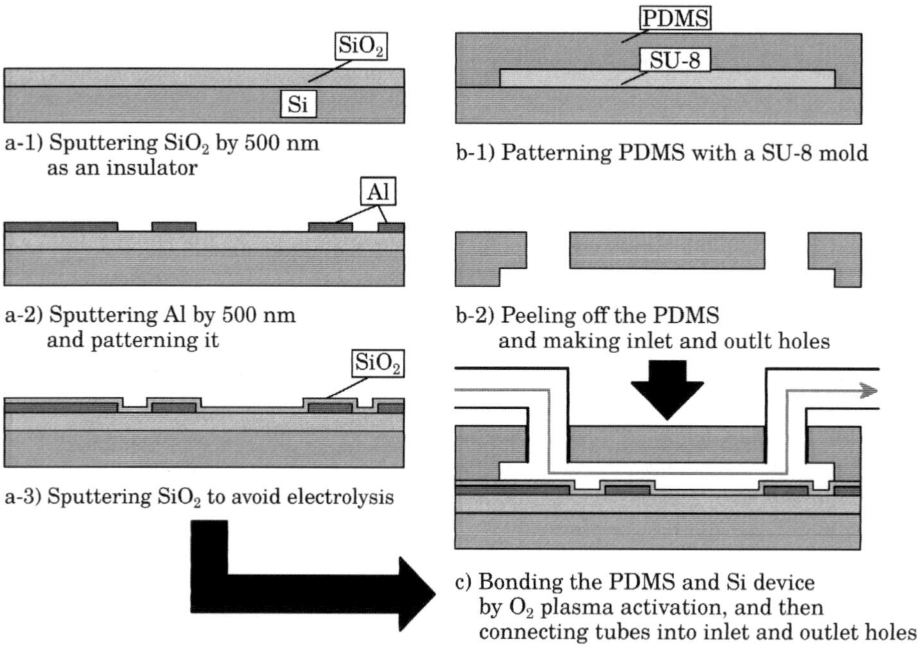

Figure 4. Process flow of the fabricated device

the temperature is measured, we calculated the temperature around a hotspot from the resister next to the hotspot as shown in Figure 6. After power consumption of a micro-heater reaches more than about 1 W, the temperature around the hotspot under the vertical flow becomes smaller than that of under the horizontal flow. It means that the heat which cannot spread to Si plate is spread by the water flow, and that the vertical flow can carry the heat more efficiently than the horizontal flow. When the power input to the hotspot is about 2.3 W, the difference of the temperature between with the horizontal flow and with the vertical flow reaches 10 ℃. This result is consistent with the simulated result, and the efficiency of vertical flow for cooling a hotspot in LSI is thereby confirmed.

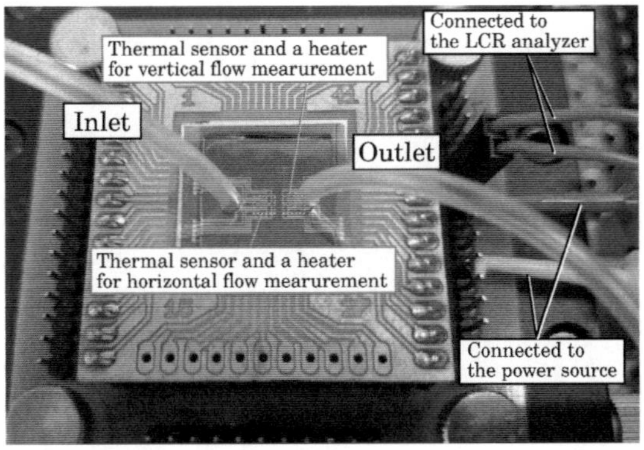

Figure 5. Setup of the experiment. The channel is connected to the external pump (450 μL/min). the resistor for a thermal sensor is connected to an LCR parameter analyzer and the resistor for a micro-heater is connected to a power source.

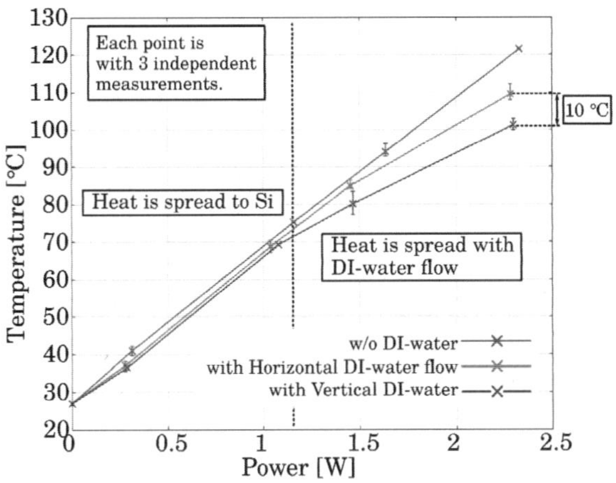

Figure 6. The temperature when the power is input to a micro-heater. When the input power is around 2.3 W is, the difference of the temperature between around the horizontal flow and with the vertical flow is 10 ℃

3. Conclusion

We propose a novel cooling method for a hotspot in an LSI with out-of-plane vertical cooling instead of in-plane horizontal cooling. The simulation results present that the vertical flow can involve much more liquid to cooling a hotspot. According to the measurement of the real fabricated thermal sensor and the micro-heater, we confirmed the higher cooling performance of vertical flow scheme for a hotspot.

Acknowledgments

The CAD patterns were designed with Cadence Virtuoso, accessible through the University of Tokyo, VLSI Design and Education Center (VDEC)'s academic program. The patterning for the device was directly performed by the VDEC's 8-inch EB writer F5112+VD01 donated by Advantest Corporation. The post-process was done using the open facilities of VDEC accessible through MEXT's Nanotechnology Platform Program. Part of this work is supported by KAKENHI No. 16H04345 and the three-dimensional integration development program of the WOW Alliance. The simulations were done with COMSOL Multiphysics^{TM}.

References

[1] Mahajan R, Chiu C P and Chrysler G 2006 *Proceedings of the IEEE* **94** 1476–1485
[2] Dang B, Bakir M S, Sekar D C, King C R and Meindl J D 2010 *IEEE Transactions on Advanced Packaging* **33** 79–87
[3] Al-Rjoub M F, Roy A K, Ganguli S and Banerjee R K 2011 *International Journal of Heat and Mass Transfer* **54** 4560–4569
[4] Green C, Fedorov A G and Joshi Y K 2009 *Journal of Electronic Packaging* **131** 025002
[5] Eng P F, Nithiarasu P and Guy O J 2010 *Microfluidics and Nanofluidics* **9** 787–795
[6] Hamann H F, Weger A, Lacey J A, Hu Z, Bose P, Cohen E and Wakil J 2007 *IEEE Journal of Solid-State Circuits* **42** 56–64

Journal of Physics: Conference Series **773** (2016) 012067 doi:10.1088/1742-6596/773/1/012067

Shared inductor hybrid topology for weight constrained piezoelectric actuators

A D T Elliott[1], A Caccia[2], A Thomas[2], A Astolfi[1], P D Mitcheson[1]

1 AMCC Consulting Ltd, London, UK

2 Animal Dynamics, Oxford, UK

E-mail: alwyn.elliott@amcc-consulting.co.uk

Abstract. This paper presents a new circuit topology designed to minimise the weight of the control circuit required to actuate multiple piezoelectric actuators. It can independently set the phase and bias voltage on each piezoelectric actuator through the use of a single inductor. This is highly desirable in weight constrained applications such as unmanned aerial vehicles as the ferroelectric material required for the inductor is heavy. Furthermore, the circuit topology can also use the same inductor to generate the high bias voltage required to drive the actuators. The full system has been verified in PSpice and a pair of piezoelectric actuators have been successfully driven using off the shelf components.

1. Introduction

Unmanned aerial vehicles (UAVs) have dramatically increased in popularity since the combined development of lightweight high density batteries, small efficient DC brushless motors and improved control and stability algorithms. However as the desired size and weight of UAVs continues to decrease, electromagnetic become increasingly difficult to efficiently operate, thus biologically inspired flapping lift mechanisms (e.g. flies) have been analysed in order to generate the required lift [1]. This has led to the development of flapping based UAV designs (Figure 1) driven by piezoelectric actuators [2, 3, 4].

Piezoelectric actuators are extremely lightweight and can be designed to mechanically resonate in the hundreds of hertz frequency range, however they require large bias voltages. Power electronics are therefore required to both step-up the voltage from a typical lightweight battery used in UAVs [5] and efficiently bias the piezoelectric actuators with the high voltage.

In [6], two approaches to meet the requirements of the power electronics were considered. The first is to use a two-stage design which steps the voltage converter to produce a constant DC voltage then switches between multiple actuators to apply the desired bias. The second is to use a single-stage design where the voltage is both simultaneously stepped up and applied to a single actuator. The two-stage design is capable of sharing it's high voltage bias between multiple piezoelectric actuators, however [6] notes that this would have a high component count thus negatively impacting on the weight. The single-stage design approach should use less components however the bias voltage cannot be shared which will increase inefficiencies in driving the actuations [6].

In this paper, a new biasing circuit called the Shared Inductor Hybrid Topology (SIHT) is presented which combines the low component count of a single-stage design with the ability to

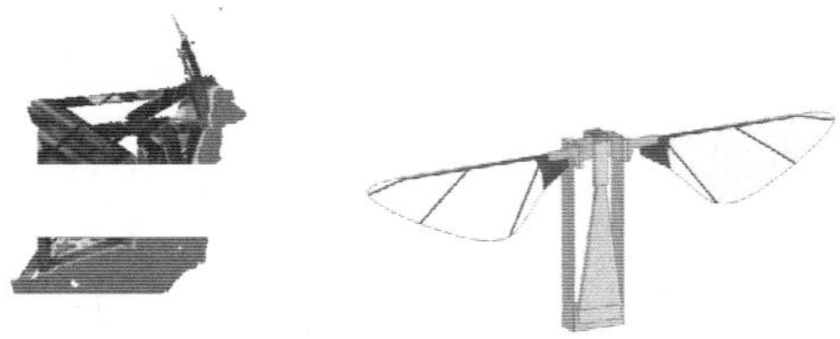

Figure 1. (Left) Structure to drive two wings using four piezoelectric actuators to achieve 4 degrees of freedom (130 mg) [2]. (Right) Two wing flapping UAV driven by a single piezoelectric actuator with passive wing rotation (60 mg) [3, 4].

share the same generated bias voltage of a two-stage design, all with the use of a single inductor. Therefore it will improve both the drive efficiency and minimise the weight associated with the power electronics. This work first analyses the current approaches to the power electronics which drive piezoelectric actuators then presents the design and implementation of the new SIHT circuit. Computer simulation and experimental verification are then presented and further recommendations to improve the design are given.

2. Previous biasing techniques

Two types of biasing drive circuits were evaluated in terms of size, control and power consumption. The first was use of a linear haptic drive for piezoelectic actuators from Texas Instruments, DRV8662 [7]. The second was a bidirectional switching amplifier circuit by [6] which uses resonant charge transfer to achieve high biasing efficiency.

2.1. Linear haptic driver

The linear haptic driver (Texas Instruments DRV8662) is a piezoelectric drive with an integrated boost converter. It samples an analog differential input voltage signal, amplifies it by a set gain then drives a differential output to bias the piezoelectric actuator. Thus an arbitrary waveform can be applied to the actuator. The output can vary between $\pm$100 V at 300 Hz into a 100 nF load. A typical implementation of the DRV8662 can be seen [7].

2.2. Dual stage switching amplifier driver

Figure 2 is the bidirectional switching amplifier technique from [6]. The circuit operates as follows. Switch S_1 is closed resulting in the voltage in between the two piezoelectric actuators, V_{mid}, being connected to the biasing voltage, V_{bias}, via an inductor. This results in the voltage bias across P_1 decreasing and whilst the voltage bias across P_2 increases and energy being stored in the inductor. When the S_1 is then opened, D_2 provides a free wheel path for the energy stored in the inductor. The inverse operation occurs when S_2 is closed instead. The result generates a square wave bias voltage across two piezoelectric actuators. For more details on the operation refer to [6].

Generation of the bias voltage is assumed to be achieved by a boost converter circuit, which could be integrated in a similar fashion to the boost converter in the Texas Instruments DRV8662 but would require an external inductor. They typically achieve around 50% efficiency [8], therefore power consumption measurements should be doubled to for an estimation of total power consumption required to drive actuators from a battery.

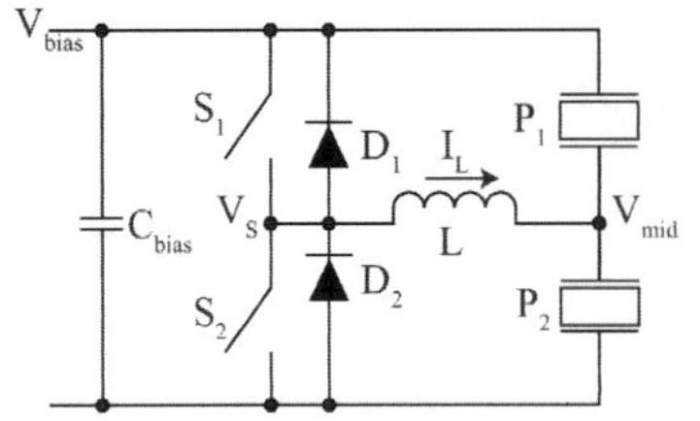

Figure 2. Bidirectional switching amplifier circuit for piezoelectric actuators [6].

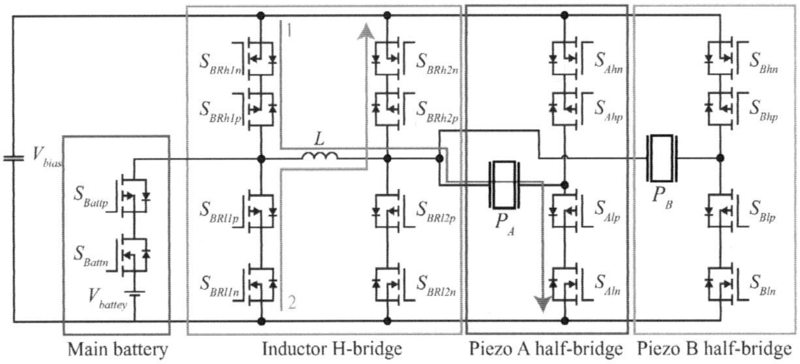

Figure 3. Circuit diagram for the SIHT used to bias multiple piezoelectric actuators and generate the bias voltage from a battery through the a single inductor.

2.3. Experimental comparison
Both driver techniques were tested with stripe actuators from APC International (350/025/0.70SA stripe actuator) [9]. The actuator's measured capacitance was 96 nF and a maximum bias deflection voltage of 200 V_{DC}. It was clamped at one end and the mechanical resonant frequency was measured at 200 Hz.

The haptic driver was connected to a single actuator and supplied with 5 V. The gain was set to maximum and a 140 mV_{p-p} square wave at 200 Hz from a signal generator was applied at the input, generating 80 V_{pk-pk} across the actuator. The power consumption was measured using a Yokogawa WT310 power meter as 1.604 W.

The switching driver circuit was connected to two actuators and a 120 V bias voltage was supplied by a laboratory power supply. The power consumption for one actuator was 87 mW, thus with 50% inefficiency for the boost circuit, the power is likely to be nearer 173 mW [8]. In comparison with the haptic driver this is still much better however the boost circuit would require an inductor, and for every pair of actuators required, another inductor would be necessary.

3. Shared inductor hybrid topology

The experimentation and analysis of drive circuits for piezoelectric actuators highlights the need to reduce the number of inductors required in order to keep the weight of a UAV to a minimum. A new circuit topology is therefore proposed which shares the inductor between multiple piezoelectric actuators by reconfiguring the connections between the inductor and the actuators in real time. The proposed scheme is called Shared Inductor Hybrid Topology (SIHT).

3.1. SIHT operation
Figure 3 shows the circuit for the SIHT. It is comprised of a bidirectional H-bridge [10] with an inductor at the centre. One side of the inductor is then connected to one side of an actuator

Journal of Physics: Conference Series **773** (2016) 012067 doi:10.1088/1742-6596/773/1/012067

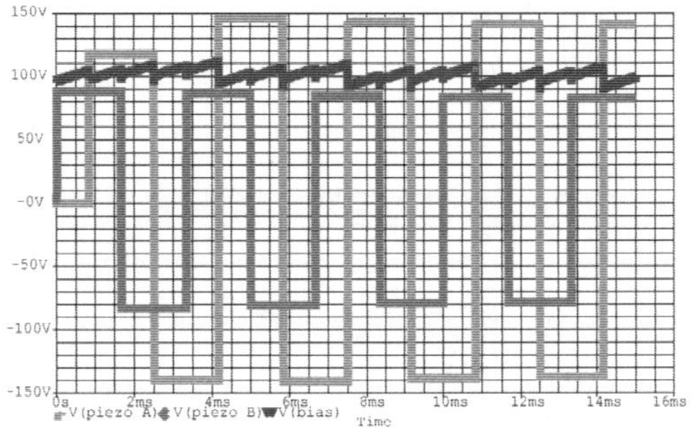

Figure 4. PSPICE simulation showing SIHT driving two actuators with different bias voltages and phases (green and red), whilst also operating a boost circuit (blue).

which itself is connected to a bidirectional half-bridge. Multiple actuators can be connected by joining them at the same inductor-piezoelectric actuator node. The SIHT operation is as follows. To bias actuator, P_A, switches S_{BRh1n}, S_{BRh1p}, S_{Alp} and S_{Aln} are closed, providing a current path from the biasing voltage through the inductor to P_A. The final voltage on P_A is then determined by the proportion of half the current path's resonant period, the gates are held on for [10]. If the switches are opened before half the resonant period, the energy stored in the inductor will be freewheel to the supply by closing switches S_{BRl1p}, S_{BRh2p} and S_{BRh2n} and opening switches S_{BRh1n}, S_{BRh1p}, S_{Alp} and S_{Aln}. Once all the energy has been removed all the switches can be opened ready for the next biasing event.

Actuator B, P_B, could also have been biased by closing switches on the low side of P_B's half bridge, so that the P_A and P_B are operated in parallel. Alternatively P_B could be biased after P_A enabling different biasing phases to be applied. To generate a negative voltage across P_A, switches S_{BRl1n}, S_{BRl1p}, S_{Ahp} and S_{Ahn} are closed first, then any freewheel current is routed through switches S_{BRl2p}, S_{BRh1p} and S_{BRh1n}. Therefore a single inductor can be used to bias multiple actuators with different biases and phases.

The number of piezoelectric actuators that can be driven in this way is limited by the electrical current path resonant frequency as this must be short enough that all switching events have been completed before the half the mechanical resonance period of the actuators. However through careful selection of the inductor, switches and piezoelectric actuators, the electrical resonance frequency can easily be in excess of 10 kHz.

Further utilisation of the inductor is also possible as the inductor can be reconfigured to form a boost converter circuit from a low voltage battery to generate the high voltage bias (Figure 3). In between biasing the piezoelectric actuators, switches S_{Battp} and S_{Battn} can be closed, then the inductor with switches S_{BRh2n}, S_{BRh2p}, S_{BRl2p} and S_{BRl2n} can be used to form a synchronous boost converter thus eliminating the need for another inductor.

3.2. SIHT verification

The functionality of the SIHT was verified fully in PSpice. The simulation used 400 nF for the piezoelectric actuator capacitance, 100 μH inductor, 1.5 Ω on-state resistance for the switches and 7.2 V battery. Two piezoelectric actuators were biased with different gate conduction times to show different final bias voltages. The biasing of each actuator was also performed out of phase. Figure 4 shows the results of the simulation which verify the functionality of SIHT. The voltage across the two actuators are shown in red and green and the bias capacitor voltage is blue. Note that the biasing capacitor was assumed to have been pre-charged and that both actuators initially started with zero bias, hence a few cycles are taken to reach steady state.

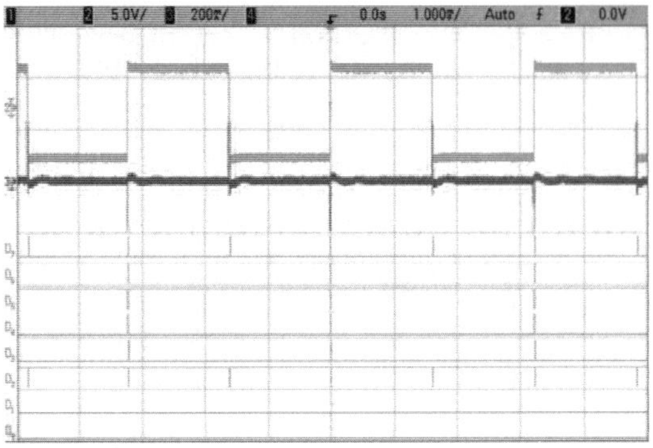

Figure 5. Measured piezoelectric voltage (green), inductor current (purple) and switch timing (blue) when driven by the SIHT.

A low voltage implementation of the piezoelectric actuator biasing was implemented using high voltage p-type and n-type MOSFETs for the switches (BSP317P and BSP126), a 100 μH inductor (VLP8040T-101M), a MSP430FR1433 microprocessor, a 350/025/0.70SA stripe actuator and lower power gate drive circuits based on [11]. Figure 5 shows the measured results where the green line is the voltage across the actuator, the purple line is the current through the inductor, and the blue lines show various gates signals used to trigger the switches.

4. Conclusion

A new circuit topology capable of both driving multiple piezoelectric actuators and stepping up a voltage through a single inductor has been presented. This new topology will enable a large weight saving in small flapping UAVs over previous piezoelectric actuator driving techniques. The topology has been fully simulated in PSpice and a low voltage bench demonstration of the actuators being driven has been presented. Further work is required to experimentally demonstrate the technique for multiple piezoelectric actuators with a high bias voltage being generated, before a custom integrated chip should be developed to reduce the size and weight of the complete power electronics.

References

[1] Balint C N and Dickinson M H 2004 *Journal of Experimental Biology* **207** 3813–3838 ISSN 0022-0949

[2] Steltz E, Avadhanula S and Fearing R S 2007 High lift force with 275 hz wing beat in mfi *2007 IEEE/RSJ Intelligent Robots and Systems* pp 3987–3992 ISSN 2153-0858

[3] Wood R J 2007 Design, fabrication, and analysis of a 3dof, 3cm flapping-wing mav *IROS*

[4] Fearing R S and Wood R J 2009 Challenges for 100 milligram flapping flight *Flying Insects and Robots* (Springer) pp 219–229

[5] Hyperion 2010 *Hyperion G3 LiPo - Specifications*

[6] Karpelson M, Wei G Y and Wood R J 2009 Milligram-scale high-voltage power electronics for piezoelectric microrobots *Robotics and Automation, 2009. ICRA '09.* pp 2217–2224 ISSN 1050-4729

[7] Texas Instruments 2014 *DRV8662 Piezo Haptic Driver with Integrated Boost Converter*

[8] Maxim Integrated 2002 *Small, High-Voltage Boost Converters* app note 1199

[9] APC International 2014 *Stripe Actuators APC 40-2040 Datasheet*

[10] Elliott A D T and Mitcheson P D 2012 Implementation of a single supply pre-biasing circuit for piezoelectric energy harvesters *Eurosensors XXVI* (Krakow, Poland)

[11] Elliott A D T 2015 *Power electronic interfaces for piezoelectric energy harvesters* Ph.D. thesis Imperial College London

Electrostatic vibrational energy harvester with ionic liquid and potassium-ion-electret

C Sano[1], H Mitsuya[1, 2], K Ishibashi[2], S Ono[3], K Miwa[3], M Ataka[1], G Hashiguchi[4], H Toshiyoshi[1] and H Fujita[1]

[1] Institute of Industrial Science, The University of Tokyo, Meguro, Tokyo, 153-8505, Japan
[2] Saginomiya Seisakusho, Inc., Sayama, Saitama, 350-1395, Japan
[3] Central Research Institute of Electric Power Industry, Yokosuka, Kanagawa, 240-0196, Japan
[4] Shizuoka University, Hamamatsu, Shizuoka, 432-8561, Japan

E-mail : chiko727@iis.u-tokyo.ac.jp

Abstract. This paper introduces an electrostatic vibrational energy harvester with an effective electrode distance of ~1 μm by utilizing the 1-nm-thick insulating gap that an ionic liquid forms, followed by a proof-of-concept with an AC current of 100 nA (peak-to-peak) and a voltage of 100 mV (peak-to-peak) generated from our prototype.

1. Introduction

Electrostatic vibrational energy harvesters attract increasing interest for their potential in miniaturization and low frequency operation towards application to IoT sensor nodes. MEMS technology and favorable scaling properties enable the miniaturization of the device, while uncoupled design freedom between the energy generating region and the mechanical suspension allows low frequency operation. The remaining issue is creating and maintaining an extremely narrow gap between moving and stationary electrodes to enlarge output power.

In this research, we have developed an energy harvester featuring ionic liquid between the electrodes. Ionic liquids are known to automatically form an extremely thin insulating gap called an electrical double layer, on the surface of charged electrodes. This thin gap may ultimately realize an effective electrode distance of ~1 nm without any specific fabrication, leading to a mechanically robust harvester with large output power.

2. Concept

2.1. Ionic liquid

An ionic liquid is a fluid salt composed only of organic cations and anions, around room temperature. Ionic liquids have unique chemical and physical properties, featuring (1) thermal and chemical stability, (2) negligible volatility, and (3) high ionic conductivity [1]. In this work, we focus on the electrochemical properties of ionic liquids, specifically the interactions that occur at the interface of ionic liquids and charged electrodes. A potential window is a voltage range within which cations or anions do not get reduced or oxidized, or in electrical terms, do not transfer charges to or from the

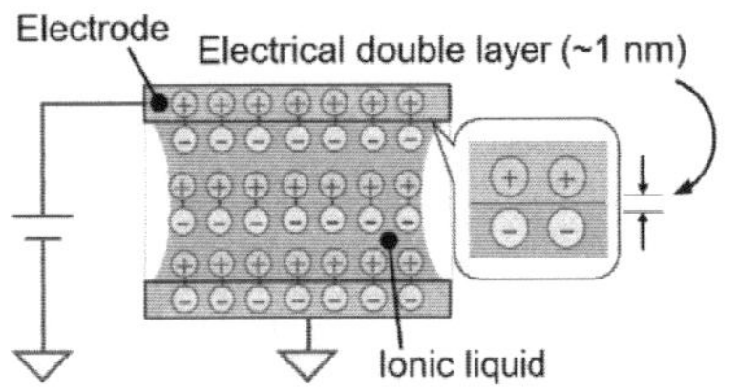

Figure 1. Schematic of electrical double layer formed on electrode surface.

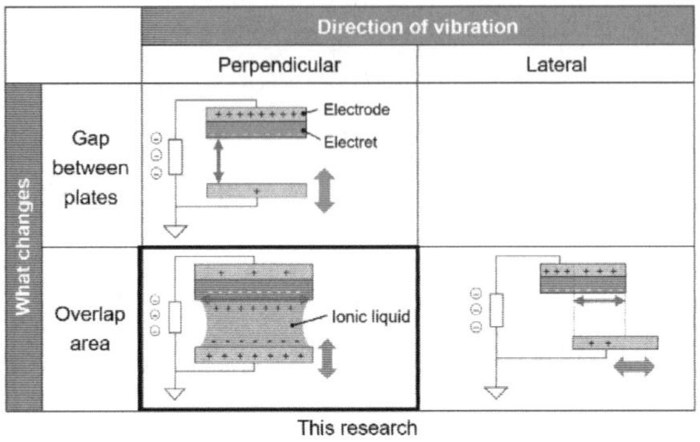

Figure 2. Comparison chart of electrostatic vibrational energy harvesters.

electrodes. Within this voltage range, electronic charges at the surface of the electrode attract counterpart ions in the ionic liquid (**figure 1**). These ions are not firmly anchored to the surface of the ionic liquid, but distributed near the surface keeping a gap of about 1 nm from the electrode surface. This structure, called an electrical double layer, provides very large capacitance of ~10 μF/cm^2 and high stored charge density [2].

2.2. Electrostatic vibrational energy harvester with ionic liquid

Conventional electrostatic vibrational energy harvesters are composed of a pair of electrodes, one of which is an electret with immobilized charge. They can be categorized into two main types : perpendicular or lateral displacement harvesters (**figure 2**). Perpendicular displacement harvesters feature a changing gap between the parallel electrodes, with the electrodes moving perpendicularly with respect to each other. Lateral displacement harvesters, however, feature overlapping surface area changes, with the electrodes moving laterally with respect to each other. In either case, charges flow through the circuit in order to compensate for the change in the capacitance between the electrodes. Another point in common is that the current flow can be increased by decreasing the width of the electrode gap during fabrication. However, there has been a limit in MEMS technology to fabricate increasingly narrow gaps while maintaining separation.

Therefore, in our device, ionic liquid is inserted between the electrodes (**figure 3**). Utilizing the overlapping surface area change with the electrodes moving in the perpendicular direction represents a novel method to generate a capacitance change for electrostatic vibrational energy harvesters (**figure 2**). The electrical double layer that the ionic liquid automatically forms leads to an effective electrode distance of ~1 nm, which is mechanically robust since the liquid is continuously touching the electrode. Furthermore, this thin insulating gap provides a large capacitance of ~10 μF/cm^2 which enlarges the output power in respect to the capacitance change.

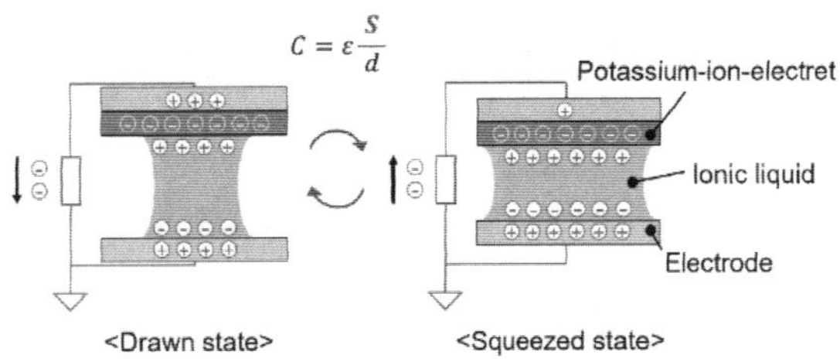

Figure 3. Schematic of our energy harvesting concept. Increase and decrease changes in contact area leads to appearance of extra holes on both electrodes.

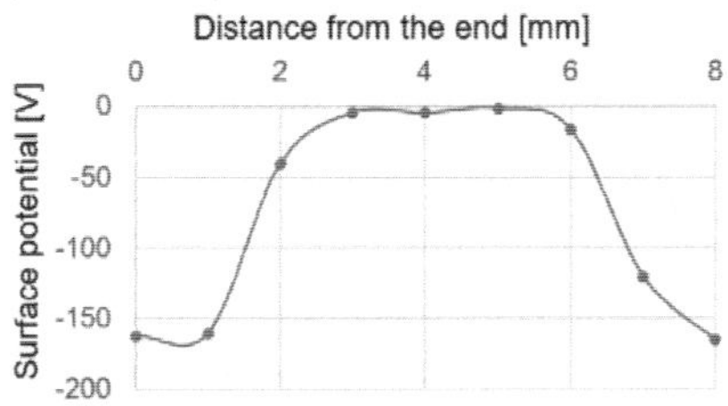

Figure 4. Measurement of SiO_2 electret surface potential after IL was removed from surface. IL was positioned around 2 mm ~ 6 mm, where charged voltage of ~-160 V diminished to ~0 V.

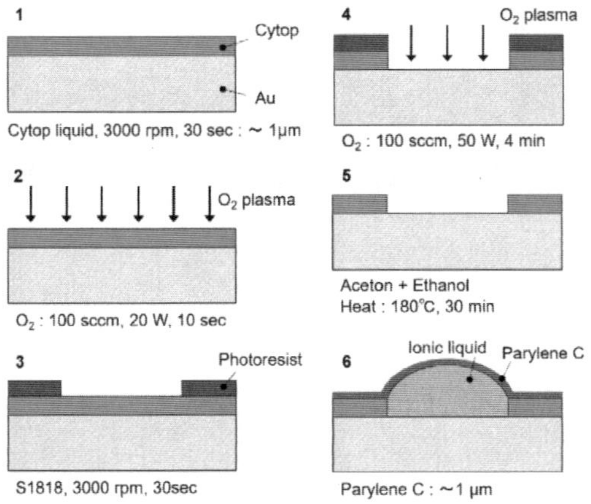

Figure 5. Process flowchart of parylene covered ionic liquid.

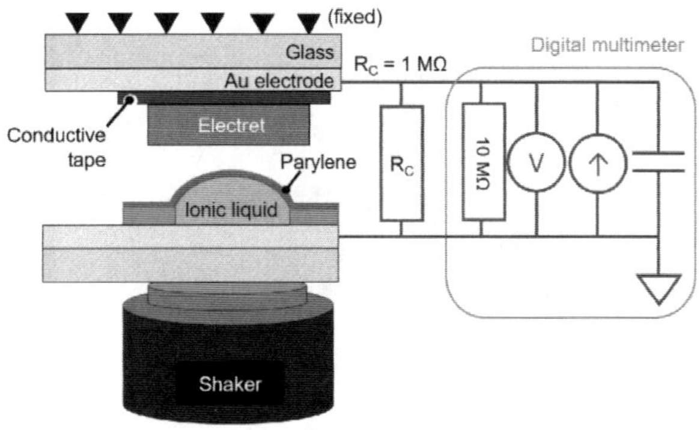

Figure 6. Schematic of measurement setup.

3. Method

3.1. Components

3.1.1. Potassium-ion-electret. In order to initiate the formation of the electrical double layer within the ionic liquid, a potassium-ion-electret is used to induce fixed charges. The fabrication of a potassium-ion-electret was developed by Hashiguchi et al [3]. First, through wet oxidization, K^+ is doped in the ~1-μm-thick SiO_2 layer formed on top of a silicon wafer. After a bias voltage is applied under high temperature (500-600℃) across the oxide layer, the redistribution of SiO^- and positive charges converts the material into a permanent electret with a surface voltage of ~-200 V. A SAM layer of HMDS covers the electret in order to prevent the gradual decay of the electret due to water vapor.

3.1.2. Protective layer covered ionic liquid. Initially, we placed the ionic liquid directly on the electret surface. The charges, however, disappeared in the area where the ionic liquid touched the surface (**figure 4**). Hence, a layer of parylene is deposited on the ionic liquid for protection (**figure 5**). To maintain the hemispherical shape of the ionic liquid during the deposition, a mold is made with a 1-μm-thick hydrophobic Cytop layer. Subsequently, a 1-μm-thick layer of parylene C is deposited on the top of the ionic liquid droplet. The diameter of the bottom area of the droplet is designed to be $\varphi = 5$ mm.

3.2. Experimental setup.

After the components of our energy harvesting structure is assembled, vibration is applied to examine the power generation. While the upper electrode with potassium-ion-electret is fixed, the lower electrode with parylene covered ionic liquid is actuated by a frequency tuneable shaker (**figure 6**). The output current and output voltage is measured using a Digital multimeter (Keysight, 34410A) with a load resistance of 1 MΩ.

4. Experimental result and discussion

Potassium-ion-electret and parylene covered ionic liquid was fabricated according to the procedures mentioned in 3.1.1 and 3.1.2. When the amount of ionic liquid was 10 μL, the height of the droplet became 1.5 mm. An Au electrode with electret attached was fixed to a support, while an Au electrode with parylene covered ionic liquid was fixed to a shaker which provides the vibration (**Figure 7**).

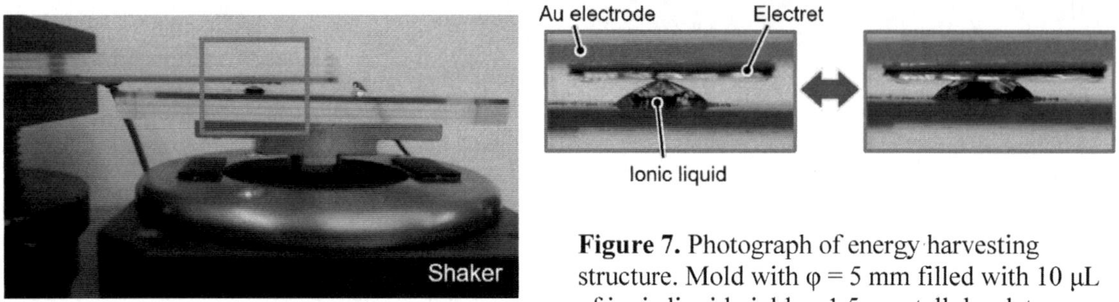

Figure 7. Photograph of energy harvesting structure. Mold with φ = 5 mm filled with 10 μL of ionic liquid yields a 1.5 mm tall droplet.

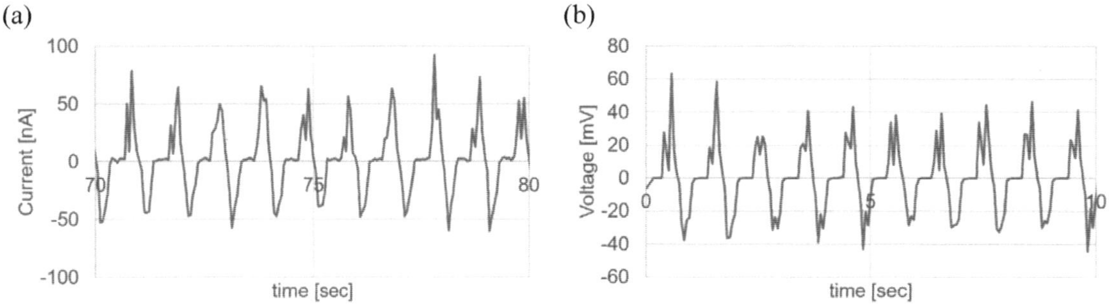

Figure 8. (a) Measurement result of AC current generated from vibration of f = 1 Hz.
(b) Measurement result of voltage generated from vibration of f = 1 Hz.

As a result, we successfully generated an AC current of 100 nA (peak-to-peak) and a voltage of 100 mV (peak-to-peak) from gently squeezing and relaxing the droplet with the electrode driven by the shaker at 1 Hz (**Figure 8**). This result verifies possible application of ionic liquid to electrostatic energy harvesting. Also, this result suggests that the SiO_2 electret kept its charge.

5. Conclusion
We introduced a novel vibrational energy harvester with an ionic liquid droplet inserted between the electrodes, followed by a proof-of-concept measurement with a calculated output power of 10 nW. We are now conducting precise measurements with the aim of device miniaturization. In addition, we are working on the improvement of the protective layer between the ionic liquid and the electret to fully utilize the 1-nm-thick electrical double layer.

6. Acknowledgements
This work was supported by CREST, JST (15664859) and the JSPS Core-to-Core Program. Also, we would like to thank Dr. Atsushi Takei for the support in parylene covered liquid.

References
[1] M Armand, F Endres, D R MacFarlane, H Ohno and B Scrosati 2009 *Nat. Mater.* **8** 621-29
[2] S Ono, S Seki, R Hirahara, Y Tominari and J Takeya 2009 *Appl. Phys. Lett.* **92** 103313
[3] T Sugiyama, M Aoyama, Y Shibata, M Suzuki, T Konno, M Ataka, H Fujita and G Hashiguchi 2011 *Appl. Phys. Expr.* **4** 114103

Study of Triboelectric Micromechanism for Three Dimensional Energy Harvesting

Lokesh Dhakar[1,2], Francis Eng Hock Tay[2,3] and Chengkuo Lee[1,2]*

[1] Department of Electrical and Computer Engineering, National University of Singapore, 4 Engineering Drive 3, Singapore 1175762
[2] NUS Graduate School for Integrative Sciences and Engineering, Centre for Life Sciences (CeLS), 28 Medical Drive, Singapore 117456
[3] Department of Mechanical Engineering, National University of Singapore, 9 Engineering Drive 1, Singapore 117575
*E-mail: elelc@nus.edu.sg

Abstract. Capturing mechanical energy from the surroundings is a potential strategy to power different sensor nodes and devices. In past few years, Triboelectric effect has been recently utilized in energy harvesting devices as a method to convert mechanical energy into electrical energy. One of the major limitations of the triboelectric mechanism based devices is that they are only responsive to stimulation in a single direction. This limitation hinders the application of triboelectric mechanism devices in various practical situations. In this paper, we present a novel design for triboelectric mechanism based devices which is sensitive to excitation in multiple directions.

1. Introduction
With the growing energy needs and energy requirements, generating green and sustainable energy is an important challenge. To power sensors and various devices, harvesting mechanical energy from various sources in environment is a potential solution. Recently, triboelectric mechanism has become a popular approach for mechanical energy harvesting owing to its high power output, easy fabrication and wide choice of materials [1, 2]. It has also been utilized to realize sensors for different applications including vibration sensing [3], chemical [4] and tactile sensing [5]. One of the major limitations of the triboelectric mechanism based devices is that they are only responsive to stimulus in a single direction. This limitation hinders the application of triboelectric mechanism devices in various practical situations. To improve the utility of triboelectric based energy harvesting devices and sensors, it is important that these devices they are responsive to the mechanical stimulations. This paper presents a novel device design for triboelectric mechanism based devices which are sensitive to mechanical stimulations in three dimensions. This design will enable the triboelectric mechanism based devices to harvest energy from multiple directions and hence increasing their applicability in different practical scenarios.

2. Device design
2.1. Device configuration
For three dimensional (3D) energy harvesting, the device should be able to generate electrical output for both in-plane and out-of-plane mechanical stimulations. For 3D energy harvesting using

triboelectric effect, a device design is proposed and fabricated as shown in figure 1. The triboelectric energy harvesting device comprises of two parts: top and bottom. The bottom part comprises of Si substrate etched with through holes. The top part consists of PDMS micropillars which are then assembled to be inserted through the holes in Si substrate. PDMS layer and the copper film coated Si substrate comes in contact with each other when mechanical force is applied on the device.

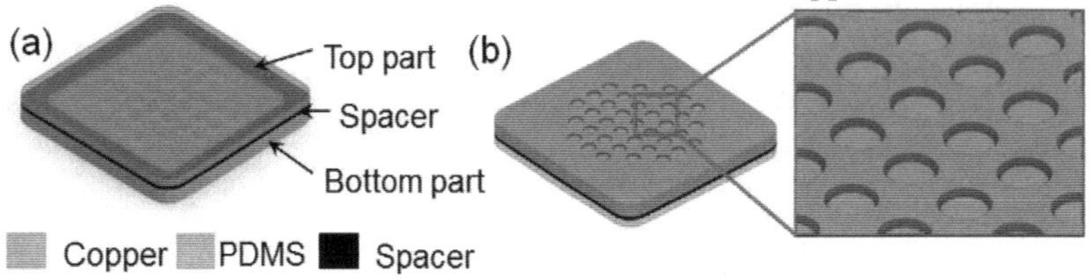

Figure 1. (a) Schematic of the device. (b) View of the device with pillars inserted inside holes.

2.2. Working principle

The device works on the principle of generating surface charges using triboelectric effect as the triboelectric layers come in contact with each other. The schematic for the working mechanism 3D energy harvester is shown in figure 2. The surface area of the PDMS layer can be divided in two parts: flat surface area and pillar surface area as shown in figure 2a. The device design results into difference in interaction between PDMS layer and Cu film, for different mechanical stimulus. When normal force is applied on the top elastic PDMS film, it presses the film towards the Cu film in the normal direction leading to the deformation of the PDMS layer. It comes in contact with Cu film in both flat area and pillar area resulting in generation of surface charge in the contacted regions as shown in figure 2b. When the shear force is applied on the PDMS film, the film undergoes shear deformation. PDMS pillars come in contact with inner sidewalls of the etched holes. Due to the contact area between PDMS pillars and Cu film on inner sidewalls, triboelectric surface charges are generated as shown in Figure 2c. The aforementioned mechanism enables the device to harvest energy from both in-plane and out-of-plane mechanical stimulation.

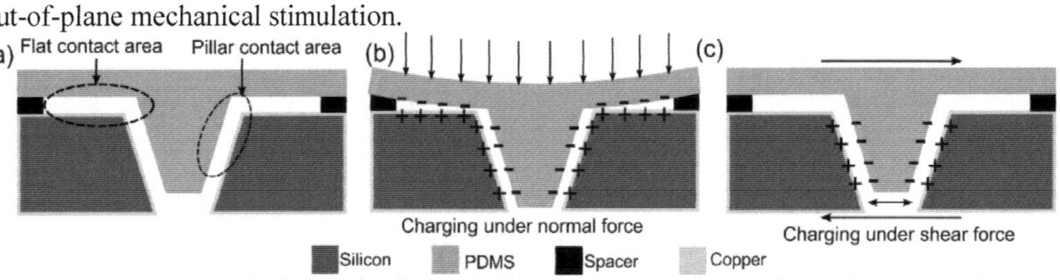

Figure 2: (a) Schematic depicting the flat and pillar contact area. (b) and (c) Shows the triboelectric charging when normal and shear forces are applied, respectively.

3. Fabrication

3.1. Bottom part fabrication

For the bottom part fabrication Si wafer was deposited with 2 μm thick silicon dioxide (SiO2) using PECVD (figure 3a). The wafer was then coated with photoresist and patterned using lithography followed by etching of SiO_2 layer using reactive ion etching (RIE). Thereafter, deep reactive ion etching (DRIE) is used to etch away Si. As the device is DRIE etched, the sidewall angle is a tilted at an angle. The photoresist layer is then stripped and SiO_2 is etched away. To create through holes, the samples are background up to 300μm. Then the samples are coated with 100nm thick Cu film from both the sides of the samples as shown in figure 3a. The Cu layer on Si serves as a triboelectric layer to generate surface charges and also an electrode to power the external load connected to device.

3.2. Top part fabrication

The top part comprises of the PDMS micropillars. For PDMS micropillar fabrication, the Si samples with etched through holes are attached to a glass slide (figure 3c). These Si chips attached to a glass slide were used as a mold to fabricate PDMS micropillar structures as shown in figure 3d. The shape of the PDMS pillar replicates the tilted the sidewalls of the Si holes and thus results into circular conical frustum shape as shown in the enlarged view in figure 3d. For the assembly of device, 100μm spacers are cut out to create gap between the top and bottom part. The spacer between the top and bottom part and aligned using manually controlled multi-axis stage to create a gap.

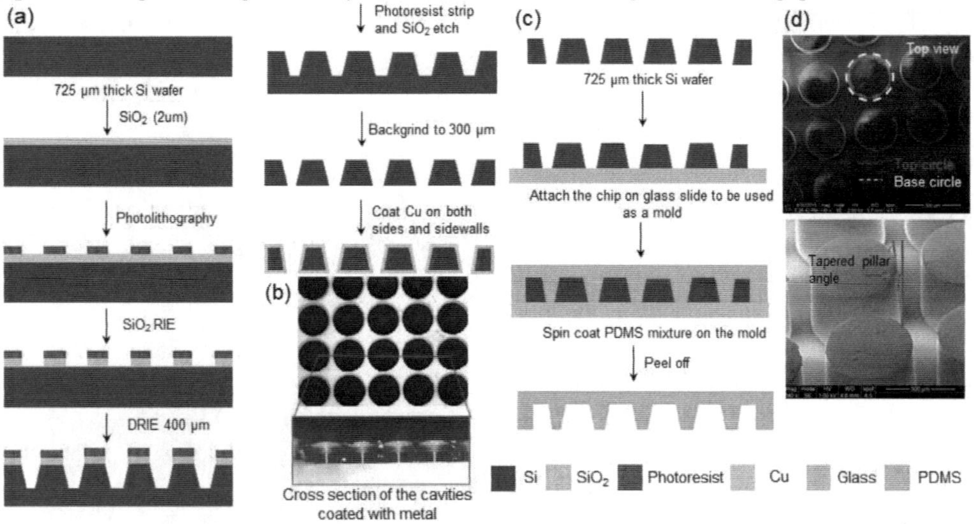

Figure 3: (a) Fabrication of bottom part. (b) Optical image of the bottom part with the cross sectional image. (c)Fabrication of top part. (d) Optical image of the pillars fabricated.

4. Results and discussion

4.1. Output under normal and shear force

As discussed in the previous sections, the device design is responsive to the normal and shear force in different ways. The device was tested for normal and sliding forces applied using human finger. The results due to press and release (normal force) is shown in figure 4a. As the device is pressed, PDMS layer and Cu film both come in contact in the flat contact area and pillar contact area resulting in generation of surface charges due to triboelectric effect. The PDMS surface gets negatively charged whereas the Cu film gets positively charged. As the normal force is released, the PDMS layer starts separating from the Cu film and electric potential at the electrode starts increasing. Thereafter, the PDMS layer again approaches the Cu film leading to decrease in electric potential. The peak voltage measured due to normal force was measured to be 16V. The working mechanism in the sliding mode is similar as the normal mode with the difference that only pillar surface area comes in contact with Cu coated sidewalls due to lateral relative motion. The peak voltage generated using in the sliding mode was measured to be nearly 2V.

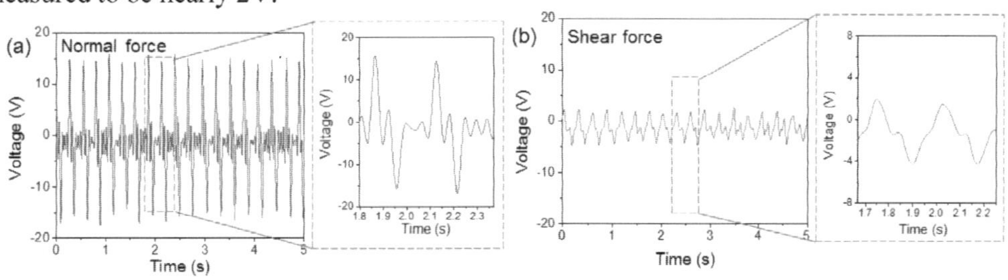

Figure 4: Voltage characteristic of the device under (a) normal and (b) shear force applied.

4.2. Energy harvesting

The device was tested for energy harvesting from force in different directions. One of the applications of the small sized device can be in harvesting energy from typing motion on the keyboards or pressing of switches. The peak voltage and current generated due to the normal force application was measured to be 90V and 0.57μA respectively, by typing motion of the finger as shown in figure 5a. Through the sliding motion, the peak voltage and current generated were measured to be 5V and 0.08μA, respectively. The generated energy was stored in a 1μF capacitor using a full wave rectifier circuit as shown in figure 5e. The capacitor was charged up to 4V using the typing motion of finger for about 40 sec. The energy was then used to light up a red commercial LED as shown in figure 5f. The power output of the devices is limited due to the small size (1cm x 1.2cm of the complete device) but proposed device design can also be used for scaled up devices to improve the power output.

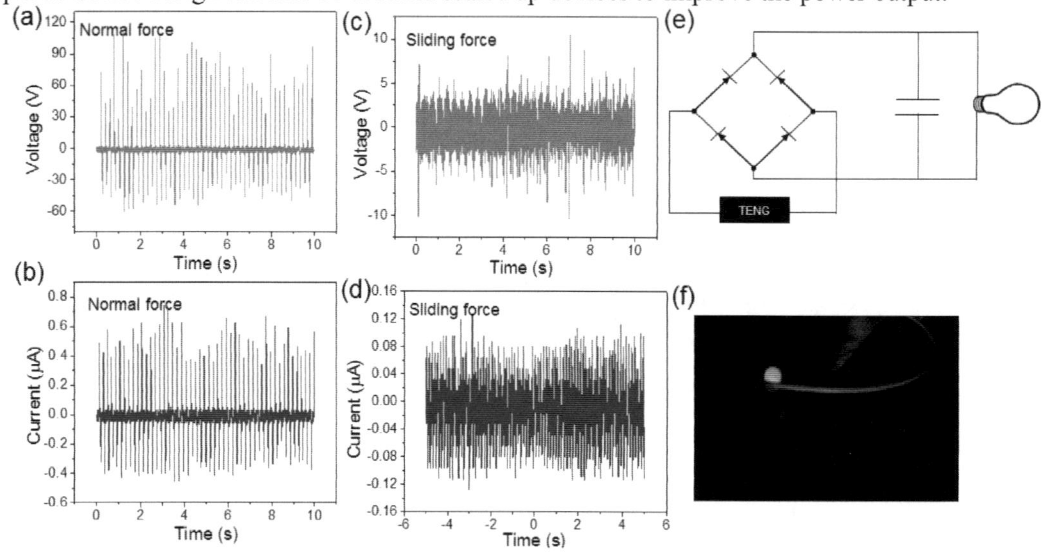

Figure 5: (a) Voltage and (b) current generated from the normal force applied on the device. (c) Voltage and (d) current generated due to sliding force applied on the device. (e) Full wave rectifier circuit used to store energy in a 1 μF capacitor. (f) A lit up LED using the harvested energy.

5. Conclusion

This paper presents a novel mechanism based device was designed and fabricated for three dimensional energy harvesting. The mechanism can be used to differentiate between the mechanical interactions in the normal and sliding directions. The device was also demonstrated to harvest mechanical energy from mechanical force in multiple directions. The device was demonstrated to generate a maximum voltage and current of 90V and 0.57μA respectively.

Acknowledgements

This work is supported NRF2011 NRF-CRP001-057 Program 'Self-powered body sensor for disease management and prevention-orientated healthcare' (R-263-000-A27-281) from the National Research Foundation (NRF), Singapore.

References

[1] F.-R. Fan, L. Lin, G. Zhu, W. Wu, R. Zhang, and Z. L. Wang 2012 *Nano letters* **12** 3109-3114.
[2] F.-R. Fan, Z.-Q. Tian, and Z. L. Wang 2012 *Nano Energy* **1** 328-334.
[3] J. Chen, G. Zhu, W. Yang, Q. Jing, P. Bai, Y. Yang, T.C. Hou and Z.L. Wang 2013 *Advanced Materials* **25** 6094-6099.
[4] Z. L. Wang 2013 *ACS Nano* **7** 9533-9557.
[5] L. Lin, Y. Xie, S. Wang, W. Wu, S. Niu, X. Wen and Z.L. Wang 2013 *ACS Nano* **7** 8266-8274.

Self-Powered energy harvester strain sensing device for structural health monitoring

A. Álvarez[1], M. Bafleur[2], J-M. Dilhac[2], J. Colomer[1], D.Dragomirescu[2], J.Lopez[1], M.Zhu[3] and P. Miribel[1]

[1] Engineering Department: Electronics, Physics Faculty, Martí i Franqués, 1, 08028 Barcelona, Spain
[2] LAAS-CNRS, Université de Toulouse, CNRS, INSA, Toulouse, France
[3] College of Engineering, Mathematics and Physical Sciences, Harrison Building, University of Exeter, Exeter, EX4, 4QF, UK

aalvarez@el.ub.edu

Abstract. This paper presents an envisaged autonomous strain sensor device, which is dedicated to structural health monitoring applications. The paper introduces the ASIC approach that replaces the discrete approach of some of the main modules.

1. Introduction

The proposed device is developed in the frame of the SMARTER (Smart Multifunctional Architecture & Technology for Energy-Aware wireless sensor) Project [1].

The device is self-powered, based on a piezoelectric transducer as an energy-harvesting source, enabling a battery-free solution. The system would be capable to sense the mechanical strain, which is an interesting topic in structural health monitoring applications, particularly in the aerospace field, which is the present field of application, defining the frequencies available for the transducer, harvesting the energy provided by the in-flight vibration of the aircraft's wing [2].

The nice approach of the device, which defines its Multifunctional concept, is that the same transducer used to harvest energy also allows extracting the mechanical strain. On the one hand, based on a piezoelectric transducer, the mechanical vibrations are harvested to generate electrical energy, in order to start-up and supply the system, but on the other hand, the system is capable to extract the open voltage (VOC) of the piezoelectric harvester, which has a relationship with the mechanical strain [3]. Then, based on the open circuit voltage concept to adapt the maximum power transferred condition between the piezoelectric mechanical to electrical energy [4,5,6], a detection of the open voltage of the piezoelectric transducer is derived. Based on these approaches, initial fully discrete electronic solutions have been implemented [7][8] as a proof of concept.

The system is based on four main modules, figure 1: a) the energy transducer as a multifunctional device (MFC) based on a piezoelectric generator (PZT); b) the Power Management, Control and Sensing Circuitry Unit, to be implemented in an ASIC solution (ASIC1); c) the Dynamic Storage Module to store the generated energy (SCap), and d) the Communication (RF) and Data Processing Unit (ASIC2). Full electronics that define the Smarter Device would be placed on a flexible PCB.

In this paper we focus the work regarding the design of ASIC1. This implementation has the role to manage the power harvested, the measurements and the management of the other modules.

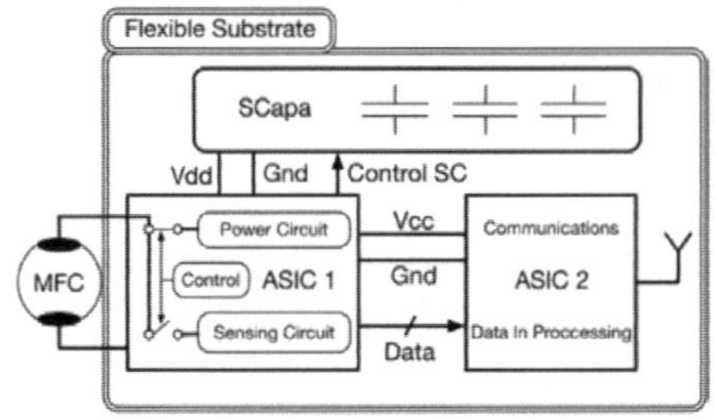

Figure 1. Main blocs for the Smarter Device.

In section 2 is introduced the ASIC1 approach. The initial discrete solutions [7][8] are translated to a microelectronic one, for a more compact solution.

Finally in section 3 are derived some conclusions and future perspectives.

2. Integrated approach

Thanks to the discrete approaches [7,8], the contemplate architecture is validated and with the gained experience we proceeded with the ASIC integration. A first ASIC is going to be implemented based on an AMS 0.35 technology. In figure 2 is depicted the ASIC1 at the functional block level and its layout design in figure 3, in an AMS 0.35μm technology, that will be placed in a QFN32 package.

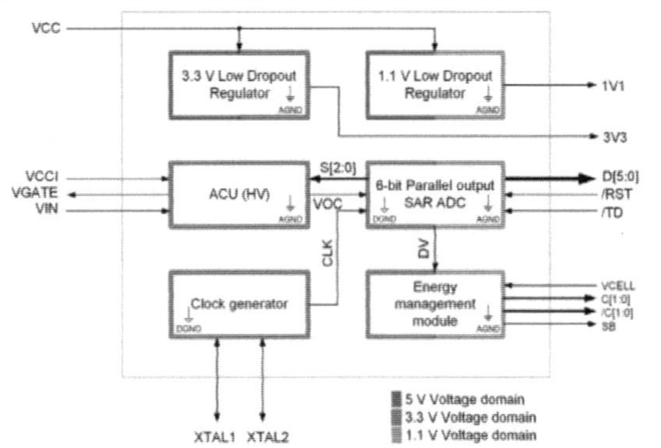

Figure 2. ASIC1's block diagram conception.

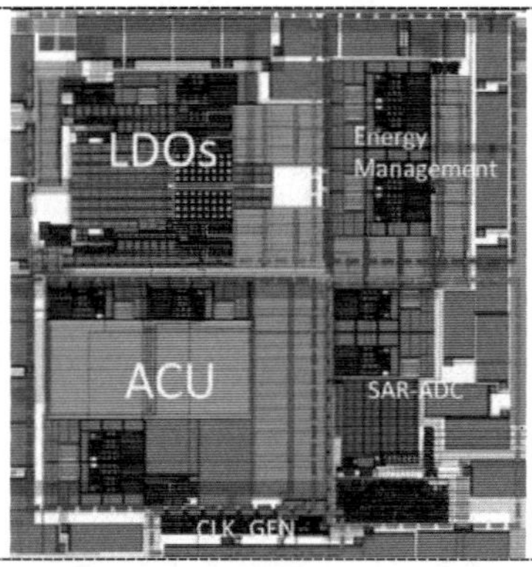

Figure 3. ASIC1 layout implementation.

To interface ASIC1 and ASIC2, figure 1,a communication protocol has been implemented, working with 8 bits: Two bits of control, and 6-bits of data. This protocol is implemented in [8] and validated. In t1 data is available on ASIC1. In t2 ASIC activates Start Bit (SB) when data is ready of transmitting. In ta SB is activated during all operation. In t3 ASIC2 is ready to operate. In tb, the bit Transmission Done (TD) is activated during UR-UWB transmission. In t4 TD is grounded when the transmission is completed. In t5 SB is turned off after communication is done. Finally, in t6 the system waits till new data is available on ASIC1.

The system is focused into the extraction of the strain through VOC, and the validation of the architecture that will combine the electronics envisaged in ASIC1 with those of ASIC2. The wireless solution implemented in [7] will be replaced by ASIC2 in the future Smarter prototype, and ASIC1 has a Data Processing Unit, based on a 6-bits SAR-ADC converter, which prepares the data to be sent by the RF module (ASIC2) replacing the TI Microcontroller in [8]. The strain is measured and related with the open circuit voltage, and the measurement is outputted by 6- parallel bit ADC, which translates the measurement to the Communication. The main blocs are defined by the Power Circuit Unit, which combines linear regulators and the Energy Management module. The system needs a 1.1V regulated voltage source and a 3.3V regulated voltage source. From the AC/DC rectifier two LDOs will generate these voltages. The 1.1V LDO is designed to bias ASIC2, and the 3.3V LDO to bias the rest of the electronics. Then, the Energy Management module has the role to control the Dynamic Storage Element or SCap [9]. Finally, the ACU module and the 6-bit SAR ADC Converter define the Control Circuit and Sensing Circuitry.

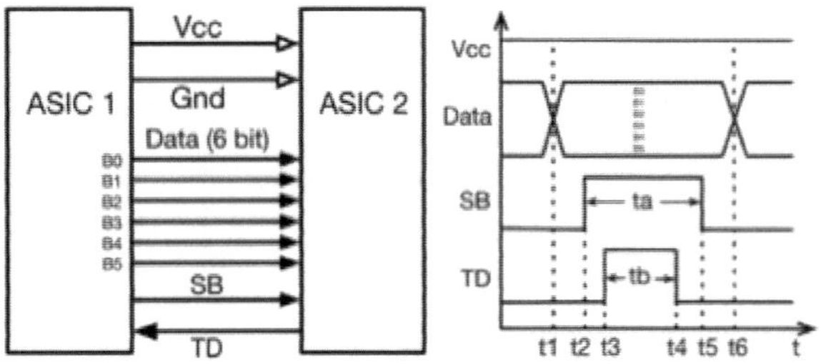

Figure 4. Communication protocol between ASICS.

The Power Circuit Unit is described more in detail. A specific low-power bandgap reference and current reference circuit has been designed; with a current consumption lower than 9μA, nominally and a nominal power consumption of 30μW. It has the capability to work from a lower voltage of 2V till 5V, the maximum operating voltage, with an output voltage of 1.23V with a Temperature Coefficient (TC) lower than 60ppm. The current reference is 105nA with a TC lower than 250ppm, in the temperature range of -40°C to 125 °C.

Then, based on the architectures presented in [5], the LDO circuits have been designed. The 3.3V LDO is designed to work with a current up to 1mA, and the 1.1V LDO with a maximum DC current of 10mA. The main parameters for the LDO are introduced. In the case of the 3.3V LDO the Linear Regulation is $68.7 \cdot 10^{-3}$ (V/V). In the case of the 1.1V LDO is $665 \cdot 10^{-9}$(V/V). The ability to maintain the output voltage with variations of the input voltage is reasonable for both designs. The measurement of the capability of the circuitry to keep the regulated output voltage to the desired value against changes of the load, that is, between light and high load conditions (in our case between few μA that define the quiescent operating point of the electronics and the load condition of 1mA for the 3.3V LDO and 10mA for the 1.1V LDO), is the load regulation, which is $81.6 \cdot 10^{-3}$ (V/A) in the case of the 3.3V LDO and $99 \cdot 3 \cdot 10^{-6}$ (V/A) for the 1.1V LDO regulator.

Full Monte Carlo analyses have been carried out for the bandgap reference circuit and LDOs. In the particular case of the Linear Dropout Regulators, it is the 1.1V LDO that has the greater limitation, with an accepted range between 1.05 and 1.18V, with 200 samples. The average regulated voltage is 1.12V with a standard deviation of 34,19 mV (Yield of 95%).

The ACU module, which is quite complex, and the Energy management module, are out of the scope of the present work. In figure 5 is depicted the caption of the simulation, at the extracted ASIC view level, of the performance of the ACU control stage though the GATE signal (VGATE) [7][8].

3. Conclusions

This paper presents the ASIC approach for the power management, control and sensing for the SMARTER project. Future full test and characterization will be carried out before its implementation in the final prototype on flexible substrate and reported in future publications.

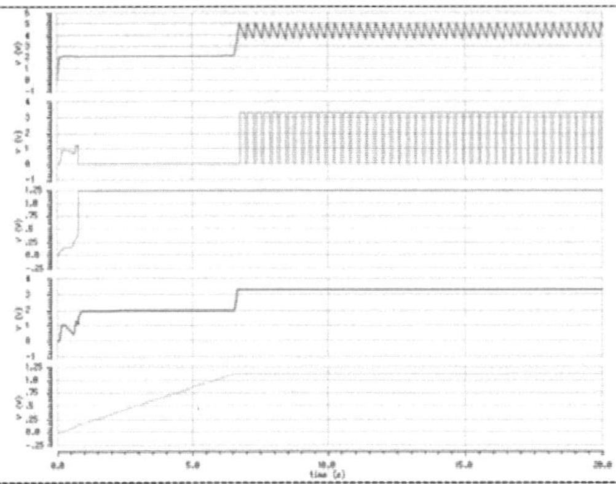

Figure 5. Caption of the transient simulation, from the ASIC1 extracted view, of the GATE control signal. Generation of the scenario operating phases. From the Top view to the bottom caption: (Top) Start-Up and harvest/measuring phases; VGATE signal that control the phases; Bandgap reference voltage; 3V3 LDO regulated voltage; (bottom) 1V1 LDO regulated voltage.

References

[1] Bafleur M and Dilhac J.M 2013 *Proc. IEEE International Conference on Green Computing and Communications and IEEE Internet of Things and IEEE Cyber, Physical and Social Computing* (*Beijing*) (Los Alamitos, California/Washignton/Tokio Conference Publishing Services IEEE Computer Society) p 1668-72

[2] Giuliano A Marsic V and Zhu M 2013 *Proc. IEEE International Conference on Green Computing and Communications and IEEE Internet of Things and IEEE Cyber, Physical and Social Computing* (*Beijing*) (Los Alamitos, California/Washignton/Tokio Conference Publishing Services IEEE Computer Society) p 681-684

[3] Chen X, Xu S, Yao N and Shi Y 2010 *Nanoletters* **10** 2133-37

[4] Ottman G.K, Hofmann H.F, and Lesieutre G.A 2003 *IEEE Tran. On Power Electronics* **18** 696-703.

[5] Colomer-Farrarons J, Miribel-Català P, Saiz-Vela A, Puig-Vidal M and Samitier J 2008 *IEEE Tran. On. Industrial Electronics* **55** 3249-57

[6] Lafuente Brinquis J 2008 Final Project Degree, (Barcelona: Electronics Department, University of Barcelona).

[7] Álvarez-Carulla A Colomer-Farrarons J López-Sánchez J Miribel-Català P 2015 *Proc. Conference on Design Circuits and Integrated Systems* (*Estoril*) (IEEE Conference Publications)

[8] Álvarez-Carulla A Colomer-Farrarons J López-Sánchez J Miribel-Català P 2016 *Proc. IEEE International Symposiun on Industrial Electronics* (*Santa Clara*) (IEEE Conference Publications)

[9] El Mahboubi F 2016 *Proc. Powermens 2016 (Paris). (To be published: Journal of Physics:Conferene Series. JPCS)*

Acknowledgements

The SMARTER project is supported by European Union under the ERA-Net funding scheme of the FP7 (CHIST-ERA), and in the case of the spaniard partner through MINECO, reference PCIN-2013-069.

Optimization of dielectric matrix for ZnO nanowire based nanogenerators

Santhosh Kannan, Mitesh Parmar, Ran Tao, Gustavo Ardila and Mireille Mouis

Univ. Grenoble Alpes, CNRS, IMEP-LaHC, F-38000 Grenoble, France

E-mail: miteshparmar17@gmail.com

Abstract. This paper reports the role of selection of suitable dielectric layer in nanogenerator (NG) structure and its influence on the output performance. The basic NG structure is a composite material integrating hydrothermally grown vertical piezoelectric zinc oxide (ZnO) nanowires (NWs) into a dielectric matrix. To accomplish this study, three materials - poly methyl methacrylate (PMMA), silicon nitride (Si_3N_4) and aluminium oxide (Al_2O_3) are selected, processed and used as matrix dielectric in NGs. Scanning electron microscopy (SEM) analysis shows the well-aligned NWs with a diameter of 200±50 nm and length of 3.5±0.3 µm. This was followed by dielectric material deposition as a matrix material. After fabricating NG devices, the output generated voltage under manual and automatic bending were recorded, observed and analyzed for the selection of the best dielectric material to obtain an optimum output. The maximum peak-to-peak open-circuit voltage output for PMMA, Si_3N_4 and Al_2O_3 under manual bending was recorded as approximately 880 mV, 1.2 V and 2.1 V respectively. These preliminary results confirm the predicted effect of using more rigid dielectrics as matrix material for the NGs. The generated voltage is increased by about 70% using Si_3N_4 or Al_2O_3, instead of a less rigid material as PMMA.

1. Introduction

Considering the present energy crisis, the scientific world is in search of alternate technological solutions to generate energy from the ambiance. Biomass, wind, solar, thermal and mechanical energies are the potential sources. Out of these, mechanical vibrations and impacts are omnipresent when compared to other sources. These mechanical inputs can be implemented for energy harvesting by typical approaches such as electromagnetic induction, electrostatic, magnetostrictive as well as piezoelectric effects [1]. The piezoelectric nanowires (NWs) based energy harvesters are widely researched considering their compatibility with the silicon fabrication technology and flexible substrates. The vertical integrated nanogenerator (VING) is one of the most common modes of harvesting energy using piezoelectric NWs [1, 2]. The basic VING structure consists of piezoelectric NWs inside dielectric matrix is shown in Fig. 1. Besides preventing the electrical leakage and short-circuit between the electrodes, the dielectric material influence the mechanical robustness and stability of the nanogenerator (NG). Moreover, Young's modulus of the dielectric material also affects the mechanical energy transfer and therefore influences overall device performance.

According to the basic VING structure in Fig.1, the dielectric medium can be categorized into two layers –matrix layer and top dielectric layer. The previous FEM analysis reported from our group described the use of polymethyl methacrylate (PMMA), silicon nitride (Si_3N_4) and aluminium oxide (Al_2O_3) as top or matrix dielectric materials in the NG fabrication [3–5]. According to these studies, the use Al_2O_3 and Si_3N_4 instead of PMMA improves the device performance under compression when used as top dielectric layer and under bending when used as whole matrix material. The present paper reports the experimental validation of these works by studying the role of suitable dielectric matrix in

NG and its influence on the output performance. To accomplish this, three dielectric materials – PMMA, Si_3N_4 and Al_2O_3 are selected, processed and used in NGs. The output performance is studied by applying strain through manual bending and automatic bending using an actuator controlled by a stepper motor.

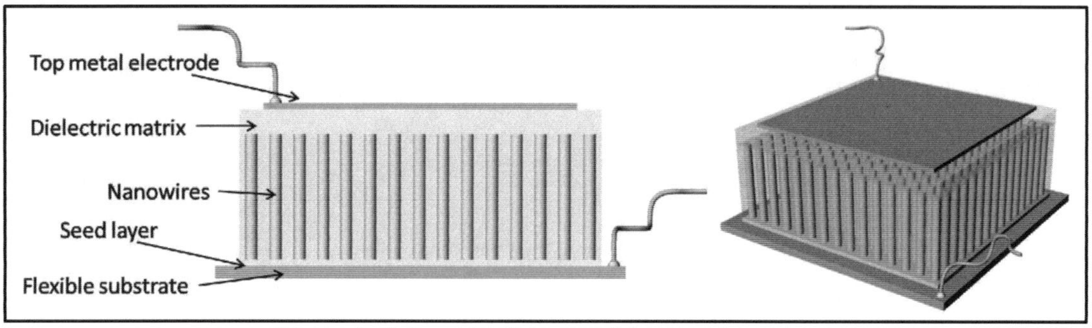

Figure 1. Schematic diagram of the nanogenerator

2. Device preparation

In the present work, piezoelectric ZnO NWs were grown using a solution growth method (Chemical Bath Deposition) on cleaned flexible stainless steel substrates with 40nm thick Atomic Layer Deposition (ALD) coated ZnO seed layer. The growth solution was comprised of a 1:1 ratio of zinc nitrate hexahydrate and hexamethylenetetramine (HMTA).The growth temperature was kept at 85 °C for the growth of 15 hours. Scanning electron microscopy (SEM) analysis shows the well-aligned ZnO nanowires grown with a diameter of 200±50 nm and length of 3.5±0.3 μm (Fig. 2(a)). This was followed by dielectric material deposition as a matrix material on three different samples. The PMMA matrix layer was deposited by spin coating consisting of a double coating of 495PMMA A2 and A6 resists using the parameters from the datasheet provided by MicroChem®. The resulting estimated thickness of the top layer is about 1 μm. The other dielectrics were deposited using sputtering for Si_3N_4 (250 nm) and ALD for Al_2O_3 (240 nm). Aluminium was deposited as a top electrode over the dielectric matrix by evaporation. The cross-sectional SEM images of these samples are shown in Fig 2(b). The active area (11 mm × 13 mm) was based on the surface area of the top electrode.

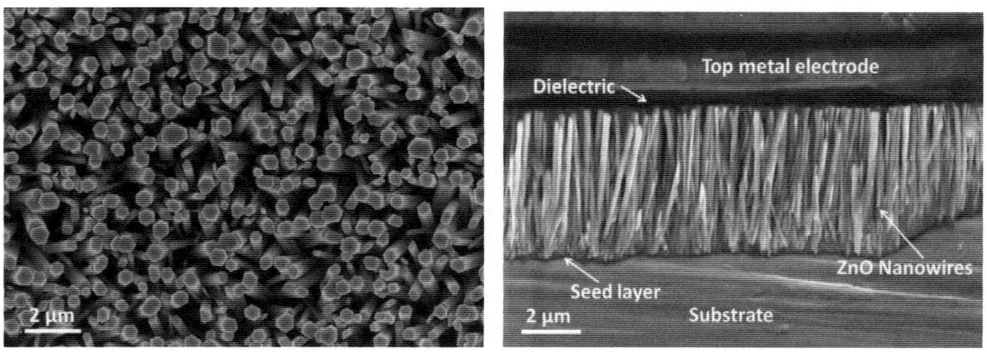

(a) (b)

Figure 2.(a) SEM image of ZnO nanowires before coating of dielectric material (b) Cross-sectional view of nanowires after dielectric (PMMA in this case) coating

3. Performance analysis

After fabricating NG devices, the performance analysis of piezoelectric NWs based NG devices were carried out by measuring open-circuit potential generation when subjected to strain. For this, the NG devices were tested under manual bending and automatic bending for minimum 3 different sets of readings. The data were recorded using digital oscilloscope (LeCroy WaveSurfer 424) and high impedance passive probe (LeCroy PP007-WS). The manual bending provides the first-hand working of the device test. However, it lacks the repeatability of the data as the bending amplitude varies every time. The automatic bending fulfils this drawback. Nevertheless, the amount of maximum strain applied to the device is limited.

3.1. Manual bending

Under manual bending tests, the peak-to-peak voltage (V_{p-p}) generation for PMMA, Si_3N_4 and Al_2O_3 were recorded as approximately 880 mV, 1.24 V and 2.07 V respectively (Fig. 3(a-c)). The bending frequency was measured to be in the range of 2–2.4 Hz. These preliminary results confirmed the predicted effect of using more rigid dielectrics as matrix material for the NGs. The generated voltage is increased by about 41% using Si_3N_4 and 135% Al_2O_3, instead of a less rigid material as PMMA.

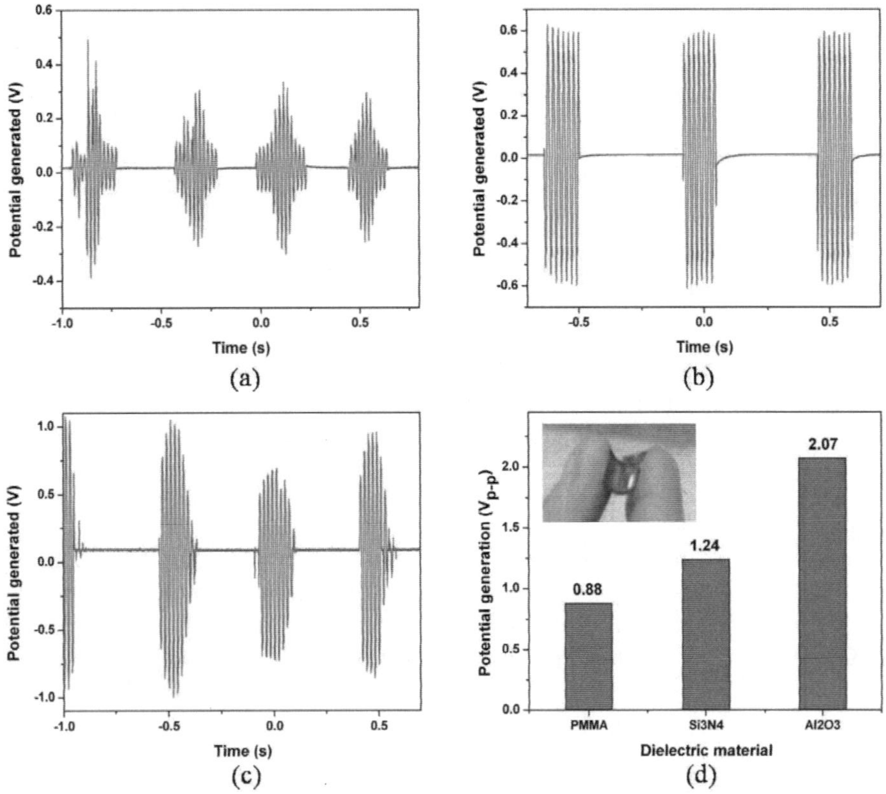

Figure 3. Generated voltage from the NGs where (a) PMMA (b) Si_3N_4 and (c) Al_2O_3 material was used as dielectric matrix, respectively. (d) Comparison of NGs performance (V_{p-p}) with different dielectric materials (inset photograph shows normal bending direction for the analysis)

3.2. Automatic bending

Although the manual results were promising, it lacked repeatability and quantification of the applied strain. Hence, a cantilever setup was made and the deflection of the cantilever was controlled using and Arduino-controlled stepper motor. The NG device was placed near the fixed-end where the strain in the cantilever is the maximum [6]. Here, the shaft deflected the cantilever by approximately 4 mm (0.4% strain) consistently to ensure the same strain in the NG. The deflection frequency was kept to be approximately 2–2.8 Hz. Si_3N_4 and Al_2O_3 based NG device generated a maximum V_{p-p} of 123 mV and 271 mV respectively (Fig. 4 (b - c). However, no output was observed in the PMMA based NG device at 4 mm cantilever deflection (Fig. 4(a)). The deflection of 7 mm (0.65% strain) generated V_{p-p} of 85 mV only (Fig. 4(d)). Al_2O_3 based NG device generated 220 % more compared to PMMA based NG with almost half the cantilever deflection.

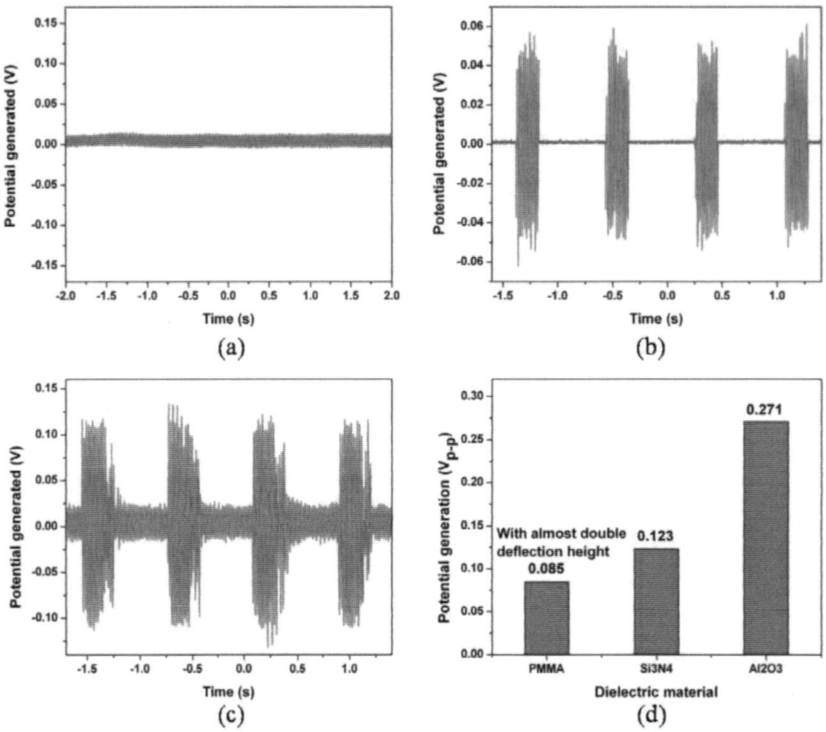

Figure 4. Generated voltage from the NGs under automatic bending where (a) PMMA (b) Si_3N_4 and (c) Al_2O_3material was used as a dielectric matrix, respectively (d) Comparison of NGs performance (V_{p-p}) with different dielectric materials

4. Conclusions

From the above study, it can be concluded that the dielectric matrix in the NG device plays significant role in the device performance as predicted by theoretical studies. The performance of the device have been studied under manual bending as well as automatic bending using an Arduino-controlled stepper motor setup. Out of the three materials tested in the present work, Al_2O_3 based NG generated 135% more potential compared with those based on PMMA. These results confirmed the predicted effect of using more rigid dielectrics as matrix material for the NGs. The detailed analysis is ongoing on these samples on several aspects: (I) to better quantify and to increase the controlled applied strain and (II) to understand the origin of the oscillation on the output potential. Further work will include the estimation of the total generated power using variable resistive loads.

References

[1] Wang Z L 2012 *Adv. Funct. Mater.* **24** 280-285.

[2] Xu S, Qin Y, Xu C, Wei Y, Yang R and Wang Z L 2010 *Nature nanotech.* **5** 366-373.

[3] Hinchet R, Lee S, Ardila G, Montès L, Mouis M and Wang Z L 2014 *Adv. Funct. Mater.* **24** 971-977.

[4] Tao R, Hinchet R, Ardila G and Mouis M 2013 *J. Phy.* **476** 012006.

[5] Tao R, Ardila G, Montes L and Mouis M 2015 *Nano energy* **14** 62-76.

[6] Zheng Y, Wu X, Parmar M and Lee D W 2014 *Rev. Sci. Instrum.* **85** 026101.

Acknowledgments

We would like to thank Paulo Oliveira for his valuable help on the testing set-up. This work has been performed with the help of the "Plateforme Technologique Amont" from Grenoble, with the financial support of the "Nanosciences aux limites de la Nanoélectronique" Fundation and CNRS Renatech network. This work is also supported by funding from the national Nano2017 project.

Air-spaced PDMS piezo-electret cantilevers for vibration energy harvesting

A Kachroudi[1,2,3], S Basrour[1,2], L Rufer[1,2] and F Jomni[3]

[1]Univ. Grenoble Alpes, TIMA Laboratory, F-38031 Grenoble, France.
[2]CNRS, TIMA Laboratory, F-38031 Grenoble, France.
[3]Université de Tunis El Manar, Faculté des Sciences de Tunis, Laboratoire Matériaux, organisation et propriétés (LMOP), 2092, Tunis, Tunisie.

E-mail: achraf.kachroudi@imag.fr

Abstract. This paper reports a design of a new prototype of air-spaced cantilevers made from a micro-structured PDMS piezo-electret material for accelerometer and energy harvesting applications. The test performed on these cantilevers in a sensor mode exhibits a stable sensitivity of 385 mV/g for a frequency ranging from 5 Hz to 200 Hz that encompass most macro-scale vibrations. In the energy harvesting mode, the cantilever generates a power of 103 nW with a load resistance of 217 MΩ.

1. Introduction

Piezoelectric materials play an important role in micro-sensors and actuators field. An important progress has been made in the development of piezoelectric materials: the study of aging with respect to external constraints (temperature, mechanical pressure, electric field ...), adjusting proportions in the preparation of conventional piezoelectric materials (ceramics and semiconductors), adjusting of geometrical dimensions of the material to be adapted to the desired applications. However, the malleability constitutes a limit for the integration of these conventional materials in certain applications. During the past fifteen years, high deformable materials (referred as piezo-electrets) were developed and tested to show a significant piezoelectric behavior. These heterogeneous materials consist of a polymer matrix containing micro-cavities. To ensure the piezoelectric behavior, electric charges of opposite signs must be implanted in the inner surfaces of the micro-cavities. The piezoelectric response of this type of material depends on the size, shape and distribution of the micro-cavities introduced into the polymer matrix. These structures are usually used in their thickness mode characterized by a large piezoelectric coefficient d_{33}. In this paper, we propose the development of a micro-structured piezo-electret PDMS material with an enhanced piezoelectric response. This piezo-electret PDMS material is integrated in an air-spaced cantilevers prototype for accelerometer and energy harvesting applications.

2. Sample and prototype preparation

The polydimethylsiloxane (PDMS) used in this work was prepared using the Sylgard 184 silicone Elastomer Kit of Dow Corning. The pre-polymer and the curing agent are mixed at *1:10* proportions. The mixed uncured PDMS was degassed in a vacuum desiccator. The micro-structured PDMS material is obtained by the molding process to be able to control the size and the shape of the introduced micro-cavities. The obtained optimized structures are composed of two micro-structured layers both *40 μm* thick, interposed with three bulk layers each *55 μm* of thickness. The micro-structured layers contain cylindrical micro-cavities of *40 μm* in height and *100 μm* in diameter that are regularly distributed over the structure area with a pitch of *150 μm*. The obtained material has the thickness of *250 μm* and the area of $1\times4\ cm^2$ (Figure 1) [1,2]. To enhance the piezoelectric response, the final structures are poled with a triangular quasi-static voltage with amplitudes between *1 kV* and *8 kV* and at a frequency of *0.5 Hz* for *15* minutes [3]. To demonstrate a possible application, the material was integrated in a mechanical vibration energy harvester. The harvester is composed of two air-spaced cantilevers attached on one side to a fixed base and on the other side to a seismic mass. The top cantilever is made of the PDMS piezo-electret material, the other cantilever, necessary to insure the system stiffness is made of aluminum (Figure 2) [4,5]. Such configuration allows a higher bending of the piezo-electret material than in a case of a simple composed beam. The design has been optimized to obtain the maximum output energy during the mechanical solicitation of the system. The geometric dimensions of the optimized energy harvester are summarized in Table 1.

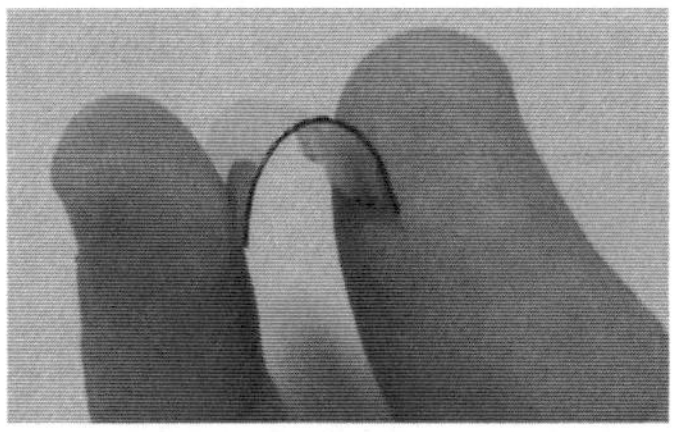

Figure 1. A photo image of the obtained flexible micro-structured PDMS piezo-electret material.

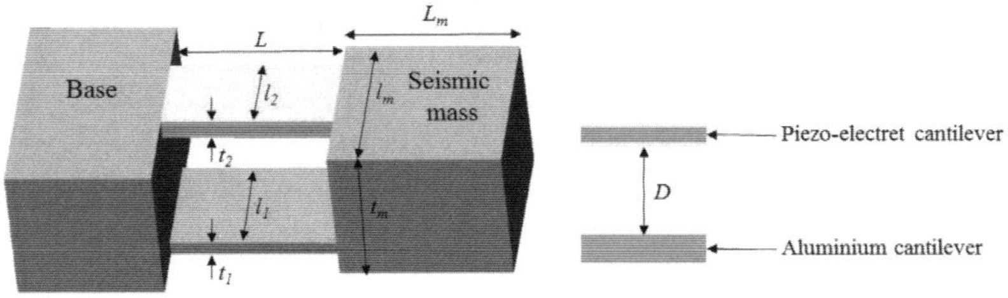

Figure 2. Schematic structures of the air-spaced cantilevers.

Table 1. Geometric parameters of the air-spaced cantilevers.

L(mm)	l_1(mm)	l_2(mm)	t_1(mm)	t_2(mm)	l_m(mm)	L_m(mm)	t_m(mm)	D(mm)
30.59	20.55	10.34	0.25	0.5	50.59	50.59	2	2

3. Results and discussion

3.1. Resonance frequency and sensitivity of the system

To determine the resonance frequency and the sensitivity, the piezo-electret PDMS material of the prototype is connected to a charge amplifier with a wide bandwidth and large dynamic range [Brüel & Kjær 2634] and placed on an electrodynamic shaker [Data Physics V20]. The shaker acceleration was controlled by an accelerometer [PCB Piezotronics 355B04]. Since we use a very soft material, the acceleration was varied between *0.06 g* to *0.1 g* with steps of *0.01 g* to avoid twisting and buckling phenomena during the system excitation. In addition, the frequency was swept in a range from *3.7 Hz* to *200 Hz,* which is compatible with the frequency range of vibrations that are available in our everyday life. A laser vibrometer [Polytec OFV 3001] is set to the seismic mass to determine the resonance frequency of the system. Figure 3 represents the displacement variations of the system in the frequency range previously specified. In Figure 3, we show separately the displacement in two frequency bands in order to observe the system resonant frequency of *4.3 Hz* for all accelerations (Figure 3a). The displacement is stable over the frequency range from *5 Hz* to *200Hz* (Figure 3b).

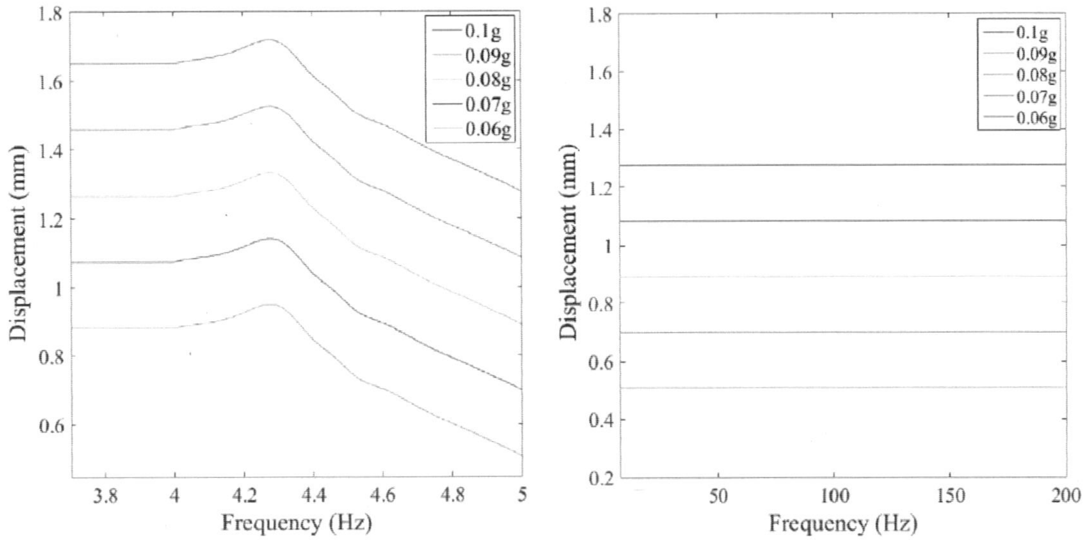

(a) (b)

Figure 3. Displacement variations for the frequency ranging from *3.7 Hz* to *200Hz*.

Figure 4 shows the normalized sensitivity of the piezo-electret PDMS material. The system presents a sensitivity of *420 mV/g* at the resonance frequency and a stable sensitivity of *385 mV/g* for the frequency ranging from *5 Hz* to *200 Hz* for various accelerations encompassing most macro-scale vibrations.

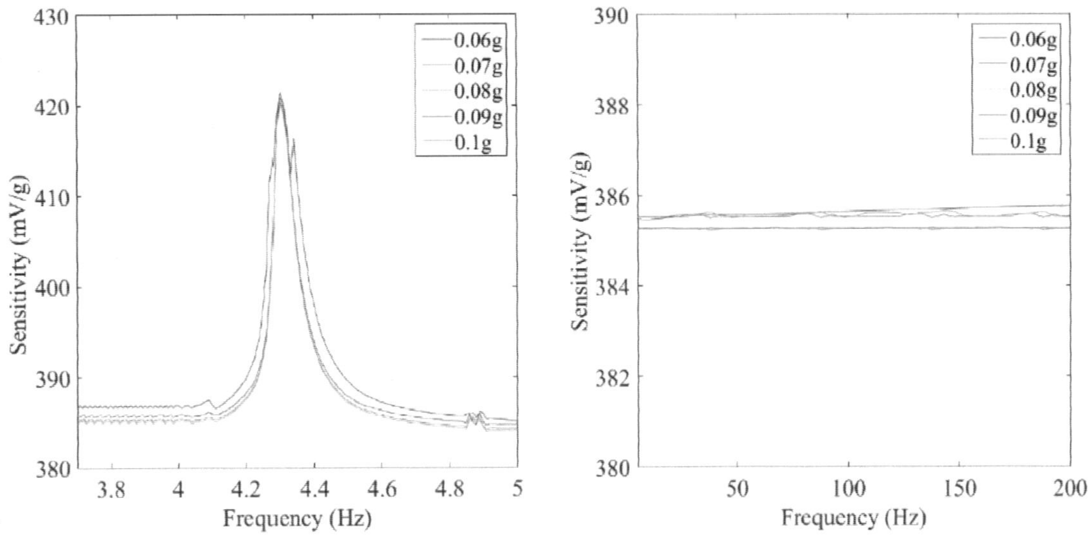

Figure 4. Sensitivity variations for the frequency ranging from *3.7 Hz* to *200Hz*.

3.2. Energy harvesting feasibility study

To study the energy to be harvested from mechanical vibrations applied to the system, the piezo-electret material was connected to variable resistors in parallel with an instrumentation amplifier [INA 116]. The system was excited with an acceleration of *0.1 g* at frequencies ranging from *3.7 Hz* to *200 Hz*. The resistance varied from *130.2 MΩ* to *295.3 MΩ*. The power shows an optimum for a resistance of *217 MΩ* for a capacitance of the material of *171 pF*. The generated power reaches its maximal value of *136 nW* and has a constant value of *103 nW* for frequencies ranging from *5 Hz* to *200 Hz* (Figure 5).

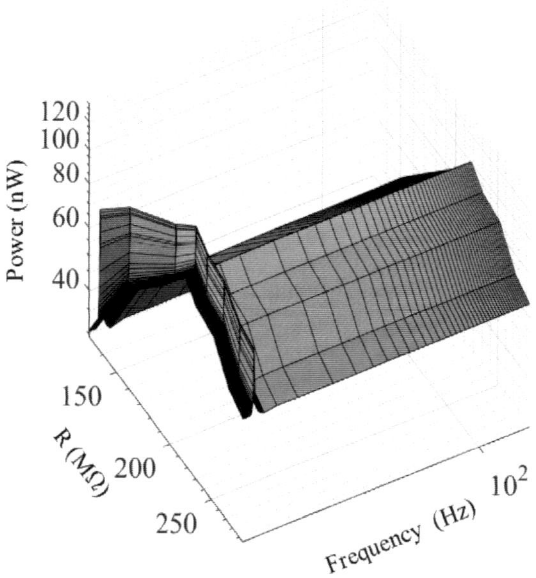

Figure 5. Power generated by the piezo-electret material for the frequency ranging from *3.7 Hz* to *200 Hz*.

4. Conclusion

In this work, a pototype of air-spaced cantilevers made from PDMS piezo-electret material is implemented. In the sensor mode, the system exhibits high stable sentivity of 385 mV/g for frequency ranging from 5 Hz to 200 Hz. In the energy-harvesting mode, the system delivers power of 103 nW to a load resistance of 217 MΩ. This frequency range covers the frequencies of the vibrations existing in our everyday life. Next work will focus design of high-capacitance structure configurations that will better match typical load resistances used in energy harvester systems.

References

[1] Kachroudi A, Basrour S, Rufer L, Sylvestre A and Jomni F 2015 *Smart Mater. Struct.* **24** 125013.
[2] Kachroudi A, Basrour S, Rufer L and Jomni F 2015 *J. Phys. Conf. Ser.* **660** 12040.
[3] Kachroudi A, Basrour S, Rufer L, Sylvestre A and Jomni F 2016 *Smart Mater. Struct.* **25** 105027.
[4] Zheng Q and Xu Y 2008 *Smart Mater. Struct.* **17** 55009.
[5] Kubba AE, Jiang K 2013 *Sensors* **14**(1) 188-211.

Thermal Energy Harvesting Using Pyroelectric and Piezoelectric Effect

Miwon Kang and Eric M. Yeatman

Department of Electrical and Electronic Engineering, Imperial College London, Exhibition Road, London, SW7 2AZ, UK

E-mail: k.miwon14@imperial.ac.uk

Abstract. This paper presents a prototype of a thermal energy harvesting mechanism using both pyroelectric and piezoelectric effect. Thermal energy is one of abundant energy sources from various processes. Waste heat from a chip on a circuit board of the electronic device involves temperature differences from a few degrees C to over 100 °C. Therefore, 95 °C of a heat reservoir was used in this study. A repetitive time-dependant temperature variation is applied by a linear sliding table. The influence of heat conditions was investigated, by changing velocity and frequency of this linear sliding table. This energy harvesting mechanism employs Lead Zirconate Titanate (PZT-5H), a bimetal beam and two neodymium magnets. The pyroelectric effect is caused by a time-dependent temperature variation, and the piezoelectric effect is caused by stress from deformation of the bimetal. A maximum power output 0.54 μW is obtained at an optimal condition when the load resistance is 610 kΩ.

1. Introduction

Energy harvesting is promising as a replacement for batteries [1]. Waste heat is produced by many processes and is, therefore, a promising source for energy harvesting technology, and various methods have been proposed. Thermoelectric generators based on the Seebeck effect have been widely researched [2]. The efficiency of these generators is limited by the figure of merit of the thermoelectric materials, ZT.

Due to high conversion efficiency, the pyroelectric effect with spontaneous polarization has attracted research attention as an alternative [3, 4]. However, one of the challenges of pyroelectric energy harvesting is to establish a temporal temperature oscillation. S. K. T. Ravindran *et al.* developed an engine chamber with sealed air and a bistable membrane [3]. Gilhwan Cha *et al.* used a macro-scale mechanical actuator [4], and Marcel Gueltig *et al.* adapted magnetic shape memory alloy films [5]. In this paper, a thermal energy harvesting mechanism is investigated, which consists of PZT-5H, a bimetal beam and two neodymium circular disc magnets with a simple structure.

2. Design and Operating Principle

The system is composed of PZT-5H, a bimetal beam, two neodymium circular disc magnets, a heat reservoir, and a linear sliding table. The schematic of the harvester is shown in Figure 1. The linear sliding table is used for time-dependent temperature variation; in a complete system an oscillating structure will be used to obtain temporal variation from a fixed heat source.

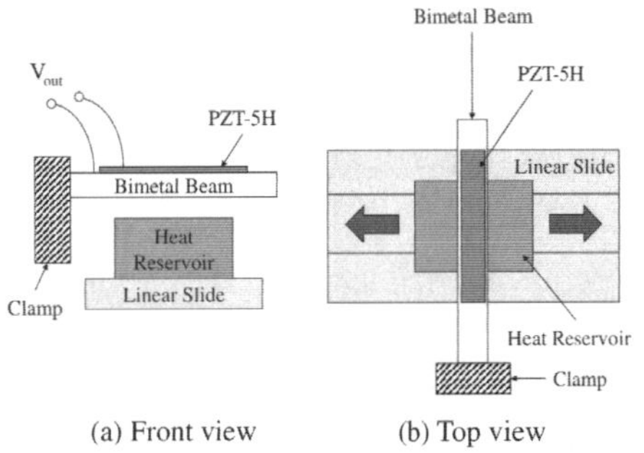

Figure 1. Schematic of the structure.

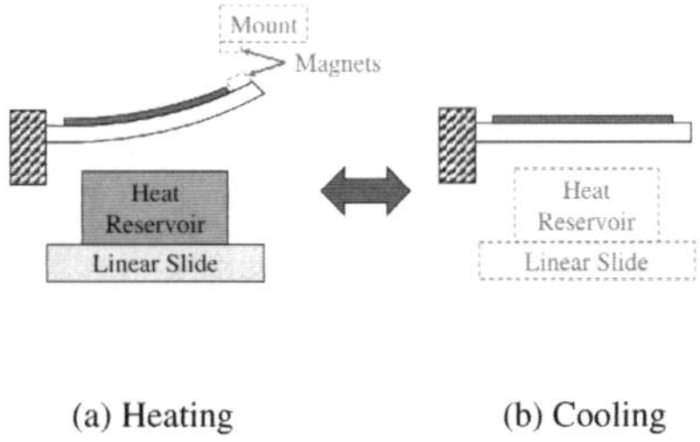

Figure 2. Schematic of the heat cycle.

Figure 2 shows the schematic of a heat cycle. Grey dotted lines show removable parts, that is the heat reservoir, the linear slide and magnets. When the heat reservoir is under the bimetal beam, the beam gets hotter and bends up as shown in Figure 2(a). When the heat reservoir is away from the bimetal beam, the beam loses heat and returns to its initial position as shown in Figure 2(b). Consequently, a reciprocating motion of the sliding table enables to convert thermal energy into electric energy.

3. Simulation and Experimental Set-Up

The bimetal beam is composed of Fe-Ni and Fe-Ni-Cr, with total thickness of 0.5 mm. The PZT-5H supplied by Piezo Systems, Inc. was cut into a beam of $50{\times}0.9{\times}0.127\ mm^3$. Figure 3 shows

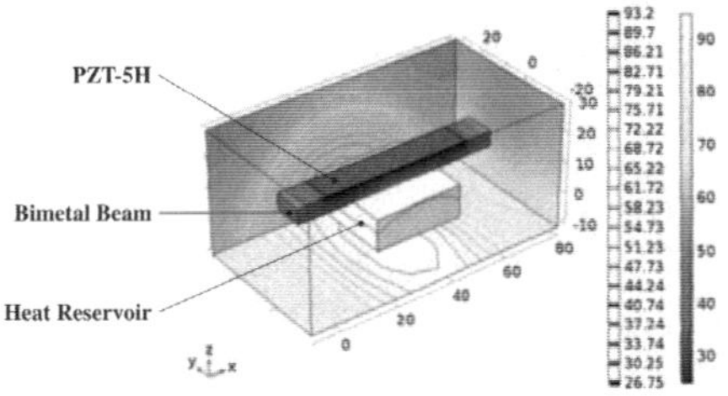

Figure 3. Simulation of temperature distribution.

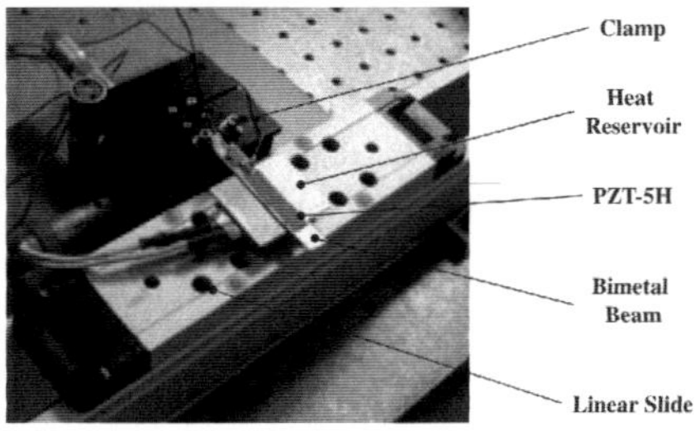

Figure 4. Experimental set-up.

a simulation of temperature distribution using COMSOL Multiphysics. This result shows heat transfer from 95 °C of the heat reservoir to 25 °C of the bimetal beam through air convection and conduction at a stationary condition. Figure 4 shows the prototype of this harvesting mechanism. The heat reservoir was mounted on the linear sliding table that allowed the heat reservoir to move under the beam.

4. Results

Figure 5 shows the open-circuit output voltage of PZT-5H as a function of velocity of the sliding table at the same table displacement frequency of 0.08 Hz. This frequency was used due to limitation of the sliding stroke at a comparative high speed of 48 mm/s and over. As velocity is decreased output voltage is slightly increased. This shows the bimetal beam obtained more thermal energy from the heat reservoir due to the longer heating period. Figure 6 shows the output voltage at different beam conditions. Measurements were made in both fixed end and free end conditions in order to help distinguish between piezo- and pyro-electric effects. The results indicate that pyroelectric effect dominates, since there is little effect of preventing beam motion; however the pyroelectric effect shows sensitivity to heating rate as expected. The output is dominated by pyroelectric effect in these cases. The voltage amplitude shows little change with frequency, which is consistent with pyroelectric effect since the slider velocity, and thus heating rate, are constant. Figure 7 shows a time-dependent simulation using COMSOL Multiphysics.

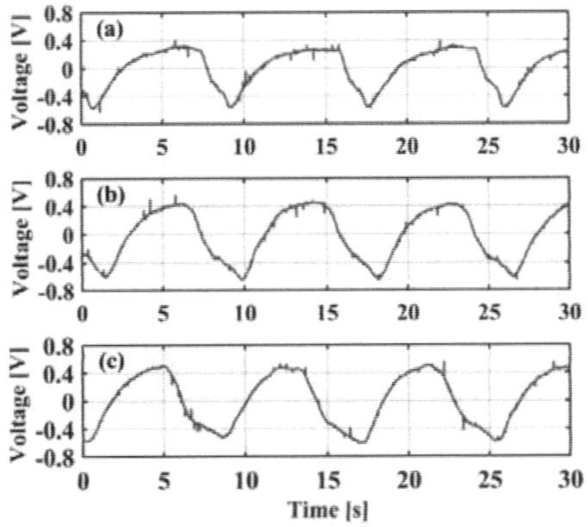

Figure 5. Measured output voltage as a function of velocity at 0.08 Hz. (a) 48 mm/s (b) 24 mm/s (c) 12 mm/s

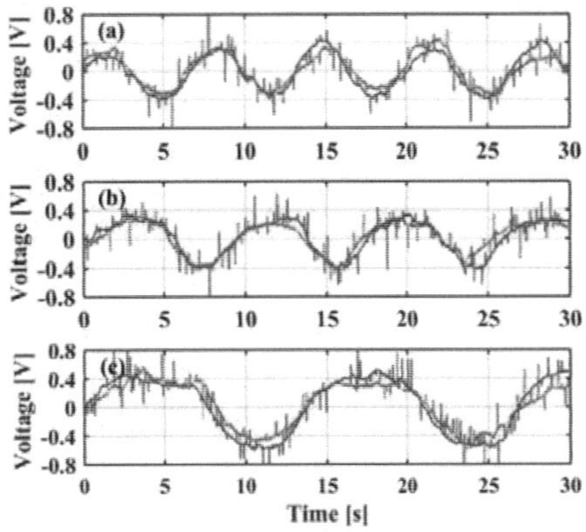

Figure 6. Measured output voltage for a fixed end (blue) and a free end (red) at 12 mm/s. (a) 0.1 Hz (b) 0.08 Hz (c) 0.05 Hz

Temperature of the beam is simulated as the temperature of the heat reservoir is oscillated from 25°C to 95°C at 0.05 Hz. Figure 8 shows measurement of temperature distribution using a thermal imaging camera. These images indicate temperature at a cross target focusing symbol, as the sliding table moves with a velocity of 12 mm/s for 0.05 Hz. When the heat reservoir is away from the beam as shown in Figure 8(a), temperature of the beam is 38.3 °C. When the heat reservoir is under the bimetal beam as shown in Figure 8(b), it is 39.1 °C.

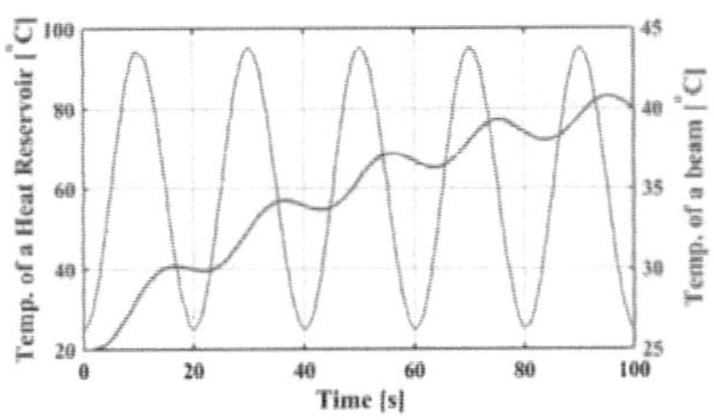

Figure 7. Simulation of time-dependant change in temperature.

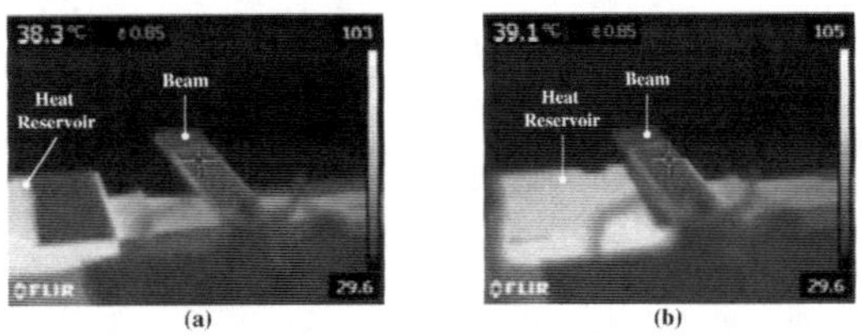

Figure 8. Measured temperature with a thermal imaging camera.

To improve the influence of piezoelectric effect, two neodymium circular disc magnets with the diameter of 5 mm and thickness of 3 mm are applied as shown by grey dotted lines in Figure 2(a). Magnets are shown as an effective method of "plucking" of a piezoelectric beam [6]. Figure 9 shows the open-circuit output voltage with the force of attraction of two magnets. Sharp peaks are seen, when two magnets are attached or released. The peaks decay due to the internal resistance of PZT-5H. Temperature changes on the beam cause broad curves between peaks. When the sliding table moves with 0.1 Hz, adjustment of the gap between the two magnets

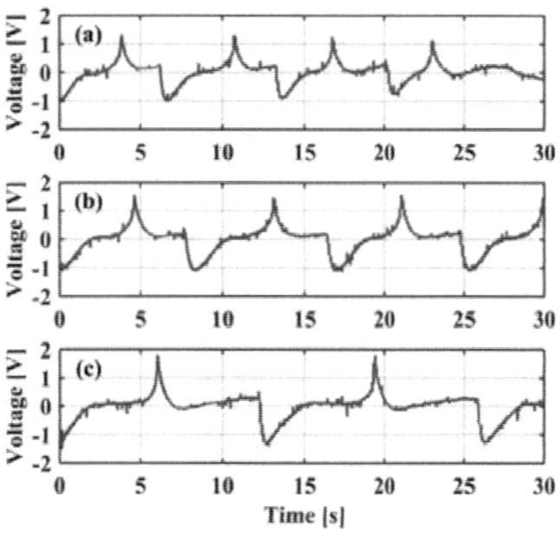

Figure 9. Measured output voltage applied with two magnets at 12 mm/s. (a) 0.1 Hz (b) 0.08 Hz (c) 0.05 Hz

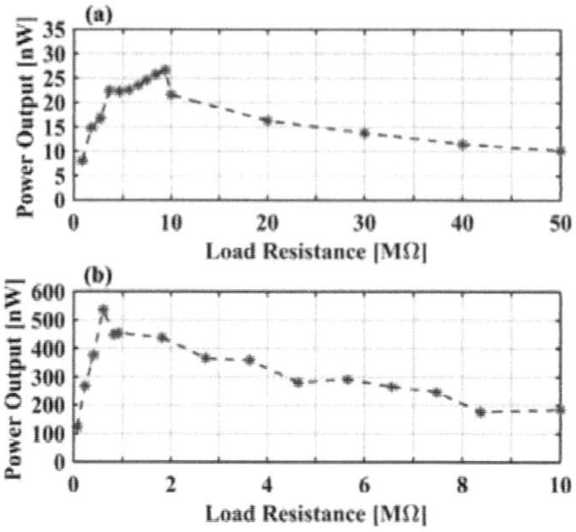

Figure 10. Measured peak power output versus load resistance. (a) without magnets (b) with magnets

is demanding due to a small deformation of the bimetal beam. Magnets were stuck easily as shown after 23 seconds of Figure 9(a). Figure 10 shows peak power output at a velocity of 12 mm/s for 0.05 Hz. Without magnets, the maximum power output 26.7 nW is obtained at an optimal condition when the load resistance is 9 $M\Omega$ as shown in Figure 10(a). With the force of attraction of two magnets, the maximum power output 0.54 μW is obtained at an optimal condition when the load resistance is 600 kΩ as shown in Figure 10(b).

5. Conclusions and Future Work

In this paper, a new simple thermal energy harvesting mechanism is shown. This mechanism consists of PZT-5H, a bimetal beam, and two neodymium magnets, enables both piezo- and pyro-effect. A prototype was built and then tested under different conditions using a heat reservoir on a linear sliding table. A maximum power output 0.54 μW is obtained at the optimal condition when a load resistance is 610 kΩ. This result is achieved when the linear sliding table works with a velocity of 12 mm/s for 0.05 Hz. Without the use of magnets, pyroelectric effect appears to dominate. Using magnets to achieve greater motion range, a stronger piezoelectric response is obtained with higher power output.

Future study will aim to improve power output and make a complete system.

6. Acknowledgments

The work is supported by the Electrical and Electronic Engineering Department, Imperial College London. The authors would also like to thank Hailing Fu and Dr. Mario D'Auria for valuable suggestions, and thank Prof. Stepan Lucyszyn and Jingye Sun for use of apparatus.

References

[1] Mitcheson P D, Yeatman E M, Rao G K, Holmes A S, Green T C, Proc. IEEE 96 (9), 2008, 1457-1486
[2] Kiziroglou M E, Wright S W, Toh T T, Mitcheson P D, Becker Th, and Yeatman E M, IEEE Transactions on Industrial Electronics, 61, 1, 2014, pp. 302-309
[3] Ravindran S K T, Huesgen T, Kroener M, and Woias P, Applied Physics Letters 99, 104102 (2011)
[4] Cha G, Ju Y S, Sensors and Actuators A 189 (2013), pp. 100107
[5] Gueltig M, Ossmer H, Ohtsuka M, Miki H, Tsuchiya K, Takagi T, and Kohl M, Adv. Energy Mater., 4, 2014, 140075
[6] Fu H and Yeatman E M, Applied Physics Letters 107, 243905 (2015)

A novel vibrational energy harvester with electric double layer electrets

S Ono[1], K Miwa[1], J Iori[1], H. Mitsuya[2], K Ishibashi[2], C Sano[3], H Toshiyoshi[3], and H Fujita[3]

[1] Central Research Institute of Electric Power Industry, Yokosuka, Kanagawa, 240-0196, Japan
[2] Saginomiya Seisakusho INC. Sayama, Saitama, 350-1395, Japan
[3] Institute of Industrial Science, The University of Tokyo, Meguro, Tokyo, 153-8505, Japan

E-mail: shimpei@criepi.denken.or.jp

Abstract. We propose a new type of vibrational energy harvester with an electric double layer (EDL) electrets. Instead of using any external bias-voltage source or dielectric layer on top of the metal electrode to sustain EDL, we succeed to anchor the ions to polymer network to form the EDL electrets. By changing contact area between the EDL electrets and the electrode, large electric current is generated in the circuit. Owing to extremely large capacitance of the EDL electret, vibrational energy harvesters have the unique capability to leverage the high-density charge accumulation to the electrode and obtained current density becomes as high as 200 μA/cm^2 with output voltage of 1V even with low frequency vibrations as low as 1 Hz.

1. Introduction

The vibrational energy harvesters have been extensively investigated in recent decade as an alternative energy source to the batteries. There have been several methods to convert vibration to electric energy such as piezoelectric, electromagnetic and electrostatic methods. Particular attention has been devoted for the electrostatic energy harvesters since they are suitable to obtain current from low-frequency and low-amplitude vibration source and can be integrated in the microsystem. However, it has been yet regarded as a serious technological challenge to enhance output power to operate autonomous sensors and wearable devices.

Recently new types of energy harvester using the microfluidics techniques have been proposed[1-3]. It is well known that when a voltage is applied to liquids, electric double layers (EDL) are formed at the interface between electrode and liquids and they work as two serially connected capacitors (fig.1). By mechanically modulating interfacial area between vibrating electrode and the EDL, the electrodes are continuously charged and discharged at different phases and AC electric current is generated. For this type of energy harvesters, it is of importance to increase carrier density accumulated at the interface of electrode when the electrodes are in contact with EDL since the difference in number of carrier over the contacted area changing with the contacting/releasing motion of the electrode directly convert into electric current. To enhance the number of carrier at the electrode, we need to use the liquids with higher capacitance and apply higher voltage between two electrodes. Among all liquids, ionic liquids (ILs), which known as room-temperature molten salts is suitable for this application since the capacitance of ILs is known to be as high as 170 μF/cm^2[4]. However, it has been reported that the external

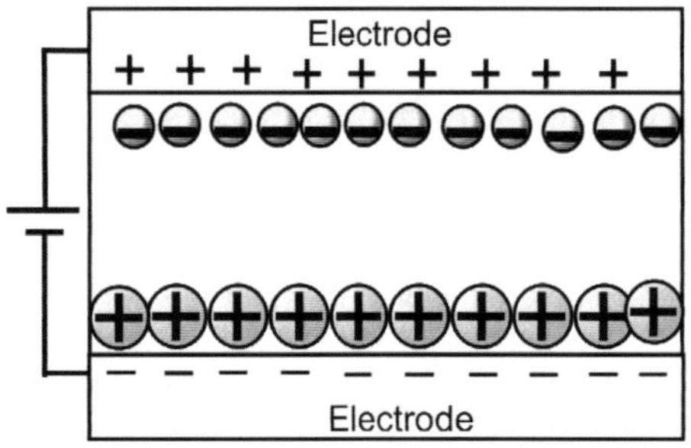

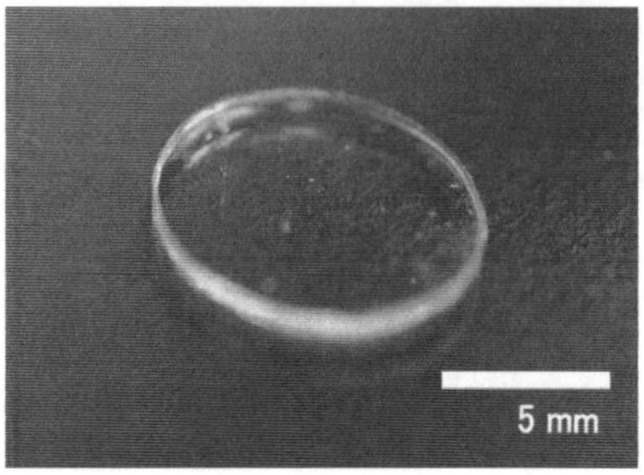

Figure 1. Schematic picture of electric double layers with electrolyte and photograph of electric double layers electrets.

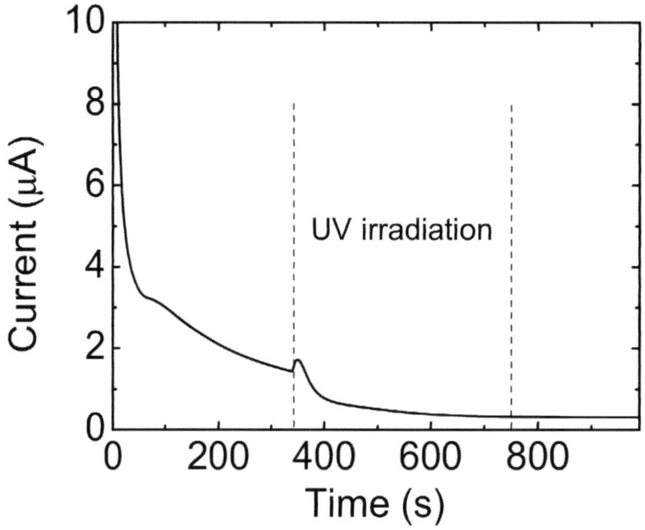

Figure 2. Current between two electrodes during polymerization of the EDL electrets with a bias voltage of 3 V. After irradiation of UV light, there is little change in current.

bias-voltage source[5] or the use of dielectric materials on the surface of electrode[6] is necessary to sustain EDL.

To overcome these problems, we are going to show a new method of electrostatic energy harvesters using an EDL electrets made with ILs. The EDL electrets are made by anchoring motion of ions inside ion gel after formation of EDL and can sustain EDL without any external bias-voltage source. By mechanically changing contact area between the EDL electrets and the electrode with the contacting/releasing motion, large electric current is generated even with low frequency vibrations.

2. Methods

The EDL electrets are made with polymers and ILs which have unsaturated bonds in molecular structure. When ILs have unsaturated bonds in the structure, we can anchor the ILs to the networks of the polymer by polymerization. The IL we use in this experiments has a cation with unsaturated bond in combination with an standard anion. It means that only cation can be anchored to the network of the polymer. We first dissolve the polymers with solvent and add the

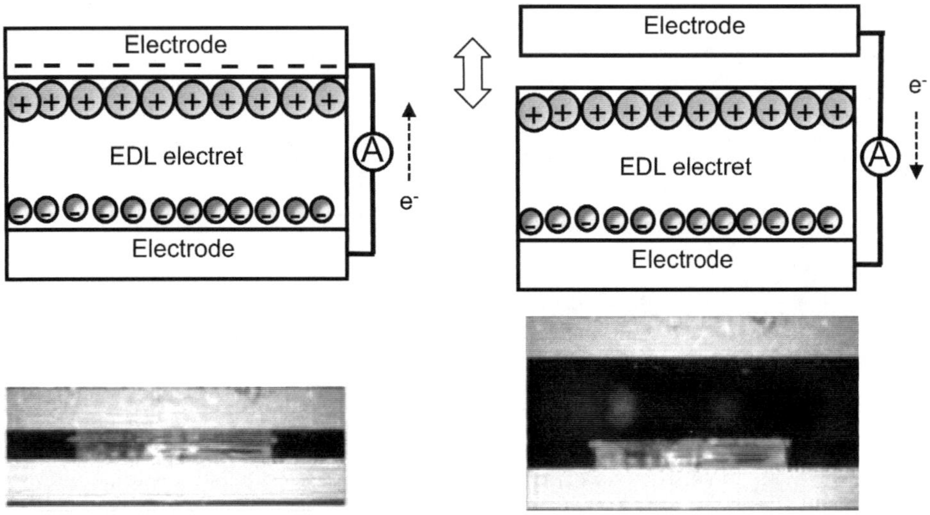

Figure 3. Mechanism of vibrational energy harvester with the electric double layers electret. Large current is generated when the electrode in contact with electrode through electrostatic induction.

IL and a polymerization initiator. This mixture is sandwiched with two electrodes where a bias voltage of 3V is applied to form EDL. Then, we polymerize EDL by exposing UV light to make the EDL electrets. The bias voltage is limited by chemical windows of the ILs and depends on the combination of cations and anions. We have measured the current between two electrodes after application of the bias voltage and waited until the current starts to show the saturation to confirm the formation of EDL (fig.2). After the exposure of UV light, this current is started to decrease and show saturation. Even after we stop the irradiation of UV light, there is little change in current. This is due to the reduction of mobile ions that will enhance the resistance of the mixture of polymer and ILs. Once we solidify ILs with polymer, we can sustain EDL without applying voltage. During polymerization process we use spacers to define the height as well as obtain flat surface of the EDL electrets. Typical size of the EDL electrets is 5mm in diameter and 500 μm in height (fig. 1).

Figure 3 illustrates the structure and the mechanism of our contacting/releasing type of energy harvester. The EDL electrets are placed between two parallel electrodes made with thin layer of Au deposited on top of glass, where distance between the electrodes is mechanically modulated by vertically vibrating the top electrode in accordance with sinusoidal wave driven by a shaker. It should be noted that the upper layer of the EDL electrets is the cations side where EDL is expected to be sustained. In this configuration we expect that larger current is generated in the contacting process. By mechanically modulating contacting area between electrode and EDL electrets, mechanical work can be converted into electrical current. The electrical current is measured with a digital multimeter with internal resistance of 10 MΩ. All the performance of the devices is measured inside glove box filled with nitrogen to avoid the water which cover the surface of the EDL electrets.

3. Results and discussion

Figure 4 shows typical current obtained with the energy harvester with the EDL electrets. Subsequent contact and release between EDL electrets and Au electrode with 1 Hz vibrations generates large current density as high as 200 μA/cm^2 and stabilized around 50 μA/cm^2 with

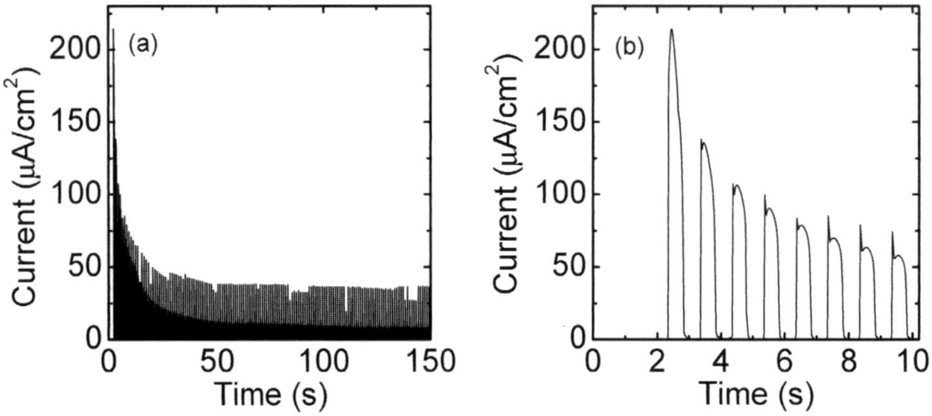

Figure 4. (a) Output current of vibrational energy harvester with the electric double layers electret generated from 1 Hz vibrations. (b) Close up of output current.

output voltage of 1V. We have measured same materials without formation of EDL and found that the obtained current density is less than $\sim$10 nA/cm^2, which indicate that the EDL is indeed sustained inside the polymer gels. The obtained current density with EDL electrets is highest value ever reported for energy harvester using EDL, since we can have the benefit of large capacitance of EDL with the EDL electrets. It has been reported that with standard the energy harvesters with the EDL, the current is electrostatically generated at the moment when the electrode touch to or release from the EDL and both positive and negative current is obtained. However, the energy harvester with the EDL electrets, we only obtain positive current during the time when the electrode is in touch with EDL electrets and negative current is not observed. This is potentially due to the water inside polymer which act as electron trap at the interface or mobile anions inside of the EDL electrets canceled out the current. Further experiments are necessary to elucidate the mechanism of energy harvester with the EDL electrets.

4. Conclusion

In this study, a novel energy harvester with the EDL electrets is proposed and experimentally verified. These results indicates that energy harvester with the EDL electrets can be well suited because (1) relatively large output currents and voltage from low frequency vibrations can be obtained. (2) the optimal advantage of the simple and low-cost production processes can be used. Therefore, it would be possible to apply this energy harvester with the EDL electrets as an eco-friendly replacement for a battery or auxiliary power source.

5. Acknowledgements

This work was partly supported by CREST, JST.

6. References

[1] Krupekin T and Taylor J A 2011 *Nat. Commun.* **2** 448
[2] Moon J K, Jeong J, Lee D, and Pak H K 2013 *Nat. Commun.* **4** 1487
[3] Kwon S H, Park J, Kim *et al.*, 2014 *Energy Environ. Sci.* **7** 3279
[4] Ono S, Miwa K, Seki S, and Takeya J 2009 *Appl. Phys. Lett.* **94** 063301
[5] Yamada S 2015 *IEEE Proc. MEMS, 18-22 Jan. 2015 Estoril, Portugal* 118
[6] Helseth L E and Guo X D 2015 *Langmuir* **31** 3269

Combinatorial approach to MgHf co-doped AlN thin films for Vibrational Energy Harvesters

H. H. Nguyen[1], H. Oguchi[2] and H. Kuwano[1]

[1] Department of Robotics, Tohoku University, Sendai, Japan
[2] AIMR, Tohoku University, Sendai, Japan

hung@nanosys.mech.tohoku.ac.jp

Abstract. In this report, we studied MgHf co-doped AlN ($(Mg,Hf)_xAl_{1-x}N$) aiming for developing an AlN-based dielectric material with the large piezoelectric coefficient. To rapidly screen the wide range of composition, we applied combinatorial film growth approach. To get continuous composition gradient on a single substrate, films were deposited on Si (100) substrates by sputtering AlN and Mg-Hf targets simultaneously. Crystal structure was investigated by X-ray diffractometer equipped with a two-dimensional detector (2D-XRD). Composition was determined by Energy Dispersive Spectroscopy (EDS). These studies revealed that we successfully covered the widest ever composition range of $0 < x < 0.24$ for this material. In addition, these studies found that we succeeded in realizing largest ever c-axis expansion of 2.7% at $x = 0.24$, which will lead to the highest enhancement in the piezoelectric coefficient. The results of this study opened the way for high-throughput development of the dielectric materials.

1. Introduction

Recently due to the restriction of lead in industry, lead-free materials have received considerable interest from many scientists for applications such as vibrational energy harvesters (VEHs) [1]-[4]. Among the lead-free piezoelectric materials that have been studied, c-axis oriented AlN film is a good candidate for the piezoelectric thin film for VEHs because of its high breakdown voltage [4] and high output power density [1]. However compared with the traditional lead zirconate titanate (PZT), piezoelectric coefficient of the AlN (d_{31} ~ 5.5 pC/N for AlN and ~ 100 pC/N for PZT[5]) is very low [1], [4]-[10]. Therefore, one of the important topics is to improve AlN piezoelectricity. The most effective approach developed so far is modifying crystal structure of AlN using substitution of Al^{3+} site by other elements.

Among the systems studied so far, MgHf co-doped AlN ($(Mg,Hf)_xAl_{1-x}N$) showed the highest increase rate of d_{31} per dopant concentration (at%). For example, Y. Iwazaki *et al.* reported MgHf doping and d_{31} of AlN showed increase rate of 11.5% [9]. This rate is the largest so far. However, in the previous study of $(Mg,Hf)_xAl_{1-x}N$, dopant concentration was limiter only to 13% although the piezoelectric coefficient tends to increase more at higher concentration of dopants. Moreover, the important crystal characteristics including lattice constant, crystal orientation *et al* were not reported, though these parameters can be dominant on the d_{31}.

In this study, to cover a wider range of the dopant concentration and systematically investigate the effect of the dopant concentration on the lattice expansion, combinatorial approach was employed. The combinatorial approach is a high-throughput material investigation method preparing films with various composition on a single substrate. Moreover, the method enables us to precisely tune the concentration of dopants.

2. Experimental

AlN targets and Mg-Hf targets (Mg target with pieces of Hf on top) were sputtered simultaneously to grow films on Si (100) substrates. Number of Hf piece was adjusted in order to get Mg: Hf ~ 1: 1. Figure 1 shows an illustration of the sputtering set up. A substrate holder and the targets were arranging to obtain composition gradient. Plasma gases used for sputtering were Ar and N_2. The latter was used to compensate the nitrogen deficiency in thin films. On the other hand, at a certain sputtering pressure, the presence of N_2 will reduce probability of collision between sputtered particles with a heavy atom of Ar. Hence, sputtered particle could reach to the substrate with higher energy, which assist the formation of higher crystallinity films [11]. Thin films were deposited at 600°C to get high crystallinity [4]. Other sputtering parameters were shown in Table 1.

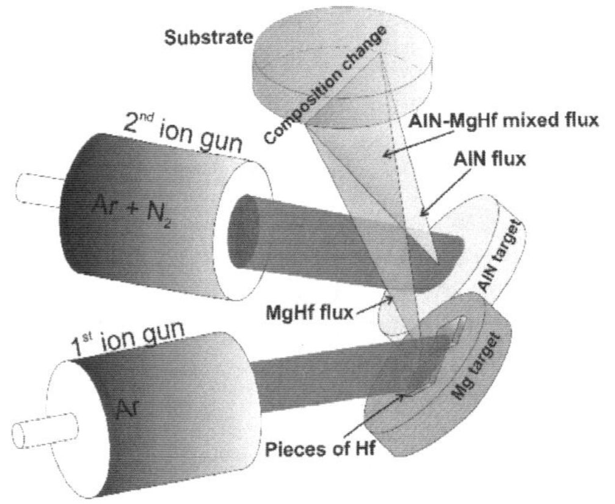

Figure 1. Schematic of combinatorial deposition using AlN target and Mg target with pieces of Hf on top

Table 1. Deposition conditions

Substrate temperature	600°C
RF power	AlN gun: 140 W MgHf gun: 100 W
Accelerated voltage	AlN gun: 1200 V MgHf gun: 1000 V
Beam current	AlN gun: 38 mA MgHf gun: 25 mA
Sputtering pressure	3.8 - 4 mTorr
Film thickness	~ 1 μm
Ar gas flow rate	15 sccm
N_2 gas flow rate	15 sccm
Deposition time	2 h

Prior to each thin film deposition, Si substrates were cleaned by standard cleaning processes of Radio Corporation of America (RCA1 and RCA2). Base pressure of the sputtering chamber was keep at less than 1×10^{-7} Torr. The AlN-target was pre-sputtered before each deposition for 20 minutes in Ar and then for 10 minutes in the mixture of Ar and N_2. Thicknesses of the films were controlled by sputtering time and measured by a surface profiler. The composition was confirmed by using the EDX spectroscopy. An atomic force microscopy (AFM, Nikon TE2000-U) was used to examine thin film surface roughness. X-ray diffractometer equipped with a two dimensional detector (2D-XRD) was employed for a measurement of crystallinity. Diameter of the focused X-ray beam was fixed at 1 mm.

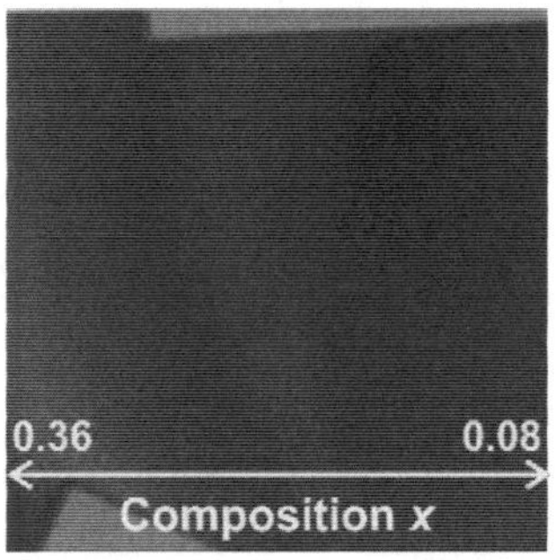

Figure 2. An example of combinatorial film, on which the composition of dopants change from MgHf rich side to AlN rich side

3. Results and discussions

Figure 2 shows a typical picture of the film surface prepared by the combinatorial approach. Lateral color change indicates change in the thickness ratio of AlN to MgHf, which corresponds to dopant concentration. For the 2 cm × 2 cm sample, typical composition difference is 2.8 at.%. Focused X-ray beam with spot diameter of 1 mm allowed for measurement of more than 15 points along the composition gradient.

XRD and EDS measurements confirmed that we succeeded in making solid solution of $(Mg,Hf)_xAl_{1-x}N$ at the highest concentration of dopant ever. Figure 3a shows a typical 2D-XRD pattern of a $(Mg,Hf)_{0.09}Al_{0.91}N$ film. Except a diffraction peak from Si substrate, only a peak from (0002) and (0004) plane of AlN were observed. Because we did not observe neither Mg nor Hf peaks, the pattern indicated that (Mg,Hf) dissolved in the AlN lattice. The solid solution was also indicated by the continuous peak shift from the (0002) peak of the pure AlN film [9], [10]. The integrated *2θ-θ* patterns obtained from 2D-XRD patterns for different compositions of $(Mg,Hf)_xAl_{1-x}N$ thin films were shown in figure 3b. The continuous peak shift in the *2θ* position of the (0002) peak indicated that c-axis lattice constant of the AlN was gradually expanded with increasing the dopant concertation.

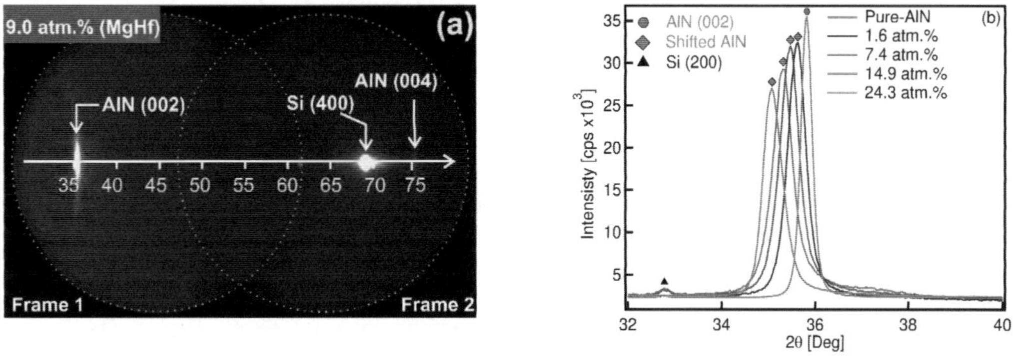

Figure 3. 2D-XRD patterns of $(Mg,Hf)_{0.016}Al_{0.984}N$ films deposited at 600°C (a) and the integrated 2θ-θ patterns at different concentration of dopants (b)

The c-axis lattice constants determined from the XRD patterns are shown in Figure 4 as a function of the dopant concentration. The *c* value was determined from the following formula:

$$\frac{1}{d_{hkl}^2} = \frac{4}{3}\left(\frac{h^2 + hk + k^2}{a^2}\right) + \frac{1}{c^2}$$

where, *h, k* and *l* are Miller indices, *a* and *c* are lattice constants, *d* is inter-planar distance for (*hkl*) plan (hear is (002)). It clearly indicated that c-axis expanded with increase in the (Mg,Hf) concentration. Note that there is difference between the reference value and experimental value for $x = 0$, which can be attributed to nitrogen deficiency.

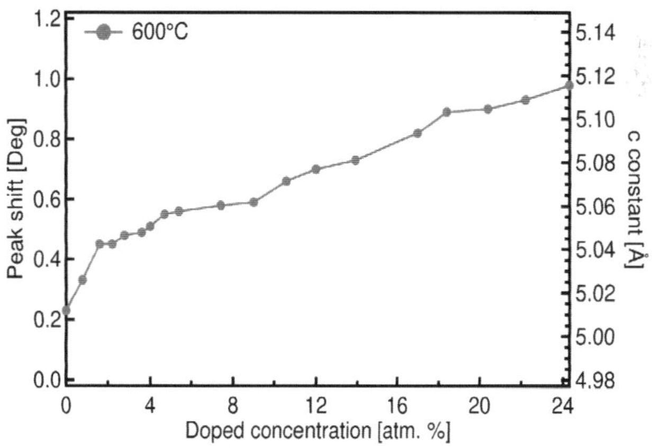

Figure 4. c-axis lattice constant of co-doped $(MgHf)_xAl_{1-x}N$ thin films as function of dopants concentration

The effect of the MgHf dopants on the surface morphologies of $(Mg,Hf)_xAl_{1-x}N$ films was investigated by atomic force microscopy (AFM). The results were shown in figure 5. AFM observation shows that root mean square (RMS) roughness of the thin films increase dramatically from 1.1 nm to 9.0 nm with increase in the dopant concentration *x* from 0 to 24.3 at%.

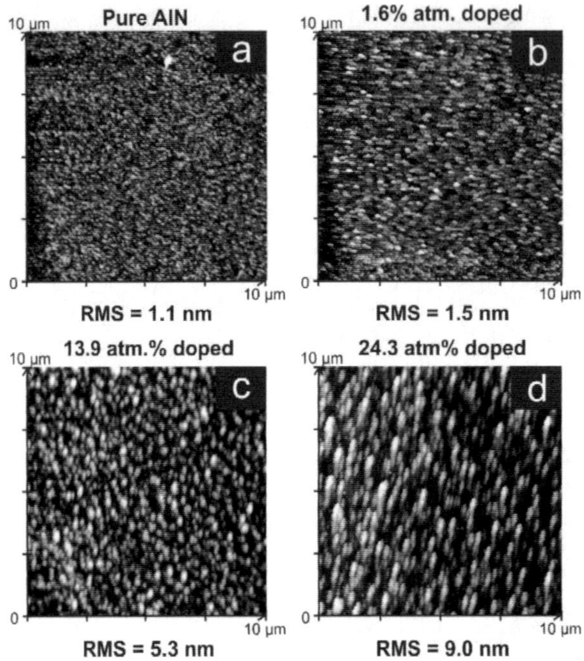

Figure 5: AFM image of MgHf doped AlN at different compositions

4. Conclusion

In summary, we successfully deposited $(Mg,Hf)_xAl_{1-x}N$ films with composition gradient covering the widest range of composition ($0 < x < 0.24$) so far reported. For these films the c-lattice of AlN was expanded almost linearly with the composition, and we obtained c-axis expansion of 2.7% at $x = 0.24$, which is the largest value for this system. We believe future study will find the largest piezoelectric coefficient for $(Mg,Hf)_xAl_{1-x}N$ among the AlN-based materials due to the largest c-axis expansion, prove the importance of this material for the vibrational energy harvesters.

References

[1] L. V. Minh, M. Hara, T. Yokoyama, T. Nishihara, M. Ueda, and H. Kuwano, IEEE Transactions on Ultrasonics, Ferroelectrics, and Frequency Control 62 (11), 2005-2008 (2015).
[2] Z. Vashaei, T. Aikawa, M. Ohtsuka, H. Kobatake, H. Fukuyama, S. Ikeda, and K. Takada, J. Crys. Growth, 311 (3), 459-462 (2009).
[3] M. Akiyama, T. Kamohara, K. Kano, A. Teshigahara, Y. Takeuchi, and N. Kawahara, Adv. Mater. 21, 593 (2009).
[4] H. H. Nguyen, H. Oguchi, and H. Kuwano, J. Phys.: Conference Series 557 (1), 012047 (2014).
[5] S. Trolier-McKinstry, P. Muralt, Journal of Electroceramics 12(1-2) 7-17 (2004).
[6] F. Tasnádi, B. Alling, C. Höglund, G. Wingqvist, J. Birch, L.Hultman, Abrikosov, Phys. R. Lett., *104* (13), 137601 (2010).
[7] H. Liu, F. Zeng, G. Tang, F. Pan, Applied Surface Science, *270* (0), 225-230 (2013).
[8] R. Matloub, M. Hadad, A. Mazzalai, N. Chidambaram, G. Moulard, C. S. Sandu, T. Metzger, P. Muralt, Applied Physics Letters 102 (15), 152903 (2013).
[9] I. Yoshiki, Y. Tsuyoshi, N. Tokihiro, U. Masanori Appl. Phys. Exp. 8:061501 (2015).
[10] T. Yokoyama et al. Proc. Ultr. Symp. (IUS) IEEE International 1382-5 (2013)
[11] Z. Vashaei, T. Aikawa, M. Ohtsuka, H. Kobatake, H. Fukuyama, S. Ikeda, K. Takada, J. Crys. Gr. 311(3) 459-462 (2009).

Comparative Analysis of Wrist-worn Energy Harvesting Architectures

R. Rantz[1], T. Xue[1], Q. Zhang[2], L. Gu[2], and K. Yang[2], S. Roundy[1]

[1]University of Utah, 1495 E 100 S (1550 MEK), Salt Lake City, UT 84112, USA
[2]Analog Devices, Inc., 804 Woburn St, Wilmington, MA 01887, USA

E-mail: robert.rantz@utah.edu

Abstract. This paper reports the simulation-based analysis of six dynamical structures with respect to their wrist-worn vibration energy harvesting capability. This work approaches the problem of maximizing energy harvesting potential at the wrist by considering multiple mechanical substructures independently of any specific transduction mechanism; rotational and linear motion-based architectures are considered. The addition of a linear spring element to the structures has the potential to improve power output. The analysis concludes that a sprung rotational harvester architecture outperforms a sprung linear architecture by 58% when real walking data is used as input to the simulations. The power output of a rotational prototype device was measured for various inputs and compared against simulation in order to corroborate the rotational device model.

1. Introduction

This paper reports the analysis of six dynamical structures with respect to their wrist-worn vibration energy harvesting capability. This work approaches the problem of maximizing energy harvesting potential at the wrist by considering multiple mechanical substructures independently of any specific transduction mechanism; this is accomplished by virtue of a modification to the standard Velocity Damped Resonant Generator (VDRG) model [1]. This approach serves as means to converge upon the device architecture that can best approach the theoretical maximum energy harvesting potential.

2. Structures

The first structure considered in the analysis is comprised of an eccentric seismic mass that rotates about an axis, as in [2]. Structure two is identical to structure one, except that a torsional spring acts on the mass, causing it to rest in the upper (with respect to gravity) semicircle when subject to no external input. See **Figure 1**. The third structure is a one-dimensional linear slide, comprised of a seismic mass that is free to move in a single dimension up to the length of the device where impact occurs. Structure four is structure three with a linear spring element. See Figure 2. Additionally, two-dimensional analogs of the third and fourth structures were considered that are composed of independent linear elements for the two available degrees of freedom, and a single seismic mass.

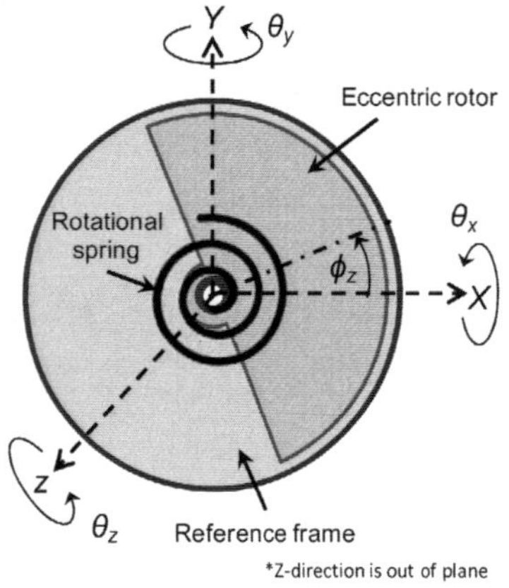

Figure 1. Rotational harvester architecture, pictured with a torsional spring (as in structure two).

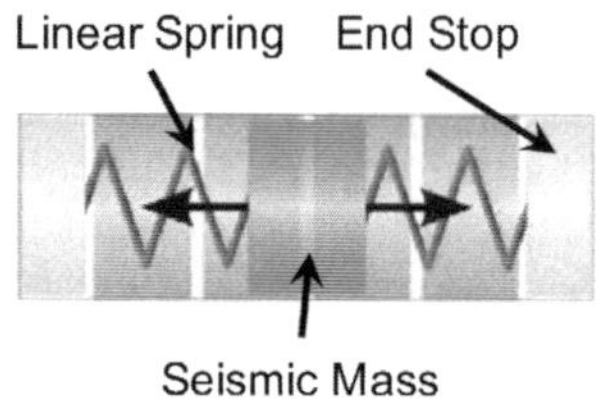

Figure 2. One-dimensional linear slide architecture, pictured with spring (as in structure four).

3. Comparative Analysis

Mathematical models of the various structures were used for simulations in order to determine the average power output of each architecture in response to various inputs. In all cases, the models assume that electrical power transduction may be modelled as an ideal viscous damper with a viscous damping coefficient (known henceforth as an electrical damping coefficient) to be determined via optimization. Power dissipated in the electrical damper is considered to be the power output of the architecture.

In the case of all structures, an arbitrary constraint of 1 cm^3 total device volume (as determined by the swept volume of the seismic mass as it is displaced through its full range of motion) is applied. The spring and damper structures are assumed to consume no volume.

The rotational structures made use of the system model described in [2], which was modified to include a rotational spring term for structure two.

The linear systems were assumed to be well-modeled as a base-excited mass spring damper system, as in the VDRG model [1], except that the seismic mass is displacement-limited, and contact with an end stop results in a reversal of velocity that is reduced via a coefficient of restitution. In the case of the unsprung architectures, the spring constant is assumed to be zero. Mechanical friction is modelled as viscous for all structures considered in the analysis.

3.1. Optimization. The value of the electrical damping coefficient and, in the case of architectures with springs, the spring constant greatly impact the nature of the dynamic response of a given architecture. For any particular input signal, the goal is to select values for the electrical damping coefficient and, when applicable, the spring constant such that maximum power is dissipated in the electrical damper. The power output of each structure may then be compared for a given input signal.

Because the input signals can be complex in nature and, in the case of structures one and two, the differential equation describing a structure highly nonlinear, a numerical Pattern Search (PS) algorithm was used in MATLAB to determine optimal system parameters for a given input signal [3]. An objective function was formed using the output of a numerical differential equations solver; the design parameters were the input to the objective function, and the output is the average power dissipated in the electrical damper over the duration of the input signal.

Some geometric optimization was performed on the linear structures. The additional design variables included the length and width of the swept volume, and the dimensions of the seismic mass.

3.2. *Inputs*. Two types of inputs were considered for the analysis. Firstly, a "pseudo-walking" input comprised of a single sinusoid derived from the motion of a driven pendulum that approximates the swing of a human arm during vigorous walking. Secondly, a collection of 6-axis inertial data collected from the wrists of five subjects during a casual walking motion. For the collection of real walking data, the optimal design parameters of each device were determined for each walking signal individually by virtue of the PS algorithm, and the maximum, minimum, and average power output of each signal found for every device architecture.

4. Model Validation

A rotational prototype device, representing a physical implementation of structure one, was constructed for the purpose of empirically validating the rotational model used in the analysis. See Figure 3.

An electromagnetic transducer was constructed for the purpose of measuring power output which consisted of an array of twenty magnets embedded in a dual-rotor design and ten corresponding circular planar copper coils. A load resistance which matched the 240 Ω coil resistance was placed across the output of the device and the voltage across the load was measured in order to compute average power dissipation. In order to reproduce a pseudo-walking input, a stepper motor-driven swing arm setup was constructed and was fed the appropriate angular position commands to reproduce the ideal signal. See Figure 4.

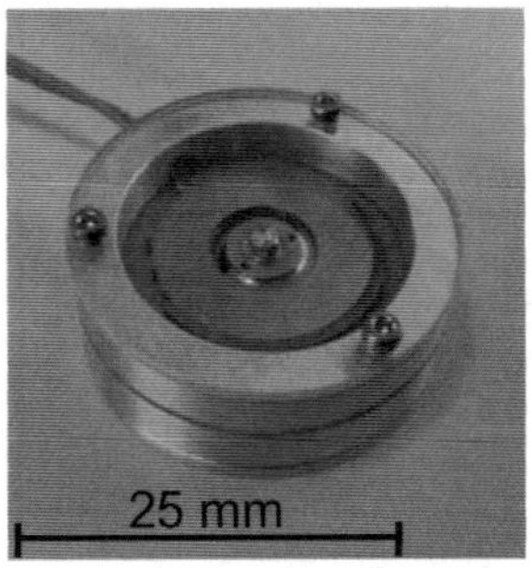

Figure 3. Rotational prototype device with an electromagnetic transducer.

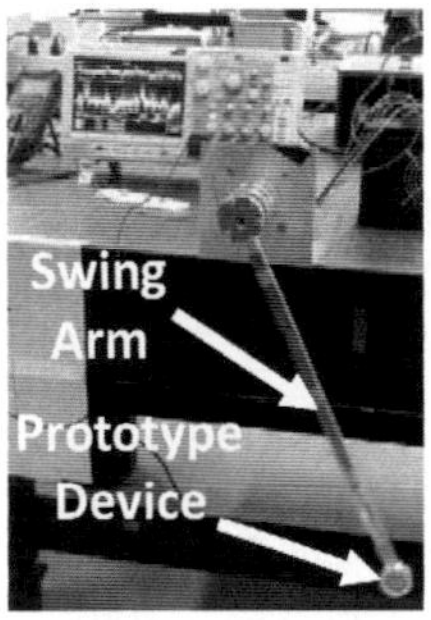

Figure 4. Swing arm experimental setup during operation.

The results of the experiment are summarized in Figure 5, which show general agreement between simulation and experimental measurement.

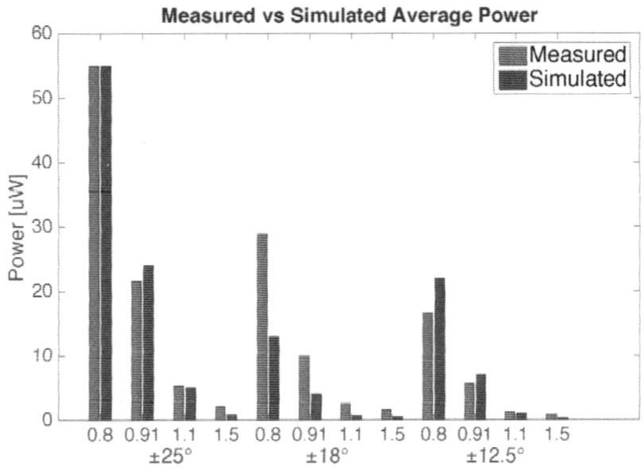

Figure 5. Comparison of measured vs simulated test data for pseudo-walking inputs. Three amplitudes and four periods were tested.

5. Results

The optimal geometry of the two-dimensional slide structures was found to converge to the optimal one-dimensional slide structure geometries for all provided inputs, indicating that the additional dimension does not provide any benefit in terms of power output. For this reason, the two dimensional linear structures (structures five and six) were no longer considered after the optimization step.

The addition of springs to all devices generally improved performance for both pseudo-walking and real walking data inputs in the simulation-based comparative analysis. In the case of sprung vs unsprung rotational harvesters (structures one and two), the addition of a spring improved performance by 365% under pseudo-walking input, and an average of 68% under real walking inputs. In the case of the linear slide structures (structures three and four), the addition of a spring did not significantly improve performance under pseudo-walking input, but improved performance by an average of 197% under real walking inputs. See Figure 6. The average power output of the sprung rotational architecture (structure two) was 58% greater than the average for the sprung linear slide (structure four) using real walking data as inputs. A consequence of adding a spring to the rotational structure is increased sensitivity of power output to electrical damping coefficient. See Figure 7.

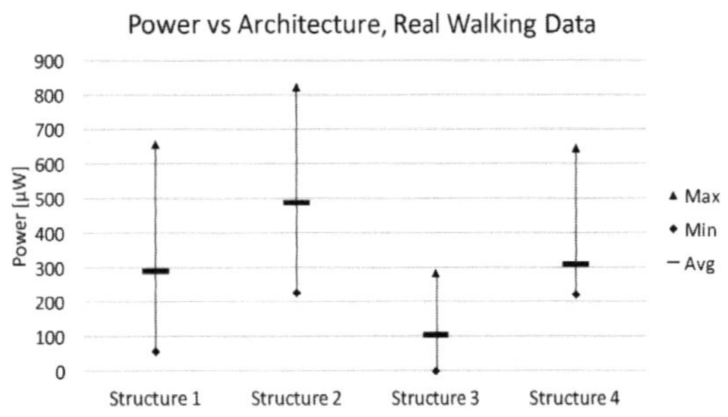

Figure 6. Maximum, minimum, and average power output of structures subject to five real walking data signals.

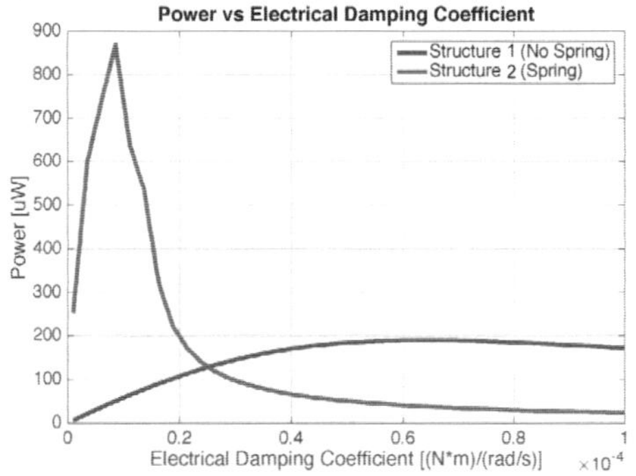

Figure 7. Power vs electrical damping coefficient for the two rotational structures. Notice sharp drop of power output when electrical damping constant is not near the optimal value.

6. Conclusions

A simulation-based comparative analysis of six vibration energy harvesting structures was performed. The addition of a spring to all structures considered in the study resulted in improved simulated power output for real walking data. The average power output of the sprung rotational structure was 58% greater than the average for the sprung linear slide using real walking data as inputs.

Acknowledgements

Funding for this research was provided by the National Science Foundation under Award Number ECCS 1342070, and a generous grant from Analog Devices Inc.

References

[1] P. Mitcheson, T. C. Green, E. M.Yeatman, and H. A. S., "Architectures For Vibration Driven Micropower Generators," *J. Microelectromechanic Syst.*, vol. 13, no. 3, pp. 429–440, 2004.

[2] T. Xue, X. Ma, C. Rahn, and S. Roundy, "Analysis of Upper Bound Power Output for a Wrist-Worn Rotational Energy Harvester from Real-World Measured Inputs," *J. Phys. Conf. Ser. 557 012090*, vol. 557, pp. 3–7, 2014.

[3] Mathworks, "Global Optimization Toolbox: User's Guide (R2016b)," 2016. [Online]. Available: http://www.mathworks.com/help/pdf_doc/gads/gads_tb.pdf. [Accessed: 15-Sep-2016].

An implantable fluidic vibrational energy harvester

S Inoue, T Takahashi, M Kumemura, H Fujita and H Toshiyoshi

Institute of Industrial Science, The University of Tokyo, 4-6-1, Komaba, Meguro-ku, Tokyo, 153-8505, Japan

E-mail: hiro@iis.u-tokyo.ac.jp

Abstract. Targeting implantable medical devices such as respiratory pace-maker, we have developed a proof-of-concept level energy harvester device that could earn electric power of 0.44 μW/cm^2 by the fluidic motion in a PDMS microchannel placed on a silicon substrate with built-in permanent electrical charges or so-called electrets. The motion of the working fluid will be operated by the heart beat or breathing as a final shape of the energy harvesting system.

1. Introduction

In this paper, we propose a fluidic type MEMS energy harvester that is under development for implantable medical electronics such as diaphragm pacemaker for ALS (amyotrophic lateral sclerosis). Such electronics is thought to consume electrical power of more than 1 mW despite today's low-power electronics technology. A piezoelectric type energy harvester has been reported that is driven by the heart beat but the total power is limited to a few tens of μW [1]. The device under development in this work, on the other hand, is scalable to a mW-class power output.

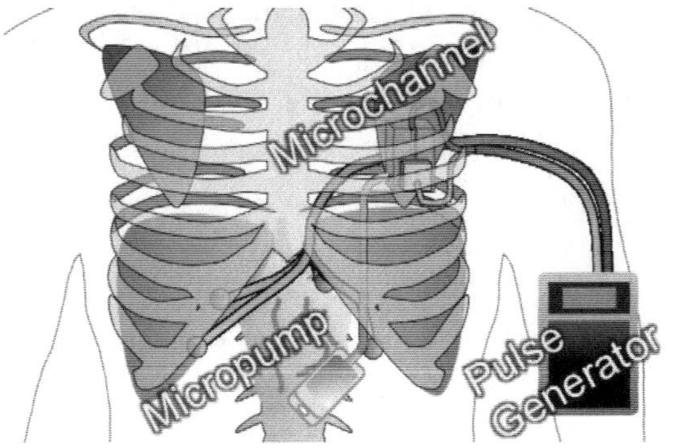

Figure 1. A schematic illustration of implantable medical electronics that assists spontaneous respiration by stimulating the diaphragm muscle. A micropump type energy harvester is placed near by the heart to drive the fluids in a microchannel to electrostatically induce current by periodically altering the capacitance with respect to the electret plate.

2. Power generation mechanism

Figure 2 shows the device mechanism, where a microchannel with upper and lower electrodes is formed with a layer of electrical insulation material that stores permanent electrical charges or so-called electrets. In this work, we used potassium-doped silicon oxide film [2]. The channel is partially filled with a liquid that is periodically displaced by the applied pressure to cover the electret region, thereby changing the effective dielectric constant in the capacitor to cause electrical current by electrostatic induction. In this work, we used air (ε_r=1) and DI water (ε_r=80, @25°C) for fluids 1 and 2, respectively.

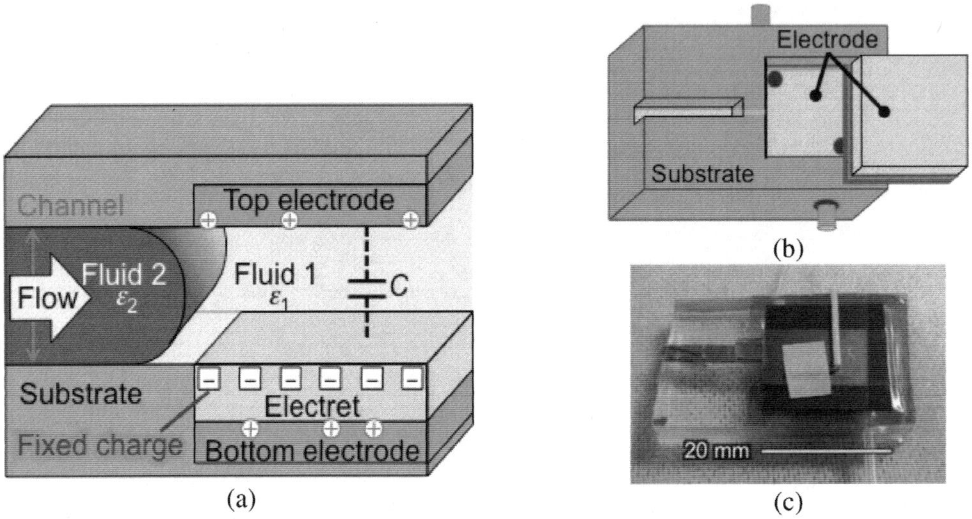

Figure 2. Device structures and operation principle. (a) Side view of the electrical micro fluidic channel. A water flow displaced by the external pressure covers the electret to change the effective dielectric constant in the capacitor ($\varepsilon_1 \leftrightarrow \varepsilon_2$) to cause electrostatic inductive current. (b) Top view schematic. (c) Photograph of the device after assembling a PDMS micro fluidic channel, Au electrodes, and silicon substrate with built-in electret layer.

3. Result of experiment

A PDMS (polydimethylsiloxane) micro fluidic channel plate was formed by using a SU8-3050 mold on a silicon substrate. After substrate removal, an electrode was formed on the PDMS by vacuum evaporation of Cr/Au. On a separate substrate, a silicon oxide electret layer was formed by K^+-doping method [2]. To improve the electrochemical resistance to water, a CYTOP (880 nm) was spun on the wafer. Finally, these two pieces are attached with the electret surface inside, and fixed with a tape.

After connecting tubes, water was injected with a motor-controlled syringe pump. As a result, an output voltage of 6 V was observed with a 10 MΩ load resistance, yielding a first result of 0.9 μW for a 2.25 cm^2 electrode device as shown in figure 3. Theoretical model predicts a 1 mW output with a smaller gap of 50 μm, an electret density of 32 mC/m^2, and a total electrode area of 31 cm^2.

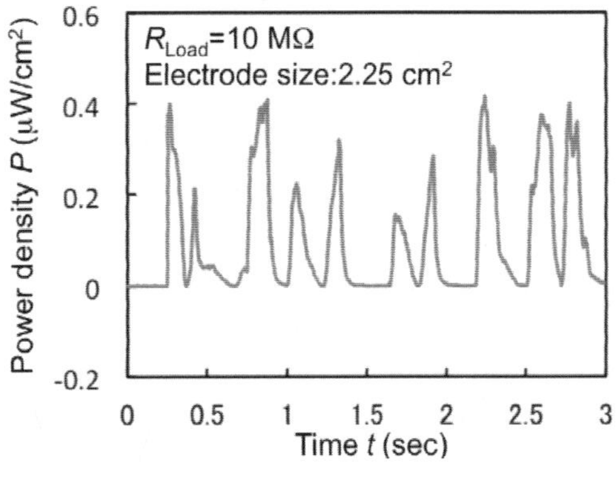

Figure 3. Power output per area for a 10 MΩ load resistor when the working fluid is displaced by a syringe pump at 1.8 Hz.

References

[1] Pfenniger A *et al* 2013 *Ann. Biomed. Eng.* **41** pp 2248-2263
[2] Hashiguchi G *et al* 2016 *AIP Adv.* **6** 035004

Hybrid nanocone forests with high absorption in full-solar spectrum for solar cell applications

Yudong Yang[1,2], Haiyang Mao[1,*], Jijun Xiong[2,*], Anjie Ming[1], Weibing Wang[1]

[1]Institute of Microelectronics of Chinese Academy of Sciences, Beijing, P. R. China
[2]National Key Laboratory for Electronic Measurement Technology, North University of China, Taiyuan, P. R. China

*E-mail: maohaiyang@ime.ac.cn, xiongjj@263.net

Abstract. In this work, hybrid nanocone forests (HNFs) with high absorption in full-solar-spectrum are fabricated based on a plasma repolymerization technique. The HNFs combine light trapping effect of the nanocone forests with surface plasmon resonance effect of the metallic nanoparticles, thus can achieve an optimized absorption larger than 80% in the full-solar spectrum (i.e. 200-2500nm). Besides, with the hybrid nanostructures, the absorption decrease around the Si bandgap width can be narrowed greatly, while the normalized utilization efficiency of solar radiation can be increased. Therefore, usage of the HNFs as a texture structure in solar cells to obtain higher conversion efficiencies is foreseeable.

1. Introduce

To realize solar cells with high conversion efficiency, textures with high optical absorption are of importance, as they can motivate a large number of optical-induced carriers in the full-solar spectrum[1]. Due to light trapping effect, nanocone forests have been regarded as a promising texture usable in solar cells. However, most of the reported nanocone forests only can enhance absorption in a very limited solar spectral range and their generation methods are usually not compatible with conventional MEMS process[2-4]. In this work, based on a plasma polymerization (PP) technique proposed to fabricate a novel type of nanocone forests, HNFs are formed by sputtering Ag nanoparticles on surfaces of the nanocone forests. As detected, the HNFs can achieve an average absorption larger than 80% in the full-solar spectral range of 200-2500nm. By integrating the HNFs in solar cells, high conversion efficiency of the cells would be further predicted.

2. Fabrication

The process for fabricating the HNFs based on the PP technique is sequentially depicted in figure 1. Firstly, a polyimide (PI) layer is spin-coated on a Si substrate. Then, plasma bombardment is adopted to self-generate nanomasks. Later on, a reactive-ion-etching (RIE) step is employed to form Si nanocones, which is then followed by buffered oxide etching (BOE) to remove the nanomasks. Finally, Ag nanoparticles are sputtered on surfaces of the nanocone forests. Scanning electron microscope (SEM) images corresponding to the fabrication steps of the HNFs are demonstrated in figure 2.

3. Experiment and Discussion

The absorption spectra of a Si wafer with planar surface, nanocone forests before Ag sputtering and the HNFs were measured using an UV-vis spectrometer. Within the wavelength range of 200-2500nm, as shown in figure 3, the average absorption of the HNFs reaches 80.6%, which is much larger than that of the nanocone forests (70.1%) and the planar Si (32.4%). As solar radiation is asymmetrically distributed along the full-solar spectrum, among which, the visible

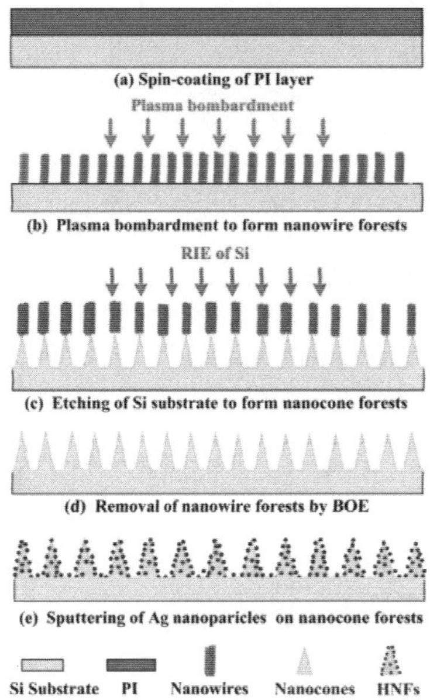

Figure 1. Fabrication process of the HNFs based on the PP technique.

light contributes slightly more than the infrared light, thus the normalized energy utilization ratio for the nanocone forests increases the most. Even though, a ratio of 84.2% can be realized by the HNFs (inset in figure 3). Compared with the nanocone forests, the larger absorption and utilization ratio of the HNFs can be attributed to the surface plasmon resonance induced by the Ag nanoparticles and the nanocones.

To demonstrate the absorption features of the three Si-based structures around the Si bandgap width, a spectral range of 1000-1200nm is chosen for the analysis. As illustrated in figure 4, the decrease of planar Si within this wavelength range is 59%, and that for the nanocone forests is 43%, while the HNFs can reduce the decrease to 6%. In other words, the HNFs have an ability to break absorption limitations caused by the Si bandgap width. Meanwhile, the absorption difference of the HNFs with Ag nanoparticles of different sizes is studied. As shown in figure 5, the absorption decrease of HNFs at the bandgap width becomes smaller when the nanoparticle sizes increase. Therefore, higher average absorption can be expected by sputtering Ag nanoparticles with larger sizes.

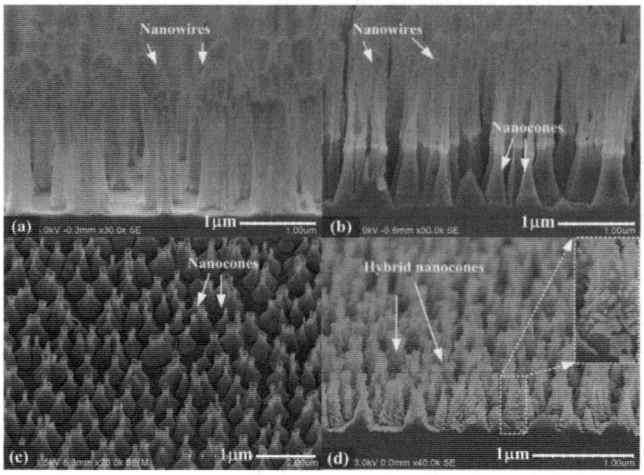

Figure 2. SEM images of nanostructure forests demonstrating formation process of the HNFs.

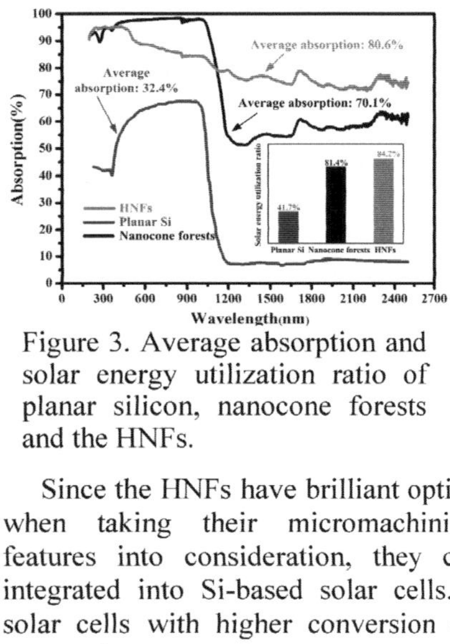

Figure 3. Average absorption and solar energy utilization ratio of planar silicon, nanocone forests and the HNFs.

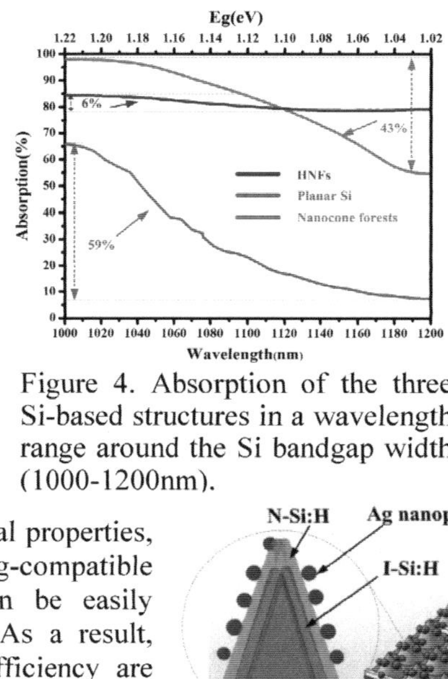

Figure 4. Absorption of the three Si-based structures in a wavelength range around the Si bandgap width (1000-1200nm).

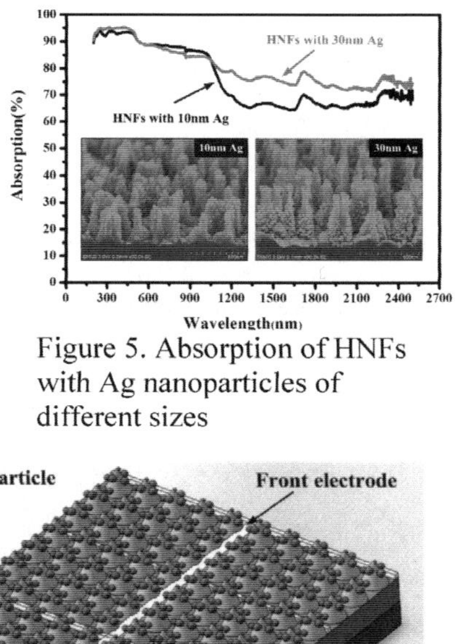

Figure 5. Absorption of HNFs with Ag nanoparticles of different sizes

Since the HNFs have brilliant optical properties, when taking their micromachining-compatible features into consideration, they can be easily integrated into Si-based solar cells. As a result, solar cells with higher conversion efficiency are expected.

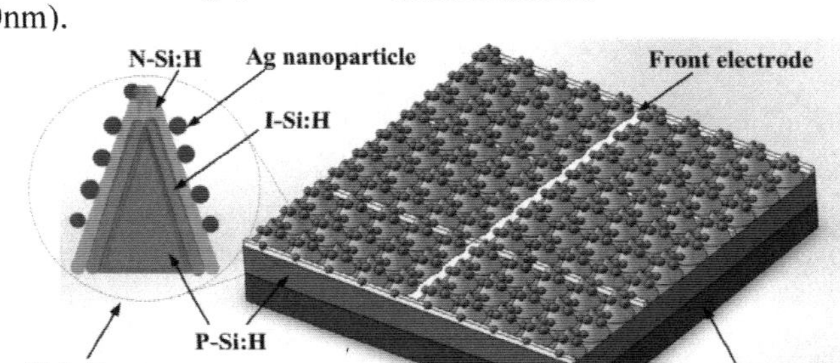

Figure 6. Schematic diagram of solar cells integrated with the HNFs.

Acknowledgement

This work was supported by National Natural Science Foundation of China (Grant No. 61401458, and 61335008), and National High Technology Research and Development Program of China (No.2015AA042605).

References

[1] Wang K, Yu Z, Liu V, Cui Y and Fan S 2012 *Nano lett.* **12** 1616-9.
[2] Qiu Y, Hao H, Zhou J, Lu M 2012 *Opt. Exp.* **20** 22087-94.
[3] Wang Y, Tang L, Mao H, Lei C, Ou W, Xiong J, Wang W and Hu J 2016 *IEEE MEMS* 2016 1185-8.
[4] Yang Y, Luo F, Kao T, Li X, Ho G, Teng J, Luo X and Hong M 2014 *Light: Sci. Appl.* **3** e185:1-8.

Multi-axis piezoelectric energy harvesting using MEMS components with slanted beams

Pontus Johannisson, Fredrik Ohlsson, Cristina Rusu

Acreo Swedish ICT AB, Arvid Hedvalls Backe 4, SE-411 33 Gothenburg, Sweden

E-mail: pontus.johannisson@acreo.se

Abstract. During the manufacturing of MEMS components, slanted beams can be produced in the etching process. We show that this can be used to produce skew motion that causes deflection of a proof mass out of the device plane also when the excitation is confined to the device plane. This allows construction of an energy harvester that uses a planar manufacturing process and produces power also with in-plane excitation. To obtain this with traditional methods it would be necessary to manufacture separate components and then mount them with their sensitive axes orthogonal to each other.

1. Introduction

The increasing need to sense and communicate is driven, e.g., by networked devices and sensors in the internet of things. It is often difficult to arrange wired power supply and the use of batteries necessitates regular replacement and disposal. Energy harvesting from vibrations is often considered, converting kinetic energy into electric energy using electromagnetic, electrostatic, or piezoelectric transduction [1]. Energy harvesters (EHs) utilizing piezoelectric materials are most popular and this approach is used in this study.

One common way to implement an EH is as a cantilever with a proof mass at the tip and with PZT in a uni- or bimorph configuration [2]. Regardless of the design choices, a traditional EH generates negligible power when the excitation is in the device plane. One obvious solution is to use more than one EH. Combining three orthogonal EHs it is possible to obtain power regardless of the excitation direction but this requires an extra assembly step, reduces the device compactness, and increases the cost.

In this paper, we consider the use of slanted beams, which can be fabricated by wet etching of silicon. As the etching rate depends on the crystal orientation, planes inclined at $\arctan(\sqrt{2}) \approx 54.74°$ can be produced. Due to this slantedness, an in-plane excitation is partly converted to an out-of-plane deflection and the axis corresponding to the maximum power is rotated from the out-of-plane direction. This method has previously been used to design an accelerometer [3], but to the best of our knowledge this approach is novel in the EH context. Using numerical simulations in COMSOL Multiphysics, we compare our approach to a traditional cantilever EH with a rectangular beam and report how the output power is affected, to what axis it corresponds, and to what extent the device is a multi-axis EH.

2. Design approach

For compatibility with MEMS manufacturing it is assumed that the PZT is deposited on top of the silicon structure. The PZT material, PZT-5H, is assumed to be operated in the 31 mode, i.e., the strain is orthogonal to the electric field. A further choice is to target a relatively low

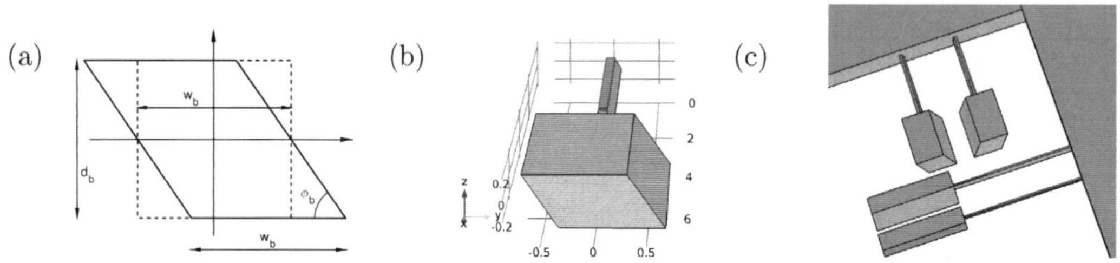

Figure 1. (a) The slanted beam with $\phi_b = \arctan(\sqrt{2}) \approx 54.74°$. In the case with a rectangular beam (dashed line), all angles are $\pi/2$. (b) The slanted beam EH with PZT (blue). A more robust design is discussed in the main text. (c) One possible combination of four EHs.

frequency at 300 Hz. The geometry of the beam is shown in Fig. 1a, both for the slanted (solid line) and rectangular (dashed line) cases, and the beam width, w_d, and thickness, d_b, are defined in the same figure. It is not possible to change only the beam angle as this changes the eigenfrequency of the structure, leading to a drastically different response to a given excitation. Furthermore, the aspect ratio, defined as $\eta \equiv d_b/w_b$, will be varied and this leads to similar problems. Therefore, the first resonance frequency of the different EHs was kept at 300 Hz by adjusting the beam length.

Focusing on the multi-axis potential, the design is kept very simple. We emphasize that other designs that preserve the qualitative features are possible. For example, one approach would be to use two beams offset in the y direction (instead of the x direction) from the proof mass. Attaching one end of the beams to the proof mass and the other to the frame, a robust S-shaped EH is obtained. Here, we investigate the cantilever structure shown in Fig. 1b. The volume of the proof mass is kept constant, although the shape depends on the etching. The proof mass is made of silicon and has the dimensions 3 mm $\times$ 1 mm $\times$ 0.5 mm, cf. Fig. 1b. The width of the beam, w_b, is selected to be 0.1 mm. When the thickness is varied, the length of the beam is adjusted to keep the first resonance at 300 Hz. The mass and the beam would have their upper surfaces aligned in practice but we here align their centroids for simplicity. The PZT thickness is 20 μm. It has the same width as the beam and we set the length of the PZT to 3 mm. The electrodes are placed on the top and bottom surfaces of the piezoelectric layer and connected by a 50 kΩ load resistance. This is approximately the average of the optimum values for the slanted (80 kΩ) and rectangular (20 kΩ) cases.

Also an EH with a slanted beam will generate negligible power when the excitation is perpendicular to the sensitive direction. Thus, the intention is to combine four of these EHs as conceptually suggested in Fig. 1c. Note that the beams are slanted in different directions.

3. Numerical simulations

As both the volume of the proof mass and the stiffness (via the resonance frequency) of the beam are kept constant, comparable motion of the proof mass is obtained in all cases. The beam length as a function of the aspect ratio is shown in Fig. 2a. As the length of the PZT is constant, it is not possible to make the slanted beam very thin. Thus, the slanted beam thickness is not made smaller than 70 μm. In the rectangular case, the beam length increases faster with the beam thickness. The reason for this difference is that when d_b increases for the slanted beam, the weakest direction of the beam is rotated.

The fact that the slanted beam has its weakest direction at an angle makes the aspect ratio an important parameter. When the aspect ratio is increased the weakest direction turns toward the in-plane direction, but at the same time the power is reduced as the strain in the PZT is reduced. The power angular dependence is exemplified in Fig. 2b. In this case, the EH is excited harmonically using body forces. The acceleration amplitude is $1g$ and the frequency is 300 Hz.

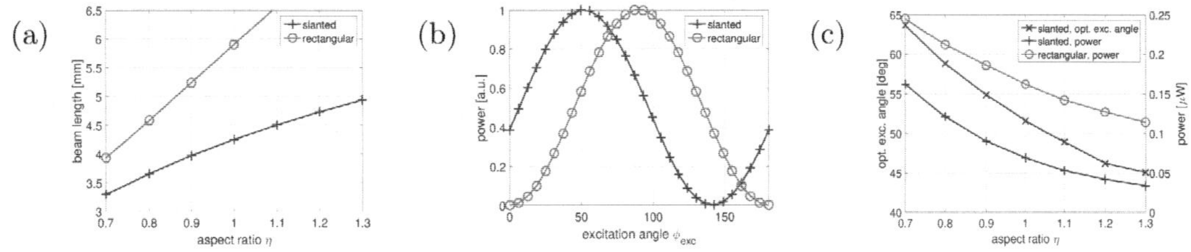

Figure 2. (a) The beam length as a function of the aspect ratio. (b) Example results of the EH power as a function of the excitation direction. (c) The optimal excitation angle and the EH power as a function of the aspect ratio.

The direction of the acceleration is rotated ϕ_{exc} in the y-z plane and the power, normalized to its peak value, is plotted. The rectangular beam has its maximum power for $\phi_{\text{exc}} = \pm 90°$, while the maximum for the slanted beam depends on the aspect ratio. In Fig. 2b, $\eta = 1$ has been used. The optimal excitation angle has been plotted as a function of the aspect ratio for the slanted beam (using $\times$ markers) in Fig. 2c. It is seen that for the thinnest slanted beam ($\eta = 0.7$), the optimal excitation angle is largest (64°). If the beam would be made even thinner, the optimal angle would approach 90°. On the other hand, for thick beams the optimal angle is reduced, but the reduction rate is not fast. For very thick beams, with $\eta \gg 1$, the weakest direction of the beam corresponds to $\phi_{\text{exc}} = 90° - 54.74° \approx 35°$, cf. Fig. 1a.

In Fig. 2c, the EH power is plotted as a function of the aspect ratio for the slanted and rectangular beams. However, these values were *not* obtained from steady-state solutions with harmonic excitation. The reason is that the damping is very small as the PZT layer is thin. This leads to low electrical losses and very high quality factors. Thus, the power is very sensitive to the exact choice of excitation frequency. In practice, the damping would be dominated by phenomena such as drag forces [4] or squeeze film damping [5]. However, these effects depend on parameters that are not part of the current investigation. Therefore, the time-dependent simulations used an acceleration pulse in the form of a sine wave at 300 Hz truncated after the first period. The energy spectral density of this pulse is obtained from the convolution of the Fourier transform of the sine function and a rectangular window. It has its maximum close to 300 Hz but will not have its energy concentrated at a single frequency. The pulse amplitude is $10g$, but the power with this short excitation is limited to the sub-μW regime.

Simulations have been performed with this pulsed excitation using the optimal excitation angle according to Fig. 2c. (For the rectangular beam, the excitation angle is $\phi_{\text{exc}} = 90°$.) After the pulse, the proof mass is oscillating and, due to the choices in Section 2, the maximum kinetic energy is equal in all cases. The maximum power has been plotted in Fig. 2c. It is seen that the slanted beam gives approximately 66 % of the power with $\eta = 0.7$ and that the relative difference increases with the aspect ratio. This is expected since the rectangular beam gives maximum strain in the PZT.

4. Discussion

To see to what extent the power generation can be made independent of the excitation direction, it is instructive to plot the EH power as a function of the excitation direction, see Fig. 3. As shown in Fig. 2b, the power varies harmonically with the excitation angle. The amplitude varies between zero and the maximum value and the period is π. Thus, the normalized power can be plotted as a function of the excitation angle in an angular plot, where the distance from the origin to the curve corresponds to the normalized power. In Fig. 3a, the case for a rectangular beam is shown. When the angle is zero, the power is zero and maximum power occurs for $\phi_{\text{exc}} = \pm 90°$. The slanted beam case can be visualized using Fig. 2c. Assuming that there are

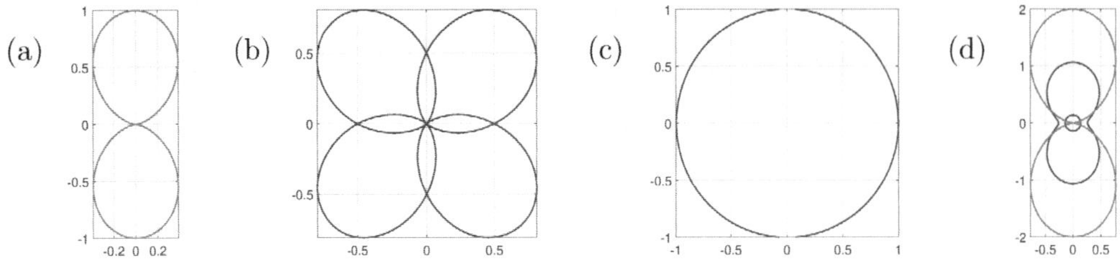

Figure 3. (a) Normalized power lobe for a rectangular EH. (b) Normalized power lobes for two EHs using $\eta = 1.3$ with beams slanted in different directions. (c) The total power for the two EHs in b. (d) Power lobes for two square EHs (red) and two slanted beam EHs (blue) using $\eta = 0.7$ and $\eta = 1.3$, respectively.

two slanted beam EHs (cf. Fig. 1c), there are two lobes as shown in Fig. 3b. The relative angle between the lobes and the absolute power is selected, via the aspect ratio, from Fig. 2c. The optimal excitation angle is $\pi/4$ when the slanted beam $\eta = 1.3$. Using this value would give an isotropic two-dimensional (2D) EH, as the sum of the power from the two EHs is then a circle, see Fig. 3c. However, this is sub-optimal as the power decreases rapidly with increasing aspect ratio. Instead maximizing the power for an in-plane excitation, $\eta = 0.7$ is optimal and the optimal excitation angle is 64°. This leads to the power lobes in Fig. 3d. In red, the power from two rectangular beams is plotted. The inner blue curve is the circle from Fig. 3c with correctly scaled relative power and the outer blue curve corresponds to $\eta = 0.7$. The difference compared to the red curve is due to that (i) the slanted beam EHs redistribute the sensitivity angles and (ii) the slanted beam EHs give 66 % of the power.

The above selection leads to maximum power for an in-plane excitation. Unfortunately, the difference in power comparing the in-plane and out-of-plane excitations is close to a factor of four. To achieve a three-dimensional (3D) system, two additional EHs are needed, cf. Fig. 1c. As the out-of-plane excitation will yield another identical contribution to the power, the power ratio for the out-of-plane and in-plane excitations will then increase to a factor of eight.

5. Conclusion

The potential of using slanted beams in MEMS EHs to obtain power also when the excitation is in the device plane has been investigated. With the current approach, the slanted beam EH gives 66 % of the power but is superior in the sense that it generates power for in-plane excitation. Combining two slanted beam EHs into a 2D harvester, it is possible to obtain constant output power, but when optimizing the EH power for in-plane excitation, there is a factor of four power difference comparing out-of-plane and in-plane excitation. For the cases where the power specification is met also considering this difference, the objective of constructing a planar multi-dimensional EH can be obtained.

Acknowledgments

The Smart-MEMPHIS project has received funding from the European Union's Horizon 2020 research and innovation programme under grant agreement No. 644378.

References

[1] Harb A 2011 *Renewable Energy* **36** 2641–2654
[2] Roundy S and Wright P K 2004 *Smart Materials and Structures* **13** 1131–1142
[3] Rödjegård H, Andersson G I, Rusu C, Löfgren M and Billger D 2005 *J. Micromech. Microeng.* **15** 1989–1996
[4] Blom F R, Bouwstra S, Elwenspoek M and Fluitman J H J 1992 *J. Vac. Sci. Technol. B* **10** 19–26
[5] Bao M and Yang H 2007 *Sens. Actuators* **136** 3–27

Equivalent-circuit model for slot-effect configuration of an electret-based generator

Cuong Phu Le[1], Michael Renaud[2] and Einar Halvorsen[1]

[1]Univ. College of Southeast Norway, Norway, [2]IMEC/Holst Centre, The Netherlands

E-mail: Einar.Halvorsen@hbv.no

Abstract. This paper presents a completely analytical method to build a lumped-model for an electret-based energy harvester. The harvester consists of two patterned electrodes facing movable electret patterns in a slot-effect electrical scheme. A Chebyshev expansion with orthogonal functions to capture singularity effects and Galerkin's method are used to determine fundamental parameters of the equivalent circuit. All important fringing fields in the microscale device are thereby taken into account. For a device example, the calculated parameters of the model agree well with those obtained by finite element modelling. The dynamic behaviour is reproduced by the established model. The advantage of the approach is to allow a fast and full exploration of the design parameters for optimization of the device structure.

1. Introduction

Biasing of electrostatic vibration energy harvesters is an essential feature for their operation. For a full MEMS integration, internal bias by an electret, which is a dielectric with trapped charges [1], is favourable over external bias by circuitry. A number of recent investigations on MEMS electret-based vibration energy harvesters consider devices configured as a set of metal-strip electrodes and a patterned electret on two separate substrates [2, 3, 4, 5]. Such a device concept has potential applications in car-tire pressure monitoring [6, 7, 8].

Modelling of electret-based harvesters is challenging when electrostatic fringing field effects are significant. Several groups have developed models and guidelines for device optimization. These models can be made either by using an ideal field approximation, which is inaccurate, or by time-consuming finite element computations [9, 10, 11]. Conformal mapping techniques can be a solution, but are limited by assumption on geometry and material [12]. An improved model by [13] included closed-form parasitic capacitances using an effective permittivity to represent the dielectric media surrounding the electrodes. The effect of the electret substrate was ignored even when the gap between the two substrates was small.

In the present work, the aforementioned limitations are overcome by developing an accurate and fully analytic method to build an equivalent lumped-model for the harvester. The method used here is adapted from our previous work [14] that includes electrostatic nonlinearities and fringing fields. The harvester parameters reported in [15] are used for example calculations that illustrate the method.

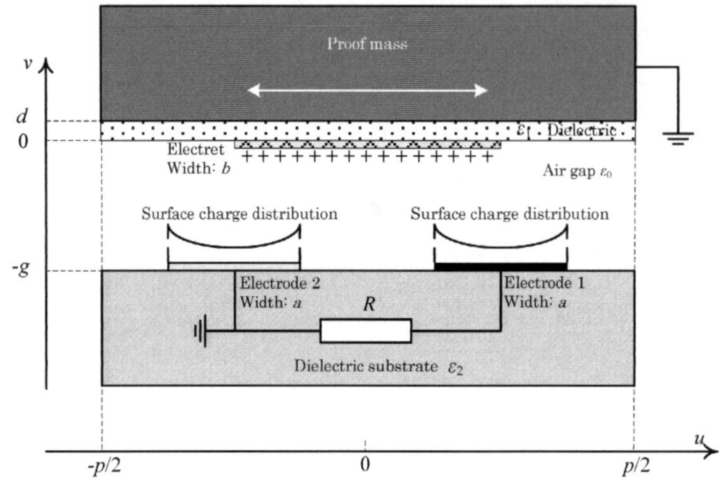

Figure 1. Cross-section of a single period for an electret-based vibration energy harvester.

2. Transducer description

Figure 1 shows a simplified cross-section of one period p for the electret-based transducer. The transducer consists of N pairs of metal electrodes 1 & 2 on a dielectric substrate with a permittivity ε_2 that face electret patterns deposited on a dielectric layer with a permittivity ε_1 and a thickness d. The dielectric layer ε_1 is supported by a conductive substrate functioning as a proof mass m suspended by mechanical springs (not shown) with a total stiffness k. The medium gap g between two substrates has a permittivity ε_0. The metal electrode and electret patterns have a width of a and b respectively. Each pattern has a length L. The electret has a total fixed charge $Q_e = \frac{\varepsilon_1 bNL}{d}V_e$, where V_e is an equivalent average surface voltage over the electret patterns.

There are several schemes for power extraction based on the arrangement of connections between the three electrical terminals, i.e the two electrodes and the conductive proof mass. A typical configuration is the slot-effect scheme [12, 15, 16] where a load, e.g. a resistor R, is connected between the two electrodes 1 & 2 and one of them is grounded together with the conductive proof mass as shown in figure 1. When the proof mass oscillates in the lateral direction, the charged electret induces varying charges on the counter electrodes and then generates an electrical current through the resistor R.

3. Analysis and modelling

In the analysis, we assume that all electrode and electret patterns are sufficiently thin that they can be treated as zero thickness. The number N of periods is large enough that end-effects can be ignored. To capture fringing fields, the surface charges induced on the electrodes are mathematically represented by an orthogonal expansion in Chebyshev polynomials multiplied by a reciprocal square root form, i.e.

$$\sigma_{1,2}(u) = \begin{cases} \sum_{n=0}^{N} C_n^{1,2}\phi_n^{1,2}(u) = \sum_{n=0}^{N} C_n^{1,2}\dfrac{T_n\left(\frac{2(u\mp\frac{p}{4})}{a}\right)}{\sqrt{1-4\frac{(u\mp\frac{p}{4})^2}{a^2}}} & -\frac{a}{2} \le u \mp \frac{p}{4} \le \frac{a}{2} \\ 0 & \text{elsewhere} \end{cases} \quad (1)$$

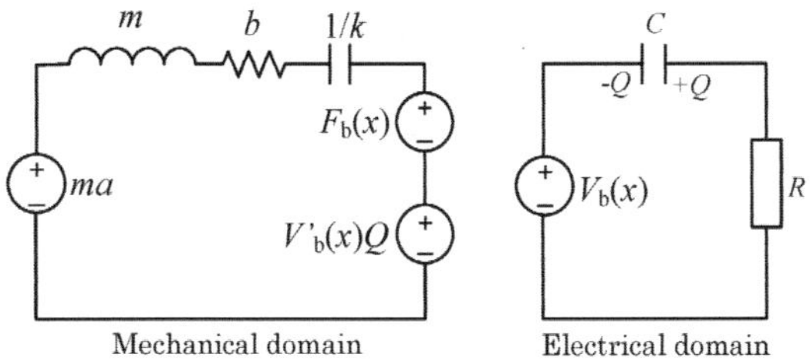

Figure 2. Lumped-model of electret-based transducer.

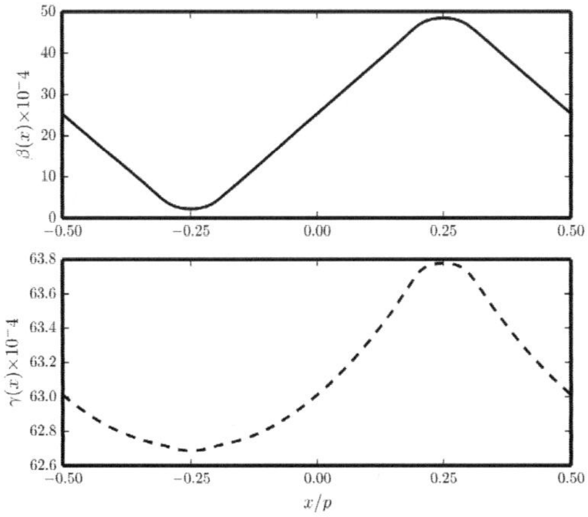

Figure 3. Induced-charge coefficients as function of proof mass displacement.

where N is a maximum Chebyshev index, $T_n(u)$ are Chebyshev polynominals of degree n=0, 1, ..., N, $\{C_n^1\}$ and $\{C_n^2\}$ are expansion coefficients. The total charges on electrodes 1 & 2 are

$$Q_{1,2} = NL\int_{-p/2}^{p/2} \sigma_{1,2}(u)du = NL\frac{\pi a}{2}C_0^{1,2}. \tag{2}$$

From here, the analytical approach to establish parameters for the equivalent circuit model can be briefly described as follows. First, Laplace's equation $\nabla^2\varphi(u,v) = 0$ is used to find the potential distributions on two substrate surfaces $\varphi(u,0)$ and $\varphi(u,-g)$. The Galerkin method with the reciprocal root form as a weight function in Eq. (1) is then employed to enforce constant potential boundary on two electrodes $V_1 = V_R$ and $V_2 = 0$ V. Now, the expansion coefficients $\{C_n^1\}$ and $\{C_n^2\}$ can be found by solving the resulting set of equations. Further details on procedure to determine the expansion coefficients can be found in [14]. As a result, the total electrostatic energy is

$$W_e(x,Q) = \frac{1}{2}\begin{bmatrix} Q_1 & Q_2 \end{bmatrix}\begin{bmatrix} V_1 \\ V_2 \end{bmatrix} + \frac{1}{2}NL\int_{-\frac{p}{2}+x}^{\frac{p}{2}+x} \varphi(u,0)\sigma_e(u)du \tag{3}$$

where the last term in the above equation is the electret energy with $\sigma_e(u) = \frac{Q_e}{bNL}$ for $-b/2 \le u - x \le b/2$. With the charge $Q = Q_1$ as the generalized displacement in the electrical domain, the energy W_e can be written on the form

$$W_e(x,Q) = \frac{\alpha}{4\pi\varepsilon_0 NL}Q^2 + \frac{\beta(x)}{2\pi\varepsilon_0 NL}QQ_e + \frac{\gamma(x)}{4\pi\varepsilon_0 NL}Q_e^2 \tag{4}$$

where the coefficients α, $\beta(x)$, and $\gamma(x)$ are dimensionless and dependent on the geometry, the transducer materials and the proof mass displacement. The corresponding electrostatic force acting on the proof mass and the voltage across the resistance R are

$$F_e = \frac{\partial W_e(x,Q)}{\partial x} = \frac{\beta'(x)}{2\pi\varepsilon_0 NL}QQ_e + \frac{\gamma'(x)}{4\pi\varepsilon_0 NL}Q_e^2 = V_b'Q + F_b(x) \tag{5}$$

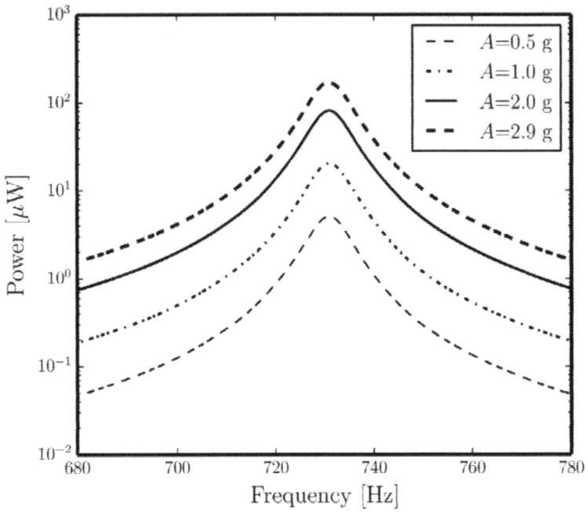

Figure 4. Frequency responses of output power for various acceleration amplitudes at $V_e = 160$ V and $R = 5.0$ MΩ.

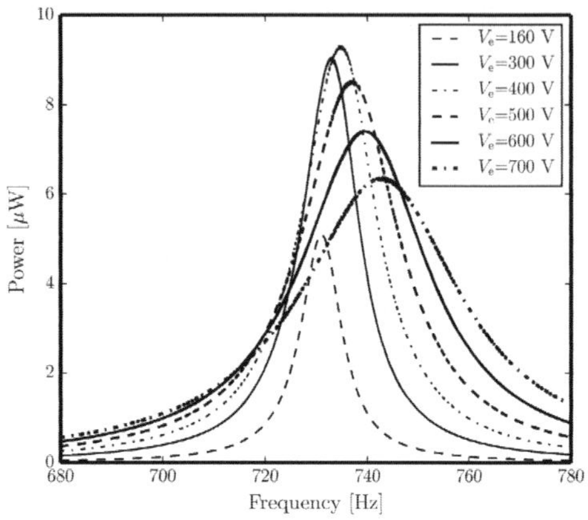

Figure 5. Effect of electret voltage V_e on frequency response at $A = 0.5$g and $R = 5.0$ MΩ.

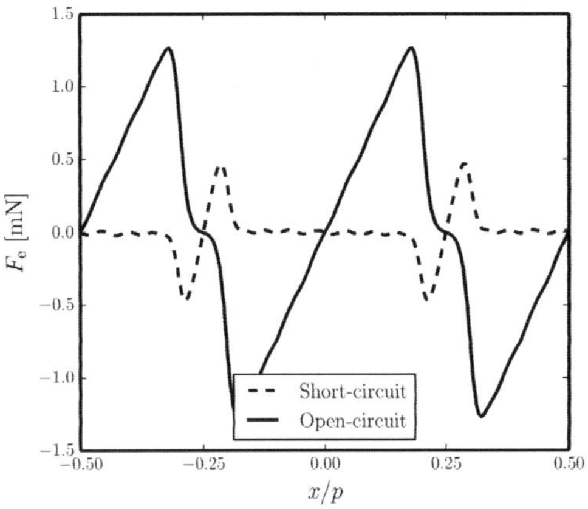

Figure 6. Electrostatic force for short-circuit and open-circuit conditions at $V_e = 160$ V.

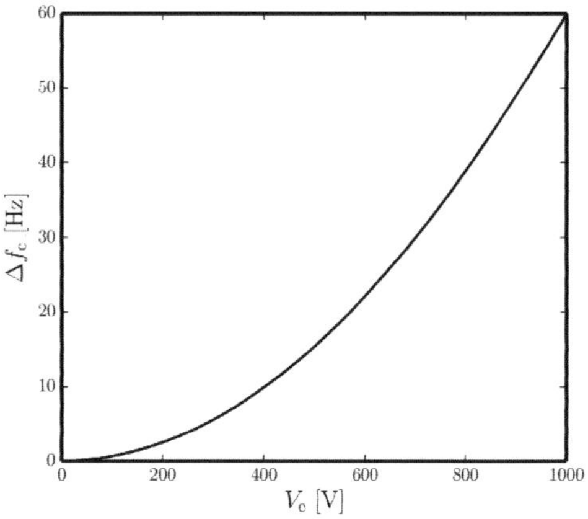

Figure 7. Center frequency shift at open circuit when increasing electret voltage V_e.

and

$$V_R = \frac{\partial W_e(x,Q)}{\partial Q} = \frac{\alpha}{2\pi\varepsilon_0 NL} Q + \frac{\beta(x)}{2\pi\varepsilon_0 NL} Q_e = \frac{1}{C} Q + V_b(x) \tag{6}$$

where $C = \frac{2\pi\varepsilon_0 NL}{\alpha}$, $F_b(x) = \frac{\gamma'(x)}{4\pi\varepsilon_0 NL} Q_e^2$ and $V_b(x) = \frac{\beta(x)}{2\pi\varepsilon_0 NL} Q_e$.

By including the inertial force $F_{ext} = ma$ and mechanical damping b, the dynamics of the harvester can be represented by the equivalent circuit as shown in figure 2.

4. Device example and simulation results

We use the parameters of the device in [15] as an example. The calculation gives $\alpha = 0.2147$ and displacement-dependent functions $\beta(x)$ and $\gamma(x)$ as shown in figure 3. It can be seen that the transducer has the maximum and minimum stored energy when the proof mass displacement are at two extremes $\pm\frac{p}{4}$. The found parameters result in $C = 51.8$ pF and $Q_0 = -3.26$ nC while the corresponding values obtained by finite elemement computation in [15] were $C^{\mathrm{FEM}} = 52.5$ pF and $Q_0^{\mathrm{FEM}} = -3.2$ nC, where Q_0 is the equilibrium charge of Q. This agreement validates the calculation approach that takes all effects of the fringing fields into account.

Figure 4 shows frequency responses of the harvester for different acceleration amplitudes at $V_{\mathrm{e}} = 160$ V and $R = 5.0$ MΩ. The behaviour reproduced by the model is in agreement with experimental results, but the power obtained from the simulation is smaller by about 15% than the measured power in [15]. Apart from the accuracy of the experimental model parameters, the deviation can be explained by the measured device having a corrugated electret while the analysis is based on a patterned flat electret.

The effect of the surface voltage V_{e} on the output power is shown in figure 5. An optimal voltage around $V_{\mathrm{e}} = 400$ V is found when increasing from 160 V while holding the acceleration amplitude fixed. Further increase of V_{e} reduces the power due to the stronger electrostatic force that limits the proof mass displacement. In addition, the electrostatic nonlinearity caused by increasing V_{e} shifts center-frequency upwards. This can be explained by analysis of the electrostatic force at short- and open-circuit conditions as shown in figure 6. There is almost no stiffness contribution from the electrostatic force at shorted circuit. With open circuit, the system stiffness is higher because of the additional force $F_{\mathrm{e}}^{\mathrm{open}} \approx 14.1$ N/m $\times\, x$ around the equilibrium position. Thus, the nonlinearity moves the center frequency by Δf_{c} as shown in figure 7 and can be used to adapt the harvester to the ambient vibration frequency.

5. Conclusion

A fully analytical method that provides induced-charge coefficients for electret-based energy harvesters in the slot-effect configuration was presented. From this method, an equivalent circuit model that takes into account effects of fringing fields and electrostatic transducer nonlinearities followed. The key model parameters can be simply calculated from the induced-charge coefficients. Variations on the proposed approach for other configurations of electret-based harvesters and further optimization on transducer geometry are left for future work.

Acknowledgments

This work was supported by the Research Council of Norway through the Grant no. 229716/E20.

References

[1] Suzuki Y 2011 *IEEJ Trans. Elec. Electron Eng.* **6** 101--111
[2] Sakane Y, Suzuki Y and Kasagi N 2008 *J. Micromech. Microeng.* **18** 104011
[3] Edamoto M, Suzuki Y et al. 2009 *Proc. IEEE MEMS2009 (Sorrento, January)*, 1059–1062
[4] Narsure Y, Matsubara N et al. 2009 *J. Micromech. Microeng.* **19** 094002
[5] Chiu Y and Lee Y C 2013 *J. Micromech. Microeng.* **23** 015012
[6] Westby E R and Halvorsen E 2012 *IEEE/ASME Trans. Mechatronics* **17** 995–1005
[7] Renaud M, Altena G et al. 2013 *Proc. Transducer2013, (Barcelona, June)*, 693–696
[8] Elfrink R, Van Acht V et al. 2014 *Proc. SSI2014* 69-76
[9] Boland J, Chao Y H et al. 2003 *Proc. IEEE MEMS2003 (Kyoto, January)*, 538–541
[10] Bartsch U, Sander C et al. 2009 *Proc. PowerMEMS2009 (Washington DC, December)* 332–335
[11] Boisseau S, Despesse G and Sylvestre A 2010 *Smart Mater. Struct.* **19** 075015
[12] Okamoto H, Hamate Y and Kuwano H 2012 *Smart Mater. Struct.* **21** 065001
[13] Chen R and Suzuki Y 2013 *J. Micromech. Microeng.* **23** 125015
[14] Le C P and Halvorsen E 2013 *J. Intell. Mater. Syst. Struct.* **24** 1176–1186
[15] Renaud M, Altena G et al. 2015 *Smart Mater. Struct.* **24** 085023
[16] Perez M, Boisseau S et al. 2016 *Smart Mater. Struct.* **25** 045015

Folded Spring and Mechanically Switching SSHI for High Performance Miniature Piezoelectric Vibration Energy Harvester

H Asanuma[1], H Okubo[1], T Komatsuzaki[1] and Y Iwata[1]

[1] Institute of Science and Engineering, Faculty of Mechanical Engineering, Kanazawa University, Kakuma-machi, Kanazawa, 920-1192, Japan

Corresponding author's e-mail: h-asanuma@se.kanazawa-u.ac.jp

Abstract. To downsize the clamp area and increase the output power of the harvester, we developed a miniature piezoelectric vibration energy harvester with combining a Z-shaped folded spring and a mechanically-switching SSHI (synchronized switch harvesting on inductor). The overall harvester size is $4\times2\times3$ cm^3. The FEM analysis revealed that the output power increases and the value of the 1st and 2nd resonance frequencies move closer as the angle of the Z-shaped spring decreases, therefore, the smaller angle would be more promising. The experimental results showed that the maximum output power of our harvester for the 1st (20.2 Hz) and 2nd (53.0 Hz) resonance frequencies at the applied acceleration of 4.9 m/s^2 are 0.088 and 0.98 mW, respectively. The reason for a marked enhancement of the output power for the 2nd resonance frequency is attributed to the vertical movement of the 2nd vibrational mode which applies larger mechanical stress to the piezo ceramic and achieves better electrical contact between the tip of the Z-shaped spring and the spring plunger.

1. Introduction

A piezoelectric vibration energy harvester (P-VEH), which typically consists of a piezoelectric material, beam, and additional mass, represents one promising power source for internet of things (IoT) modules because of the high output power density. However, in most cases, volume for clamping the beam accounts for a large portion of the harvester, which leads to decrease in the actual power density. In addition, we need to further increase the output power of P-VEHs to shorten the sensing period of the wireless sensor module.

Recently, SSHI (synchronized switch harvesting on inductor) interface incorporating mechanical switch was developed to increase the output power and simplify the complexity of the circuit [1, 2]. However, the clamp jig for their clamped-free beam structure increase the harvester's volume. To downsize the clamp volume and increase the output power, we propose a miniature P-VEH with combining a folded spring and a mechanically-switching SSHI.

2. Harvester Structure and Interface Schematic

Figure 1 shows a photograph and interface schematic of the proposed harvester. The overall harvester size is $4\times2\times3$ cm^3 and the clamp area of the bottom base is 1.7×2 cm^2. We employed a fine-grained stainless steel (Nippon Steel & Sumitomo Metal, SUS304 H-SR3, $67\times20\times0.15$ mm^3) to ensure mechanical reliability [3, 4], and formed the Z-shaped spring by folding the steel thin film. As shown in Figure 1(a), the screwed ball spring plunger was employed to achieve soft contact with the tip of the Z-shaped spring, as well as to adjust the gap between the tip and the plunger. The gap distance between the upper and lower plunger was set to 4.2 mm. The piezo ceramic (Fuji Ceramics, C-6, 12×20×0.2 mm^3) and the copper mass (26.7 gram) are attached on the Z-shaped spring. As shown in Figure 1(b), the lower electrode of the piezo ceramic is electrically connected with the conductive Z-shaped spring. When the tip of the Z-shaped spring contacts with the plunger under large enough acceleration, the mechanical switch closes and thus the SSHI works.

Firstly, we used the FEM analysis (Comsol Multiphysics) to investigate the dependence of the output power on the angles of the Z-shaped spring. The applied acceleration and the resistive load are 4.9 m/s^2 and 270 kΩ, respectively. In this FEM analysis, the output power is evaluated when the piezo ceramic is only connected to the load. Figure 2 shows output power P_{out} vs frequency with different five angles of the Z-shaped spring and vibrational mode. The analysis revealed that the output power increases and the value of the 1st and 2nd resonance frequencies move closer with reducing the angle. Therefore, the smaller angle would be more promising. However, for a technical difficulty attaching the piezo ceramic onto the Z-shape spring with the angle of 15°, the angle of 30° was chosen in the experiment.

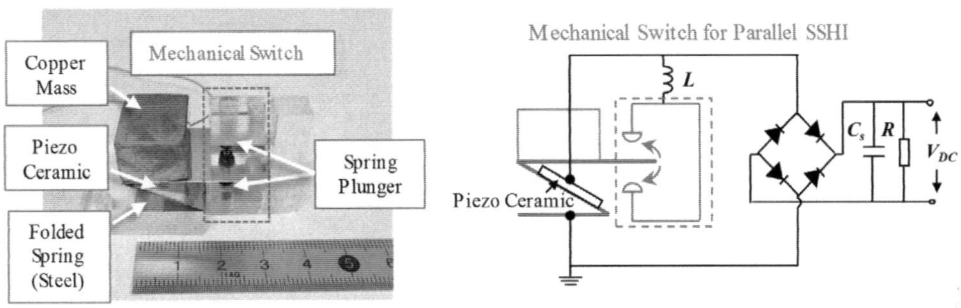

Figure 1. (a) Photograph and (b) interface schematic of the proposed harvester. The overall harvester size is 4×2×3 cm^3.

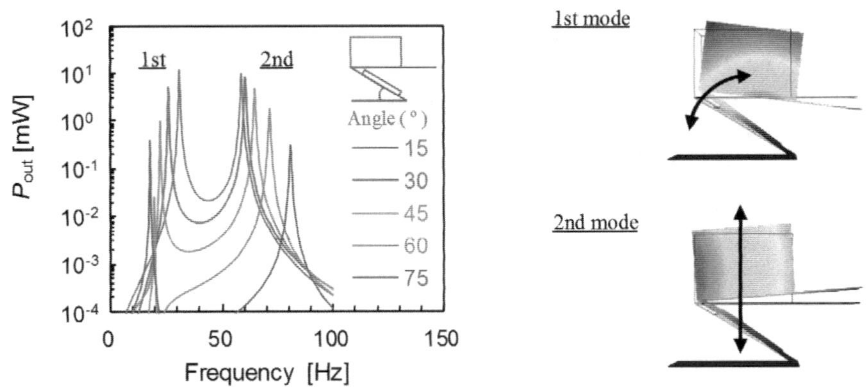

Figure 2. Output power P_{out} vs frequency with different five angles of the Z-shaped spring.

3. Experimental

Figure 3 shows frequency sweep of the output power P_{out} and of the peak-to-peak tip displacement $Disp._{p\text{-}p}$ of the Z-shaped spring for the mechanically-switching SSHI (MSS) and the standard circuit. $Disp._{p\text{-}p}$ was measured by a laser displacement meter (Keyence LC-2450 and LC2400). The standard circuit is equal to the MSS circuit shown in Figure 1(b) without the inductor L. The output power is calculated from V_{DC}^2/R. The applied acceleration a, load R, inductor L, capacitor C_s, are 4.9 m/s², 270 kΩ, 100 mH, and 47 μF, respectively. The harvester for the MSS circuit showed 1.2-fold increase in the power at the 1st resonance frequency (20.2 Hz) and 2.1-fold increase at the 2nd resonance frequency (53.0 Hz) over the standard circuit. The maximum output power for the MSS at 1st and 2nd resonance frequencies are 0.088 and 0.98 mW, respectively. The reason why the output power for the 2nd resonance frequency was much higher than that for the 1st one is attributed to the vertical movement of the 2nd mode which applies larger mechanical stress to the piezo ceramic and achieves better electrical contact between the tip and the plunger. For the standard circuit removing the mechanical switch structure shown in Figure 1(a), the maximum output power for the 2nd resonance frequency was 0.72 mw, however, the piezo ceramic was broken during the measurement due to the large displacement. Therefore, the mechanical switch also works as a stopper for improving mechanical reliability.

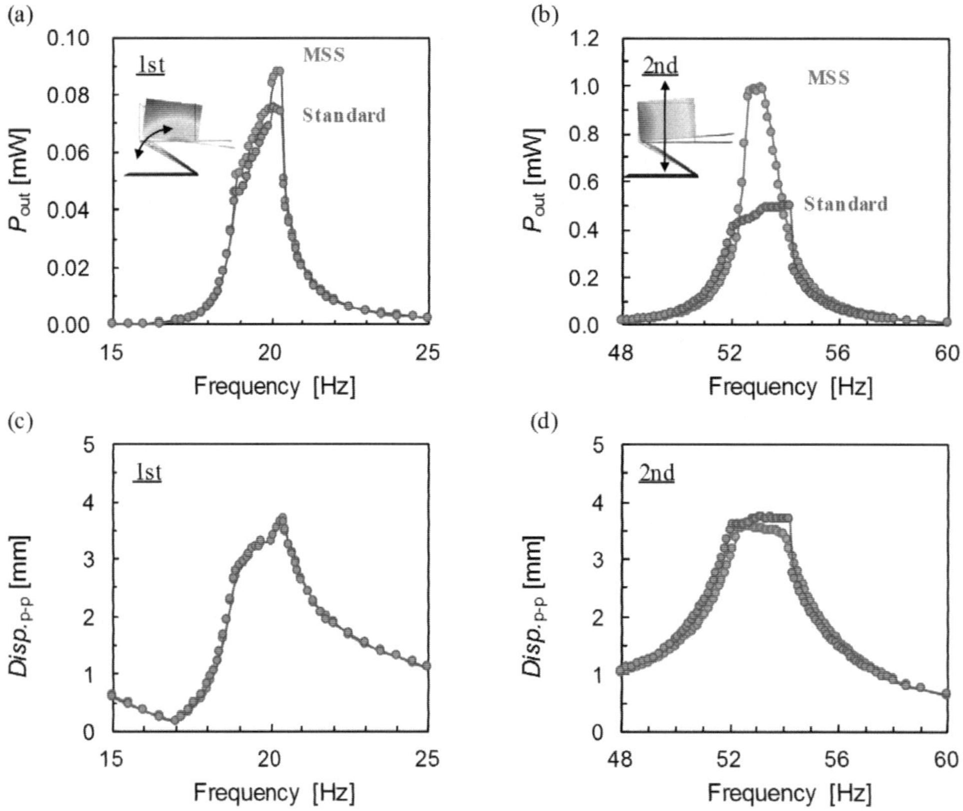

Figure 3. Frequency sweep of the output power P_{out} and of the peak-to-peak tip displacement $Disp._{p\text{-}p}$ of the Z-shaped spring for the mechanically-switching SSHI (MSS) and the standard circuit at the applied acceleration a of 4.9 m/s².

Figure 4 shows the output power P_{out} vs the acceleration a at (a) the 1st and (b) 2nd resonance frequencies. For the 2nd mode, after the tip of the Z-shaped spring start contacting the plunger under large enough acceleration, the output power for the MSS continues to increase, while the power for the standard circuit showed saturation. The increase in power may be attributed to longer contact duration. The too long contact duration at the acceleration greater than 5 m/s^2 may lead to the decrease in the power for the MSS.

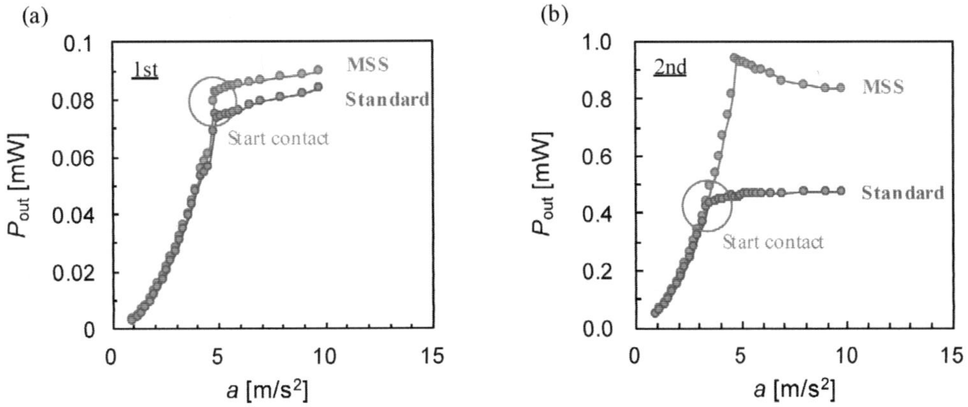

Figure 4. Output power P_{out} vs acceleration a at (a) 1st and (b) 2nd resonance frequencies.

4. Conclusion

We developed a miniature piezoelectric vibration energy harvester with combining a Z-shaped folded spring and a mechanically-switching SSHI (synchronized switch harvesting on inductor). The FEM analysis revealed that the output power increases and the value of the 1st and 2nd resonance frequencies move closer as the angle of the Z-shaped spring decreases, therefore, the smaller angle would be more promising. Under the applied acceleration of 4.9 m/s^2, the proposed harvester with the size of 4×2×3 cm^3 exhibited the maximum output power of 0.088 mW at the 1st resonance frequency and of 0.98 mW at the 2nd resonance frequency. The reason for a marked enhancement of the output power for the 2nd resonance frequency is attributed to the vertical movement of the 2nd vibrational mode which applies larger mechanical stress to the piezo ceramic and achieves better electrical contact between the tip of the Z-shaped spring and the spring plunger.

Acknowledgements

We would like to thank the Japan Society for the Promotion of Science (JSPS). This work was supported by Grant-in-Aid for Young Scientists (B) (No. 16K18039).

References

[1] Giusa F *et al.*, *Sensor and Actuators A: Physical* 2013 **198** pp 35-45
[2] Wu Y *et al.*, *Journal of Intelligent Material Systems and Structures* 2014 **25** pp 1658-1663
[3] Asanuma H *et al.*, *Proc. PowerMEMS 2014 Journal of Physics: Conference Series* **557** 012124
[4] Asanuma H *et al.*, *Journal of Micromechanics and Microengineering* 2015 **25** 104013

Investigation of Interface between Ge Electrodes and Ionic Liquid Electrolytes for Electric Double Layer Capacitors

R.M Abeysinghe[1], H. Oguchi[2] and H. Kuwano[1]
[1]Division of Nanosystems, Department of Robotics, Tohoku University, 6-6-01, Aramaki-Aoba, Aoba-ku Sendai, Miyagi 980-8579, Japan.
[2] AIMR, Tohoku University, 2-1-1 Katahira, Aoba-ku, Sendai, Miyagi 980-8577, Japan.

E-mail : maduka@nanosys.mech.tohoku.ac.jp

Abstract. This study discusses novel way of use of ionic liquids to develop Ge-based electrodes for electric double layer capacitors (EDLC). We found that ionic liquids change their electrochemical properties depending on the amount of the absorbed water. Wet ionic liquids work as solvents to dissolve Ge and make porous structures, whereas dry ones work as electrolytes of the EDLCs. The former property was used to increase surface area of the electrodes which is desired to increase the capacity of EDLCs. This method showed another advantage in contrast to the dry ionic liquids; wet ones could fill the complex Ge pores in parallel to porous structure formation. Finally, after porous formation, we dried the ionic liquid at 100 °C and prepared the EDLCs composed of Ge porous electrodes. Cyclic voltammetry and impedance measurements indicated that the obtained devices can work as EDLCs.

1. Introduction

Semiconductor-based electric double layer capacitors (EDLC) have been collecting attention as energy storage devices for integrated micro devices consisting of sensors, diodes, actuators *et al* [1]. In this context we have challenged to fabricate the world first Germanium (Ge)-based EDLCs [2] compatible with micro-fabrication processes. Moreover, Ge-based EDLCs will enable integration with Ge diodes which has lower rectification threshold voltage than Silicon (Si).

In the previous research we succeed in obtaining large surface area Ge electrodes with micro porous structure preferable to increase capacity of the EDLCs by etching Ge in acid solutions. Then we fabricated the EDLCs using ionic liquid (IL) electrolytes [2] to take advantage of its higher decomposition voltage, which leads to higher energy storage. However large current observed in a cyclic voltammetry (CV) indicated unfavourable chemical reactions at the interface of Ge/IL (Fig.1). In addition, we noticed that IL did not fill in the porous structure as indicated by contact angle of 48.8° (Fig.2(a)) and Scanning Electron Microscope (SEM) images (Fig.2(b)).

The primary purpose of this study is to stop chemical reactions at the interface of Ge/IL based on understanding of the chemical reaction origin, and the secondary purpose is to develop proper method to fill in the pores by the IL. For these purposes firstly we studied the importance of the absorbed water contained in the IL. After that by uniquely controlling water amount dissolved in the IL electrolytes, we solved the both issues mentioned above.

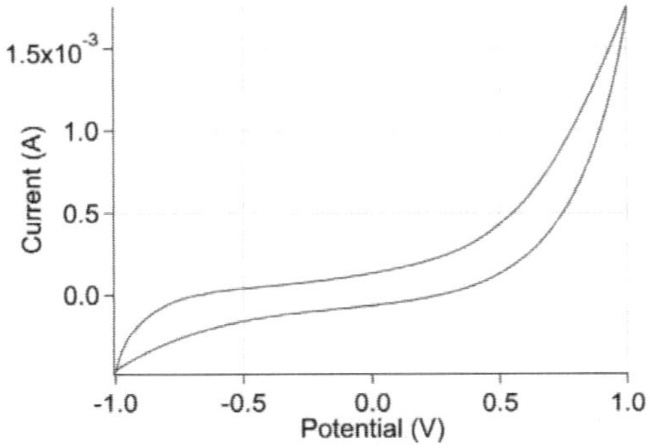

Figure 1. A CV curve obtained with scanning rate of 0.01V/sec

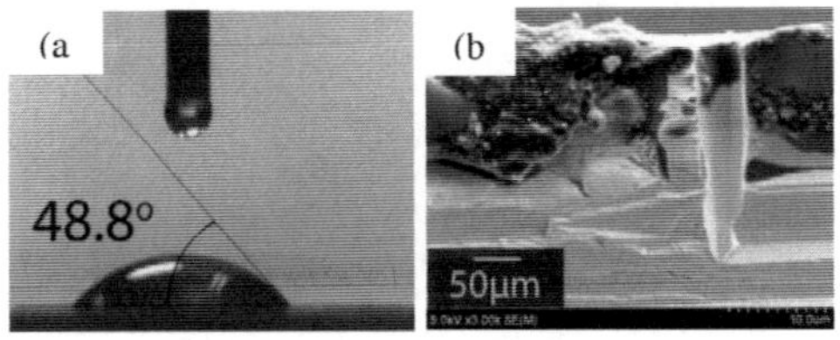

Figure 2. (a) Contact angle measurement of an IL droplet on a porous Ge substrate. (b) A cross-sectional SEM image of the Ge micro porous structure. Black part corresponds to the IL. A pore on the right was only partially filled by IL.

2. Experiment

In this study we used one-side polished Sb-doped n-type (100)-oriented Ge wafers with resistivity of 5 - 20 mΩcm. The wafers were rinsed using Isopropyl Alcohol (IPA) in ultraviolet cleaner for 20 mins to remove contamination. On the rough side of the wafer, we deposited Ti adhesion layer (thickness: 20nm) and Pt (thickness: 200nm) electrode using Sputtering (EIKO Sputtering) method. Then samples were diced into 9.5×9.5 mm^2. For these samples, we tried to form Ge porous structure using experimental setup shown in Fig.3. By applying constant voltage between two Pt electrodes, chemical reactions occurred at the interface between the ionic liquid and Ge substrate. The distance between two electrodes was set to be 1mm. Electrochemical reactions were studied using electrochemical measurement system (VersaSTAT, version 4.0, Princeton Applied Research). Surface morphology was observed using scanning electron microscopy (HITACHI SU-70).

3. Results and Discussion

3.1. Investigation of the chemical reactions origin

Guessing from reported reaction between Ge and water [3], the chemical reaction observed in our previous experiments can be attributed to the water absorbed in the IL.

$$Ge + 2H_2O \longleftrightarrow GeO_2 + 4H^+ + 4e^- \quad (1)$$

To confirm our guess, for ionic liquid EMIM-IM with different amount of absorbed water, we implemented CV measurements using the setup shown in Fig.3. The amount of the absorbed water was changed by pumping and heating the as-supplied EMIM-IM in the vacuum chamber with back pressure of 2×10^{-4} Pa. As-supplied EMIM-IM showed the current (at 1V, 10×10^{-6} A) attributed to the reaction between the ionic liquid and Ge (Fig.4). On the other hand, current decreased after pumping (at 1V, 3×10^{-6} A). Pumping and heating treatment further decreased the current value (at 1V, 1.5×10^{-6} A). These observations clearly indicated that water is the dominant factor for the chemical reactions.

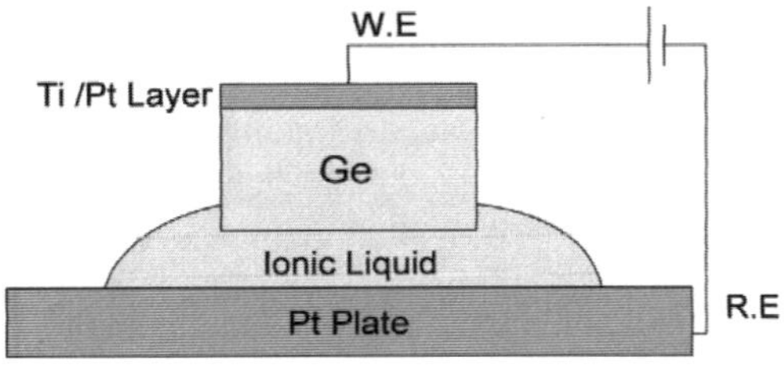

Figure 3. Schematics of electrode – electrolyte – electrode three layer structure.

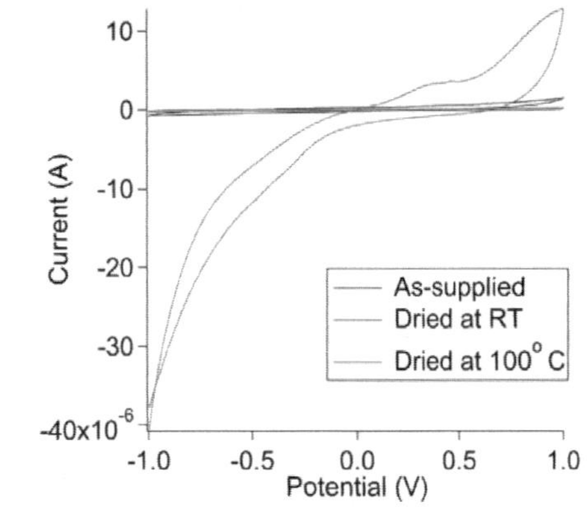

Figure 4. Comparison of CV curves for differently prepared BMIM-PF_6.

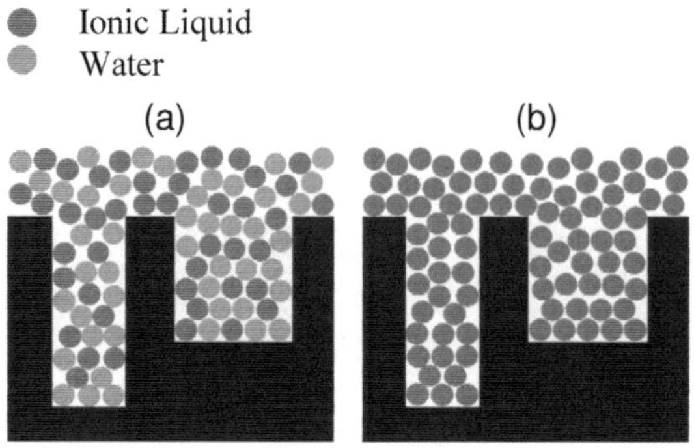

Figure 5. IL drying procedure (a) IL with water (As-supplied), (b) After completely dried up at 100°C in vacuum

3.2. Development of new porous formation method

Based on the understanding that absorbed water in the IL reacts with Ge, we proposed an interesting method to form and fill in the porous structures in parallel. The method forms porous structure by use of IL with water (Fig.5(a)). Then the IL is dried up at 100°C in vacuum (Fig.5(b)). In this way, porous structure is obtained and filled by IL without water, which works as an electrolyte for EDLCs.

To verify our idea, we conducted Ge porous formation in air by using the as-purchased ILs. In addition, to find the appropriate conditions, we controlled three variables; IL species, voltage, and elapsed time. SEM observations (not shown) indicated that among the three different ILs which were used in the experiment (BMIM-PF_6, BMIM-AC and EMIM-IM), EMIM-IM has the highest capability to form the porous structure. Moreover, larger voltage and longer elapsed time increase penetration depth of the pores.

Figure 6. SEM images of the Ge after porous formation at 3V for 60hrs, (a) Surface morphology and Cross section of (b) before clean up (c) after clean up.

Based on the above results we formed the largest and uniform porous structures by applying 3V for 60 hours in air (Fig.6(a)). Furthermore the cross-sectional SEM image showed that the ionic liquid appeared as black part penetrated into the deepest part of the pores (Fig. 6(b), (c)), indicating that the porous structure can be filled by the ionic liquids using this method. Note that the black part was confirmed as the EMIM-IM from the fact that it was removed by cleaning up the sample using ethanol (Fig.6(c)).

To get EDLC operation, we dried the EMIM-IM remained in the pores at 100 °C for more than 24 hours. The cyclic voltammetry (Fig.7(a)) and impedance plots (Fig.7(b)) before and after the drying treatment clearly changed the behaviour. After the drying treatment, CV current reduced and Impedance plot was elongated along the vertical axis (imaginary part of the impedance Z_{im}), indicating that the chemical reaction was almost stopped at the Ge/IL interface.

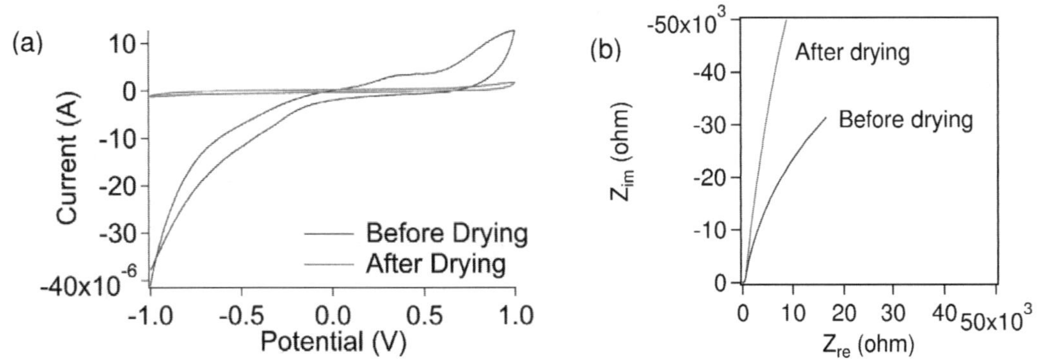

Figure 7. (a) CV curves and (b) Impedance plots obtained before and after drying treatments.

Finally, capacity of the EDLCs was evaluated for the Ge electrode with porous structures filled by dry IL. Increase in the current value of CV curves (Fig. 8) obtained before and after Ge porous formation found that we succeeded in increasing the capacity of the Ge-based EDLCs by using the Ge electrode with micro porous structures. The increase is, however, not so much as we expected. Possibly straight and shallow pores could not increase surface area of the Ge effectively. In addition, water could be trapped in the pore and drying of the IL was not complete.

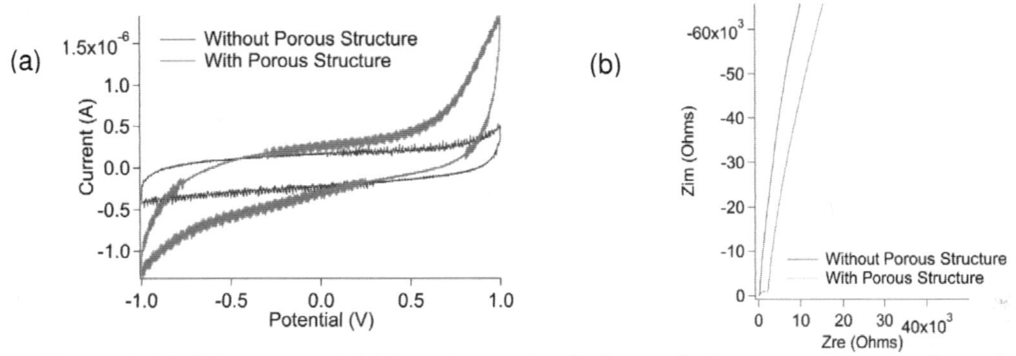

Figure 8. (a) CV curves and (b) Impedance plots before and after porous structure formation.

4. Conclusion

We verified that water amount contained in the IL is one of the dominant factor of the chemical reactions at the Ge/IL interface. In addition, we revealed that Ge micro porous structure need special treatment to be filled by the IL electrolyte. Based on these facts, we proposed a new method; forming and filling the micro porous structure simultaneously by controlling the absorbed water in IL. Using this method we succeeded in preparing Ge electrodes with a porous structure filled by IL. Moreover, by drying the IL filling the porous structures, chemical reactions were almost suppressed, indicating the promise of the proposed method for Ge-based EDLCs.

References

[1] Beom-Jin Yoon, Soo-Hwan Jeong, Kun-Hong Lee, Hyung Seok Kim, Chan Gyung Park and Jong Hun Han, 2004 "Electrical properties of electrical double layer capacitors with integrated carbon nanotube electrodes," *Chemical Physics Letters,* Vol **388** pp 170–174.

[2] A. Sekiguchi, M. Hara, H. Oguchi and H. Kuwano, 2014 "Porous Ge based Electric Double Layer Capacitors with High Compatibility for Low Threshold Voltage Diode Rectifiers," *Journal of Physics: Conference Series,* Vol **557**, *Conf.*1.

[3] Tatsuya Kawase, Atsushi Mura, Katsuya Dei, Keisuke Nishitani, Kentaro Kawai, Junichi Uchikoshi, Mizuho Morita and Kenta Arima, 2013 "Metal-assisted chemical etching of Ge(100) surfaces in water toward nanoscale patterning," *Nanoscale Research Letters,* chapter **8** p 151.

20000G shock energy harvesters for gun-fired munition

J. Willemin[1,2], S. Boisseau[1,2], L. Olmos[1,2], M. Gallardo[1,2], G. Despesse[1,2], T. Robert[3]
[1] Univ. Grenoble Alpes, F-38000 Grenoble, France
[2] CEA, Leti, Minatec Campus, 17 rue des Martyrs - 38054 Grenoble Cedex 9, France
[3] Nexter Munitions, La Chapelle, Route de Villeneuve - 18570 La Chapelle Saint Ursin, France

E-mail: jerome.willemin@cea.fr

Abstract. This paper presents a 20000G shock energy harvester dedicated to gun-fired munitions and based on a mass-spring resonant structure coupled to a coil-magnet electromagnetic converter. The 20000G shock energy is firstly stored in the spring as elastic potential energy, released as mass-spring mechanical oscillations right after the shock and finally converted into electricity thanks to the coil-magnet transducer. The device has been modeled, sized to generate 200mJ in 150ms, manufactured and tested in a gun-fired munition. The prototype sizes 117cm³ and weighs 370g. 210mJ have been generated in a test bench and 140mJ in real conditions; this corresponds to a mean output power of 0.93W (7.9mW/cm³) and a maximum output power of 4.83W (41.3mW/cm³) right after the shock.

1. Introduction - Shock energy harvester concept

The use of electromagnetic converters in energy harvesters dates back to the beginning of the energy harvesting era. Yet, they still offer great innovation and optimization opportunities in various fields such as biomechanical energy harvesting [1], water flow energy harvesting [2] or high-G shock energy harvesting, which is the topic of this paper. The concept of the energy harvesters proposed here is close to the one developed by Rastegar et al. [3-4]; but, because of cost issues, the choice has been made to use an electromagnetic converter instead of a piezoelectric one. The basic concept of our high-G energy harvesters is presented in Figure 1a: during the firing phase, the energy harvester is subjected to a 20000G acceleration, the mass is projected towards the bottom of the device and elastic potential energy is stored in the spring. Right after the shock, this potential energy is released as mechanical oscillations (mass-spring) which are eventually converted into electricity thanks to a coil-magnet transducer. The use of mechanical oscillations enables to convert energy in several periods, which has the great benefit of limiting the size of the electromagnetic transducer. The acceleration as a function of time of the shock to harvest is presented in Figure 1b: the shock lasts $T_{shock}\approx 10$ms and reaches exactly 22700G at $t\approx 5$ms.

2. Shock energy harvester – models and equations

The energy harvester's model is presented in Figure 2a. The mechanical part is a mass-spring resonator damped by mechanical (b_m) and electrical forces; the electric part a coil-resistor circuit. This energy harvester is ruled by the system of equation presented in Figure 2b, deduced from Newton's second law and Kirchhoff's laws, where m is the mobile mass, k, the spring stiffness, b_m, the mechanical damping ratio, z(t), the mass position, Bl, the electromechanical coupling of the coil-magnet converter,

L, the coil inductance, r_l, the coil resistance and R, the load. The energy harvester can be modeled by an equivalent electric circuit with a mechanical and an electric part (Figure 2c) [5-6]. The electromechanical coupling is modelled by a transformer with a $Bl{:}Bl$ conversion ratio and with $\dot{z}$ flowing on the primary and i on the secondary.

(a) (b)

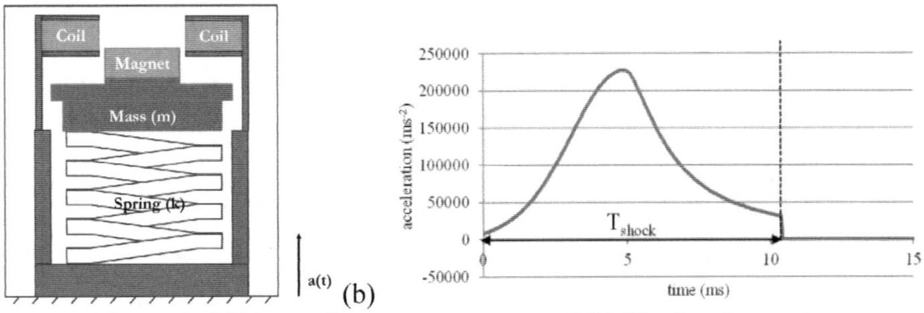

Figure 1. (a) Energy harvester concept and (b) Shock to harvest

(a)

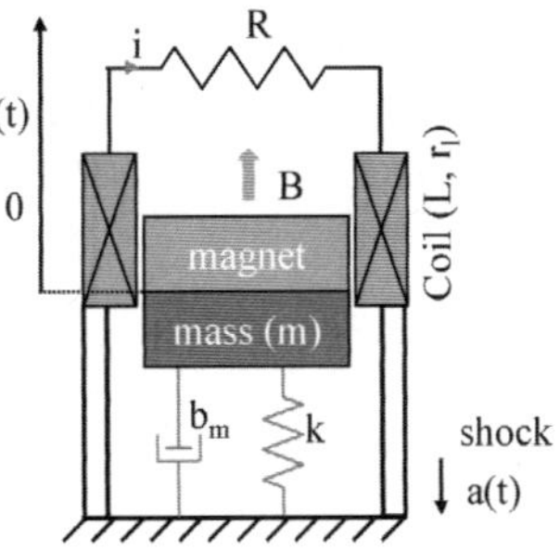

(b)

$$\begin{cases} m\ddot{z} + b_m\dot{z} + kz = -\frac{Blu_R}{R+r_l} - ma \\ \frac{L}{R}\dot{u_R} + \frac{R+r_l}{R}u_R = Bl\dot{z} \end{cases}$$

(c)

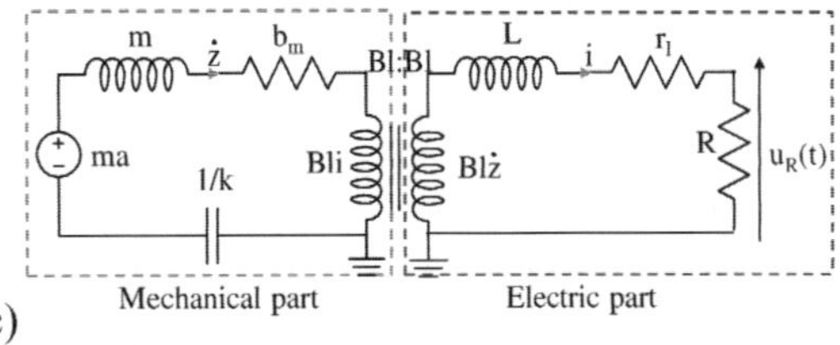

Figure 2. (a) Energy harvester model, (b) system of differential equations ruling the energy harvester's behavior, and (c) equivalent electric model

3. Simulations and energy harvester conception

3.1. Parameters and simulated output energies

The energy harvester's equations have been implemented in Matlab/Simulink to determine the parameters maximizing the output energy. The time acceleration (Figure 1b) is directly used as the input excitation; the output energy targeted is E=200mJ in 150ms. The most important parameter is the resonant frequency of the mass-spring system (f_{mk}). Actually, so as to ensure the mechanical oscillations, the 20000G acceleration must be perceived as a shock by the mass-spring resonator. Then $T_{mk}=1/f_{mk}$, the mechanical period of the mass-spring oscillator, must be small compared to T_{shock}; then, on the one hand, f_{mk}>>100Hz. On the other hand, f_{mk} should be small enough to induce a sufficient oscillation amplitude z. Simulations have been performed and have shown that the output energy is maximized for f_{mk}=299Hz. The maximum displacement of the mobile mass has been limited by stoppers and fixed to z_{max}=±8mm to limit the size of the device and to avoid buckling issues. Simulations have shown the necessity of using soft stoppers to prevent rebounds and to ensure that $z(t)=z_{max}$ right after the shock. Moreover, experimental trials have shown the necessity of using guiding parts in order to keep oscillations on the vertical axis, avoiding the energy to be transferred in lateral resonant modes.

The mobile mass and the spring stiffness have also been optimized by simulations and should be respectively equal to m=130g and k=460kN/m. The mechanical quality factor is taken equal to 50. The electromagnetic converter is made of a NdFeB magnet and a 600-turn coil. The Bl conversion coefficient is imposed by the characteristics of the magnet, the coil and the distance between them, and has been experimentally measured at Bl=3.44. The simulated output voltage, output power and mass displacement as a function of time are presented in Figure 3.

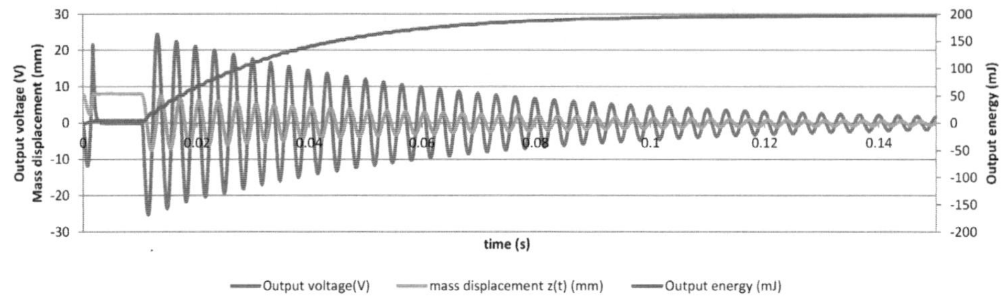

Figure 3. Simulated output voltage, output power and mass displacement as a function of time

3.2. Spring design

A double-helix spring (Figure 4) has been chosen to ensure a good stability and to better withstand buckling during the shock phase. The spring parameters have been determined by using the system of equations $k = Gb^4/4D^3n$ and $\tau(z_{max}) = 4K_wkz_{max}D/\pi b^3$, where G is the shear modulus, D the mean diameter of the spring, b the wire diameter, n the number of active coils and K_w the wahl factor, to keep $\tau(z_{max})$, the spring stress when $z=z_{max}$, below 1GPa and to get k=460kN/m. Springs have been made from round steel wires, soldered on two supports (Figure 4b) and characterized in a Mecmesin 25-i traction bench (Figure 4c). The experimental spring stiffness is k=430kN/m.

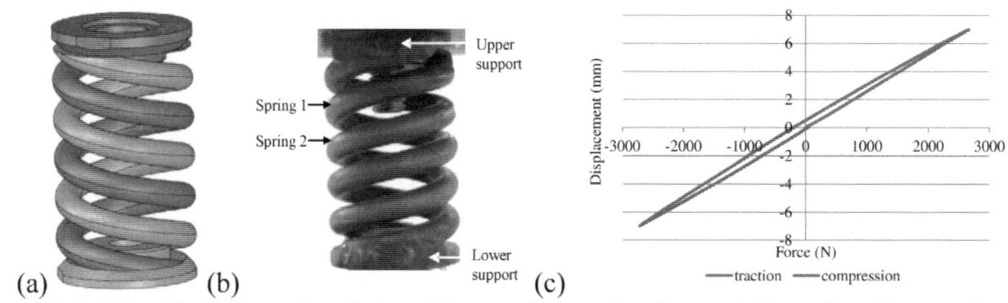

Figure 4. (a) Double helix spring design, (b) manufactured spring and (c) spring characterization in a traction bench (Mecmesin 25-i).

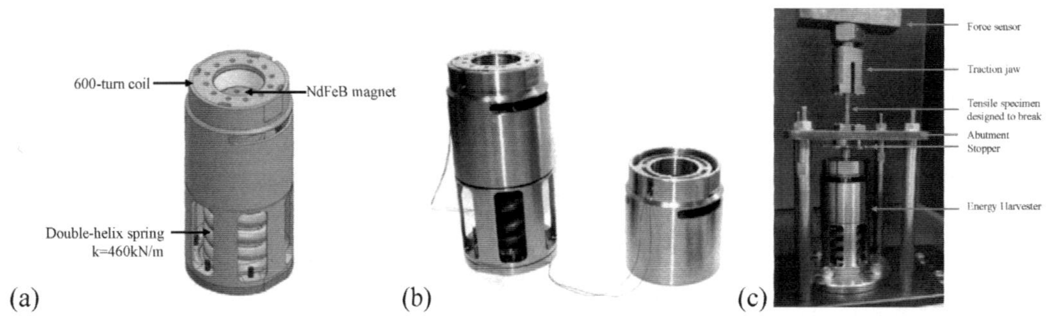

Figure 5. (a) Energy harvester CAD conception, (b) Prototype and (c) test bench

4. Manufacturing, tests and experimental results

The energy harvester has been designed in Solidworks (Figure 5a) and fabricated with standard manufacturing tools, integrating the double-helix 430kN/m spring, stoppers and guiding parts. The complete device (Figure 5b) sizes 117cm³ and weighs 370g.

The system has firstly been tested in a traction bench (Mecmesin 25-i): the energy harvester is connected to a tensile jaw with a tensile specimen (Figure 5c). A stopper is used to impose the energy harvester's elongation at startup ($z=z_{max}$=8mm). The tensile specimen is designed to break after the energy harvester abutted to liberate the device which enters in oscillations. The experimental resonant frequency is f_{mk}=289Hz (Figure 6a); output voltages reach 23 volts, as predicted by simulations (Figure 3). Finally, 210mJ have been generated in 150ms (vs 200mJ targeted).

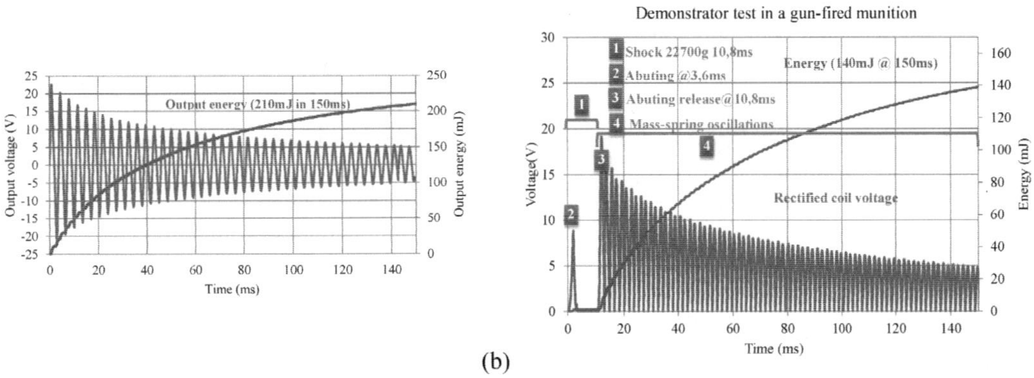

(a) (b)

Figure 6. (a) Experimental output voltage and energy in the test bench – (b) In-situ tests – output voltage and output energy

The prototype was finally installed and tested in a gun-fired munition in November 2013. Experimental output voltages and energies are presented in Figure 6b: 140mJ were generated in 150ms after a 22700G shock. This corresponds to a mean output power of 0.93W and a maximum output power of 4.83W, right after the shock. The mechanical quality factor Q=1/(2b_m) was finally evaluated from experimental data and varied between Q=20 right after the shock and Q=180 at the end of the oscillations. Low Q is due to peripheral high stress on the springs at large displacements. Further improvements are targeted to reduce these stress and thus to increase output powers.

5. Conclusions

We have reported on a 20000G shock energy harvester for gun-fired munition, designed to generate 200mJ in 150ms. Prototypes have been manufactured, tested in lab and in real conditions. 210mJ have been reached in a test bench and 140mJ in a gun-fired munition, validating, for the first time, the proof of concept of 20000G shock energy harvesters for gun-fired munition exploiting an electromagnetic conversion. Output powers have reached 4.83W (41.3mW/cm^3) right after the shock, one of the highest densities of power of the state of the art [7]. Research actions are now targeted towards the size reduction and the optimization of the power densities of these energy harvesters.

References

[1] Geisler M et al. 2016 Scaling effects of a non-linear electromagnetic energy harvester for wearable sensors, *Proc. PowerMEMS*
[2] Boisseau S et al. 2016 Water flow energy harvesters for autonomous flowmeters, *Proc. PowerMEMS*
[3] Rastegar J et al. 2011 Energy-Harvesting Power Sources for Gun-Fired Munitions, *Proc. SPIE 8035*
[4] Rastegar J et al. 2006 Piezoelectric-based energy-harvesting power sources for gun-fired munitions, *Proc. SPIE 6174*
[5] Saha C 2011 Modelling Theory and Applications of the Electromagnetic Vibrational Generator, Chapter 3: Sustainable Energy Harvesting Technologies – Past, Present and Future, *Intech*
[6] Cheng S et al. 2007 Modeling of magnetic vibrational energy harvesters using equivalent circuit representations, *Journal of Micromechanics and Microengineering*
[7] Moss S D et al. 2015 Scaling and power density metrics of electromagnetic vibration energy harvesting devices, *Smart Materials and Structures*

Model Based Optimization of Integrated Low Voltage DC-DC Converter for Energy Harvesting Applications

H M P C Jayaweera[1], Ali Muhtaroğlu[1,2]

[1]Sustainable Environment and Energy Systems
[2]Dept. of Electrical-Electronics Engineering
Middle East Technical University Northern Cyprus Campus
Kalkanli, Guzelyurt, Mersin 10, Turkey

E-mail: {pradeep.jayaweera, amuhtar}@metu.edu.tr

Abstract. A novel model based methodology is presented to determine optimal device parameters for the fully integrated ultra low voltage DC-DC converter for energy harvesting applications. The proposed model feasibly contributes to determine the maximum efficient number of charge pump stages to fulfill the voltage requirement of the energy harvester application. The proposed DC-DC converter based power consumption model enables the analytical derivation of the charge pump efficiency when utilized simultaneously with the known LC tank oscillator behavior under resonant conditions, and voltage step up characteristics of the cross-coupled charge pump topology. The verification of the model has been done using a circuit simulator. The optimized system through the established model achieves more than 40% maximum efficiency yielding 0.45 V output with single stage, 0.75 V output with two stages, and 0.9 V with three stages for 2.5 kΩ, 3.5 kΩ and 5 kΩ loads respectively using 0.2 V input.

1. Introduction

The continuous urban growth with sustainable development thrust demands low cost, low power electronics with small footprint for wireless sensor networks. Low voltage output (around 0.2 V) from thermoelectric or small photovoltaic energy harvesters utilized in such applications needs to be stepped up to support small loads. Researchers proposed several DC-DC step-up circuit topologies and stratergies to improve the conversion efficiency. Body biasing technique proposed by [1] and [2] required an additional power supply and resulted in losses from added wiring. Tanzava [3] modelled the effect of resistance on switching devices in an effort to optimize performance, which guided further developments of such models. The equivalent half circuit model presented by [4] and [5] is limited to the Dickson [6] charge pump topology, and does not extend to more recent designs. An integrated LC tank was proposed as a good alternative to traditional charge-pump clock generators in our previous work [7] for reduced losses. The LC tank half circuit model proposed by [8] and [9] investigates characteristics of such circuits in VCOs. Recent papers [10, 11] present optimization models for the charge pump design, but without linking these to the VCO efficiency. This paper reports a novel model based methodology to determine the optimal device parameters for the full DC-

DC converter [7] in Figure 1 with varying charge pump stages, depending on the required output voltage range for a given load.

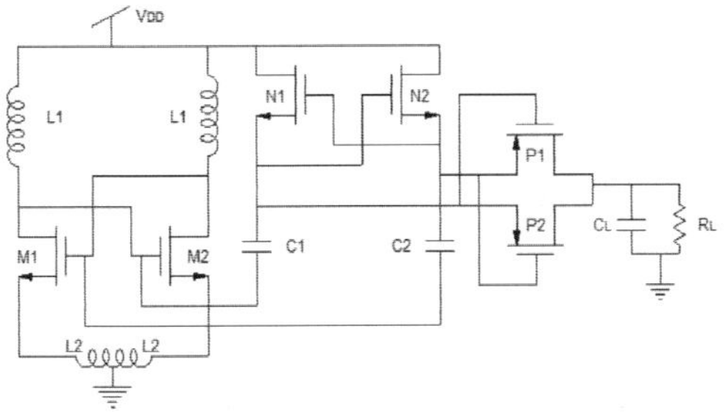

Figure 1. Single stage cross-coupled DC-DC converter with LC tank based oscillator.

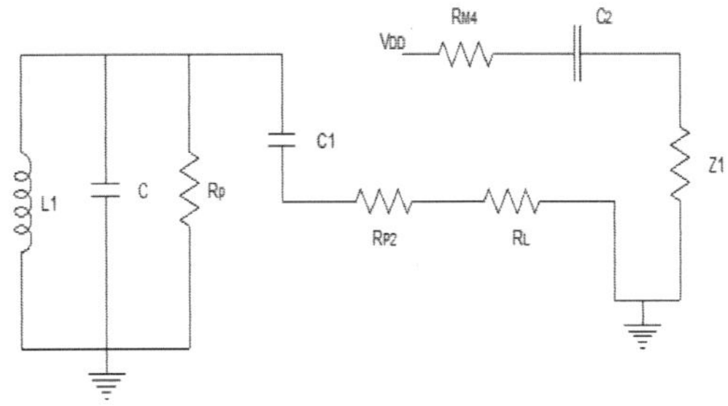

Figure 2. Equivalent half circuit model for Single stage DC-DC converter.

2. DC-DC converter model

The lumped element half circuit for the LC tank based oscillator, shown in Figure 2, consists of the parallel identical inductor (*L1*), capacitor (*C*) and resistor (R_p). The parallel capacitance necessary to sustain oscillation in the LC tank is provided by the switching NMOS (*M1* and *M2* in Figure 1) gate. Charge pump MOSFETs (*N1*, *N2*, *P1*, *P2*) in Figure 1 are modeled using series resistors. MOSFET characteristics have been extracted from the simulation using the target 180nm CMOS process under operation conditions as depicted in Figure 3 and 4 for gate capacitance and resistance respectively.

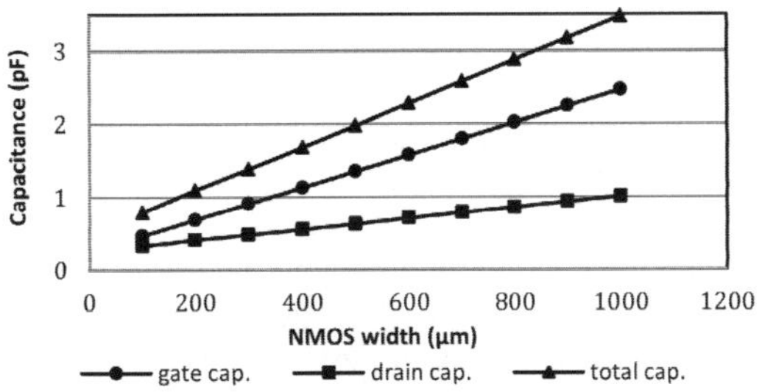

Figure 3. The variation of the NMOS capacitance with MOSFET width, W (length, L=240 nm).

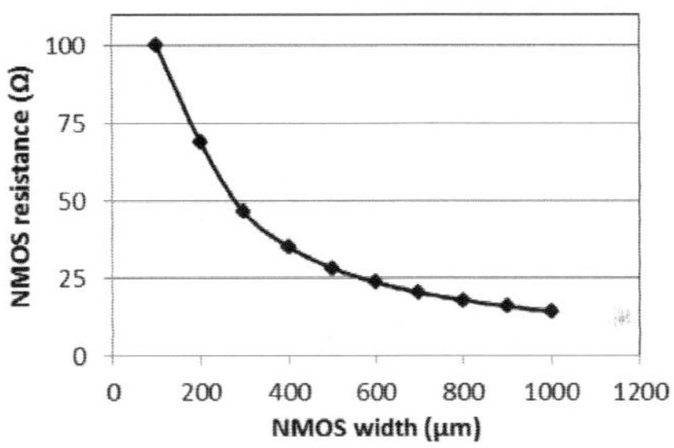

Figure 4. The variation of the NMOS resistance with MOSFET W (L=240 nm).

Equation 1 provides the power output for the single stage DC-DC converter based on the lumped power consumption model. This enables the analytical derivation of the charge pump efficiency when utilized simultaneously with the known LC tank oscillator behavior under resonant conditions, and voltage step-up characteristics of the cross-coupled charge pump topology. The extended equations for two, three, and more stages are not shown here for brevity.

$$P_{out} = \left(\frac{2V_{DD}\, e^{\frac{-t}{C_1(R_L+R_{P2})}} + V_0}{R_L+R_{P2}} \right)^2 \times e^{\frac{-2t}{C_1(R_L+R_{P2})}} \times R_L\,, \tag{1}$$

where, V_0 is a constant (voltage across the *C1* capacitor), t is the half period of the clock signal. V_{DD}, R_{R2} and R_L are input voltage, charge pump NMOS resistance and load resistance respectively. Equation 2 provides an analytical model for the total power in the single stage DC-DC converter, which mainly consists of the switching power consumption in the LC tank, switching power consumption due to the charge pump capacitors, and the power output.

$$P_{total} = P_{out} + V_{rms}^2 \left(\frac{1}{R_p} + \sqrt{C/_{L1}}\ \left(1 + {}^1/_{2\pi}\right) \right) + 2C_1 V_{rms}^2 \sqrt{C/_{2\pi L1}}\ . \tag{2}$$

Relationship of LC tank capacitance to inductance (*L1*), series (R_S) and parallel (R_P) resistances is as shown in Equation 3. Equation 4 and 5 provide fitted models for Figure 3 and Figure 4 respectively.

$$C = L1/(R_P R_s)\,, \tag{3}$$

$$C = 0.003x + 0.4912\,, \tag{4}$$

$$R_N = 3 \times 10^{-10}x^4 - 9 \times 10^{-7}x^4 + 0.001x^2 - 0.5757x + 148.33\,, \tag{5}$$

where, x is the NMOS width in μm and C is the NMOS capacitance in pF. R_N is the NMOS drain-to-source ohmic resistance.

3. Optimization methodology

The area of the on-die inductors is a critical parameter in the optimization process. The first step is to select the maximum possible size of drain inductors (*L1*), and set the source inductors at least 10% of the drain inductors (the purpose of the source inductors is oscillation with 50% duty cycle [7]). LC tank capacitance for the desired average oscillator load range R_P is calculated using Equation 3, and LC tank NMOS size is then determined from Equation 4. The NMOS and PMOS sizes in the charge pump are balanced, based on technology characteristics, to set $R_{PI} = R_N$. Next, $R_{PI}+R_L = R_L$ ($R_{pI} \ll R_L$) is substituted into equation $\frac{dP_{out}}{dC} = 1$ to calculate the optimum charge pump capacitance for the maximum output power. Similarly, input-to-output power transfer efficiency equation can be used to determine the optimal capacitance for the maximum efficiency, since the main losses are modelled in the equations in Section 2. The optimized capacitor value can then be substituted into Equation 1 to calculate the corresponding PMOS resistance (R_{PI}), which is converted to the device size using the characteristic curves in the chosen technology. Similarly, models with two or more charge-pump stages can be used to determine the optimal parameters for the corresponding stages, and also the minimum number of stages to fullfil the required power/voltage output with maximum efficiency.

4. Results and conclusion

The simulated LC tank power consumption is depicted in Figure 5 along with the model results. The power efficiency curves for one- and three-stage DC-DC converter are depicted in Figure 6, as example, for varying L1 inductor sizes. Figure 7 depicts the output power variation of an optimized converter with the load resistance for different number of stages. The optimized system through the established model achieves more than 40% maximum efficiency yielding 0.45 V output with single stage, 0.75 V output with two stages, and 0.9 V output with three stages for 2.5 kΩ, 3.5 kΩ, and 5 kΩ loads respectively using 0.2 V input. The optimizations presented results in efficiency improvement of more than 50% for the unlimited input power scenario with unregulated output, as compared to the alternatives [7], [12].

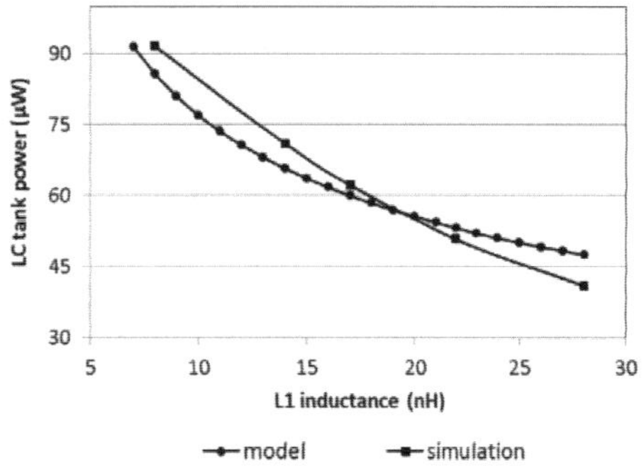

Figure 5. LC tank oscillator power consumption vs. L1 inductance.

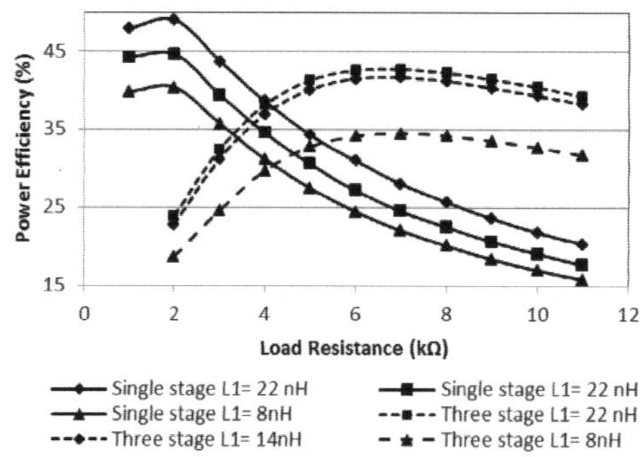

Figure 6. Converter power efficiency vs. load resistance and L1 size for 1 & 3 stages.

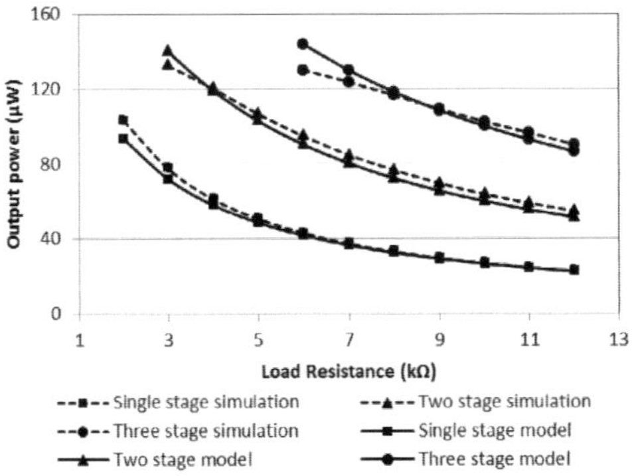

Figure 7. The variation of converter output power with load resistance for 1-3 stages.

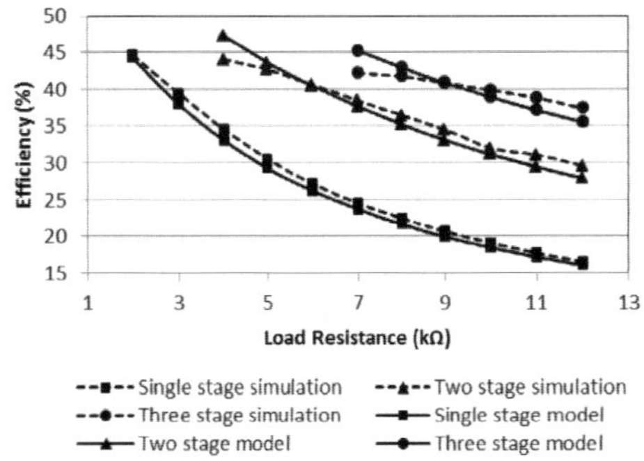

Figure 8. The variation of converter efficiency with load resistance for 1-3 stages.

The developed model can be utilized for circuit optimization in a variety of energy harvesting scenarios. Die area is a primary design constraint for fully integrated (low cost) solutions. After determining the application requirements including input power capacity, load voltage and power targets, either efficiency or output power can be optimized using the provided method.

5. References

[1] Kim J, Mok P K T and Kim C 2014 23.1 A 0.15V-input energy-harvesting charge pump with switching body biasing and adaptive dead-time for efficiency improvement *Dig. Tech. Pap. - IEEE Int. Solid-State Circuits Conf.* **57** 394–5

[2] Chen P H, Ishida K, Zhang X, Okuma Y, Ryu Y, Takamiya M and Sakurai T 2010 0.18-V input charge pump with forward body biasing in startup circuit using 65nm CMOS *Proc. Cust. Integr. Circuits Conf.* 2–5

[3] Tanzawa T 2011 A switch-resistance-aware dickson charge pump model for optimizing clock frequency *IEEE Trans. Circuits Syst. II Express Briefs* **58** 336–40

[4] Tanzawa T and Tanaka T 1997 A dynamic analysis of the Dickson charge pump circuit *IEEE J. Solid-State Circuits* **32** 1231–40

[5] Witters J S, Groeseneken G and Maes H E 1989 Analysis and Modeling of On-Chip High-Voltage Generator Circuits for Use in EEPROM Circuits *IEEE J. Solid-State Circuits* **24** 1372–80

[6] Dickson J F 1976 On-Chip High-Voltage Generation in MNOS Integrated Circuits Using an Improved Voltage Multiplier Technique *IEEE J. Solid-State Circuits* **11** 374–8

[7] Jayaweera HM, Pathirana WP, Muhtaroğlu A 2015 Fully Integrated Ultra-Low Voltage Step-up Converter with Voltage Doubling LC- Tank for Energy Harvesting Applications *Journal of Physics: Conference Series* **660** 012017

[8] Kwok K and Luong H C 2005 Ultra-low-voltage high-performance CMOS VCOs using transformer feedback *IEEE J. Solid-State Circuits* **40** 652–60

[9] Hsieh H, Member S and Lu L 2007 A High-Performance CMOS Voltage-Controlled **55** 467–73

[10] Tanzawa T 2015 A comprehensive optimization methodology for designing charge pump voltage multipliers *Proc. - IEEE Int. Symp. Circuits Syst.* **2015–July** 1358–61

[11] Cabrini A, Gregori S and Torelli G 2016 Integrated charge pumps : a generalised method for power efficiency optimisation **10** 12–9

[12] Bassi G, Colalongo L, Richelli A and Kovacs-Vajna Z 2012 A 150 mV-1.2 V fully-integrated DC-DC converter for Thermal Energy Harvesting *Int. Symp. Power Electron. Electr. Drives, Autom. Motion* 331–4

INTEGRATED FLEXIBLE SOLID-STATE SUPERCAPACITOR FABRICTED IN A SINGLE FABRIC LAYER

S Yong[1], J R Owen[2], M J Tudor[1], S P Beeby[1]

[1]Electronic &Electrical Engineering Group, Electronics and Computer Science, University of Southampton, Southampton, SO17 1BJ, UK
[2]Electrochemistry, Chemistry, University of Southampton, Southampton, SO17 1BJ, UK

E-mail: sy2g11@soton.ac.uk

Abstract. This paper presents the design, fabrication and characterization of a flexible solid-state electrical double layer supercapacitor fabricated in a single fabric layer. The proposed supercapacitors were based on fabric electrodes fabricated with low cost carbon materials via a spray coating technique. The single layer solid state supercapacitors achieved a specific capacitances of 40.5 $F.g^{-1}$, area capacitance of 40.5 $mF.cm^{-2}$.

1. Introduction

Wearable electronics also known as electronic textile or e-textile is the integration of electrical system in clothing and accessories. Powering such system could benefit from a flexible solid-state energy storage device with high energy and power density. A supercapacitor is an electrochemical device used for storing electrical energy, it can act as a standard energy storage and buffer devices with high capacitance while having a fast cycling rate like a conventional capacitor. Traditional energy storage devices such as electrochemical secondary batteries are limited in several areas such as poor cycling stability, lack of flexibility and low power density that not always compatible with wearable application like electronic textile. Therefore, new lightweight and flexible energy storage devices like fabric supercapacitors are highly of great interests as energy reservoir for smart fabric applications and desirable for e-textile approaches [1], but these have not yet been fully developed [2].

To date there are many examples of multilayer solid–state supercapacitor implemented on different flexible substrate, most of them attempt used either expansive carbonized materials like carbon nanotube (CNT) [3] and reduced graphene oxide [4], or pseudo-capacitive material such as manganese oxide [3] and conductive polymer [5] that suffered relative short device life time. H Sun et.al [4] implemented a solid-state flexible supercapacitor with polyester textile electrodes and gel electrolyte, the textile substrate was coated by reduced graphene oxide and conductive polymer(polyaniline), this multilayer device achieved an aerial capacitance of 781.8 $mF.cm^{-2}$. D V Lam et al. [6] demonstrated a flexible and all solid state supercapacitor with calligraphic ink coated hybrid cotton textile electrode and lithium salt based gel electrolyte. This device achieved area capacitance of 36 $mF.cm^{-2}$.

An electrical double layer supercapacitor is constructed using two electrical double layer interfaces, each having opposite polarity with respect to the electrolyte solution. The supercapacitor used in wearable electronics needs to be lightweight, flexible, solid-state, easy to be integrated into the system and made with non-hazardous materials. Previously, researchers have used hazardous substances in their solid-state supercapacitor, for example acid [7] in the gel electrolyte or corrosive oxide material [8] in the electrode, also these devices are consisting of two layers of fabric electrode and possibly a paper separator. In wearable electronics application, these supercapacitor requires extra care when packaging reduces the potential for skin irritation and compress/seal the multilayer device. Therefore, although such supercapacitor may demonstrate electrochemical performance the use of hazardous materials and multilayer device structure mean they are not the best opinion for wearable electronics.

This paper demonstrates an energy storage device for wearable electronics by fabricating a solid-state fabric electric double layer supercapacitor with non-hazardous materials in one piece of cotton fabric. The fabric electrodes were achieved via the spray coating process with various inexpensive carbon powders and cotton fabric substrates. The supercapacitor was fully tested with a non-hazardous gel electrolyte in order to study their operation and stability as a function of test rate and cycle number.

2. Device fabrication

Firstly, the conductive carbon ink for the coating process is prepared, the ink contains the commercial available inexpensive carbon powders, 1, 2, 4- Trichlorobenzene (1, 2, 4 TCB) solvent, a surfactant and polymer binder Ethylene-co vinyl acetate (EVA) that provides the adhesion between the carbon powders and fabrics. In this work the carbon powders used are activated carbon powder Kuraray Yp-80F and carbon black powder Shawinigan Black. Kuraray Yp-80F, has high purity, a particle size between 5-25 μm in diameter and an effective surface area of 2200 $m^2.g^{-1}$. Shawinigan Black has a mean particle size of 42 nm in diameter and an effective surface area of 75 $m^2.g^{-1}$.

In the next step the carbon ink was spray coated on both side of the cotton fabric substrates with a stainless steel mask and pneumatic spray system. During the spray coating process, carbon vapor coats the selected part of the fabric substrate and adheres to the yarns of fabrics uniformly, the penetration depth of the carbon vapor in the fabric substrate and was controlled by the pressure and the distance of the spray coating process, it allows carbon vapor get into the cotton substrate but did not penetrate all the way through the fabric substrate. After curing process, the carbon powder bonds to the fabric substrate via the polymer binder. The photograph of the spray coated carbon fabric electrode on the cotton fabrics can be seen in figure 1.a. As shown in figure 1.b, the SEM photograph of cross section view of cotton electrode indicates the top and bottom carbon coated cotton electrode yarns are physically separated by the cotton fibres.

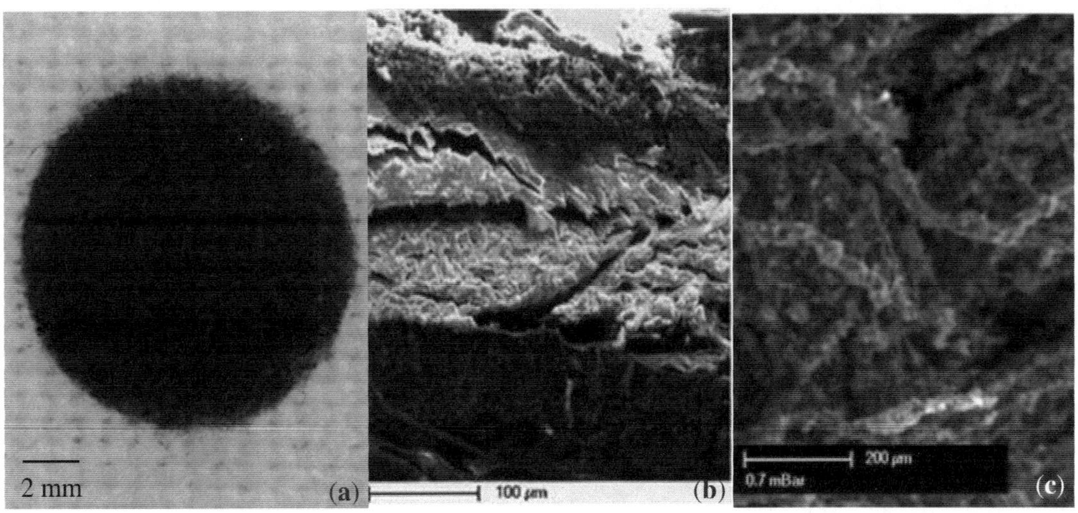

Figure 1. (a) Black carbon fabric electrode on original white cotton fabric substrate. (b) SEM photograph of cross section view of cotton electrode. (c) SEM photograph of carbon cotton electrode

Figure 1 (c) shows the SEM micrograph of the carbon particles adhered to the fibre to make a conductive network in the fabric substrate. The circular fabric electrode shown in figure 1 (a) will be cut out to assemble the supercapacitor cell. In this study the area of each cotton fabric electrode is 0.785

cm^2 with thickness about 400 μm, each single piece of fabric electrode has a weight of 69.8 mg.cm^{-2} before coating and carbon materials account for 1.8% of the final electrode weight.

The gel electrolyte was prepared by mixing ammonium salt, vinyl- alcohol polymer and water. The gel electrolyte solution was put under vacuum for 2 hours to reduce the amount of air bubbles inside the solution. In the next step, the cotton electrode was dipped into the gel electrolyte under vacuum for 20 minutes. Finally, the cotton electrodes containing the wet gel electrolyte were compressed and cured in oven for 2 hours at 60 °C. As shown in figure 2 the assembled supercapacitor device is compressed from the top to bottom electrodes using spring loaded stainless steel current collectors and spring housed within a Swagelok PFA tube fitting for testing.

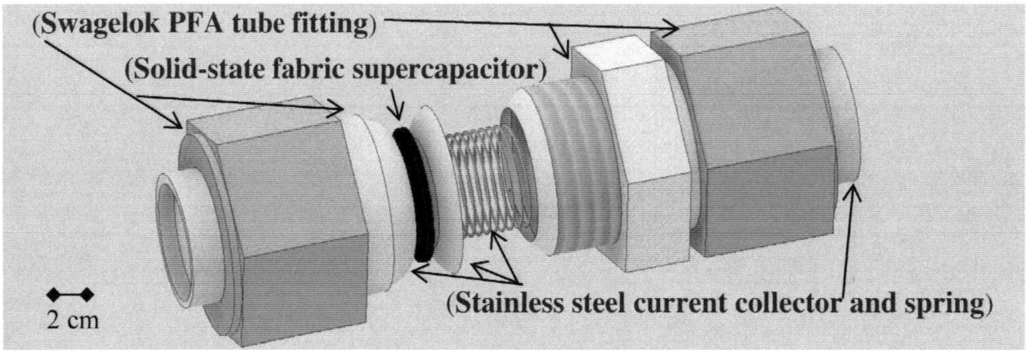

Figure 2. Swagelok PFA tube fitting used to testing before closing the two fitting caps

3. Results and discussion

The Solid-state fabric supercapacitor was tested using a VMP2 potentiostat/galvanostat (Biologic, France). The encapsulated supercapacitor was characterized by impedance spectroscopy (EIS), cyclic voltammetry (CV) at different scan rates and galvanostatic cycling (GC) at different scan voltage.

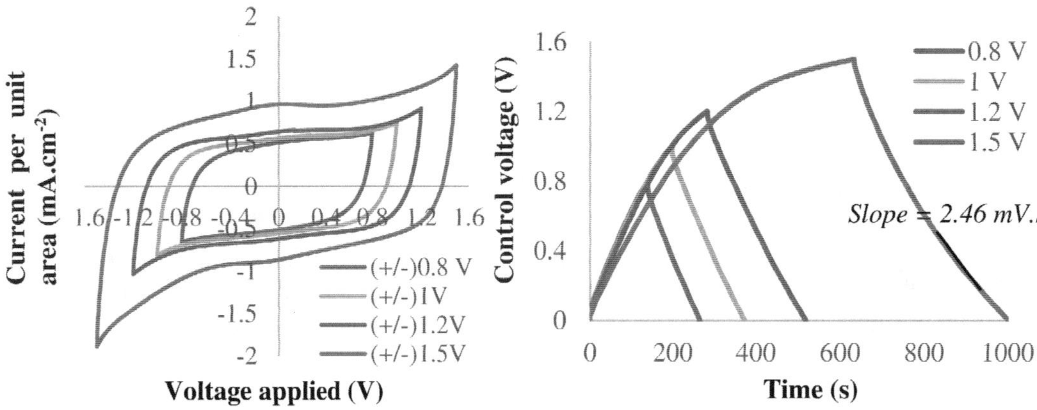

Figure 3. (a) Cyclic voltammetry tests (25 mV.s^{-1}) between +/-0.8 V to +/-1.5 V. (b) Galvanostatic cycling tests (0.127 mA.cm$^{-2)}$ between 0.8 V to 1.5 V

As shown figure 3(a) The cyclic voltammetry curve (CV) shows the device did not exhibit any chemical reaction or instability on both positive and negative scan directions. Figure 3 (b) shows the Galvanostatic cycling (GC) test results (0.127 mA.cm^{-2}) with different test voltages. The encapsulated single layer fabric supercapacitor achieved an equivalent series resistance of 54.1 Ω.cm^{-2} and an area capacitance of 40.5 mF.cm^{-2} at a test voltage of 1.5 V.

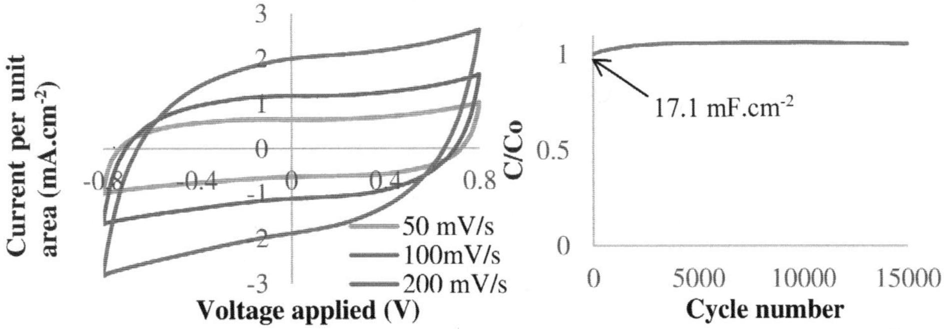

Figure 4. (a) CV test of the device between +/- 0.8 V at the scan rate of 200, 100, 50 mV.s^{-1}, (b) stability of device over 15000 cycles. C_o is the initial area capacitance (17.1 mF.cm^{-2}) of the device measured from cycle 1 of the CV test between +/- 0.8 V at the scan rate of 200 mV.s^{-1}.

According to figure 3 (a) and figure 4 (a), the CV curves show that the device is electrochemically stable at the scan rates from 25 to 200 mV, and area capacitance values vary from about 17.1 to 25.7 mF.cm-2 depending on the different scan rate. Figure 4 (b) indicates the device stability with a CV response. The overall capacitance increased about 5% in the first 2500 test cycles, it is correlates with the devices were not fully immersed by the gel electrolyte solution. In the rest of 12500 test cycles the device demonstrates a good cycling stability of less than 1% capacitance variation, it is correlates with the small temperature changes in the test lab. However the high device stability indicates excellent adhesion of the carbon powders forming a continuous conducting network, which is due the use of the spray coating technique and vacuum impregnation process with EVA polymer binder and gel electrolyte. The spray coating process results in activated carbon and carbon black powder being spread uniformly over the fabric samples and infiltrate into but not penetrate the cotton fabrics. The vacuum impregnation process ensures the gel electrolyte get in contract with most of the electrode materials to forms an effective double layer interface during characterisation.

4. Conclusion

This report presents a single layer supercapacitor on cotton fabrics. The non-hazardous solid-state fabric supercapacitor implemented in this work achieves a mass specific capacitance of 40.5 F.g^{-1}, area specific capacitance of 40.5 mF.cm^{-2}, a low normalized ESR of 54.1 Ω.cm^{-2} and achieves good cycling stability of less than 1% capacitance variation over 12500 cycles. In comparison with other works, these result are not as good as formal multilayer supercapacitor, however in this work both spray coated carbon electrodes and gel-electrolyte reinforced charge separator are integrated into just a single layer of cotton fabrics, the proposed supercapacitor is non-hazardous, fully wearable, scalable, inexpensive. This work demonstrates a reliable approach for obtaining a low cost and appropriated energy storage supercapacitors in fabrics for wearable applications. Future work will include improving the formulation of the carbon and gel electrolyte ink so the single layer solid-state fabric supercapacitor can achieve better electrochemical performances. The final device could see use applications in a wide range of wearable electronic systems.

References

[1] Zhu D, Beeby S, Tudor J and Harris N. 2011 Sensors and Actuators A Physical 169 **2** 317-25
[2] Beeby, S and Zhu D. 2011 Materials Science Foundations 197-218.
[3] Choi C and Kim S et al. 2015 SCIENTIFIC REPORTS, 5, 9387.
[4] Sun H and Peng H et al. 2016 Adv. Mater, 10.1002.
[5] Yuksel Rand Unalan H.2015 Int. J. Energy Res, 39,2042–2052
[6] Lam D and Lee S et al. 2016 J. Mater. Chem. A, 4, 4082
[7] Zang X and Zhu H et al. 2015 Nanoscale **7** 7318
[8] Jin H and Chana H. et al. 2014 RSC Advanes **4** 33022

Enhancing the power conversion efficiency of solar cells employing down-shifting silicon quantum dots

R Lopez-Delgado[1, 2], H J Higuera-Valenzuela[2], A Zazueta-Raynaud[1, 2], A Ramos[3], J E Pelayo[1, 4], D Berman[2, 3], M E Álvarez-Ramos[2] and Arturo Ayon[1].

[1]MEMS Research Lab, Department of Physics and Astronomy, University of Texas at San Antonio, San Antonio, TX, 78249, USA.
[2]Departamento de Física, [3]Departamento de Investigación en Física, Universidad de Sonora, Hermosillo, Son, 83000, México.
[4]Centro de Ciencias Exactas e Ingenierías, Universidad de Guadalajara, Guadalajara, Jal, 44430, México.

Corresponding author email: Rosendo.lopezdelgado@utsa.edu

Abstract. We report the synthesis and characterization of silicon quantum dots that exhibit down-shifting, photo luminescent characteristics. We also discuss the fabrication and characterization of single crystal Silicon (c-Si) Solar cells with and without the influence of the previously mentioned QDs. The incorporation of these nanostructures triggers improvements in the performance of the fabricated photovoltaic devices, especially in the open circuit voltage (V_{oc}) and short circuit current density (J_{sc}). Specifically, the experimental results showed increments in the V_{oc} from 532.6 to 536.2 mV and in the J_{sc} from 33.4 to 38.3 mA/cm^2. The combined effect of those improved V_{oc} and J_{sc} values led to an increment in the power conversion efficiency (PCE) from 11.90 to 13.37%. This increment represents an improvement of the order of 12.4% on the power conversion efficiency of this type of solar cells. The observed results could be conducive to promoting the proliferation of photovoltaic structures.

1. Introduction.

Even though the wavelength distribution of the solar spectrum extends from the infrared (2500 nm) to the ultraviolet (280 nm) one of the major problems in solar energy conversion is the relatively small absorption range of the materials employed in solar cells [1-4]. The underlying reason is that in principle only photons with energy greater than the bandgap can be absorbed by the material leading to the loss of relatively low energy photons. But even photons with energies much larger than the bandgap have a non-negligible probability of interacting with crystal phonons, in a process referred to as thermalization, rather than producing electron-hole pairs [3].

These fundamental losses can be ameliorated by the utilization of luminescent materials. To promote the absorption of low energy photons, near and below the bandgap, particles with up-conversion properties can be employed [5, 6]. On the other hand, in the case of energetic photons that contribute to thermalization, the alternative is to employ down-shifting materials, such as silicon quantum dots (Si-QDs) [7-9], able to emit lower energy photons that fall within the range of absorption by the underlying solar cell.

2. Experimental.

2.1. Silicon Quantum Dots Synthesis.

Silicon quantum dots (Si-QDs) were synthetized following the method described by Wang et al. [10] to achieve water-dispersible Si-QDs at room temperature. The Si-QDs were prepared by adding (1.5, 1.75 and 2.0) mL of (3-Aminopropyl) triethoxysilane 99% (APTES) to 4 mL of deionized (DI) water and stirred for 10 minutes. After that, 1.25 mL of 0.1 M (+)-Sodium L-ascorbate ≥ 98% (SA) was added to the previous mixture and stirred by an additional 30 minutes. The synthesis of Si-QDs was carried out using different amounts of APTES to achieve a batch of samples in order to characterize their photo-luminescence under UV radiation and the performance of solar cells in the presence of the aforementioned QDs.

2.2. Silicon Solar Cell Fabrication.

The silicon solar cells under study were fabricated from 4-inch, n-type, <100> silicon wafers with resistivity of 10-20 Ω-cm employing a spin on dopant technique (SOD) that has been described elsewhere [11].

The wafers were cut into pieces of approximately 30 x 30 mm; subsequently the samples were cleaned employing an extended 4-steps RCA cleaning process. To create the p-type emitter and the n+ back surface field (BSF) for charge collection, Boron and Phosphorous spin-on-dopant (SOD) solutions were prepared by the sol–gel method [11]. The Boron (SOD) solution was spin coated on the window side of the silicon sample and the Phosphorous (SOD) was deposited on the opposite side. Then, the samples were baked at 120°C for 10 min to remove the organic solvents. Subsequently, an annealing of 10 minutes at 1000°C was performed to diffuse the dopants and create the p–n junction and the BSF. To make the electrical contacts, a layer of 200 nm of aluminum was deposited on each side of the device by thermal evaporation. A shadow mask was used on the window side to make a pattern of finger electrodes, while a blanket deposition was performed in the back side. Then, the samples were annealed at 585°C for 10 min to promote the formation of an ohmic Al/Si contact.

Subsequently, Si-QDs films were spin cast on the window side of the solar cell. To this end, 200 μl of Si-QDs solution was dispersed with an angular velocity of 4000 rpm during 60s. All the samples were characterized by the same method in order to compare the performance of the device with and without the influence of the Si-QDs. Figure 1 show the flowchart of the fabrication and deposition process.

Figure 1. Flowchart of the fabrication of c-Silicon solar cell and Si-QDs deposition: 1) Extended RCA cleaning, 2) Diffusion of the emitter layer and BSF, 3) Front and back electrode deposition and 4) spin casting of a Si-QDs layer.

3. Results and Discussion.

3.1. Characterization of Silicon Quantum Dots.

The UV-Vis absorption spectra of the synthesized Si-QDs were measured using a Varian Cary 5000 UV-Vis spectrometer, while the photoluminescence effects were recorded using an Ocean Optics Flame-S-UV-VIS spectrometer. The absorption and photoluminescence spectra of the different synthesis process of Si-QDs are shown in Figure 2. All the synthesized Si-QDs show an absorption spectrum tapering off beyond ~500 nm. The photo-luminescence spectra consist of broad bands extending from 450 to 650 nm in all cases with variations in the maximum intensity and the peak position. Figure 2 shows that the variation in the amount of APTES used in the synthesis has a relatively minor effect on peak position that fluctuates from 521 nm to 526 nm, for 1.5 ml and 2.0 ml of APTES, respectively.

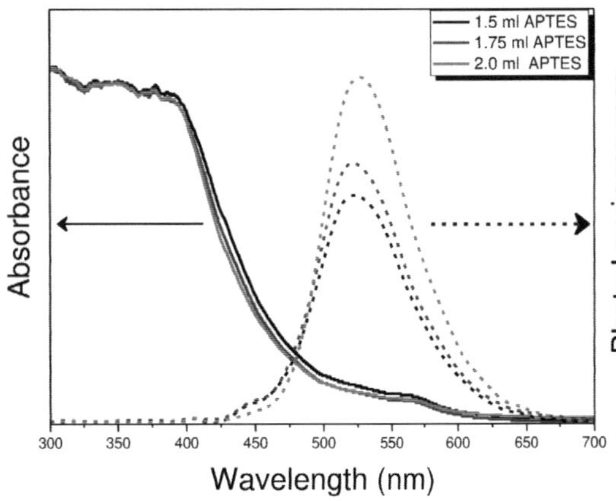

Figure 2. Absorption (continuous line) and photo-luminescence (dashed line) spectrum of Si QDs varying the amount of APTES.

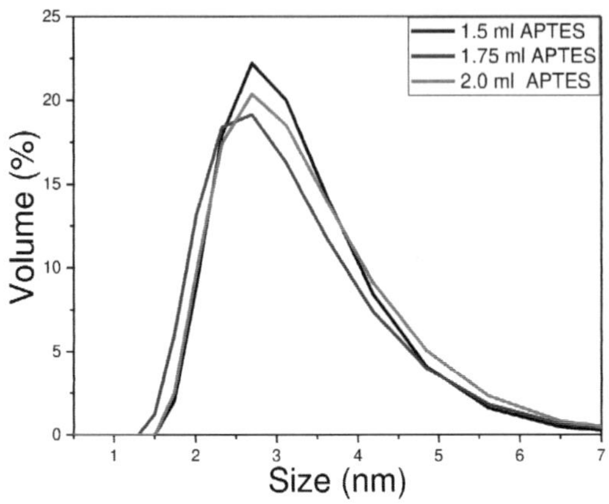

Figure 3. Dynamic light scattering (DLS) size distribution of Si-QDs for the different synthesis methods.

A Zetasizer Nano Series using a dynamic light scattering method (DLS) was employed to perform measurements of the average size of the synthesized silicon nanoparticles. As shown in Figure 3, the amount of APTES has a relatively minor influence on the size of the QDs whose diameter is centred around 2.6 nm. For subsequent characterization efforts, the samples with the highest photoluminescence intensity were selected, namely, 2.0 ml APTES (with 0.1M SA and 30 min of reaction).

3.2. Solar Cell Characterization.

A set of three c-Si solar cells were fabricated with the previously described approach. Si-QDs with 2.0 ml APTES were deposited on the set of solar cells to determine the influence of synthesis variable. I – V characteristics of the devices were measured before and after the deployment of the Si-QDs. In all cases, the utilization of Si-QDs improved the performance of the solar cells. In general the short circuit current had a more pronounced improvement than the open circuit voltage as shown in Figure 4.

The averages of the solar cell performance parameters are listed in Table 1. The application of these nanostructures on the c-Si solar cells lead to increases in two of the three main parameters that contribute to solar cell efficiency, namely, the open circuit voltage (V_{oc}) and the short circuit current density (J_{sc}). An improvement was observed in all the solar cells after the deposition of the Si-QDs synthesized, where the solar cell V_{oc} increased from an average of 532.57 to 536.20 mV, the J_{sc} went from 33.42 to 38.28 mA/cm^2 and although the fill factor (FF) decreased from 66.87 to 65.20%, the combination of all these parameters led to an improvement of the power conversion efficiency (PCE) from 11.90 to 13.37% for an overall PCE improvement of 12.4%.

Table 1. C-Silicon solar cell performance parameters before and after application of Si-QDs.

Solar cell performance	Voc (mV)	Jsc (mA/cm^2)	FF (%)	PCE (%)
c-Si Solar cell set	532.57	33.42	66.87	11.9 ± 0.27
c-Si Solar cell set + Si-QDs	536.2	38.28	65.20	13.37 ± 0.19

Figure 5 shows the spectral response of the c-Si solar cell with the application of Si-QDs. EQE measurements indicate that the QDs are improving the photocurrent generation specially in two segments, namely, (i.) below 600 nm, in agreement with the photo-luminescence spectra, and (ii.) a broad contribution with maxima ranging from 850 to 950 nm which could be attributed to a change in the reflectance due to the Si-QDs layer. These EQE improvements are thought to be responsible for the current increase measured during the current – voltage analysis.

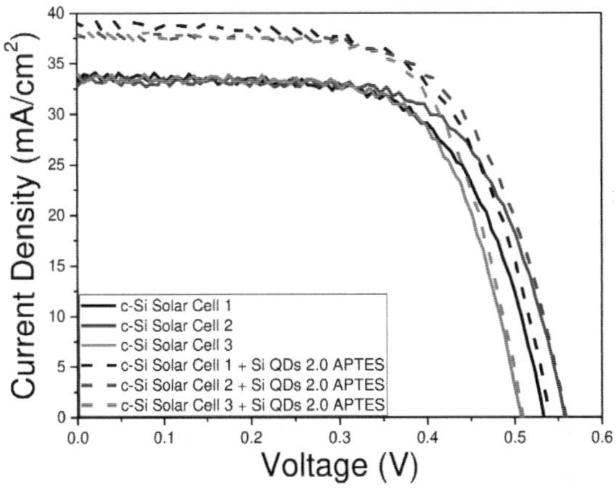

Figure 4. Current – Voltage characteristics for c-Si Solar Cells before (continuous lines) and after (dashed lines) application of Si-QDs.

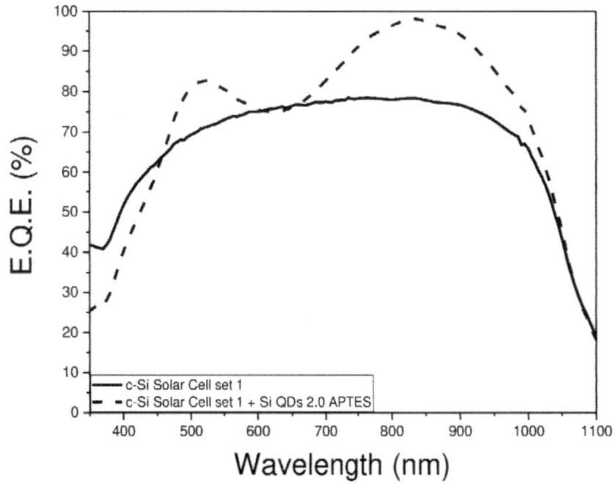

Figure 5. External Quantum Efficiency (EQE) of c-Si Solar Cells before (continuous lines) and after (dashed lines) deploying the Si-QDs.

4. Conclusions.

In the present work it was studied the influence of Si-QDs in c-Si solar cells as photovoltaic response improvers. The Si-QDs were synthetized using a water-dispersible method. Different amount of APTES were used in order to obtain different emission spectra. Experimental results showed that the incorporation of Si-QDs as down-shifting material on the window side of a solar cell improves the photocurrent generation which can be seen from the current density and open circuit voltage increase. Even though the FF was observed to decrease in all cases, the overall power conversion efficiency achieved after the application of Si-QDs was increased from 11.90% to 13.37 %. This result represents an improvement of 12.4% in the photovoltaic performance. The experimental results evince that this strategy represents a promising way to improve the efficiency of existing solar cells, which could promote the proliferation of photovoltaic energy harvesting.

Acknowledgements

The authors would like to acknowledge the U.S. Army Research Office (Grant W911NF-13-1-0110), CONACYT, the Physics and Astronomy Department of the University of Texas at San Antonio and the Physics Department of University of Sonora, for the financial support.

References

[1] Hirst L C and Ekins-Daukes N J, 2011. *Progress in Photovoltaics: Research and Applications*. **19**(3): p. 286-293.
[2] Scholes G D, et al., 2011. *Nature chemistry*. **3**(10): p. 763-774.
[3] Huang X, et al., 2013. *Chemical Society Reviews*. **42**(1): p. 173-201.
[4] Richards B, 2006. *Solar energy materials and solar cells*. **90**(15): p. 2329-2337.
[5] Qu B, et al., 2016. *Journal of Alloys and Compounds*. **658**: p. 848-853.
[6] Van Der Ende B M, Aarts L, and Meijerink A, 2009. *Physical Chemistry Chemical Physics*. **11**(47): p. 11081-11095.
[7] Trupke T, Green M, and Würfel P, 2002. *Journal of Applied Physics*. **92**(3): p. 1668-1674.
[8] Zhang Q and Huang X, 2010. *Progress in Materials Science*. **55**(5): p. 353-427.
[9] Stupca M, et al., 2007. *Applied Physics Letters*. **91**(6): p. 063107.
[10] Wang J, et al., 2014. *Journal of Materials Chemistry B*. **2**(27): p. 4338-4345.
[11] Pudasaini P R and Ayon A A, High performance nanopillars array silicon solar cells in *Design, Test, Integration and Packaging of MEMS/MOEMS (DTIP), 2013 Symposium on*: IEEE.

Influence of photo-luminescent CdSe/CdS core shell quantum dots in solar cell efficiency

R Lopez-Delgado[1, 2], Y Zhou[4], A Zazueta-Raynaud[1, 2], H Zhao[4], J E Pelayo[1, 3], A Vomiero[5], M E Álvarez-Ramos[2], F Rosei[4], and A Ayon[1].

[1]MEMS Research Lab, Department of Physics and Astronomy, University of Texas at San Antonio, San Antonio, TX, 78249, USA.
[2]Departamento de Física, Universidad de Sonora, Hermosillo, Son, 83000, México.
[3]Centro de Ciencias Exactas e Ingenierías, Universidad de Guadalajara, Guadalajara, Jal, 44430, México.
[4]INRS-EMT, Varennes, QC J3X 1P7, Canada.
[5]Luleå University of Technology, 971 87 Luleå, Sweden

Corresponding author email: Rosendo.lopezdelgado@utsa.edu

Abstract. We report the synthesis and characterization of CdSe/CdS core-shell quantum dots (CdSe/CdS-QDs) that exhibit absorption in the UV range of the solar spectrum and emit photons with wavelengths centered around 625 nm, a wavelength that is well suited for silicon absorption and electron-hole pair generation. We also report the fabrication and characterization of single crystal silicon (c-Si) solar cells with and without the aforementioned photo luminescent, down-shifting CdSe/CdS-QDs. The incorporation of these nanostructures triggered improvements in the performance of the devices, particularly in the open circuit voltage (V_{oc}) and short circuit current density (J_{sc}) for which the measured values showed an increase from 543 to 546 mV and from 32.5 to 37.0 mA/cm^2, respectively. The combined effect of the improved values led to an increment in the power conversion efficiency (PCE) from 12.01 to 13.54%. This increase represents a 12.7% improvement in the PCE of the fabricated devices. The effort described herein is considered a good fit to the generalized trend to improve the efficiency of solar cells with mass-compatible techniques that could serve to promote their widespread utilization.

1. Introduction.

In recent years, silicon solar cells have been researched with the aim of improving their power conversion efficiency by different methods. However, one of the biggest problems in the energy conversion from sunlight is the absorption range of the material constructing solar cells, since they produce energy only from a small part of the total solar spectrum [1, 2]. The wavelength distribution of the solar spectrum extends from the infrared to the ultraviolet but only a fraction is absorbed by common materials in solar cells[3, 4]. The main reasons are that only photons with energy greater than the bandgap can be absorbed by the material, and photons with energies much larger than the bandgap do not produce electron-hole pairs but only contribute to thermalization of the material, leading to losses of energy from low and high energy photons[3].

Improvements in the absorption range could be achieved by the utilization of luminescent materials, using up-conversion materials for low energy photons [5, 6] and down conversion materials for high energy photons [7, 8]. In this report, we focus on down-shifting CdSe/CdS quantum dots.

2. Experimental.

2.1. CdSe/CdS Quantum Dots Synthesis.

Core CdSe QDs were synthesized by using the hot injection approach [9]. To this end, trioctyl phosphine oxide (TOPO) (1g) and Cd-oleate (0.38 mmol, 1mL) in 8 mL of octadecene (ODE) were purged by N_2 at room temperature for 30 min. The reaction system was evacuated for 30 min at 100 °C, and then the temperature was raised to 300 °C. A mixture of trioctyl phosphine (TOP)–Se (4 mmol, 4 mL), 3 mL of oleylamine and 1 mL of ODE at was quickly injected into the Cd-oleate suspension under vigorous stirring. Ethanol was added, then the suspension was centrifuged and the supernatant was removed. The Core CdSe QDs were dispersed in toluene. Deposition of CdS layers (shell) on CdSe (core) QDs followed the procedure described in Ghosh et al [9]. Oleylamine (OLA) (5 ml), ODE (5 mL) and CdSe QDs (~2×10^{-7} mol in hexane) were degassed at 110°C for 30 min. Then the temperature was raised to 240°C with stirring. Cd-oleate dispersed in ODE (0.25 mL, 0.2 M) was added dropwise and the mixture was allowed to react for 2.5 h, followed by dropwise addition of 0.2 M sulfur in ODE with same volume and a reaction time of 1 h. The shell was further annealed for 10 min. All subsequent shells were annealed at 240 °C for ~10 min. The reaction was cooled to room temperature using ice water (0°C). Ethanol was added, the suspension was centrifuged and the supernatant was removed. The QDs were then dispersed in toluene for further characterization.

2.2. Silicon Solar Cell Fabrication.

Single crystal silicon solar cells were fabricated employing 4-inch, n-type, <100> silicon wafers with a resistivity of 10-20 Ω-cm employing a spin on dopant technique (SOD) that has been described elsewhere [10].

The silicon samples were cleaned using an extended 4-steps RCA cleaning process. In order to create the p-type emitter layer and the n+ back surface field (BSF) for charge collection, Boron and Phosphorous spin-on-dopant (SOD) solutions were prepared by the sol–gel method [10]. The Boron (SOD) solution was deposited on the front side of the silicon sample and the Phosphorous (SOD) on the opposite side. Then, the samples were annealed at 120°C for 10 min to remove the organic solvents. Subsequently, the samples were annealed at 1000°C for 10 min to diffuse the dopants and create the p–n junction and the BSF. To make the electrical contacts, 200 nm of aluminum were deposited on each side of the device by thermal evaporation. A shadow mask was used on the front side to make a pattern of finger electrodes, while a blanket deposition was performed in the back side. Finally, the samples were annealed at 585°C for 10 min to promote the formation of an ohmic Al/Si contact.

After a set a solar cells were fabricated and characterized, CdSe/CdS-QDs were spin cast on the window of the solar cells. To this end, 200 µl of CdSe/CdS-QDs solution was dispersed with an angular velocity of 5000 rpm during 60s. All the samples were characterized by the same method in order to compare the performance of the device with and without the influence of the CdSe/CdS -QDs.

Figure 1. Flowchart of the fabrication of c-Silicon solar cell and CdSe/CdS-QDs deposition: 1) Extended RCA cleaning, 2) Diffusion of the emitter layer and BSF, 3) Front and back electrode deposition and 4) spin casting of CdSe/CdS-QDs.

3. Results and Discussion.

3.1. Characterization of CdSe/CdS Quantum Dots.

The UV-Vis absorption spectra of the synthesized CdSe/CdS-QDs were measured using a Varian Cary 5000 UV-Vis spectrometer, while the photoluminescence effects were recorded using an Ocean Optics Flame-S-UV-VIS spectrometer. The absorption and photoluminescence spectra of the synthetized CdSe/CdS-QDs are shown in Figure 2. It is observed that the CdSe/CdS-QDs have absorption of photons below ~525 nm while the photo-luminescence spectra consist of a broad band extending from 525 to 675 nm with a maximum around 625 nm, confirming the down-shifting effect, which is desired to improve the absorption by the silicon solar cell.

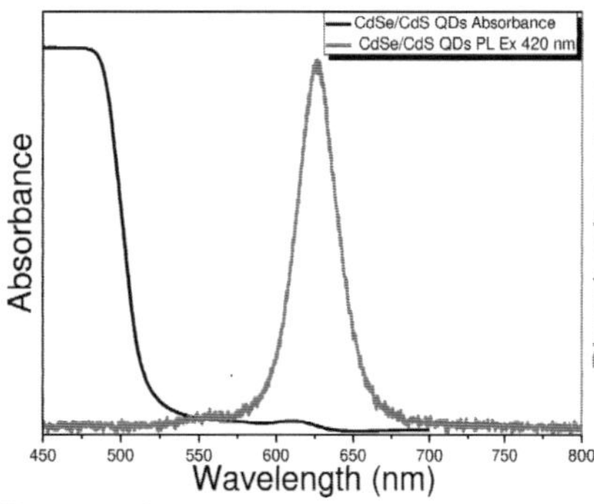

Figure 2. Absorption and photo-luminescence spectrum of CdSe/CdS-QDs varying the amount of APTES.

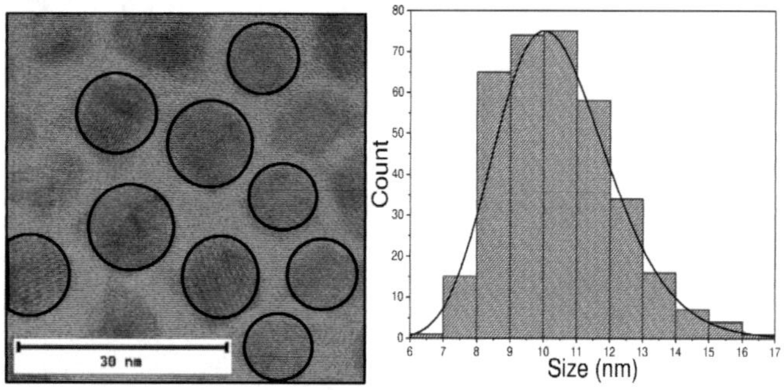

Figure 3.TEM images of CdSe/CdS QDs and size distribution histogram.

Transmission electron microscopy (TEM) images were collected employing a JEOL 2010-F transmission electron microscope. Figure 3 show TEM images of the synthetized CdSe/CdS-QDs as well as the size distribution histogram where it was found that the size average is around 10 nm.

3.2. Solar Cell Characterization.

A set of three c-Si solar cells were fabricated with the previously described approach. CdSe/CdS-QDs were deposited on the set of solar cells to determine their influence. The I – V characteristics of the devices were measured before and after the deployment of the CdSe/CdS-QDs. In all cases, the utilization of these nanostructures improved the performance of the solar cells as shown in Figure 4.

The averages of the solar cell parameters are listed in Table 1. The application of the QDs on c-Si solar cells lead to increases in two of the three main parameters that contribute to solar cell efficiency, namely, the open circuit voltage (V_{oc}) and the short circuit current density (J_{sc}). An improvement was observed in all the solar cells after the deposition, where V_{oc} showed an increase from an average of 543.38 to 545.90 mV, the J_{sc} went from 32.49 to 37.03 mA/cm^2 and although the fill factor (FF) decreased from 68.04 to 66.96%, the combination of all these parameters led to an improvement of the power conversion efficiency (PCE) from 12.01 to 13.54% for an overall PCE improvement of 12.7%.

Table 1. C-Silicon solar cell performance parameters before and after application of CdSe/CdS-QDs.

Solar cell performance	Voc (mV)	Jsc (mA/cm^2)	FF(%)	PCE (%)
c-Si Solar cell set	543.38	32.49	68.04	12.01 ± 0.2
c-Si Solar cell set +CdSe/CdS QDs	545.90	37.03	66.96	13.54 ± 0.2

The External Quantum Efficiency was measured in order to study the spectral response of the fabricated c-Si solar cells with and without the deployed CdSe/CdS-QDs. The observations indicate that the photovoltaic response improved in a wavelength segment extending from the UV region to the visible with the maximum influence between 400 and 600 nm as shown in Figure 5. These EQE improvements are thought to be responsible for the current increase measured during the I–V analysis.

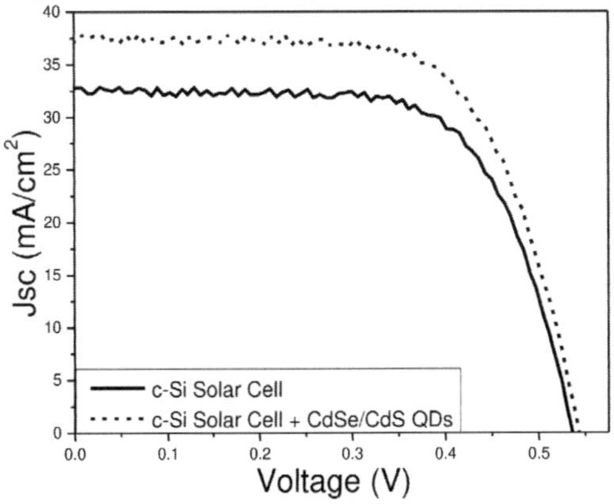

Figure 4. Current – Voltage characteristics for c-Si Solar Cells before (continuous lines) and after (dashed lines) application of CdSe/CdS-QDs.

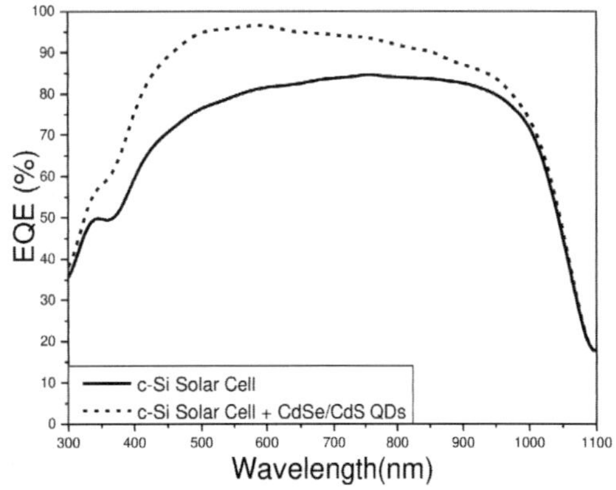

Figure 5. External Quantum Efficiency of c-Si Solar Cells before (continuous lines) and after (dashed lines) deploying the CdSe/CdS-QDs.

4. Conclusions.

In the present work it was studied the influence of CdSe/CdS-QDs in c-Si solar cells as photovoltaic response improvers. Experimental results showed that the incorporation of CdSe/CdS-QDs as down-shifting material on the window side of a solar cell improves the photocurrent generation, which can be seen from the current density and open circuit voltage increase as well as the EQE improvement in the segment in accordance to the absorbance and photoluminescent results. Even though the FF was observed to decrease in all cases, the overall power conversion efficiency achieved after the application of CdSe/CdS-QDs was increased from 12.01% to 13.54 %. This result represents an improvement of 12.7% in the photovoltaic performance. The experimental results evince that this strategy represents a promising way to improve the efficiency of existing solar cells, which could promote the proliferation of photovoltaic energy harvesting.

Acknowledgements

The authors would like to acknowledge the U.S. Army Research Office (Grant W911NF-13-1-0110), CONACYT, the Physics and Astronomy Department of the University of Texas at San Antonio and the Physics Department of University of Sonora, for the financial support.

References

[1] Hirst L C and Ekins-Daukes N J, 2011. *Progress in Photovoltaics: Research and Applications*. **19**(3): p. 286-293.
[2] Scholes G D, et al., 2011. *Nature chemistry*. **3**(10): p. 763-774.
[3] Huang X, et al., 2013. *Chemical Society Reviews*. **42**(1): p. 173-201.
[4] Richards B, 2006. *Solar energy materials and solar cells*. **90**(15): p. 2329-2337.
[5] Qu B, et al., 2016. *Journal of Alloys and Compounds*. **658**: p. 848-853.
[6] Van Der Ende B M, Aarts L, and Meijerink A, 2009. *Physical Chemistry Chemical Physics*. **11**(47): p. 11081-11095.
[7] Zhang Q and Huang X, 2010. *Progress in Materials Science*. **55**(5): p. 353-427.
[8] Trupke T, Green M, and Würfel P, 2002. *Journal of Applied Physics*. **92**(3): p. 1668-1674.
[9] Ghosh Y, et al., 2012. *Journal of the American Chemical Society*. **134**(23): p. 9634-9643.
[10] Pudasaini P R and Ayon A A, High performance nanopillars array silicon solar cells in *Design, Test, Integration and Packaging of MEMS/MOEMS (DTIP), 2013 Symposium on*: IEEE.

A miniaturized human-motion energy harvester using flux-guided magnet stacks

M A Halim[1] **and J Y Park**
Department of Electronic Engineering, Kwangwoon University, 447-1 Wolgye-dong, Nowon-gu, Seoul 139-701, South Korea
[1] He is currently with the Department of Mechanical Engineering, University of Utah, 50 S. Central Campus Drive, Salt Lake City, Utah 84112, USA

E-mail: halim.miah@utah.edu

Abstract: We present a miniaturized electromagnetic energy harvester (EMEH) using two flux-guided magnet stacks to harvest energy from human-generated vibration such as hand-shaking. Each flux-guided magnet stack increases (40%) the magnetic flux density by guiding the flux lines through a soft magnetic material. The EMEH has been designed to up-convert the applied human-motion vibration to a high-frequency oscillation by mechanical impact of a spring-less structure. The high-frequency oscillator consists of the analyzed 2-magnet stack and a customized helical compression spring. A standard AAA battery sized prototype (3.9 cm^3) can generate maximum 203 μW average power from human hand-shaking vibration. It has a maximum average power density of 52 μWcm^{-3} which is significantly higher than the current state-of-the-art devices. A 6-stage multiplier and rectifier circuit interfaces the harvester with a wearable electronic load (wrist watch) to demonstrate its capability of powering small-scale electronic systems from human-generated vibration.

1. Introduction

Due to the increasing popularity of the multifunctional and low-power consuming wearable electronic devices and systems, energy harvesting from human-generated (e.g., walking, running, shaking limbs etc.) vibration is recently being interesting as an alternative to the commonly used electrochemical batteries. However, generating significant power from human-generated vibration is challenging because these vibrations have extremely low frequency, random vibration characteristics [1]. Coupling such human-generated, low-frequency random vibrations into the electromechanical transducer of an energy harvester needs a clever design choice. Among a number of design approaches proposed by the researchers, mechanical frequency up-conversion has become the mainstream approach which is executed either by mechanical (direct or transverse) impact or by plucking (mechanical or magnetic) [2-5]. It allows the transducer element to generate voltage/power at high frequency while excited by a low-frequency vibratory system in response to the human-motion. Due to convenience in spring-mass suspension, relatively low source resistance, and high power density in macro-scale, an electromagnetic generator best suits the frequency up-conversion mechanism for energy harvesting from human-generated motion. The output power of an EMEH depends on the coil characteristics (number of turns, height, resistance, relative position to magnet, etc.), relative motion between the magnet and coil, and magnetic flux linkage with the coil. The first and second factors can be optimized by proper electro-mechanical design of the harvester, the third factor can be improved either by an array of magnets, or by Halbach arrays [6-8]. However, using a number of magnets (more than two) in an array increases the size of the harvester for specific application, which, in turn, affects its power density.

The goal of this work is to improve the output performance of our recently reported proof-of-concept prototype [9] to achieve higher power density. In order for that, we choose to increase the magnetic flux densities using a flux-guided magnet stack that we have analyzed and optimized by finite element method (FEM) analysis. Other parameters (e.g., spring-mass design, damping, coil characteristics etc.) were optimized through power spectral density analysis of a vibrating body, magnet displacement within the boundaries of electromagnetic interaction as well as the magnet-coil equilibrium position. Then, output responses of a fabricated prototype were carried out for human hand-shaking motion. An interface circuit was designed and was also analyzed its charging characteristics while connected to a wearable electronic load.

2. Harvester structure and its operation

Figure 1 shows the schematic structure and basic working mechanism of the proposed EMEH. It consists of two front-facing frequency up-converted electromagnetic generators which are struck by a freely-movable non-magnetic ball placed between the ball. Each generator consists of a cylindrical shaped flux-guided magnet stack placed on a helical compression spring and a coil wound around the magnet-stack over a cylindrical housing. The magnet stack (Type-3) was formed by assembling two magnets with a spacer in a repulsive manner. This particular type of stack was chosen by analyzing a number of 2-magnet stacks using Finite Element Method Magnetic (FEMM) program, as shown in Fig. 2. FEMM simulations were done to determine the magnetic flux density distributions of the different 2-magnet stacks. Results show that among those magnetic structures, Type-3 magnet-stack has the largest magnetic flux density (increased by 40% than a single magnet). The flux lines for the first three magnetic structures interact with the coil windings along the axial direction whereas those of the Type-3 stack are guided by the ferromagnetic spacer that allows the flux lines to interact with the coil windings along the radial direction, as shown in Fig. 3. Type-3 stack significantly increases the flux linkage through the coil windings which are aligned with the flux-guided magnet stack.

On low frequency excitation (in the axial direction) generated by human-limb motion with sufficiently large acceleration amplitude, the freely-movable ball oscillates within its displacement limit between the two front facing high-frequency oscillatory systems and impacts on the magnet stacks consecutively with a certain force $F(t)$ at its peak displacement. This allows each magnet stack to vibrate with higher frequency (at its resonance). Following basic mechanical vibration theory [10], the resultant response is an oscillatory motion with exponential decay between two consecutive impacts

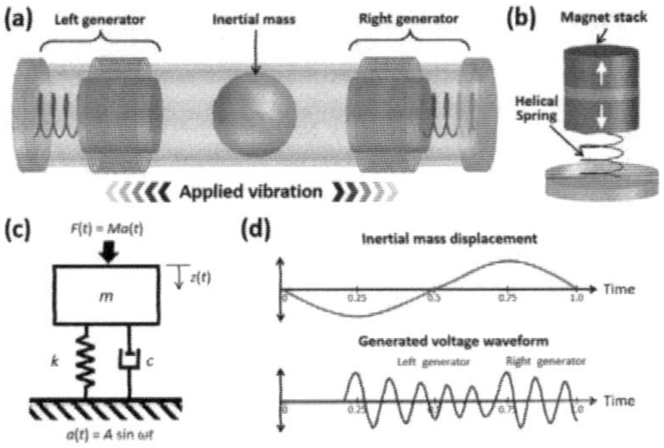

Figure 1. Schematics of the (a) proposed energy harvester, (b) flux-guided magnet stack, (c) equivalent mechanical model of one generator, and (d) basic operation principle.

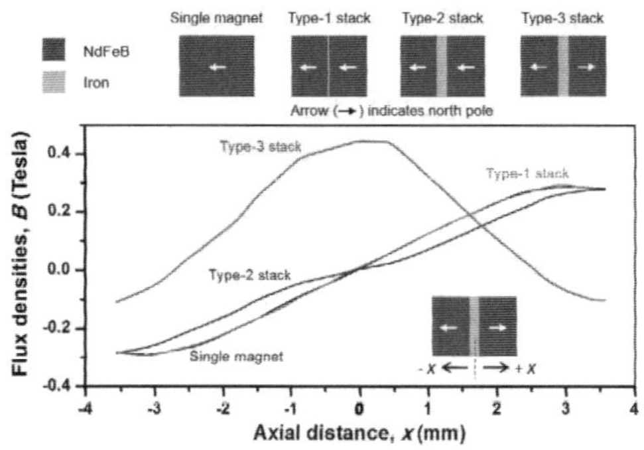

Figure 2. Change of magnetic flux densities with axial distance of the magnetic structures (sketched at the top of the graph) at 1.25 mm radial distance from the magnet wall.

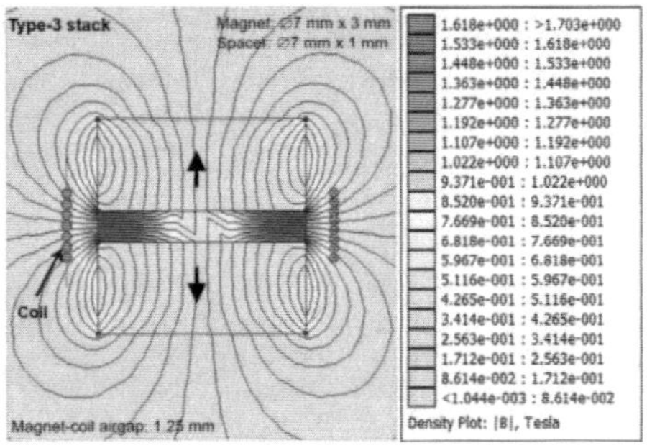

Figure 3. Magnetic flux distribution of the Type-3 magnetic structure obtained from FEMM simulation. The flux lines interact with the coil windings along the radial direction.

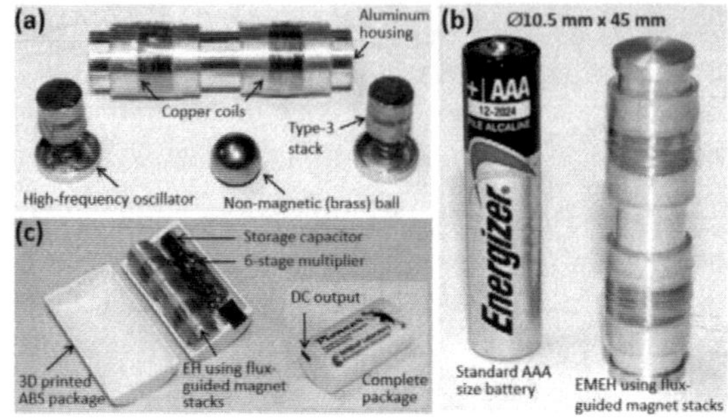

Figure 4. Photographs of the (a) harvester components before assembling, (b) assembled prototype alongside a standard AAA battery, and (c) complete package including multiplier and rectifier circuit for AC to DC conversion.

(by the ball). The kinetic energy generated by human-limb motion is transferred to the helical compression spring, which converts it to electrical energy by the relative motion between the flux-guided magnet stack and the coil. The maximum power delivered to a matched load resistance is

$$P_{\max} = \frac{(NBl)^2}{8R_l}\left[\frac{MA\omega_r e^{-\zeta\omega_r t}}{k\sqrt{1-\zeta^2}}\sin\left(\omega_r\sqrt{1-\zeta^2}\right)\right]^2 \tag{1}$$

where N is the number of coil turns, B is the magnetic flux density, l is the coil height, M is the mass of the ball, A is the applied acceleration, k is the stiffness of the helical spring, ω_r is the resonant frequency of the high-frequency oscillator, ζ is the total damping, and R_l is the load resistance. We fabricated, characterized and demonstrated the newly designed prototype to drive a wearable electronic load (wrist watch). The fabricated prototype, along with a 6-stage multiplier (and rectifier) circuit and a storage capacitor, was accommodated in a 3D printed ABS package, as shown in Fig. 4.

3. Experimental results and discussion

The performances of the fabricated EMEH were evaluated by applying hand-shaking vibration. It generates 1.54 $V_{p\text{-}p}$ (123 mV_{rms}) open circuit voltage when both the generators were connected in series, as shown in Fig. 5(a). The maximum average power delivered to 43 Ω matched load resistance is 203 μW, as shown in Fig. 5(b). To use the generated power for driving an electronic load, we built a multiplier and rectifier circuit with a 22 μF storage capacitor by soldering surface-mount devices on a

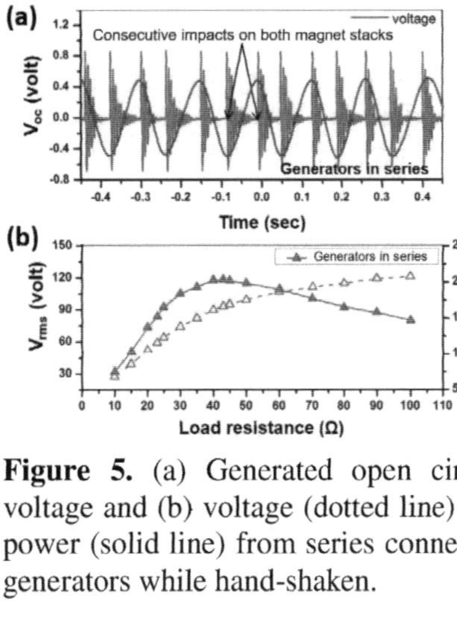

Figure 5. (a) Generated open circuit voltage and (b) voltage (dotted line) and power (solid line) from series connected generators while hand-shaken.

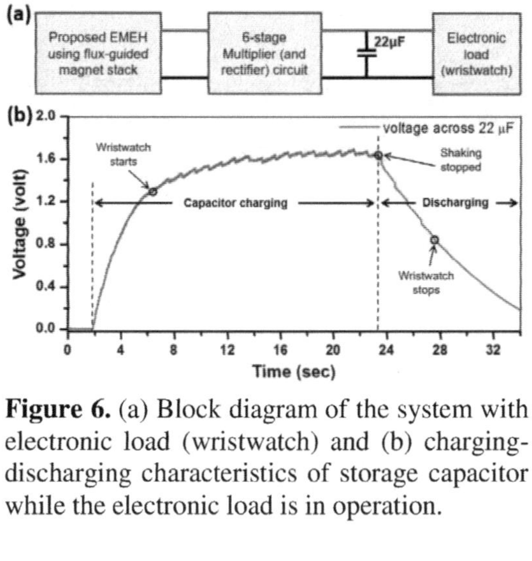

Figure 6. (a) Block diagram of the system with electronic load (wristwatch) and (b) charging-discharging characteristics of storage capacitor while the electronic load is in operation.

Table 1: Comparison of recently reported energy harvester operated by human-motion vibration.

Reference	Tested by	Prototype volume (cm^3)	Operating condition	Max. average power (μW)	Avg. power density ($\mu W cm^{-3}$)
[3]	Handy motion	19.2	5.8 Hz, 2 g	103.55	5.4
[5]	Running motion	5	2 Hz, 2 g	43[a]	8.6[a]
[9]	Hand-shaking	7.16	4.6 Hz, 2 g	110	15.4
[11]	Walking motion	51.84	5 kmh^{-1}	51	0.74
This work	*Hand-shaking*	*3.9*	*4.9 Hz, 2 g*	*203*	*52*

[a] Peak value

self-designed printed circuit board (PCB). Figure 6(b) shows that the wrist watch starts running when the capacitor voltage rises to 1.3 V_{dc} after 6 sec and kept running as long as the harvester was being excited. Until then, the capacitor voltage increased up to 1.66 V_{dc} at about 24 sec. The wrist watch kept running even after the excitation was stopped until the capacitor voltage dropped to 0.86 V_{dc} at 28 sec. Note that, the storage capacitor discharges very fast. However, a proper choice of energy regulation process and a suitable storage unit (e.g., supercapacitor) could store the energy for a long time to use the stored energy later. The performances of vibration energy harvesters are generally compared using the power density. Table 1 presents a performance comparison of this work with similar works (human-motion driven) recently reported.

4. Conclusions

We have proposed, designed and demonstrated a miniaturized EMEH using flux-guided magnet stacks for harvesting energy from human hand-shaking motion to power wearable electronics. Flux-guided magnet stacks were designed by analyzing electromagnetic fields characteristics of several magnetic structures by FEMM simulation. It not only increases the resultant flux densities but also improves the flux linkage with the coil windings in the radial direction, rather than in the axial direction for other magnetic structures. We built a standard AAA battery sized prototype by connecting two high-frequency generators in series which were struck consecutively by a freely movable ball during human hand-shaking. The device generates 203 μW average power and offers maximum 52 $\mu W cm^{-3}$ average power density which is significantly higher than the current state-of-the-art devices designed for human-motion operation. A complete harvesting system was built by incorporating the harvester unit, an ac/dc converter (6-stage multiplier and rectifier circuit), and a storage capacitor in a single package to drive a wearable electronic load (wrist watch).

Acknowledgements

The authors are grateful to acknowledge the Pioneer Research Centre Program (2010-0019313) funded by National Research Foundation (NRF) of Korea. Special thanks to Shad Roundy of the University of Utah for his support to present this work at PowerMEMS-2016.

References

[1] Ylli K, Hoffmann D, Willmann A, Becker P, Folkmer B and Manoli Y, 2015 *Smart Mater. Struct.* **24** 025029
[2] Renaud M, Fiorini P, van Schaijk R and van Hoof C, 2009 *Smart Mater. Struct.* **18** 035001
[3] Halim M A and Park J Y, 2015 *Sens. Actuators A* **229** 50-58
[4] Pozzi M and Zhu M, 2012 *Smart Mater. Struct.* **21** 055004
[5] Pillatsch P, Yeatman E M and Holmes A S, 2014 *Sens. Actuators A* **206** 178-185
[6] Wang Y-J, Chen C-D, Sung C-K and Li C, 2012 *J. Intel. Mat. Sys. Struc.* **23** 933-943
[7] Roundy S and Takahashi E, 2013 *Sens. Actuators A* **195** 98-104
[8] Salauddin M, Halim M A and J Y Park, 2016 *Smart Mater. Struct.* **25** 095017
[9] Halim M A and Park J Y, 2014 *J. Appl. Phy.* **115** 094901.
[10] Jung S-M, Yun K-S, 2010 *Appl. Phys. Lett.* **96** 111906.
[11] Wei S, Hu H and He S, 2013 *Smart Mater. Struct.* **22** 105020.

Airflow energy harvesting with high wind velocities for industrial applications

Z J Chew, S B Tuddenham and M Zhu

College of Engineering, Mathematics and Physical Sciences, University of Exeter, Exeter, EX4 4QF, UK

E-mail: m.zhu@exeter.ac.uk

Abstract. An airflow energy harvester capable of harvesting energy from vortices at high speed is presented in this paper. The airflow energy harvester is implemented using a modified helical Savonius turbine and an electromagnetic generator. A power management module with maximum power point finding capability is used to manage the harvested energy and convert the low voltage magnitude from the generator to a usable level for wireless sensors. The airflow energy harvester is characterized using vortex generated by air hitting a plate in a wind tunnel. By using an aircraft environment with wind speed of 17 m/s as case study, the output power of the airflow energy harvester is measured to be 126 mW. The overall efficiency of the power management module is 45.76 to 61.2 %, with maximum power point tracking efficiency of 94.21 to 99.72 % for wind speed of 10 to 18 m/s, and has a quiescent current of 790 nA for the maximum power point tracking circuit.

1. Introduction

Airflow is one of the most promising sources to be harvested in industrial environments since the velocity of air in ventilation systems or induced by fast moving objects is quite high, which can be more than 10 m/s [1]. So far, miniaturized wind turbines [2, 3] or flapping airflow generators [4, 5] are often used to harvest energy from airflow. Wind turbine generators generally consist of vaned wheels to convert direct airflow into electrical energy via electromagnetic transducers. On the contrary, flapping airflow generators are usually composed of elastic cantilevers which require vortexes to oscillate the cantilevers. Oscillating motion of the cantilever can be converted into electrical energy via piezoelectric or electromagnetic transductions. These airflow energy harvesters are capable of producing high output power of around 10 mW at air speed of around 10 m/s [2, 5]. The harvested power can be used to power wireless sensors to monitor physical quantities or environment phenomena, which is important in many industries [6-10] so that appropriate measures can be taken for the safety and well-being of the lives in the environment that is being monitored. However, these airflow energy harvesters are mainly designed to operate under airflow coming at a certain angle and/or direction, which greatly constrained their applications because airflow can come in any direction in the real-world. Also, obstructions and cavities present in the real-world environment prevent airflow from smoothly passing through, creating micro-climates of swirling air vortexes. Although flapping airflow generators can capture vortexes for energy harvesting, the operation of flapping airflow generators is limited to a certain airflow speed. The flapping airflow generator will stall and therefore could not generate any power if the air speed is too high and also be in a risk of breaking due to excessive stress sustained by the cantilever which is bended to one side [5]. This paper herein presents an airflow energy harvesting system that can harvest swirling airflow at high speed.

2. System descriptions

The implemented system consists of a modified helical Savonius to capture the swirling airflow, a DC motor as the electromagnetic generator and a power management module for the power conditioning. The DC motor used is a brushed motor with dimensions of Ø7 × 16mm (Hubsan, Shenzen, China).

2.1. Modified helical Savonius turbine

A Savonius turbine which is a drag-type device is capable of harvesting both the direct and swirling airflows. However, standard Savonius turbine is a vertical-axis wind turbine which captures airflow in the direction normal to the rotating shaft [11]. Therefore, based on a standard Savonius turbine, a modified helical Savonius turbine with horizontal axis is implemented as shown in figure 1. The prototyped modified helical Savonius turbine has a diameter of 15 mm and length of 20 mm, which is small so that it can fit into small gaps or spaces. The modified helical Savonius turbine is connected to the shaft of an electromagnetic generator to convert the captured airflow in the form of rotational kinetic energy into electrical energy as the shaft is rotated by the turbine due to the airflow.

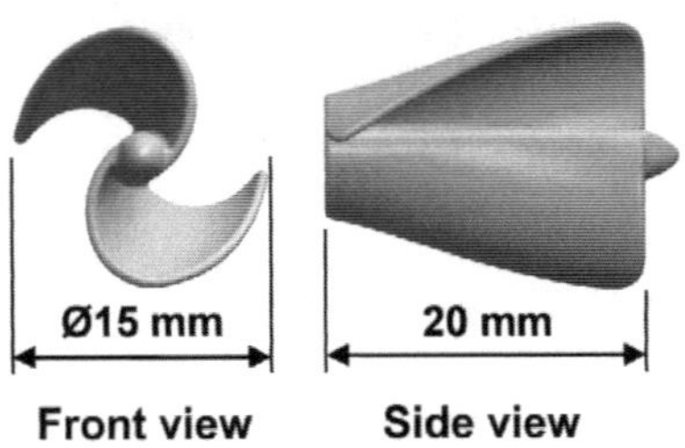

Figure 1. The designed turbine.

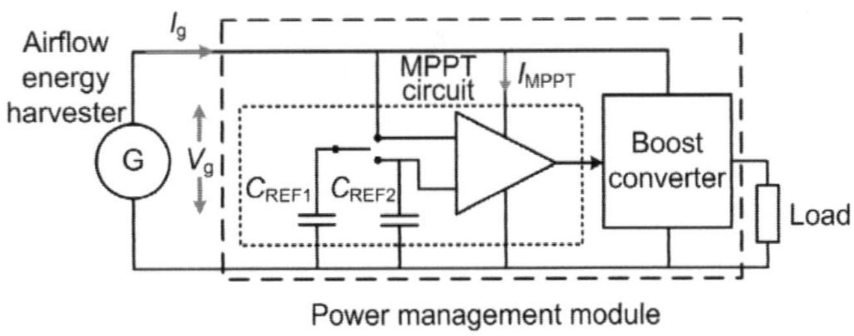

Figure 2. Schematic of the implemented system.

2.2. Power management module

The power management module consists of a maximum power point tracking (MPPT) circuit and a commercially available boost converter (bq25504, Dallas, USA). A boost converter is used because the output voltage from the electromagnetic generator is typically lower than 1 V, which is insufficient to power up a wireless sensor node. The maximum power point finding is based on maximum power transfer that occurs at half of the open-circuit voltage of the electromagnetic generator [12]. Therefore, the circuit is implemented using low power analog components without the need of a power hungry microcontroller as shown in figure 2 since the operation is simple. The circuit momentarily disables the boost converter. The boost converter appears as open-circuited when disabled so that open-circuit voltage of the generator can be obtained and sampled using a sampling capacitor C_{REF1}. The sampling capacitor is then disconnected from the generator output and connected in parallel with another capacitor C_{REF2} of similar value. Voltage across the capacitors becomes half due to charge sharing between the capacitors. The half-open-circuit voltage is used as the reference voltage to harvest energy from the generator and is sampled every second to obtain a new reference voltage as the air speed in actual environment is dynamic, causing voltage from the electromagnetic generator to change.

3. Experiment

The airflow energy harvester was positioned behind a vortex inducing plate in a wind tunnel as shown in figure 3. Vortex is produced when the air stream hits the plate as shown in figure 4. By using the aircraft environment as a case study, the wind tunnel was set to produce an air speed of 17 m/s. This air speed was obtained based on the boundary layer airflow at a distance of 20 mm from the wing surface at the trailing edge, where the wing has a chord length of 13 m and when the aircraft is travelling at the take-off free stream air speed of 77 m/s. The generator output was connected to a variable resistor to determine its power generation capability by manually tuning the resistor.

To determine the maximum power point finding capability of the power management module, tests for the output power of the generator when connected with a variable resistor were carried out at different air speeds of 10 to 18 m/s. Then, the tests were repeated with the electromagnetic generator connected to the power management module. Output power from both tests were compared to obtain the maximum power point tracking efficiency of the implemented circuit based on an open-circuit voltage method. Overall system efficiency of the power management module and the current consumption of the MPPT circuit were also measured using Keithley 2612B.

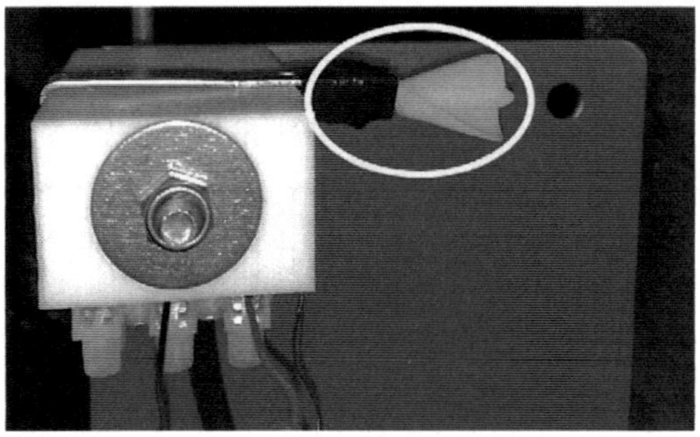

Figure 3. Prototyped airflow energy harvester behind a vortex inducing plate as indicated by the yellow line.

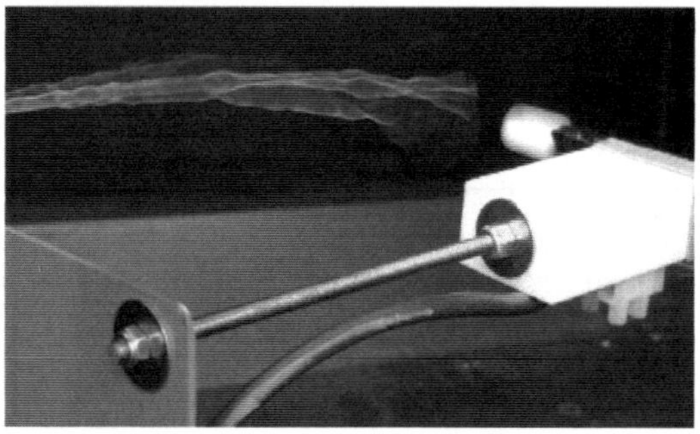

Figure 4. Airflow energy harvester rotating within single velling .

4. Results

Figure 5 shows the power, voltage and current generated by the airflow energy harvester using different resistance load at air speed of 17 m/s. Results indicate that maximum power of 126 mW can be achieved with optimal load resistance of 3.1 Ω. Figure 6 shows the power generated by the airflow energy harvester when connected to its optimal load resistance at different air speed from 10 to 18 m/s. The power generated is 25.5 mW to 140 mW in the range of tested airspeed where the power generated at 10 m/s is double of previously reported airflow energy harvester [2, 4].

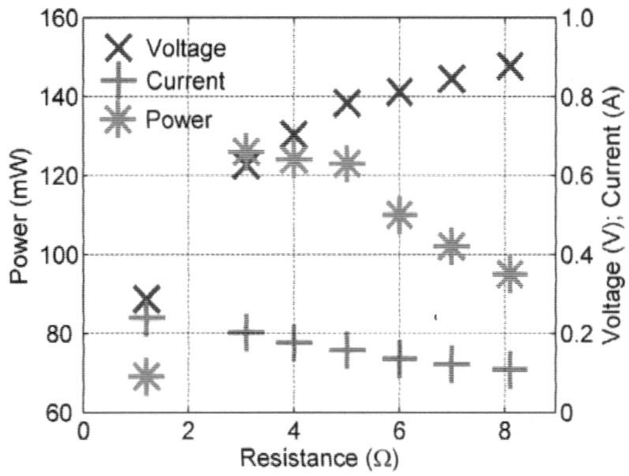

Figure 5. Power, voltage, and current generated by the vortex shedding energy harvester with different resistive load at air speed of 17 m/s.

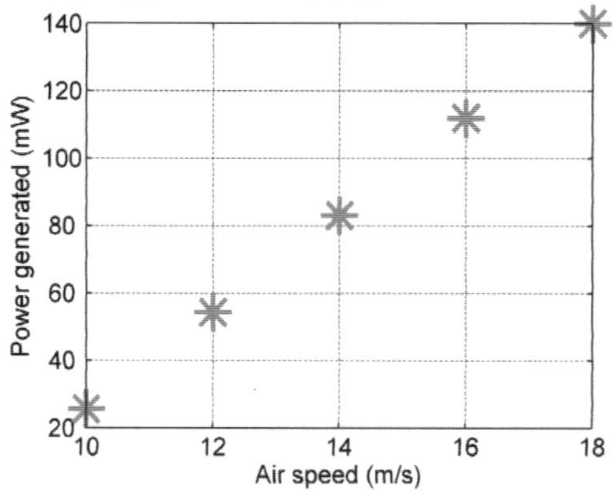

Figure 6. Power generated by the airflow energy harvester subjected vortexes induced by airflow with speed of 10 to 18 m/s.

Overall system efficiency is 45.76 to 61.2 %, which is mainly dictated by the efficiency of the boost converter which has to boost voltages below 0.7 V to 3.1 V. The maximum power point finding efficiency using the MPPT circuit is 94.21 to 99.72 % as shown in figure 7 while consuming 790 nA of quiescent current consumption with short peaks of no more than 4 μA as shown in figure 8. This indicates the effectiveness of the implemented circuit in harvesting energy close to maximum power point of the airflow energy harvester without the need of power hungry components, which is essential in energy harvesting applications so that most of the harvested energy can be delivered to the wireless sensors to perform the intended tasks instead of consumed by the power management module and minimizes the energy loss to prevent further decrease in the system efficiency.

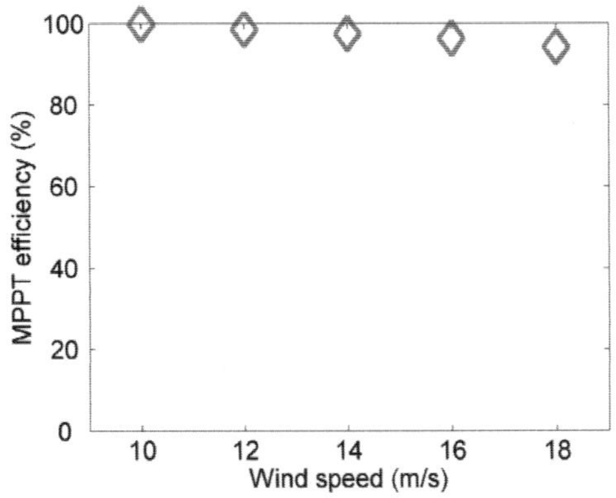

Figure 7. MPPT of the power management module at different wind speeds.

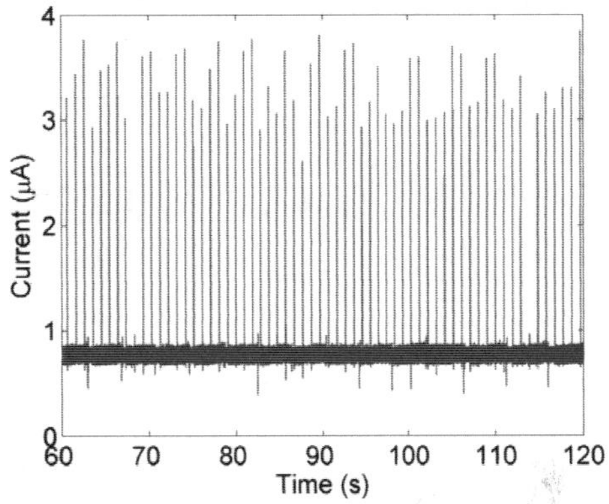

Figure 8. Current consumption of the maximum power point finding circuit.

5. Conclusion

An airflow energy harvester which is able to operate under swirling air at high speed and its associated power management module are presented and demonstrated. The airflow energy harvester produces up to 126 mW of power under vortex induced by airflow of 17 m/s. Overall system efficiency is 45.76 to 61.2 %. The maximum power point finding based on open-circuit voltage has high efficiencies between 94.21 and 99.72 % and the MPPT circuit has a low quiescent current of 790 nA. The capability of the implemented airflow energy harvester to run by vortex at high speed and generated high power of several hundreds milliwatts makes it suitable for industrial applications.

Acknowledgments

The authors would like to thank EPSRC for the sponsorship of this project.

References

[1] Fei F, Zhou S, Mai J and Li W 2014 *Energies* **7** 2985-3003
[2] Holmes A S, Howey D A, Bansal A and Yates D C 2010 *Proc. PowerMEMS* 115-8
[3] Brunelli D 2016 *Sensors* **16** 327 (19 pp)
[4] Erturk A, Vieira W G R, De Marqui C and Inman D J 2010 *Appl. Phys. Lett.* **96** 184103 (3 pp)
[5] Zhu D, Beeby S P, Tudor M J, White N M and Harris N R 2013 *IEEE Sensors J.* **13** 691-700
[6] Jurik A D and Weaver A C 2009 *IEEE Softw.* **26** 71-3
[7] Morais R, Matos S G, Fernandes M A, Valente A L G, Soares S F S P, Ferreira P J S G and Reis M J C S 2008 *Comput. Electron. Agr.* **64** 120-32
[8] Kafi M A, Challal Y, Djenouri D, Doudou M, Bouabdallah A and Badache N 2013 *Procedia Comput. Sci.* **19** 617-26
[9] Gilbert J M and Balouchi F 2008 *Int. J. Autom. Comput.* **5** 334-47
[10] Chew Z J, Ruan T and Zhu M to be published *Procedia Eng.*
[11] Mahmoud N H, El-Haroun A A, Wahba E and Nasef M H 2012 *Alexandria Eng. J.* **51** 19-25
[12] Tan Y K and Panda S K 2011 *IEEE Trans. Power Electron.* **26** 38-50

A frequency up-converting harvester based on internal resonance in 2-DOF nonlinear systems

Yipeng Wu, Jinhao Qiu and Hongli Ji

Nanjing University of Aeronautics and Astronautics, Nanjing, 210016, China

yipeng.wu@nuaa.edu.cn

Abstract. This paper reports the design and experimental testing of a novel frequency up-converting piezoelectric energy harvester. The harvester is firstly approximated as a 2-degree-of-freedom cubic nonlinear system instead of the general Duffing systems. A 1:3 internal resonance innovatively applied in the frequency up-conversion approach is thoroughly investigated. Finally, the theoretical dynamic model confirmed by the experimental results clearly shows the effect of the frequency up-conversion.

1. Introduction

Vibration energy harvesting technology describes the process of converting ambient mechanical energy into useful electrical energy through the use of electromechanical structures. It has the ability to offer the prospect of powering autonomous electronic devices such as wireless sensor nodes without the use of conventional batteries. Piezoelectric cantilevers are the simple and typical electromechanical structures, however, real-word applications of such linear cantilevers are not practical due to the random nature of vibrations. For this reason, resonant frequency tuning harvesters, multimodal harvesters, nonlinear harvesters, etc. are proposed [1, 2]. The design of these harvesters usually follows the principle of the resonant-frequency-matching strategy.

More recently, an increasing number of researchers realize that low-frequency excitations are the most often available from the ambient environment. In this case, the utilization of the above matching strategy may unacceptably lead to increased encumbrance [3]. Consequently, the need for a technique capable of bridging between the high-frequency response and the low-frequency excitation is clearly needed. This technique is usually called the frequency up-conversion effect.

The most common frequency up-conversion approaches are implemented by mechanical plucking or impact. For example, Pozzi and Zhu proposed a knee-joint pizzicato harvester which consisted of several plectra and four piezoelectric bimorphs [3]. Gu and Livermore proposed the similar frequency up-converting harvester using vibrations of beam structures at high resonant frequencies that were excited by mechanical impact [4]. Besides the direct contact approaches, Han and Yun implemented the frequency up-conversion effect by a high impulse-like acceleration without physical contact, the harvesters apply the impact force to the high-frequency oscillators through the instantaneous state transition of a bucked bridge beam [5]. However, it is worthy of note that the above frequency up-converting harvesters usually operate at the high-level excited accelerations.

In this paper, we proposed a novel frequency up-converting harvester based on the principle of 1:3 internal resonance. The harvester can be considered as 2-Degree-Of-Freedom (DOF) cubic nonlinear system, when the internal resonance happens, part of the vibration energy can be transferred from the low-order vibration mode to the high-order one, which realizes the frequency up-conversion without physical contact and energy losses. Finally, the simulation results validated by the experiment clearly explained the principle of this frequency up-conversion effect.

2. Theoretical model

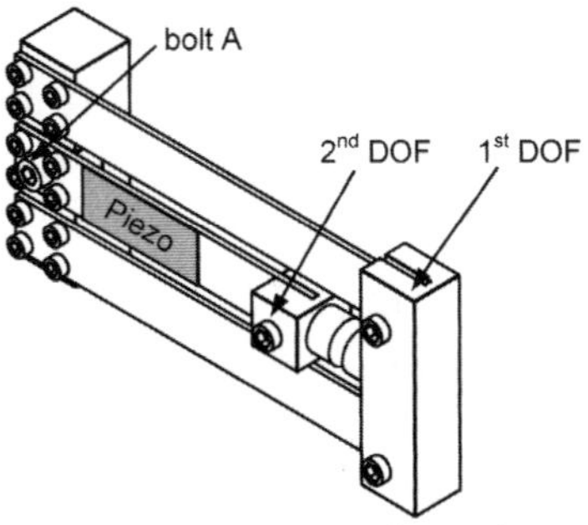

(a) Structure schematic

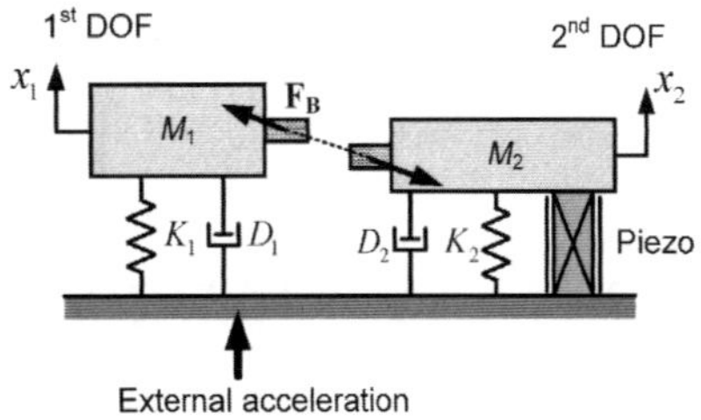

(b) Dynamic model

Figure 1. The novel frequency up-converting piezoelectric harvester

Figure 1 shows a compact prototype of the proposed energy harvester, it consists of two asymmetrical cantilevers and a pair of NdFeB permanent magnets. It is worthy of note that the base frame was composed of 2 sub-bases which are combined by the bolt A, so the separated distance between the two magnets can be adjusted manually by the relative position of the 2 sub-bases.

According to the previous literature [6], the proposed harvester can be considered as 2 spring-mass-damper models which are coupled by the magnetic force $\mathbf{F_B}$. The parameters K_i and D_i (i = 1, 2) are the global equivalent stiffness and the damping coefficient, respectively. An external acceleration γ is applied to the model, induces a relative displacements x_1 and x_2 between the masses and the base frame. The governing equations of the system is then obtained as shown in equation (1), where M_i are the proof masses of the corresponding cantilevers, F_B is the magnitude of the magnetic force $\mathbf{F_B}$, α is the piezoelectric force-voltage coefficient of the cantilever, V is the output voltage of the piezoelectric element.

$$\begin{bmatrix} M_1 \\ M_2 \end{bmatrix} \gamma = \begin{bmatrix} M_1 & 0 \\ 0 & M_2 \end{bmatrix} \begin{bmatrix} \ddot{x}_1 \\ \ddot{x}_2 \end{bmatrix} + \begin{bmatrix} D_1 & 0 \\ 0 & D_2 \end{bmatrix} \begin{bmatrix} \dot{x}_1 \\ \dot{x}_2 \end{bmatrix} + \begin{bmatrix} K_1 & 0 \\ 0 & K_2 \end{bmatrix} \begin{bmatrix} x_1 \\ x_2 \end{bmatrix} + \begin{bmatrix} -F_B \\ \alpha V + F_B \end{bmatrix} \quad (1)$$

Because the shape and size of the magnets are not so important compared with the whole harvester device, the permanent repulsive magnets can be modeled as a dipole-dipole magnetic interaction in this article [7]. Finally, the simplified magnetic force F_B is expressed in equation (2), where a_i, b_i, c_i are the constant values for a given PEG.

$$F_B = a_1 x_1 - a_2 x_2 - b_1 x_1^3 + b_2 x_2^3 + c_1 x_1^2 x_2 - c_2 x_1 x_2^2 \quad (2)$$

From the above 2 equations, it is clearly found that the dynamic system has two DOFs (x_1 and x_2), the equivalent nonlinear stiffness only has linear and cubic terms. Hence, the proposed harvester can be considered as a 2-DOF cubic nonlinear system. A typical nonlinear vibration phenomenon called 1:3 internal resonance will be excited, provided that the nature frequency of the 2nd cantilever is the triple-frequency of the 1st one. This phenomenon has been thoroughly explained in multi-DOF nonlinear dynamics theory.

3. Simulation performances and experimental validation

Since it is very complicated to solve the above dynamic equation analytically, a numerical integration techniques using MATLAB software will be discussed in this section. All simulation parameters are set according to the experimental prototype of the harvester.

Figure 2 shows the photo of the experimental setup and the proposed frequency up-converting harvester. The experimental platform is mainly composed of a vibration controller, a power amplifier, a vibration table and a feedback accelerometer. The experimenter can easily establishes a PID feedback loop to make sure the table vibrate at a constant acceleration amplitude. The harvester is fixed on the vibration table by the screws, one Macro-fiber-composite (MFC) is attached on the inner cantilever for energy generation, the other is attached on the outer cantilever for comparison. It is worthy of note that in this article, we mainly focus on the frequency up-conversion effect of the piezoelectric harvester, the optimization of the electromechanical coupling coefficient will not be discussed here.

Based on the measured data, the key parameters of the model utilized in the simulation are given in table 1. Some of the parameters are identified using the equation given in reference [6].

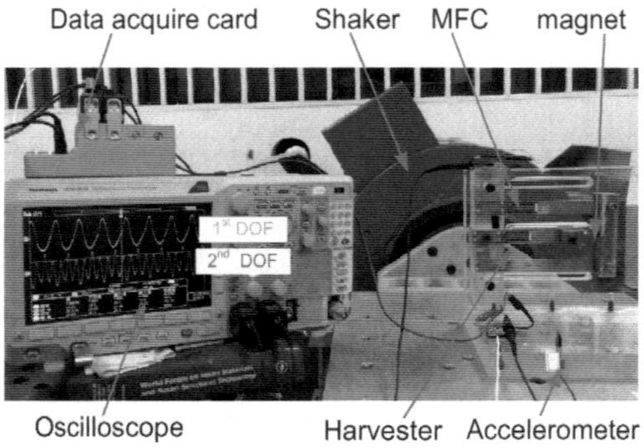

Figure 2. Experimental setup and the photo of the proposed harvester

Table 1. Parameters used for numerical computation

	Parameters	1st DOF	2nd DOF
M_i	Effective mass (g)	782.9	104.6
l_i	Equivalent length (mm)	125	100
f_i	Resonant frequency (Hz)	8.03	23.27
ξ_i	Damping ratio	0.011	0.007
α	Force-voltage coefficient (N/V)		8.4×10^{-5}
C_0	Clamped capacitance (nF)		47.58

3.1. Simulation performances of the frequency up-conversion

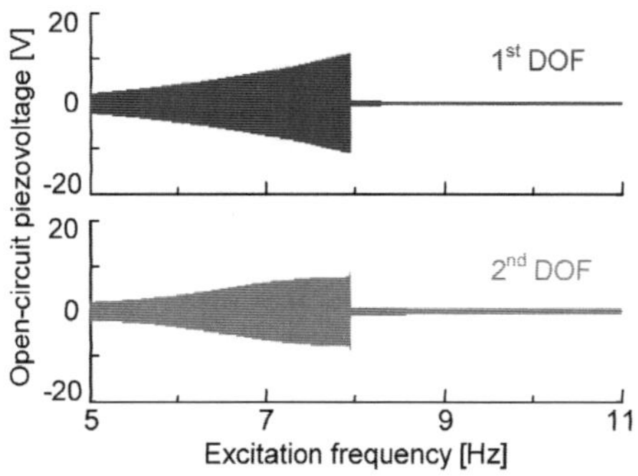

Figure 3. Experimental setup and the photo of the proposed harvester

To study the performances of the frequency up-converting harvester, a forward sweep signal with a rate of 0.04 Hz/min is used as an excited acceleration, two magnets are placed with the distance of 23 mm.

The range of the excited frequency is from 5 Hz to 11 Hz, the acceleration amplitude is a constant value of 1 m/s^2. Figure 3 shows the open-circuit piezoelectric output voltages of the two DOFs, because the piezoelectric outputs are proportional to the corresponding vibration displacements of the cantilevers, it can be observed when the excited frequency is near the resonance of the 1st beam, this outer beam can amplify the vibration displacement at its DOF, leads to the drastic alteration of the magnetic coupling force in the nonlinear structure. Consequently, although the excited frequency is much lower than the resonance of the 2nd cantilever, the 2nd cantilever also vibrates strongly at its DOF.

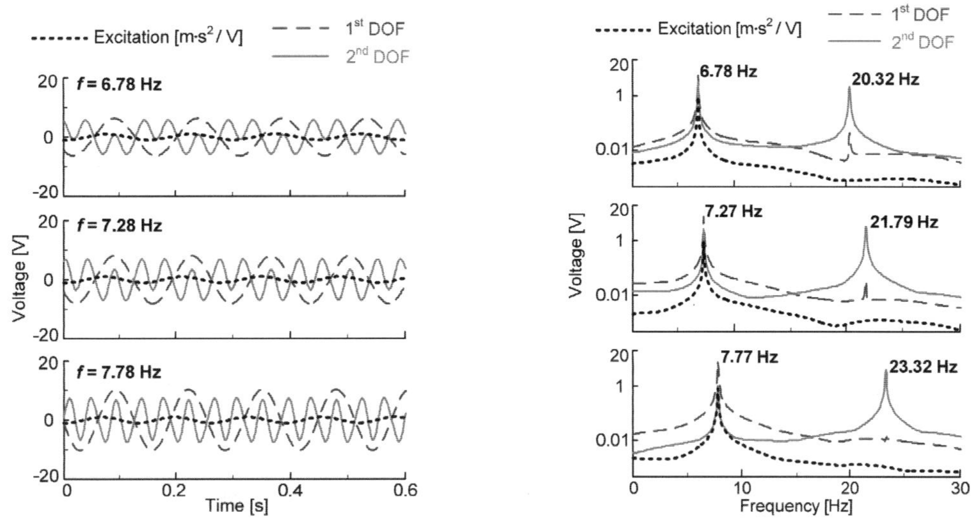

(a) Waveforms of the piezoelectric voltages and the excitation signal

(b) Frequency response of the piezoelectric voltages and the excitation signal

Figure 4. Simulation results of the frequency up-conversion effect under the different excited frequencies

The zoomed waveforms of the piezoelectric output voltages are plotted in figure 4(a), where the excited acceleration signal is also plotted in the same coordinate axes. From the figure, it is seen that the vibration of the 2nd DOF has two major vibration modes: low-order vibration mode whose frequency corresponds to the excited frequency, and high-order vibration mode whose frequency is three times increased. Obviously, the higher vibration value of the high-order mode corresponds to the better performance of the frequency up-conversion.

The performances of frequency up-converting harvester in the frequency domain are obtained by fast Fourier transformation, for several sweep excitations (around 6.78 Hz, 7.28 Hz and 7.78 Hz), as shown in Figure 4(b). It can be observed that the high-order vibration mode of the 2nd cantilever is excited, and the vibration frequency is exactly three times higher than the excited frequency. In addition, the excited frequency is more close to 7.78 Hz, the vibration amplitude of the 1st beam is larger, the energy percentage of the high-order vibration mode is also enlarged.

3.2. Experimental results and validation

The results in this sub-section were obtained from the experiment setup, shown in figure 2. As presented in the simulation sub-section, the experimental results of a typical frequency up-conversion effect is shown in figure 5. A forward sweep signal with a rate of 0.05 Hz/min is used as an excitation, two magnets are placed with the distance of 22.65 mm. The range of the excited frequency is 5~11 Hz and the acceleration amplitude is a constant value of 1 m/s^2.

However, due to the large vibration amplitude of the outer beam and the nonlinear magnetic force, it seems that there exists the harmonic distortion in the 1st beam, and the third harmonic component may cause the vibration of the 2nd cantilever. Based on the simulation analysis, the authors still think that the 1:3 internal resonance existed in the 2-DOF cubic nonlinear system is the main reason. In addition, when the excited frequency is converging with the natural frequency of the 1st beam, the fundamental component of the 1st DOF is drastically enhanced, while the corresponding component in the 2nd DOF is decreased, which also validates the above conclusion on the other hand.

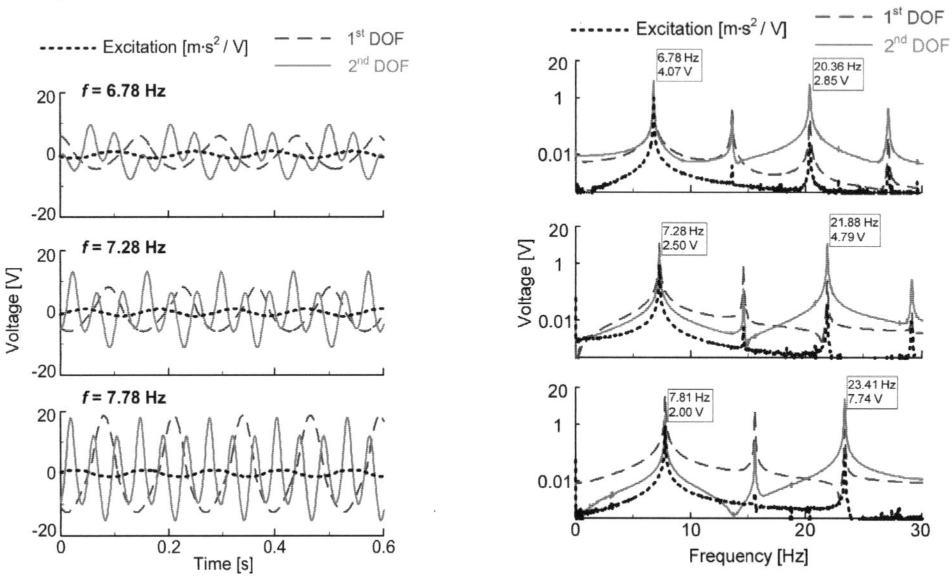

(a) Waveforms of the piezoelectric voltages and the excitation signal

(b) Frequency response of the piezoelectric voltages and the excitation signal

Figure 5. Experimental results of the frequency up-conversion effect under the different excited frequencies

4. Conclusion

In summary, the proposed compacted harvester design, which is comprised of the magnets and the simple cantilevers, offers the capability of the frequency up-conversion effect which is benefit to low-frequency vibration energy harvesting. Ongoing studies can aim at researching the frequency up-conversion effect as functions of the different design parameters and optimizing the coupling coefficient of the piezoelectric harvester.

Acknowledgment

This work has been supported by Jiangsu Youth Found (BK20160790), China Postdoctoral Science Foundation (2015M581786), Jiangsu Postdoctoral Science Grants Program (1501025C) and the Fundamental Research Funds for the Central Universities (NS2016016).

References

[1] Tang L, Yang Y and Soh C K 2013 *Advances in Energy Harvesting Methods* ed N Elvin and A Erturk (New York: Springer) chapter 2 pp 17-61

[2] Daqaq M F, Masana R, Erturk A and Quinn D D 2014 *Applied Mechanics Reviews* **66** 040801

[3] Pozzi M and Zhu M 2013 *Advances in Energy Harvesting Methods* ed N Elvin and A Erturk (New York: Springer) chapter 5 pp 119-140

[4] Gu L and Livermore C 2011 *Smart Materials and Structures* **20** 045004

[5] Han D and Yun K S 2015 *Microsystem Technologies* **21** 1669-1676

[6] Guyomar D, Badel A, Lefeuvre E and Richard C 2005 *IEEE Transactions on Ultrasonics, Ferroelectrics, and Frequency Control* **52** 584-595

[7] Wu H, Tang L, Yang Y and Soh C K 2014 *Journal of Intelligent Material Systems and Structures* **25** 1875-1889

Influence of nonlinear constitutive relations in unimorphs piezoelectric harvesters

C D Gatti[1], J M Ramirez[1], S P Machado[1] and M Febbo[2]

[1]Grupo de Investigación en Multifísica aplicada, Universidad Tecnológica Nacional FRBB and CONICET, 11 de Abril 461, 8000 Bahía Blanca, Argentina

[2]Instituto de Física del Sur, CONICET and Departamento de Física. Universidad Nacional del Sur. Av Alem 1253, 8000 Bahía Blanca, Argentina.

Email: smachado@frbb.utn.edu.ar

Abstract. This paper presents the influence of nonlinear terms of a previously proposed constitutive piezoelectric equation on the dynamics of a cantilever aluminium beam with a piezoelectric unimorph PZT (MIDE QP16N) attached to it. The system is subjected to different levels of base acceleration with the intention to evidence the limits of the linear model. To carry out the analysis, a one-dimensional model is applied and solved employing a single-term solution of the harmonic balance method to compare with the experiments. A model identification of linear and nonlinear parameters such as dissipation, stiffness, and electromechanical coupling were then performed. From the results, it is possible to observe the departure of the linear model even for very low acceleration levels (0.1G). It can be concluded that the nonlinearity plays an unavoidable roll in predicting electric generation for the considered systems.

1. Introduction

Nonlinearity in piezoelectric beams is a challenging and complex phenomenon. Although it is commonly supposed that a linear description captures the principal features of the mechanical and electrical response of a piezoelectric material under usual practical situations, its behavior is far from being linear even for low values of mechanical and electrical excitation. There are several factors that make the dynamics to be nonlinear. Between them, it is possible to mention strong electric fields [1, 2], damping, influence of adhesive layer, nonlinear beams theories, nonlinear constitutive equations under nonlinear elastic and electromechanical coupling, among others [3]. The observed discrepancy between linear model and experiments is evident even at weak electric fields.

Non-linear theory of dielectrics has been developed by the pioneering works of Toupin [4] and Tiersten [5]. Joshi [6] employed the thermodynamic Gibbs potential to derive constitutive equations including higher order effects up to second order terms in the linear formulation. Aurelle et. al. [7] studied the effect of a nonlinear electromechanical coupling parameter and nonlinear elasticity on the response of a piezoelectric transducer. Wagner and Hagedorn [2] studied the nonlinear elastic and coupling behavior of piezoelectric beam actuators, proposing an enthalpy density with quadratic and cubic nonlinearities of strain and coupling. The first approach to study nonlinear piezoelectricity in a harvester was published by Hu [8]. His results show a hardening response of a piezoelectric plate due to a shear mode vibration. Triplett and Quinn [9] proposed a strain dependence of the piezoelectric constant as responsible for nonlinearity in the electromechanical equations. In a first work Stanton et. al. [10] used the nonlinear constitutive equations of ref. [8] in a bimorph PZT 5H and proposed a nonlinear quadratic damping to account for the theoretical overshoot of experimental data. In a second work, the same author and collaborators [11] extended the nonlinear constitutive relations to include non-conservative stress and electric displacement to introduce dissipation within electroelastic media. Abdelkefi et. al. [12] performed a nonlinear analysis of a multi-layered piezoelectric beam with a tip mass taking into account geometric and piezoelectric nonlinearities into the electromechanical model. The constitutive relations of the piezo are derived using the approach of Joshi [6]. Goldshmidtboeing et. al. [13] attributed to ferroelastic hysteresis the source of nonlinearities in PZT cantilever beams. They modelled the constitutive nonlinear equations and nonlinear dissipation effects with the simplest hysteresis model, the Rayleigh model, to obtain their theoretical predictions. More recently, Leadenham and Erturk [14] based on the work of refs [2,13] proposed their own enthalpy density expression which predicts a linear backbone curve in the response amplitude of a bimorph piezoelectric sheet. Finally, Yang and Upadrashta [15] considered a nonlinear electromechanical model and damping nonlinearities to model a cantilever macro-fiber composite for energy harvesting. They proposed to identify nonlinear elastic and damping coefficients by matching analytical responses and experimental results. As a consequence, a high-order stress/strain relationship is obtained which contains seven polynomial terms to fit the experimental data up to very high strains.

In this paper we used the model of Leadenham and Erturk [14] to investigate the validity of their model applied to an energy harvesting system (EH) containing a unimorph piezoelectric sheet. In most of the works mentioned previously, high resonant frequencies were considered to study the behaviour of piezoelectric materials [10 - 14]. Only Yang and Upadrashta [15] investigated a harvester working at low frequencies with large deformation but with another model of dissipation, not considering the ferroelastic hysteresis.

2. Design and Modelling

The model of the proposed EH system comprises an aluminum cantilever beam with a piezoelectric unimorph PZT sheet (MIDE QP16N) attached at its root. The system is excited by its base with constant acceleration level, observed schematically in figure 1. The mathematical model is based on the Bernoulli-Euler theory for flexural beams. The electric field is considered uniform and linear over the thickness of the piezo sheet and the constitutive relations (stress/strain+electric field/electric displacement) contain nonlinear elastic terms, linear and nonlinear electromechanical coupling and dissipation effects due to ferroelastic hysteresis as stated in [13].

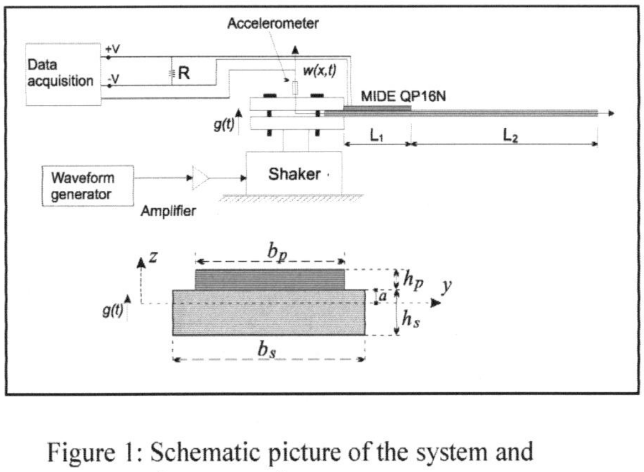

Figure 1: Schematic picture of the system and its geometric parameters.

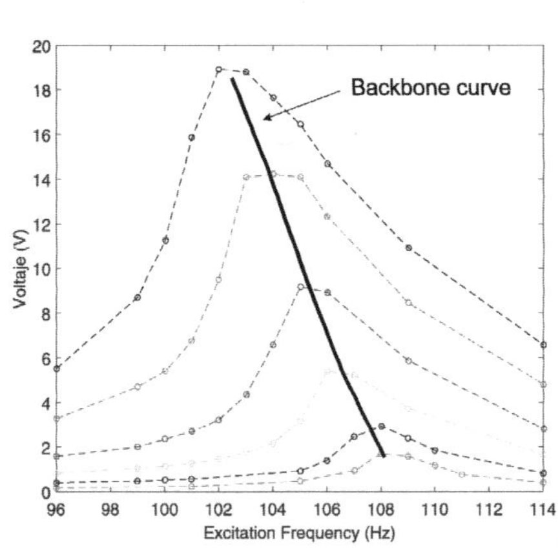

Figure 2: Experimental response curves at various excitation amplitudes.

2.1. Mathematical formulation

Following the conclusions of Leadenham and Erturk [14], we also observed in the experimental results that the backbone curve changes linearly with the response amplitude (see figure 2). Therefore, to address the nonlinear problem of the constitutive model, we consider the enthalpy density for the substrate H_s and for the piezo beam $H_p$as [14]:

$$H_s = \frac{1}{2} c_{11}{}^{s} S_1{}^2 \tag{1}$$

$$H_p = \frac{1}{2} c_{11}{}^{p} S_1{}^2 + \frac{1}{3} c_{111}{}^{p} S_1{}^3 sgn(S_1) - e_{31} S_1 E_3 - \frac{1}{2} e_{311} S_1{}^2 sgn(S_1) E_3 - \frac{1}{2} \epsilon_{33} E_3{}^2 \tag{2}$$

where $c_{11}{}^{s}$ y $c_{11}{}^{p}$ are the Young modulus of the aluminum and the piezo sheet respectively, $c_{111}{}^{p}$ is the third order nonlinear elastic modulus, e_{31} is the linear piezoelectric constant, e_{311} is the third order nonlinear piezoelectric constant, ϵ_{33} is the piezo permittivity and E_3 is the electric field in the transverse direction given by $E_3 = v(t)/h_p$ ($v(t)$ is the voltage).

As mentioned previously, to account for the dissipation, the model of ferroelastic hysteresis of [13] provides a dissipation energy as $D_p = \frac{1}{3} b_{111} S_1{}^3$, whereas the damping for the substructure is modelled as simple proportional damping. Kinetic and potential energy of the piezo and the substructure can be obtained in a straightforward manner following [14] with the only difference that here, we consider a unimorph piezoelectric beam [16].

2.2. Governing equations

Two governing electromechanical equations have been derived using Lagrange equations, with a single-mode solution for the first bending mode:

$$\begin{aligned} m\ddot{q}[t] + (bq[t]sgn(q[t]) + b_n q[t]^2) sgn(\dot{q}[t]) + kq[t] + k_n q[t]^2 sgn(q[t]) \\ - \left(\theta + \theta_n q[t] sgn(q[t])\right) v[t] = -m_a \ddot{g}[t] \end{aligned} \tag{3}$$

$$C_p \dot{v}[t] + \frac{v[t]}{R_l} + (\theta + \theta_n q[t]) \dot{q}[t] = 0 \tag{4}$$

Where $m, b, b_n, k, k_n, \theta, \theta_n, m_a, C_p, R_l$ are the modal mass, linear and nonlinear damping, linear and nonlinear electromechanical coupling, modal forcing mass, internal capacitance of the piezoelectric unimorph sheet and load resistance. These lumped parameters are functions of the device dimensions and material properties and are given in [14]. The nonlinear differential equations are further solved analytically using a single-term solution in the harmonic balance method, so that the dynamic response of the system is obtained. From this, it is possible to find the voltage $v[t]$ to compare it with the experiments.

3. Experiments and Results

The experiments were carried out using the setup of figure 3. We considered a large range of accelerations ranging from 0.1-3.0 G (G=gravity =9.8m/sec^2). The sweeps for each acceleration amplitude were performed at constant amplitude. Material and geometrical constants of the EH system are given in table 1.

Journal of Physics: Conference Series **773** (2016) 012093 doi:10.1088/1742-6596/773/1/012093

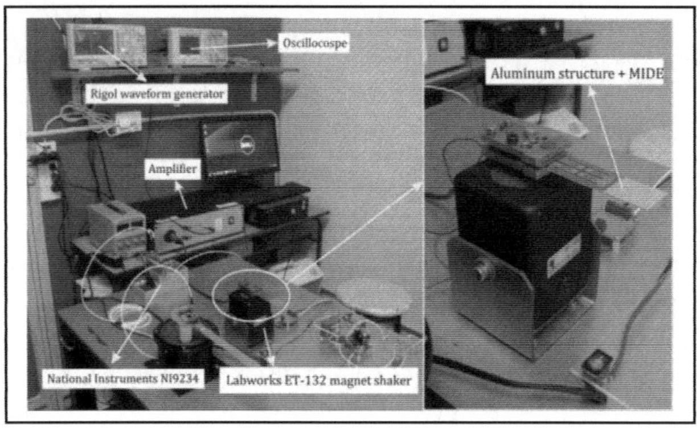

Figure 3. Experimental setup (left), zoom in of the EH system (right)

Table 1. Geometrical and material constants.

Parameter	**Alum.**	**QP16N**
Length (mm)	80	45.9
Width (mm)	21	20.57
Thickness (mm)	0.5	0.25
Density (kg/m^3)	2700	7800
Young`s modulus (GPa)	67	67
Coupling d_{31}(pm/V)		-190
Capacitance (nF)		90.78
Permittivity		1500 ε_0

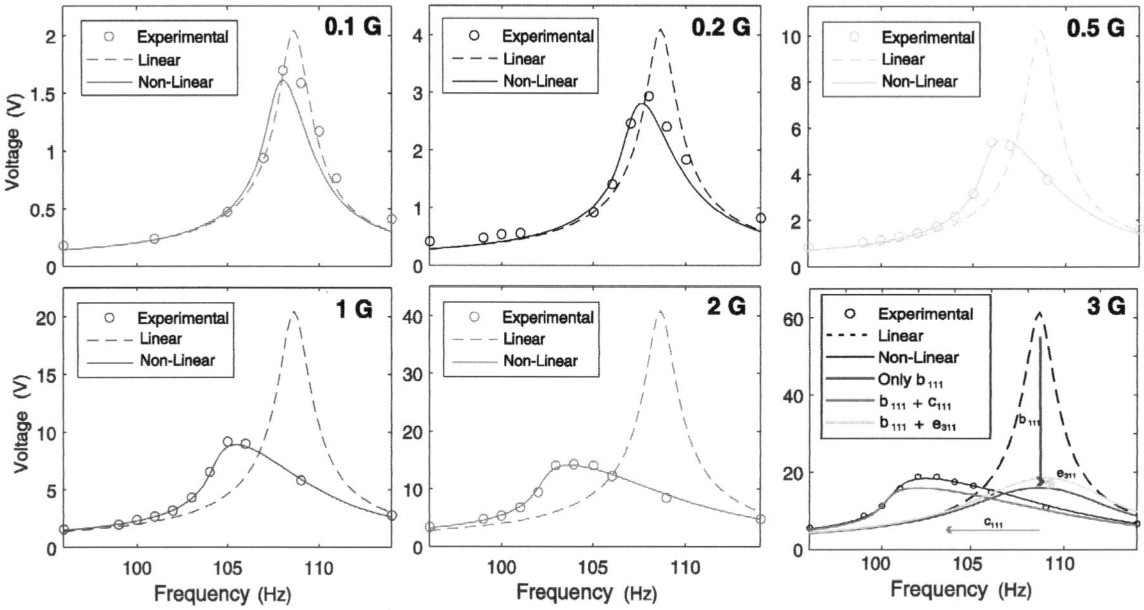

Figure 4: Comparison between experimental and linear/nonlinear responses for accelerations ranging from 0.1 to 3 G

In order to apply and solve the model of the proposed device, it was necessary to make an identification of linear and nonlinear parameters $(b, b_{111}, c_{111}, , e_{311})$. To this end, the experimental tests for an acceleration level of 3 G were used. After that, we employ a least squares technique (*lsqnonlin* command in Matlab) along with the proposed analytical model to extract the more approximate values in a least squared sense. The obtained values are: $b_{11} = 3.59\ 10^9\ \mathrm{N/m^2}$, $b_{111} = 6\ 10^{13}\ \mathrm{N/m^2}$, $c_{111} = -303.44\ \mathrm{TPa}$ and $e_{311} = -68.15 \frac{\mathrm{kC}}{\mathrm{m^2}}$. These values are fed into the analytical model and plot together with the experimental results. The comparison can be observed in figure 4 where both linear and nonlinear responses are shown along with the experiments for accelerations ranging from 0.1-3G. As a first singular feature, it is possible to observe that the curves bend to the left (softening response) as the amplitude of the base acceleration increases. This clearly represents a significant departure of the linear regime. Even for low accelerations levels (0.1G) the linear prediction is poor. For larger accelerations, instead, the nonlinear model represents an accurate description of the voltage generation. A very good agreement can be appreciated for all the performed tests including the maximum value of 3G. In this case, the voltage peak amplitude diminishes four times compared with its linear counterpart and the system's natural frequency (frequency peak) decreases up to 102 Hz, almost 6 Hz less than the linear frequency of 108.6 Hz. An interesting analysis can be observed for the 3G case. There, we plot the contribution of the nonlinear parameters (b_{111}, c_{111} and e_{311}) on the FRF curves. The linear response amplitude is reduced considerably by means of nonlinear damping b_{111} while the nonlinear elastic modulus c_{111} is responsible for the softening behavior of the EH system. The third order nonlinear piezoelectric constant e_{311} plays a secondary role in this case.

4. Conclusions

In this paper, the dynamic behavior of a cantilever beam with a piezoelectric unimorph sheet (MIDE QP16N) attached to an aluminum beam subjected to different levels of acceleration, was studied for low frequency energy harvesting purposes. A non-linear constitutive model for the piezoelectric sheet was used, which was previously proposed by Leadenham and Erturk [14] for a bimorph EH system. The model takes into account the effect of ferroelastic hysteresis resulting in terms of nonlinear stiffness and nonlinear dissipation. The results evidence a very good agreement between the nonlinear analytical model and the experiments, even though for low levels of base acceleration. Therefore, it is concluded that the model correctly predicts the dynamic behavior of the studied system. This contrasts definitively with the application of the linear model for these cases. Future efforts will be conducted to explore the effect of ferroelastic hysteresis in macro-fiber composite EH, considering the piezoelectric 33 and 31-effects.

Acknowledgments

The authors wish to thank CONICET, UNS, UTN FRBB, ANPCyT under grant PICT 2013-2065 for their financial support.

References

[1] Beige H 1983 *Ferroelectrics* **51** 113-119
[2] Tiersten H 1993 *J. Appl. Phys.* **74** 3389–3393
[3] Von Wagner U and Hagendorn P 2002 *J. Sound Vibration* **256** 861–872
[4] Toupin R A 1983 *Int.Jj. Eng. Sci.* **1** 101-26
[5] Tiersten H F 1993 *J. Acoust. Soc. Am.* **70** 1567-1570
[6] Joshi S P 1992 *Smart Mater. Struct.* **1** 80-83
[7] Aurelle N *et al* 1996 *Ultrasonics* **34** 187–191
[8] Hu Y *et al* 2006 *IEEE Trans. Ultrason. Ferroelect. Freq. Control* **53** 1387–1391
[9] Triplett A and Quinn D 2009 *J. Intel. Mat. Syst. Str.* **20** 1959–1967
[10] Stanton S C *et al* 2010 *J. Appl. Phys.* **108** 074903
[11] Stanton SC *et al* 2012 *J. Intel. Mat. Syst. Str.* **23** 183–199
[12] Abdelkefi A *et al* 2012 *Nonlinear Dyn.* **67** 1147–1160
[13] Goldschmidtboeing F *et al* 2011 *Proc. Power MEMS, Seul Korea* 114-117
[14] Leadenham S and Erturk A 2014 *Nonlinear Dyn.* **79** 1727-1743
[15] Yang Y and Upadrashta D 2016 *Nonlinear Dyn.* **84** 2487-2504
[16] Machado S P *et al* 2015 *Smart Mater. Struct.* **24** 115011

Butterfly micro bilayer thermal energy harvester geometry with improved performances

E. Trioux[1,2], S. Monfray[3] and S. Basrour[1,2].

[1]Université Grenoble Alpes, TIMA Laboratory, F-38031 Grenoble, France
[2]CNRS, TIMA Laboratory, F-38031 Grenoble, France
[3]STMicroelectronics, Crolles, France

E-mail: emilie.trioux@imag.fr

Abstract. This paper reports the recent progress of a new technology to scavenge thermal energy, implying a double-step transduction through thermal buckling of a bilayer aluminum nitride / aluminum bridge and piezoelectric transduction. A completely new scavenger design is presented, improving greatly its final performance. The butterfly shape reduces the overall device mechanical rigidity, which leads to a decrease of buckling temperatures compared to previously studied rectangular plates. In a first time we compared performances of rectangular and butterfly plates with an equal thickness of Al and AlN. In a second time, with a thicker Al layer than AlN layer, we will study only butterfly structure in terms of output power and buckling temperatures, and compare it to the previous stack.

1. Introduction

The interest in thermal energy harvesters has grown steadily in the past few years [1, 2]. Indeed thermal gradients are present in every environment, from large building facades to small electronic circuits. Harvesters based on Seebeck thermoelectric materials are the most commonly used devices to scavenge thermal energy. A lot of efforts have been invested in order to downscale them, and also to propose alternative ways to harvest thermal gradients at micro scale [3]. We previously presented an original way of harvesting thermal gradients by combining in a two-step transduction both thermal buckling and piezoelectricity [4, 5]. The harvesters are fabricated using standard CMOS processes.

The working principle of the device is based on a two-step transduction. A bilayer plate, initially curved downward (by the effect of residual stresses from fabrication), is put in contact with a hot source. The difference in thermal expansion coefficient of the two layer materials, aluminum (Al) and aluminum nitride (AlN) generates huge thermal stresses in the films that are leading to a mechanical instability called buckling. The plate is inverting suddenly its curvature from a downward to an upward position. The large deformations occurring during buckling are converted into usable electric charges by the piezoelectric AlN layer. By changing the sign of curvature, the plate is not anymore in contact with the hot source. It begins to cool down by natural convection and for the same reasons as before, the change in temperature in the plate is generating thermal stresses that are leading to the inverse buckling of the harvester. In the same time, it goes back in contact with the hot source. Then we can have regular oscillations of the device with regular electrical peaks at each snap up and down of the plate. The harvested signal could be used later on by some low power-consuming device.

The previously studied harvesters were rectangular, doubly clamped bilayer plates, with equal thickness of AlN and Al layers. This kind of structure was very rigid, especially if width and length of the plate are comparable, so the buckling temperatures were relatively high (over 100°C). Moreover, the neutral plane was situated near the middle of the piezoelectric layer, which was not favorable. In this contribution we will present the advantage of butterfly design compared to rectangular one in terms of buckling temperatures and output power for the equal AlN/Al thickness stack. Then we will present the performances of butterfly structure having a thicker Al layer than AlN layer, and compare its performances to the previous stack.

2. Butterfly structure improvements

The structure previously presented were rectangular plates, clamped on the short sides and free on the long sides. Their fabrication process is described in details in [6]. They were composed of 200nm of AlN seedlayer, 100nm of bottom Pt electrodes and 0.5µm of AlN and 0.5µm of Al as the two active layers with respectively $4.5.10^{-6}K^{-1}$ and $24.10^{-6}K^{-1}$ thermal expansion coefficients. A harvester of this type measuring 1400µm long by 700µm wide could scavenge an output energy density of $230pJ.mm^{-3}$ with a thermal hysteresis (temperature difference between snap up and down) of 30°C, and a relatively high snap up temperature of 114°C.

The values of thermal hysteresis as well as snap up temperature are not compatible with a future application as thermal switch, which requires a more reduced thermal hysteresis for a better detection of the threshold temperature, or with harvesting applications under 100°C. The rectangular shape of the bridge shows also an aging issue at the anchor, where the stresses are maximized. In order to solve these two issues, we designed a new shape of harvester. It is based on the fact that tapered beam shape, largely studied for piezoelectric energy harvesting [7, 8], is homogenizing the stress along the length generated during vibrations. So we designed a bridge having a butterfly shape as shown in Figure 1. The central, squared part is measuring 400µm, the total length of the device is 1400µm (as the rectangular plate studied before) and the clamped side is measuring 1400µm. This shape decreases the overall mechanical rigidity of the device and also decreases the stress at the clamps.

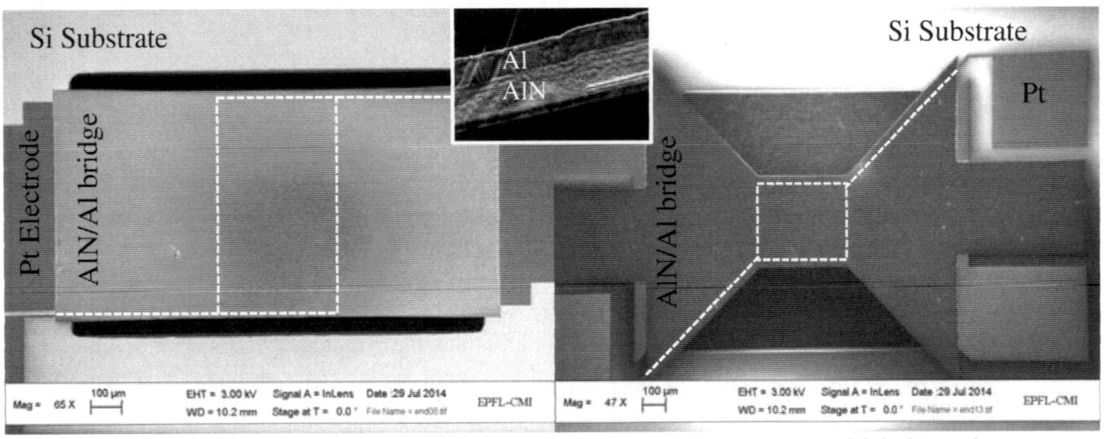

Figure 1. SEM pictures of a rectangular structure and butterfly harvester, with in inset the cross-section of the bilayer bridges. The bottom Pt electrodes are indicated with dashed lines.

We tested the butterfly structure with the same procedure as rectangular plate: the whole wafer is put on a hotplate where the temperature is ramped until reaching the structure snap up. The bottom Pt electrode and top Al electrode are connected to an instrumentation amplifier with a load resistance of

1MΩ (which is not optimized). Then the hotplate is switched off and cooled down until reaching the snap down temperature of the butterfly harvester. Both electrical output signals are shown in Figure 2 (a). Two peaks are observed for each snap and are of opposite sign.

The snap up temperature for one particular structure is 100°C, while snap back occurs at 90°C. The buckling temperature is a bit lower than the rectangular plate, but the great improvement lies in the reduction by three of the thermal hysteresis. The only drawback of the butterfly harvester is its smaller active area compared to rectangular plate, which results in a lower output energy.

3. Non-symmetric stack butterfly structure

Optimizing the thickness ratio between AlN and Al active layers can increase the efficiency of the harvester. Indeed in the previous studied case, the neutral plane was situated near the middle of the active AlN layer (90nm above the mid-thickness). This is not a favorable situation as the stresses at the neutral plane are zero. Moreover the neutral plane being near the mid-thickness of the piezoelectric layer, one part will undergo compressive stress while the other part tensile stress, and the charges generated by the AlN will compensate.

A device with a thicker Al layer present several advantages: it will shift the neutral plane toward the Al layer and the thermal behavior will be enhanced: the thicker Al layer, which has the higher thermal expansion coefficient, acting as a "thermal leader". With a thicker Al layer we expect also a lower snap up temperature, as it will generate higher thermal stresses. The position of the neutral plane with 1µm of AlN and 2µm of Al is 70nm under the Al layer, which is much more favorable. We will study only butterfly geometry having this non-symmetric stack and compare the results to the same butterfly geometry having symmetric stack.

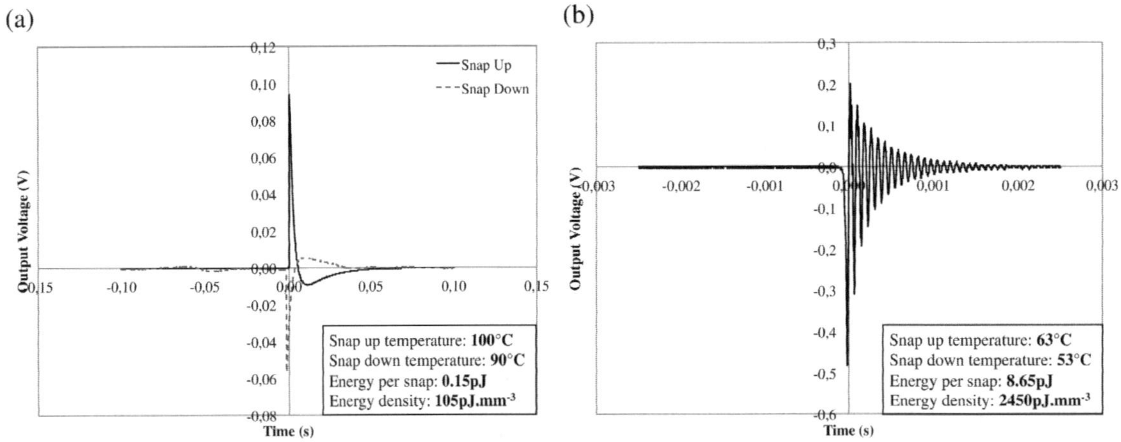

Figure 2. Output voltage during thermal buckling of a butterfly structure with symmetric stack (a) and non-symmetric stack (b)

In a first time, we studied the output signal of the butterfly with the new stack. As before, we put the whole wafer on a hotplate and increased the temperature until reaching the snap up. The output signal is measured through an instrumentation amplifier with a load resistance of 1MΩ. The output voltage is shown in Figure 2 (b). The buckling temperatures are much lower than with symmetric stack, and the thermal hysteresis is still 10°C. But in this case, the structure, when buckling, is oscillating at its natural frequency. Moreover, the output energy is twenty times higher than with symmetric stack.

In a second time, we will study more precisely the harvester behavior during heating up and cooling down. The impedance phase of the device is measured at different temperatures of the hotplate, which can be considered as device temperature, as shown in Figure 3 for the first resonance mode. We can see that during heating, the resonance frequency is decreasing until a threshold temperature where there is a jump in resonance frequency. This threshold temperature corresponds to the snap up temperature. The resonance frequency is increasing continuously if the temperature is further increased.

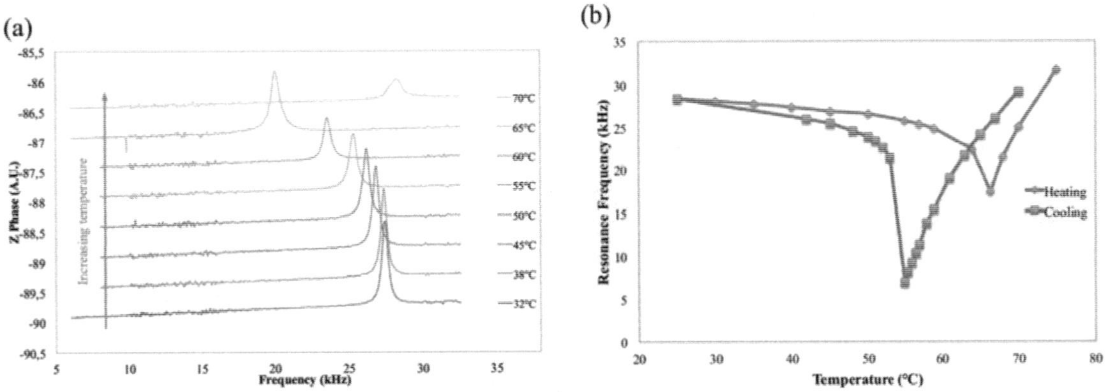

Figure 3. Impedance phase during heating (a) and resonance frequency during heating and cooling (b) of butterfly structure with 1µm of AlN and 2µm of Al.

When the device is cooled down, the resonance frequency is decreasing, and as during heating, a gap in resonance frequency is observed at a specific threshold temperature, lower than the first one, and corresponding to the snap down temperature. This change of resonance frequency is due to the shape change of the structure. Indeed during heating up, the structure is first deforming linearly, until the stress level reaches the buckling limit, where the plate is suddenly changing its curvature to an upward position. The resonance frequency is getting lower during cooling down because the structure being in an upward position, the Al layer can deform more than in the downward position. The measurement of the device impedance constitutes a reliable means to track the structure buckling.

4. Conclusion

This paper present an innovative structure, butterfly shape for micro thermal energy harvesting. The thermal and electrical performances of this new design are compared to the previous, rectangular harvester. The buckling temperature, and especially the thermal hysteresis are decreased in order to adapt to a future commercial application. In a second time, a different, non-symmetric AlN/Al stack is tested with an improvement of the output energy and a decrease of the buckling temperature. This new design increases up to 20 times the output energy. A new technique in order to track precisely the structure buckling consists in the measurement of the electrical impedance for various temperatures.

References

[1] E. Hourdakis and A. G. Nassiopoulou, *Sensors 2013*, **13**, pp. 13596 - 608
[2] A. Moser, M. Erd, M. Kostic, K. Cobry, M. Kroener and P. Woias, *J. Elec. Mat.*, 2012, **41**, pp. 1653 – 61
[3] T. Huesgen, J. Ruhhammer, G. Biancuzzi and P. Woias, *J. Micromech. Microeng.*, 2010, **20**, pp. 104004 – 13
[4] E. Trioux, S. Monfray, T. Skotnicki, S. Basrour and P. Muralt, *Proc. PowerMEMS 2014*, **557**
[5] E. Trioux, L. Rufer, S. Monfray, T. Skotnick, P. Muralt and S. Basrour, *Proc. PowerMEMS 2015*, **660**
[6] E. Trioux, S. Monfray, T. Skotnick, P. Muralt and S. Basrour, *IEEE Sensors'14*, pp. 2171 – 74, Valencia, Spain
[7] D. Benasciutti, L. Moro, S. Zelenika and E. Brusa, *Microsyst. Technol.*, 2010, **16**, pp. 657 – 68
[8] M. Defosseux, M. Allain and S. Basrour, *Proc. PowerMEMS 2010*, pp. 339 – 46

Improvement of vibration energy harvesters mechanical Q-factor through high density proof mass integration

A Dompierre[1,2] **and L G Fréchette**[1,2]

[1]Institut Interdisciplinaire d'Innovation Technologique (3IT), 3IT.nano, Université de Sherbrooke, Sherbrooke, QC, J1K 0A5, Canada
[2]Laboratoire de Nanotechnologie et Nanosystème (LN2), CNRS UMI-3463, Université de Sherbrooke, Sherbrooke, QC, J1K 0A5, Canada

andre.dompierre@usherbrooke.ca

Abstract. This paper reports on improvement of the mechanical Q-factor of resonant energy harvesters at ambient pressure via the use of tungsten proof masses by evaluating the impact of the mass size and density on the squeeze film damping. To this end, a simplified model is first proposed to evaluate cantilever beams deflection and the resulting fluid pressure build up between the mass and a near surface. The model, which accounts for simultaneous transverse and rotational motion of very long tip masses as well as for 2D fluid flow in the gap, is used to extract a scaling law for the device fluidic Q-factor Q_f. This law states that Q_f can be improved by either increasing the linear mass density of the tip mass or by reducing the side lengths compared to the gap height. The first approach is validated experimentally by adding a tungsten proof mass on a silicon based device and observing an improvement of the Q-factor by 103%, going from 430 to 871, while the resonance frequency drops from 457 to 127 Hz. In terms of fluidic Q-factor, this represents an increase from 562 to 1673. These results successfully demonstrate the benefits of integrating a tungsten mass to reduce the fluid losses while potentially reducing the device footprint.

1. Introduction

As demonstrated by previous research [1], energy losses due to fluidic damping can substantially reduce the performance of MEMS vibration energy harvesters in ambient pressure. Unpackaged devices are typically subject to viscous friction with air, while squeeze film damping is observed in packaged harvesters due to fluid confinement in the cavity [1,2]. The latter actually leads to worse performance in terms of damping and efficiency, but packaging is nonetheless necessary to protect the fragile structure. The typical solution to this issue is to use a vacuum sealed package [1,3], but hermiticity now becomes a new concern. The mass can also be perforated to improve fluid circulation, although it comes at the expense of its effective density.

High density materials improve the power density of vibration energy harvesters due to an increased sensitivity to input vibrations [4,5]. In this paper, we furthermore highlight how their integration can also be an effective way to reduce the effects of fluidic damping on the device mechanical Q-factor. We first propose a simplified model to analytically verify the impact of a mass density variation from silicon (Si) to tungsten (W) on squeeze film damping. This effect is then shown experimentally by characterization of the Q-factor of Si-AlN based piezoelectric microcantilevers fabricated using MEMSCAP's PiezoMUMPs process.

2. Beam and fluid damping modelling

2.1. General assumptions and definitions

Several factors affect the interaction of the harvester with the surrounding fluid. For this work, we assume isothermal conditions, ideal venting (P=0 at the boundaries), small displacement and that air is an ideal gas. The variables are therefore the ambient pressure P_a, geometry and frequency ω.

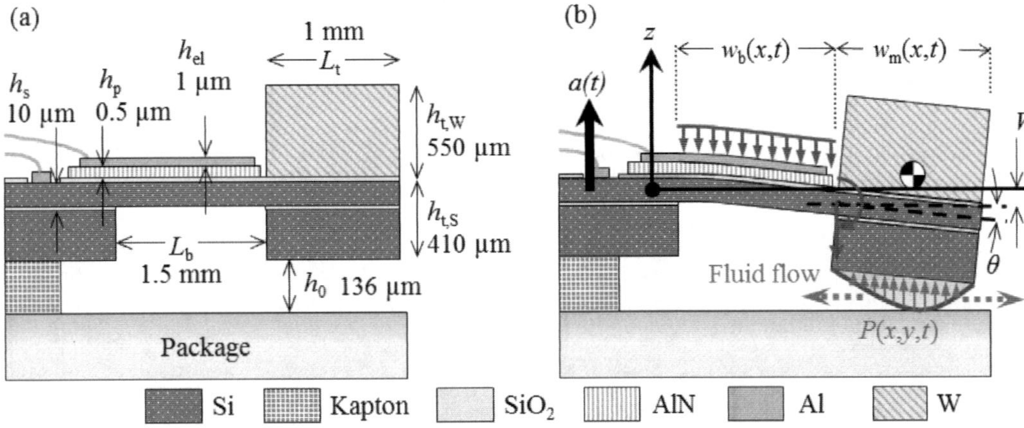

Figure 1. Diagram of the packaged piezoelectric cantilevers, unstrained with the dimensions (a) and deflected under the combined load with the squeeze film pressure in the gap (b).

The configuration of the packaged piezoelectric cantilevers studied is depicted in figure 1. The first mode of the beam $\hat{w}_b(x)$, excited by the harmonic base acceleration $a(t)$, is approximated to the static deflection of a cantilever subject to a combined load. The inertia of the beam of length L_b and width b is considered as a constant distributed load while the tip mass applies a punctual force and a moment to the tip of the cantilever. The tip mass rotary inertia is also assumed negligible. The beam deflection $w_b(x,t)$, defined over the domain x=[0, L_b], can therefore be expressed as in equation (2.1) [6]

$$w_b(x,t) = \hat{w}_b(x) W_0 \sin(\omega t) = \frac{\left[6 + 12\lambda\left(\frac{L_t}{L_b} + \frac{1}{2}\left(\frac{L_t}{L_b}\right)^2\right)\right]\left(\frac{x}{L_b}\right)^2 - 4\left(1 + \lambda\frac{L_t}{L_b}\right)\left(\frac{x}{L_b}\right)^3 + \left(\frac{x}{L_b}\right)^4}{3 + 8\lambda\frac{L_t}{L_b} + 6\lambda\left(\frac{L_t}{L_b}\right)^2}\sin(\omega t) \tag{2.1}$$

where W_0 is the beam tip amplitude, L_t is the length of the proof mass and λ is the ratio of the linear mass density of the proof mass to the beam which depends on the material's densities ρ and thicknesses h,

$$\lambda = \left(\rho_s h_{t,S} + \rho_W h_{t,W}\right) / \left(\rho_s h_s + \rho_p h_p + \rho_{el} h_{el}\right). \tag{2.2}$$

The displacement of the mass $w_m(x,t)$, defined over x=[L_b, L_b+L_t], is then the combined result of the beam tip transverse motion W_0 and rotation due to the tip angle θ. Noting $\hat{\theta} \approx \mathrm{d}\hat{w}_b(L_b)/\mathrm{d}x$ as the normalized tilt angle,

$$w_m(x,t) = \left[1 + (x - L_b)\hat{\theta}\right] W_0 \sin(\omega t). \tag{2.3}$$

Equation (2.4) finally provides the definition of the Q-factor,

$$Q_{tot} = \left(Q_{int}^{-1} + Q_f^{-1}\right)^{-1} = 2\pi \frac{\text{Stored Energy}}{\text{Dissipated Energy per Cycle}} = \pi M_{eq} \omega^2 W_0^2 / U_D \,, \tag{2.4}$$

where Q_{int} and Q_f are respectively the intrinsic and fluidic Q-factor. Meanwhile, U_D is obtained by integrating over the structure area and over one period the product of the local drag force induced by the pressure field $P(x,y,t)$ with the local velocity $\dot{w}(x,t)$. The problem therefore lies in properly evaluating the distribution of this force. This section quickly presents the basic model for 2D squeeze film flow to showcase the impact of the tip mass density and its geometry.

2.2. *Squeeze film damping*

Due to the fluid flow restriction in the tight h_0 thick gap between the mass and the package, squeeze film damping will only occur in this area. The linearized differential equation of the normalized pressure field P^* for a compressible 2D squeeze film is described by the classic Reynolds equation [7]

$$\frac{\partial^2 P^*}{\partial x^2}+\frac{\partial^2 P^*}{\partial y^2}-\frac{12\mu_{\text{eff}}}{h_0^2 P_\text{a}}\frac{\partial P^*}{\partial t}=-\frac{12\mu_{\text{eff}}}{h_0^2 P_\text{a}}\frac{\partial \tilde{h}(x,t)}{\partial t} \tag{2.5}$$

with $\tilde{h}$ the normalized gap function given by equation (2.6)

$$\tilde{h}(x,t)=1-w_\text{m}(x,t)/h_0 \,. \tag{2.6}$$

Equation (2.5) is solved using the approach presented in [8] which yields equation (2.7), the pressure field expressed as a sum of eigenmodes

$$P(x,y,t)=P_\text{a}\frac{W_0}{h_0}\tilde{P}(x,y)\cos(\omega t)=P_\text{a}\frac{W_0}{h_0}\sum_{mn}\phi_{mn}C_m C_n \begin{matrix}\cos\\ \sin\end{matrix}\left(\frac{m\pi(x-L_\text{b})}{L_\text{t}}-\frac{m\pi}{2}\right)\begin{matrix}\cos\\ \sin\end{matrix}\left(\frac{n\pi y}{b}\right)\cos(\omega t), \tag{2.7}$$

where cos and sin are respectively used for odd and even indices m, n while C_m, C_n and ϕ_{mn} are computed from equations (2.8) and (2.9).

$$\phi_{mn}=\frac{\sigma_{mn}}{1+\text{j}\sigma_{mn}}, \qquad C_m=\begin{cases}\frac{2(L_\text{t}\theta+2)}{\pi m}(-1)^{\left(\frac{m-1}{2}\right)}\\ \frac{2L_\text{t}\theta}{\pi m}(-1)^{\left(\frac{m}{2}+1\right)}\end{cases}, \qquad C_n=\begin{cases}\frac{4}{n\pi}(-1)^{\left(\frac{n-1}{2}\right)} & \text{odd}\\ 0 & \text{even}\end{cases} \tag{2.8}$$

$$\sigma_{mn}=\frac{12\mu_{\text{eff}}\,\omega}{h_0^2 P_\text{a}}\left[\pi^2\left(\frac{m^2}{L_\text{t}^2}+\frac{n^2}{b^2}\right)\right]^{-1} \tag{2.9}$$

Here σ_{mn} are equivalent to modal squeeze numbers. For large σ_{mn}, the fluid is compressible and contributes to an increased stiffness. Conversely, the pressure field is mainly dissipative for smaller σ_{mn} which should be the case here ($\sigma_{11}\approx$ 5E-6 at 1 atm) and therefore $\text{Re}[P]\approx P$. The energy dissipated, U_D, is derived in equation (2.10) by integrating the pressure field over the mass surface and one period,

$$U_\text{D}=\int_{L_\text{b}}^{L_\text{b}+L_\text{t}}\int_{-b/2}^{b/2}\int_{0}^{2\pi/\omega}P(x,y,t)\dot{w}_\text{m}(x,t)\text{d}t\text{d}y\text{d}x=W_0^2\frac{P_\text{a}}{h_0}\int_{L_\text{b}}^{L_\text{b}+L_\text{t}}\int_{-b/2}^{b/2}\tilde{P}(x,y)\left[1+(x-L_\text{b})\hat{\theta}\right]\text{d}y\text{d}x \,. \tag{2.10}$$

Replacing U_D in equation (2.4) by equation (2.10) finally yields the fluidic Q-factor,

$$Q_\text{f}=M_\text{eq}\omega^2 h_0\left(P_\text{a}\int_{L_\text{b}}^{L_\text{b}+L_\text{t}}\int_{-b/2}^{b/2}\tilde{P}(x,y)\left[1+(x-L_\text{b})\hat{\theta}\right]\text{d}y\text{d}x\right)^{-1} \tag{2.11}$$

An exact simplified expression for Q_f that covers all geometry range is not straightforward to obtain, but we can nonetheless extract a general trend by developing equation 2.11 for the case of a sizeable tip mass. Upon simplification, this trends is expressed as equation (2.12),

$$Q_\text{f}\propto\frac{\omega(\rho_\text{t}h_\text{t}b)}{\mu_{\text{eff}}}\left(\frac{h_0}{l}\right)^3, \tag{2.12}$$

where l is the smallest planar feature size of the mass. The first term therefore states that Q_f will improve proportionally with the linear mass density. This is either done by increasing the material density, as we propose here, or by increasing the thickness. This second option may however reduce the gap h_0 if the space is constrained. Meanwhile, the second term is affected by a cubic power and shows potential for drastic improvement if the gap to side length ratio is increased.

3. Case study application and experimental validation

3.1. Experimental setup and procedure

The piezoelectric devices, shown on figure 2, were fabricated via MEMSCAP's PiezoMUMPs process. The process uses a n-type SOI wafer as the starting substrate with a 10 µm thick device layer and 400 µm handle layer. The device layer acts as the beam support material and is phosphorous doped to also serve as the bottom electrode. The piezoelectric layer consists of 0.5 µm thick sputtered AlN and the top electrode is a metal stack of 20 nm Cr and 1 µm thick Al defined by lift-off. The beam and mass geometries are machined by DRIE through the device and handle layer respectively.

The nominal width of the beam and mass of the tested design is 1 mm while the other dimensions are illustrated in figure 1(a). A shaker setup was used to excite the cantilever at its resonance frequency and was suddenly turned off to measure the decay time. A first measurement was done with the Si proof mass (ρ_s=2.33 g/cm^3) in vacuum (P_a=5E-4 mbar) to evaluate Q_{int} and a second one was conducted at atmospheric pressure to evaluate Q_{tot} and extract Q_f from equation (2.4). A W mass (ρ_W=19.3 g/cm^3) was subsequently glued on top of the Si mass and the same methodology was repeated. The displacement was measured via the piezoelectric open circuit voltage and the Q-factors were computed via the logarithmic decrement method using a curve fitting algorithm. Four waveforms were recorded and analyzed for each condition.

3.2. Results and discussion

The results of the simulation (max m,n=10) are compared with our experiments in table 1 and show good agreement despite some of the nonlinearities observed. As anticipated, the tungsten mass reduces the resonance frequency ω_n significantly from 457.49 Hz to 126.27 Hz while a 103% increase of Q_{tot} is observed. The model follows the same trends, although with a small offset which could be explained by errors on the dimensions or by residual stresses (the value of Q_f is especially sensitive to h_0). Acknowledging for the measured Q_{int}, the derived Q_f shows a 3 fold increase which is expected considering the frequency drop (see equation 2.12). For comparison's sake, we also considered an additional modelling scenario where the mass length is adjusted to conserve the frequency. A significant reduction of fluid damping is again observed. However, viscous flow would probably more aptly represent this scenario considering the reduction of h_0/L_t. These results successfully demonstrate the benefits of integrating a tungsten mass to simultaneously reduce the footprint and the fluid losses.

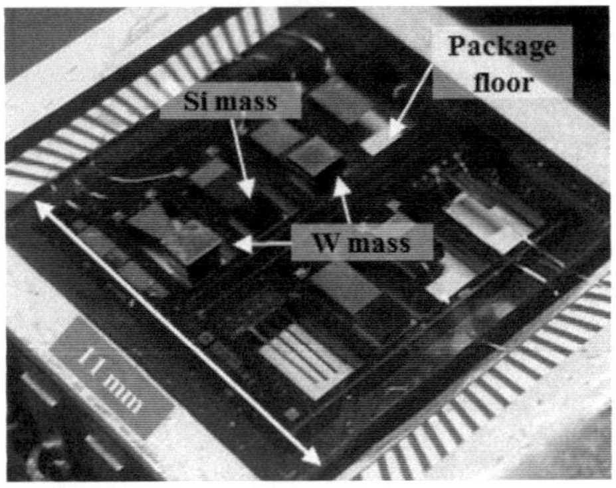

Figure 2. Close up view of a PiezoMUMPs die, featuring designs with silicon and glued tungsten masses.

Table 1. Simulation and experimental results

		ω_n [a] (Hz)	Q_{int}	Q_{tot} [a]	Q_f [a]
Exp	Si	457.49	1836	430	562
	Si + W	126.27	1817	871	1673
Model	Si	483.18	1825	437	575
	Si + W	136.68	1825	962	2033
	Si + W[b]	484.18	1825	1812	5E5

[a] at 1 atm
[b] L_t = 106 µm

4. Conclusion

We have successfully shown analytically and experimentally the impact of the tip mass density and geometry on the fluidic damping and device Q-factor. A simplified analytical model was first introduced to describe the 2D squeeze film and the beam behaviour. The model mainly shows that Q_f can be improved by increasing the linear mass density, by using tungsten instead of silicon for instance, or by reducing the mass area if the frequency is fixed. The first approach was verified experimentally and has shown that it is possible to double the total Q-factor at ambient pressure. Beside the improvement in efficiency, the harvester also benefits from an improved sensitivity to input vibration which can drastically improve its power density.

5. References

[1] Elfrink R, *et al.* 2010 Vacuum-packaged piezoelectric vibration energy harvesters: damping contributions and autonomy for a wireless sensor system *J. Micromech. Microeng.* **20** 104001

[2] Hosaka H, Itao K and Kuroda S 1995 Damping characteristics of beam-shaped micro-oscillators *Sensors and Actuators* A **49** 87–95

[3] Aktakka E E and Najafi K 2014 A micro inertial energy harvesting platform with self-supplied power management circuit for autonomous wireless sensor nodes *IEEE J. of Solid-State Circuits* **49** 2017–29

[4] Mitcheson P D, Green T C, Yeatman E M and Holmes A S 2004 Architectures for vibration-driven micropower generators *J. Microelectromech. Syst.* **13** 429–40

[5] Dompierre A, Vengallatore S and Fréchette L G 2011 Theoretical and practical limits of power density for piezoelectric vibration energy harvesters *Proc. of PowerMEMS 2011 (Seoul)* 249–52

[6] Bazergui A, Bui-Quoc T, Biron A, McIntyre G and Laberge C 2002 *Résistance des Matériaux, troisième édition* (Montreal: Presses Internationales Polytechnique) chapter 5 p 118

[7] Bao M and Yang H 2007 Squeeze film air damping in MEMS *Sensors and Actuators* A **136** 3–27

[8] Darling R B, Hivick C and Xu J 1998 Compact analytical modeling of squeeze film damping with arbitrary venting conditions using a Green's function approach *Sensors and Actuators* A **70** 32–41

Acknowledgment

We would like to acknowledge the financial support of NSERC and the technical support of the 3IT staff members. We'd especially like to thank Daniel Blackburn for his help in setting up the vacuum chamber used in these tests.

MINIATURE FUEL CELL WITH MONOLITHICALLY FABRICATED Si ELECTRODES - Au-Pd-Pt MULTILAYER CATALYST -

Ryo Shirai[1], N.Vasiljevic[2], and Masanori Hayase[1]

[1]Tokyo Unversity of Science, Japan
2641 Yamazaki, Noda, Chiba 278-8510, Japan
[2]University of Bristol, United Kingdom
Tyndall Avenue, Bristol BS8 1TL, UK

E-mail: mhayase@rs.noda.tus.ac.jp

Abstract. A novel catalyst layer structure is proposed for our miniature fuel cells. In our fuel cells, conventionally porous Pt was used as a catalyst layer. In order to reduce the Pt amount, instead of porous Pt, porous Pd was formed on a Si chip and Pt was deposited atomically on the Pd by UPD-SLRR(Under Potential Deposition – Surface Limited Redox Replacement). The Pd-Pt catalyst showed satisfying performance, besides high CO tolerance was observed. Though the Pd-Pt catalyst is quite promising, Pd is also a rare metal and reduction of Pd amount is necessary. In this study, a novel Au-Pd-Pt catalyst formation strategy is proposed by UPD-SLRR, and the layered structure is preliminary fabricated.

1. Introduction

We have developed miniature fuel cells with monolithically fabricated Si electrodes (Fig.1). In the fuel cells, two Si electrodes with porous Pt catalyst layer, which Pt usage was 3.6mg/cm^2, were hot-pressed onto either side of a PEM (polymer electrolyte membrane), and the prototype fuel cells showed promising performances of over 500mW/cm^2 output at 313K with H_2-O_2 supply[1-3]. However, recent improvements of Li-ion battery are lowering expectation to the miniature fuel cells, and miniature fuel cell products are not available yet commonly. The advantage of the fuel cell, i.e. clean and efficient energy source, should be strengthened.

Bio-mass is an ultimate energy source, and effective utilization of the bio-mass is desired. Hydrogen can be produced by gasification of the bio-mass, but gasification is performed in reducing environment, and high CO concentration in the bio-mass derived hydrogen is inevitable. To utilize the bio-mass derived hydrogen, catalyst poisoning by CO is problematic. In the PowerMEMS 2013, we reported that our Pd-Pt catalyst had promising feature [3]. Though the catalyst layer used only 6μg/cm^2 of Pt, high tolerance to CO was obtained. But the catalyst layer was based on the porous Pd, and more than 1.5mg/cm^2 of Pd was used. Pd and Pt are rare metals, and both amounts should be minimized.

In this study, a novel Au-Pd-Pt multilayer catalyst is proposed as shown in figure 2. Though Au is also a precious metal, reserve of Au is much larger than Pd and Pt, and porous Au can be formed on a Si substrate by similar process to porous Pt [4].

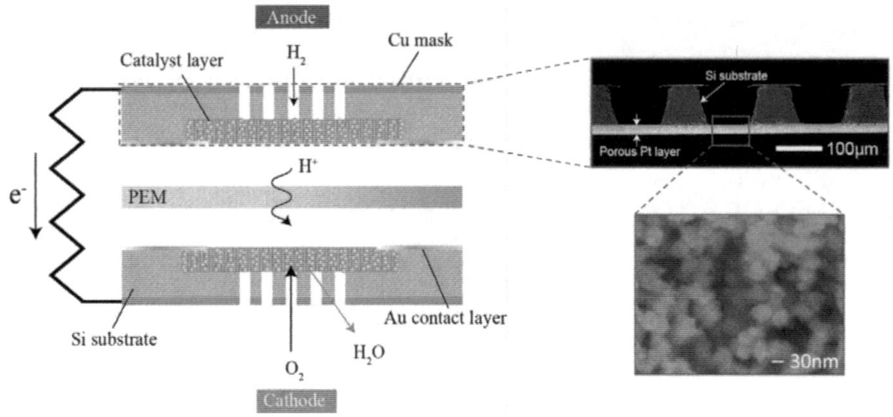

Figure 1. Miniature fuel cell with monolithically fabricated Si electrodes. Porous Pt catalyst layer was formed on a Si substrate, and fuel channels were opened by plasma etching.

2. Au-Pd-Pt Multilayer Catalyst

There is no clear evidence, but it is assumed that high CO tolerance of the Pd-Pt catalyst is realized with sub-monolayer Pt on Pd, and release of absorbed hydrogen from Pd has important role. Hydrogen absorption into Pd is sensitive to the number of atomic layer of Pd, and fine control of Pd and Pt deposition on Au is crucial. In our previous study, sub-monolayer Pt was deposited with Cu-UPD-SLRR (Cu – Under Potential Deposition - Surface Limited Redox Replacement). In this process, formation of a monoatomic layer of Cu on porous Pd was performed by UPD. Then, the deposited Cu was replaced to Pt galvanically, and a sub-monolayer Pt was expected to deposit on porous Pd. But we needed to separate Cu plating bath and Pt plating bath, because simultaneous deposition of Cu and Pt occurs. Because of the two baths limitation, it is difficult to repeat the Cu-UPD-SLRR for Pd multilayer formation. Recently, instead of Cu or Pb, it was reported that UPD hydrogen can be used for SLRR. Hydrogen also absorbs on some metal surfaces and form atomic monolayer at slightly under the equilibrium potential [5]. The H-UPD-SLRR can be performed just in an acidic plating bath containing Pd^{2+} or Pt^{4+}. There is no need to replace the plating bath, and repetition of the atomic layer deposition process will be easy. In this study, basic fabrication strategy of the Au-Pd-Pt multilayer catalyst using H-UPD-SLRR process was examined.

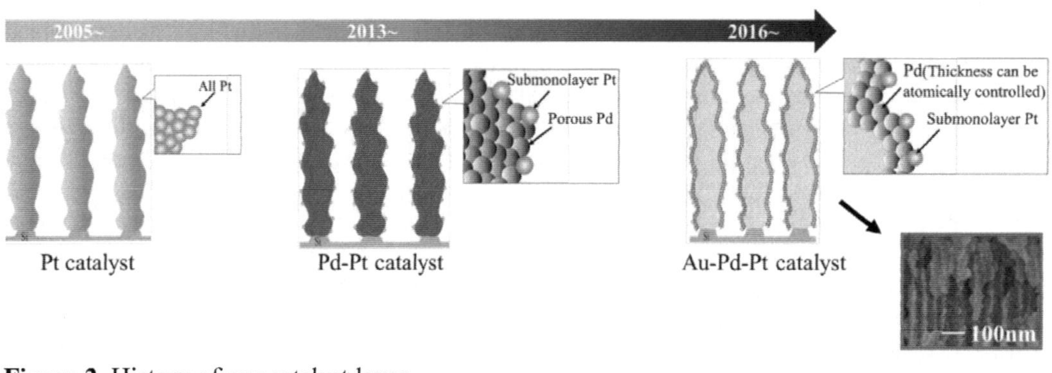

Figure 2. History of our catalyst layer.
Pt catalyst is original porous catalyst. Pt amount was 3.6mg/cm^2. Pd-Pt catalyst is sub-monolayer Pt deposited on porous Pd. Pt amount was reduced to 6μg/cm^2 and high CO tolerance was obtained. Au-Pd-Pt catalyst will realize that amount of platinum group can be reduced, while expected performance can be maintained. Porous Au can be formed on Si substrate by similar process to porous Pt.

3. Experimental

Conditions for the H-UPD-SLRR Au-Pd-Pt catalyst formation process are shown in Table 1. Figure 3 shows the experimental setup. Au wire was masked by polyester film tapes, and 1cm of Au wire was exposed to the solution. A mercury-mercurous sulfate electrode (MSE) and a Pt wire were used as the reference and the counter electrodes, respectively. An electrolyte solution of 0.1M $HClO_4$+10μM $PdCl_2$ was prepared, and the solution was bubbled with nitrogen gas for at least 30 min prior to the experiment to eliminate dissolved oxygen and carbon dioxide. Before the UPD process, the Au wire was cleaned by a piranha solution, then the Au wire was immersed in a 0.1M $HClO_4$ solution, and the electrode potential was cyclically varied from -1000 to 1500 mV vs. MSE more than 100 times. After the cleaning process, the wire was immersed into the electrolyte solution in a vessel filled with nitrogen gas.

Figure 4 shows the schematic of the H-UPD-SLRR process. H-UPD was performed by applying -600mV vs. MSE. Because the process bath contains Pd ions, bulk electrodeposition also proceeds during the UPD process. To minimize the Pd electrodeposition, UPD process was limited for 1 s. The amount of Pd electrodeposition for 1 s UPD period was estimated to be less than 1% of the electrode surface assuming a monolayer, and hydrogen UPD was assumed to be dominating. Then, the electrical circuit was opened until the electrode potential reached 0 mV vs. MSE. During the circuit was opened, adsorbed hydrogen was replaced by novel metals. This process is called as the SLRR, and sub-monolayer Pd or Pt deposition was expected. Pd multilayer formation was expected by repeating the H-UPD-SLRR process in the Pd process bath. Finally, Pt sub-monolayer deposition was performed by identical way in the Pt process bath.

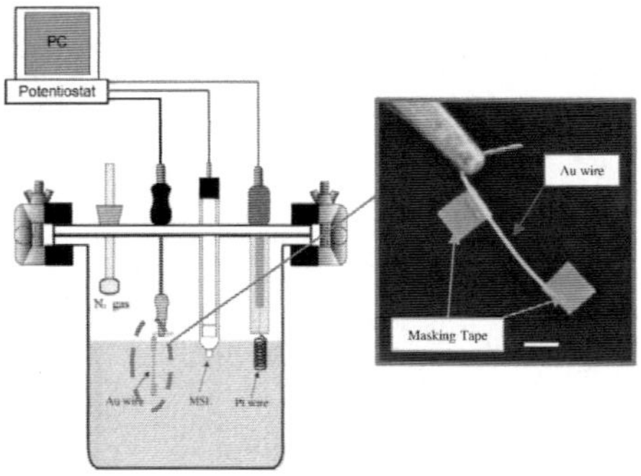

Figure 3. Experimental model of H-UPD-SLRR.
In this time, three-electrode cell was used, and during experiment, nitrogen gas was filled in a vessel because oxygen and carbon dioxide were concerned.

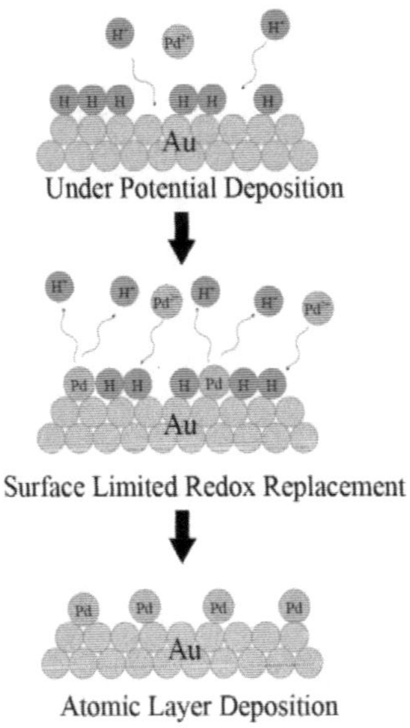

Figure 4. Schematic of the H-UPD-SLRR. In H-UPD-SLRR process, Electrochemical sub-monolayer metal is completed.

Table 1. Condition for atomic layer Pd or Pt deposition on Au wire by H-UPD-SLRR.

Experimental condition	
Pd deposition	
Working Electrode	Au wire
Counter Electrode	Pt
Composition of solution	0.1M $HClO_4$ + 10μM $PdCl_2$
Negative potential(E_1)	-600mV vs. MSE
UPD time	1s
Positive potential(E_2)	0mV vs. MSE
Pt deposition	
Working Electrode	50R Pd + Au wire
Counter Electrode	Pt
Composition of solution	0.1M $HClO_4$ + 10μM H_2PtCl_6
Negative potential(E_1)	-600mV vs. MSE
UPD time	1s
Positive potential(E_2)	0mV

4. Results and discussion

Figure 5 shows Pd film thickness estimated from H-UPD charges. Figure 6 shows variation of CV (cyclic voltammograms) along Pd deposition cycles. The CV was obtained in a solution of 0.1M $HClO_4$. The hysteresis became larger along the repetition of the deposition process. Negative current suggests hydrogen adsorption or absorption into Pd, and it is assumed that amount of hydrogen absorption can be modulated.

Figure 7 showed CVs after Pt deposition on the Au wire with 50 Pd deposition cycles. Surprisingly, the hysteresis loop became larger significantly with only one Pt deposition cycle, and Pt catalytic activity was expected. Typical Pt electrode behavior with several peaks was observed with increasing the Pt deposition cycles, but no significant enlargement of the hysteresis loop was obtained.

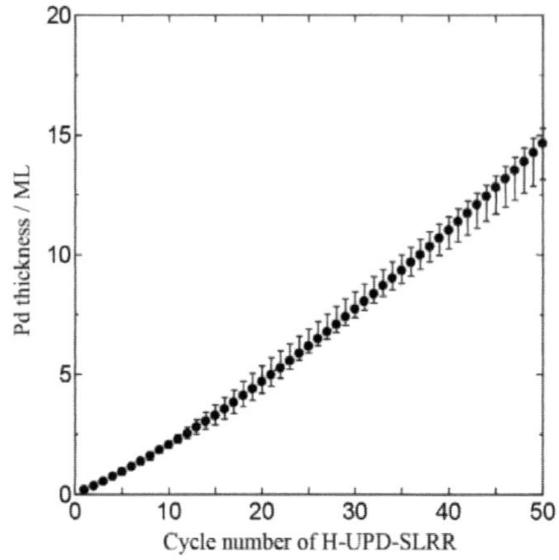

Figure 5. Estimated palladium thickness was evaluated by multiplying current and time during H-UPD.

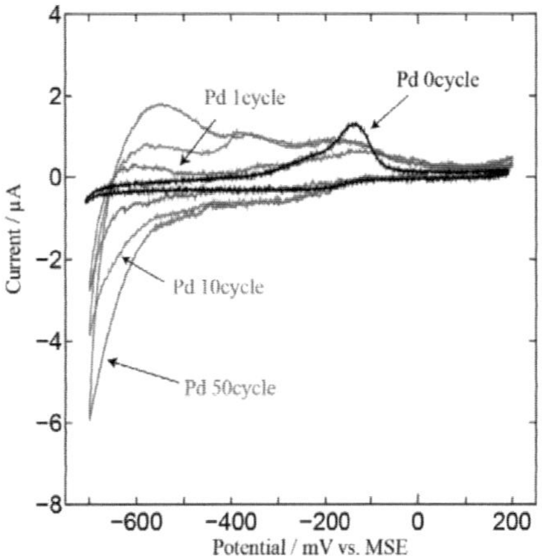

Figure 6. Variation of cyclic voltammogram along the Pd deposition cycles.

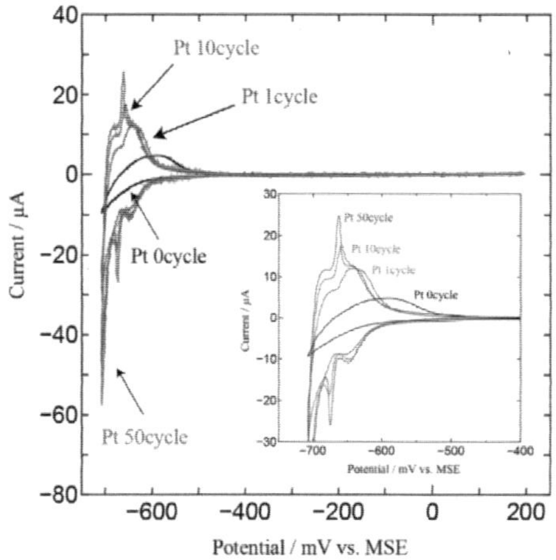

Figure 7. Cyclic Voltammogram after Pt deposition on Au-Pd wire.

5. Conclusion

In this study, we proposed a novel Au-Pd-Pt multilayer catalyst structure, and fabrication process using H-UPD-SLRR was preliminary examined with Au wire. Electrochemical analyses suggested that amount hydrogen absorption in the Pd layer, which associates to CO tolerance, can be controlled by the cycle of Pd deposition process. Pt catalytic activity was also observed with only one cycle of the H-UPD-SLRR deposition process. Further study with porous Au will be performed.

References

[1] M. Hayase, K. Horikoshi, T.Tajima, *Proc. PowerMEMS2009*, pp. 13-16.
[2] M. Kobayashi, T. Suzuki, M. Hayase, *J. Power Sources*, **267** (2104), pp. 622-628.
[3] D.Ogura, T. Suzuki, N. Katayama, K. Dowaki, M. Hayase, *Proc. PowerMEMS2013*, 012066.
[4] Jose. G. A. Brito-Neto, K.Kondo, and M.Hayase, *J. Electrochem. Soc.*, **155**(2008), pp. D78-D82.
[5] J. Nutariya, M. Fayette, N. Dimitrov, N.Vasiljevic, *Electrochim. Acta,* **112** (2013), pp.813-823.

Fully casted soft power generating triboelectric shoe insole

Rubaiyet I. Haque[1,2], Pierre-André Farine[2] and Danick Briand[1]

[1] Microsystems for Space Technologies Laboratory, École Polytechnique Fédérale de Lausanne, Neuchâtel, Switzerland
[2] Electronics and Signal Processing Laboratory, École Polytechnique Fédérale de Lausanne, Neuchâtel, Switzerland

E-mail: rubaiyet.haque@epfl.ch

Abstract. Power generating soft triboelectric based shoe insole fully elastomeric and compatible with large-scale fabrication technique has been developed. During the process, film casting and stencil printing techniques were implemented to deposit/pattern elastomeric and soft/flexible materials, such as, polydimethylsiloxane (PDMS) and polyurethane (PU). Carbon-based elastomeric materials were used as electrodes, which were also film casted. The developed triboelectric generator (TENG) was capable of harnessing electrical power effectively from mechanical deformation of the system during walking or running activities. The performance of the device was tested for walking with frequency of 0.9±0.2 Hz. The power (rms value) of 0.25 mW was achieved for load resistance of 100 MΩ, which corresponded to the power density (rms value) of 1.9 μW/cm^2.

1. Introduction

The limitation regarding lifetime, immobility and toxic hazards, of traditional power supply, like, batteries, encourage researchers to explore the alternative sources for energy to develop maintenance-free/ autonomous electronics system. With the fast development of wireless sensors and consumer electronics, the urge for self-powered autonomous system is even stronger. Various ambient kinetics motions and mechanical vibrations, that is generally wasted in everyday life, can be used as renewable source to harvest energy and to develop green technologies.

Nanogenerator technology, that is capable of producing electricity from small variation of mechanical deformation, can be used to power portable electronics [1]. For many years, electromagnetic, electrostatic and piezoelectric mechanism are mainly implemented to harvest mechanical energy [2]–[4]. Triboelectric generator (TENG), which uses the contact electrification and electrostatic induction between materials having opposite electron affinity [5], can be used to convert kinetic energy to electrical energy efficiently. Since its first discovery by Prof. Wang's group in 2012 [1], TENGs have demonstrated the possibilities of exciting applications in mechanical energy harvesting. To date, several TENGs based energy harvesters have been reported [6]–[9]; however, most of their fabrication process is complex and none of them are truly compatible with the large scale fabrication method.

We are reporting the development of highly soft triboelectric based shoe insole using elastomeric materials as triboelectric and electrodes, which is compatible with large scale fabrication process and capable of generating power from human movement while walking or running.

2. Experimental procedure

2.1. Materials and design

The reported triboelectric generator, which uses cycled vertical contact-separation mechanism, was fabricated with casting technique. Elastomers and their composites were used both as triboelectric and electrodes materials during the fabrication process. Both triboelectric layers were facing each other separated by a spacer between them (Figure 1). The dimension/ geometric parameters of the TENG-shoe insole are listed in Table 1.

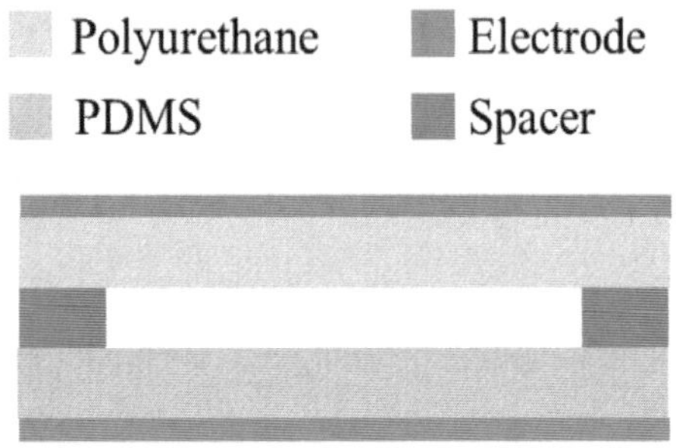

Figure 1. Schematic diagram of triboelectric based shoe insole.

Polydimethylsiloxane (PDMS, Sylgard 186) was purchased from DowCorning and polyurethane (PU, MM4520) was supplied by SMP Technology Inc. Electro-conductive carbon black pellets (Ketjenblack EC-300J & Ketjenblack EC-600JD) were purchased from Akzo-Nobel, which were used to prepare conductive elastomers and used as electrode for the devices. Solvents were purchased from Sigma-Aldrich. All products were used as received. Casting on flexible substrate was performed on commercially available polyethylene terephthalate (PET) film.

Table 1: Dimensions of TENG-based shoe insole

Total area	:	133 cm^2
Gap between two dielectric layers	:	2 mm
Thickness of PDMS layer	:	60 μm
Thickness of PU layer	:	50 μm

2.2. Experimental process

The fabrication process was composed of three parts, namely, preparation of first dielectric layer with electrode, preparation of second dielectric layer, and assembly. Initially, conductive carbon-black based PDMS ink was deposited on flexible PET substrate using film casting technique and baked at 80 °C for 2 hours. PDMS layer was then film casted on conductive layer and cured for 2 hours at 80 °C. On the other hand, conductive carbon-black based PU ink was film casted on thin PET film and cured on hot plate at 80 °C for 2 hours. Thereafter, PU dissolved in solvent was deposited on electrode layer using casting method, which acts as second triboelectric layer, and cured on hot plate. The surface of the substrate was rinsed with isopropanol and dried using blown nitrogen, prior to casting process. Finally, layers were assembled together having dielectric layers facing each-other and separated by a spacer between them. Figure 2 illustrates the process flow of TENG based shoe insole. Figure 3 shows the photograph of the top-view of the developed TENG-based shoe insole structure. Optimization was performed based on the effect of physical parameter, such as, layer stiffness, on output responses.

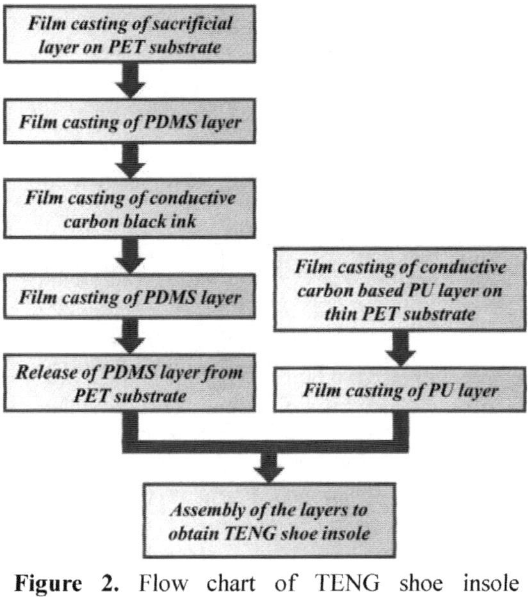

Figure 2. Flow chart of TENG shoe insole fabrication.

Figure 4(a) presents the experimental setup that has been used to record the electrical responses of the TENG-based shoe insole using oscilloscope while walking, and the amount of electrons in a single charge-transfer process was measured using experimental setup showed in Figure 4(b), where a capacitor (C_L) of 2.2 μF was used. Walking experiments were performed for the frequency of 0.9±0.2 Hz.

Figure 3. Top view of the TENG-based shoe-insole.

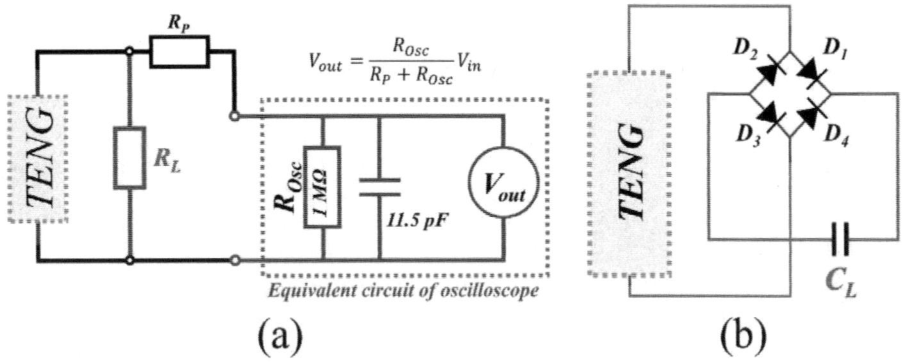

Figure 4. Schematic diagram of the measurement circuits (a) to characterize the electrical response and (b) to store energy generated by TENG-based shoe insole during walking.

3. Results and discussion

Fully casted elastomeric TENG-based shoe insole was tested while walking using electronics described above. Figures 5(a) to 5(c) illustrate the cycle of electrical responses generated by TENG-based shoe insole while walking. At initial state, the two dielectric layers were fully in contact with each other. Since PDMS is much more triboelectrically negative and has higher affinity towards negative charge than PU, electrons were injected from PU into PDMS, leading to the positive triboelectric charges on PU layer and equal amount of negative charges on PDMS layer.

As observed, the developed triboelectric generator was capable of generating open circuit voltage (rms value) of 312.1 V, and closed circuit current (rms value) of 1.14 μA for 100 MΩ load resistance, which corresponded to the power of 0.25 mW. The power density of the developed shoe insole was 1.9 μW/cm^2.

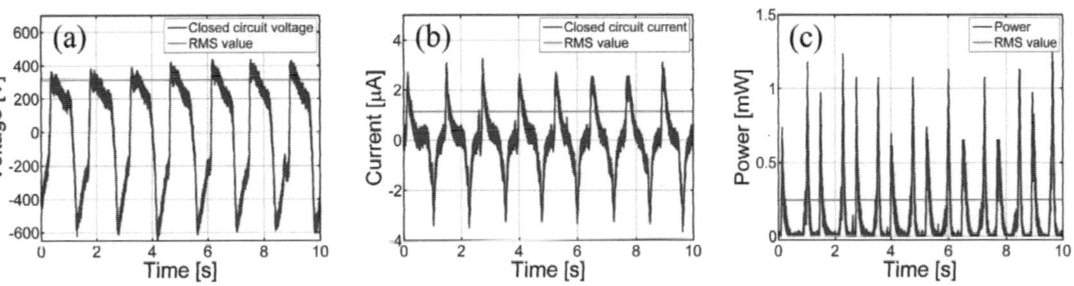

Figure 5. (a) Open circuit voltage; (b) closed circuit current for the load resistance of 100 MΩ; (c) power generated for closed circuit measurement for the load resistance of 100 MΩ.

The total charge transfer (Q_C) was obtained by measuring voltage (V_C) of the capacitor after walking for 10 steps and using relating Q_C=C_LV_C. Thereafter, the charge transfer in a single step was measured by dividing the Q_C by number of steps. The amount of electrons in a single charge-transfer process reached 0.89±0.03 μC, which corresponded to the surface charge density of 6.9±0.22 nC/cm^2 generated by TENG-based shoe insole while walking. Moreover, the corresponding energy stored in a 2.2 μF capacitor for single step was 0.18±0.01 μJ.

4. Conclusions

In summary, we have fabricated fully elastomeric, vertical contact separation mode TENG-based shoe insole using elastomeric PDMS and PU as triboelectric layer and carbon based elastomeric composites as electrodes. The developed TENG was capable of producing electrical energy from the kinetic energy/ movements while walking and/or running, and its fabrication process was simple, compatible with large scale fabrication method, and economical. High flexibility increased the durability of the device and capacity to withstand large deformation. Due to the mechanical deformation during walking, the fabricated TENG-based shoe insole provided power (rms value) of 0.25 mW for load resistance of 100 MΩ, which corresponds to the power density (rms value) of 1.9 μW/cm^2, and charge density of 6.9±0.22 nC/cm^2/cycle.

Further work will be focused on the test of different materials and designs, and on the improvement of the processing steps to develop TENGs with higher capability. In addition, encapsulation method and procedure will also be investigated to protect the TENG from humidity.

Acknowledgements

The authors would like to acknowledge Swiss National Science Foundation (SNSF) and NanoTera.ch for the funding and support of this work within the framework of BodyPowerSenSE project.

References

[1] Wang L Z, Chen J and Lin L 2015 Progress in triboelectric nanogenerators as a new energy technology and self-powered sensors *Energy & Environmental Science* **8(8)** 2250
[2] Kong L B, Li T, Hng H H, Boey F, Zhang T and Li S 2014 Waste mechanical energy harvesting (I): piezoelectric effect (*Waste Energy Harvesting* vol 24) Berlin: Springer Berlin Heidelberg pp. 19–133
[3] Yusuf S T, Yatim A H M, Samosir A S and Abdulkadir M 2013 Mechanical energy harvesting devices for low frequency applications: revisited *ARPN Journal of Engineering and Applied Sciences* **8(7)** 504
[4] Janicek V, Husak M, Jakovenko J and Formanek J 2012 Design and fabrication of 3D electrostatic energy harvester *Radioengineering* **21(1)** 231
[5] AlphaLab Inc. 2009 The triboelectric series *Retrieved from https://www.trifield.com/content/tribo-electric-series/* [Accessed: 2016]
[6] Zhu G, Bai P, Chen J and Wang Z L 2013 Power-generating shoe insole based on triboelectric nanogenerators for self-powered consumer electronics *Nano Energy* **2(5)** 688
[7] Hou T -C, Yang Y, Zhang H, Chen J, Chen L -J and Wang Z L 2013 Triboelectric nanogenerator built inside shoe insole for harvesting walking energy *Nano Energy* **2(5)** 856
[8] Zhang X -S, Han M -D, Meng B and Zhang H -X 2015 High performance triboelectric nanogenerators based on large-scale mass-fabrication technologies *Nano Energy* **11** 304
[9] Cheng X, Meng B, Zhang X, Han M, Su Z and Zhang H 2015 Wearable electrode-free triboelectric generator for harvesting biomechanical energy *Nano Energy* **12** 19

Utilising Nonlinear Air Damping as a Soft Mechanical Stopper for MEMS Vibration Energy Harvesting

Shao-Tuan Chen[1], Sijun Du[1], Emmanuelle Arroyo[1], Yu Jia[1,2] and Ashwin Seshia[1]

[1] Nanoscience Centre, University of Cambridge, Cambridge, CB3 0FF, UK
[2] Department of Mechanical Engineering, University of Chester, Chester CH1 4BJ, UK
E-mail: **stc33@cam.ac.uk**

Abstract. This paper reports on the theory and experimental verification of utilising air damping as a soft stopper mechanism for piezoelectric vibration energy harvesting to enhance shock resistance. Experiments to characterise device responsiveness under various vibration conditions were performed at different air pressure levels, and a dimensionless model was constructed with nonlinear damping terms included to model PVEH response. The relationship between the quadratic damping coefficient ζ_n and air pressure is empirically established, and an optimal pressure level is calculated to trade off harvestable energy and device robustness for specific environmental conditions.

1. Introduction

Microelectromechanical systems (MEMS) technology has enabled the construction of chip-scale vibration energy harvesters with the potential for close in-package or monolithic co-integration with standard CMOS technology. Despite the demonstration of useful power output from MEMS vibration energy harvesters [1], the lack of robustness of MEMS devices due to the brittleness of silicon has hindered commercialisation and real-world application, as large amplitude vibrations in operational environments often cause cantilevered MEMS devices to fracture [2], as shown in Fig.1. Nonlinear air damping has been observed by Jia et al. [3] when the PVEH operates under atmospheric conditions, which limits the maximum displacement of the PVEH when subjected to large amplitude excitation. While this effect reduces the overall power output, it can also serve as a 'soft' mechanical stopper to constrain the vibration amplitude of the harvester. This in turn helps the harvesters to survive at high acceleration levels whereas PVEHs operating in vacuum environments fracture due to excessive displacement under identical drive conditions. Therefore, nonlinear air damping can potentially serve as a soft stopper and prevent devices from failure due to shock or other excessive dynamic loading.

Figure 1. Picture of MEMS PVEH [1] The top left corner reveals fractured PVEH due to shock excitation.

2. Nonlinear model

This paper investigates the variation of maximum power output by PVEH with varying air pressure inside the MEMS package. The PVEH is designed by Jia et al. [3] as a cantilever with proof mass at the free end. The dimensions of the cantilever are 3.5mm by 3.5mm, and the proof mass is comprised of an un-etched silicon substrate (400 μm). The particular device considered in this study has a proof mass to cantilever length ratio of 70%.

A dimensionless model was constructed with nonlinear damping to analytically model the time response of the cantilever and to verify the effect of air damping ζ_n on the displacement of the MEMS PVEH. Nonlinear dissipation of vibrating cantilevers can have several origins, such as geometric nonlinearities or viscous damping by the ambient medium. The equation of motion with nonlinear damping has been previously investigated by Ravindra [4] on soft Duffing oscillators, with the nonlinear damping term taken to be proportional to the pth power of the velocity of the system, in the form of :

$$a_p \dot{x} |\dot{x}|^{p-1} \tag{1}$$

where $p \geq 1$ is the nonlinear damping exponent, and α_p is the corresponding damping coefficient.

The precise value for the nonlinear damping exponent can be determined by considering the physical origin of the damping effect. For MEMS devices operating in air, the cantilever is oscillating in a fluid, and the fluid drag force accounts for the nonlinear air damping effect in the physical model. The amplitude of the fluid drag force can be expressed as a quadratic function of velocity [5]. As a result, the response of a PVEH oscillating in air is modelled by a nonlinear-damped Duffing equation with a quadratic damping term as:

$$\ddot{x} + 2\zeta w_n \dot{x} + \zeta_n \dot{x}|\dot{x}| + \mu x^3 + w_n^2 x = A w^2 \cos(wt) \tag{2}$$

where x is the displacement, ζ is the viscous damping, ζ_n is the quadratic damping representing nonlinear air drag, μ is the Duffing coefficient, w_n is the natural frequency, A is the excitation displacement amplitude, w is the excitation frequency and t is the time domain.

The maximum displacement of the cantilever in air as a function of the nonlinear air damping coefficient ζ_n can be numerically obtained through the dimensionless model, and is shown in Fig.2. As seen from the figure, the maximum displacement flattens off and amplitude saturation is observed as the excitation increases with higher nonlinear air damping.

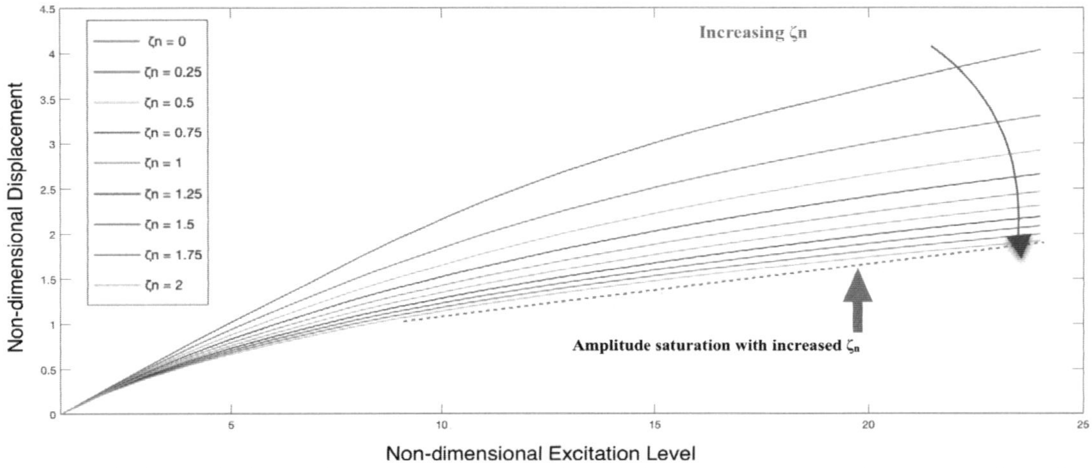

Figure 2. Dimensionless model demonstrating excitation-displacement relationship for various values of nonlinear air damping ζ_n. As ζ_n increases, amplitude saturation is observed

3. Experimental characterisation

3.1 Experimental setup

Characterisation experiments are performed to obtain the relationship between output power and input excitation for the PVEH with different air pressure levels. The experimental apparatus is shown in Fig. 3. The PVEH is inserted onto a chip carrier, mounted to a shaker with a custom PCB, and the entire setup is placed in a vacuum chamber, evacuated using a vacuum pump to adjust the air pressure level inside the chamber. Sinusoidal excitation operating at the device resonance frequency with mean amplitude varying from 0.5 to 10 g is supplied to the PVEH via a function generator (Agilent 33220A), and the operating air pressure is set to be 1, 3, 10, 200, 500 and 1000 mbar, respectively.

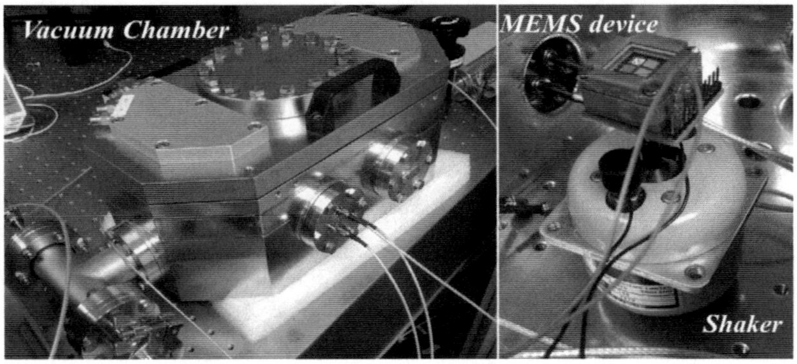

Figure 3. Experimental setup.

3.2 Sinusoidal excitation

The maximum power output of the PVEH under varying excitation levels recorded at different air pressure levels is shown in Fig.4. For larger excitations, lower air pressure cases have steeper slope between excitation and power output, whereas the slope decreases as air pressure increases. This is due to the diminishing displacement of the PVEH due to nonlinear damping, which results in the device describing amplitude saturation.

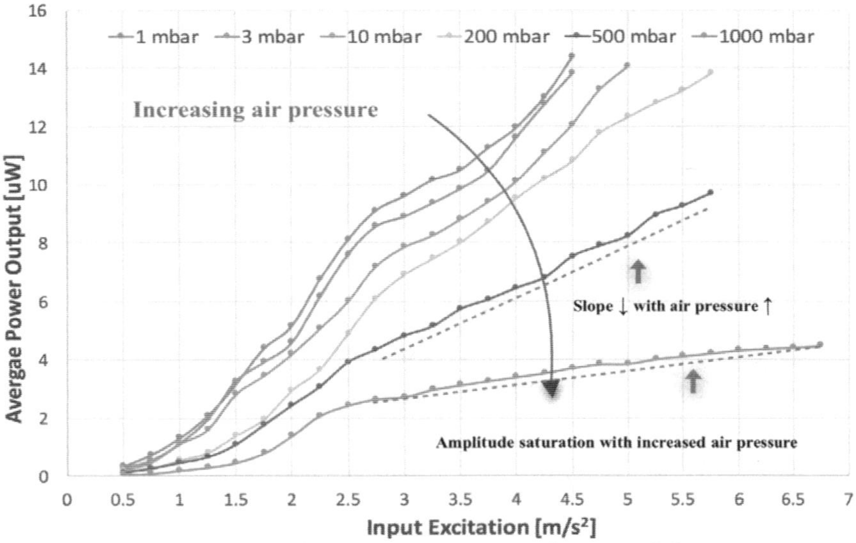

Figure 4. Excitation-Power output curve with sinusoidal excitation.

3.3 Parameters extraction

The relationship between power output under different air pressure levels and the quadratic damping coefficient in the nonlinear Duffing model can be established by nonlinear curve fitting using the Levenberg-Marquardt (LM) algorithm, described in [6]. The linear parameters of the cantilever in this study are measured and reported in [3]. The nonlinear duffing stiffness μ is derived from the experimental results at 1 mbar, considering the effect of nonlinear damping in rarefied regime negligible [7]. The parameters of the cantilever are listed in Table 1.

Table 1. Parameters of the PVEH

	w_n	ζ	μ
Value	215	0.0035	3.02×10^{-4}
Unit	Hz	1/s	N/m

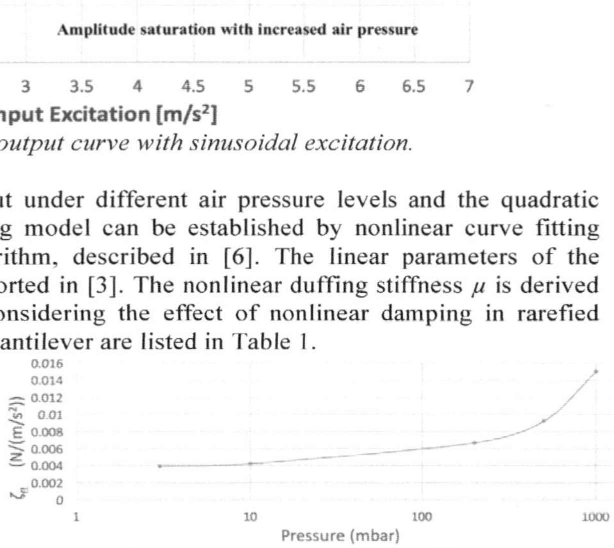

Fig.5 Quadratic damping coefficient VS air pressure

Fig.5 demonstrates the relationship between quadratic damping coefficient and air pressure level by fitting the excitation-power output response of the cantilever using the LM algorithm. As seen from the figure, the quadratic damping coefficient increases as the ambient pressure in the vacuum chamber increases, which in turn reduces the maximum displacement of PVEH for the same excitation level conditions.

3.4 Random vibration testing

Apart from the sinusoidal resonant response, Fig.6 further compares the average power output for PVEH under different air pressure subjecting to band-limited white noise vibration. Random excitation experiments are performed with band-limited white noise vibration input (10 Hz to 2 kHz) under varying air pressure from 1 to 1000 mbar to validate utilising nonlinear damping as a soft stopper.

Fig.6 (a) illustrates the time response of the cantilever when subjected to 7 g of band-limited white noise (0.025 g^2Hz^{-1}). The average power response for different air pressure were 76.63 nW, 64.48 nW, 35.07 nW, 28.85 nW and 13.39 nW, for air pressure at 1 mbar, 3 mbar, 10 mbar, 200 mbar, 500 mbar and 1000 mbar respectively, as shown in Fig.6 (b). The experiment demonstrates that increasing air pressure level can effectively reduce the maximum displacement of PVEH, reducing the impact of high frequency shock in a stochastic excitation scenario, which prevents onset of failure by fracturing and serves as a protective soft stopper for PVEH employed in harsh environments.

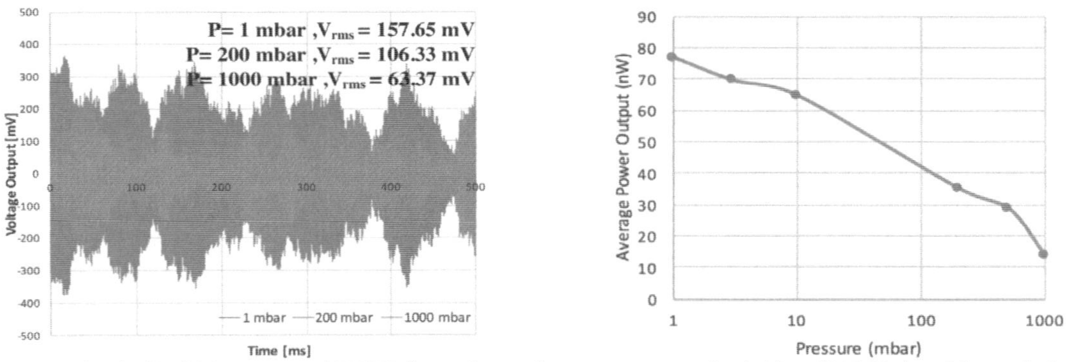

Figure 6(a) & (b). (a) Response of PVEH driven by random excitation with 0.15 g ,1000 Hz band-limited white noise. (b)Average power output with varying air pressure. When pressure increased, the output voltage diminished from 76.63 nW to 13.39nW, verifying air damping as soft stopper for random excitation environment

4. Design Guidelines

In order to prevent device from fracturing, the failure limit in terms of maximum displacement of PVEH is derived based on the fracture strength of silicon and PVEH geometry.

For a simple rectangular cantilever with length l subjecting to base excitation, the static maximum stress and deflection of the cantilever can be derived with Euler-Bernoulli beam equation [8]. However, in dynamic loading scenarios, the deflection and stress fluctuates with time, also nonlinear stiffness and damping of the cantilever contribute to the system dynamics.

If a sinusoidal excitation $P=Aw^2cos(wt)$ with amplitude $|Aw^2|$ is applied to the cantilever, the maximum deflection δ_l under different excitation amplitudes and air pressure levels can be solved numerically with the value of both linear and nonlinear parameters derived from prior experimental characterisation of the devices. The time response of the cantilever with nonlinear air damping is described in Eqn. 2, where $|Aw^2|$ is the amplitude of the fluctuating load, and x is the deflection of the cantilever.

Also under dynamic loading, the maximum bending stress of the cantilever as a function of maximum deflection of the cantilever can be approximated as [9]:

$$\sigma_{max} = \frac{EI}{Z}\delta''(0) = \frac{3EI}{Z}\frac{\delta_l}{l^2} \tag{3}$$

where E is the Young's Modulus of the cantilever, I is the moment of inertia and Z is the section modulus of the cantilever. As seen from Eqn. 3, the bending stress is a function of maximum deflection of the cantilever when it is subjected to dynamic loading. The failure limit is then defined as the maximum sustainable deflection of the cantilever before the maximum bending stress reaches the fracture strength of silicon, thus for a given PVEH, the failure limit can be determined by the geometry of the PVEH.

For a given environment, the excitation range can be defined as the maximum excitation amplitude of the cantilever before the cantilever reaches the failure limit. Based on the dimensionless model, the excitation range can be expressed as a function of nonlinear damping ζ_n when the failure limit is known, as shown in Fig.7. As seen from the graph, for the same excitation level, as nonlinear damping increases, the deflection of the cantilever decreases. As a result, the excitation range for the device can be extended as nonlinear damping is increased, meaning the device with higher nonlinear damping can be employed into environment with higher excitation range without the device reaching the maximum deflection and fractures with excessive dynamic stress.

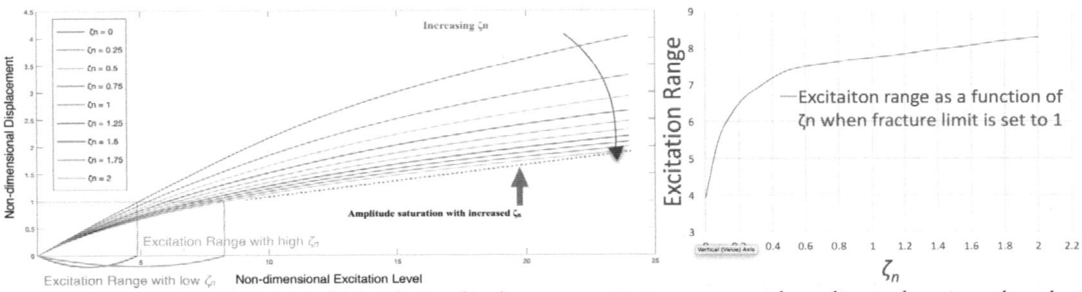

Figure 7(a)&(b). Schematic of (a) relationship between excitation range with nonlinear damping when the failure limit of PVEH is set to be 1 (b) Excitation range as function of ζn in the range between 0 to 2 calculated with the dimensionless model when the failure limit of PVEH is set to be 1, and failure limit of a PVEH based on the geometry of the device and fracture strength of silicon

5. Conclusion and Future work

While vacuum packaging is typically used in MEMS applications to minimise air damping, this paper presents a novel application of utilising air damping as a soft stopper for PVEHs. The results show that the excitation-power characteristic of the PVEH can be altered by varying the air pressure level. For larger excitation, higher air pressure reduces the power output of the PVEH, and the bounded PVEH displacement can prevent device failure by fracturing. As a result, the device is able to withstand higher excitation level before reaching the failure limit with higher air pressure, thus the range of input excitation can be extended at higher pressures, facilitating suitability for deployment in high excitation environments. The failure limit and excitation range of a rectangular cantilever with air damping subjecting to sinusoidal excitation is demonstrated in this study, as future work aims to calculate the optimized level of nonlinear damping based on device fracture limits and the excitation condition for a given device and deployment environment.

References:

[1] Cook-Chennault K, Thambi N, and Sastry AM 2008 *Smart Materials and Structures* **17(4)** p. 043001.
[2] Wang Z, Elfrink R, Rovers M, Matova S, van Schaijk R and Renaud M 2014 *Journal of Microelectromechanical Systems* **23(3)** p. 539-548.
[3] Jia Y and Seshia A A 2016 *Journal of Microelectromechanical Systems* **25(1)** p. 108-117.
[4] Ravindra B and Mallik A K 1994 *Physical Review E* **49(6)** p. 4950-4954.
[5] Hoerner S F 1965 *Fluid-dynamic drag: practical information on aerodynamic drag and hydrodynamic resistance*, Hoerner Fluid Dynamics.
[6] Lourakis M A 2005 *Foundation of Research and Technology* **4** p. 1-6.
[7] Sumali H 2007 *Journal of Micromechanics and Microengineering* **17(11)** p. 2231.
[8] Truesdell C and Noll W2004 *The non-linear field theories of mechanics*, in *The non-linear field theories of mechanics*, Springer. p. 1-579
[9] Suhir E 1997 *IEEE Transactions on Components, Packaging, and Manufacturing Technology* **20(4)** p.513-517.

MEMS ion source for mass spectrometer integrated on a chip

P Szyszka, T Grzebyk, A Górecka-Drzazga and J A Dziuban

Faculty of Microsystem Electronics And Photonics, Wroclaw University of Science and Technology, 11/17 Janiszewskiego Str., 50-372, Wroclaw, Poland

piotr.szyszka@pwr.edu.pl

Abstract. The paper describes silicon-glass MEMS electron impact ion source developed for miniature mass spectrometer (MS) integrated on a chip. The device consists of the field emission electron source with an electrophoretically deposited carbon nanotube cathode and ion beam formation electrodes. Ion source structure has been fabricated using MEMS technology. A complete manufacturing process of the test structures has been successfully elaborated and implemented.

1. Introduction

Miniaturization of mass spectrometers (MS) is a complex and difficult technical problem. So far, different individual modules of MEMS mass spectrometer have been presented [1-4], but up to now only one on chip integrated MS has been reported [5].

This paper focuses on design, fabrication and first tests of MEMS ion source that works on the base of electron impact process. The device is designed as a part of a concept of portable MEMS mass spectrometer (Fig. 1a). According to this concept, all components of the spectrometer: ion source, mass separator, detector and high vacuum micropump are made from silicon and glass, and are integrated on a single, hermetically sealed chip (Fig. 1b). Sample is introduced into ionization microchamber using microleak channels that are able to suppress ambient pressure to vacuum level [6] and is ionized in ion-source chamber. Created ions are introduced to the mass separator, where are sorted. In our concept all parts of MS are made as MEMS, including high vacuum (10^{-6} mbar) MEMS pump, recently developed by us [7].

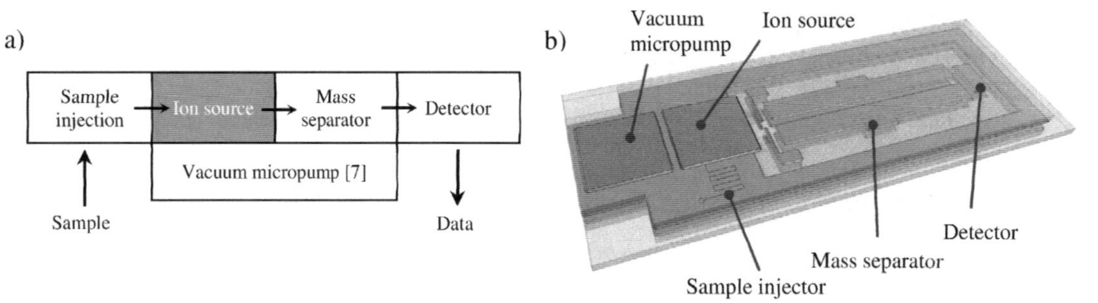

Figure 1. Author's concept of the MEMS mass spectrometer: a) block diagram of the device, b) schematic view

2. Ion source construction and technology

The ion source consists of several silicon/glass micromachined layers bonded anodically into a form of a sandwich (Fig. 2). Electrons field emitted from the carbon nanotube cathode are extracted by the silicon gate and collected by the silicon anode [8]. Electrons collide with gas molecules and ionize them. Ionized molecules are focused by repeller electrode and are intercepted by the ion collector, which directs them perpendicular in a relation to the electron beam.

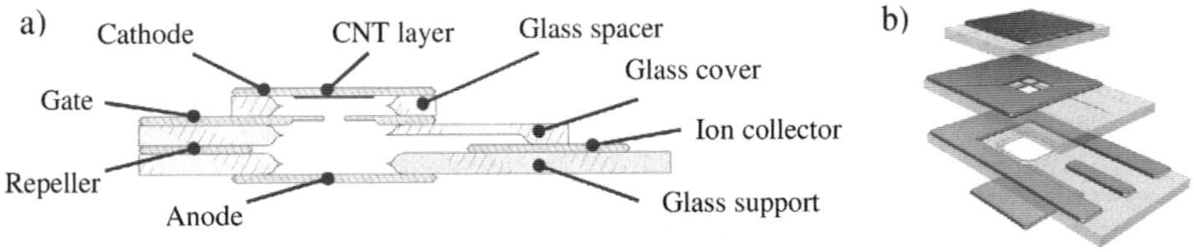

Figure 2. Ion source: a) schematic cross-sectional view, b) schematic extended view

Silicon parts of ion source are made from double-side polished, 400 μm thick (100) oriented monocrystalline wafer delivered by ITME, Poland. Glass components are micromachined in 1.1 mm thick Borofloat glass double side polished wafer of Shott, Germany. CNT emissive layer is formed electrophoretically from the material delivered by Cheap Tubes, USA.

Fabrication process (Fig. 3) starts from fabrication of glass elements, 3-inch Borofloat wafer is sewed into smaller parts (a). Next via holes are selectively etched in 40% HF solution through polymer laser-cut mask (b, c).

Following that a preform of ion beam formation electrodes is double-side micromachined in 10M KOH solution in 80 °C. Etch process forms 50 μm thick membrane (d). In the separate wet KOH etch process anode, cathode and gate are formed (e). The next step includes anodic bonding (380 °C, 1200 V) of ion beam formation electrodes and glass support (f), followed by final wet etch of the silicon membrane in 10M KOH in 80 °C (g). The silicon-glass sandwich is now bonded to glass cover (h). After that silicon gate is anodically bonded (i). Simultaneously 3x3 mm^2 CNT layer is deposited on the cathode (j) and finally all parts are bonded anodically in air (k).

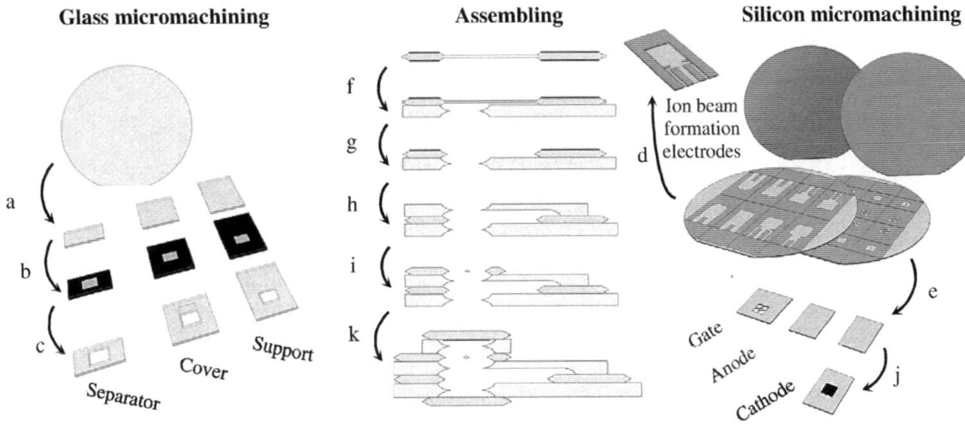

Figure 3. Basic steps of the ion source fabrication process

The ion source test structures with various ion formation electrodes geometries have been fabricated (Fig. 4). Overall dimensions of the fabricated structures are 21x13x5 mm^3, volume of ionization microchamber is approx. 0.04 cm^3. The distances cathode–gate and cathode–anode were 1.1 mm and 3.7 mm, respectively. Ion collector is placed 6 mm away from the center of the ionization microchamber.

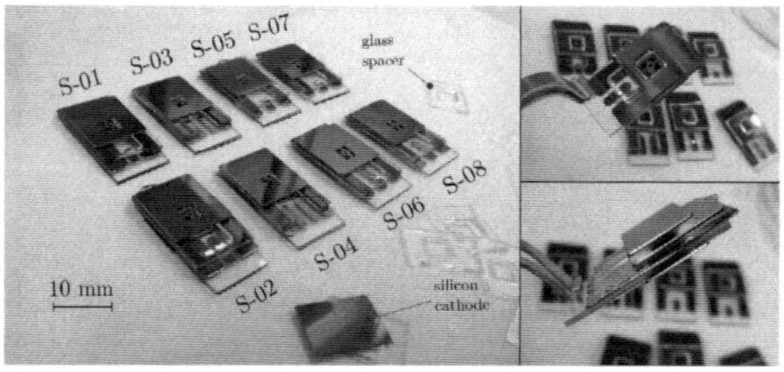

Figure 4. Fabricated ion source structures

3. Measurements

Tests of the ion source have been carried out inside reference vacuum chamber (up to 5×10^{-6} mbar). Firstly, field emission of electrons from CNT cathode was verified and *I-V* characteristics were plotted (Fig. 5a). Electron current of 0.1 µA starts at threshold voltage of 600 V, current of 400 µA was obtained for 1500 V. To inject electrons into ionization chamber the anode potential must be higher than the gate one (Fig. 5b), otherwise it acts as an ion collector and introduction of ions to the next module would be impossible.

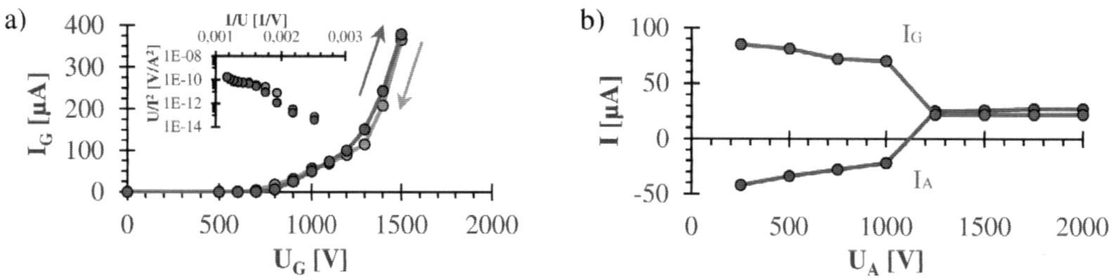

Figure 5. Results of the electron source measurements, current-voltage characteristics of: a) emission of electrons from the CNT cathode, in insert F-N curve, b) anode and gate currents vs. anode voltage, $U_G = 1200$ V, $p = 5\times10^{-6}$ mbar

As observed, the electron beam ionize gases inside microchamber and generates an ion current which is successfully drawn by the collector (Fig 6a). Ions start to be intercepted (1 nA) when collector voltage is at least −600 V. Collector current reaches value of 50 nA at potential of −1000 V.

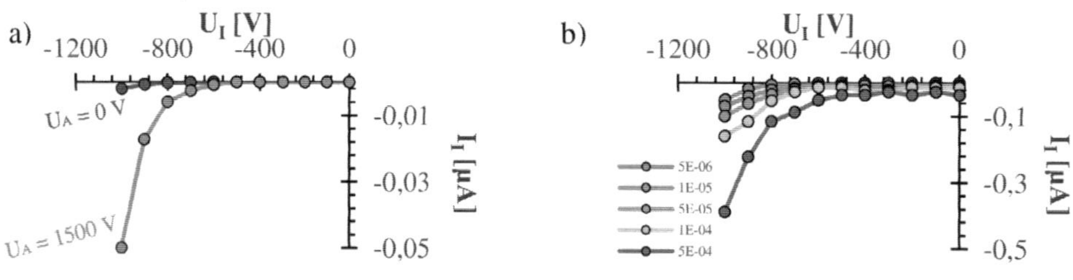

Figure 6. Results of the ion source measurements, current-voltage characteristics of ions intercepting by collector: a) for different anode voltage, $p = 5\times10^{-6}$ mbar, b) for different pressure levels

When bias polarization is applied on an anode, interception of ions by collector is very limited (up to 1 nA), all charged molecules are drawn by anode. Lower pressure inside chamber causes increase of ions that reach collector, up to 390 nA at pressure of 5×10^{-4} mbar (Fig. 6b).

Simulations of 3D ion source model, performed in COMSOL Multiphysics software confirm experimental results. Figure 7a showing electric field distribution documents that under these conditions repelling of ions is possible (Fig. 7b).

a) b)

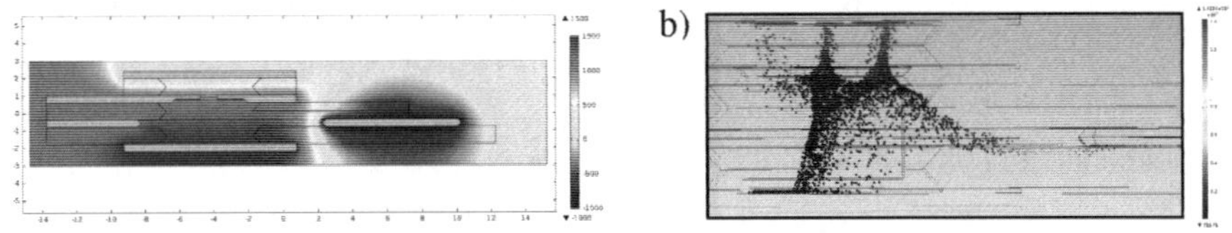

Figure 7. Computer simulations of ion source: a) equipotential lines (scale – [V]), b) ion interception by collector electrode (scale – ion velocity [m/s])

4. Conclusion

MEMS ion source for integrated mass spectrometer has been successfully elaborated. The source contains field electron emitter and ion formation electrodes. Satisfactory electron field emission (I_G = 400 μA for U_G = 1500 V, p = 5×10^{-6} mbar) and ionization of gases were obtained. Created ions were successfully directed into collector electrode (I_I = 390 nA for U_I = −1000 V, p = 5×10^{-4} mbar).

Acknowledgements

The work was partially financed by the Statutory Grant of Wroclaw University of Science and Technology and project number UMO-2013/09/B/ST7/01602 of the Polish National Science Centre.

References

[1] Wright S, Syms R R A, Moseley R, Hong G, O'Prey S, Boxford W E, Dash N and Edwards P 2010 MEMS-based nanospray-ionization mass spectrometer *J. Microelectromech. Syst.* **19** 1430–43

[2] Geear M, Syms R R A, Wright S and Holmes A S 2005 Monolithic MEMS quadrupole mass spectrometers by deep dilicon etching *J. Microelectromech. Syst.* **14** 1156–66

[3] Tassetti C, Mahieu R, Daniel J, Peyssonneaux O, Progent F, Polizzi J, Machuron-Mandard X and Duraffourg L 2013 A MEMS electron impact ion source integrated in a microtime-of-flight mass spectrometer *Sensors Actuators* B **189** 173–8

[4] Velasques-Garcia L F, Gassend B and Akinwande A 2010 CNT-based MEMS/NEMS gas ionizers for portable mass spectrometry applications *J. Microelectromech. Syst.* **19** 484–93

[5] Wapelhorst E, Hauschild J-P and Muller J 2007 Complex MEMS: a fully integrated TOF micro mass spectrometer *Sensors Actuators* A **139** 22–7

[6] Jamieson B G, Lynch B A, Harpold D N, Niemann H B, Shappirio M D and Mahaffy P R 2007 Microfabricated silicon leak for sampling planetary atmospheres with mass spectrometer *Rev. Sci. Instrum.* **78** 065109-8

[7] Grzebyk T, Górecka-Drzazga A, Dziuban J A 2014 Glow-discharge ion sorption micropump for vacuum MEMS *Sensors Actuators* A **208** 113–9

[8] Grzebyk T, Dziuban A J and Górecka-Drzazga A 2015 MEMS-type self-packaged field-emission electron source *IEEE Trans. Electron Devices* **62** 2339–45

A DESIGN STUDY OF A WIRELESS POWER TRANSFER SYSTEM FOR USE TO TRANSFER ENERGY FROM A VIBRATION ENERGY HARVESTER

N J Grabham[1], C Harden[2], D Vincent[2] and S P Beeby[1]

[1] Electronics and Computer Science, University of Southampton, Southampton, UK
[2] Perpetuum Ltd, Epsilon House, Southampton Science Park, Southampton, UK

njg@ecs.soton.ac.uk

Abstract. A wirelessly powered remote sensor node is presented along with its design process. The purpose of the node is the further expansion of the sensing capabilities of the commercial Perpetuum system used for condition monitoring on trains and rolling stock which operates using vibration energy harvesting. Surplus harvested vibration energy is transferred wirelessly to a remote satellite sensor to allow measurements over a wider area to be made. This additional data is to be used for long term condition monitoring. Performance measurements made on the prototype remote sensor node are reported and advantages and disadvantages of using the same RF frequency for power and data transfer are identified.

1. Introduction
This paper presents a case study into the integration of RF power transfer [1] with vibration energy harvesting which is being performed by the University of Southampton and Perpetuum Ltd. In the prototype integrated system energy is harvested using a Perpetuum vibration energy harvester and condition monitoring device [2] which is attached to the wheel bearing mount on a train, the existing Perpetuum sensor node is able to harvest more energy than it requires for its sensing and radio communications. This surplus harvested energy can then be used to wirelessly power a remote satellite sensor to increase the sensing capabilities of the system as a whole, with the remote measurements being transmitted back to the Perpetuum sensor node for onward transmission. The motivation for this novel approach is that in many deployment cases the optimum locations for both energy harvesting and sensing purposes are not coincident. In the rail bogie environment a cabled interconnection is impractical and an alternative route to provide power to the remote sensor is required. In this work this is achieved using RF power transfer at 915 MHz to transfer power over a distance of up to 1 m. The sensor measurements made by the remote satellite node are transferred back to the main sensor node over a single direction RF link operating at 915 MHz which shares an antenna with the RF energy transfer system.

2. System Specification
The first stage in the design was to draw up a list of key specifications of the wirelessly powered sensor node as these requirements will drive the design process, these specifications are as follows:

- Power to the remote sensor node is to be from harvested RF energy.

- Measured data from the remote sensor node will be transmitted to a local data sink over a unidirectional RF link from where it will be relayed over the existing data infrastructure.
- The power transmitter and local data sink will be connected to the existing Perpetuum vibration energy harvesters, and will be required to operate from harvested energy surplus to their needs.
- Power and data are to be transferred using a single antenna at each end of the power and data link over a distance of up to 1 m, with both power and data operating at circa 915 – 920 MHz.
- The wirelessly powered satellite sensor node is to store the harvested RF energy until it has sufficient to perform the desired measurements and then transmit the results to any available data sink.
- Target measurement and reporting rate is to transmit from the satellite sensor node at least once an hour to permit tracking of long term changes over time.
- Initially the satellite sensor is to measure the local temperature, with scope for adding further analogue sensors for other monitoring purposes.

From these requirements it can be seen that, as with most energy harvesting applications, the first consideration must be the total energy requirements of the satellite sensor node. Within the sensor node the highest energy consuming item is expected to be the wireless data communications, therefore the device selected should be as low power as possible whilst still able to achieve the anticipated communications range. Other components such as microprocessors and sensors should be selected to be low power, ideally with very low power sleep modes so that their power consumption during inactive periods is minimized. To operate the node from the harvested RF energy it is necessary to store the harvested energy until sufficient is available to operate the node. This energy store, typically a supercapacitor, should be sized such that it can contain sufficient energy to meet the node's requirements but is not excessively sized. Too much capacity will cause the node to take longer to harvest sufficient energy to reach the required start-up voltages.

3. System Design and Implementation

Given the outline requirements that have been established from the initial specification the design and implementation of the remote sensor node can be performed. Initially a block level approach is taken to designing the system, with the actual functional implementation following on from this stage.

3.1. System Block Level Design

The block level design of the wirelessly powered remote sensor is shown in Figure 1, this design consists of seven main function blocks whose functions are as follows:

Patch Antenna: This is used both to harvest the transmitted RF energy and for communication back to the data sink attached to the main Perpetuum wheel vibration energy harvester.

RF Switch: The purpose of the RF switch is to isolate the RF energy harvester from the patch antenna during data transmission so that the RF energy associated with the transmission is not directly re-absorbed by the harvester but is radiated out to provide the communications link.

RF Energy Harvester: This section of the design takes the received RF energy, converts it back to electrical energy and then manages the storage and voltage regulation of this harvested energy to provide a suitable power supply to operate the rest of the sensor node's electronics.

Energy Storage: The role of the energy storage is to store the energy harvested from the RF source until a sufficient quantity has been harvested to enable operation of the sensor node.

Low Power Microprocessor: The microprocessor serves to control the operation of the sensor node, perform the measurements from the sensors and format the data ready to transmit using the RF transceiver module at the appropriate time.

Sensors: The initial sensing capability for the node is a single temperature sensor which measures the environmental temperature in the vicinity of the node. The node also has the capability to add two further, off-board, analogue sensors to add further sensing capacity.

RF Transceiver: The RF transceiver serves to take the data package prepared by the microprocessor and transmit it to the data sink attached to the RF energy source transmitter and main Perpetuum wheel-mounted vibration energy harvester.

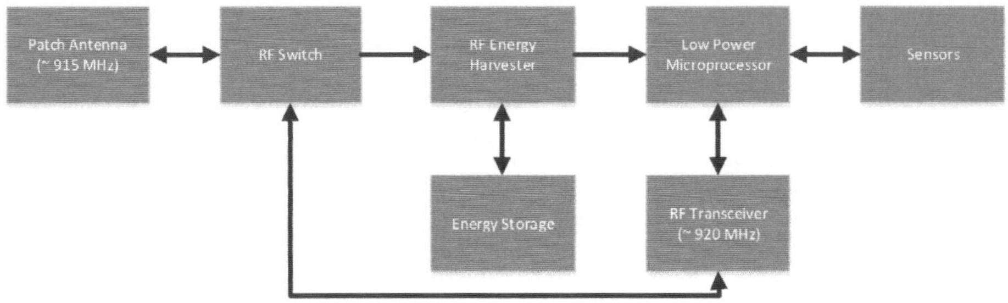

Figure 1. Block level design of remote sensor node.

3.2. *Hardware Implementation*

To implement the prototype system design it was decided to use pre-manufactured modules for the RF energy harvester and radio communications. The RF to DC harvesting and conversion being achieved with a Powercast P2110B module, which also provides a DC-DC boost convertor to boost and regulate the voltage from the energy storage element to the level required by the rest of the node. The radio data communication link is realised using an LPRS eRA900TRS Easy Radio module which takes serial data from the microprocessor and encodes and transmits it.

The RF switch is implemented as a normally closed switch, so when it is not powered it directs the incoming RF energy to the harvesting circuity, when the RF data transmission is desired it is then powered so that it opens and isolates the harvesting circuitry from the radio module and antenna.

The energy storage is implemented using a low-ESR double layer supercapacitor.

The low power microprocessor used is an XLP device from Microchip which has a very low power consumption sleep mode and the ability to gate the power to all of the internal function modules, such as the ADC, so that they are only powered when required. The microprocessor also controls the power supply to the radio module and the sensors so that they are only powered when they are needed.

The realised prototype wirelessly powered sensor node is shown in Figure 2, the patch antenna is not shown but connects to the SMA connector on the board and provides both harvesting and communication functionality.

Figure 2. Prototype satellite sensor.

3.3. *Software Implementation and Satellite Sensor Operation*

The operation of the remote sensor node is controlled by the software programmed into the microprocessor, the steps in the node operation are as follows:

The node harvests sufficient RF energy until it has sufficient stored to enable full operation of the sensor node. The microprocessor is first powered up (sensors and radio module remain off) and the microprocessor sleeps in a low power mode until there is a pause in the RF power transmission. The microprocessor then wakes up, powers up the sensors and measures their values, after which they are then powered down again. A measurement is also made of the internal band-gap reference to scale the measured sensor values. The measured data is then formed into a data packet, with a node identification string and a checksum. The RF energy harvester is isolated from the antenna and the radio powered up and the data packet transmitted. After transmission, the radio is powered down again, the RF energy harvester is reconnected to the antenna, the microprocessor shuts itself down and the node returns to harvesting and storing energy ready for the cycle to repeat.

4. Satellite Sensor Node Performance

A Powercast RF energy 3W source was used to provide the RF energy to test the operation of the prototype satellite sensor node and a local data sink consisting of an Arduino Due and an LPRS Easy Radio module connected to a terminal window on a PC to display the received data. Using the experimental setup it was verified that the sensor node would harvest energy from the RF source and charge the energy storage supercapacitor. Once the supercapacitor was sufficiently charged the DC-DC boost converter commenced operation and the microprocessor was powered up. Once powered up the microprocessor was observed to remain in the low sleep mode and periodically check if the RF energy source was still present. When the RF source was switched off the microprocessor commenced its measurement cycle, followed by transmitting the data package which was received and decoded by the local data sink.

During the node's operation its total energy consumption was measured, a full measurement cycle consisting of the three ADC input channels and the band gap reference, followed by transmitting the RF data packet twice with a 2 s interval between the transmissions, requires 143 mJ of energy over 2.25 s. This figure includes all the operational overheads such as the DC-DC convertor loses.

The DC-DC boost convertor operates with the supercapacitor voltage between 1.02 V and 2.3 V. With a 3 W source the time taken for the supercapacitor to charge over this range was 52 s. Based on the energy used by the RF transmitter a source of ~50 mW should be able to fully charge the supercapacitor in around 37 minutes. In use the supercapacitor does not get fully discharged before the measurement and transmission cycle completes, so the charging time will be further reduced.

5. Discussion and Conclusions

The use of the same frequency for both the energy transfer and data communications has both advantages and disadvantages. The advantages include reduced node size and cost due to only needing one antenna and the ability to harvest energy from data communication between other nodes and sensors operating at the same frequency. The main disadvantage is that the RF energy transmission can block the data transmission and therefore requires the energy broadcast to be paused periodically to permit the remote node to transmit its data. In a system with multiple RF energy sources these pauses may need to be synchronised. In this work the target application requires that the satellite node be as compact as possible so the advantage of a single antenna outweighs the inconvenience of having to the pause the RF energy transmission.

Ongoing work in the project is aiming to further reduce the energy usage of the satellite sensor node through the use of an integrated microprocessor and RF transceiver with enhanced power control over the RF system, the revised system will then be tested in its target deployment environment and the effects of this environment on the propagation of RF energy investigated.

References

[1] Visser H J and Vullers R J M 2013 RF Energy Harvesting and Transport for Wireless Sensor Network Applications: Principles and Requirements, *Proc. IEEE* **101** 1410–1423

[2] Perpetuum Ltd *Condition Monitored Maintenance using self-powered sensors & vibration based analytics* Online: www.passenger-rolling-stock-maintenance.com/media/downloads/inline/perpetuum-case-study.1449847586.pdf Accessed September 2016

Acknowledgements

The work presented in this paper was performed as part of the Innovate UK funded project "Wireless Autonomous Rail Network Sensing Systems", Project No. 102157.

Effect of packaging pressure on energy harvesting from vibration source

Tianyang Wang[1], Yulong Zhang[1,2], Xingwei Chen[1], Anxin luo[1] and Fei Wang[1,2]

[1] Dept. Electrical and Electronic Engineering, Southern University of Science and Technology, Shenzhen 518055, China

[2] Shenzhen Key Laboratory of 3rd Generation Semiconductor Devices, Shenzhen 518055, China

E-mail: wangf@sustc.edu.cn, wangty@mail.sustc.edu.cn

Abstract. In this paper, we have studied the effect of air pressure damping on the vibration-based energy harvesting devices. The device we used is an electrostatic energy harvester with an out-of-the-plane gap closing scheme. A broadened bandwidth is observed at 4 Pa when the acceleration is increased from 0.5 m/s^2. A power output of 2.2 μW is achieved at the air pressure of 4 Pa when a low acceleration of 0.5 m/s^2 is applied, while the power output is only 0.02 μW at atmosphere with an acceleration of 1.2 m/s^2. Detailed study is made to investigate the waveform of the displacement and the voltage output. Nonlinearity is observed at low air gas pressure about 239 Pa and it is more significant at 4 Pa.

1. Introduction

Recently, energy harvesters has been developed a lot because of the incomparable highlights as an alternative choice to replace the button cells in the wireless sensor networks, micro systems and in-body health monitoring devices [1-2]. Among various energy harvesters, the vibration-based energy harvesting devices (VEHD) use the vibration-to-electricity conversion methods to convert kinetic energy to electric energy. VEHD based on the piezoelectric [3-6], electromagnetic [7-9] and electrostatic methods [10-17] have been developed during the past years. For most of the VEHD, the performance of the devices has been generally studied with a few governing equations [18]. However, there is not much attention has been paid for the damping force in the energy harvesters. In this paper, we will study how the air pressure affects the performance of VEHD.

To evaluate the air pressure effect, we have designed an electrostatic VEHD with an out-of-the-plane gap closing scheme, as shown in Figure 1. The device can be simply regarded as a variable capacitor with parallel electrodes on two separate plates as shown in Figure 2. When the proof mass is driven by an ambient vibration source, induced charge caused by the pre-charged CYTOP electret moves back and forth between the two electrodes, which generates a current through an external load. The overall size of the device we used is about 13×18 mm^2, and the surface potential of the electret is charged to about -400 V. During the vibration, the squeeze air damping force would affect the mechanical performance of the proof mass and therefore influence the output power.

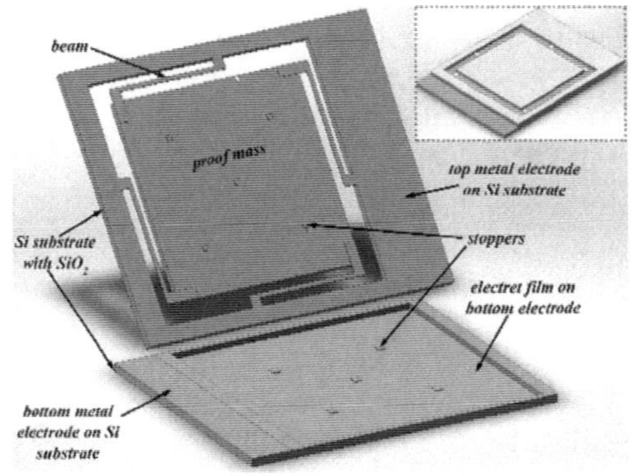

Figure 1. 3D schematic of the out-of-plane electrostatic energy harvesting device.

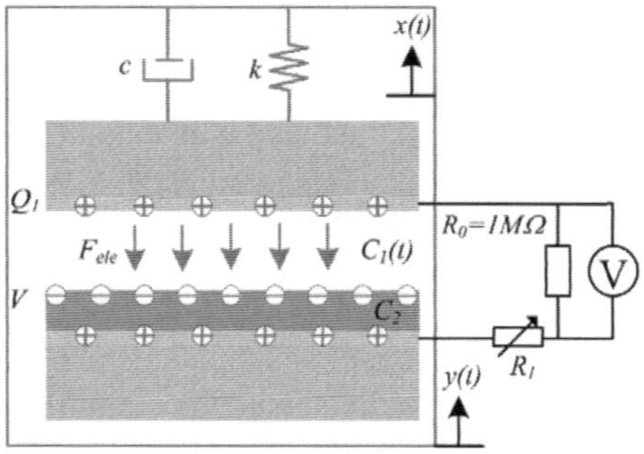

Figure 2. Schematic of the energy transduction between the mechanical system and the electric circuit with external

2. Characterization

The effect of air pressure on the performance of the VEHD is characterized with the equipment setup shown in Figure 3. It consists of a chamber where the pressure can be controlled with a pump and monitored with a pressure sensor. A shaker is put in the chamber to mimic the vibration source as described in [19]. A laser detector is used to monitor the displacement of the proof mass during the vibration. By tuning the air pressure, we can characterize the device performance under various damping conditions.

The performance test circuit is shown in Figure 2. The output power can be calculated by measuring the voltage across a fixed resistance $R_0=1\ M\Omega$ while the overall optimal load resistance is $17\ M\Omega$. The whole experiment to characterize the air pressure effect of VEHD can be divided into two parts to achieve convincing results.

Figure 3. The measurement setup for the air damping effect.

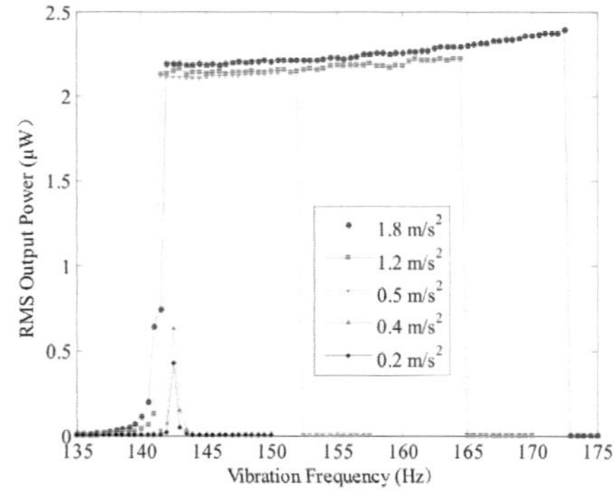

Figure 4. The relationship between the vibration frequency and the output power of the energy harvester at a low air pressure of

2.1. Bandwidth at low air pressure

This part of the experiment starts with pumping down the ambient gas pressure in the chamber. Then a sinusoidal signal is given from the signal generator (Brüel&Kjær, LAN-XI 3160) and a power amplifier (Brüel&Kjær, 2719) to excite the shaker and the device.

The result is shown in Figure 4. Five different accelerations are applied on the VEHD at a low air pressure of 4 Pa. It is clear that a maximal power of 2.2 μW is achieved at low acceleration of 0.5 m/s^2 and a bandwidth is continuously broadened when the vibration amplitude is increased from 0.5 m/s^2 to 1.8 m/s^2, which could be interesting for energy harvesting from random vibration sources. A broad bandwidth of about 30 Hz (from 142 Hz to 172 Hz) has been observed at an acceleration of 1.8 m/s^2. This is because large vibration amplitude could be easily reached when low air pressure is applied in the chamber. As it is shown in Figure 7(a), a maximum displace of the proof mass is measured up to 250 μm, which is exactly the gap between the stoppers of the two plates.

2.2. Output power with changing air pressure

The energy loss caused by air pressure damping can be clearly seen in this part. When the chamber pressure is increased, higher vibration amplitude is needed to achieve the same power, as shown in Figure 5.

The air damping effect can be more clearly seen in Figure 6. With a fixed vibration amplitude of 1.2 m/s^2, the RMS power of the device at resonance decreases gradually from 2.2 μW to about 0.02 μW when the chamber pressure increases from 4 Pa to atmosphere.

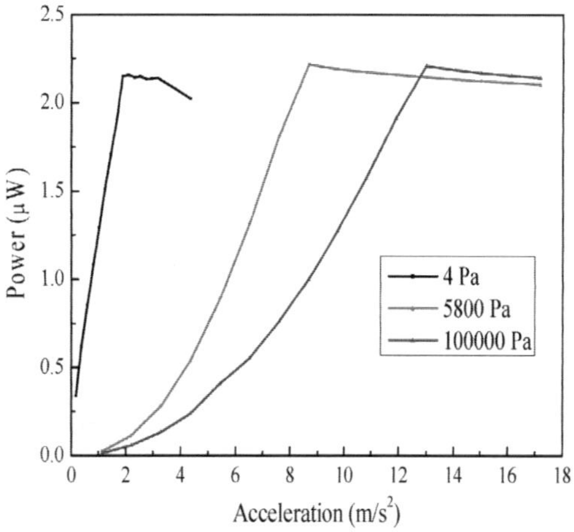

Figure 5. The output power of the energy harvester with increased acceleration amplitude under different air pressure of 4 Pa, 5800 Pa and 10^5 Pa, respectively.

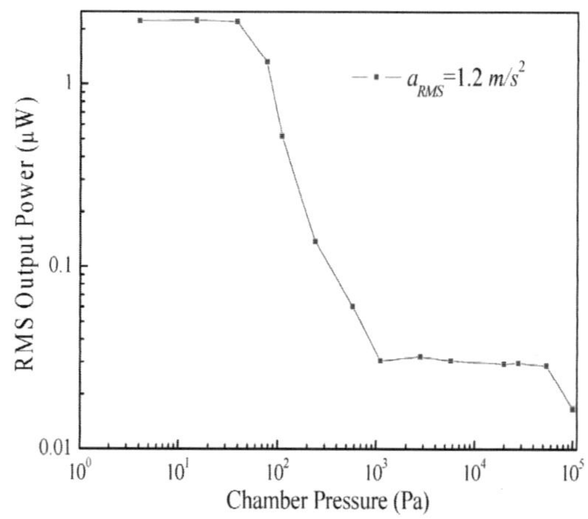

Figure 6. The output power of the energy harvester at resonance with chamber pressure increased from 4 Pa to atmosphere.

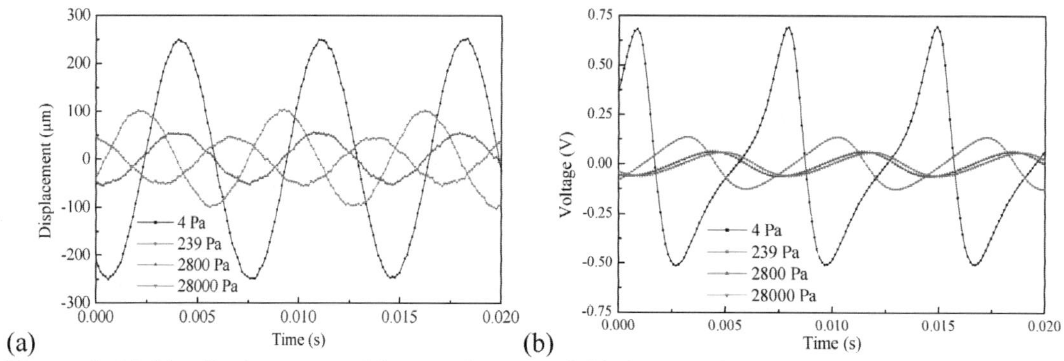

Figure 7. (a) The displacement of the proof mass and (b) the waveform of the voltage output on R_0 at the same vibration amplitude of 1.2 m/s^2 when the chamber pressure is tuned to 4 Pa, 239 Pa, 2800 Pa and 28000 Pa, respectively.

A more detailed study has been done to investigate the waveform of the displacement and the voltage output when the chamber pressure is increased, as shown in Figure 7. The displacement amplitude of the proof mass decreases/stabilizes to 55 μm under a chamber pressure of 2800 Pa, and the voltage output amplitude increases from 0.1 V to 0.65 V. Nonlinearity can be observed at the ambient gas pressure of 239 Pa, and it is more obvious at 4 Pa. The nonlinearity effect is mainly due to the fact that the capacitance between the two electrodes is inversely proportional to gap distance [16]. Furthermore, when the two plates of device get close enough, the electrostatic force will influence the output voltage.

3. Conclusion
A primary research was made to find the effect of air pressure damping on VEHD. A broadened bandwidth can be observed when the acceleration is increased from 0.5 m/s^2. At a low air pressure of 4 Pa, maximum power output of 2.2 μW can be achieved with a low acceleration of 0.5 m/s^2. When the air pressure increases up to atmosphere, the power output drops down to 0.02 μW. When the ambient gas pressure is low enough, the waveform of the voltage output will become nonlinear for the variation of the capacitance.

Acknowledgement
This work is supported by National Natural Science Foundation of China (Project No.: 51505209) and University Research Committee of SUSTech (Project No.: FRG-SUSTC1501A-31). Fei Wang is also supported by Guangdong Natural Science Funds for Distinguished Young Scholar (Project No.: 2016A030306042).

References
[1] S. P. Beeby, M. J. Tudor, and N. M. White, Meas. Sci. Technol., 17 (2006), pp. R175-R195.
[2] P. D. Mitcheson, E. M. Yeatman, G. K. Rao, A. S. Holmes, and T. C. Green, P. IEEE, 96 (2008), pp. 1457-1486.
[3] L. Tang and Y. Yang, Appl. Phys. Lett., 101 (2012), 094102.
[4] S. Li, Z., A. Zhang, and F. Wang, AIP Adv., 6 (2016), pp.015019.
[5] Q. Shi, T. Wang and C. Lee, Sci. Rep., 6 (2016), pp. 24946.
[6] Q. C. Tang and X. X. Li, IEEE/ASME Trans. Mech., vol. PP, no. 99, pp. 1-7, Jan. 2014.
[7] S. P. Beeby, et.al., J. Micromech. Microeng., 17 (2007), pp. 1257–1265.
[8] M. Wischke, M. Masur, F. Goldschmidtboeing, and P. Woias, J. Micromech. Microeng.,, Volume 20, Number 3, 035025 (2010).
[9] I. Sari, T. Balkan, and H. Kulah, Sensor Actuat. A-Phys, 145–146(2008), 405.
[10] T. Takahashi, M. Suzuki, T. Nishida, Y. Yoshikawa, and S. Aoyagi, in Proc. MEMS 2015, pp. 1145–1148, 2015.
[11] T. Tsutsumino, Y. Suzuki, N. Kasagi, and Y. Sakane, Proc. MEMS 2006, pp. 98–101, 2006.
[12] Y. Xu, et.al., Micro & Nano Letters, 2016, DOI: 10.1049/mnl.2016.0336.
[13] F. Wang, C. Bertelsen, G. Skands, T. Pedersen, and O. Hansen, Microelectronics Engineering, 97 (2012), pp. 227–230.
[14] A. Crovetto, F. Wang, and O. Hansen, J. Micromech. Microeng., 23 (2013), 114010 (10pp).
[15] F. Wang, and O. Hansen, Sensor Actuat. A-Phys, 211 (2014), pp.131–137.
[16] S. Boisseau, G. Despesse, T. Ricart, E. Defay, and A. Sylvestre, Smart Mater. Struct., 20 (2011), p. 105013.
[17] Y. Suzuki, IEEJ T. Electr. Electr., 6 (2011), pp. 101–111.
[18] A. Crovetto, F. Wang, and O. Hansen, J. Microelectromechan. Sys., 23 (2014), pp. 1141-1155.
[19] S. Li, et.al., Sensor Actuat. A-Phys, 247 (2016), pp. 547–554.

TIMING A PULSED THIN FILM PYROELECTRIC GENERATOR FOR MAXIMUM POWER DENISTY

A.N. Smith[1,3], B.M. Hanrahan[2], C.J. Neville[2] and N.R Jankowski[2]

[1]U.S. Naval Academy, Annapolis, Maryland, USA, 21402
[2]U.S. Army Research Laboratory, Adelphi, Maryland, USA, 20783

Abstract. Pyroelectric thermal-to-electric energy conversion is accomplished by a cyclic process of thermally-inducing polarization changes in the material under an applied electric field. The pyroelectric MEMS device investigated consisted of a thin film PZT capacitor with platinum bottom and iridium oxide top electrodes. Electric fields between 1-20 kV/cm with a 30% duty cycle and frequencies from 0.1 – 100 Hz were tested with a modulated continuous wave IR laser with a duty cycle of 20% creating temperature swings from 0.15 - 26 °C on the pyroelectric receiver. The net output power of the device was highly sensitive to the phase delay between the laser power and the applied electric field. A thermal model was developed to predict and explain the power loss associated with finite charge and discharge times. Excellent agreement was achieved between the theoretical model and the experiment results for the measured power density versus phase delay. Limitations on the charging and discharging rates result in reduced power and lower efficiency due to a reduced net work per cycle.

1. Introduction

Pyroelectric thermal-to-electric energy conversion is accomplished by a cyclic process of thermally-inducing polarization changes in the material under an applied electric field. The energy density of this process depends on the strength of the applied field and the amount of polarization change realized from a given temperature change, described by the pyroelectric coefficient of the material. High power density is achieved by frequent cycling, which is easy to realize with a low thermal mass, thin film active material. Only recently have these devices been operated at the frequencies required to produce reasonable power densities [1]. The synchronization of heating/charging becomes critical when thermodynamic cycles are realized on real systems [2]. In this work, we examine the timing of the various processes within the pyroelectric energy conversion (PEC) cycle in order to maximize the power density of a thin-film pyroelectric generator.

The PEC cycle consists of charging the pyroelectric film, heating the material under an applied field, discharging the material while hot, then cooling the material back to the original state (figure 1). There are variations in the specifics of these processes that will strongly influence cycle efficiency and energy densities. For example, the difference between an Ericsson and Brayton cycle is whether the charging processes are performed isothermally or adiabatically, respectively. It has been hypothesized that the Brayton cycle, which includes instantaneous adiabatic charging is ideal for thin film systems [3], while the Ericsson cycle is commonly accepted as the best cycle for bulk pyroelectrics [4-6]. Practically, there

[3] Corresponding Author <ansmith@usna.edu>

are resistive losses during the charge/discharge processes that scale deleteriously with cycle frequency and applied field. The charge/discharge losses will ultimately limit the rise time of the electric field, preventing the realization of an adiabatic processes and increasing the duty cycle of the electric field. The ideal duty cycle for both the heat addition and electric field is very short for two reasons (1) thermal losses continue during the heating event since the device remains in contact with the substrate and (2) leakage losses increase when the system is held charged [7].

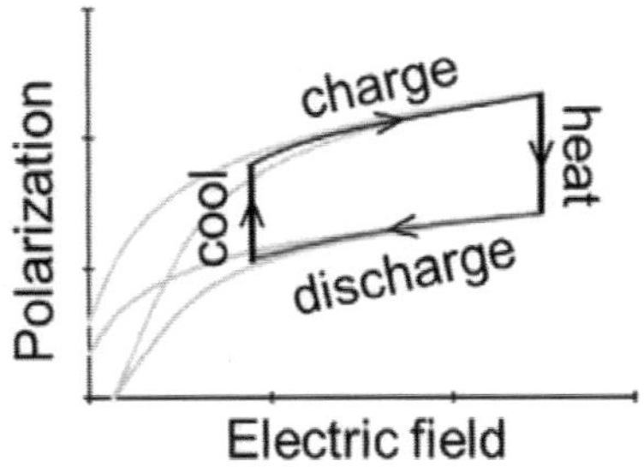

Figure 1. Pyroelectric energy conversion cycle where the enclosed area represents the energy density of the cycle.

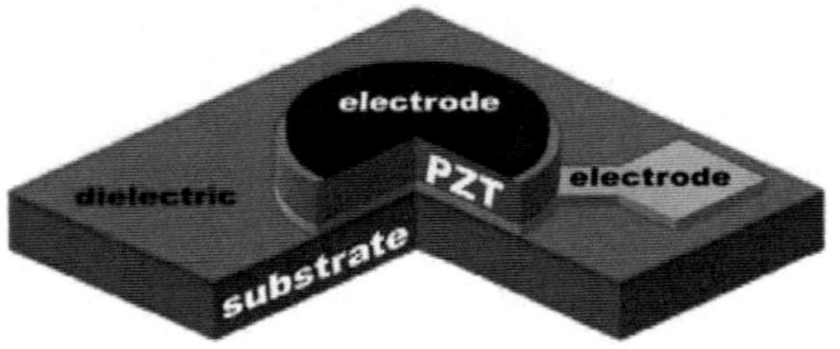

Figure 2. The pyroelectric MEMS device utilizing thin film 60:40 PZT capacitors with platinum (bottom) and IR-absorbing iridium oxide (top) electrodes.

2. Experimental Results

The pyroelectric MEMS device used in this investigation consisted of thin film lead zirconate titanate (PZT) with a 60:40 ratio of Zr:Ti cations in a capacitor structure with platinum bottom and iridium oxide top electrodes (figure 2). Figure 3 shows experimental data that illustrates the cycle timing, showing both the applied electric field responsible for charging and discharging the device along with the transient temperature response to the heating event. The temperature response is determined from the pyroelectric current where the pyroelectric coefficient has been independently measured.

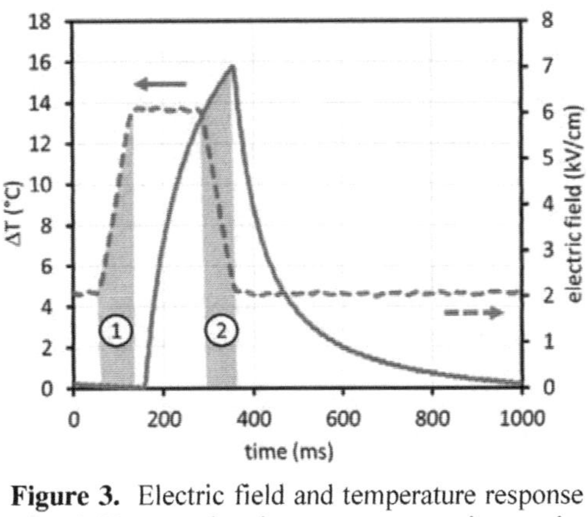

Figure 3. Electric field and temperature response of a 1 Hz pyroelectric energy conversion cycle. Regions 1 and 2 represent the times when the cycle deviates from ideal operation.

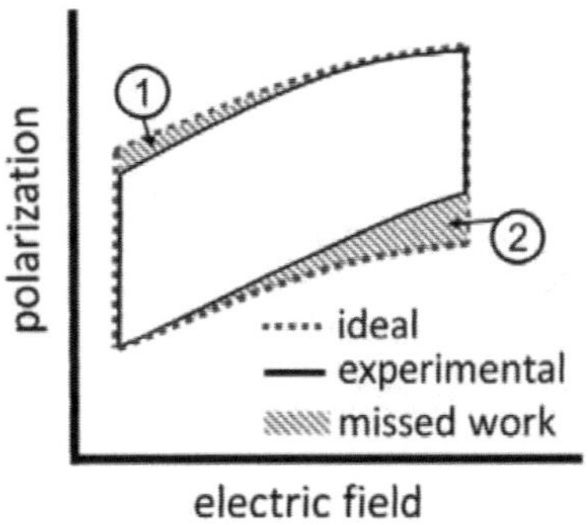

Figure 4. The time required for charging and discharging lowers the specific work of the cycle. Regions 1 and 2 represent deviations from ideal, also shown in figure 3.

The electric field had a duty cycle of 30% and was applied with a rise time that represented 7% of the cycle period while the laser heating duty cycle was 20%. Practically the substrate remains in thermal contact with the device. During the laser heating period the system exponentially approaches a steady state "on" value, and once the heat source is removed the system exponentially approaches the substrate temperature. The thermal time constant of the device is a function of the thermal mass and the thermal impedance between the device and the substrate; and governs the performance of the device at high frequencies [3]. The thermal time constant for the device in this study was determine to be 100 ms by comparing the measured temperature response shown in figure 3 to a thermal model for the heating and cooling. Figures 3 & 4 show how the finite rise leads to a reduction in net work per cycle. Ideally cooling would only occur at the minimum electric field and heating would only occur at the maximum electric field.

Figure 5 shows experimental results with phase delays of 25°, 30° and 45° between the initiation of charging and heating. The pyroelectric current, applied electric field, and laser power are shown along with a plot of the measured polarization versus electric field for the adjacent timing, where the lost work is illustrated by the shaded area. At a 25° delay the lower left corner is shaded because the system temperature is allowed to decrease before the electric field has been fully lowered. At a 45° phase delay heating continues after the field has been removed, therefore less heating has occurred while the field was raised reducing the length of the heating line and cooling lines on the right and left side respectively. The optimal power density occurs around 30° phase delay, which roughly aligns the peak temperature with the end of the electric field pulse due to the 10% difference in duty cycles.

3. Thermal Model

The thermal model based on a simple heating/cooling lumped capacitance model was useful in explaining the behavior observed in the polarization plots as shown in Figure 5. The predicted transient temperatures were combined with the electric field values and used to predict the polarization based on the assumption of a constant pyroelectric coefficient, $p = -0.03$ μC/cm^{2}K and dielectric constant, $\varepsilon = 0.2$ μC/kV·cm. Figure 6 shows the predicted polarization versus electric field at different phase delays, where the same shaded regions seen in Figure 5 are evident and a phase delay of 36° has the highest net work per cycle.

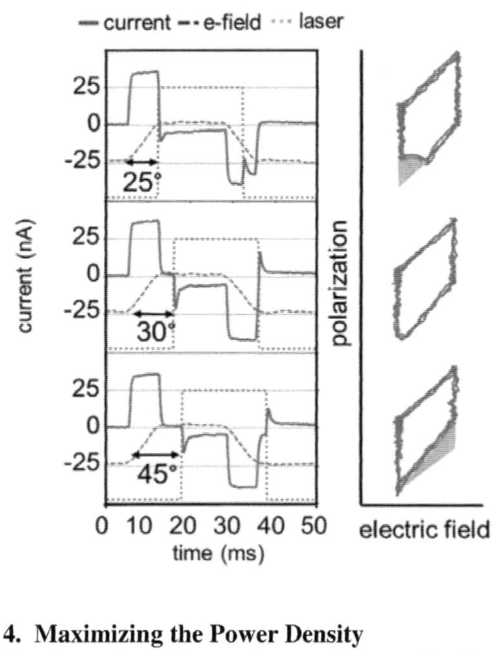

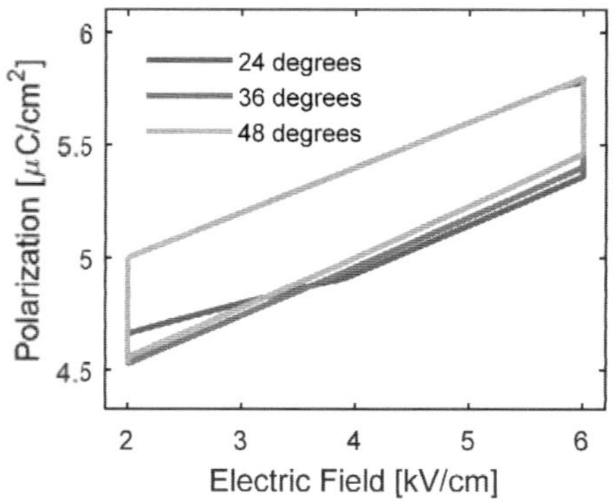

Figure 6. (Above) Polarization versus electric field for the 24°, 36° and 48° phase delays where 36° has the highest energy density.

Figure 5. (Left) 10 Hz pyroelectric energy conversion cycles with various phase lag between charging/discharging and heating/cooling cycles

4. Maximizing the Power Density

The thermal model was then used to examine the impact of the charging rates on the net work per cycle. Figure 7 shows the optimal phase delay of the heating pulse where the solid line represents the phase delay that would align the end of the laser and electrical field pulses. Given a laser and electric field duty cycle of 20%, instantaneous charging requires a 0% rise time while the minimum charging rate is given by a triangular pulse for a 20% rise time. Increasing the rise time reduces the net work per cycle.

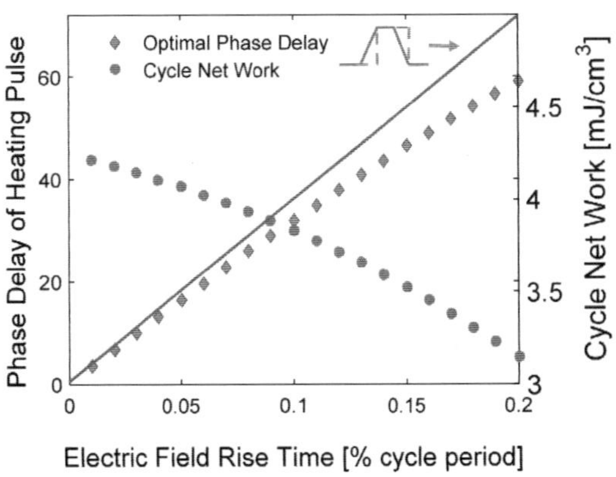

Figure 7. Optimal Phase delay of heating pulse and net cycle work over a range of electric field rise times (charging rates) for a 1 Hz cycle

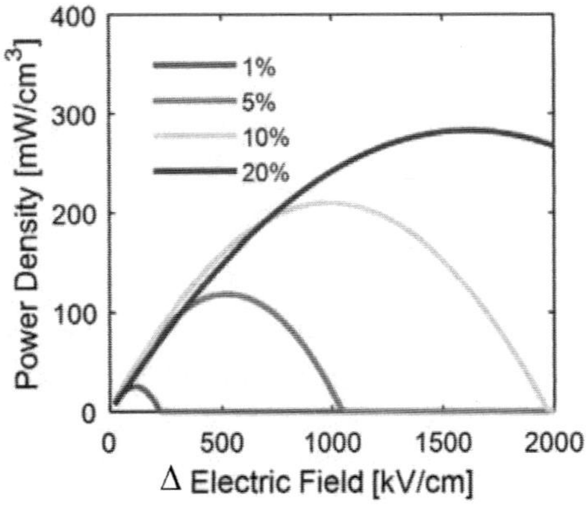

Figure 8. Calculated power density for 1 Hz cycles over a range of applied fields at different charging rates, where C=3 nF and $\tan\delta = 0.1\%$

With the optimum phase for each rise time established, the electrical power loss during the charge and discharge cycles can be added to the model to find a maximum power density for a give applied field and charging rate. The electrical power loss is calculated as,

$$P_{loss} \approx \frac{C\Delta V^2 f \tan\delta}{\pi(\%t_r)} \qquad [1]$$

where C is the capacitance, ΔV is the change in voltage, f is the frequency, $\tan\delta$ is the dissipation factor, and $\%tr$ is the faction of the cycle period used to linearly increase the electric field. Depending on the values of $\tan\delta$, it was unexpectedly found that longer rise times can be advantageous despite a reduction in the net work per cycle.

5. Conclusions

Dissipative losses that occur during charging and discharging can place a practical limit on the rise time of the electric field and hence the charging current. Finite rise times lead to a situation where the phase delay must be optimized to produce the maximum net work per cycle and power density. If the heating pulse terminates too early then the temperature drops while the electric field in still rising and potential work is lost. If the heating pulse terminates too late, heating occurs while the electric field has been reduced which is wasteful and lessens the portion of the heating pulse that is actually utilized. Longer rise times may have the potential to be more power dense due to reduced charge/discharge losses.

References

[1] B. Bhatia, A. R. Damodaran, H. Cho, L. W. Martin, W. P. King, *J. of App. Phys,* **116**, 194509, 2014

[2] B. Hanrahan, C. Neville, A. Smith, N. Ter-Gabrielyan, N. Jankowski, and C. Waits, *Advanced Mat. Tech.*, **1**, October 2016

[3] N. Jankowski, A. Smith, B. Hanrahan, *Proc. Of ASME SHTC 2016*, Washington, D.C. , July 10-14, 2016.

[4] S. P. Alpay, J. Mantese, S. Trolier-McKinstry, Q. Zhang, R. W. Whatmore, *MRS Bulletin*, **39**, 1099, 2014

[5] R. B. Olsen, D. A. Bruno, and J. M. Briscoe, *J. of App. Phys,* **58**, p. 4709, 1985

[6] G. Sebald, D. Guyomar, and A. Agbossou. *Smrt Mat. and Struct.,* **18**, p. 125006, 2009

[7] B. Hanrahan, L. Sanchez, C. M. Waits, R. G. Polcawich, *Smrt Mat. and Struct.,* **25**, 015025, 2015

Self-Powered Adaptive Switched Architecture Storage

F. El Mahboubi[1], M. Bafleur[1], V. Boitier[1], A. Alvarez[2], J. Colomer[2], P.Miribel[2] and J-M. Dilhac[1]

[1]LAAS-CNRS, Université de Toulouse, CNRS, INSA, UPS, Toulouse, France
[2]University of Barcelona, Barcelona, Spain

E-mail: felmahbo@laas.fr

Abstract. Ambient energy harvesting coupled to storage is a way to improve the autonomy of wireless sensors networks. Moreover, in some applications with harsh environment or when a long service lifetime is required, the use of batteries is prohibited. Ultra-capacitors provide in this case a good alternative for energy storage. Such storage must comply with the following requirements: a sufficient voltage during the initial charge must be rapidly reached, a significant amount of energy should be stored and the unemployed residual energy must be minimised at discharge. To answer these apparently contradictory criteria, we propose a self-adaptive switched architecture consisting of a matrix of switched ultra-capacitors. We present the results of a self-powered adaptive prototype that shows the improvement in terms of charge time constant, energy utilization rate and then energy autonomy.

1. Introduction

Energy harvesting is necessary to ensure energy autonomy of networked wireless sensors. Generally, this type of system incorporates a device to store energy, in order to cope with ambient energy intermittency. The device used for storage, is either a battery or an ultra-capacitor. In some cases the ultra-capacitor is an interesting alternative to batteries: on the one hand, to benefit from an almost infinite lifetime compared to battery and on the other, in the case of extreme temperatures (in an aircraft, for ex.), to get rid of safety issues. Moreover, ultra-capacitors can deliver much larger power peaks than batteries. However, ultra-capacitors have three major disadvantages. Firstly, the output voltage varies with the amount of stored energy, preventing the direct supply of a load and therefore requiring a DC-DC regulator. The second disadvantage is relative to start-up with empty ultra-capacitors: enough energy has to be stored to reach a sufficient voltage to operate the associated electronics. If storage of a large amount of energy is needed, the capacitance should be high, and so will be the charge time constant. If a fast start-up is required, the value of the capacitance could be too small for efficient storage. A compromise must be found between these two constraints. The third disadvantage is that the DC-DC converter stops working below a certain voltage threshold. Thus, during the discharge, part of the stored energy in the ultra-capacitors is lost. To partially solve these problems, we propose in this paper a self-adaptive architecture of energy storage.

2. Self-adaptive switched architecture and prototype

The self-adaptive structure consists of four identical ultra-capacitors (with a capacitance of C=100mF). The cells are judiciously interconnected by switches, thus allowing for three configurations. The first one is a configuration « all-series » (S) which allows a fast start-up by having an equivalent capacitance four times smaller than the final one. Then, when sufficient voltage is reached for proper operation, the architecture moves to a « series-parallel » (SP) configuration, and finally to an « all-parallel » (P) configuration to maximize energy storage. The process is reversed towards the SP step and finally the S configuration when discharging the ultra-capacitor to use as much as possible the stored energy.

The use of adaptive storage topologies is not new. There are many solutions: one using DC-DC converters [1], another requiring many ultra-capacitors [2], and a third one exhibiting heavy losses due to the number of used switches [3]. This study is carried out within the European project SMARTER [4], dedicated to the development of adaptive storage architecture for a wireless sensor node that is self-powered by piezoelectric energy harvesting. In this project, the whole node will be implemented on a flexible substrate and ultra capacitors integrated onto silicon [4].

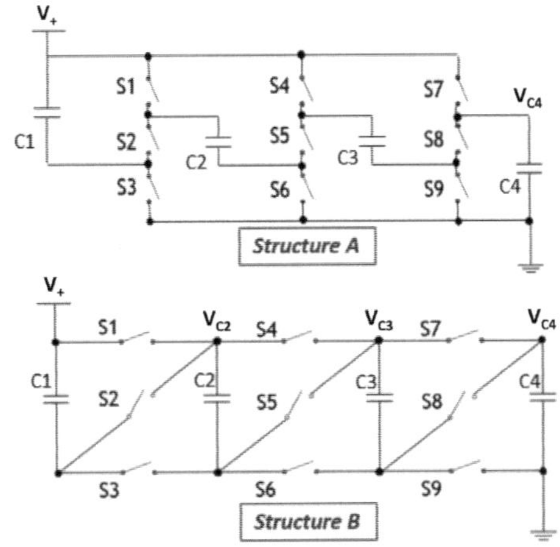

Figure 1. Configuration of the two studied self-adaptive switched architectures.

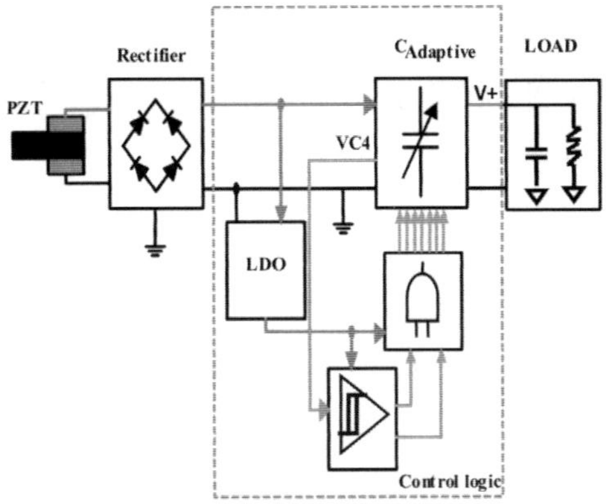

Figure 2. Block diagram of the self-powered adaptive storage architecture.

Two different architectures (Figure 1) have been studied [6][7] and compared with regard to losses to assess if a balancing circuitry is needed. Table 1 summarizes the simulation results, for both structures A and B: it provides the losses related to the variability of the capacitance value for a tolerance range of ±20%. We found out that to minimize losses, a judicious placement of the capacitors allows limiting them to an acceptable level and balancing circuitry is then useless. This information is quite important since ultra-capacitors integrated onto silicon could exhibit such large variability. For the implementation of a self-powered adaptive storage prototype, we then used structure B that is slightly more efficient. In addition, this structure has an advantage regarding the associated control circuitry. Instead of monitoring the voltage V+ that is applied to the load (see Fig. 1 and 2), the control circuitry only monitors the voltage across the ultracapacitor C4 that is connected to ground, assuming that it provides a good image of V+ voltage. Compared to the approach of [6] that uses a complex sequential logic, for this latter architecture, a simpler combinational logic can be used that is beneficial in terms of energy consumption.

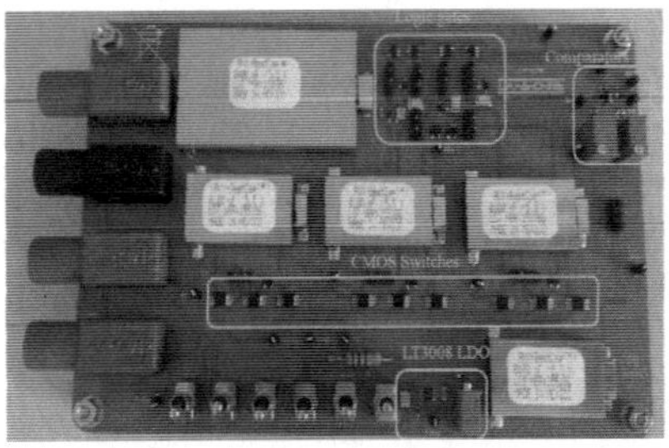

Figure 3. Prototype of the self-powered adaptive storage.

Tolerance range ±20%	E_{max} loss (%)	C1 (F)	C2 (F)	C3 (F)	C4 (F)
Structure A	2.08	0.12	0.08	0.08	0.12
Structure B	2.16	0.08	0.08	0.12	0.12

Table 1. Impact of the dispersion in capacitance values on losses (worst case capacitor placement).

A preliminary validation of this concept was already presented [6] but it was not autonomous, i.e. an external power supply was used for the control circuitry. The schematic architecture of the self-powered adaptive storage architecture is given in Figure 2 and was validated using commercial components (Figure 3). The matrix of ultracapacitors is made up of four AVX-BestCap supercapacitors of 100mF that are interconnected by nine ADG801 normally-off CMOS switches. An LT3008 LDO allows powering the control circuitry that is composed of two comparators and four logical gates driving the switches.

The comparison of the charge voltage of the structure (or of a cell) to predetermined voltage thresholds (V_{high} and V_{low}) allows switching from one configuration to another. During the charge, when the measured charging voltage exceeds V_{high}, the control circuitry changes the topology of the structure towards a higher capacitance value. Similarly, during the discharge, when the measured voltage reaches V_{low}, the configuration switches to a lower capacitance value. The choice of the threshold depends on the desired goal (maximizing harvested power [6], high voltage…). For an optimized behavior, V_{high} and V_{low} should respect:

$$2V_{min} < V_{high} < V_{max} \quad \text{and} \quad V_{min} < V_{low} \tag{1}$$

where V_{min} and V_{max} are the minimum and maximum operating voltages of the electronics to supply, respectively.

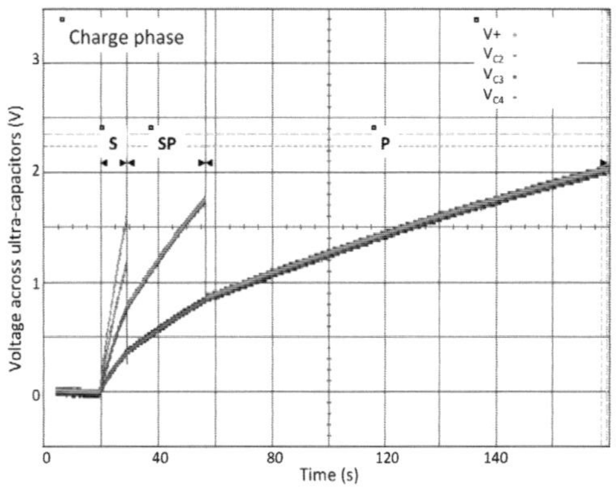

Figure 4. Measurement of the self-adaptive storage prototype in the charge phase. S, SP and P indicate ultra-capacitor configuration: all series, series/parallel and all parallel.

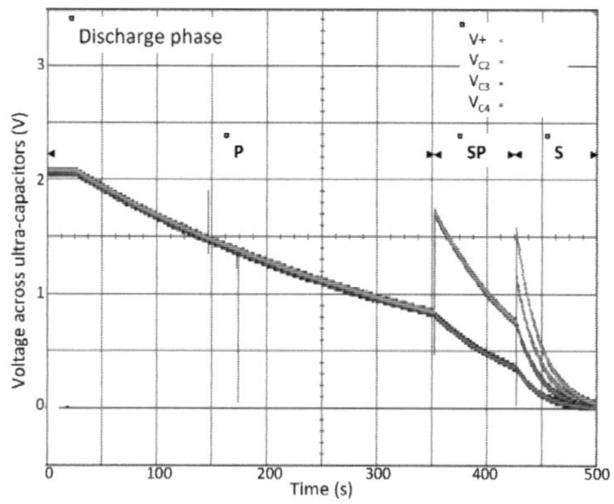

Figure 5. Measurement of the self-adaptive storage prototype in the discharge phase. S, SP and P indicate ultra-capacitor configuration: all series, series/parallel and all parallel.

3. Experimental results

We have tested the proposed architecture. For the charging phase, we used a Thevenin generator made up of 1kΩ resistor and 4.5V voltage source to simulate an energy harvester source. For the discharge phase, we connected a resistive load of 1kΩ between V+ and ground to simulate a constant power load. During these two phases, in addition to V+, we monitored the voltages at each intermediate node (VC4, VC3 and VC2) with reference to ground.

3.1. Charge phase

Figure 4 presents the experimental waveforms obtained during the charging phase. At starting, the ultra-capacitors are empty and they are configured in series (S). This allows reaching very rapidly, in 5 s, the voltage threshold (0.8V) required for operating the electronics. This has to be compared with a configuration with a single storage capacitor that requires a starting time of 55 s. We can notice that voltage V_{C4} used to monitor the switching to another configuration is exactly proportional to one fourth (400 mV) of V+ voltage when switching from S to SP configuration. Then the adaptive storage switches to a series-parallel (SP) configuration and finally to an all parallel (P) one.

We have measured the total energy stored in the adaptive storage and compared it to a single capacitor architecture when the storage element is fully charged up to 4.5V. The losses related to the required control electronics result in 91% energy storage efficiency.

3.2. Discharge phase
Figure 5 depicts the voltage evolution across the self-adaptive matrix during the discharge phase. In this case, the system stops operating at a minimum voltage corresponding to the one required to operate the electronics, while a significant portion of the energy is still stored in the capacitors. Using the adaptive configuration that switches back to SP and finally to S configurations, it is possible to minimize this residual energy. Compared to single storage capacitor architecture, the adaptive approach allows saving 5% of this residual energy, which is in this case, does not completely compensate the energy required to power the control electronics. However, this performance should be greatly improved by moving to a dedicated power-management integrated circuit.

4. Conclusion
We have validated the concept of an adaptive storage for improving the dynamic behaviour of a battery-free storage architecture based on ultra-capacitors. It allows significantly boosting the start-up time and the energy management requires some more optimization to get the full benefits of this approach in terms of energy usage efficiency that an integrated solution should provide.

5. References
[1] M. Uno, K. Tanaka, Single-Switch Cell Voltage Equalizer Using Multistacked Buck-Boost Converters Operating in Discontinuous Conduction Mode for Series-Connected Energy Storage Cells, *IEEE Trans. Veh. Technol.*, vol. 60, no 8, p. 3635-3645, oct. 2011

[2] J.-F. Nie, X. Xiao, Z. Nie, P. Tian, and R. Ding, A novel three-level changeover circuit of super-capacitor bank for energy storage systems, in *IECON 2012 - 38th Annual Conference on IEEE Industrial Electronics Society*, 2012, p. 144-149.

[3] M. Uno, H. Toyota, Equalization technique utilizing series-parallel connected supercapacitors for energy storage system, in *IEEE International Conference on Sustainable Energy Technologies*, 2008. ICSET 2008, 2008, p. 893-897

[4] M. Bafleur, J.M. Dilhac, Towards Energy Autonomy of Wireless Sensors in Aeronautics Applications: SMARTER Collaborative Project, *Proc. IEEE International Conference on Green Computing and Communications and IEEE Internet of Things and IEEE Cyber, Physical and Social Computing*, 2013, pp.1668-1672.

[5] T.M. Dinh, A. Achour, S. Vizireanu, G. Dinescu, L. Nistor, K.Armstrong, D. Guay, D. Pech, Hydrous RuO2/carbon nanowalls hierarchical structures for all-solid-state ultrahigh-energy-density micro-supercapacitors, *Nano Energy, vol. 10 (2014) p. 288-294.*

[6] R. Monthéard, M. Bafleur, V. Boitier, J.-M. Dilhac and X. Lafontan, Self-adaptive switched ultra-capacitors: a new concept for efficient energy harvesting and storage, in *PowerMEMS Proceedings*, 2012, pp. 283-286.

[7] P.Miribel-Català, J.Colomer-Fararons, J.Lafuente Brinquis, J.López-Sánchez, Self-Powered Adaptive Circuit Sampling for a Piezoelectric Harvester, *2014 DCIS Conference on Design of Circuits and Integrated Systems*, 26-29 Nov. 2014, Madrid, Spain.

Acknowledgments
This work is carried out within the framework of the European project SMARTER funded by the CHIST-ERA program, "Green ICT, towards Zero Power ICT".

Self-Pumped Coolant Circulation Driven by Thermal Gradient over Microstructures

Nicolas Lobato-Dauzier[1], Anthony J. Genot[2] and Hiroyuki Fujita[1]

[1]CIRMM-Institute of Industrial Science, The University of Tokyo,
[2]LIMMS/CNRS-Institute of Industrial Science, The University of Tokyo

hfujita@iis.u-tokyo.ac.jp

Abstract. We propose a passive cooling device which converts thermal energy into fluid flow via Marangoni effect: a gradient of temperature induces a gradient of surface tension. This in turn triggers a fluid flow at the air/liquid interface which runs transversally to the heat flow. We show how to amplify the global fluid flow thanks to geometrical optimization. We also show how the addition of an asymmetric small wall across the channel reduces the backflow and improves our device.

1. Introduction

Electronic devices produce heat which needs to be dissipated - often with a fluid flow -, which implies the availability of an extern supply of energy. The idea behind this work is to use the dissipated heat to propel the cooling fluid. Our aim is to power a micro-pump without applying an external pressure gradient (**Figure 1**), which enables closed loop flows that cycle between hot and cold sources Recycling the dissipated heat to propel the fluid itself would not only reduce energy consumption, but would boost the autonomy of ambient devices, such as those needed for « the Internet Of Things ». Such devices are very small and therefore a coolant circulation based on MEMS technology is very suitable. At the micrometer scale surface tension effects become very strong, the Marangoni effect – which depends on the variation of surface tension with temperature – is therefore suitable to propel a fluid at such scale.

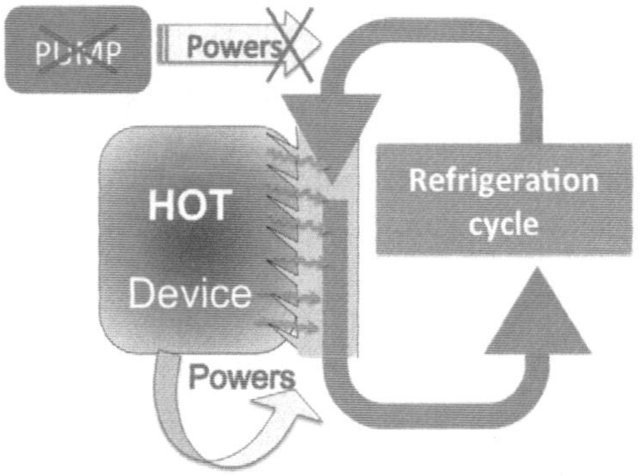

Figure 1. Scheme of the concept: the dissipated heat powers the refrigeration cycle without the need of a pump

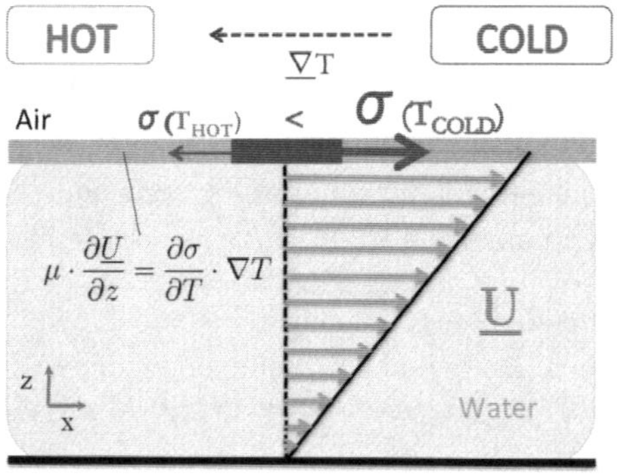

Figure 2. Schematic side view of Marangoni flow: the fluid at the surface is pulled by the cold side

At interface air/fluid, the fluid is pulled stronger on the cold side. For an interface in the xy-plane, this can be expressed by :

$$\mu \cdot \frac{\partial \boldsymbol{U}}{\partial z} = \frac{\partial \sigma}{\partial T} \cdot \boldsymbol{\nabla T} \tag{1}$$

The flow triggered by such effect- also called Marangoni flow- is a couette flow (**Figure 2**). We show how to amplify this effect thanks to geometrical optimization -first developed for autophoretic pumping [1] - which engineers temperature gradients between the heat and cold source. Our proposed device differs from most Marangoni-based pumps, which often considered the effect from a 2 dimensional perspective [2], or applied a longitudinal gradient (**Figure 3A**) of temperature with respect to fluid flow [3]. Although a transverse heat flow (**Figure 3B**) should only create a transverse rotating motion, microstructures allow the fluid to move in the $x > 0$ direction. (**Figure 3C**)

Crucially, because in our device the fluid flow runs transversally to the heat flow, we can decouple the hydrodynamic and thermal scale: this means we can propel fluid over centimetres thanks to the large, but localized thermal gradients that MEMS fabrication enable.

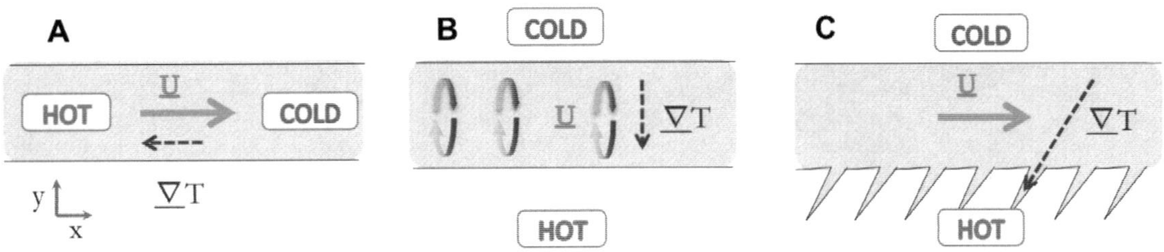

Figure 3. Schematic top view comparison of the Marangoni flow induced by different channels. **A:** Flat channel + parallel gradient/ **B:** Flat channel + transverse gradient / **C:** Patterned channel + transverse gradient. Patterning allows the formation of a fluid flow transverse to the original gradient of temperature.

2. Preliminary study and modelization

2.1. Device concept and characteristics

Our device consists on a succession of cavities in which the fluid is pumped out at the surface and pumped in the lower layers. The angle between the cavity and the channel enables the outgoing flow to trigger a surface flow to the right. A wall prevents the pumped-in fluid to come from the same side, which would suppress the global flow in the $x > 0$ direction.(**Figure 5)**

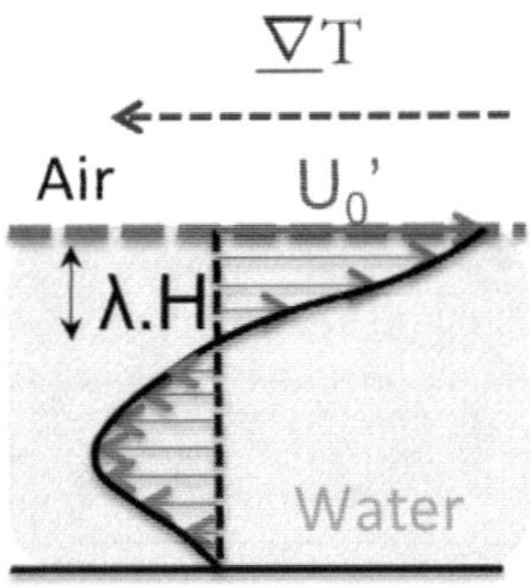

Figure 4. Schematic side view of flow inside a cavity.

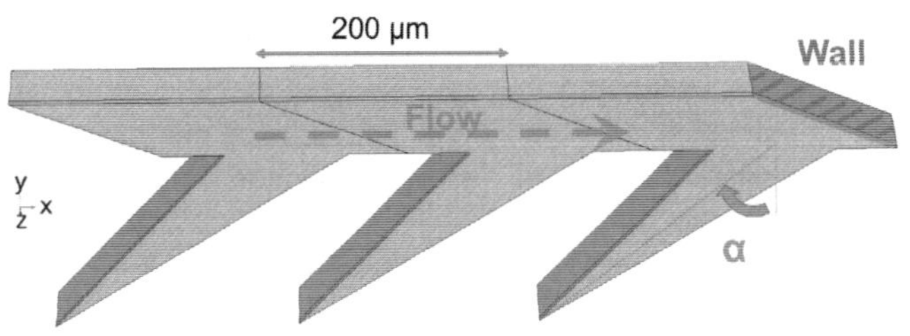

Figure 5. COMSOL model of our device: the red wall stops the backflow from the right, while the fluid on the top can still go over.

2.2. Simulation parameters

We used COMSOL to simulate the flow field inside our device. We only considered diffusive heat transfer, and a Stokes flow. For the heat transfer, we used periodic boundaries for the narrow upper left and right side, set the temperature on the upper wall and modeled the device to be cooled by a heat source. The rest of the boundaries were thermally insulated. For solving the Stokes flow, we used equation (1) as a boundary condition on the top plane, periodic flow condition on the narrow upper left and right side, while considering all other boundaries as "no slip" walls. The fluid chosen is water, and we considered heat source of 65 W/m^2, which corresponds to a gradient of 1 K/mm for a hot source made of copper for instance.

3. Results and discussion

The simulations show how the micro cavities create a temperature gradient that is not orthogonal to the channel (**Figure 6**). The angle α is therefore the key parameter.

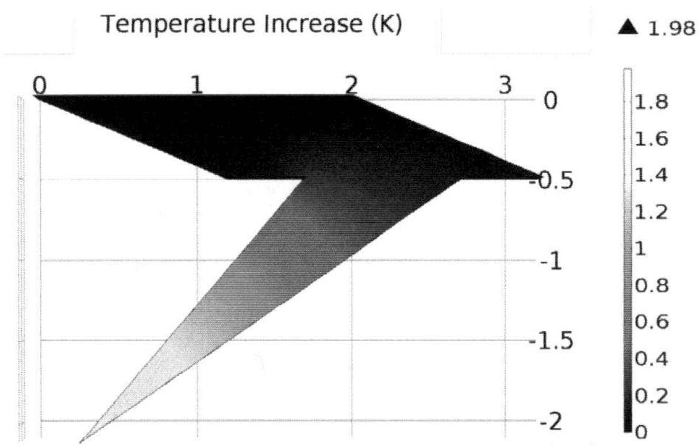

Figure 6. Simulation result for temperature distribution in the channel, for an angle of 50°

As predicted, at the surface the fluid is pumped out of the cavity at a relatively high velocity(**Figure 7**), whereas it is pumped in in lower layers (**Figure 8**). The addition of the wall works properly : it only allows the top layers to pass while preventing backflow. Preventing the backflow from the right side allows us to have a wide opening for the cavity, which can propel more fluid then. However the cost is quite high, as having such a wall decreases the velocity.

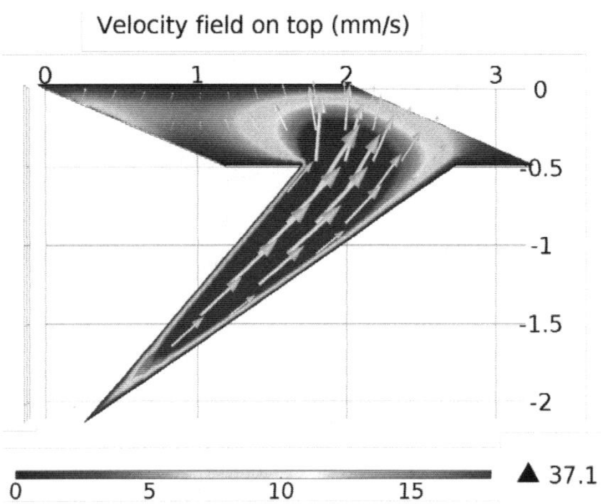

Figure 7. Simulation result for the velocity field on top : the fluid is pumped out at high velocity

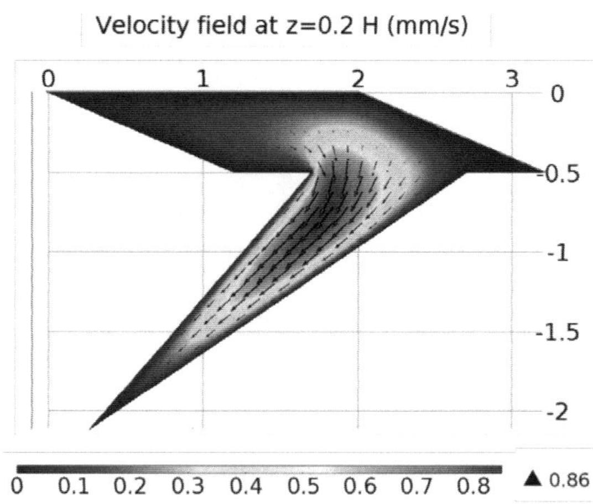

Figure 8. Simulation result for the velocity field near the bottom: the fluid is pumped in, mainly from the left side, at a lower velocity

The variation of the angle of the cavity reveals a trade-off relationship: high angles are more efficient to propel the fluid in the right direction. However, when the angle increases the cross section of the cavity decreases and therefore the volumic flow rate also decreases. There is therefore an optimal angle which measures around 50°(**Figure 9)**

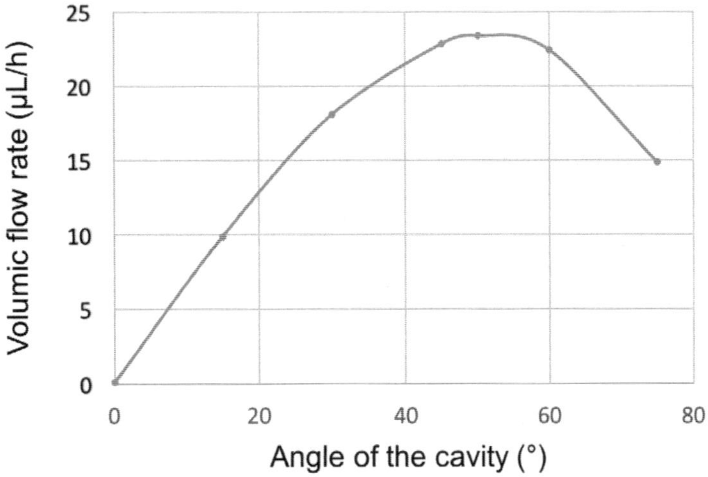

Figure 9. Evolution of the volumic flow rate with the variation of the cavity angle. There is an optimal angle that maximizes the flow rate around 50°

4. Conclusion

We showed that thanks to micro cavities on a channel walls, it was possible to use Marangoni effect to propel a fluid in a direction orthogonal to the temperature gradient with flow rates that however small they are, correspond to what is used in some microfluidic devices. As we can decouple the hydrodynamic and thermal scale this device is suited for propelling a fluid over distances that can reach the centimeter while having the benefits of the large thermal gradients that MEMS fabrication allows.

5. References

[1] S. Michelin et al. , Soft Matter, 2015,**11**, 5804-5811
[2] C. Maggi, et al.Nature Communications, 2015, **6**, 7855
[3] Michael Grunze, Sciencemag 1999, **283**, 5398

A Smart Load Interface and Voltage Regulator for Electrostatic Vibration Energy Harvester

Mohammed Bedier[1], Philippe Basset[2] and Dimitri Galayko[1]

[1]Sorbonne Universite's, UPMC Univ Paris 06, UMR 7606, LIP6, F-75005, Paris, France
[2]Université Paris Est ESYCOM, ESIEE Paris, France

E-mail: mohammed.bedier@lip6.fr

Abstract. This paper presents a new implementation in ams 0.35μm HV technology of a complete energy management system for an electrostatic vibrational energy harvester (e-VEH). It is based on the Bennet's doubler architecture and includes a load voltage regulator (LVR) and a smart Load Interface (LI) that are self-controlled with internal voltages for maximum power point tracking (MMPT). The CMOS implementation makes use of an energy harvester that is capable of producing up to 1.8μW at harmonic excitation, given its internal voltage is kept within its optimum. An intermediate LI stage and its controller makes use of a high side switch with zero static power level shifter, and a low power hysteresis comparator. A full circuit level simulation with a VHDL-AMS model of the e-VEH presented was successfully achieved, indicating that the proposed load interface controller consumes less than 100nW average power. Moreover, a LVR regulates the buffer and discharge the harvested energy into a generic resistive load maintaining the voltage within a nominal value of 2 Volts.

1. Introduction

The electrostatic vibrational energy harvesters (e-VEH) is a category of energy harvesters of which the vibrational energy is transduced into electric energy through an electrostatic transducer. The electrostatic transducer is usually a variable MEMS capacitor. In most of the implementations, such as [1, 2], the transducer is usually accompanied by a conditioning circuit (CC) where its role is to assist the energy conversion process and can be regarded as an autonomous form of maximum power point tracking algorithm (MPPT). In recently published e-VEH implementations, only the transducer and the CC are considered, with almost no form of energy management for the harvested energy to be interfaced with a load. In this paper we present a smart load interface that manages the extraction of the harvested energy from the CC and ensures the CC operating within its optimum region.

This paper is arranged as follows, first the e-VEH system blocks are introduced, then the CMOS implementation of each block is proposed, finally the simulations and conclusion are discussed.

2. E-VEH System Blocks

A smart e-VEH system is proposed that is formed of three main blocks - shown in Figure 1: a Bennet's doubler CC, followed by an intermediate load interface (LI), and load voltage regulator block (LVR). The load interface is a DC/DC converter like architecture. The load could be a low power sensor - such as in body sensor network (BNS) or a motion sensor. The requirements of the load can vary from one application to another, yet it is not the main concern of this work. However, some load voltage regulation is proposed in this work for a generic load.

2.1. Bennet's Doubler CC

Conditioning circuits based on Bennet's doubler are one of the promising CC that were first introduced by A. Querioz et al in in 2014 [3]. Since then, it has gained the interest of the electrostatic energy harvesting community such as [4] , [5] and [6]. It implements a rectangular QV cycle, with two main advantages on other QV cycles: First, It allows the accumulation of energy through self-increasing of the internal voltage. This allows a build up of energy even from small initial external biasing. Second, It allows synchronization of the charge flow with the movement of the MEMS capacitor plates through a network of diode, thus remove the need of extra controlling blocks.

However, Bennet's doubler CC has a fundamental characteristic of which the power of conversion is function of the instantaneous value of its internal energy. An energy management should achieves a smart control of this internal energy by setting the CC internal voltage within an interval maximizing the power of conversion. This is vital for circuits with self-increasing biasing, as Bennets doublers. Thus a load interface (LI) that separates the load from the CC while controlling that smart control is proposed. Moreover, high voltages are easily achieved with e-VEHs (up to tens of volts) while the harvested energy are typically within few micro watts, leaving $1 - 2\mu W$ available for controller overhead at most. In consequence a mix of high voltage interface and low voltage low power controller is a must for proper operation of the e-VEH. The next subsection discuss in details this load interface.

2.2. Load Interface and Load interface Controller

The LI is formed of an inductor (L), freewheeling diode (D_{LI}), buffer load capacitor (C_{buffer}) and a high voltage switch (SW_{LI}). The C_{load} is a buffer low voltage storage capacitor. It is necessary for interfacing the harvester with the load, since V_{res} is usually too large to interface (up to 30V for some e-VEH) with the load. Moreover, V_{res} may experience slow variations, whereas the load voltage should be stabilized.

In general, the reservoir voltage (V_{res}) has an optimum interval in which the harvesting power is maximized [7, 8]. If the voltage V_{res} exceeds this interval, then it should be forced back. The optimum interval is defined where V_{res} is within the two thresholds, V_H and V_L. The energy to be removed from C_{res} to drop its voltage from V_H to V_L is,

$$\Delta W = \frac{1}{2} C_{res}(V_{rH}^2 - V_L^2) \tag{1}$$

Extracting ΔW is achieved by LI and is controlled by sensing V_{res} and controlling the switching order for SW_{LI}.

However, the technology of which the switch is implemented usually sets a fundamental upper limit on the allowed inductor current ($I_{L_{max}}$). In an effort to transfer the energy with a constrained current, the energy is transferred through number of energy-shots inspired from the well know burst-mode control. This burst-mode energy transfer process is repeated until ΔW is extracted from C_{res}.

The proposed load interface controller is achieved by applying a scaled divided version of V_{res} to a comparator with a hysteresis as shown in Figure 3. The hysteresis gap is adjusted to the scaled version of the two thresholds V_L and V_H. The comparator output - COMP - controls an enable clock that precisely generates the switching decision for SW_{LI}. The transistor level implementation of the LI controller is discussed in the next section.

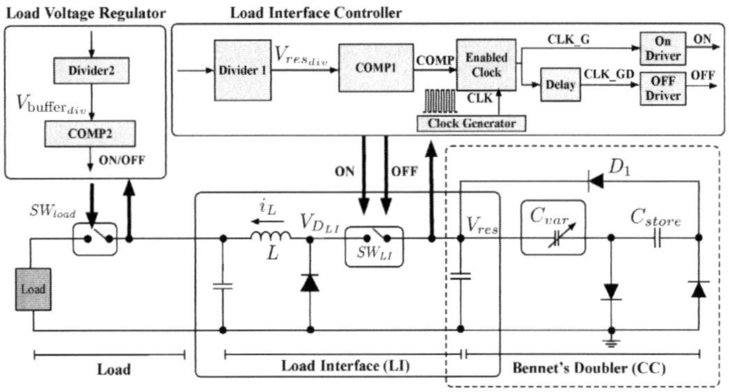

Figure 1. E-VEH with Bennets doubler CC, LI and LVR.

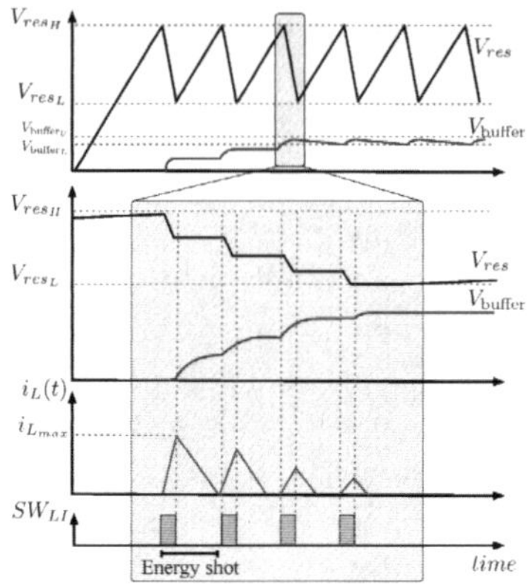

Figure 2. Multiple energy-shot transfer illustration of $V_{res}(t)$, $V_{\text{buffer}}(t)$ and $i_L(t)$.

2.3. Output Voltage Regulator

The LVR regulates V_{buffer} low voltage to maintain it between V_{buffer_U} and V_{buffer_L} while supplying the load with thae required energy thus maintaining V_{buffer} within its average nominal value. The LVR controller senses a divided voltage of V_{buffer} using a second comparator (COMP2) as shown in Figure 4. It maintains V_{buffer} between $2V < V_{\text{buffer}} < 2.2V$ by allowing the discharge of C_{buffer} into a load modeled by a resistor.

3. CMOS implementation and simulation

The load interface and the voltage regulator are implemented in ams 0.35 μm HV (50V) technology. It makes use of the energy harvester presented in [9], which is capable of producing up to $1.8\mu W$ at harmonic excitation, given V_{res} is kept within its optimum. The intermediate LI stage and its controller, shown in Figure 3, makes use of a high side switch with zero static power level shifter [10] , and a low power hysteresis comparator (COMP1). The LI controller was previously presented in details in [11] with average power consumed of less than $100nW$ (excludeding the switch and the voltage divider). The optimum V_{res} interval is set between 15V and 13.5V. The C_{buffer} is allowed to discharge into a light resistive load of (525Ω) by a LVR , shown in and Figure 4. It makes uses of a low power hystersis comparator (COMP2), similar to the one used for the LI controller. The system simulation depicting the behavior of the load interface proposed system is shown in Figure 5 . With the reservoir capacitor is initially charged with 20V, simulation shows the evolution of both the harvester internal voltage (V_{res}), LI buffer voltage (V_{buffer}) and Load voltage (V_{Load}) with successful regulation of load voltage and maintaining V_{res} within its optimum interval.

4. Conclusion

This paper presents a complete energy management interface for vibrational energy harvesters. It is capable of regulating high internal voltage of the conditioning circuit while maintaining maximum energy harvested. This achieved by a mix of high and low voltage design. The controller design takes into account low energy constrain, as the whole system should be able to operate within low power generated from the harvester. The proposed e-VEH system also offer an output load voltage regulation of a resistive load and is capable of periodically supplying a generic load with a 2V for period of 1ms.

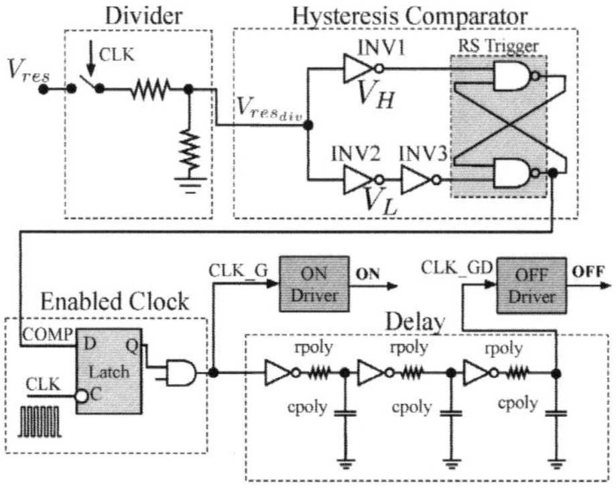

Figure 3. Load Interface Controller.

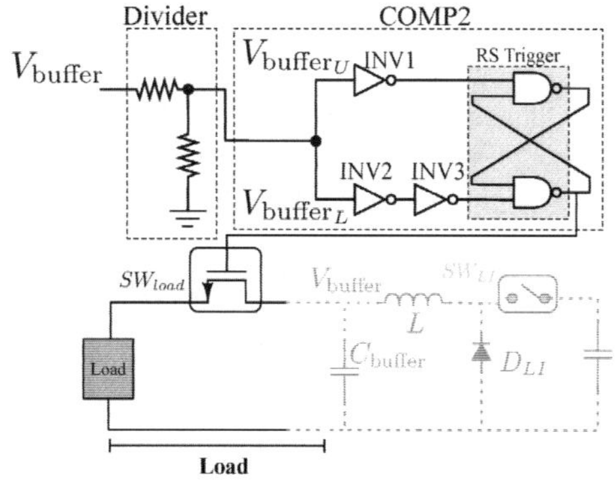

Figure 4. Load Voltage Regulator.

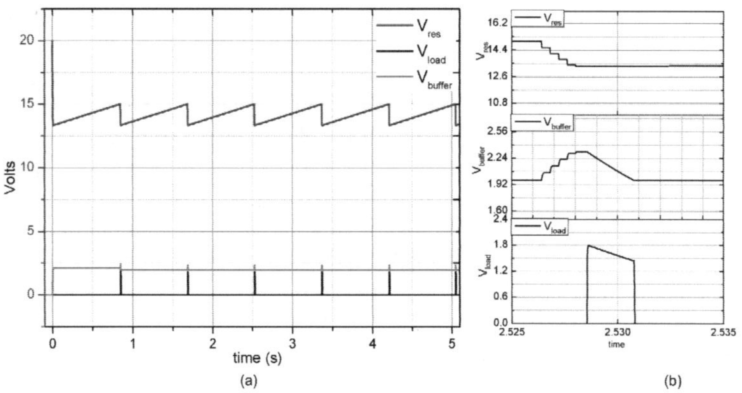

Figure 5. E-VEH LI simulation $0s < t < 5s$ (a) zoom-in at $2.525s < t < 2.535s$(b).

References

[1] Torres E O and Rincon-Mora G A 2009 Energy budget and high-gain strategies for voltage-constrained electrostatic harvesters *2009 IEEE International Symposium on Circuits and Systems* pp 1101–1104 ISSN 0271-4302

[2] Kempitiya A, Borca-Tasciuc D A and Hella M M 2012 *IEEE Transactions on Industrial Electronics* **59** 456–463 ISSN 0278-0046

[3] de Queiroz A C M and de Souza M S 2014 Batteryless electrostatic energy harvester and control system *2014 IEEE International Symposium on Circuits and Systems (ISCAS)* pp 1977–1980 ISSN 0271-4302

[4] Dorzhiev V, Karami A, Basset P and Galayko D 2014 *Journal of Physics: Conference Series* **557** 012126

[5] de Queiroz A C M 2014 Electrostatic energy harvesting without active control circuits *Circuits and Systems (LASCAS), 2014 IEEE 5th Latin American Symposium on* pp 1–4

[6] Lefeuvre E, Risquez S, Wei J, Woytasik M and Parrain F 2014 *Journal of Physics: Conference Series* **557** 012052

[7] Dorzhiev V, Karami A, Basset P, Marty F, Dragunov V and Galayko D 2015 *IEEE Electron Device Letters* **36** 183–185 ISSN 0741-3106

[8] Basset P, Blokhina E and Galayko D 2016 *Electrostatic Kinetic Energy Harvesting: Nanotechnologies for Energy Recovery Set* (Wiely)

[9] Karami A, Basset P and Galayko D 2015 *Journal of Physics: Conference Series* **660** 012025

[10] Dudka A, Galayko D and Basset P 2012 Design of controller ic for asynchronous conditioning circuit of an electrostatic vibration energy harvester *Green Computing and Communications (GreenCom), 2012 IEEE International Conference on* pp 660–663

[11] Bedier M and Galayko D 2016 Multiple energy-shot load interface for electrostatic vibrational energy harvesters *The 4th IEEE International NEWCAS Conference 2016*

Improved Energy Management System for Low-Voltage, Low-Power Energy Harvesting Sources

D Newell[1] and M Duffy[1].

Power Electronics Research Centre, NUI Galway, Ireland.

d.newell2@nuigalway.ie

Abstract. This paper focuses on improving the energy conversion process for low-voltage energy harvester powered wireless sensors by optimising the conversion stages for pulsed sensor operation. The proposed circuit has been designed to operate efficiently with both a low-voltage low-power energy harvester source and a low-power pulsed load. This ensures that continuous conversion losses are kept to a minimum and power is only delivered to the sensor when required. This has shown an increase in energy delivered to a sensor of up to 10% versus that of the best existing solution.

1. Introduction

Energy harvesting (EH) sources, including dye-sensitised solar cells (DSSCs), thermoelectric generators (TEGs), microbial fuel cells (MFC's) and piezoelectrics, provide a low-cost, reliable and convenient solution to powering wireless sensors compared to both batteries and mains connections [1]. By utilising EH sources, wireless sensors can become autonomous and may be fitted in a larger range of locations when compared with conventional power sources. However, to date, power levels provided by EH sources have proven insufficient to enable fully autonomous sensor operation; this is explained by the combination of limited power production by EH sources and low power conversion efficiency due to low source voltage or current levels.

This paper focuses on improving the energy conversion process for low-voltage EH sources by optimising system operation for the pulsed nature of wireless sensor loads. The best performing step-up converter found for a solar cell powered wireless sensor is a boost converter operating under burst mode, with an efficiency of between 70% to 80%, depending on input lux [2]. Due to the low source power levels, efficiency is found to be limited by baseline losses associated with monitoring and control circuitry. In terms of sensor operation, many low-power wireless sensors operate in a pulsed manner consuming high levels of power (relative to the EH source) for short bursts of time. This means that conversion of the EH source voltage may only be needed during the time intervals when the wireless sensor is operating.

The circuit solution proposed in this work has been developed to exploit the pulsed nature of typical EH loads by reducing the operating on-time of voltage step-up converters and therefore the associated losses. It consists of a bank of energy storage supercapacitors connected directly to the EH source in a switched capacitor configuration. In this way, the circuit provides dual functionality of energy storage and voltage step-up conversion; it can be connected to supply a wireless sensor load directly or it can be fed into another converter stage, e.g. a boost converter, where its higher voltage

Journal of Physics: Conference Series **773** (2016) 012106 doi:10.1088/1742-6596/773/1/012106

can result in increased converter efficiency. For the latter, connection to the next conversion stage may be restricted to time intervals when the wireless sensor needs to complete an operation, thereby reducing baseline converter losses.

2. Switched Capacitor Energy Buffer

As discussed, wireless sensors operate in a pulsed manner to perform periodic sense and data transmission/store actions. For example a sensor might perform an action once every 15 minutes, remaining in low-power, sleep mode between actions. This operating mode is compatible with low-power EH sources whose energy can be built up to a sufficient level between each sense operation. However, many existing step-up conversion circuits for EH sources operate continuously to deliver energy to the load, with excess energy stored either in a capacitor, battery or a combination of both. These storage devices ensure that the wireless sensor has sufficient energy when the source is not available. However, due to ultra-low source power levels, baseline converter losses are significant and therefore a more optimum system design would allow the converter to deliver power only when required by the wireless sensor, while also delivering excess energy to a long-term storage element only during the sensor operating time.

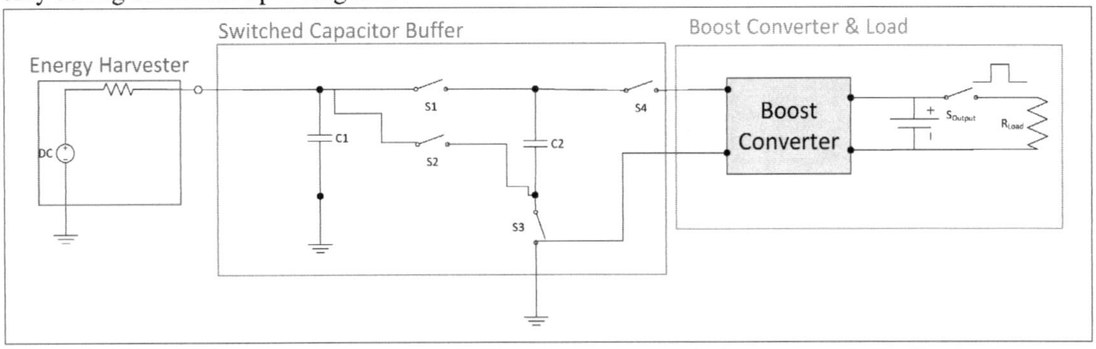

Figure 1: EH energy management system including a 2-stage switched capacitor energy buffer circuit.

Figure 1 shows the circuitry involved in a 2-stage version of the proposed switched capacitor energy buffer block that supports such operation. The circuit is similar in operation to that of a series-parallel charge pump, [3] and [4], but supercapacitors [5] rather than regular capacitors are utilised (for C1, C2) to provide sufficient energy storage capacity, while their robustness under repeated charge-discharge cycling is superior to that of batteries. In comparison with regular charge pumps, the proposed circuit operates at a very low frequency, as seen in Figure 2, to match the long time periods between wireless sensor operations. Note that the number of switched capacitor stages can be varied depending on output voltage requirements.

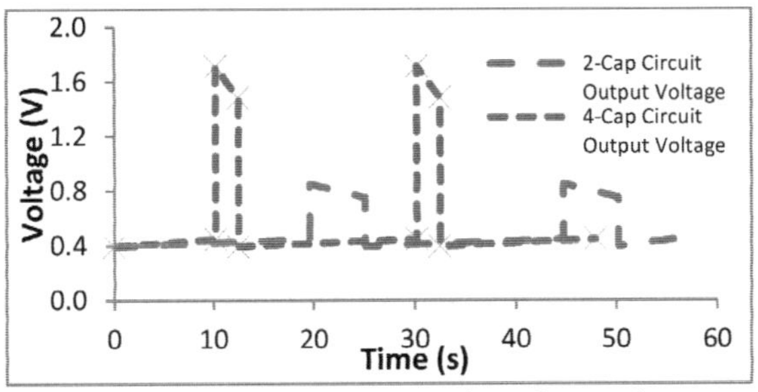

Figure 2: Switched capacitor output voltage for 2 and 4-capacitor circuits.

Table 1: Different states of operation the circuit seen in Figure 1.

State	Conditions	Action	Switch States
Charging	V_{C1} & V_{C2} < $V_{MPP+10\%}$,	Charge C_1 & C_2	S_1 & S_3 are on, S2 & S4 are off
Discharging	V_{C1} & V_{C2} >/= $V_{MPP-10\%}$	Power load	S_2 & S_4 are on, S1 & S3 are off

The circuit operates in two states, charging and discharging, to maintain the EH source at its maximum power point (MPP), as outlined in Table 1. During the charging stage, supercapacitors C_1 and C_2 are in a parallel configuration and are charged directly from the energy harvesting source. Once the voltage on the supercapacitors reaches a predetermined value set as the maximum power point (MPP) voltage + 10%, V_{MPP+}, the supercapacitors stop charging. At this point, S_1 and S_3 are open while S_2 and S_4 are closed, and the supercapacitors discharge to the load in a series connection. In this way, the output voltage is increased, doubled in the case of the circuit in Figure 1. The supercapacitor voltage is again monitored until the value reaches a predetermined minimum, MPP voltage – 10%, V_{MPP-}, when the switches are changed to the charging state and the cycle repeats. This hysteretic control method ensures that the energy harvesting source is kept at its MPP [6]. Figure 2 shows the output voltages of 2- and 4-cap implementations of the switched capacitor block of Figure 1.

3. Loss Analysis

Losses for the proposed switched capacitor block comprise of MOSFET conduction for each of the switches, S1-S4:

$$P_{MOS\_cond} = I_{RMS}^2 R_{DS(on)} \quad , \tag{1}$$

where I_{rms} is the rms current through each MOSFET over a complete charge-discharge cycle and $R_{DS(on)}$ is their on-resistance. In addition there are MOSFET switching losses:

$$P_{SW} = \frac{t_{sw,on} V_{out} I_{Peak} F_{SW}}{2} + \frac{t_{sw,off} V_{out} I_{Peak} F_{SW}}{2} \quad , \tag{2}$$

where F_{sw}, is the switching frequency and $t_{sw,on}$ and $t_{sw,off}$ are the MOSFET turn-on and turn-off times respectively. Note that due to the large capacitance values applied, F_{sw} is very low (approximately 30 mHz in this case), which reduces switching losses significantly compared to the other switched capacitor circuits [7].

The circuit also encounters significant capacitor related losses. As the circuit utilises supercapacitors, an accurate model is required to understand the losses within the supercapacitor structure. Figure 3 shows an equivalent circuit model of a supercapacitor [8], which shows that there are losses through multiple branches, each of which contributes losses of:

$$P_{CAP\_x} = I_{xrms}^2 R_x \quad , \tag{3}$$

where x relates to the rms current and resistance of each branch over a complete switched capacitor cycle. Table 2 shows the equivalent circuit values that were obtained by characterising the supercapacitor according to methods described in [9]. Here, R_{ESR} accounts for the usual internal interconnect and dielectric material losses, while $R_{LEAKAGE}$ accounts for dielectric leakage losses. In addition, there is a component of $R_{REDISTRIBUTION}$ which is explained by the double layer structure of supercapacitors. By adding a third branch the long-term characteristics can be modelled more accurately [10].

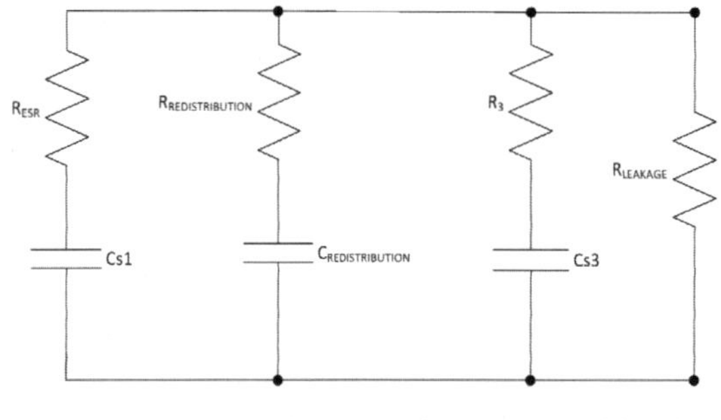

Figure 3: Supercapacitor equivalent circuit model and equivalent circuit values.

Table 2: Equivalent supercapacitor circuit values.

C_1	0.195 F
$C_{Redistribution}$	16.5 mF
C_3	27.5 mF
R_1	6 Ω
$R_{Redistribution}$	4 kΩ
R_3	45 kΩ

Following a charging period of the supercapacitor, the voltage on capacitors Cs_1 is charged to the source voltage, V_{MPP+} in the proposed circuit. However, due to a higher time constant in the redistribution branch, it takes a longer time for the voltage on $C_{REDISTRIBUTION}$ to reach V_{MPP+}. Therefore, during the discharge cycle, some energy flows from branch 1 (Cs_1), to branch 2,

($C_{REDISTRIBUTION}$), incurring losses in R_{ESR} and $R_{RESISTRIBUTION}$. This effect may be minimised by charging the supercapacitor sufficiently to V_{MPP} before connecting it into the circuit, but some losses remain due to the variation in voltage applied to enable hysteretic control. These losses are included in capacitor conduction losses in Figure 4 (b).

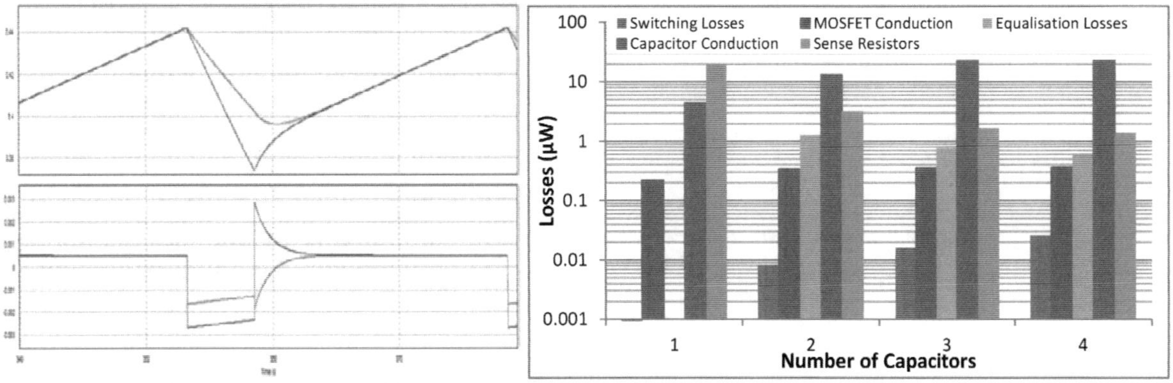

Figure 4: (a) Switched capacitor buffer operation waveforms and (b) loss breakdown versus number of capacitors.

Similarly equalisation losses, [11], occur between each of the supercapacitors, C_1 & C_2, in Figure 1. This imbalance is caused by the fact that the EH source always remains connected in parallel with the first supercapacitor, C_1. Figure 4 (a) shows this voltage imbalance between C_1, red, and C_2, blue, and the current difference between the supercapacitors during a discharge cycle. It is clear to see that C_2 supplies more current to the load in comparison with C_1. These losses are detailed as equalisation losses in Figure 4, where it is clear that both they and capacitor conduction losses are the most significant.

In a bid to reduce equalisation losses, the EH source was disconnected from the circuit during discharging but, mainly due to reduced input power, the efficiency was found to be significantly lower,. Tests were also performed with the EH source connected in series with the switched supercapacitor stages to overcome the imbalance. In this case the DSSC severely limited the output current and therefore had a major impact on the circuit performance and efficiency. Sense resistors are required to measure the current flowing into and out of the switched capacitor circuit. These losses would not feature in a final version of the circuit but for the prototype these losses play a significant part due to the overall low-power being supplied by the source.

4. Measurements/Results

A prototype of the proposed switched capacitor block has been built and initial testing performed with a DSSC source and a variable load resistance. Figure 5 (a) shows measured output waveforms of a 2-capacitor version, while Figure 5 (b) includes measured efficiency for 1- and 2-capacitor implementations. Figure 5 (b) also shows the predicted efficiency of the switched capacitor stage, of a compatible boost converter stage [2] (during discharging) and the overall combined system efficiency.

The overall trend is a decrease in efficiency of the switched capacitor circuit with an increase in number of capacitors. However due to the sense resistors the efficiency of the 1-capacitor circuit is reduce significantly. There is also a significant difference between the measured and predicted efficiency of the 2-capacitor circuit. This requires a review of the equalisation model as well as examining new methods to increase the accuracy of the efficiency measurements. This is due to the low current values, mA, that are being measured on both the input and output of the switched capacitor circuit. When the switched capacitor circuit is combined with a boost converter to deliver the required 3.3 V output, the overall step-up conversion efficiency is improved versus the standard boost converter.

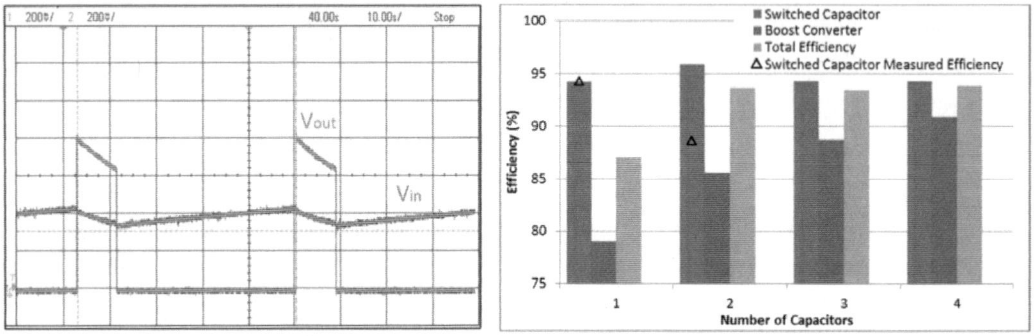

Figure 5: Oscilloscope waveform of circuit operation for 2-cap circuit.

The higher total conversion efficiency is mostly due to the increased boost converter efficiency which is enabled by the switched capacitor buffer stage. At higher number of capacitors the overall step-up conversion efficiency remains at a value around 15 % higher than that of the best existing solution conversion solution.

5. Conclusions & Future Work

Results for the proposed switched-capacitor buffer block show a very promising increase of up to 10% in efficiency of power delivered from low-voltage low-power EH sources over an existing boost converter. This increases the energy available to wireless sensor loads which can increase their data transfer rate or functionality/accuracy. This increase is due to the switched capacitor circuit enabling operation in a similar pulsed manner as the wireless sensor, thus reducing the continuous conversion losses encountered by many converters. It is also provides a voltage step-up function that increases the boost stage efficiency. Future work will involve determining the optimum number of capacitors and boost converter configuration to give the highest possible overall step-up efficiency.

Acknowledgements

This work is funded by the National University of Ireland, Galway (NUIG) under the college of engineering postgraduate scholarship.

References

[1] S. Roundy, D. Steingart, L. Frechette, P. Wright, and J. Rabaey, "Power sources for wireless sensor networks," *Eur. Conf. Wirel. Sens. Networks*, pp. 1–17, 2004.

[2] D. Newell, M. Duffy, and R. Twohig, "Effects of DSSC configuration on power conversion efficiency," in *Applied Power Electronics ...*, 2014, pp. 3155–3159.

[3] G. Palumbo and D. Pappalardo, "Charge pump circuits: An overview on design strategies and topologies," *Circuits Syst. Mag. ...*, 2010.

[4] T. Tanzawa, "On two-phase switched-capacitor multipliers with minimum circuit area," *IEEE Trans. Circuits Syst. I Regul. Pap.*, vol. 57, no. 10, pp. 2602–2608, 2010.

[5] Cooper Bussmann, "B-Series Supercapacitors," 2011. [Online]. Available: http://www.cooperindustries.com/content/dam/public/bussmann/Electronics/Resources/product-datasheets/Bus_Elx_DS_4390_B_Series.pdf.

[6] A. S. Weddell, G. V. Merrett, and B. M. Al-Hashimi, "Photovoltaic Sample-and-Hold Circuit Enabling MPPT Indoors for Low-Power Systems," *IEEE Trans. Circuits Syst. I Regul. Pap.*, vol. 59, no. 6, pp. 1196–1204, Jun. 2012.

[7] W. Jung, S. Oh, S. Bang, Y. Lee, Z. Foo, G. Kim, Y. Zhang, D. Sylvester, and D. Blaauw, "An ultra-low power fully integrated energy harvester based on self-oscillating switched-capacitor voltage doubler," *IEEE J. Solid-State Circuits*, vol. 49, no. 12, pp. 2800–2811, 2014.

[8] H. Yang and Y. Zhang, "Analysis of supercapacitor energy loss for power management in environmentally powered wireless sensor nodes," *IEEE Trans. Power Electron.*, vol. 28, no. 11, pp. 5391–5403, 2013.

[9] R. Faranda, "A new parameters identification procedure for simplified double layer capacitor two-branch model," *Electr. Power Syst. Res.*, vol. 80, no. 4, pp. 363–371, 2010.

[10] A. Weddell and G. Merrett, "Accurate supercapacitor modeling for energy harvesting wireless sensor nodes," *Circuits Syst. ...*, vol. 58, no. 12, pp. 911–915, 2011.

[11] M. Uno and H. Toyota, "Equalization Technique Utilizing Series–Parallel Connected Supercapacitors for Energy Storage System," *Sustain. Energy Technol. 2008. ICSET 2008. IEEE Int. Conf.*, 2008.

Self-powered In-plane Accelerometer Using Triboelectric Mechanism

Rahul Kumar Gupta [1], **Lokesh Dhakar** [1] **and Chengkuo Lee** [1,2]

[1] Department of Electrical and Computer Engineering, National University of Singapore, Singapore
[2] NUS Graduate School for Integrative Sciences and Engineering, Singapore

E-mail: elelc@nus.edu.sg

Abstract. This paper presents a self-powered triboelectric based accelerometer to detect wide range of in-plane acceleration utilizing the triboelectric mechanism. The freestanding sliding mode was utilized to realize the in-plane sensing. The fabricated device consists of soft polymer spring which displays wide detection range from ±1g to ±6g ($g=9.8m/s^2$) in *x* and *y* directions with sensitivity of 21mV/(g). The proposed device can be utilized for self-powered shock sensing in various future applications.

1. Introduction

Self- powered sensing has recently gained among modern researchers in view of increasing demand of sensors with development of Internet of Things (IoT) applications. In this trend, self-powered strain, pressure and motion sensors have been investigated in past years. Acceleration sensing is important for applications in various fields as portable electronics, healthcare and industrial applications. Over the years, researchers has developed accelerometers with different mechanisms which piezoelectric, electromagnetic, thermoelectric, capacitive and piezo-resistive. Some of these mechanisms such as piezo-resistive and capacitive need external bias to realize sensing. Conventionally, piezoelectric and electrostatic mechanisms have been commonly used for accelerometers [1-2]. Self-powered accelerometers also have been developed with use of piezoelectric mechanisms. Recently, triboelectric mechanism based on the coupling of contact-electrification and electrostatic induction has been widely investigated for energy harvesting and self-powered sensing applications since it was first demonstrated in 2012 [3]. The triboelectric mechanism using as motion sensor has been reported in few papers. A spherical triboelectric based accelerations sensor was illustrated by Zhang *et. al* which can detect acceleration in three axes. Another approach for motion sensing was presented by Yi *et. al,* which uses time difference obtained between the voltage peaks from pixel arrays to calculate accurate velocity and acceleration of moving object. The time interval information is required for such approach and further mathematical operation is complex to realize without any processing units. A much simpler device was demonstrated as three dimensional acceleration sensor [6] which can achieve wide detection range in three directions by integration of three standalone acceleration sensors. But integration of three individual devices makes it to be bulky and non-robust. In this paper, we present a self-powered triboelectric based accelerometer to detect wide range of in-plane acceleration utilizing the triboelectric mechanism. The freestanding sliding mode was utilized to realize the in-plane sensing. The fabricated device consists of soft polymer spring which displays wide detection range from ±1g to ±6g ($g=9.8m/s^2$) in x and y directions with sensitivity of 21mV/(g). The proposed device can be utilized for self-powered shock sensing in various future applications.

2. Device configuration and working principle

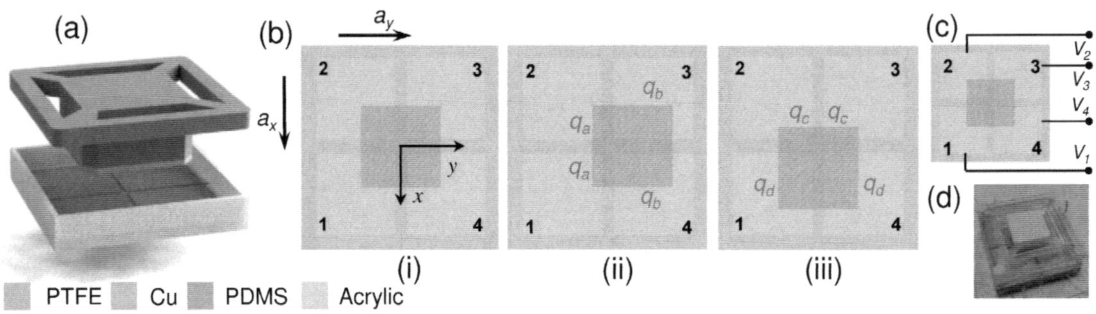

Figure 1. Schematic illustration of the Triboelectric accelerometer, 1(b) Working mechanism of the device in x and y-axis; equal charge generated with similar sliding area (c) Voltage representation of four pixels (d) fabricated accelerometer

Herein we fabricated a novel device utilizing the freestanding mode [7] to realize the accelerometer. The working mechanism for freestanding sliding mode is shown in Fig. 1(a-b). As shown in Fig. 1(a), the proposed accelerometer comprises of top and bottom parts with two triboelectrically opposite materials. The top part consists of proof mass of solid acrylic (2cm×2cm×1cm) attached with a PTFE layer (2cm×2cm×200μm). The proof mass is suspended through PDMS polymer springs (Fig.1a, d) enabling it to move in-plane with respect to fixed bottom triboelectric layer when subjected to external mechanical vibration. The bottom part consists of four copper pixels (1cm×1cm×35μm), which serve dual purpose of charge generation by triboelectrification and also function as electrode to collect the generated charge. Fig. 1(b) represents the working mechanism of the device in x and y directions; Voltage V_1, V_2, V_3 and V_4 are electrical output from four pixels as shown in Fig. 1(c). The two metal electrodes as bottom layers are stationary, whereas upper polytetrafluoroethylene (PTFE) dielectric layer can move freely from one end to other end. With relative motion between upper and lower triboelectric layers, positive and negative charge is generated on metal electrode and dielectric layer respectively such that total charge is zero. As PTFE dielectric layer slides with copper surface, electrons will be injected into PTFE from copper electrode according to triboelectric series [8]. Thus, voltage between two electrodes can be expressed by a $V = \sigma\Delta A/C$, where σ is surface charge density of PTFE layer, ΔA is change in area from initial to final state and C is capacitance between stationary electrodes.

Fig. 2(a-d) shows comparison between the two possible approaches namely metal TENG and dielectric TENG method [9]. As shown in Fig. (2d) the simulated V_{oc}-x curve exhibit linear relation due to which this configuration was adopted to implement the accelerometer. For dielectric TENG, since there is no top metal electrode, the generated charge cannot be redistributed. Due to absence of charge re-distribution, capacitance for dielectric mode remains constant irrespective of any displacement. For metal TENG, overall capacitance is maximum at the centre and minimum at two edges leading to non-linear behaviour of open circuit voltage derived from bottom electrodes. Due to constant capacitance between the two electrodes for dielectric mode, it results in linear V_{oc}-x curve which is desirable for acceleration sensing. With above analysis, dielectric TENG is found to be suitable mode for realizing the accelerometer. Two freestanding sliding modes based on only-dielectric and with metal-electrode has been simulated using analytical methods. Both structures look very similar, but due to presence of metal electrode in Fig. 2(a) results in charge re-distribution whereas in Fig. 2(b) charge re-distribution won't take place. We have simulated expressions for both cases using basic electrostatic principle. For metal-electrode mode, when PTFE moves in positive *x*-direction by displacement *x*, overlap area between upper and bottom layers can be given as $l(l-x)$ and $l(l+x)$ respectively where $2l$ is length of moving and stationary layers.

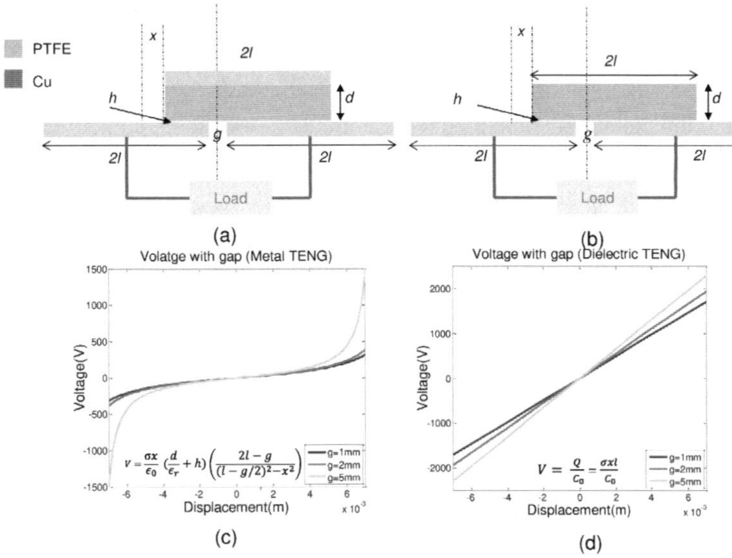

Figure 2. (a, c) Representation of Metal TENG method and dependence of open circuit voltage with displacement (x) showing non-linear behaviour, 2(b, d) Dielectric TENG method and dependence of open circuit voltage with displacement (x) displays linear relationship.

3. Experiment and results

The assembled device (Fig.1d) was mounted on electromagnetic shaker to characterize the acceleration sensing functionality of device. The output voltage from four pixels were recorded from digital oscilloscope for different applied accelerations in *x* and *y* directions. Fig.3 and 4 present the time-dependent voltage signals from four pixels for *x* and *y* directions taken at 1g and dwell frequency of 183Hz. In Fig.3, voltage generated due to *x*-axis acceleration is shown; it can be noted that the pair voltages (V_1, V_2) and (V_3, V_4) are 180° out of phase whereas the voltages (V_2, V_3) and (V_1, V_4) are in phase. Similarly for y-axis acceleration, the pair voltages (V_2, V_3) and (V_1, V_4) are 180° out of phase and voltages (V_1, V_2) and (V_3, V_4) are in phase as shown in Fig.4. Thus, the calculated voltage $V_x=V_{12}+V_{34}$ and $V_y = V_{23}+V_{14}$ can be correlated to represent *x* and *y*-axis accelerations. The linear relation between voltage signal and applied acceleration can be observed from the *V-a* curve shown in Fig. 5(a-b). When the device was subjected to *x*-axis acceleration, it can be noted that voltage V_y is close to zero and similar behaviour was observed when *y*-axis acceleration was applied. From Fig. 5(a) and 5(b), we can conclude that the fabricated device has wide detection range from ±1g to ±6g (g=9.8m/s2) in *x* and *y* directions with sensitivity of 21mV/(g). The triboelectric output is low due to impedance mismatch between device and testing setup. Further, improvements can be done by eliminating the noise from external connections and optimizing the structure.

4. Conclusion

In this paper, a self-powered triboelectric based accelerometer was fabricated to detect wide range of in-plane acceleration utilizing the triboelectric mechanism. The fabricated device consists of soft polymer spring which displays wide detection range from ±1g to ±6g ($g=9.8m/s^2$) in *x* and *y* directions with sensitivity of 21mV/(g). The device performance was analytically studied and dielectric TENG was found suitable to obtain linear voltage acceleration relation.

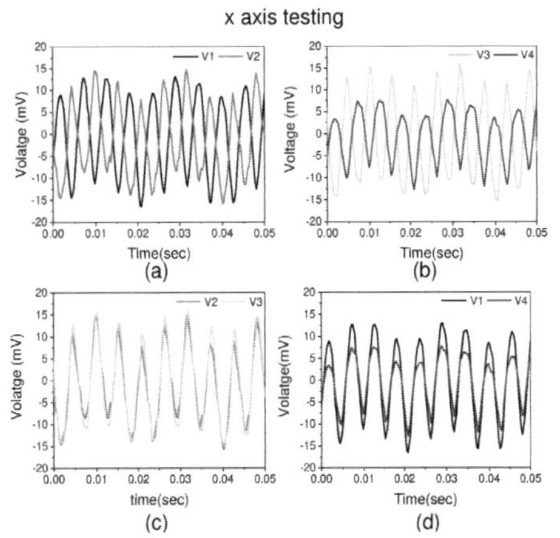

Figure 3. Time dependent voltage signals from four pixels for x-axis acceleration measured at 1g, 183Hz, (a-d) presents different pair voltages

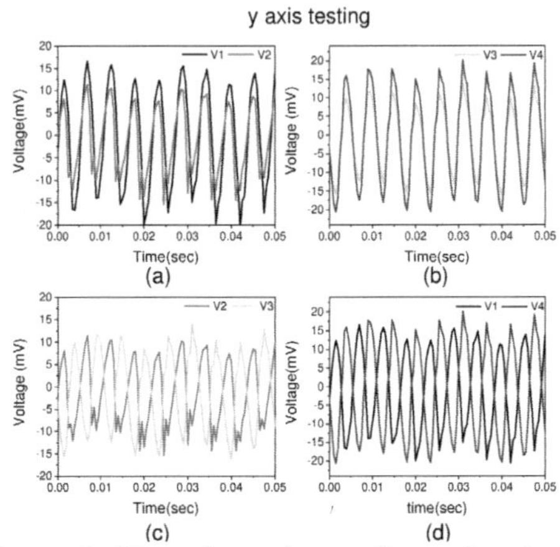

Figure 4. Time dependent voltage signals from four pixels for y-axis acceleration measured at 1g, 183Hz, (a-d) presents different pair voltages

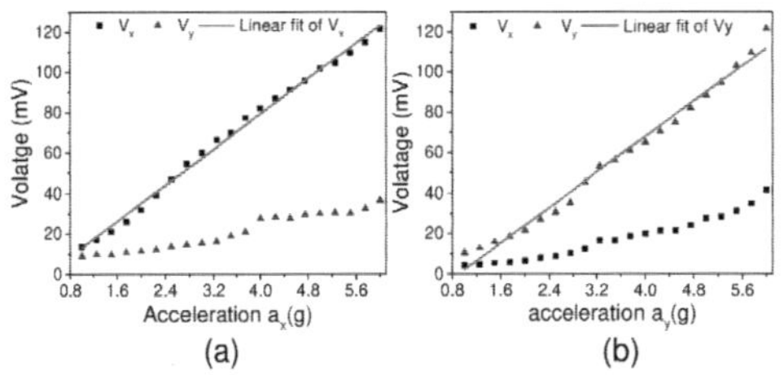

Figure 5. Linear relationship between output voltage and applied acceleration for (a) x-axis and (b) y-axis

Acknowledgements
This work is supported NRF2011 NRF-CRP001-057 Program 'Self-powered body sensor for disease management and prevention-orientated healthcare' (R-263-000-A27-281) from the National Research Foundation (NRF), Singapore.

References

[1] N. Yazdi, F. Ayazi, and K. Najafi 1998 *Proc. of the IEEE* **86(8)** 1640-1659.
[2] G.A. MacDonald 1990 *Sensors and Actuators A: Physical* **21(1-3)** 303-307.
[3] Feng-Ru Fan, Zhong-Qun Tian, and Zhong Lin Wang 2012 *Nano Energy* **1(2)** 328-334.
[4] Hulin Zhang, Ya Yang, Yuanjie Su, Jun Chen, Katherine Adams, Sangmin Lee, Chenguo Hu, and Zhong Lin Wang 2014 *Advanced Functional Materials* **24(10)** 1401-1407.
[5] Yi, Fang, Long Lin, Simiao Niu, Jin Yang, Wenzhuo Wu, Sihong Wang, Qingliang Liao, Yue Zhang, and Zhong Lin Wang 2014 *Advanced Functional Materials* **24(47)** 7488-7494.
[6] Yao Kun Pang, Xiao Hui Li, Meng Xiao Chen, Chang Bao Han, Chi Zhang, and Zhong Lin Wang 2015 *ACS applied materials & interfaces* **7(34)** 19076-19082.
[7] S. Wang, Y. Xie, S. Niu, L. Lin, and Z. L. Wang 2014 *Advanced Materials* **26(18)** 2818-2824.
[8] Zhong Lin Wang, 2013 *ACS nano* **7(11)** 9533-9557.
[9] Simiao Niu, Ying Liu, Xiangyu Chen, Sihong Wang, Yu Sheng Zhou, Long Lin, Yannan Xie, and Zhong Lin Wang 2015 *Nano Energy* **12** 760-774

An integrated microcombustor and photonic crystal emitter for thermophotovoltaics

Walker R. Chan[1], Veronika Stelmakh[1,2], William R. Allmon[3], Christopher M. Waits[3], Marin Soljacic[4], John D. Joannopoulos[1,4] and Ivan Celanovic[1]

[1] Institute for Soldier Nanotechnologies, Massachusetts Institute of Technology, 77 Massachusetts Avenue, Cambridge, Massachusetts 02139, USA
[2] Department of Electrical Engineering, Massachusetts Institute of Technology, 77 Massachusetts Ave., Cambridge, Massachusetts 02139, USA
[3] Sensors and Electron Devices Directorate, US Army Research Laboratory, 2800 Powder Mill Rd., Adelphi, Maryland 20783, USA
[4] Department of Physics, Massachusetts Institute of Technology, 77 Massachusetts Ave., Cambridge, Massachusetts 02139, USA

E-mail: `wrchan@mit.edu`

Abstract. Thermophotovoltaic (TPV) energy conversion is appealing for portable millimeter-scale generators because of its simplicity, but it relies on a high temperatures. The performance and reliability of the high-temperature components, a microcombustor and a photonic crystal emitter, has proven challenging because they are subjected to 1000–1200°C and stresses arising from thermal expansion mismatches. In this paper, we adopt the industrial process of diffusion brazing to fabricate an integrated microcombustor and photonic crystal by bonding stacked metal layers. Diffusion brazing is simpler and faster than previous approaches of silicon MEMS and welded metal, and the end result is more robust.

Power demand for mobile electronics and robotics continues to rise but current Li-ion battery chemistry is approaching its thermodynamic limit [1]. Thermophotovoltaic (TPV) energy conversion is promising as a high specific energy, millimeter-scale portable power source capable of harnessing the energy content of hydrocarbon fuels (12.6 kWh/kg compared to 180 Wh/kg for batteries) without moving parts or working fluids. In TPV, a microcombustor heats a photonic crystal emitter to incandescence and the resulting spectrally confined thermal radiation drives low-bandgap photovoltaic (PV) cells to generate electricity, as shown in Fig. 1(a). The microcombustor reacts propane and oxygen in a catalyst-loaded serpentine channel to heat the photonic crystal to 1000–1200°C. The microstructured tantalum 2D photonic crystal is engineered to only emit thermal radiation that is convertible by the PV cell, leading to high efficiency [2–4]. In this paper, we report a simple and stable method for fabricating an integrated microcombustor and photonic crystal (hot side assembly) by diffusion brazing.

Designing a stable TPV hot side has been challenging because of the high temperatures and the thermo-mechanical stresses arising from thermal expansion mismatch between the microcombustor and photonic crystal. Microcombustors (for TPV and other applications) have been fabricated from silicon by MEMS techniques [5–7], from laminated metal layers by diffusion bonding [8], and from welded metal components [9, 10]. These methods are difficult

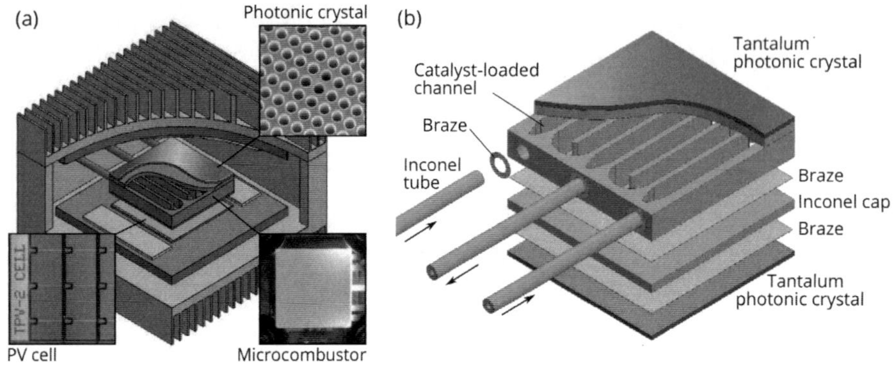

Figure 1. (a) Rendering of a millimeter-scale TPV generator with integrated microcombustor and photonic crystal hot side and a PV cell. (b) Exploded view of the hot side.

and unreliable. In the past, to integrate the photonic crystal, we directly deposited a multilayer silicon/silicon dioxide stack (1D photonic crystal) onto a MEMS microcombustor [7], or welded a 2D tantalum photonic crystal to an Inconel microcombustor [11]. The optical performance offered by the multilayer stack and the thermal contact offered by welding were not ideal. Diffusion brazing can be used to both fabricate the microcombustor and integrate the tantalum photonic crystal without any of these limitations.

Diffusion brazing is an industrial process similar to brazing, however addition and subsequent diffusion of melting point depressants out of the brazing alloy increases its remelt temperature—allowing the assembly to be reliably operated *above* the original brazing temperature. In this work, we chose BNi-2 (Lucas-Milhaupt) for the brazing alloy, with a solidus temperature of 971°C and a liquidus temperature of 999°C, and the following composition: 7% chromium, 3% boron, 4.5% silicon, 3.0% iron, and balance nickel. The braze alloy is subjected to a prolonged anneal above its liquidus, during which the silicon and boron diffuse out and the molten alloy undergoes isothermal solidification [12–14]. Assuming the silicon and boron are completely removed, the remelt temperature exceeds 1400°C [15]. The increase in remelt temperature enables us to use the same braze alloy for multiple brazing steps, to avoid exposing the photonic crystal to a higher temperature than absolutely required, and to perform the brazing in a low-cost furnace which was limited to 1200°C even through the target operating temperature is 1000–1200°C.

We were concerned that significant stresses, and therefore deflections, could occur in the final brazed assembly owing to the differential thermal expansion between Inconel and tantalum. Inconel was required for the microcombustor for its high-temperature oxidation resistance, low cost, and machinability; tantalum was required for the photonic crystal for its low vapor pressure, low optical loss, and etchability. We modeled the deflection of an Inconel-tantalum bimetallic strip in SolidWorks Simulation using available material properties and found significant deflection. To test the validity of the simulation, two test coupons with 1.02 mm and 3.18 mm thick Inconel and 0.51 mm thick tantalum were prepared, and deflections of 1.1 and 0.2 mm were measured, 2–3 times greater than simulated values possibly because of inconsistent material properties or plastic deformation. To prevent deflection, we balanced the stress with a symmetric design and eliminated long spans by brazing each channel wall to both the top and bottom Inconel caps, as shown in Fig. 1(b).

The microcombustor channels were designed as described in Ref. 9 and fabricated by abrasive water jet cutting from Inconel sheet stock. The holes for the tubes were reamed over-sized to ensure a consistent 25 µm gap between the tube and hole so that the braze would reliably flow by capillary action. Tubing was commercially available and was simply cut to length. We

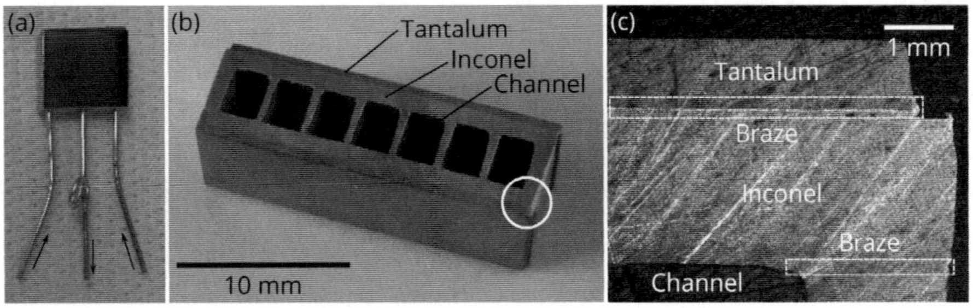

Figure 2. (a) Completed hot side with a diffraction pattern visible on the photonic crystal. Arrows indicate direction of gas flow. (b) Cross section with the channels visible. (c) A micrograph of the indicated corner of the hot side assembly with the tantalum-Inconel braze joint visible and the Inconel-Inconel braze indicated but not visible.

bent a loop in the center tube to relieve stress arising from differing thermal expansion between inlet and outlet tubes, visible in Fig. 2(a). The photonic crystal was fabricated separately, as described in Ref. 4. Braze preforms were fabricated from foil by photochemical machining using dry film photoresist and ferric chloride etching solution commonly used for printed circuit board fabrication. Preforms were dimensioned to deliver a slight excess of braze alloy to the joint.

Brazing was conducted in three steps: tubes were brazed to the channels, caps were brazed to seal the channels, and the photonic crystals were brazed to the completed microcombustor. Jigs were used to hold components in place for each of the steps. The jigs used for the first and second brazing operations were machined from Inconel. The one for the final brazing operation was machined from tantalum to avoid contamination of the photonic crystal, as some outgasing of the Inconel was observed.

Brazing was performed a quartz tube furnace evacuated by a turbomolecular pump. We relied on the high temperature and high vacuum to shift the chemical equilibrium to favor the dissociation of surface oxides before the braze alloy melted [12–14]. Flux and reactive atmospheres (e.g. hydrogen) were avoided to prevent contamination of the photonic crystal. After pump-down, the furnace was ramped at 10°C/minute, with one hour stops at 350°C and 500°C for degassing, to a final brazing temperature of 1100°C. When the brazing temperature was reached, furnace pressure would initially spike to $\sim 5 \times 10^{-5}$ Torr then reduce to $\sim 3 \times 10^{-6}$ Torr. The temperature was held at 1100°C for two hours to ensure full diffusion before returning to room temperature.

During process development, we fabricated ten hot side assemblies with bare tantalum substituted for the photonic crystal, and experienced high fabrication yield as defined by visual inspection and helium leak detection. A cross section of one is shown in Fig. 2(b) and (c). One microcombustor without a photonic crystal was operated for one week without failure or visual damage. Once we had perfected the brazing process, we used a photonic crystal instead of bare tantalum. The finished assembly is shown in Fig. 2(a). Reflectance measurements of the photonic crystal before and after brazing confirmed no degradation of optical properties of the photonic crystal.

In summary, we adopted the industrial process of diffusion brazing and fabricated an integrated Inconel microcombustor and tantalum photonic crystal to serve as the hot side of a millimeter-scale TPV generator. In diffusion brazing, fast-diffusing elements contained in the alloy diffuse out of the joint during heating, thus increasing the remelt temperature of the braze above the original brazing temperature. Table 1 compares this method to prior methods of silicon MEMS and welded Inconel. This approach represents a breakthrough for TPV because it integrates the microcombustor and photonic crystal, and is fast, simple, and reliable. Diffusion

	Silicon MEMS	Welded Inconel	Brazed Inconel
Channels	KOH etched silicon, sealed by wafer bonding	Machined Inconel, sealed by welding	Machined Inconel, sealed by brazing
Tubes	Glass soldered borosilicate tubes	Welded Inconel tubes	Brazed Inconel tubes
Emitter	Five layer Si/SiO_2 stack deposited directly on microcombustor	Tantalum 2D fabricated separately and welded to microcombustor	Tantalum 2D fabricated separately and brazed to microcombustor
Results	2.7% at 300 mW_e but unreliable and limited lifetime	2.4% at 1.6 W_e limited by poor thermal contact	Experiments in progress

Table 1. Comparison of the silicon MEMS hot side [7], welded Inconel hot side [9, 11], and the novel brazed Inconel hot side.

brazed Inconel is applicable to microcombustor fabrication for other applications, and to other millimeter-scale devices operating at high temperatures.

Acknowledgments

Photonic crystal fabrication was supported as part of the Solid-State Solar-Thermal Energy Conversion Center (S3TEC), an Energy Frontier Research Center funded by the U.S. Department of Energy (DOE), Office of Science, Basic Energy Sciences (BES), under Award No. DE-SC0001299 / DE-FG02-09ER46577. This work was additionally supported by the Army Research Office through the Institute for Soldier Nanotechnologies under Contract No. W911NF-13-D-0001 and the Micro Autonomous Systems and Technology Collaborative Technology Alliance under Contract No. 892730.

References

[1] Zu C X and Li H 2011 *Energy & Environmental Science* **4** 2614
[2] Celanovic I, Jovanovic N and Kassakian J 2008 *Applied Physics Letters* **92** 193101
[3] Yeng Y X, Ghebrebrhan M, Bermel P, Chan W R, Joannopoulos J D, Soljacic M and Celanovic I 2012 *Proceedings of the National Academy of Sciences* **109** 2280–2285
[4] Stelmakh V, Rinnerbauer V, Geil R D, Aimone P R, Senkevich J J, Joannopoulos J D, Soljačić M and Celanovic I 2013 *Applied Physics Letters* **103** 0–4
[5] Peck J, Jacobson S A and Waitz I A 2010 *Journal of Engineering for Gas Turbines and Power* **133** 1–10
[6] Marton C H, Haldeman G S and Jensen K F 2011 *Industrial & Engineering Chemistry Research* **50** 8468–8475
[7] Chan W R, Bermel P, Pilawa-Podgurski R C N, Marton C H, Jensen K F, Senkevich J J, Joannopoulos J D, Soljacic M and Celanovic I 2013 *Proceedings of the National Academy of Sciences of the United States of America* **110** 5309–14
[8] Matson D W, Martin P M, Stewart D C, Tonkovich A, White M, Zilka J and Roberts G 1999 *3rd International Conference on Microreaction Technology 3rd International Conference on Microreaction Technology*
[9] Chan W R, Wilhite B A, Senkevich J J, Soljacic M, Joannopoulos J and Celanovic I 2013 *Journal of Physics: Conference Series* vol 476 p 12017
[10] Tolmachoff E D, Allmon W and Waits C M 2014 *Applied Energy* **128** 111–118
[11] Chan W R, Stelmakh V, Waits C M, Soljacic M, Joannopoulos J D and Celanovic I 2015 *Journal of Physics: Conference Series* vol 660 p 012069
[12] Pattee H E 1980 *Soucebook on Brazing and Brazing Technology* ed Schwartz M M (Metals Park, OH: American Society for Metals) pp 17–63
[13] Schwartz M M 1980 *Soucebook on Brazing and Brazing Technology* ed Schwartz M M (Metals Park, OH: American Society for Metals) pp 109–132
[14] Schwartz M 2003 *Brazing* (Materials Park, OH: American Society for Metals)
[15] Villars P 1995 *Handbook of ternary alloy phase diagrams* (Materials Park, OH: ASM International)

Modelling and fabrication of a compliant centrally supported meandering piezoelectric energy harvester using screen-printing technology

E Fernandes[1], S Zarabi[1], H Debéda[2], C Lucat[2], D Nairn[1], L Wei[1] and A Salehian[1]

[1] University of Waterloo, Waterloo, Canada
[2] Université de Bordeaux, Laboratoire IMS, Talence, France

Email: salehian@uwaterloo.ca

Abstract. This paper reports the design, simulation, fabrication, and experimental testing of a low-frequency piezoelectric energy harvester fabricated using screen-printing technology. A centrally supported meandering geometry is chosen to reduce the torsional mode effects during the unit vibrations and to achieve better power efficiency. The design experiences alternating stress along the beams and a strain-matching polarization technique is used to minimize voltage cancellation. The test results are validated against the finite element solution for which there is a good agreement.

1. Introduction

The use of piezoelectric cantilever beams for energy harvesting has been studied extensively and used for wireless sensing applications [1]. One application is the electromagnetic harvesting from current carrying wires. Previous work by [2, 3] demonstrated this capability with a cantilever design with piezoelectric layers and a permanent tip magnet. The maximum power output of such structures occurs when the fundamental frequency of the structure matches the dominant frequency available through the electric current. The major drawback in using cantilevers for MEMS vibrational energy harvesters is their relatively high natural frequency resulting from their short lengths. Work done by [4] for the same application produced a quad-folded cantilevered design to fit within a 100mm^2 area. This design, however, is susceptible to some torsional vibrations and may include several strain nodes throughout the harvester that reduces the overall harvesting efficiency from the bending modes for which more complex electrode configurations have to be used.

PZT components (T_c ~300°C) [5] are often considered for energy harvesting because of their outstanding electromechanical properties compared to those of PVDF [6] or AlN [4]. Although PZT ceramics (>100 μm thick) demonstrate high intrinsic coupling factors and are commercially available, they are relatively expensive and their geometry is often limited to simple shapes which may necessitate additional expensive micromachining steps for specific applications [7]. PZT thin films (<1 μm) are often deposited onto silicon or metallic supports, but are not suitable when large power is required due to their weak electromechanical coupling. PZT thick-films have been found to make good candidates for MEMS applications due to their higher coupling, and can bridge the 1-100 μm gap between thin film and bulk components [8]. They can be formed by screen-printing through a mask. Free-standing thick-film structures can be mass produced and do not need to be assembled manually, like with ceramics. It is therefore possible to create quite complex structures with a series of relatively simple fabrication steps. They can also be integrated with other thin or thick-film layers and microelectronic components, thereby offering an appealing approach for this project.

Free-standing piezoelectric thick-films fabricated on a silicon substrate have been shown to produce ~3 μW from an acceleration of 2.8 m/s^2 at 250 Hz and only up to ~33 μW at about five times the acceleration [9]. For the same 2.8 m/s^2 acceleration thick-films on stainless steel substrates have been shown to deliver power of 240 μW harvested at 66.2 Hz [10].

This paper reports the design, simulation, fabrication, and experimental testing of a compliant centrally supported meandering piezoelectric energy harvester (EH) for current monitoring applications. Future work involves employment of a tip magnet to both reduce the natural frequency of the present unit, and to harvest from the electromagnetic flux around a wire carrying an AC current.

2. Energy harvester design, modelling, and simulation

Compared to a simple cantilever of similar area, a zigzag design reduces the fundamental natural frequency of the structure and helps maintain a high power density [11]. A so-called meandering design uses symmetry about the free end of a zigzag and contains two fixed ends [12]. A centrally supported meandering geometry, used in this paper, is created by mirroring the zigzag design about the clamped beam. This design achieves better power efficiency than a zigzag due to the reduction of the torsional mode effects in the clamped beam, which results in bending modes being dominant during harvesting. The geometry of the harvester is shown in figure 1. Using COMSOL, the simulated strain contour plot for the fundamental frequency is obtained and presented in figure 2.

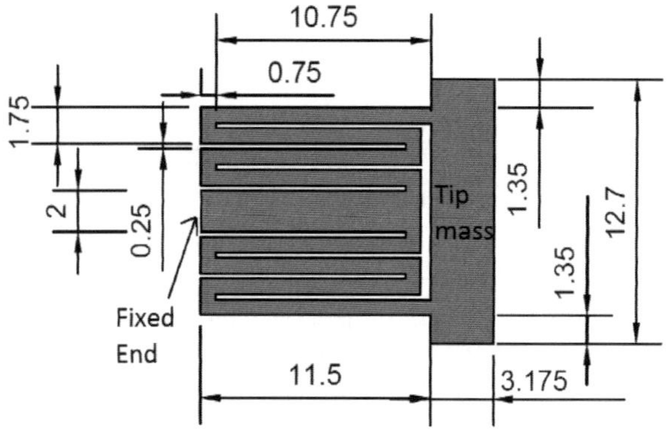

Figure 1. Schematic of the harvester (mm).

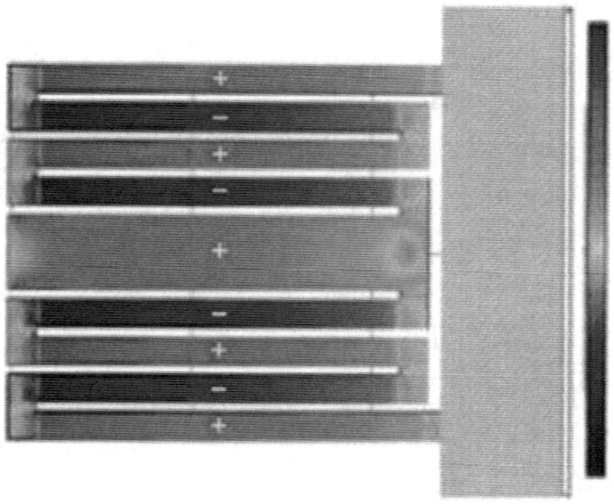

Figure 2. Strain contour plot showing tension (red) and compression (blue).

As shown in figure 2 the adjacent beams experience alternating strains and each individual beam experiences either tension or compression along its entire length during the cyclic motion at its fundamental frequency. This is one of the most attractive features of the proposed design and quite critical as any strain node within the length of the beam would require segmented electrodes to avoid a charge cancellation. The segmented electrode was adopted as a solution in [4].

3. Fabrication process

This screen-printing method is a low-cost alternative process to classical silicon manufacturing. The most valuable advantages of screen-printing are the low investment costs for equipment and choice of materials including the substrate and paste. Screen printing also allows for thicker PZT layers which can be in the range of 1-100 μm.

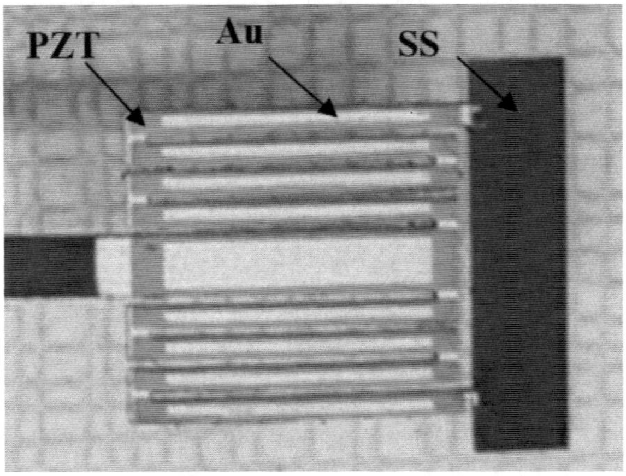

Figure 3. PZT energy harvester fabricated using screen-printing technology.

Figure 3 shows the EH fabricated by screen-printing techniques. It consists of a PZT layer sandwiched between two gold electrodes deposited on a stainless steel substrate. The substrate material is 301 stainless steel (SS) with a thickness of 254 μm and was laser cut according to the design geometry. The SS was selected for its compatibility with the firing temperature required by the screen-printable process and for its low Young's Modulus of 193GPa. The gold ink used for the electrode layer is commercially available (ESL8836 from ElectroScience Laboratories), whereas the PZT is prepared at the IMS laboratory at the University of Bordeaux. The piezoelectric paste is prepared from a commercial piezoelectric PZT powder (PZ26 from Ferroperm) mixed with 3wt% LBCu (25wt% Li_2CO_3, 40wt% Bi_2O_3, 35%wt CuO), blended with ESL4000 organic vehicle from ElectroScience Laboratories [13].

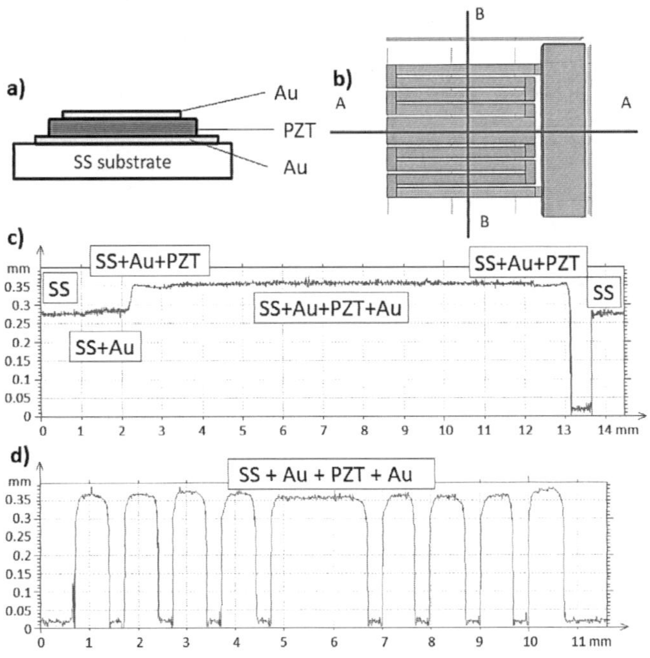

Figure 4. a) Cross section of stainless steel substrate with screen printed layers b) Thickness measurement lines across the sample for optical profiling c) Optical profile along A-A d) Optical profile along B-B

After aligning the substrate for printing, the bottom electrode is printed, dried for 20 minutes at 120°C then fired for 10 minutes at 850°C. Next, a layer of PZT and its top electrode are printed successively, and dried for 20 minutes at 120°C between each deposition. Then, the dried samples are isostatically pressed for one minute at 40 MPa to improve the densification of the layers. Afterwards, the samples are fired for 2 hours at 900°C in air atmosphere. The densification and firing reduced the thickness of the PZT layer to approximately 55 μm. The thickness profiles of the dried Au/PZT/Au layers on the SS substrate are shown in figure 4.

Polarization is required for the PZT layer to exhibit piezoelectric properties. Polarization of the PZT occurs close to the measured Curie temperature (~280°C) for five minutes at an electric field of 50kV/cm under a dry nitrogen environment. A strain-matched polarization (SMP) technique is used to minimize voltage cancellation across the harvester. As described in [12] the piezoelectric material of each beam is poled such that each adjacent member has the opposite poling direction. By employing this approach all of the electrodes will have the same voltage polarity. This allows for a single continuous electrode to be used across the bottom layer of the EH. A ball-bonding technique was used to make micro-wire connections between the top layers of electrodes. As a preliminary step, only four beams that experienced the same type of strain were poled.

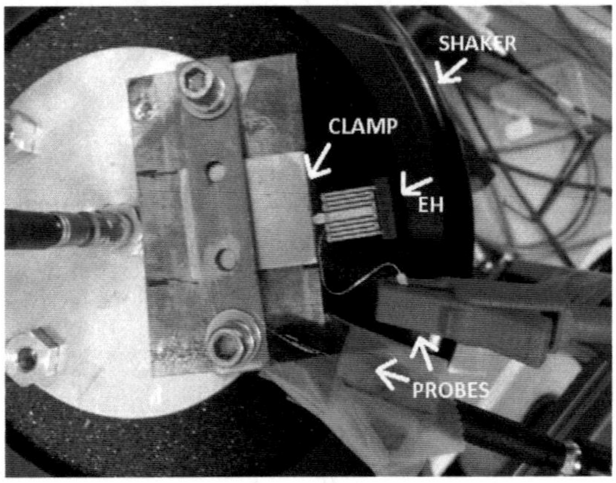

Figure 5. Experimental test setup of the fabricated EH without a tip mass.

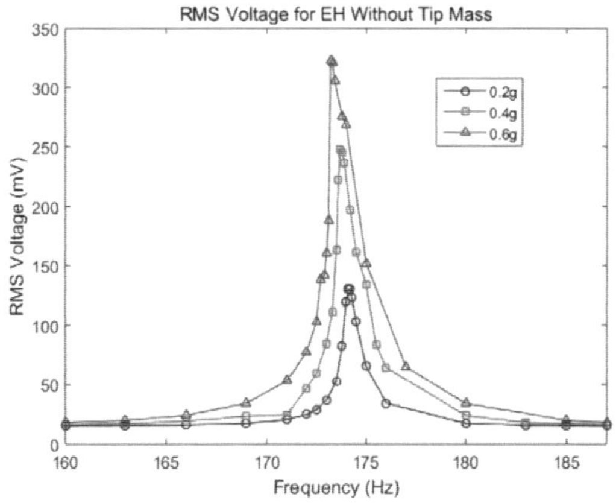

Figure 6. Experimental open-circuit voltage of the harvester without tip mass for various accelerations.

4. Experimental procedure and results
To characterize the PZT parameters, the EH was mounted on an electromagnetic shaker and the voltage readings were taken across the top and bottom electrodes, as shown in figure 5. Figure 6 shows the frequency response of the output voltages for the EH in an open-circuit state for multiple base accelerations.

The maximum RMS voltage produced by the harvester without a tip mass was 322 mV at 0.6 *g* with a natural frequency of 173 Hz. Using COMSOL simulations an eigenfrequency analysis was performed on the same unit. The thickness profile used for the simulation is shown in figure 4. The simulations results predict a fundamental frequency of 171 Hz that is in great agreement with the test results. Future work involves adding the tip magnetic mass to both reduce the natural frequency for the unit and to help harvest from the electromagnetic flux. Further calculation using the model shows that adding a neodymium permanent magnet tip mass of 0.70 g would tune the overall natural frequency to 60 Hz, conforming to the frequency of AC power in the North American power grid.

5. Conclusion
In summary, a piezoelectric EH of MEMS scale has been successfully fabricated using low-cost screen-printing technology on a stainless steel substrate. The centrally supported meandering design experiences alternating strain in the subsequent beams. The PZT is poled according to the strain matched polarization technique to minimize voltage cancellation. The simulation and test results are shown to be in good agreement.

Future work will include improving the adhesion to the substrate and the polarization of all beams. The geometry will also be optimized for greater power density and the EH will be integrated with power conditioning circuitry for the electromagnetic energy harvesting of current carrying lines.

Acknowledgements
This work was supported by grants from the Waterloo Institute of Sustainable Energy, Cisco Systems, UWUB grant and IRPG. The authors would like to thank Sofia Benouakta, Simon Grall, and Isabelle Favre for their help in the fabrication process.

References
[1] Roundy S, Wright P K and J. Rabaey 2003 A study of low level vibrations as a power source for wireless sensor nodes. *Computer Communications*, 26(11):1131–1144
[2] Pollock T 2014 Design, modelling, fabrication & testing of a miniature piezoelectric-based EMF energy harvester
[3] Paprotny I, Xu Q, Chan W W, White R M and Wright P K 2013 Electromechanical energy scavenging from current-carrying conductors. IEEE Sensors Journal, 13(1):190–201
[4] Paprotny I 2010 Modeling, design and fabrication of (10×10×4 mm^3) MEMS AC energy scavenger for smart grid applications *PowerMEMS*
[5] Beeby S *et al* 2006 Energy harvesting vibration sources for microsystems applications *Meas. Sci. Technol.*, vol. 17, no 12, pp.175-195
[6] Liu W 2014 Fabrication of spiral-shaped PVDF cantilever based vibration energy harvester, Proceedings of the 9th IEEE International Conference on Nano/Micro Engineered and Molecular Systems, Hawaii, USA, April 13-16
[7] Ibrahim M 2014 Modeling, fabrication, and experimental validation of hybrid piezo-magnetostrictive and piezomagnetic energy harvesting units *J. of Intelligent Material Systems and Structures* vol. 1 (13)
[8] Hindrichsen C 2010 Advantages of PZT thick film for MEMS sensors *Sensors and Actuators A*, vol 163, no 1, pp.9–14
[9] Xu R 2012 Screen printed PZT/PZT thick film bimorph MEMS cantilever device for vibration energy harvesting *Sensors and Actuators A* vol.188 pp.383– 388
[10] Zhu D 2011 A credit card sized self-powered smart sensor node *Sensors and Actuators A* vol.169, pp.317– 325
[11] Karami A 2010 Analytical Modeling and Experimental Verification of the Vibrations of the Zigzag Microstructure for Energy Harvesting," Journal of Vibration and Acoustics Vol **133**
[12] Berdy D 2012 Low-Frequencyy Meandering Piezoelectric Vibration Energy Harvester *IEEE Transactions on Ultrasonics, Ferroelectrics, and Frequency Control* Vol **59**
[13] Debéda H 2015 One-step firing for electroded PZT thick-films applied to MEMS *Smart Mater. Struct.* **24**

PowerMEMS 2016 IOP Publishing
Journal of Physics: Conference Series **773** (2016) 012110 doi:10.1088/1742-6596/773/1/012110

Electrical current measurement system for energy harvesting applications

S Heller, F Mueller, M Kroener and P Woias

Laboratory for Design of Microsystems, IMTEK, University of Freiburg,
Georges-Koehler-Allee 102, 79110 Freiburg, GERMANY

Email: simon.heller@imtek.uni-freiburg.de

Abstract. The measurement of the dynamic power consumption is an important task in the field of energy harvesting regarding the characterization and optimization of low power systems. This paper reports on the development and characterization of a computer-controlled measurement system for the measurement of the dynamic current consumption of low power systems in a range from 100 nA to 100 μA, with a time resolution down to 1 μs.

1. Introduction

From a deterministic point of view there is a general rule: "The better a system is understood, the more it can be improved". In energy harvesting, a system first of all is "better" if it is thoroughly optimized for minimum power consumption, which secondly leads to maximum functionality, reliability and availability. To achieve this, the power consumption, i.e. the current draw of an electronic circuit has to be well known, which typically is an easy task by measuring current with a multimeter. However, the current consumption of low power circuits and of embedded systems is not constant over time from several reasons: First, to increase power efficiency, only the circuit parts used for the current task are powered, while unneeded parts are turned off or set into a sleep mode. Second, the system itself shows a high dynamics in power consumption, with e.g. a low current draw for sensing and a significantly higher current draw for wireless communication. Such a change in power consumption will happen rapidly, i.e. within microseconds, and potentially over a large range of supply current.

Astonishingly, currently there is no commercial measurement system available that can perform a real-time current measurement analogous to an oscilloscope for small currents in the microampere range. Most suitable for this task is the picoammeter B2980A from *Keysight* with a maximum sample rate up to 20 kHz which is by far too low to detect, for example, the readout of a sensor or a short-time - but high power - wireless data transmission [1]. Another approach is the *EnergyTrace™* Technology developed by *TI* for their microcontroller platforms. By deriving the clock rate and the number of cycles of the switching signal of a buck-converter integrated in the debug-interface they calculate the power consumption for a logging period up to 5 min [2]. However, this system has to be powered by the debug interface, therefore self-sufficient systems cannot be analysed, as well as the power consumption of individual parts of the circuit cannot be measured. In this research, we present a system, which bridges all these gaps, and allows for measurement of all single components in a typical wireless energy-autonomous embedded system with a time resolution of up to 10 μs for currents from 0.1 μA to 100 μA.

2. Concept and preliminary considerations

2.1. Hardware concept

The goal of the project was the development of a low-cost measurement system, which is able to measure the dynamic power consumption of an embedded system in a range from 100 nA to 100 μA with a resolution of 1 μs. To achieve a simple system operation, the device should be controlled by a computer to enable an efficient post-processing of the data and the possibility of automated long-term measurements. The measurement principle of the system is identical to that of a multimeter, i.e, the measured current is derived from the voltage drop over a current sense resistor put in series to the investigated electronic circuit, by applying Ohm's law. To minimize the effect of a decreasing current consumption with decreasing supply voltage, respectively to maintain the proper functioning of the device under test, the measurement system is designed such, that the voltage drop over the sense resistor does not exceed a voltage of 10 mV. To evaluate such small voltage drops at the sense resistor the signal has to be amplified. The sense resistor is located at the high-side of the power supply chain, in order to prevent errors in the device under test from different ground levels. Therefore, the current sense amplifier needs a high common mode rejection ratio (CMRR) which is guaranteed by using an instrumentation amplifier.

A high resolution over the whole measurement range from 100 nA to 100 μA is achieved by splitting the measurement range into three ranges (see *figure 1*). To switch between the ranges, the total value of the current sense resistor is changed dynamically via two relays. The highest resistor value is always connected to the amplifier as a load, to protect the inputs from high voltage peaks, and further resistors are connected in parallel to decrease this value as required. As a safety feature a reverse voltage protection is implemented by a symmetric supply of the amplifier, which also allows the measurement of negative currents, but with half of the resolution.

Measurement range	Resistance
100 nA – 1 µA	$R_{ges} = R_1 = 10\ k\Omega$
1 µA – 10 µA	$R_{ges} = R_1 \parallel R_2 = 1\ k\Omega$
10 µA – 100 µA	$R_{ges} = R_1 \parallel R_2 \parallel R_3 = 100\ \Omega$

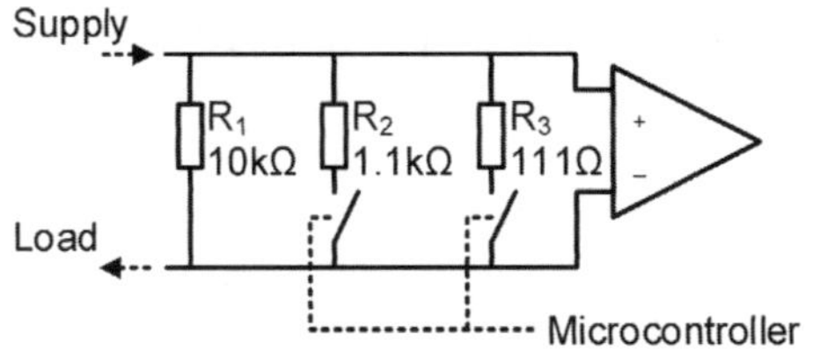

Figure 1. *Schematic of the implementation of the three measurement ranges.*

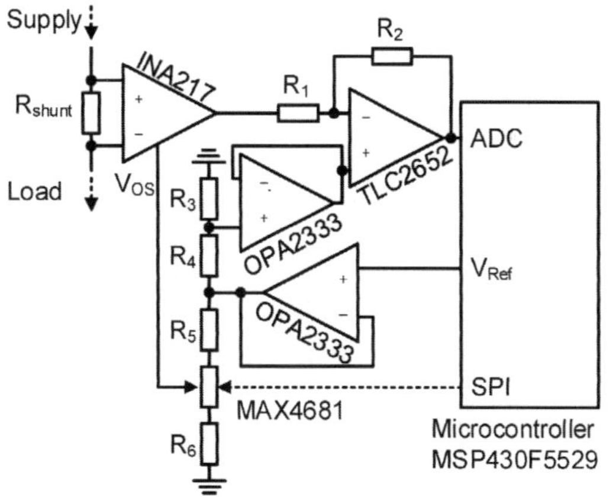

Figure 2. *Schematic of the offset compensation circuit.*

2.2. Interpretation of dynamic measurement results

One problem of dynamic current measurements is, that any amplifier typically shows a low-pass behavior, in combination with a limited slew rate. Both factors lead to a specific rise and fall time of the output signal, during which the measure of the output current is incorrect. On the other hand, an embedded system controlled by an internal microcontroller tends to show a stepped current consumption with a staircase-like behavior. Therefore, to minimize errors induced by the current sense amplifier, an averaged current approach can be used. For this, the energy present in a staircase step of the current curve can be calculated by integration of the current sense amplifier's output signal over time. A requirement to apply this method is an identical positive respectively negative slew rate of this amplifier.

3. Circuit design

The core of the measurement circuit is a low-noise instrumentation amplifier with a low offset voltage and a sufficiently high slew rate to guarantee high signal fidelity. Moreover the amplifier should have a high input impedance to minimize measurement errors introduced by the input offset currents of the amplifier. The best accordance to these requirements shows the monolithically integrated instrumentation amplifier INA217 from *TI*. At a gain of 100 the amplifier shows an appropriate bandwidth of 800 kHz, a slew rate of 15 V/μs, a CMRR of 116 dB, a moderate input offset current of 100 nA and a maximum input offset voltage of 300 μV [3].

Since an input offset voltage of 300 μV already introduces a measurement error of 3%, the offset voltage needs to be compensated. This can be achieved by adjusting the voltage at the *REF*-pin of the amplifier. In order to maintain the good ease of use, the circuit should perform the offset compensation automatically, which is realized with a microcontroller MSP430F5529 from *TI*. The controller measures the offset voltage with its internal ADC and sets an appropriate compensation resistance using a MAX5481 (*Maxim Integrated*) digital potentiometer (see *figure 2*). Thereby a theoretical resolution of 39 μV can be achieved. As the offset voltage is by an order of 100 smaller than the output voltage of the instrumentation amplifier, it is again amplified by a factor of 40 with a chopper-stabilized operational amplifier (TLC2652, *TI*), showing a maximum offset voltage of 0.5 μV, to take benefit of the whole input range of the ADC. To prevent the amplifier from saturation during a measurement, a relay is used to disconnect the offset measurement circuit.

For digitizing the instrumentation amplifier's output signal a 12-bit ADC (MAX11108, *Maxim Integrated*), is used. This ADC is controlled via an SPI interface and capable of sampling up to 3 million times per second. Due to the unipolar input of the ADC the output signal of the instrumentation amplifier must be shifted by half of the reference voltage using a precise voltage divider together with low offset operational amplifier (OPA2333, *TI*) as a buffer. The digital resolution (LSB) is therefore decreased to 610 μV at a reference voltage of 2.5 V.

4. Measurements results

4.1. DC Measurements

In a first measurement series the transfer function of the circuit was characterized. Therefore DC currents in a range from 100 nA to 100 μA generated by a sourcemeter (*Keithley* 2450) were applied and the output voltage was measured with a multimeter (*Agilent* U1252B). As expected, a linear transfer characteristic with different slopes depending on the measurement ranges (see *figure 3*) could be determined with an $R^2 > 0.999$.

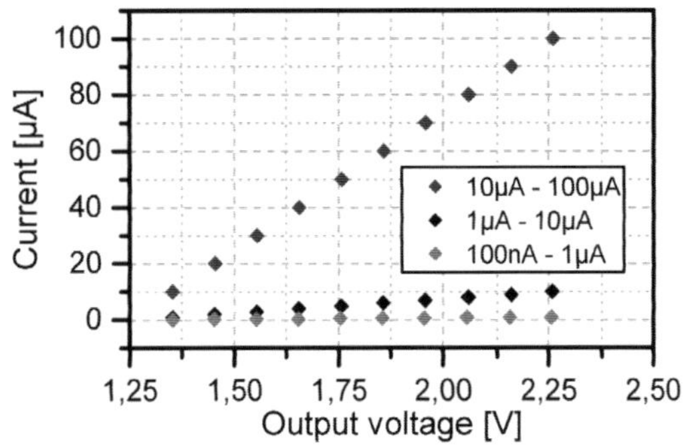

Figure 3. *Output voltage of the measurement system as a function of positive input currents for the three measurement ranges.*

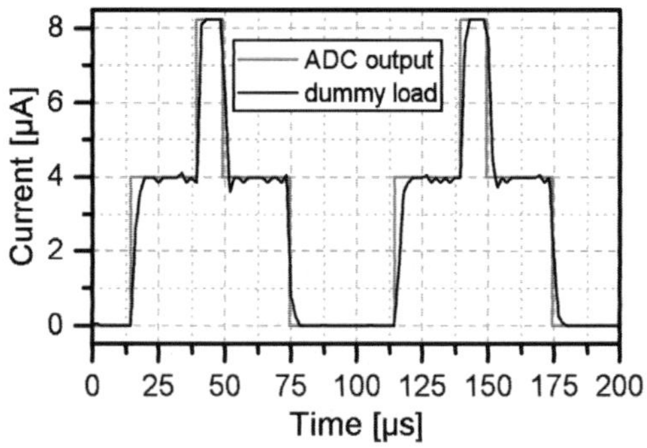

Figure 4. *Current course (black) recorded with the ADC at a sample rate of 500 kHz for a dummy load (red) toggling between 0 μA, 4 μA and 8 μA.*

To test the compensation of the offset voltage, the output voltage of the amplifier was measured five times with the multimeter before and after the automated offset compensation. Over the five measurements the initial offset voltage could be reduced from 8.48 mV to a maximum offset voltage of 84.0 μV. As this voltage represents only 13.8% of the LSB, the error caused by the offset voltage can be neglected.
A further error source is related to the input common mode voltage which cannot be perfectly suppressed by the amplifier. To determine this error, a voltage signal with a differential voltage of 5 mV and a variable common mode voltage varying between 0 V and 3 V was generated with the two outputs of a function generator (*Agilent* 33500B). With increasing common mode voltage a decrease of the amplified differential voltage could be detected. Considering the measurement at 0 V common mode as a reference, a maximum error of 1.37 % (mean 0.76 %) in the output signal is measured at a common mode voltage of 3 V.

4.2. AC Measurements
To determine the error related to the rise and fall time of the amplifier, rectangular pulses generated by the function generator, with an amplitude of 5 mV at a common mode voltage of 0 V were applied to its input. The output signal was recorded with an oscilloscope (*Tektronix* TDS2024B) and post processed in *MATLAB*. As expected, the error increases with the increasing influence of the rise and fall time. Up to a pulse length of 10 μs the error stays below 4 %, and for pulses up to 1 μs it is still below 7 %.
The function of the ADC was tested by applying a dummy load drawing a current between 0 μA and 4 μA respectively approx. 8 μA for different time intervals. The ADC is sampled with a frequency of 500 kHz. As the recorded current course in *figure 4* shows, the signal is of low noise and depicts the course of the real current consumption.

5. Conclusion and outlook
This paper reports the development of a measurement system for the measurement of the dynamic current consumption of low power circuits in a range from 100 nA - 100 μA (see *figure 5*). The implementation of an automated offset compensation with a maximum error of 13.8% of the LSB and a *LabVIEW* user interface leads to a system with a high ease of use. With a price of only 50 € for the electronic components and the housing, the system is affordable to everyone. For dynamic current measurements, an error of less then 4% is achieved for pulses of up to 10 μs, and an error less than 7% for pulses up to 1 μs.

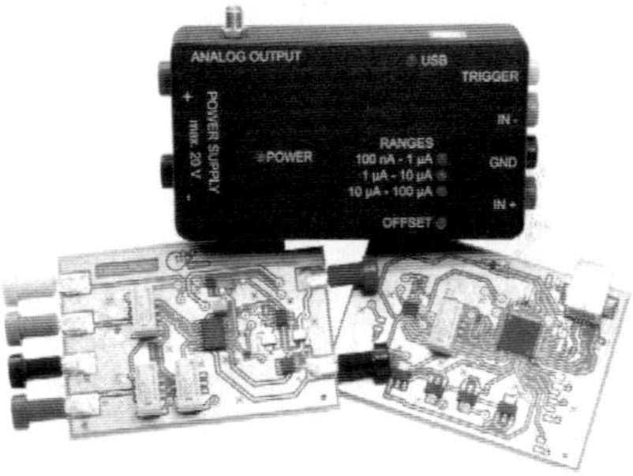

Figure 5. *Packaged measurement device (top) and the topside PCB (left) respectively underside PCB (right).*

For the future a real time measurement function is planned, which makes use of the internal USB interface of the microcontroller. However, the interface only supports USB 2.0 Full speed. The expected sample rate is therefore limited to 500 kHz for a 12-bit resolution and can probably be increased to 1 MHz by decreasing the resolution to 8-bit [3]. Furthermore, a digital compensation of the limited system bandwidth is envisaged via means of digital signal processing.

References
[1] Keysight Technologies, *B2980A Series Femto/Picoammeter and Electrometer/High Resitane Meter*, Datasheet, March 2016.
[2] Finch B, Goh W, *MSP430 Advanced Power Optimizations: ULP Advisor Software and Energy Trace Technology*, Applications Report Texas Instruments, June 2014.
[3] Quiring K, *Starting a USB Design Using MSP430 MCUs*, Application Report Texas Instruments, May 2014.

Electrostatic Vibration Energy Harvester Pre-charged Wirelessly at 2.45 GHz

Z. Saddi[1], H. Takhedmit[1], A. Karami[2], P. Basset[1] and L. Cirio[1]

[1]Université Paris-Est. ESYCOM (EA 2552), UPEM, ESIEE-Paris, CNAM, F.77454 Marne-la-Vallée, France

[2]Laboratoire d'informatique de Paris 6 (LIP6), Université Paris 6, Paris 75005, France

hakim.takhedmit@u-pem.fr

Abstract. This paper reports the design, fabrication and experiments of an electrostatic vibration harvester (e-VEH), pre-charged wirelessly for the first time by using an electromagnetic waves harvester at 2.4 GHz. The rectenna uses the Cockcroft-Walton voltage doubler rectifier. It is designed and optimized to operate at low power densities and provides high voltage levels: 0.5 V at 0.5 µW/cm² and 0.8 V at 1 µW/cm². The e-VEH uses the Bennet doubler as conditioning circuit. Experiments show 23 V voltage across the transducer terminal when the harvester is excited at 25 Hz by 1.5 g of external acceleration. An accumulated energy of 275 µJ and a maximum power of 0.4 µW are available for the load.

1. Introduction

Advances in wireless communications and low consumption electronics in recent decades had contributed to the emergence of sensors and connected objects in different fields. An exponential growth of the number of devices is expected with the advent of the Machine-to-Machine (M2M) and the Internet of Things (IoT). The energy autonomy of such devices constitutes one of the main obstacles before reaching the full mobility. Instead of traditional batteries which require periodic replacement or recharging and raises recycling issues, the energy harvesting consisting to convert the energy of ambient sources such as vibrations, electromagnetic waves, thermal, solar and wind into electrical energy, became a potentially promising solution. From these ambient sources, the mechanical vibration energy is of particular relevance due to its availability. The vibration harvesters are based on the transduction mechanism, and there are typically of three kinds: electromagnetic, piezoelectric and electrostatic. In electrostatic vibration energy harvesters (e-VEHs), the mechanical energy is converted into electricity by a mechanical attraction force due to charged variable capacitor plates. This force opposes the motion of the mobile plate. The generated power from mechanical to electrical conversion is proportional to the square of the accumulated quantity of charge in the capacitor. Therefore, an external source providing sufficient voltage is necessary to convert vibrations into electricity in a sufficient manner. One solution consists to use an electret layer [1]. Another solution consists in using a transducer's pre-charge containing a power source and a conditioning circuit which generates the bias voltage itself and then creates a force between the two plates of the variable capacitor [2].

Most conditioning circuits reported in the literature requires inductive elements and switches [3] to generate a high bias voltage. However, inductive elements are not compatible with batch manufacturing process and switches require additional power-consumption control circuits. Recently issued, a conditioning circuit based on the Bennet doubler generated high bias voltage without using any switch or inductor [4]. The outline of this paper is as follows. Section 2 describes the electrostatic vibration energy harvester and section 3 reports the experimental results of the full circuit. Section 3 concludes the paper.

2. Description of the electrostatic vibration energy harvester

The schematic drawing of the e-VEH prototype is shown in figure 1. The device includes a variable capacitor and a Bennet's doubler conditioning circuit. The top view (Fig. 1 (a)) shows the rectenna and the conditioning circuit on the same substrate, which is linked to the mobile plate of the variable capacitor with four teflon bolts (Fig. 1 (b)). The variable capacitor is made by 2 circular doped silicon wafers of 100 mm diameter and 0.5 mm thickness, pasted on square epoxy board of 1.5 mm thickness. The wafer of the mobile plate is provided with a SiO_2 insulating layer of 50 μm thickness to prevent short circuits with the second wafer. The second plate of the variable capacitor is fixed to a shaker. Four flat metal springs, of 6 mm length and 15 mm width, are used to link the two plates of the variable capacitor, one on each side of the epoxy layer.

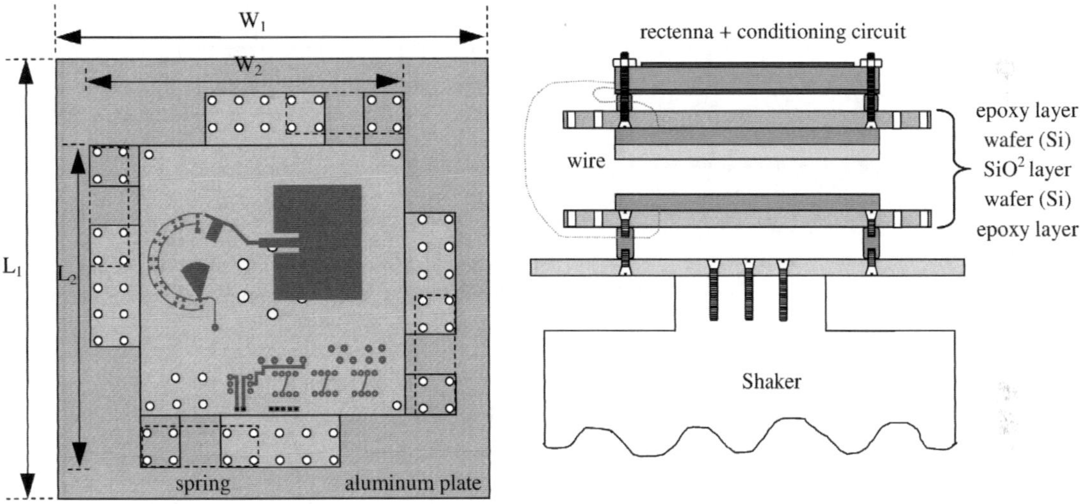

(a) (b)

Figure 1. Geometry of the proposed e-VEH: L_1= W_1 =150, L_2=120, W_2=120 (dimensions are in mm). Top view (a) and profil view (b).

3. Experiments of the e-VEH full system

The proposed harvester was achieved and measured. The photograph of the prototype is depicted on Fig. 2 (a). The experiments were carried out using the macro-scale resonant variable capacitor described in the previous section. The resonance frequency was 25 Hz. The measured unbiased transducer capacitance variation of 1.5 g amplitude at 25 Hz frequency was: C_{max}/C_{min} = 250pF / 40 pF (η = 6.25). Figure 2 (b) shows the schematic diagram of the used experimental setup. The e-VEH prototype is mounted on a shaker (Bruel & Kjaer type 7541) and placed in an anechoic chamber at a distance R = 1.5 meters from a transmitting horn antenna, where the far field condition is satisfied. An accelerometer, adhered to the shaker, is used to control and then regulate the acceleration. The experiment was carried using external vibrations at 25 Hz with acceleration amplitude of 1.5 g. The rectenna uses the Cockcroft-Walton circuit proposed by authors in [5]. It provides 0.5 V at 0.5 μW/cm², 0.8 V at 1 μW/cm² and 1.17 V at 1.75 μW/cm², as shown in Fig. 3 (a). The measured voltage evolution, over a capacitive load of 1 mF, for different power densities from 1 to 10 μW/cm² is depicted in Fig. 3 (b). The capacitor load stores energy of about 281 μJ at 1 μW/cm², 1620 μJ at 3 μW/cm² and 5445 μJ at 10 μW/cm².

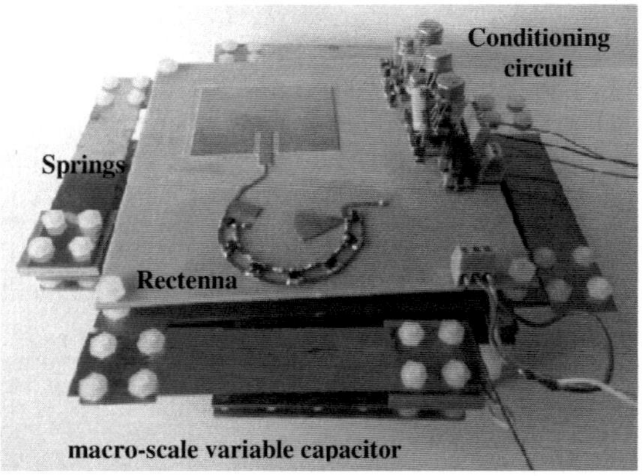

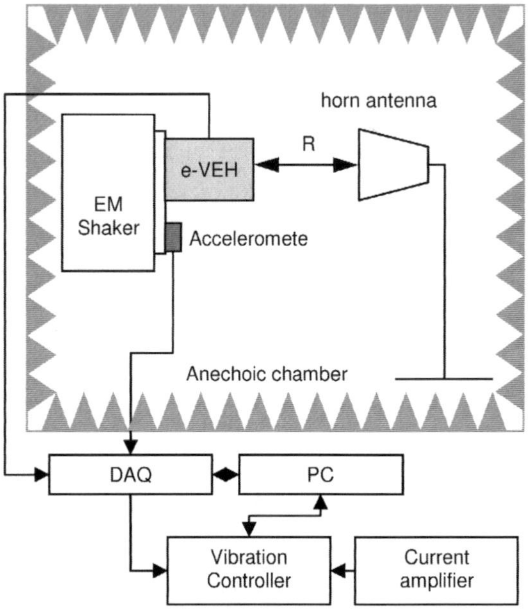

(a) (b)

Figure 2. Photograph of the prototype (a) and schematic diagram of the used experimental setup (b).

The Bennet's doubler used as conditioning circuit is shown in Fig. 4 (a). The circuit contains three capacitors, the variable capacitor C_{var}, C_{res} = 1 μF and C_{store} = 47 nF and three diodes D_1, D_2 and D_3 (JPAD5). The initial pre-charge V_0 applied to C_{res} is supplied by the Cockcroft-Walton rectenna. The operation of the Bennet's doubler is described in [4].

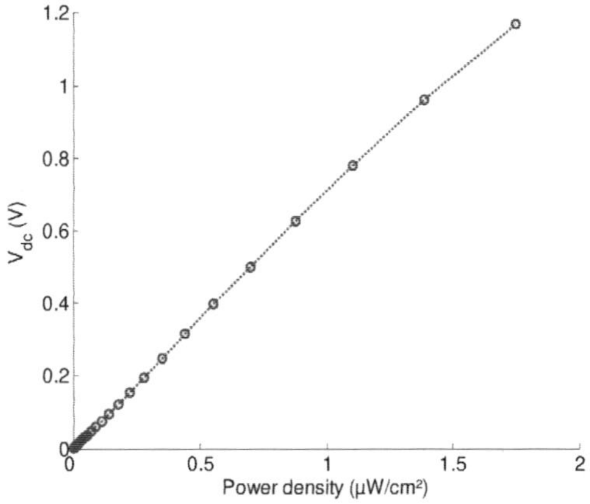

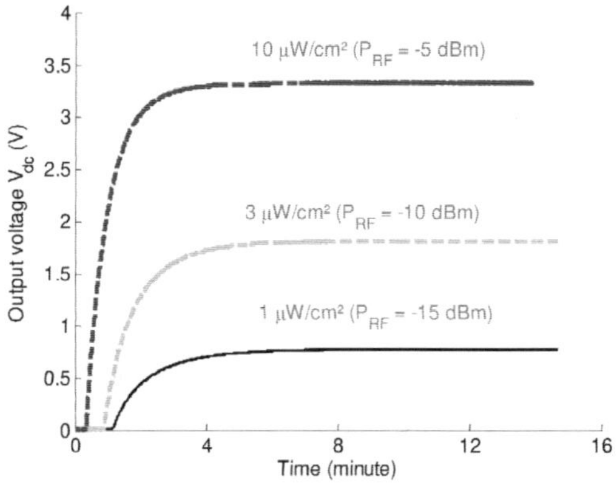

(a) (b)

Figure 3. Measured output voltage vs. power density (output load = 100 MΩ, f = 2.4 GHz) (a). Measured voltage evolution vs. time over a capacitive load of 1 mF (b).

The measured voltage evolution across the reservoir capacitor C_{res} = 1 μF at several pre-charge voltages are shown in Fig. 4 (b). The output voltage progressively increases up to 23 V, where saturation occurs. This increase of voltage across C_{res} corresponds to the accumulated energy. The saturation of the voltage across C_{res} is due to the spring-softening effect induced by the electromechanical coupling [4]. Indeed, the resonance frequency of the harvester decreases when the self biasing by the conditioning circuit increases, and the frequency becomes outside the mechanical system bandwidth.

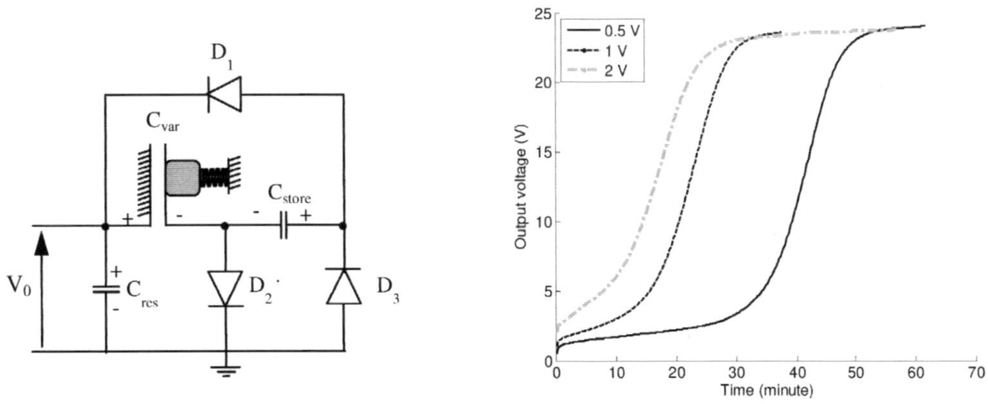

(a) (b)

Figure 4. Bennet's doubler conditioning circuit (a). Measured evolution of voltage across C_{res} = 1 µF for several bias voltages 0.5, 1 and 2 V with 25 Hz and 1.5 g external vibrations (b).

4. Conclusion

This paper presents the first experiments of an electrostatic vibration energy harvester start-up using RF waves. The RF harvester consists on a Cockcroft-Walton rectenna at 2.4 GHz. The circuit was fabricated and validated. It provides 0.8 V at 1 µW/cm² and 1.8 V at 3 µW/cm². The Bennett doubler is used as conditioning circuit. It doesn't need any inductive element or switch. Experiments show 23 V voltage across the transducer terminal, when the harvester is excited at 25 Hz by 1.5 g of external acceleration. An accumulated energy of 275 µJ and a maximum power of 0.4 µW are available for the load.

References

[1] Y. Lu, F. Cottone, S. Boisseau, F. Marty, D. Galayko and P. Basset, "A Nonlinear MEMS Electrostatic Kinetic Energy Harvester for Human-Powered Biomedical Devices", *AIP Applied Physics Letters* 107, 253902, 2015.

[2] E O'Riordan, D Galayko, P Basset and E Blokhina, "Complete electromechanical analysis of electrostatic kinetic energy harvesters biased with a continuous conditioning circuit", *Sensors and Actuators A: Physical* 247, 379-388, 2016.

[3] P. Basset, E. Blockhina and D. Galayko, "Electrostatic Kinetic Energy Harvesters", *Smart adaptive systems on silicon series*, 244 pages, March 2016, Wiley-ISTE.

[4] V. Dorzhiev, A. Karami, P. Basset, F. Marty, V. Dragunov, and D. Galayko, "Electret-free micromachined silicon electrostatic vibration energy harvester with the Bennet's Doubler as conditioning circuit," *IEEE Electron Device Letters*, vol. 36, no. 2, pp. 183-185. 2015.

[5] H. Takhedmit, H. Kilani, L. Cirio, P. Basset, and O. Picon, "Design and experiments of a 2.4-GHz voltage multiplier for RF energy harvesting," *Proc. PowerMEMS*, pp. 448-51. 2012.

Sol-gel based TiO_2 thin film deposition on frustules towards facile and scalable manufacturing

A Li[1], J Wang[1], W Zhang[1], R McNaughton[1], S Anderson[2], X Zhang[1]

[1] Mechanical Engineering, Boston University, Boston, MA 02215, USA
[2] Boston University Medical Center, Boston, MA 02118, USA

E-mail: xinz@bu.edu

Abstract. Diatom frustules have drawn a lot of attention from engineering researchers in the past decades. As a type of biomaterial, diatom frustules have been applied in a variety of areas such as biosensors and solar cells due to their excellent material and optical properties. Titanium dioxide (TiO_2), on the other hand, is also semiconductor material and photocatalyst, micro and nanoparticles of which can be found in applications such as dye sensitised solar cells (DSSC). It has been demonstrated that by using diatom frustule-TiO_2 composite particles in DSSCs, the performance of the solar cells could be increased. In this paper, we introduce a sol-gel based method to deposit TiO_2 layers on the surface of diatom frustules. TiO_2 nanoparticles were deposited on the surface of the frustules. After a subsequent annealing process, TiO_2 crystal grains were formed. The method in this paper has the potential for scalable manufacturing of frustule-TiO_2 composite materials for future solar cell applications.

1. Introduction

Diatoms, photosynthetic microalgae that are ubiquitous in aquatic systems, have attracted significant attention from researchers over the past several decades [1]. The exoskeleton structures (frustules) that enclose their cytoplasm consist of hierarchical micro/nanostructures that are composed of pure silica. The morphology of these structures gives rise to their outstanding mechanical strength, surface area and optical properties [1].

Frustules have been directly employed in applications ranging from sensors in chemical/bio-sensing [2] to light scatterers [3] in solar cells because of their advanced properties and their compatibility with existing device fabrication techniques [1]. It has been reported that diatom frustules have been used in multiple types of sensors [2], in which the large surface area of diatom frustules was used for effective sensing areas enabling improved performance. Specifically, in energy applications, it has also been reported that diatom frustules were coated with TiO_2 nanoparticles and added in dye sensitized solar cells (DSSCs) for enhanced energy harvesting performance [3]. It was believed that diatom frustules acted as "light traps" in these solar cells for scattering light and worked together with the deposited photocatalyst titanium dioxide to increase the efficiency of these devices.

It is believed, however, that the TiO_2 nanowires and nanotubes are superior to TiO_2 nanoparticles. Furthermore, in studies of nanowire/tube structures, it has been shown that the performance of those containing higher degrees of crystallinities were superior [4]. Compared with the randomly organized TiO_2, highly organized crystals of TiO_2 can provide more direct channels for electron transmission [4]. To simultaneously take advantage of the light scattering property of diatom frustules and the superior electron collection/transmission properties of TiO_2 nanocrystals, we designed a facile method for TiO_2 nanowire growth on diatom frustule's surface.

Herein, we report the method to deposit thin titanium dioxide layers on the surface of diatom frustules via sol-gel process, and to consecutively anneal the TiO_2 layers yielding crystalline structures. The reported method differs from prior work [5] in that a) diatomite was used as the starting material, which is economic and readily available, b) a one-step deposition technique with minimum chemical reaction steps was employed, and c) nanowire structures were formed within the hollow structures of the frustules. The thusly-produced composite particles are believed to have great potential in enhancing DSSC performance.

2. Experimental

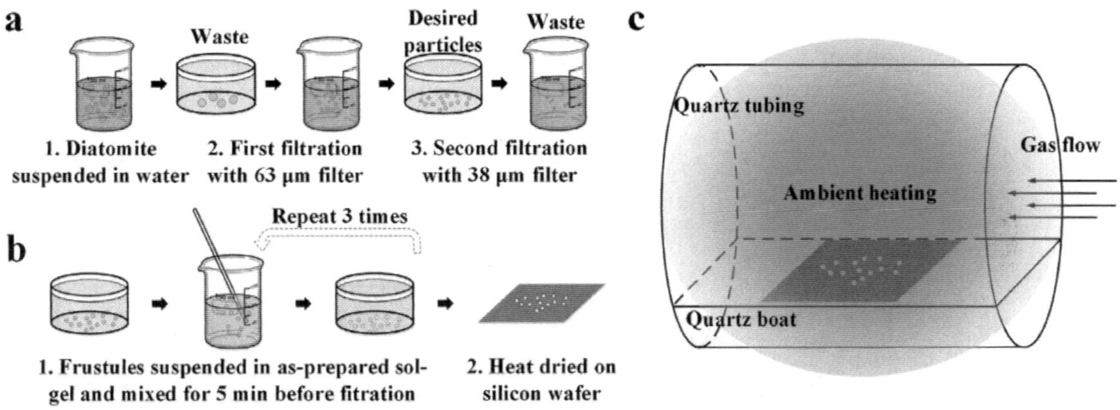

Figure 1. Fabrication of the frustule-TiO_2 composite particles. (a) Filtration of the diatomite. (b) TiO_2 precursor deposition process. (c) Schematic of the thermal treatment setup.

In this paper, diatomaceous earth (Celite 289, Almeria, Mexico) was initially suspended in deionized water (DI water) and then filtered to obtain diatom frustules (diameter 38 μm-63 μm). The frustules were then heat-dried (Figure 1(a)).

To prepare titanium oxide sol-gel, titanium isopropoxide (TIP, 99.999%), ethyl alcohol (>= 99.5%) and oxalic acid (>= 99%) were used (all purchased from Sigma Aldrich). 2.5 ml ethyl alcohol and 25 μl water were first mixed for 5 min, then 250 μl TIP was added to the solution. 2 ml of saturated oxalic acid solution was then added slowly as the solution was vigorously mixed at 700 rpm (sol-gel A).

In a slightly different process, sol-gel B was made for comparison purposes. In this process, 500 μl TIP was first mixed with 5 ml of ethyl alcohol. A mixture of 5 ml alcohol, 100 μl acetic acid, and 50 μl water was consecutively added in the as prepared TIP solution. After 5 min of mixing (700 rpm), 2 ml of water was added to the solution drop by drop to form sol-gel B.

The diatom frustules were then submerged in both types of sol-gel and mixed thoroughly for 5 min, followed by filtering and drying for further processing. This process was repeated for 3 times to achieve higher deposition thicknesses. The prepared samples were then annealed under 500 °C with a continuous O_2 flow (108 sccm) for 1 h. The heating rate was 20 °C/min. Diatoms processed by sol-gel A were heated with a high temperature (HT) of 750 °C. Diatoms processed by sol-gel B were heated with 980 °C. In both high temperature treatments, 30 min of Ar flow (40 sccm) and 40 min of O_2 flow (65 sccm) were used during the annealing process.

3. Results and discussion

Through SEM observation, it was found that most of the debris associated with the diatomaceous earth was filtered out and the remaining frustules were mostly circular micro particles, the size of which ranged from 36.5 μm to 61.9 μm. The sub-structures include nano pores with sizes of 20.1±4.2 nm and hierarchical sub-micron pores with sizes ranging from 373.3 nm to 435.8 nm (Figure 2).

Journal of Physics: Conference Series **773** (2016) 012112 doi:10.1088/1742-6596/773/1/012112

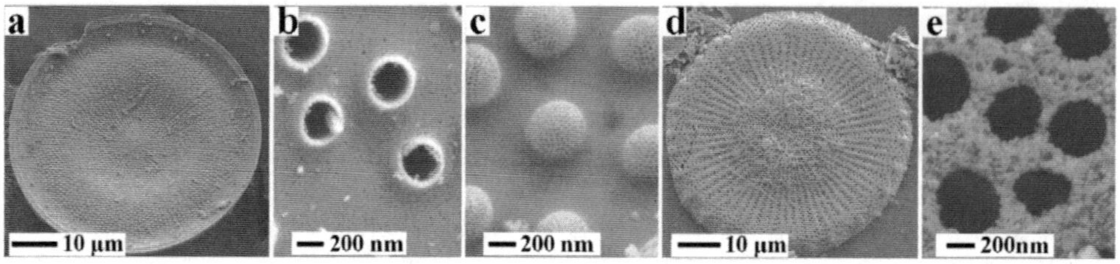

Figure 2. Morphology of the extracted frustules. The 3D structure of a frustule is dish-like. (a) Concave side of a frustule. (b, c) Sub-micron pores and nano pores on the concave side of the frustule (zoomed view of (a)). (d) Convex side of a frustule. (e) Sub-micron pores on the convex side of the frustule (zoomed view of (d)).

The prepared sol-gels differ in their final states. Sol-gel A's pH value was 1-1.5, while sol-gel B's pH value was 3. With a higher pH, sol-gel B was more viscous and opaque while A is more liquid-like and transparent.

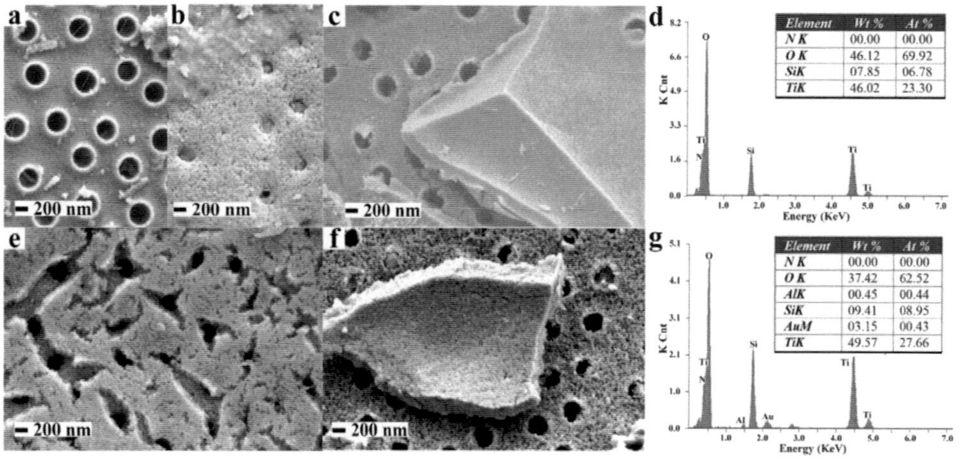

Figure 3. Deposited TiO_2 micro/nanostructures after 500 °C annealing. (a)-(d) show type A nano structures in frustule pores ((a) and (b)), thick films on frustule surface ((c)), and corresponding EDX analysis ((d)). (e)-(g) show type B films ((e) and (f)) on frustule and EDX analysis ((g)).

With both types of sol-gels, thick layers of TiO_2 were deposited on the surface of frustules after initial annealing (Figure 3). With sol-gel A, the TiO_2 films (type A films) are dense compared with films formed from sol-gel B (type B films) (Figure 3(c) and (f)). In type B films, TiO_2 nanoparticles' diameter was down to ~10 nm, while in type A films, no obvious nanoparticle structures were observed. With sol-gel A, nanowires (diameter ~ 15 nm) and films were formed in the pores (Figure 3(a)). It was also observed that some of the sub-micron pores were blocked. Although a portion of the nano pores were also blocked, they were not affected to the same degree when compared with other pore structures. The pore-filling phenomenon was not observed in the frustules treated with sol-gel B. After high temperature treatments, both types of films showed clear crystal growth (Figure 4). In type A films, crystal grain sizes range from 30 nm to 272 nm. In type B films, the sizes range from 30 nm to 360 nm.

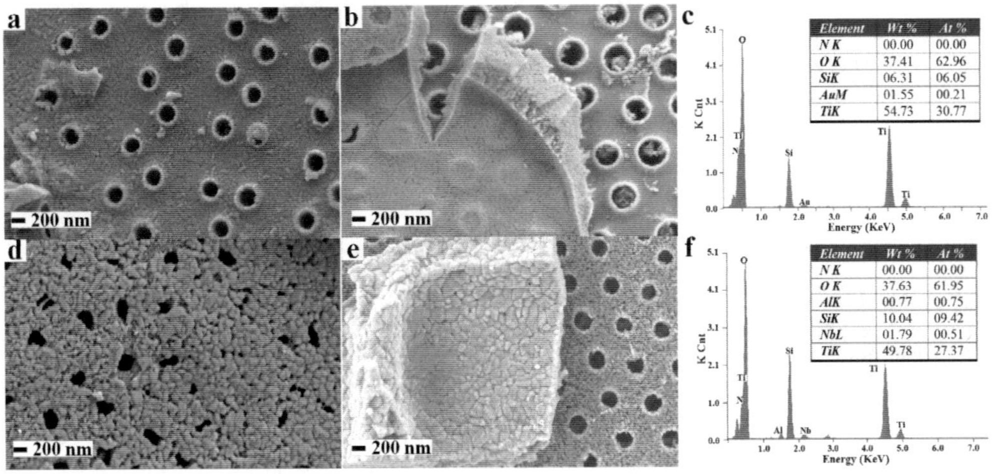

Figure 4. TiO_2 nanocrystals on frustule surface after HT treatment. (a)-(c) show crystal growth in type A films on frustule surface after HT treatment ((a) and (b)), and corresponding EDX analysis ((c)). (d)-(f) show similar crystal growth in type B films.

Energy dispersive X-ray spectroscopy (EDX) was performed for initial analysis of the resultant composite particles (Figure 3 and 4). The atomic ratios between oxygen, titanium and silicon varied in the different processes and films. For type A films, before HT treatment, atomic ratios (O:(Ti+Si)) ranged from 2.07 to 2.32, while after HT treatment, the ratio changed to 1.65-1.81, indicating that oxygen deficits may have been created during the HT treatment process. In type B films, this ratio was 1.70-1.91 before HT treatment, and remained relatively static after the HT treatment (1.68-1.88). In type A sol-gel, the excessive hydrogen ions likely inhibited the hydrolysis process between TIP and water molecules, and thusly significantly reduced the consecutive gelation. This property of the solution likely lead to increased diffusion into the nano pores of the diatom frustules, which may have contributed to the formation of nanowires/films in the pores.

4. Conclusion

In this work, a sol-gel based TiO_2 deposition technique was introduced for fabricating diatom frustule-TiO_2 composite particles. Two types of sol-gel were employed. With a lower pH, the sol-gel solution was able to diffuse into the sub-micro pores to form nanowire/film structures. TiO_2 nanoparticles and nanocrystals were formed on diatom frustules' surfaces. The material preparation method introduced herein has the potential for scalable frustule-TiO_2 nanocrystal fabrications, which may be further applicable to a broad range of research areas and novel technologies.

References

[1] Gordon R, Losic D, Tiffany MA, Nagy SS and Sterrenburg FA 2009 *Trends Biotechnol.* **27** 116–27
[2] Dolatabadi JE and de la Guardia M 2011 *Trac-Trends Anal. Chem.* **30** 1538–48
[3] Toster J, Iyer KS, Xiang W, Rosei F, Spiccia L and Raston CL 2013 *Nanoscale* **5** 873–6
[4] Lin J, Guo M, Yip CT, Lu W, Zhang G, Liu X, Zhou L, Chen X and Huang H 2013 *Adv. Funct. Mater.* **23** 5952–60
[5] Zhang Q, Chen R and Li L 2012 *J. Ocean Univ.* **11** 507–10

Influence of nonlinearities on the power output of the Self-Oscillating Fluidic Heat Engine (SOFHE)

A Tessier-Poirier[1,2] , T Monin[1,2,3,4], E Léveillé[1,2] F Formosa[3], S Monfray[4], and L G Fréchette[1,2]

E-mail: Luc.Frechette@usherbrooke.ca
[1] 3IT, Université de Sherbrooke, Sherbrooke, QC, Canada
[2] LN2, CNRS UMI-3463, Université de Sherbrooke, Canada
[3] SYMME, Université Savoie Mont-Blanc, Annecy-le-Vieux, France
[4] STMicroelectronics, Crolles, France

Abstract. In this paper, it is shown that two non-linearities drive the oscillations amplitude and the potential power density of the Self-Oscillating Fluidic Heat Engine (SOFHE). This new type of engine converts thermal energy into mechanical energy by producing self-sustained oscillations of a liquid column from a continuous heat source to power wireless sensors from waste heat. The underlying theoretical modeling shows that the pressure and the temperature nonlinearities limit the final oscillations amplitude, hence its achievable power density.

1. Introduction

The new paradigm of the Internet of things (IoT) leads to a multiplication of sensors everywhere in our environment. Powering those wireless sensors longer than a battery life can be done by harvesting nearby energy sources. In this paper, a new type of thermal energy harvester inspired by [1] is studied, the Self-Oscillating Fluidic Heat Engine (SOFHE) [2, 3], which couples an electro-mechanical transducer to a motor based on the self-oscillatory phenomenon observed in the pulsating heat pipes (PHP) [4]. The motor in SOFHE consists of a small tube closed at one end, filled with water (figure 1). By heating the closed-end, evaporation of the liquid leads to the formation of a vapor bubble which increases in length until an equilibrium position is reached. From this equilibrium, increasing the temperature of the heat source leads to an instability [5], the liquid column starts to oscillate as shown by the experimental pressure measurements in figure 2 from [2]. This paper shows that the final amplitude of the oscillations and the power output of the SOFHE are limited by the gas pressure and the wall temperature nonlinearities.

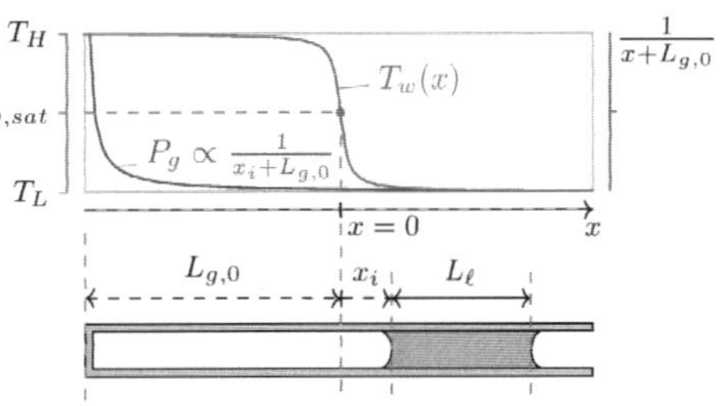

Figure 1: Tube, pressure and temperature.

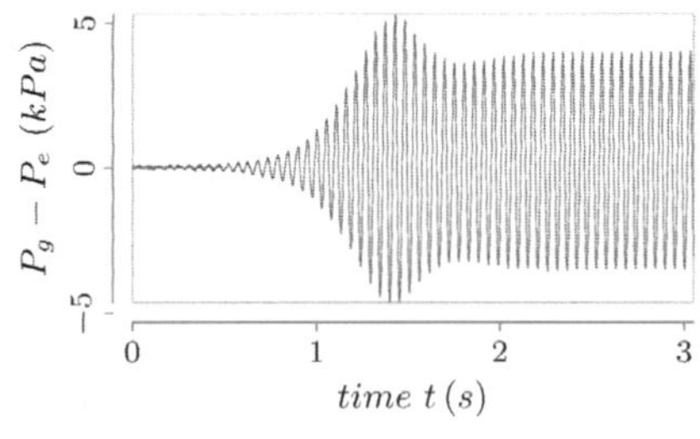

Figure 2: Start-up of the oscillations.

2. Model of the self-oscillations

Newton's second law is applied to the liquid column Eq. (1), where movement of the column (x_i) will result from the difference between the gas bubble pressure (P_g) in Eq. (2) and the external pressure P_e and be opposed by the liquid's friction (F_f) assuming Poiseuille flow (Eq. (3)) and the harvester, taken as a linear damping ($F_h = c_h \dot{x}_i$ where $c_h \leqslant 0$). In Eq. (2), the mass of vapor m_g change accordingly to the evaporation/condensation rate which is modelled from the heat conduction (with a thermal resistance R_{th} and the enthalpy of vaporization H_v) to the meniscus location, leading to the first part of Eq. (6). The value of the thermal resistance R_{th} should come from experiment or theoretical estimation. The wall temperature T_w is constant in the heat source at T_H and at the heat sink at T_L and varies linearly between those two regions. This is modelled by an arctangent as shown in figure 1 and given by Eqs. (4) and (5).

$$\ddot{x}_i = (1/m_\ell)\,(P_g\,A - P_e\,A + F_f + F_h) \quad (1) \qquad F_f = -\,8\pi\,\mu\,L_\ell\,\dot{x}_i = \,c_f\,\dot{x}_i \quad (3)$$

$$P_g = m_g\,R_g\,T_g/\left((x_i + L_{g,0})\,A\right) \quad (2) \qquad T_w = A_{th}\,\arctan(M_{th}\cdot x + B_{th}) + C_{th} \quad (4)$$

$$A_{th} \equiv (T_H - T_L)/\pi \quad (5a) \qquad M_{th} \equiv 1/A_{th}\left(B_{th}^2 + 1\right)\cdot (dT/dx)_{x=0} \quad (5c)$$

$$C_{th} \equiv (T_H + T_L)/2 \quad (5b) \qquad B_{th} \equiv \tan\left[(T_{g,sat,0} - C_{th})/A_{th}\right] \quad (5d)$$

$$\dot{m}_g = \frac{\dot{Q}}{H_v} = \frac{T_{wall}(x_i) - T_{g,sat}}{H_v R_{th}} = \frac{A_{th}}{H_v R_{th}}\arctan(M_{th}\cdot x_i + B_{th}) + \frac{C_{th} - T_{g,sat}}{H_v R_{th}} \quad (6)$$

The equilibrium solution of these equations is a constant x_i such that $T_{wall}(x_i) = T_{g,sat}$, leading to $x_i = \dot{x}_i = \dot{m}_g = 0$. It is useful to study variations of the system relative to this equilibrium and to nondimensionalized the problem with $L_{g,0}$ and the natural angular frequency ω_n (7f), leading to Eq. (8a) (nonlinear because of P_g) and Eq. (8c) (nonlinear because of the arctan T_w profile). It is useful to study the effect of the nonlinearities. In that purpose, we can compare the nonlinear equations (8a) and (8c) to their linearized counterpart given by Eqs. (8b) and (8d). Finally, the harvester extracts an adimensionalized power $\widetilde{\mathcal{P}_h}$ given by Eq. (9).

$$\widetilde{x_i} \equiv x_i/L_{g,0} \quad (7a) \qquad \dot{\widetilde{x_i}} \equiv \left(1/(\omega_n L_{g,0})\right)\dot{x}_i \quad (7e) \qquad \widetilde{\Delta m_g} \equiv (m_g - m_{g,0})/m_{g,0} \quad (7i)$$

$$\tau \equiv \omega_n\,t \quad (7b) \qquad \omega_n \equiv \left(P_{g,0}/\rho_\ell\,L_\ell\,L_{g,0}\right)^{1/2} \quad (7f) \qquad \widetilde{\omega} \equiv \omega/\omega_n \quad (7j)$$

$$\widetilde{M_{th}} \equiv L_{g,0}M_{th} \quad (7c) \qquad \widetilde{C_{th}} \equiv (C_{th} - T_{g,sat})/T_C \quad (7g) \qquad \zeta_f \equiv -c_f/(2\omega_n m_\ell) \quad (7k)$$

$$\widetilde{A_{th}} \equiv A_{th}/T_C \quad (7d) \qquad \zeta_h \equiv -c_h/(2\omega_n m_\ell) \quad (7h) \qquad T_C \equiv \omega_n\,m_{g,0}\,H_v\,R_{th} \quad (7l)$$

$$\ddot{\widetilde{x_i}} = -\left(\frac{1}{1+\widetilde{x_i}}\right)\widetilde{x_i} + \left(\frac{1}{1+\widetilde{x_i}}\right)\widetilde{\Delta m_g} - 2\,(\zeta_f + \zeta_h)\,\dot{\widetilde{x_i}} \quad (8a)$$

$$\ddot{\widetilde{x_i}} = -\widetilde{x_i} + \widetilde{\Delta m_g} - 2\,(\zeta_f + \zeta_h)\,\dot{\widetilde{x_i}} \quad (8b)$$

$$\dot{\widetilde{\Delta m_g}} = \widetilde{A_{th}}\,\arctan(\widetilde{M_{th}}\cdot\widetilde{x_i} + B_{th}) + \widetilde{C_{th}} \quad (8c)$$

$$\dot{\widetilde{\Delta m_g}} = \left(\widetilde{A_{th}}\widetilde{M_{th}}/(1 + B_{th}{}^2)\right)\,\widetilde{x_i} \quad (8d)$$

$$\widetilde{\mathcal{P}_h} \equiv \left(\pi/\omega_n{}^3 m_\ell\,L_{g,0}{}^2\right)\,\mathcal{P}_h = \widetilde{\omega}\,\zeta_h\int_\tau^{\tau+2\pi/\widetilde{\omega}} \dot{\widetilde{x_i}}^2\,d\tau \;\approx \pi\,\zeta_h\,\dot{\widetilde{X_i}}^2 = \pi\,\widetilde{\omega}^2\,\zeta_h\widetilde{X_i}^2 \quad (9)$$

3. The amplitude of the limit cycle is driven by the nonlinearities

As expected, numerical simulations of the linear equations (8b) and (8d) show an exponential growth with no saturation. Including either one or the other nonlinearity does lead to saturation of the amplitude. In the saturated regime, the variables $\widetilde{x_i}$, $\dot{\widetilde{x_i}}$ and $\dot{\widetilde{\Delta m_g}}$ varie nearly periodically and is assumed to be reached when their difference between two consecutive cycle is less then 0.1% on all the cycle for ten consecutive cycles. The saturated regime is shown by the orbits from numerical simulations in figure 3 for two cases (see table 1).

Table 1: Numerical value of the parameters for two studied cases

	ζ_f	$\widetilde{A_{th}}$	$\widetilde{M_{th}}$	B_{th}	$\widetilde{C_{th}}$	Conclusions
Case 1	0.0217	0.0043	-14.50	0.3792	-0.0016	T_N dominates over P_N
Case 2	0.0034	0.0271	-0.3625	0.3792	-0.0098	P_N dominates over T_N

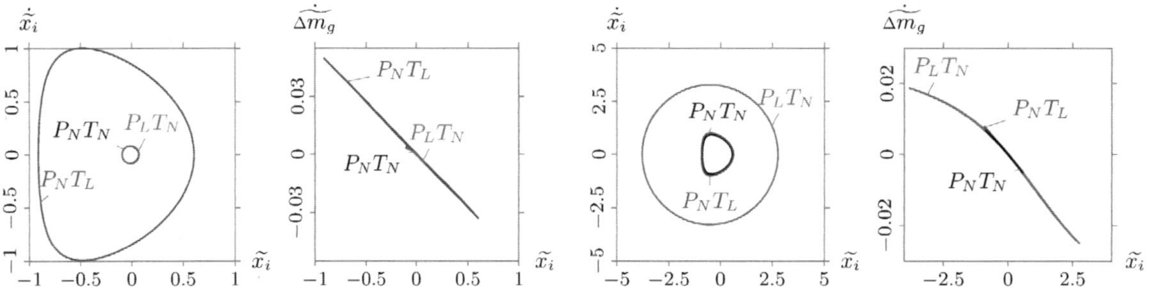

(a) case 1 (b) case 1 (c) case 2 (d) case 2

Figure 3: Saturated regime for $\zeta_h = 0$ with parameters of table 1.

3.1. *Wall Temperature Profile nonlinearity*

As the meniscus oscillates at small amplitude, the wall temperature profile seen by the meniscus is mostly linear (figure 1). As the amplitude increase, however, the wall temperature increase/decrease less than linearly (T_w saturates, (4)). This leads to a saturation of the evaporation/condensation (8c) and therefore, of the injection of energy into the system. Since the dissipation of energy is linear, the system eventually reach an amplitude where the injection of energy and the dissipation of energy equals each other over a cycle. The behavior of this saturated regime (with only the T_w nonlinearity is shown in figure 3 by the curves denoted $P_L T_N$. A saturation of $\dot{\widetilde{\Delta m_g}}$ should be seen by plotting it against $\widetilde{x_i}$, given Eq. (8c). This signature is indeed observed as shown by the curve $P_L T_N$ in figure 3d. Such behavior in experimental measurements could indicate the dominance of the T_w nonlinearity.

3.2. *Pressure nonlinearity*

When the meniscus oscillates, the compression/expansion of the gas gives rise to variations of the pressure P_g with $\Delta P_g \propto 1/(x_i + L_{g,0})$ (as shown in Eqs. (2), (8a) and in the figure 1). For oscillations of small amplitude at constant mass, P_g is proportional to the volume variation : $P_g \propto x_i$. As the amplitude increases, however, this linear approximation is not valid anymore. As the meniscus goes toward the closed end ($x_i \to -L_{g,0}$), the pressure tends to infinity (as shown in figure 1), preventing the liquid column to go touch the closed end. This nonlinear behavior limits the evaporation/condensation rate and forces the system to settle on a saturated regime as shown by the $P_N T_L$ curve on figure 3a. The saturated regime exhibit an asymmetry along $\widetilde{x_i}$ on the plane $(\widetilde{x_i}, \dot{\widetilde{x_i}})$, a signature due to the presence of the closed end. As it reaches the closed end, the system bounces back, reverses its velocity on a short distance.

3.3. Pressure and Wall Temperature nonlinearities combined - Amplitude and Power

The combination of both nonlinearities leads to a saturated regime shown by the $P_N T_N$ curves in figure 3. In case 1, figures 3a and 3b, suppressing the pressure nonlinearity (the $P_L T_N$ curve which only contains the temperature nonlinearity) has so little effect on the size of the saturated regime that the two curves $P_N T_N$ and $P_L T_N$ are almost undistinguishable. However, suppressing the temperature nonlinearity (the $P_N T_L$ curve) changes drastically the amplitude of the saturated regime. One concludes that the wall temperature nonlinearity is dominant. Incidentally, the $P_N T_N$ curve is exhibiting a wall temperature nonlinearity signature and no pressure nonlinearity signature. Now, from case 1, taking a $\zeta_h > 0$ allows to produce a power $\widetilde{\mathcal{P}_h}$ given by (9). By assuming a sinusoidal motion, we get the approximation in (9) which clearly shows that for a given ζ_h and a given $\widetilde{\omega}$, the power $\widetilde{\mathcal{P}_h}$ will increase if the size of the saturated regime increases (the peak amplitude of $\widetilde{x}_i$ and $\dot{\widetilde{x}}_i$, $\widetilde{X}_i$ and $\dot{\widetilde{X}}_i$ respectively, increase). The optimal ζ_h (which produce the maximal power) leads to $\widetilde{\mathcal{P}_h}(P_N T_N) = 23$, $\widetilde{\mathcal{P}_h}(P_L T_N) = 24$ and $\widetilde{\mathcal{P}_h}(P_N T_L) = 5250$. Just like for the amplitude, suppressing either nonlinearity increases the power but the temperature nonlinearity is clearly dominant.

Going from case 1 to case 2 is the adimensionnal equivalent of dividing $L_{g,0}$ by 40 and keeping all the other dimensional parameters of the equations constants. The equilibrium is therefore closer to the closed end and the pressure nonlinearity becomes more apparent. This is shown in figures 3c and 3d where the pressure nonlinearity is now clearly dominant (suppressing the temperature nonlinearity as much less influence on the size of the saturated regime). This could have been concluded from the $P_N T_N$ curve shape in figure 3c, which exhibit a nonlinear pressure signature. In case 2, the optimal ζ_h leads to $\widetilde{\mathcal{P}_h}(P_N T_N) = 631$, $\widetilde{\mathcal{P}_h}(P_L T_N) = 5750$ and $\widetilde{\mathcal{P}_h}(P_N T_L) = 848$. Again, just like for the amplitude, suppressing either nonlinearity increase the power but the pressure nonlinearity is now clearly dominant. One may conclude that one non-linearity can dominate over the other to set the amplitude and the power output, depending on the parameters of the system.

4. Conclusion

A nondimensional model of the SOFHE was presented which include two nonlinearities and a linear harvester. It has been shown that the final amplitude of the system without harvesting is limited by the nonlinearities. Furthermore, one nonlinearity can dominate over the other (have more impact on the final amplitude than the other). Finally, it has been shown that the nonlinearities also limits the power output. In this paper, the focus was solely on the nonlinearity but the power output also depends on the linear part. It will eventually be useful to show and understand the influence of the adimensional numbers on the amplitude and on the power output, leading to design guidelines for the SOFHE.

Acknowledgments

This research was funded by NSERC and conducted in collaboration with STMicroelectronics. We are grateful to Thomas Skotnicki from STMicroelectronics for discussions on the subject.

References

[1] Léveillé *et al.* 2012 *PowerMEMS 2012* (Atlanta)
[2] Monin T *et al.* 2016 *PowerMEMS 2016* (Paris)
[3] Skotnicki T 2013 Thermoelectric generator
[4] Das S P *et al.* 2010 *Int. J. Heat Mass Transfer* **53** 3905–3913
[5] Jenkins A J 2013 *Physics Reports* **525** 205–207

Stack air-breathing membraneless glucose microfluidic biofuel cell

J Galindo-de-la-Rosa,[1] A Moreno-Zuria,[1] V Vallejo-Becerra,[2] N Arjona,[3] M Guerra-Balcázar,[2] J Ledesma-García[2], L G Arriaga[3*]

[1] Centro de Investigación y Desarrollo Tecnológico en Electroquímica, Pedro Escobedo, Qro., C.P. 76703, México.
[2] Facultad de Ingeniería, División de Investigación y Posgrado, Universidad Autónoma de Querétaro, Querétaro, Qro., C.P. 76010, México.
[3] Centro de Investigación y Desarrollo Tecnológico en Electroquímica, Unidad Tijuana, Tijuana, B.C., C.P. 22444, México.

E-mail: larriaga@cideteq.mx

Abstract. A novel stacked microfluidic fuel cell design comprising re-utilization of the anodic and cathodic solutions on the secondary cell is presented. This membraneless microfluidic fuel cell employs porous flow-through electrodes in a "V"-shape cell architecture. Enzymatic bioanodic arrays based on glucose oxidase were prepared by immobilizing the enzyme onto Toray carbon paper electrodes using tetrabutylammonium bromide, Nafion and glutaraldehyde. These electrodes were characterized through the scanning electrochemical microscope technique, evidencing a good electrochemical response due to the electronic transference observed with the presence of glucose over the entire of the electrode. Moreover, the evaluation of this microfluidic fuel cell with an air-breathing system in a double-cell mode showed a performance of 0.8951 $mWcm^{-2}$ in a series connection (2.2822$mAcm^{-2}$, 1.3607V), and 0.8427 $mWcm^{-2}$ in a parallel connection (3.5786$mAcm^{-2}$, 0.8164V).

1. Introduction

Microfluidic fuel cells that use enzymes as catalysts to convert the chemical energy of fuels into electrical energy are of special interest due to their operational conditions such as low-temperature (20–40 °C), activity near to neutral pH, selectivity and specificity [1-3]. Additionally, designing aspects have shown also an increase in the performance of these miniaturized devices. For instance, the incorporation of an air-breathing window intake increases the oxygen concentration, and due to its superior diffusivity in air than in aqueous solution avoids the transfer limitations of dissolved oxygen [4, 5]. Stacked membraneless microfluidic hybrid or biofuel cells have been tested in both, series or parallel electrical modes enhancing the overall fuel utilization by reusing the unreacted fuels from the first cell to the others [6, 7]. In this work, we present the scanning electrochemical microscope characterization of glucose oxidase immobilized on Toray carbon paper and its use as bioanode towards the evaluation of two air-breathing membraneless microfluidic fuel cells, one operated in a single mode, and the other in the stack mode in both series and parallel.

2. Experimental

2.1 Reagents

Glucose oxidase (EC 1.1.3.4 from *Aspergillus niger*), tetrabutylammonium bromide (TBAB), glutaraldehyde solution 5% (GA), potassium phosphate dibasic dihydrate (K_2HPO_4), potassium phosphate monobasic monohydrate (KH_2PO_4), Nafion® (5% diluted in water), and potassium hydroxide (88.7%) were purchased from Sigma-Aldrich.

2.2 Electrodes fabrication

The bioanode consisted on Toray carbon slides (EC-TP1-060T) of 1.25 mm width and 30 mm length dipped for 15min in a catalytic ink composed of the following components: 7 μL Nafion, 13 μL glutaraldehyde solution, 990μL of 0.1M phosphates buffer at pH 7, 20 mg Vulcan carbon (Vulcan XC-72 from E-ETEK) and 1 mg of TBAB for each mg of enzyme glucose oxidase. This ink was prepared by sonication during 20 min followed by a vortex for other 15 min. The bioanode was then dried at room temperature. On the other side, the air-breathing cathode was prepared on a piece of carbon nanofoam (2.5×30 mm, Marketech®) and covered with an ink containing 7 μL of Nafion® 5% (Sigma–Aldrich), 73 μL of isopropyl alcohol (J.T. Baker) and 1 mg of commercial Pt/C (30 wt. % on Vulcan XC-72 from E-TEK), which was deposited using an airbrush to assure homogeneous deposition with a final metal loading of 1 $mgcm^{-2}$ over the entire surface.

2.3 Construction and evaluation of microfluidic air-breathing membraneless fuel cells.

Schematic representations of the single and stack microfluidic fuel cells are illustrated in figure 1. The significant difference between them consists in the electrode array due to the stack microfluidic design allows re-utilization of the anodic and cathodic solutions in the second cell in order to increases the fuel and oxidant utilization.

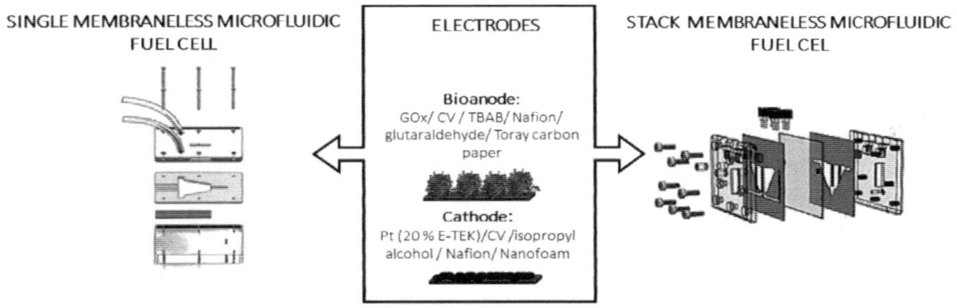

Figure 1. Schematic representations of single and stack air-breathing membraneless and electrode array.

The single microfluidic air-breathing device was previously reported [8]. The second section consisted of a home-made silicone elastomer film (Silastic®, Dow Corning, prepared using an Elcometer® Film Applicator with a final thickness of 200 μm). On the other hand, the stack air-breathing membraneless microfluidic fuel cell employs porous flow-through electrodes, but in a "V"-shape architecture. The microfluidic fuel cells were evaluated by injecting 5mM glucose N_2-saturated (4 U. P. D. Praxair) as the fuel and oxygen from the air as the oxidant within two aqueous streams consisting in 0.1M phosphate buffer (pH 7) and 0.3 M KOH injected on the anodic and the cathodic side, respectively, with a flow rate of 30 μL min^{-1} for both streams.

3. Results and discussion

3.1 Topography and localized enzymatic activity.

Scanning electrochemical microscopy (SECM) has been used widely to monitor the distribution and the enzymatic activity on surfaces [9], therefore the study of glucose oxidase on Toray paper was achieved using a BioLogic scanning system instrument model ac-SECM/SECM470. SECM measurements were performed in a four electrode cell consisting of a graphite rod and Ag|AgCl|3M KCl served as auxiliary and reference electrodes, respectively, a Platinum ultramicroelectrode (UME) of 25 μm, using 5mM of glucose in 2mM $K_3[Fe(CN)_6]$ in 0.1M phosphate buffer pH 7.0 as working solution. Activity evaluated by H_2O_2 production due to an applied potential of 0.7V, thus ensuring diffusion-controled oxidation of H_2O_2 [10]. The SECM images were obtained using software MIRA.

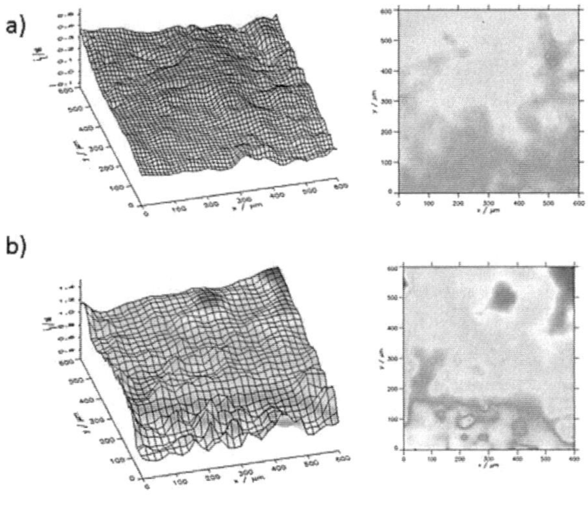

Figure 2. SECM images of catalytic ink of GOx in absence (a) and presence of 5mM glucose (b) at E_T = +0.7 V vs Ag/AgCl in pH 7.0 (0.1 M) phosphate buffer solution.

Figure 2 shows SECM images obtained following the H_2O_2 generated by influence of both, the enzyme and the applied potential. In the absence of glucose (figure 2a), currents in the order of nano-Amps are observed associated mostly to non-Faradic processes. Meanwhile, with the addition of glucose (figure 2b), yellow and red zones related to higher current are observed and related to the Faradic process involving the electron transference from the glucose oxidation to form gluconolactone and hydrogen peroxide, indicating that the enzyme is on the electrode.

3.2 Evaluation of the single and stack microfluidic fuel cells.

The single microfluidic fuel cell produced a power density of 0.4522 mW/cm^2 (figure 3a). Meanwhile, the stack cells increased at least twice the power density compared with the single cell regardless the electrical configuration (figure 3b). As expected, the highest current density was obtained in the parallel mode, with a value of 3.5786 mA/cm^2. However, the maximum cell performance (0.89 mW cm^{-2}) was obtained in series mode due to the reduction of both, the polarization and ohmic lossses.

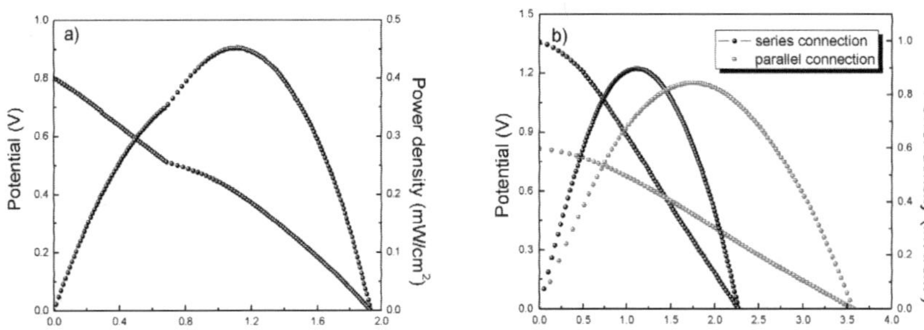

Figure 3. Polarisation and power density curves using 5Mm glucose in 0.1M buffer phosphates pH 7.0 at10 mV s^{-1} as fuel and 0.3M KOH as oxidant of a) single and b) stack microfluidic fuel cell connected in series and parallel.

Table 1. Performance of single and stack microfluidic fuel cells using 5mM glucose.

Microlfuidic fuel cell	Current density (mA/cm^2)	Open circuit potential (V)	Power density (mW/cm^2)
Single microfluidic fuel cell	1.9314	0.797	0.4522
Stack microfluidic fuel cell in series connection	2.2822	1.3607	0.8951
Stack microfluidic fuel cell in parallel connection	3.5786	0.8164	0.8427

According with table 1, the best results were obtained using the stack cell in a series mode, where the re-utilization of fuel and oxidant from the first to the second cell increased the open circuit potential and, in a lower manner the current density compared with the single cell, demonstrating the excellent energetic capability and versatility of this device.

4. Conclusions

Evaluation of two different designs of microfluidic fuel cells was performed varying the number of cells that composed them. The first design corresponds to a previously reported single membraneless microfluidic fuel cell and, in the second case, a stack cell composed of two cell in which the oxidant and fuel are re-utilized in the second cell promoting the fuel consumption. In addition, the stack cell was tested in both modes, series and parallel, increasing the current density or the voltage. The enzymatic activity was tested by cyclic voltammetry and scanning electrochemical microscope evidencing the electron transference from the glucose oxidation through the entire electrode related to a good enzymatic immobilization. Consequently, the stack membraneless microfluidic fuel cell operated in a series configuration showed the highest cell performance herein found (0.8951 mW cm^{-2}) due to the decreasing of the polarization losses associated mainly to the electrode kinetics and reduction of ohmic losses.

5. References

[1] Desmaële D, Nguyen-Boisse T T, Renaud L, Tingry S, 2016, *Microelectron. Eng.* **165** 23.
[2] Kizling M, Stolarczyk K, Tammela P, Wang Z, Nyholm L, Golimowski J, Bilewicz R, 2016, *Bioelectrochemistry* **112** 184.
[3] Narvaez Villarrubia C W, Soavi F, Santoro C, Arbizzani C, Serov A, Rojas-Carbonell S, Gupta G, Atanassov P, 2016, *Biosens. Bioelectron.* **86** 459.
[4] Zhu X, Zhang B, Ye D, Li J, Liao Q, 2014, *J. Power Sources* **247** 346.
[5] Zou H, Chen J, Fang Y, Ding Y, Peng W, Liu R, 2016, *Nano Energy* 27 619.
[6] Wang H, Gu S, Leung D Y C, Xu H, Leung M K H, Zhang L, Xuan J, 2014, *Electrochim. Acta* **135** 467.
[7] Salloum K S, Posner J D, 2011, *J. Power Sources* **196** 1229.
[8] Burchardt M, Wittstock G, 2008, *Bioelectrochemistry* **72** 66.
[9] Ortiz-Ortega E, Goulet M A, Lee J W, Guerra-Balcázar M, Arjona N, Kjeang E, Ledesma-García J and Arriaga L G 2014 *Lab Chip* **14** 4596.
[10] Schäfer D, Maciejewska M, Schuhmann W, 2007, *Biosens. Bioelectron.* **22** *1887.*

Annular screen printed thermoelectric generators for ultra-low-power sensor applications

Z T Gima[1], K Gururangan[2], J Evans[2], P Wright[1]

[1]Department of Mechanical Engineering, University of California, Berkeley, California, 94720, USA

[2]Department of Materials Science and Engineering, University of California, Berkeley, California, 94720, USA

E-mail: ztakeo@berkeley.edu

Abstract. We report on the materials preparation and device fabrication for screen printed $Bi_2Te_3/Bi_{0.5}Sb_{1.5}Te_3$ annular thermoelectric generators (TEGs) for use in ultra-low-power sensor applications. At a 20K temperature gradient, test couples — n- & p-type 1 cm^2 squares connected electrically in series with printed Ag traces — demonstrated an average power of 0.068nW at 26nA and 2.6mV. The material preparation leverages mechanical alloying for both the n- and p-type materials in order to reduce electrical resistivity and increase device power output. Screen printed thermoelectric elements were found to have substantial challenges with mechanical failure due to cracking and delamination. This work specifically describes these challenges and corresponding mitigation strategies for screen printing the TEG slurries.

1. Introduction

The need for cost-effective methods of energy harvesting continues to grow as autonomous wireless sensor networks proliferate due to the exploding interest in the Internet of Things and Industrial Internet. For applications where waste heat exists, thermoelectric energy harvesting can provide the requisite power for these sensor networks and wireless devices.

To enable cheaper thermoelectric energy harvesting, high-throughput screen printing is investigated. This research builds upon previous work with dispenser printed circular thermoelectric generators (TEGs) [1], highlighting challenges in switching manufacturing methods. Similar preparation for the n-type (Bi_2Te_3) and p-type ($Bi_{0.5}Sb_{1.5}Te_3$) materials was used with new ink formulations appropriate for screen printing. As a test application common to many commercial and industrial applications, a 50-couple annular TEG with element dimensions 3.5mm x 0.6mm x 120 μm was designed to harvest heat from a hot pipe (Figure 1).

2. Experimental Methods

2.1. Mechanical Alloy & Ink Synthesis

Compositions of the thermoelectric and epoxy materials are shown in Table 1. Mechanical alloying is leveraged to prepare the thermoelectric materials. For the n- & p-type materials, 1% Se and 8% Te are added respectively to increase electrical conductivity [2, 3]. Previous work has indicated that electrical conductivity is greater for particle sizes below 10 μm [3]. To achieve such particle size, the mechanically alloyed material is wet grinded. The ink slurry is

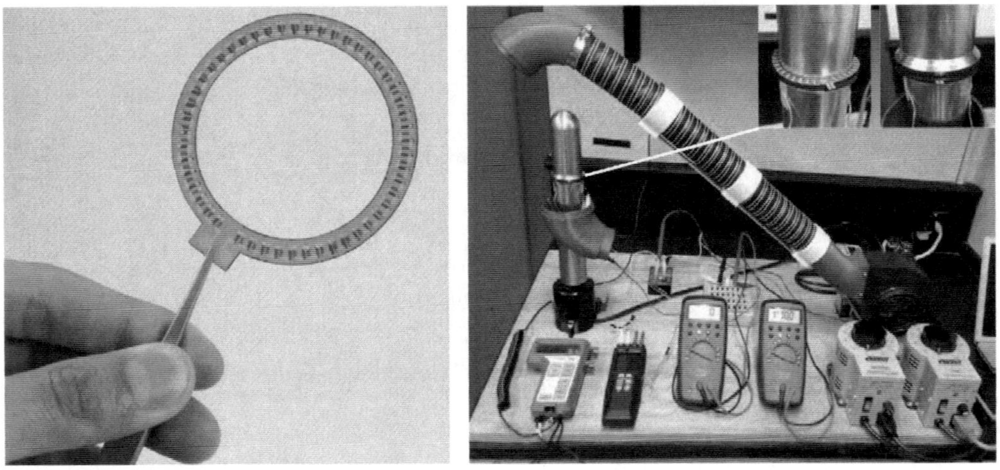

Figure 1. *Left*: 50-couple annular screen printed thermoelectric generator; *Right*: Case study application setup for testing annular TEG power output, voltage, and current. *Image courtesy of Marco Salvioli.*

Table 1. Compositions of n- & p-type mechanically alloyed materials and epoxy binder. N- & p-type inks were created by combining the active material with epoxy in a 7:1.5 mass ratio with toluene added to tailor rheology. All materials measured in wt%.

N-type (Bi_2Te_3)	P-type ($Bi_{0.5}Sb_{1.5}Te_3$)	Epoxy binder
47% - Bi	15% - Bi	45% - Epon 862 Resin
52% - Te	25% - Sb	8% - Heloxy 61
1% - Se	60% - Te	46% - MHHPA
–	–	1% - AC-8

then formed by mixing the powder with a thermoset epoxy solution (Broadview Tech./Hexion) in an 82% solid loading ratio. Toluene is added depending on the desired viscosity.

2.2. Screen Printing & Heat Treatment

The devices were screen printed on polyimide that had been roughened to increase surface contact. Screens (Dynamesh, Inc.) for the n-type, p-type, and Ag electrical contacts (Dupont 5025) were designed to screen print each material in the annular geometry. After printing, each layer was dried at 70°C for 20 minutes to prevent smearing. The fully printed device was left out overnight in air to release any leftover toluene and limit void formation within the bulk material (Figure 2). After drying, the TEGs were cured at 250°C for 8 hours to relax the printed structure and form longer percolated active material pathways through each leg.

2.3. Testing

Initial material tests were performed on doctor-blade casted test couples — n- & p-type 1cm^2 squares connected electrically in series with printed Ag traces — using a bespoke planar TEG testing setup. Full annular devices were evaluated using the setup in Figure 1. Measurements of the OCV and short-circuit current were recorded.

Figure 2. *Left*: Confirmation of optimal particle size (1-3μm) and epoxy presence in n-type material. Grain coalescence due to the curing procedure can also be seen. P-type similar. *Right*: SEM cross-section of p-type thermoelectric leg. N-type similar.

3. Results & Discussion

At a 20K temperature gradient, test couples demonstrated an average power of 0.068nW at 26nA and 2.6mV. However, screen printed thermoelectric legs frequently exhibited visible mechanical failures, which prevented the overall device from producing measurable power. Legs tended to print uniformly, but then began to crack and sometimes delaminate from the polyimide substrate during either the drying or curing phase. Ink viscosity and device geometry were evaluated as potential factors.

3.1. Ink Viscosity

Viscosity was initially thought to be the main factor in cracking. TEG inks were confirmed to be thixotropic and rheology was tuned subject to printing performance. While increasing the solvent content to lower viscosity initially helped mitigate crack formation by allowing the printed material to flow and fill in gaps and voids, it ultimately did not change the fundamental mechanical behavior of the material. Figure 3 qualitatively demonstrates the effect that lowering the viscosity of the ink has on crack formation. It is important to note, however, that all devices printed ultimately showed signs of cracking and/or delamination. Low viscosity inks tended to print well; however, cracking and delamination still appeared during drying and curing. Furthermore, lowering the viscosity of the ink resulted in practical printing challenges, such as poor feature resolution and edge fidelity.

3.2. Device Geometry

While thin film cracking can be attributed to many coupled factors including viscosity, orientation of packed particles, and inertial/gravitational forces [4], the high aspect ratio (height/width) of the thermoelectric legs proved most important. As [5] and [6] report, printed thermoelectric layers (or particle based inks in general) over 20μm are prone to cracking and delamination. Thickness of the printed layer with respect to the substrate thickness is also important for mitigating cracking. As a general rule of thumb, thickness of the printed layer should never exceed that of the substrate.

4. Future Work

Based on these results, new screen designs with lower aspect ratio thermoelectric elements will be evaluated. Improving electrical conductivity of the materials will also be targeted. The authors

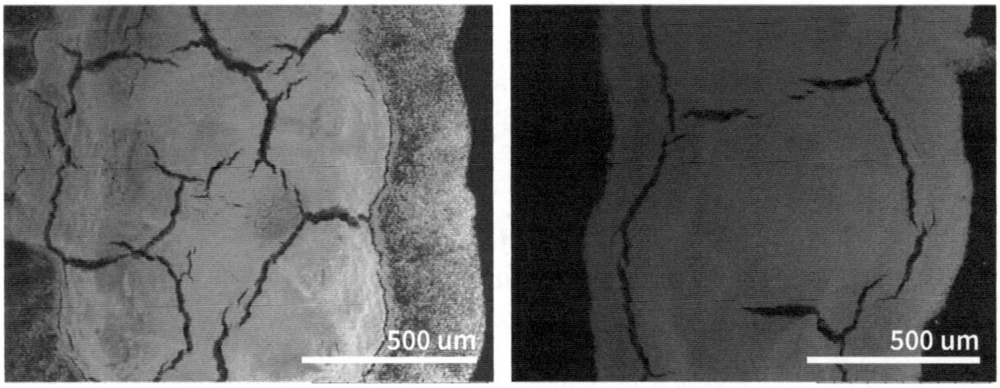

Figure 3. Influence of viscosity on cracking based on top-down SEM images. *Left*: P-type material with a viscous (lower solvent content) ink composition. N-type similar; *Right*: N-type material with lower viscosity and less visible cracking. P-type similar.

expect that the change in geometry, which increases the cross-sectional area through which electrons can flow, coupled with the decreased instance of cracking, can lower electrical resistance enough for the devices to be functional. Other ongoing efforts in printed TEGs suggest several pathways forward for overcoming the low electrical conductivity that tends to plague these devices including: 1) densification of the printed materials via mechanical or thermal processes [7, 8], 2) use of a conductive polymer binder to connect active material particles electrically and mechanically [9], 3) ink parameter optimization, and 4) avoiding printing altogether and resolving bulk material processing with flexible substrates [10].

5. Conclusion

Annular screen printed TEGs were manufactured and tested. While the ink slurries demonstrated reasonable thermoelectric properties when deposited as test couples, cracking and delamination prevented the 50-couple devices from generating power. While many factors such as ink viscosity, surface tension, solvent-substrate compatibility, solid loading ratio, and drying/curing procedure likely contributed, high aspect ratio was the most significant factor in causing the mechanical failure of the printed TEGs.

References

[1] Madan D, Wang Z, Chen A and Winslow R 2014 *Applied Physics Letters* **104** 4
[2] Madan D, Wang Z, Chen A, Juang R c, Keist J, Wright P K and Evans J W 2012 *ACS Applied Materials & Interfaces* **4** 6117–6124
[3] Madan D, Chen A Juang R, Keist J, Wright P and Evans J 2012 *ACS Applied Materials Interfaces* **4**
[4] Prosser J H, Brugarolas T, Lee S, Nolte A J and Lee D 2012 *Nano Letters* **12** 5287–5291
[5] Søndergaard R R, Hösel M, Espinosa N, Jørgensen M and Krebs F C 2013 *Energy Science & Engineering* **1** 81–88
[6] Orrill M and LeBlanc S 2016 *Journal of Applied Polymer Science* 1–15
[7] Dimitriadou I A, Fulham A, Robbins M C, Simpson K, Dorey R A, Jones P, Bernadet S, Laroche J, Guillem S P, Potyrala C and Wood J 2014 *Proceedings of the 11th European Conference on Thermoelectrics* (Cham: Springer International Publishing) pp 177–182
[8] Rudež R, Markowski P, Presečnik M, Košir M, Dziedzic A and Bernik S 2015 *Ceramics International* **41** 13201–13209
[9] Kato K, Hagino H and Miyazaki K 2013 *Journal of Electronic Materials* **42** 1313–1318
[10] Geppert B and Feldhoff A 2016 *Energy Harvesting and Systems* **3** 1–11

Fatigue of micromachined stainless steel structural materials for vibrational energy harvesting

Y Shimizu, L Van Minh, H Kitayoshi and H Kuwano

Graduate School of Engineering, Tohoku University, Sendai, Japan

y.shimizu@nanosys.mech.tohoku.ac.jp

Abstract. This work presents fatigue measurement for micromachined stainless steel (SUS304) structural substrate using resonant bending mode. Micromachined specimens for fatigue test had a cantilever structure with a proof mass. They were fabricated by $FeCl_3$ wet etching and wire-discharged cutting. The SUS specimens had Young's modulus of 198 GPa on average. The endurance limit of micromachined specimens was 213 MPa on average after 10^8 cycles under our fracture definition. The large SUS specimens had the endurance limit of 229 MPa after 10^7 cycles.

1. Introduction

Development of wireless sensor nodes for internet of things (IoTs) has needed miniaturize power source that is capable of converting on-site ambient energy resources into electricity. From a point of view of energy capacity and microfabrication compatibility, piezoelectric vibrational energy harvesters are promising for practical applications. To improve the further performance, previous researchers focused on developing piezoelectric materials for energy harvesting [1-4]. Minh *et al.* developed lead-free piezoelectric energy harvesters in which piezoelectric thin films were grown on silicon wafers [1,2]. The KNN-based harvester was able to provide the normalized power density of 1.98 $mW.g^{-2}.cm^{-3}$ [1]. Recently, Minh *et al.* provided the (Mg,Zr)-doped AlN based micro-energy harvesters at level of 4 $mW.g^{-2}.cm^{-3}$[2]. Although they provided high power density, those energy harvesters fabricated from Si structural materials are likely to be suitable for ambient vibrations with frequency above 200 Hz. Low fracture toughness and low elastic strain limit prevent silicon to be applied in the case of heavy weight and low frequency resonance. Therefore, there is a desire of seeking a new structural material for substituting silicon in those low frequency applications. Cao *et al.* had used the stainless steel substrate for fabricating piezoelectric macro-energy harvesters [3,4]. However, there is likely to have no research to consider the mechanical and fatigue properties of stainless steel substrate at micro scale. In this work, we aim at studying the mechanical and fatigue properties of micromachined stainless steel that might provide the important data for guiding a design of energy harvesters.

2. Experimental procedure

2.1. Resonant-vibration fatigue measurement

Fatigue test technique of micro-scale structural substrates for micro-energy harvesters is not matured, although it is a prerequisite for devices' design [5-7]. Most fatigue tests have been performed under tension-tension loading by using low frequency (tens of Hz) cyclic fatigue system [5]. For reducing time consumption of fatigue test and being compatible with both macro- and micro-scale fatigue specimens, resonant bending mode is preferable [6]. Therefore, in this work, we applied resonant bending mode to measure endurance limit of micromachined SUS304 specimens. We prepared both of large and microscale cantilever-like fatigue specimens. In our fatigue measurement, fatigue specimens were vibrated at their resonant frequencies. Displacement amplitudes were measured by a laser vibrometer (Onosokki LV-1710) and controlled by a shaker controller (Asahi Seisakusho G-Master 05) (Figure 1(a)). The endurance limit was determined by the corner point at which the displacement amplitude decreased 20% because of initial occurrence of the micro-fracture as shown in Figure 1(b). Applied stress values were deduced from the experimental displacement amplitudes by using FEM simulation (ANSYS 16) with SOLID186 element.

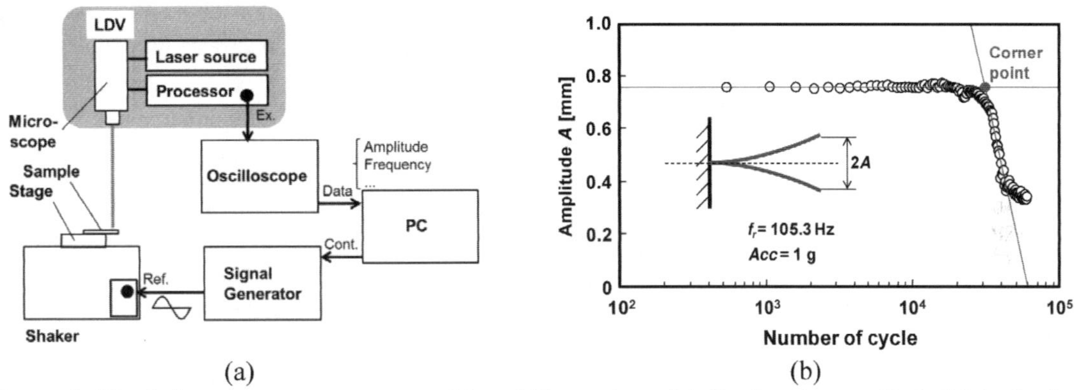

Figure 1. The fatigue measurement set-up (a) and illustration of defined endurance limit (b) for both of large and micro SUS304 cantilevers.

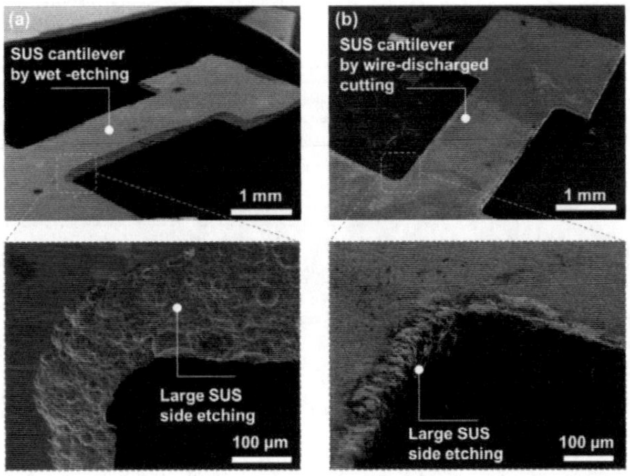

Figure 2. Scanning electron microscopic images (SEM) of micromachined SUS cantilevers: by $FeCl_3$ wet etching (a), by wire-discharged cutting (b).

2.2. Fatigue specimens

We conducted fatigue test for three types of specimens including SUS304 large cantilevers and micro-cantilevers. The large specimens were fabricated by a dicer (DISCO DAD522). The micromachined specimens were fabricated by wet etching using $FeCl_3$ etchant at 40 °C or by a wire discharge cutting (Sodick AQ360). Large cantilevers had beam geometric parameters including length of 1.9 mm, width of 9.4 mm, and thickness of 0.1 mm. The micromachined cantilevers had beam length of 3.0 mm, width of 1.6 mm, and thickness of 0.1 mm. A copper proof mass was attached at the tip of cantilever to turn the resonant frequency. Figure 2 show the scanning electron microscopic (SEM) images for both wet-etching and wire-discharge cutting SUS micro-cantilevers.

3. Results and Discussions

3.1. SUS304 Young's modulus

For determining the applied stress, at first, we determined the Young's modulus of the SUS304 materials. By using the micro-cantilevers, we measured the resonant spectra (Figure 3 (a)). The Young's modulus values were deduced from FEM simulation. The average Young's modulus of SUS304 was 198 GPa (Figure 3 (b)).

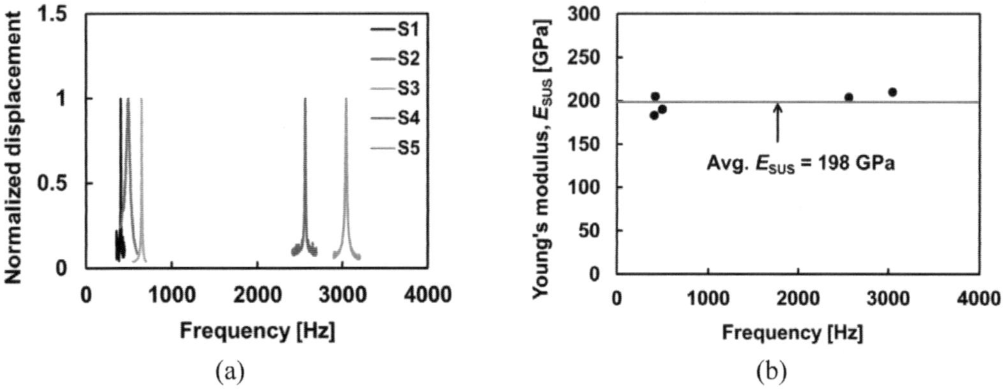

Figure 3. The resonant spectra of SUS304 based cantilevers (a), SUS304 Young's modulus deduced from FEM model (b).

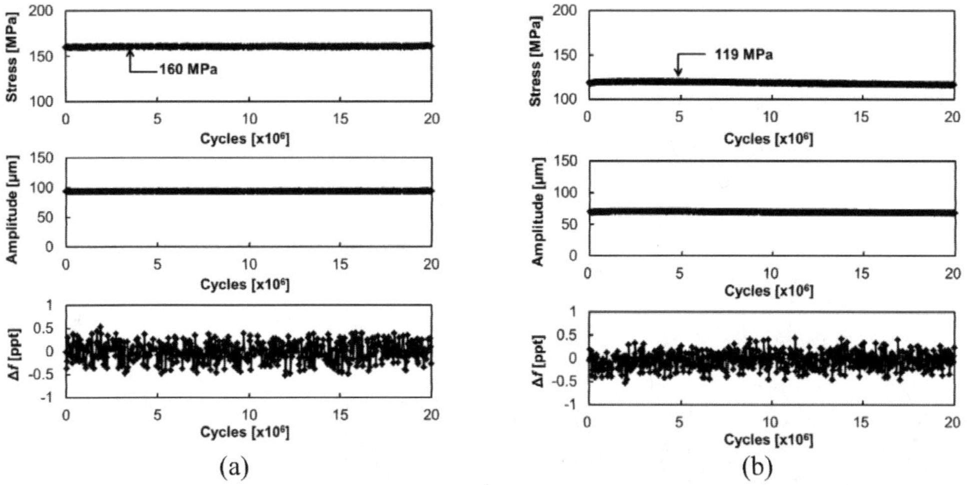

Figure 4. Stress amplitude, tip deflection amplitude, and frequency shift, Δf (Δf *(ppt)* = $10^3\times(f-f_0)/f_0$) at applied acceleration of 24.5 m.s^{-2} over 20×10^6 cycles: wet-etching micro-cantilever (f_0 = 1140 Hz, *Q*-factor = 37) (a), wire-discharge cutting micro-cantilever (f_0 = 1160 Hz, *Q*-factor = 26) (b).

3.2. Fatigue measurement

Figure 4 (a) and (b) exhibit the data sets for the micromachined cantilevers by the wet etching and the wire-discharged cutting. The root-mean-square (rms) stress values were calculated from FEM and were 160 MPa and 119 MPa for the wet-etching and the wire-discharged cutting, respectively, at the acceleration of 24.5 m/s^2. The deviation of the stress values comes from the differences in the resonant frequency, f_0, and *Q*-factor. f_0 and *Q* were 1140 Hz and 37 for the wet-etching and 1160 Hz and 26 for the wire-discharged cutting. The normalized frequency shift, Δf, defined as $ppt = 10^3 \times (f\text{-}f_0)/f_0$. Both of the micromachined cantilevers show small Δf of ±0.5 *ppt* over 20×10^6 cycles.

Figure 5 shows S-N diagram for the micromachined SUS and the large SUS cantilevers. By tuning the accelerations, various bending stress values were achieved for fatigue measurement. For the macro samples, at large stress levels, there was the instantaneous occurrence of fracture. The endurance limit approached 229 MPa after 10^7 cycles.

Further fatigue measurements were conducted on the micromachined SUS cantilevers. At the applied stress level of 311 MPa, the cantilever amplitude decreased after approximately 2 ×10^6 cycles The micromachined samples (enclosed by dot circle) had no fracture occurrence. Therefore, the endurance limit might be the average level of 213 MPa, being comparable with that of the large SUS specimen.

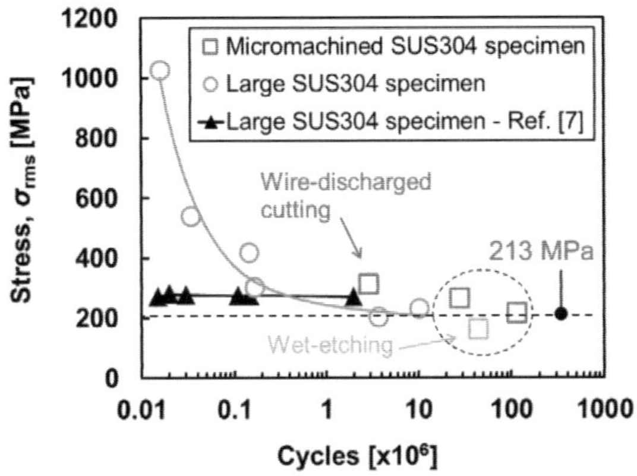

Figure 5. Stress-lifetime (S-N) for the SUS cantilevers including large cantilevers and micro-cantilevers fabricated by both wet-etching and wire-discharge cutting techniques.

4. Conclusions

Fatigue measurement for micro-machined stainless steel (SUS304) specimens was tested by using resonant bending mode. A cantilever structure with a proof mass was applied as fatigue specimen. They were micro-machined by $FeCl_3$ wet etching or wire-discharged cutting. SUS 304 had Young's modulus of 198 GPa. The endurance limit of micromachined samples was 213 MPa on average, being similar to those of the previous works on large SUS specimen.

References

[1] Van Minh L, Hara M, Horikiri F, Shibata K, Mishima T and Kuwano H 2013 *J. Micromech. Microeng.* **23** 035029
[2] Van Minh L, Hara M, Yokoyama T, Nishihara T, Ueda M and Kuwano H 2015 *IEEE Trans. Ultrason., Ferroelect., Freq. Control* **62** 11 2005-2008
[3] Cao Z, He J, Wang Q, Hara M, Oguchi H and Kuwano H 2013 *J. Phys.: Conf. Ser.* **476** 012034
[4] Cao Z, Zhang J and Kuwano H 2012 *Sensors and Actuators A* **179** 178-184
[5] Ando T, Shikida M and Kazuo S 2001 *Sens. Actuators A*, 93 1 70-75
[6] Kwak K, Otsu M and Takashima K 2010 *Sensors and Materials* **22** 1 51-59
[7] Hanby K R 1975 *Handbook on Materials for Superconducting Machinery* National Bureau of Standards p 579

DESIGN IMPROVEMENT FOR PREVENTING DISCHARGE DURING FABRICATION OF ELECTROSTATIC ENERGY HARVESTER

N Miwatani, T Fujita, Y Kitagawa, K Kanda, and K Maenaka

Graduate School of Engineering, University of Hyogo
2167 Shosha, Himeji, Hyogo 671-2280, Japan

er14q043@steng.u-hyogo.ac.jp

Abstract. This paper reports an improvement of structure design of bipolar charged electret energy harvester to prevent an electrical discharge during a corona charging process on electret. We confirmed that differential output power of 33 μW is obtained with 8.8 *g* at 350 Hz sinusoidal vibration from the developed device. By using a commercially available power management IC, regulated voltage of 1.8 V is properly obtained and can drive a load resistance more than 500 kΩ.

1. Introduction

The vibration energy harvester is expected as a maintenance-free power supply for wireless sensors, such as tire pressure monitoring system (TPMS) [1]. In our electret type vibration energy harvester (VEH), the density of counter electrode (CE) was doubled for higher output power. we confirmed that the device can generate four times (above 40 μW) as large power as previous device in the same dimension [2]. However, the output power from the fabricated device was lower than expected because of the unintended discharge during the fabrication. In this paper, the device design is revised to overcome this problem. We also described the novel device's performance by measurement.

2. Harvester with double-density counter electrode

Figure 1(a) shows the cross-sectional schematic diagram of previous device with a normal-density counter electrode (CE) and a bipolar charged electret [3]. The buried grid electrodes (BGEs) beneath positive and negative charged electret are named the BGE-P (Positive) and the BGE-N (Negative) respectively. In this structure, CE is arranged in same pitch of a pair of BGE-P and BGE-N. Relative position and capacitance between CE and BGE are changed by a horizontal vibration. Unlike the ideal value, the capacitance change is smaller than theoretical calculation because of a large fringing effect.

Figure 1(b) shows the structure of our revised harvester with double-density CE. In this structure, the CE is electrically divided to two part i.e. the CE-P and CE-N, to induce reversed-polarity charges above the bipolar charged electret. In order to double the CE density, the electrode was arranged in half pitch size of previous one. The double-density CE structure can produce larger capacitance change compare to the previous one because an additional counter electrode acts as a guard-electrode to reduce the fringing effect. The half-pith CE also increase spatial efficiency, so the output will be improved.

Figure 2 shows top view diagrams of single and double-density counter electrode. The interdigital-shaped CE-P and CE-N which are fabricated by D-RIE process and electrically divided to individual electrode by removing conductive part after the dicing process.

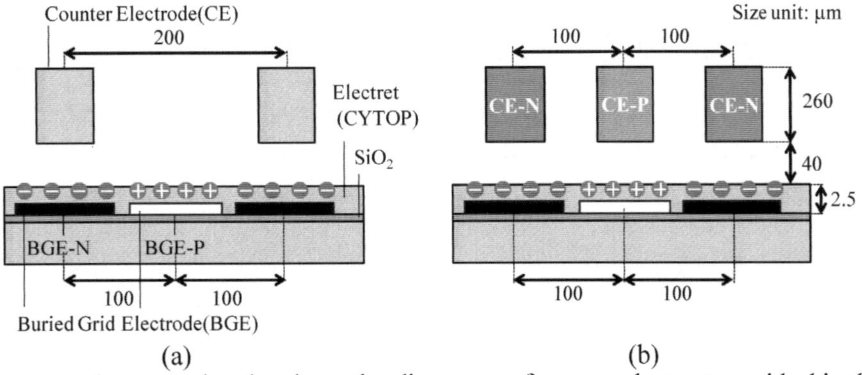

Figure 1. Cross-sectional schematic diagrams of energy harvester with bipolar charged electret for (a)normal-density CE and (b)double-density CE.

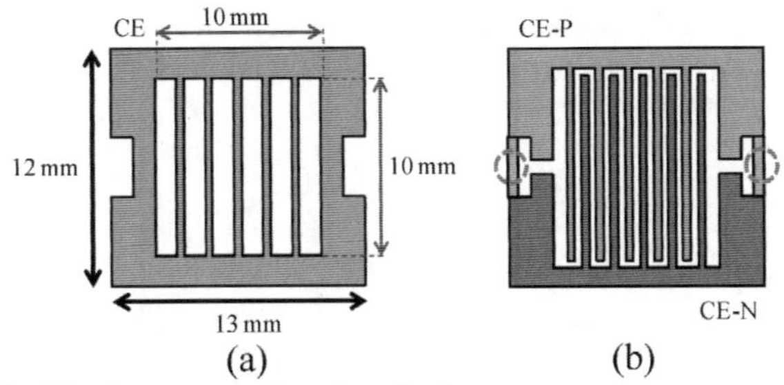

Figure 2. Top view of structural schematics of CE (a) normal-density and (b) double-density. Red circle shows conductive part which is removed after dicing process to divide between CE-P and CE-N elactrically.

3. Discharge prevention design

In our previous study, the harvester with double-density CE showed large difference between output powers from the CE-P and the CE-N. There was unintended parasitic resistance of few MΩ, R_p that was generated between CE-N and BGE [2]. The R_p brought an unbalance of output voltage from the harvester. We assumed that the R_p was generated from the carbonization of polymer by discharging on the CE tip where an electric field was highly concentrated. During the bipolar charging process as shown in Fig. 3, the charging voltage is as twice as normal monopolar electret.

To avoid the discharge problem, We modified the counter electrode to be covered by polymer film (KMPR1005; Nihonkayaku Co., ltd, Japan) for insulation as shown in Fig. 4. Electret charging voltage was reduced from ±200 V to ±150 V for considering a dielectric breakdown voltage , 9 kV/0.1 mm of electret material (CYTOP: CTL-809M, Asahi Glass Corp., Japan). Table 1 shows the R_p and the parasitic capacitance C_p for (a) before and (b) after the isolation measure. We confirmed that the unintended conduction between CE-N and BGE was drastically reduced from the previous device.

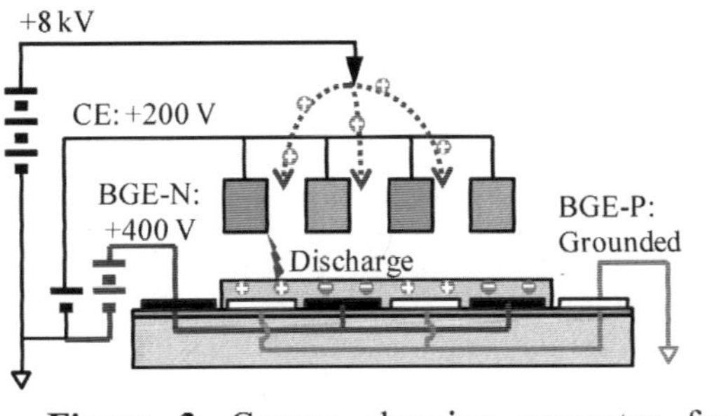

Figure 3. Corona charging apparatus for bipolar-electret.

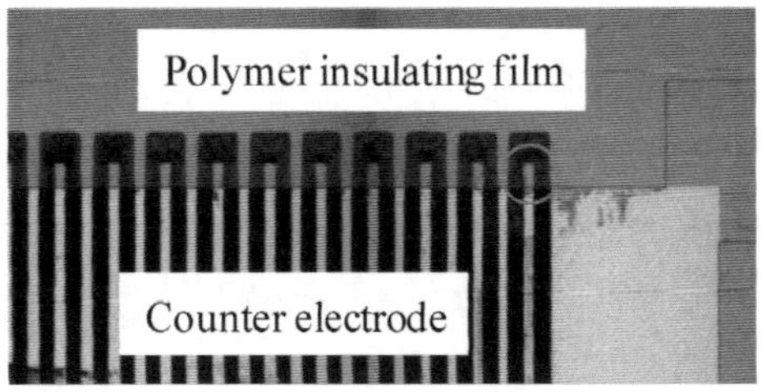

Figure 4. Covered area of the CE tip insulation polymer film.

Journal of Physics: Conference Series **773** (2016) 012117 doi:10.1088/1742-6596/773/1/012117

Table 1. Parasitic elements between each of electrodes by using a LCZ meter for (a)previous design and (b)improved design. The LCZ meter was configured with 0 V bias voltage and 1 kHz and 1 V_{pp}

(a)

Measured electrode	C_p [pF]	R_p [MΩ]
CE-P and BGE	52.5	-
CE-N and BGE	60.0	3.1
CE-N and CE-P	64.0	27.5

(b)

Measured electrode	C_p [pF]	R_p [MΩ]
CE-P and BGE	77.9	-
CE-N and BGE	80.4	28.7
CE-N and CE-P	73.0	-

4. Experiments

4.1. Waveform measurement

The novel harvester was mounted on a shaker and applied acceleration of 8.8 *g* (*g*: gravitational acceleration 9.81 m/s^2) at a frequency of 350 Hz. Figure 5 shows the schematic of the measurement apparatus. The CE-P and CE-N are individually connected to load resistances. Measuring prove was connected through the ultra-high-impedance buffer (AD549; Analog Device Inc., USA).

Figure 6 shows output waveforms from the CE-P and CE-N with load resistance of 1 MΩ. The output anti-phased waveforms obtained from CE-P and CE-N were as expected. The differential output waveform from two CEs will be approximately as doubled as individual CEs (thin black line in Fig. 6).

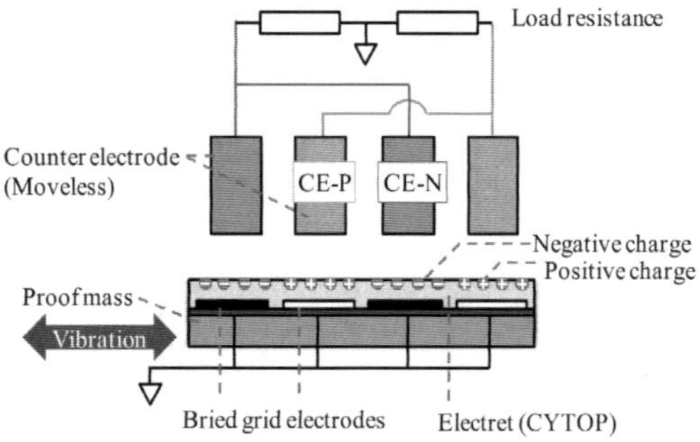

Figure 5. Connection diagram of each counter electrode and load resistance.

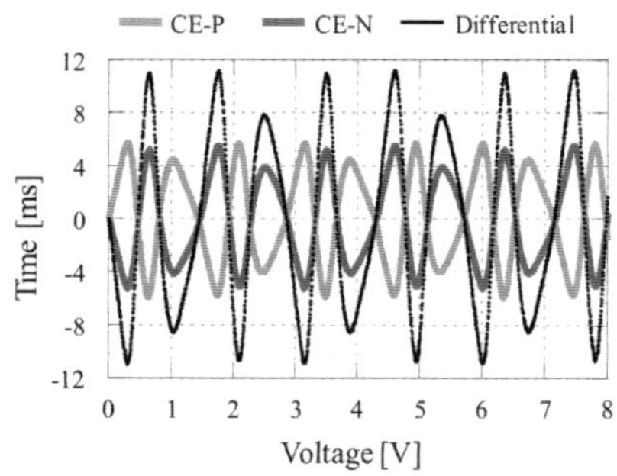

Figure 6. Output waveform of energy harvester with double-density counter electrode.

4.2. Load characteristic

Figure 7 shows measurement and simulation results for maximum output power versus load resistance. The simulated output power was calculated by an equivalent circuit model of SPICE simulator in our previous work [4]. The maximum powers from CE-P and CE-N that connected with optimal load of 0.4 MΩ were 18 μW and 15 μW, respectively. The maximum differential output from both electrodes was reached 33 μW. We confirmed the maximum output power of novel device was 2.2 times as large as the simulation result. We assumed that wafer-level adhesive bonding process causes air-gap reduction, then the output was increased above the simulation.

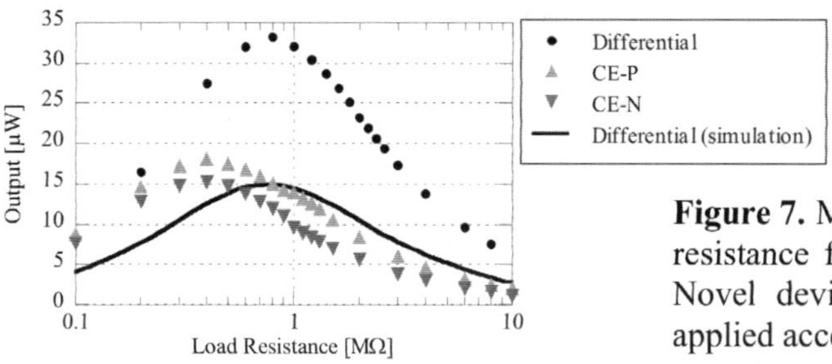

Figure 7. Maximum output power versus load resistance from 10 × 10 mm^2 of active area. Novel device was charged at ±150 V and applied acceleration of 8.8 *g* at 350 Hz.

4.3. *Connect to power management IC*

Our novel VEH was connected to commercially available power management IC (LTC3588-1: Linear Technology Co., USA), which is designed for high impedance energy harvester. Figure 8 shows input (V_{in}) and output (V_{out}) voltage on LTC3588-1. The regulated output voltage was set to 1.8 V. Figures 8(a) to (c) show start up voltage of V_{in}, V_{out} for load resistances of 1 MΩ, 500 kΩ and 100 kΩ, respectively. The specified voltage of 1.8 V was obtained after 37s case of 1 MΩ load. Due to a current from load resistance during charging, the IC never reaches specified voltage to connect under 100 kΩ.

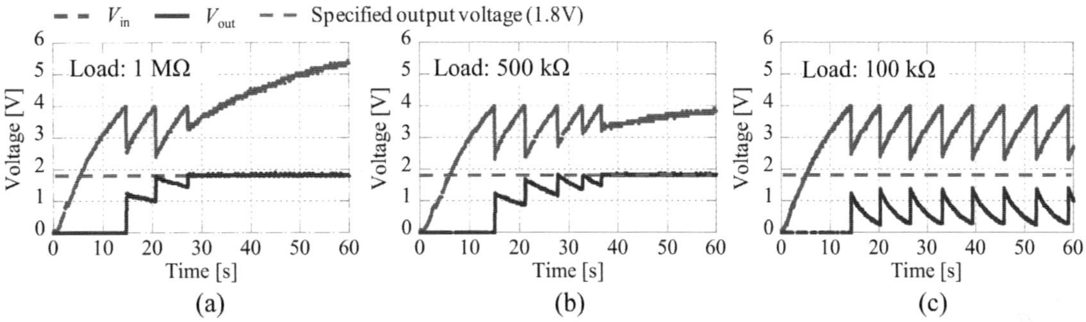

Figure 8. Input and output waveform from LTC3588-1. (a)1 MΩ, (b)500 kΩ, and (c)100 kΩ load resistance. V_{in} is charging terminal from differential output of harvester to connected capacitor. V_{out} is output terminal to supply DC power for loads.

5. Conclusion

The energy harvester with double-density counter electrode was developed and improved by polymer-insulation film. We confirmed that conduction by R_p was greatly reduced from discharge prevention device. The differential output power of 33 μW was obtained from novel energy harvester as the acceleration was 8.8 *g* at 350 Hz with 0.8 MΩ of optimum load resistance from 10 × 10 mm^2 of active area. Specified regulated voltage of 1.8 V is properly obtained with load resistance more than 500 kΩ by using a power management IC.

Acknowledgement

This work was supported by JSPS(Japan Society for the Promotion of Science) KAKENHI Grant-in-Aid for Scientific Research(C) Number 16K06308.

References

[1] Altena G, Renaud M, Elfrink R, Goedbloed M H, Nooijer C de and Schaijk R van 2013 *Proc. of PowerMEMS2013* 371-375
[2] Miwatani N, Minami K, Fujita T, Kanda K, and Maenaka K 2016 *Proc. of APCOT 2016*, 219-229
[3] Fujita T, Onishi T, Fujii K, Katsuma K, Kanda K, Higuchi K, and Maenaka K 2012 *Proc. of PowerMEMS2012* 436-439
[4] Minami K, Fujita T, Sonoda K, Miwatani N, Kanda K and Maenaka K 2014 *Proc. of PowerMEMS2014* 208-212

Journal of Physics: Conference Series **773** (2016) 012118 doi:10.1088/1742-6596/773/1/012118

A triboelectric wind turbine for small-scale energy harvesting

Matthias Perez[1,2,3], Sebastien Boisseau[1,2], Matthias Geisler[1,2], Ghislain Despesse[1,2], and Jean Luc Reboud[1,3]
[1] Univ. Grenoble Alpes, F-38000 Grenoble, France
[2] CEA, Leti, Minatec Campus, 17 rue des Martyrs, F-38054 Grenoble, France
[3] Univ. Grenoble Alpes G2ELab, F-38000 Grenoble, France

E-mail: matthias.perez1@gmail.com

Abstract. This paper deals with a rotational energy harvester including a Horizontal Axis Wind Turbine (HAWT), a cylindrical stator covered by several electrodes, and thin Teflon dielectric membranes hung on the rotor. The sliding contact of the Teflon membranes on the stator provides simultaneously large capacitance variations and a polarization source for the electrostatic converter by exploiting triboelectric phenomena. 1μW has been harvested at 4m/s; 130μW at 10m/s and 550μW at 20m/s with a 40mmØ device. In order to validate the energy harvesting chain, the airflow energy harvester has been connected to a power management circuit implementing Synchronous Electric Charge Extraction (SECE) to supply a wireless sensor node with temperature and acceleration measurements, transmitted to a computer at 868MHz.

Introduction

Converting the kinetic energy present in an airflow into electrical energy requires an intermediate mechanical energy converter turning this kinetic energy into a relative movement between two elements, for example a relative movement of rotation between a stator and a rotor. This kinetic-to-mechanic conversion can be achieved thanks to a Horizontal Axis Wind Turbine (HAWT) or a Vertical Axis Wind Turbine (VAWT). The mechanic-to-electric conversion is finally achieved thanks to an electromagnetic, a piezoelectric, an electrostatic or a triboelectric converter.

Over the past two decades, many researchers have investigated the airflow energy harvesting field and, as a consequence, many configurations have already been tested: HAWTs have been coupled to electromagnetic [1-3], piezoelectric [4] and electrostatic converters [5-6]; VAWTs have been coupled to piezoelectric [7] and triboelectric converters [8]. This paper focuses on a HAWT using a triboelectric converter implementing a free-standing triboelectric conversion scheme with flexible dielectric membranes so as to insure large capacitance variations and a self-polarization.

1. Micro wind turbines design

HAWTs are driven by the lift force F_L and limited by the drag force F_D (Figure 1). Here, we want to develop small devices (Ø=4cm) able to work at relatively low speeds (U<10m/s). These two constraints imply a Reynolds number lower than $Re<3\times10^4$. At these Reynolds numbers, the viscous drag is significant and the lift is limited by early detachments of the boundary layer. Consequently, the specific speed ($\lambda=\omega R/U$) of micro-turbines is very limited, as well as the power coefficient C_p. Yet, it appears that the best aerodynamic profiles to maximize the F_L/F_D ratio are simple thin flat plates, which greatly simplifies the fabrication of the devices.

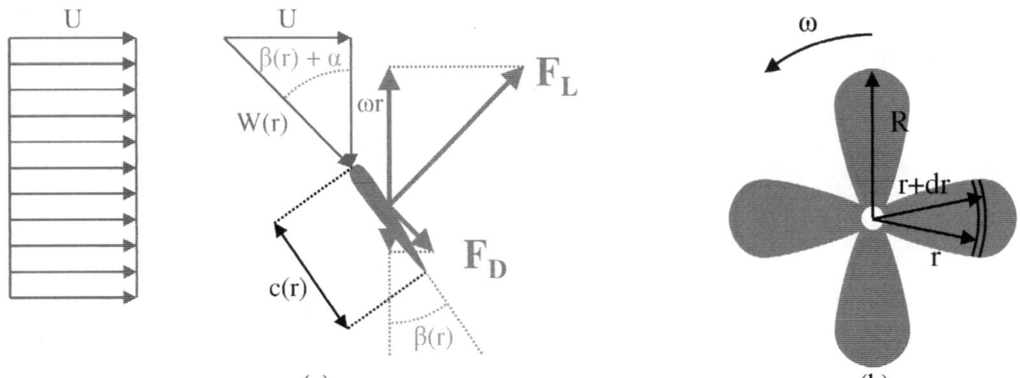

(a) (b)

Figure 1. Working principle of a HAWT: (a) lateral view and (b) front view.

2. Triboelectric conversion

Triboelectricity is a fascinating phenomenon of electric charges exchange that occurs when two materials of different triboelectric natures are brought into contact (Figure 2). The strength of the triboelectric effect depends on many parameters such as the materials' nature, surface roughness, friction conditions (intensity and speed), moisture, temperature... In most cases, triboelectricity is damaging and must be avoided (overvoltage, electric shock, fuel ignition...), but it can also offer interesting possibilities in several applications including tactile sensors [9], and electrostatic energy harvesters to ensure a stable polarization source through time [10,11].

As presented in Figure 2, representing the free-standing triboelectric conversion scheme implemented here, triboelectricity enables to generate and to store negative charges on the dielectric layer. Then, as the dielectric moves from an electrode to another, positive charges are able to move freely from a conductive electrode to another, thus creating an electric current through a resistive load.

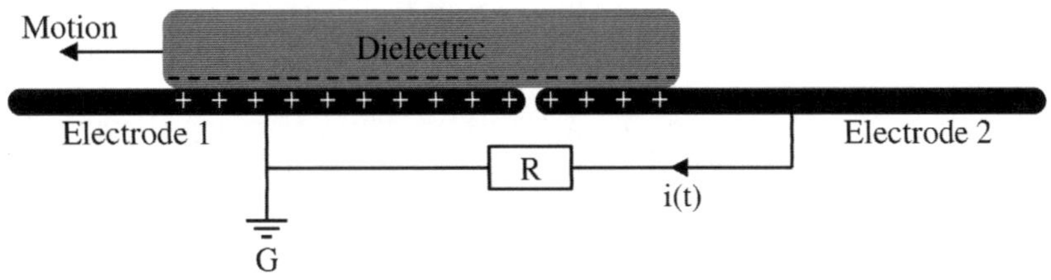

Figure 2. Physical principle of the triboelectric conversion – Free-standing triboelectric conversion scheme

3. Experimental results

The blades of the rotor are made of 750μm-thick ABS. Several films (N) of 50μm-thick Teflon FEP are attached to the rotor and slip inside the multi-electrodes copper stator. When the rotor starts to turn, the centrifugal force drags the membranes away, entering in contact with the stator's electrodes on N angular sectors θ (Figure 3a). As a result, the Teflon layers are negatively charged and the copper electrodes positively charged by triboelectric effects. By linking wisely the different electrodes (Figure 3a), a flow of electric current is generated through an external load.

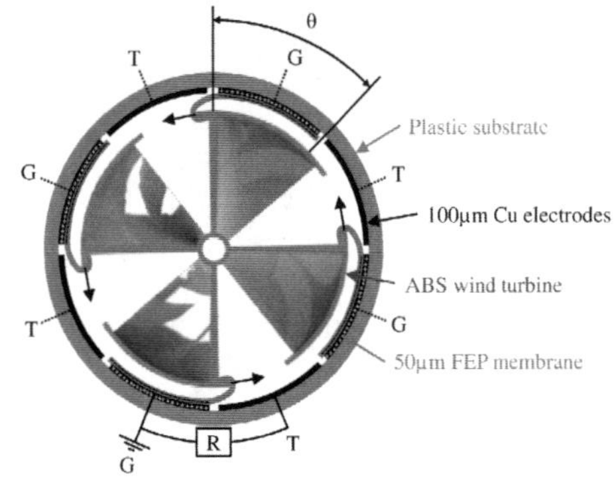

(a)

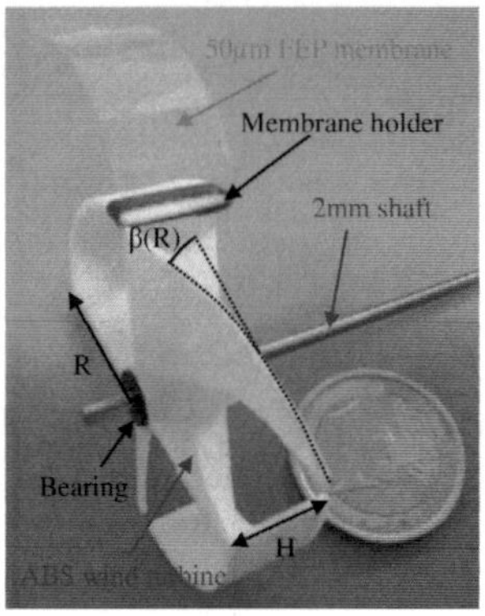

(b)

Figure 3. Triboelectric wind turbine: (a) schematic view with N=4 membranes and θ=45°, (b) Picture of prototypes with respectively β(R)=20°, θ=90° and N=1

Various rotors with a constant diameter of Ø=40mm, a depth of H=10mm and different pitch angles β(R) have been designed. Several membranes (N) have also been fixed to the rotor's rim (Figure 3b). The stator is located 3mm away from the point of connection, so there is no need to use very precise manufacturing techniques to obtain high electrostatic couplings with this concept [4]. Experiments have shown that the membranes tend to increase the start-up speed from 1m/s to 5m/s (Figure 4a). The friction of the membranes on the stator also decreases the rotational speed of the rotor from 7300rpm@10m/s without membranes to 3700rpm@10m/s with N=1 membrane and 1200rpm@10m/s with N=2 membranes (Figure 4a). Yet, the sliding contact of the membranes turns FEP layers into 300V-triboelectrets as expected, ensuring the polarization of the electrostatic converter for long. 1μW has been generated at 4m/s; 130μW at 10m/s and 550μW at 20m/s (Figure 4b).

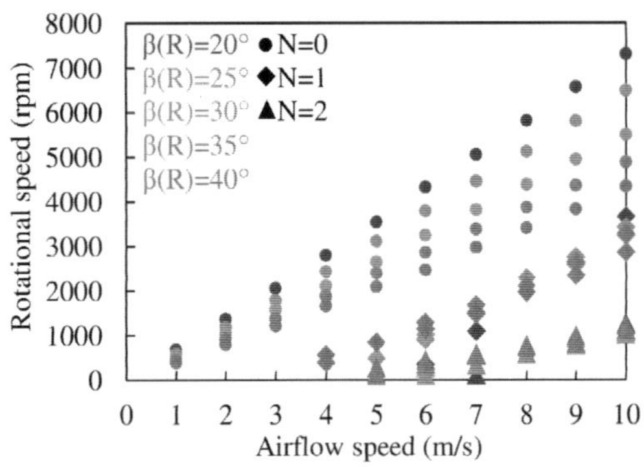

(a)

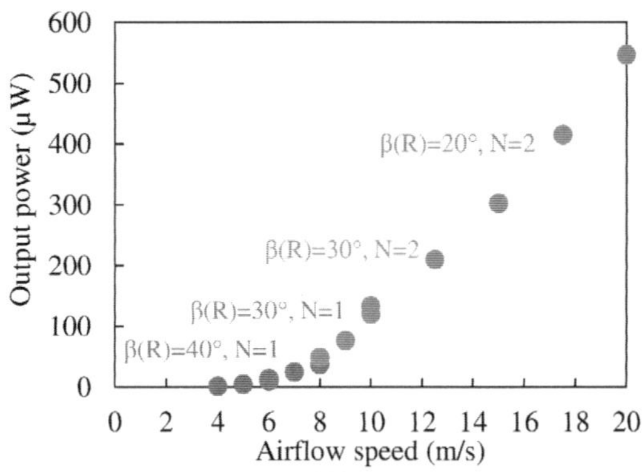

(b)

Figure 4. (a) Reduction of the rotational speed by the membranes (θ=90°). (b) Output power of the best prototypes between 0 and 20m/s (θ=90°).

4. Self starting battery-free power management circuit

So as to validate the viability of this approach, a power management circuit implementing SECE (Synchronous Electric Charge Extraction) has been designed (Figure 5a) to supply a wireless sensor node. This circuit was already presented in [12] and tested with a flutter electrostatic energy harvester in [13]. It uses two energy paths, a passive one and an active one, to start the system from scratch without battery, and then to perform a SECE conversion scheme with a flyback converter.

Our circuit has been tested with a triboelectric wind turbine at 12m/s and it turned out to be very efficient with a first measurement of temperature and acceleration and its RF emission after 30 seconds followed by other measurements and RF emissions every 6 seconds (red curve in Figure 5b). This active circuit was compared to a passive strategy with a diode bridge rectifier directly connected to the buffer capacitor C_b (blue curve in Figure 5b). In this case, the first measurement is performed after 3 minutes and 37 seconds (7 times as much time) followed by the other measurements every 1 minute (10 times as much time).

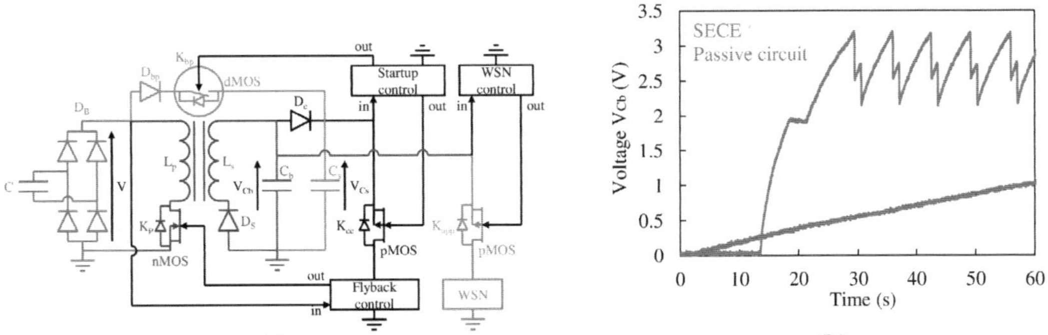

(a) (b)

Figure 5. (a) Schematic of the SECE circuit. (b) Evolution of the voltage V_{Cb} during the measurements (R=20mm, β(R)=20°, N_p=4, N_m=2 and U=12m.s^{-1}, C_b=100μF).

Conclusions

We have presented an airflow electrostatic energy harvester exploiting flexible membranes to ensure large capacitance variations and a stable polarization by taking advantage of triboelectric phenomena. Up to 550μW have been generated at 20m/s. The device has been connected to a self-starting SECE power management circuit to supply a temperature and acceleration sensor node and its RF data transmission. Research actions are now focused on the reduction of friction losses to improve the efficiency of the turbine and its cut-in speed.

References

[1] Rancourt D, Tabesh A, and Frechette L G 2007 Evaluation of centimeter-scale micro wind mills: Aerodynamics and electromagnetic power generation, *Proc. PowerMEMS*

[2] Bansal A, Howey D A, and Holmes A S 2009 Cm-scale air turbine and generator for energy harvesting from low-speed flow, *Proc. Transducers*

[3] Howey D A, Bansal A, and Holmes A S 2011 Design and performance of a centimeter-scale shrouded wind turbine for energy harvesting, *Smart Materials and Structures*

[4] Priya S, Chen C T, Fye D, and Zahnd J 2005 Piezoelectric windmill: A novel solution to remote sensing, *Japanese Journal of Applied Physics*

[5] Perez M, Boisseau S, and Reboud J L 2015 Design and performance of a small-scale wind turbine exploiting an electret-based electrostatic conversion, Proc. *Electrostatics*

[6] Perez M et al. 2016 A cm scale electret-based electrostatic wind turbine for low-speed energy harvesting applications, *Smart materials and structures*

[7] Myers R, Vickers M, and Kim H 2007 Small scale windmill, *Applied Physics Letters*

[8] Xie Y et al. 2013 Rotary triboelectric nanogenerator based on a hybridized mechanism for harvesting wind energy, *ACS Nano*

[9] Zhu G, Yang W Q, Zhang T, Jing Q, Chen T, Zhou Y S, Bai P, and Wang Z L 2014 Self-powered, ultrasensitive, flexible tactile sensors based on contact electrification, *Nano letters*

[10] Yang Y et al. 2013 Triboelectric nanogenerator for harvesting wind energy and as self-powered vector sensor system, *Nano Letters*

[11] Despesse G et al., Electrostatic Device and method for recovering mechanical energy by triboelectric effect, Patent US 2015/0035408

[12] Boisseau S et al. 2013 Self-starting power management circuits for piezoelectric and electret-based electrostatic mechanical energy harvesters, *Proc. PowerMEMS*

[13] Perez M et al. 2015 Electret-based aeroelastic harvester and its self-starting battery-free power management circuit *Proc. NEWCAS*